Methods in Enzymology

Volume 197

PHOSPHOLIPASES

METHODS IN ENZYMOLOGY

EDITORS-IN-CHIEF

John N. Abelson Melvin I. Simon

DIVISION OF BIOLOGY
CALIFORNIA INSTITUTE OF TECHNOLOGY
PASADENA, CALIFORNIA

FOUNDING EDITORS

Sidney P. Colowick and Nathan O. Kaplan

Methods in Enzymology

Volume 197

Phospholipases

EDITED BY

Edward A. Dennis

DEPARTMENT OF CHEMISTRY
UNIVERSITY OF CALIFORNIA, SAN DIEGO
LA JOLLA, CALIFORNIA

ACADEMIC PRESS, INC.
Harcourt Brace Jovanovich, Publishers
San Diego New York Boston
London Sydney Tokyo Toronto

ACADEMIC PRESS, INC.
San Diego, California 92101

United Kingdom Edition published by
ACADEMIC PRESS LIMITED
24-28 Oval Road, London NW1 7DX

LIBRARY OF CONGRESS CATALOG CARD NUMBER: 54-9110

ISBN 0-12-182098-X (alk. paper)

PRINTED IN THE UNITED STATES OF AMERICA
91 92 93 94 10 9 8 7 6 5 4 3 2 1

Table of Contents

Section I. Phospholipase Assays, Kinetics, and Substrates

Section II. Phospholipase Structure–Function Techniques

Section III. Phospholipase A_1

Section IV. Phospholipase A_2

A. Phospholipase A_2

B. Platelet-Activating Factor Acetylhydrolase

C. Lecithin–Cholesterol Acyltransferase

Section V. Lysophospholipase

A. Lysophospholipase

B. Lysoplasmalogenase

Section VI. Phospholipase C

A. Phospholipase C

B. Sphingomyelinase

C. Phosphatidate Phosphatase

Section VII. Phospholipase D

Contributors to Volume 197

Article numbers are in parentheses following the names of contributors.
Affiliations listed are current.

A. J. AARSMAN (34, 44), *Center for Biomembranes and Lipid Enzymology, University of Utrecht, 3584 CH Utrecht, The Netherlands*

KATHY A. ALDERN (11), *Department of Medicine, University of California, San Diego, La Jolla, California 92093*

MICHAEL P. ANGIOLI (51), *SmithKline & Beecham Laboratories, Philadelphia, Pennsylvania 19101*

YOSHIKO BANNO (50), *Department of Biochemistry, Gifu University School of Medicine, Gifu 500, Japan*

DAFNA BAR-SAGI (24), *Cold Spring Harbor Laboratory, Cold Spring Harbor, New York 11724*

JOHN D. BELL (21, 22), *Department of Zoology, Brigham Young University, Provo Utah 84602*

GUNILLA BENGTSSON-OLIVECRONA (32), *Department of Medical Biochemistry and Biophysics, University of Umeå, S-901 87 Umeå, Sweden*

C. FRANK BENNETT (51), *Department of Molecular and Cellular Biology, ISIS Pharmaceuticals, Inc., Carlsbad, California 92008*

RODNEY L. BILTONEN (21, 22), *Departments of Biochemistry and Pharmacology, University of Virginia, Charlottesville, Virginia 22908*

MERLE L. BLANK (14), *Medical and Health Sciences Division, Oak Ridge Associated Universities, Oak Ridge, Tennessee 37831*

DAVID N. BRINDLEY (55), *Department of Biochemistry and Lipid and Lipoprotein Research Group, Heritage Medical Research Centre, University of Alberta, Edmonton, Alberta T6G 2S2, Canada*

KAROL S. BRUZIK (23), *Department of Chemistry, The Ohio State University, Columbus, Ohio 43210*

GEORGE M. CARMAN (54), *Department of Food Science, Rutgers University, New Brunswick, New Jersey 08903*

HELEN R. CARTER (16), *Department of Biochemistry, University of Tennessee, Memphis, Memphis, Tennessee 38163*

HYEUN WOOK CHANG (36), *Department of Biochemistry, College of Pharmacy, Yeungnam University, Gyongsan 713-749, Korea*

HUGUES CHAP (28), *Departments of Membrane Phospholipids, Cell Signalling, and Lipoproteins, INSERM, Hôpital Purpan, 31059 Toulouse Cedex, France*

SUBROTO CHATTERJEE (53), *Department of Pediatrics, The Johns Hopkins University, Baltimore, Maryland 21205*

FLOYD H. CHILTON (15), *Division of Clinical Immunology, The Johns Hopkins University School of Medicine, Baltimore, Maryland 21224*

KEY SEUNG CHO (48), *Department of Biochemistry, College of National Science, Han Yang University, Ansan, South Korea*

WONHWA CHO (6), *Department of Chemistry, University of Illinois, Chicago, Illinois*

ENRIQUE CLARO (16), *Department of Biochemistry, University of Tennessee, Memphis, Memphis, Tennessee 38163*

XAVIER COLLET (40), *Department of Physiology, Cardiovascular Research Institute, University of California Medical Center, San Francisco, California 94143*

STANLEY T. CROOKE (51), *ISIS Pharmaceuticals, Inc., Carlsbad, California 92008*

R. A. DEEMS (1, 25, 43), *Department of Chemistry, University of California, San Diego, La Jolla, California 92093*

G. H. DE HAAS (19), *Department of Enzymology and Protein Engineering, State University of Utrecht, CBLE, University Center de Uithof, Utrecht NL-3584 CH, The Netherlands*

J. G. N. DE JONG (34, 44), *Laboratory of Pediatrics and Neurology, University of Nymegen, 6500 HB Nymegen, The Netherlands*

EDWARD A. DENNIS (1, 5, 25, 33, 43), *Department of Chemistry, University of California, San Diego, La Jolla, California 92093*

PETER ELSBACH (2), *Department of Medicine, New York University School of Medicine, New York, New York 10016*

YASUFUMI EMORI (49), *Department of Biochemistry, Faculty of Science, University of Tokyo, Tokyo 113, Japan*

JOHN N. FAIN (16), *Department of Biochemistry, University of Tennessee, Memphis, Memphis, Tennessee 38163*

JOSETTE FAUVEL (28), *Departments of Membrane Phospholipids, Cell Signalling, and Lipoproteins, INSERM, Hôpital Purpan, 31059 Toulouse Cedex, France*

CHRISTOPHER J. FIELDING (40), *Department of Physiology, Cardiovascular Research Institute, University of California Medical Center, San Francisco, California 94143*

MICHAEL F. GARDNER (11, 29), *Department of Medicine, University of California, San Diego, La Jolla, California 92093*

AMA GASSAMA-DIAGNE (28), *Departments of Membrane Phospholipids, Cell Signalling, and Lipoproteins, INSERM, Hôpital Purpan, 31059 Toulouse Cedex, France*

MICHAEL H. GELB (10), *Departments of Chemistry and Biochemistry, University of Washington, Seattle, Washington 98195*

NUPUR GHOSH (53), *Department of Pediatrics, The Johns Hopkins University, Baltimore, Maryland 21205*

ANTONIO GOMEZ-MUÑOZ (55), *Department of Biochemistry and Lipid and Lipoprotein Research Group, Heritage Medical Research Centre, University of Alberta, Edmonton, Alberta T6G 2S2, Canada*

O. HAYES GRIFFITH (47), *Department of Chemistry, Institute of Molecular Biology, University of Oregon, Eugene, Oregon 97403*

RICHARD W. GROSS (38, 45), *Division of Molecular and Cellular Cardiovascular Biochemistry, Washington University School of Medicine, St. Louis, Missouri 63110*

MICHAEL R. HANLEY (13), *Department of Biological Chemistry, University of California, Davis, Davis, California 95616*

SHUNTARO HARA (36), *Department of Health Chemistry, Faculty of Pharmaceutical Sciences, University of Tokyo, Tokyo 113, Japan*

PHILLIP T. HAWKINS (13), *Department of Biochemistry, AFRC Institute of Animal Physiology and Genetics, Babraham, Cambridge CB2 4AT, England*

STANLEY L. HAZEN (38), *Division of Molecular and Cellular Cardiovascular Biochemistry, Washington University School of Medicine, St. Louis, Missouri 63110*

ROBERT L. HEINRIKSON (18), *Department of Biopolymer Chemistry, The Upjohn Company, Kalamazoo, Michigan 49008*

H. STEWART HENDRICKSON (8), *Department of Chemistry, St. Olaf College, Northfield, Minnesota 55057*

ALBIN HERMETTER (12), *Department of Biochemistry, Graz University of Technology, A-8010 Graz, Austria*

Y. HIRASHIMA (7), *Department of Neurosurgery, Toyama Medical and Pharmaceutical University, Toyama 930-01, Japan*

YOSHIMI HOMMA (49), *Department of Biosignal Research, Tokyo Metropolitan*

Institute of Gerontology, Tokyo 173, Japan

KAZUHIKO HORIGOME (36), *Takarazuka Research Center, Sumitomo Chemical Co., Ltd., Takarazuka 665, Japan*

L. A. HORROCKS (7, 46), *Department of Physiological Chemistry, The Ohio State University, Columbus, Ohio 43210*

KARL Y. HOSTETLER (11, 29), *Department of Medicine, University of California, San Diego, La Jolla, California 92093*

KUO-SEN HUANG (56), *Department of Protein Biochemistry, Hoffmann La Roche, Inc., Nutley, New Jersey 07110*

TONY HUNTER (26), *Molecular Biology and Virology Laboratory, Salk Institute, La Jolla, California 92037*

KEIZO INOUE (20, 36), *Department of Health Chemistry, Faculty of Pharmaceutical Sciences, University of Tokyo, Tokyo 113, Japan*

RICHARD L. JACKSON (31), *Department of Biochemistry, Merrell Dow Research Institute, Cincinnati, Ohio 45215*

MAHENDRA KUMAR JAIN (10), *Departments of Chemistry and Biochemistry, University of Delaware, Newark, Delaware 19716*

ZAHIRALI JAMAL (55), *Department of Biochemistry and Lipid and Lipoprotein Research Group, Heritage Medical Research Centre, University of Alberta, Edmonton, Alberta T6G 2S2, Canada*

M. S. JURKOWITZ-ALEXANDER (7, 46), *Department of Anesthesiology, College of Medicine, The Ohio State University, Columbus, Ohio 43210*

JULIAN N. KANFER (57), *Department of Biochemistry and Molecular Biology, University of Manitoba, Winnipeg, Manitoba R3E 0W3, Canada*

KEN KARASAWA (41), *Faculty of Pharmaceutical Sciences, Teikyo University, Sagamiko, Kanagawa 199-01, Japan*

E. S. KEMPNER (25), *National Institute of Arthritis and Musculoskeletal and Skin Diseases, National Institutes of Health, Bethesda, Maryland 20892*

DONALD A. KENNERLY (17), *Department of Internal Medicine, University of Texas Southwestern Medical Center, Dallas, Texas 75235*

FERENC J. KÉZDY (6), *Biopolymer Chemistry Unit, The Upjohn Company, Kalamazoo, Michigan 49001*

MUTSUHIRO KOBAYASHI (57), *Department of Geriatrics, Endocrinology, and Metabolism, Shinshu University School of Medicine, Nagano 390, Japan*

RUTH M. KRAMER (35), *Lilly Research Laboratories, Eli Lilly and Co., Indianapolis, Indiana 46285*

GREGORY L. KUCERA (30), *Internal Medicine, Bowman Gray School of Medicine, Wake Forest University, Winston-Salem, North Carolina 27103*

ICHIRO KUDO (20, 36), *Department of Health Chemistry, Faculty of Pharmaceutical Sciences, University of Tokyo, Tokyo 113, Japan*

ANDREAS KUPPE (47), *Institute of Molecular Biology, University of Oregon, Eugene, Oregon 97403*

BRIAN K. LATHROP (21), *Department of Pharmacology, University of Virginia, Charlottesville, Virginia 22908*

KEE YOUNG LEE (48), *Department of Biochemistry, College of Medicine, Chonnam National University, Kwangju, Korea*

SHIRLEY LI (56), *Department of Protein Biochemistry, Hoffman La Roche, Inc., Nutley, New Jersey 07110*

YI-PING LIN (54), *Department of Food Science, Rutgers University, New Brunswick, New Jersey 08903*

LORI A. LOEB (38), *Division of Molecular and Cellular Cardiovascular Biochemistry, Washington University School of Medicine, St. Louis, Missouri 63110*

MARTIN G. LOW (56), *Rover Physiology Research Laboratories, Department of Physiology and Cellular Biophysics, College of Physicians & Surgeons of Columbia University, New York, New York 10032*

Ashley Martin (55), *Department of Biochemistry and Lipid and Lipoprotein Research Group, Heritage Medical Research Centre, University of Alberta, Edmonton, Alberta T6G 2S2, Canada*

Thomas M. McIntyre (39), *Departments of Medicine and Biochemistry, Nora Eccles Harrison Cardiovascular Research and Training Institute, University of Utah, Salt Lake City, Utah 84112*

Larry R. McLean (31), *Department of Biochemistry, Merrell Dow Research Institute, Cincinnati, Ohio 45215*

Jill Meisenhelder (26), *Molecular Biology and Virology Laboratory, Salk Institute, La Jolla, California 92037*

H. Moreau (4), *C.N.R.S. Centre de Biochimie et de Biologie Moléculaire, 13402 Marseille, Cedex 9, France*

Makoto Murakami (20), *Department of Health Chemistry, Faculty of Pharmaceutical Sciences, University of Tokyo, Tokyo 113, Japan*

Yasuhito Nakagawa (27), *Faculty of Pharmaceutical Sciences, Teikyo University, Sagamiko, Kanagawa 199-01, Japan*

Shoshichi Nojima (27, 41), *Faculty of Pharmaceutical Sciences, Teikyo University, Sagamiko, Kanagawa 199-01, Japan*

Yoshinori Nozawa (50), *Department of Biochemistry, Gifu University School of Medicine, Gifu 500, Japan*

Mitsuhiro Okamoto (37), *Department of Biochemistry and Molecular Physiological Chemistry, Osaka University Medical School, Osaka 530, Japan*

Tadayoshi Okumura (42), *Department of Medical Chemistry, Kansai Medical School, Osaka 570, Japan*

Thomas Olivecrona (32), *Department of Medical Biochemistry and Biophysics, University of Umeå, S-90187 Umeå, Sweden*

Takashi Ono (37), *Department of Biochemistry and Molecular Physiological Chemistry, Osaka University Medical School, Osaka 530, Japan*

J. C. Osborne, Jr. (25), *Advanced Development Unit, Beckman Instruments, Fullerton, California 92634*

Fritz Paltauf (12), *Department of Biochemistry, Graz University of Technology, A-8010 Graz, Austria*

R. Blake Pepinsky (35), *Biogen, Inc., Cambridge, Massachusetts 02142*

David R. Poyner (13), *MRC Laboratory of Molecular Biology, MRC Centre, Cambridge CB2 2QH, England*

Stephen M. Prescott (39), *Departments of Medicine and Biochemistry, Nora Eccles Harrison Cardiovascular Research and Training Institute, University of Utah, Salt Lake City, Utah 84112*

L. E. Quintern (52), *German Aerospace Research Establishment, Institute for Aerospace Medicine, Biophysics Division, D-5000 Köln 90, Germany*

S. Ransac (4), *C.N.R.S. Centre de Biochimie et de Biologie Moléculaire, 13402 Marseille, Cedex 9, France*

Laure J. Reynolds (1, 25, 33), *Department of Chemistry, University of California, San Diego, La Jolla, California 92093*

Sue Goo Rhee (48), *Laboratory of Biochemistry, National Heart, Lung, and Blood Institute, National Institutes of Health, Bethesda, Maryland 20892*

C. Rivière (4), *C.N.R.S. Centre de Biochimie et de Biologie Moléculaire, 13402 Marseille, Cedex 9, France*

Mary F. Roberts (3, 9), *Department of Chemistry, Boston College, Chestnut Hill, Massachusetts 02167*

Sung Ho Ryu (48), *Department of Life Science, POSTECH, Pohang, South Korea*

Kunihiko Saito (42), *Department of Medical Chemistry, Kansai Medical School, Osaka 570, Japan*

K. Sandhoff (52), *Institut für Organische Chemie und Biochemie, Universität Bonn, D-5300 Bonn 1, Germany*

Morio Setaka (27), *Faculty of Pharmaceutical Sciences, Teikyo University, Sagamiko, Kanagawa 199-01, Japan*

PATRICIA J. SISSON (30), *Department of Biochemistry, Bowman Gray School of Medicine, Wake Forest University, Winston-Salem, North Carolina 27103*

FRED SNYDER (14), *Medical and Health Sciences Division, Oak Ridge Associated Universities, Oak Ridge, Tennessee 37831*

DIANA M. STAFFORINI (39), *Department of Medicine, Nora Eccles Harrison Cardiovascular Research and Training Institute, University of Utah, Salt Lake City, Utah 84112*

JAY C. STRUM (58), *Department of Biochemistry, Wake Forest University Medical Center, Winston-Salem, North Carolina 27103*

JUNKO SUGATANI (42), *Department of Medical Chemistry, Kansai Medical School, Osaka 570, Japan*

KIYOSHI TAKAYAMA (20), *Department of Health Chemistry, Faculty of Pharmaceutical Sciences, University of Tokyo, Tokyo 113, Japan*

TADAOMI TAKENAWA (49), *Department of Biosignal Research, Tokyo Metropolitan Institute of Gerontology, Tokyo 173, Japan*

TOM Y. THUREN (30), *Department of Biochemistry, Bowman Gray School of Medicine, Wake Forest University, Winston-Salem, North Carolina 27103*

HIROMASA TOJO (37), *Department of Biochemistry and Molecular Physiological Chemistry, Osaka University Medical School, Osaka 530, Japan*

MING-DAW TSAI (23), *Department of Chemistry, The Ohio State University, Columbus, Ohio 43210*

MASATO UMEDA (20), *Department of Health Chemistry, Faculty of Pharmaceutical Sciences, University of Tokyo, Tokyo 113, Japan*

H. VAN DEN BOSCH (34, 44), *Center for Biomembranes and Lipid Enzymology, University of Utrecht, 3584 CH Utrecht, The Netherlands*

R. VERGER (4), *C.N.R.S. Centre de Biochimie et de Biologie Moléculaire, 13402 Marseille, Cedex 9, France*

H. M. VERHEIJ (19), *Department of Enzymology and Protein Engineering, State University of Utrecht, CBLE, University Center de Uithof, Utrecht NL-3584 CH, The Netherlands*

JOHANNES J. VOLWERK (47), *Institute of Molecular Biology, University of Oregon, Eugene, Oregon 97403*

MOSELEY WAITE (30), *Department of Biochemistry, Bowman Gray School of Medicine, Wake Forest University, Winston-Salem, North Carolina 27103*

MICHAEL A. WALLACE (16), *Department of Biochemistry, University of Tennessee, Memphis, Memphis, Tennessee 38163*

WILLIAM N. WASHBURN (1), *Life Sciences Research Laboratories, Eastman Kodak Company, Rochester, New York 14650*

JERROLD WEISS (2), *Department of Medicine, New York University School of Medicine, New York, New York 10016*

REBECCA W. WILCOX (30), *Department of Biochemistry, Bowman Gray School of Medicine, Wake Forest University, Winston-Salem, North Carolina 27103*

ROBERT L. WYKLE (58), *Department of Biochemistry, Wake Forest University Medical Center, Winston-Salem, North Carolina 27103*

LIN YU (5), *Department of Chemistry, University of California, San Diego, La Jolla, California 92093*

YING YI ZHANG (43), *Physiological Chemistry, Karalinska Institutet, Stockholm, Sweden*

Preface

Until recently, lipids were thought to play important roles in only energy storage and membrane structure. It is now clear that they also play many other critical roles, especially as mediators in cell activation and signal transduction. The phospholipases are key enzymes in all of these functions. Recent advances in enzymology and molecular biology have given us new tools to study how various phospholipases function and their importance in regulation and metabolism.

A few chapters on phospholipases have been included in the Lipids and Biomembranes volumes of this series. We are indeed pleased that the central role and importance of phospholipases has now been recognized and that an entire *Methods in Enzymology* volume is devoted to this subject.

In this volume, the term phospholipases has been broadly interpreted to include not only those enzymes traditionally defined as phospholipase A_1, A_2, C, D, and lysophospholipase, but also other enzymes which break down phospholipids, including sphingolipids, such as sphingomyelinase, lecithin-cholesterol acyltransferase (LCAT), platelet activating factor (PAF) acetyl hydrolase, lysoplasmalogenase, and some lipases. Chapter 1 more completely defines the phospholipase substrates and provides an overview to the special problems involved in the assay and study of the phospholipases.

We appreciate the authors' cooperation in promptly preparing manuscripts. We also thank the staff of Academic Press for their help in producing a timely volume in this important field. The counsel of Drs. Laure Reynolds and Raymond Deems is much appreciated. Special thanks go to Lynn Krebs whose secretarial assistance aided immeasurably in the compilation of this volume.

EDWARD A. DENNIS

METHODS IN ENZYMOLOGY

VOLUME I. Preparation and Assay of Enzymes
Edited by SIDNEY P. COLOWICK AND NATHAN O. KAPLAN

VOLUME II. Preparation and Assay of Enzymes
Edited by SIDNEY P. COLOWICK AND NATHAN O. KAPLAN

VOLUME III. Preparation and Assay of Substrates
Edited by SIDNEY P. COLOWICK AND NATHAN O. KAPLAN

VOLUME IV. Special Techniques for the Enzymologist
Edited by SIDNEY P. COLOWICK AND NATHAN O. KAPLAN

VOLUME V. Preparation and Assay of Enzymes
Edited by SIDNEY P. COLOWICK AND NATHAN O. KAPLAN

VOLUME VI. Preparation and Assay of Enzymes (*Continued*)
Preparation and Assay of Substrates
Special Techniques
Edited by SIDNEY P. COLOWICK AND NATHAN O. KAPLAN

VOLUME VII. Cumulative Subject Index
Edited by SIDNEY P. COLOWICK AND NATHAN O. KAPLAN

VOLUME VIII. Complex Carbohydrates
Edited by ELIZABETH F. NEUFELD AND VICTOR GINSBURG

VOLUME IX. Carbohydrate Metabolism
Edited by WILLIS A. WOOD

VOLUME X. Oxidation and Phosphorylation
Edited by RONALD W. ESTABROOK AND MAYNARD E. PULLMAN

VOLUME XI. Enzyme Structure
Edited by C. H. W. HIRS

VOLUME XII. Nucleic Acids (Parts A and B)
Edited by LAWRENCE GROSSMAN AND KIVIE MOLDAVE

VOLUME XIII. Citric Acid Cycle
Edited by J. M. LOWENSTEIN

VOLUME XIV. Lipids
Edited by J. M. LOWENSTEIN

VOLUME XV. Steroids and Terpenoids
Edited by RAYMOND B. CLAYTON

VOLUME XVI. Fast Reactions
Edited by KENNETH KUSTIN

VOLUME XVII. Metabolism of Amino Acids and Amines (Parts A and B)
Edited by HERBERT TABOR AND CELIA WHITE TABOR

VOLUME XVIII. Vitamins and Coenzymes (Parts A, B, and C)
Edited by DONALD B. MCCORMICK AND LEMUEL D. WRIGHT

VOLUME XIX. Proteolytic Enzymes
Edited by GERTRUDE E. PERLMANN AND LASZLO LORAND

VOLUME XX. Nucleic Acids and Protein Synthesis (Part C)
Edited by KIVIE MOLDAVE AND LAWRENCE GROSSMAN

VOLUME XXI. Nucleic Acids (Part D)
Edited by LAWRENCE GROSSMAN AND KIVIE MOLDAVE

VOLUME XXII. Enzyme Purification and Related Techniques
Edited by WILLIAM B. JAKOBY

VOLUME XXIII. Photosynthesis (Part A)
Edited by ANTHONY SAN PIETRO

VOLUME XXIV. Photosynthesis and Nitrogen Fixation (Part B)
Edited by ANTHONY SAN PIETRO

VOLUME XXV. Enzyme Structure (Part B)
Edited by C. H. W. HIRS AND SERGE N. TIMASHEFF

VOLUME XXVI. Enzyme Structure (Part C)
Edited by C. H. W. HIRS AND SERGE N. TIMASHEFF

VOLUME XXVII. Enzyme Structure (Part D)
Edited by C. H. W. HIRS AND SERGE N. TIMASHEFF

VOLUME XXVIII. Complex Carbohydrates (Part B)
Edited by VICTOR GINSBURG

VOLUME XXIX. Nucleic Acids and Protein Synthesis (Part E)
Edited by LAWRENCE GROSSMAN AND KIVIE MOLDAVE

VOLUME XXX. Nucleic Acids and Protein Synthesis (Part F)
Edited by KIVIE MOLDAVE AND LAWRENCE GROSSMAN

VOLUME XXXI. Biomembranes (Part A)
Edited by SIDNEY FLEISCHER AND LESTER PACKER

VOLUME XXXII. Biomembranes (Part B)
Edited by SIDNEY FLEISCHER AND LESTER PACKER

VOLUME XXXIII. Cumulative Subject Index Volumes I–XXX
Edited by MARTHA G. DENNIS AND EDWARD A. DENNIS

VOLUME XXXIV. Affinity Techniques (Enzyme Purification: Part B)
Edited by WILLIAM B. JAKOBY AND MEIR WILCHEK

VOLUME XXXV. Lipids (Part B)
Edited by JOHN M. LOWENSTEIN

VOLUME XXXVI. Hormone Action (Part A: Steroid Hormones)
Edited by BERT W. O'MALLEY AND JOEL G. HARDMAN

VOLUME XXXVII. Hormone Action (Part B: Peptide Hormones)
Edited by BERT W. O'MALLEY AND JOEL G. HARDMAN

VOLUME XXXVIII. Hormone Action (Part C: Cyclic Nucleotides)
Edited by JOEL G. HARDMAN AND BERT W. O'MALLEY

VOLUME XXXIX. Hormone Action (Part D: Isolated Cells, Tissues, and Organ Systems)
Edited by JOEL G. HARDMAN AND BERT W. O'MALLEY

VOLUME XL. Hormone Action (Part E: Nuclear Structure and Function)
Edited by BERT W. O'MALLEY AND JOEL G. HARDMAN

VOLUME XLI. Carbohydrate Metabolism (Part B)
Edited by W. A. WOOD

VOLUME XLII. Carbohydrate Metabolism (Part C)
Edited by W. A. WOOD

VOLUME XLIII. Antibiotics
Edited by JOHN H. HASH

VOLUME XLIV. Immobilized Enzymes
Edited by KLAUS MOSBACH

VOLUME XLV. Proteolytic Enzymes (Part B)
Edited by LASZLO LORAND

VOLUME XLVI. Affinity Labeling
Edited by WILLIAM B. JAKOBY AND MEIR WILCHEK

VOLUME XLVII. Enzyme Structure (Part E)
Edited by C. H. W. HIRS AND SERGE N. TIMASHEFF

VOLUME XLVIII. Enzyme Structure (Part F)
Edited by C. H. W. HIRS AND SERGE N. TIMASHEFF

VOLUME XLIX. Enzyme Structure (Part G)
Edited by C. H. W. HIRS AND SERGE N. TIMASHEFF

VOLUME L. Complex Carbohydrates (Part C)
Edited by VICTOR GINSBURG

VOLUME LI. Purine and Pyrimidine Nucleotide Metabolism
Edited by PATRICIA A. HOFFEE AND MARY ELLEN JONES

VOLUME LII. Biomembranes (Part C: Biological Oxidations)
Edited by SIDNEY FLEISCHER AND LESTER PACKER

VOLUME LIII. Biomembranes (Part D: Biological Oxidations)
Edited by SIDNEY FLEISCHER AND LESTER PACKER

VOLUME LIV. Biomembranes (Part E: Biological Oxidations)
Edited by SIDNEY FLEISCHER AND LESTER PACKER

VOLUME LV. Biomembranes (Part F: Bioenergetics)
Edited by SIDNEY FLEISCHER AND LESTER PACKER

VOLUME LVI. Biomembranes (Part G: Bioenergetics)
Edited by SIDNEY FLEISCHER AND LESTER PACKER

VOLUME LVII. Bioluminescence and Chemiluminescence
Edited by MARLENE A. DELUCA

VOLUME LVIII. Cell Culture
Edited by WILLIAM B. JAKOBY AND IRA PASTAN

VOLUME LIX. Nucleic Acids and Protein Synthesis (Part G)
Edited by KIVIE MOLDAVE AND LAWRENCE GROSSMAN

VOLUME LX. Nucleic Acids and Protein Synthesis (Part H)
Edited by KIVIE MOLDAVE AND LAWRENCE GROSSMAN

VOLUME 61. Enzyme Structure (Part H)
Edited by C. H. W. HIRS AND SERGE N. TIMASHEFF

VOLUME 62. Vitamins and Coenzymes (Part D)
Edited by DONALD B. MCCORMICK AND LEMUEL D. WRIGHT

VOLUME 63. Enzyme Kinetics and Mechanism (Part A: Initial Rate and Inhibitor Methods)
Edited by DANIEL L. PURICH

VOLUME 64. Enzyme Kinetics and Mechanism (Part B: Isotopic Probes and Complex Enzyme Systems)
Edited by DANIEL L. PURICH

VOLUME 65. Nucleic Acids (Part I)
Edited by LAWRENCE GROSSMAN AND KIVIE MOLDAVE

VOLUME 66. Vitamins and Coenzymes (Part E)
Edited by DONALD B. MCCORMICK AND LEMUEL D. WRIGHT

VOLUME 67. Vitamins and Coenzymes (Part F)
Edited by DONALD B. MCCORMICK AND LEMUEL D. WRIGHT

VOLUME 68. Recombinant DNA
Edited by RAY WU

VOLUME 69. Photosynthesis and Nitrogen Fixation (Part C)
Edited by ANTHONY SAN PIETRO

VOLUME 70. Immunochemical Techniques (Part A)
Edited by HELEN VAN VUNAKIS AND JOHN J. LANGONE

VOLUME 71. Lipids (Part C)
Edited by JOHN M. LOWENSTEIN

VOLUME 72. Lipids (Part D)
Edited by JOHN M. LOWENSTEIN

VOLUME 73. Immunochemical Techniques (Part B)
Edited by JOHN J. LANGONE AND HELEN VAN VUNAKIS

VOLUME 74. Immunochemical Techniques (Part C)
Edited by JOHN J. LANGONE AND HELEN VAN VUNAKIS

VOLUME 75. Cumulative Subject Index Volumes XXXI, XXXII, XXXIV–LX
Edited by EDWARD A. DENNIS AND MARTHA G. DENNIS

VOLUME 76. Hemoglobins
Edited by ERALDO ANTONINI, LUIGI ROSSI-BERNARDI, AND EMILIA CHIANCONE

VOLUME 77. Detoxication and Drug Metabolism
Edited by WILLIAM B. JAKOBY

VOLUME 78. Interferons (Part A)
Edited by SIDNEY PESTKA

VOLUME 79. Interferons (Part B)
Edited by SIDNEY PESTKA

VOLUME 80. Proteolytic Enzymes (Part C)
Edited by LASZLO LORAND

VOLUME 81. Biomembranes (Part H: Visual Pigments and Purple Membranes, I)
Edited by LESTER PACKER

VOLUME 82. Structural and Contractile Proteins (Part A: Extracellular Matrix)
Edited by LEON W. CUNNINGHAM AND DIXIE W. FREDERIKSEN

VOLUME 83. Complex Carbohydrates (Part D)
Edited by VICTOR GINSBURG

VOLUME 84. Immunochemical Techniques (Part D: Selected Immunoassays)
Edited by JOHN J. LANGONE AND HELEN VAN VUNAKIS

VOLUME 85. Structural and Contractile Proteins (Part B: The Contractile Apparatus and the Cytoskeleton)
Edited by DIXIE W. FREDERIKSEN AND LEON W. CUNNINGHAM

VOLUME 86. Prostaglandins and Arachidonate Metabolites
Edited by WILLIAM E. M. LANDS AND WILLIAM L. SMITH

VOLUME 87. Enzyme Kinetics and Mechanism (Part C: Intermediates, Stereochemistry, and Rate Studies)
Edited by DANIEL L. PURICH

VOLUME 88. Biomembranes (Part I: Visual Pigments and Purple Membranes, II)
Edited by LESTER PACKER

VOLUME 89. Carbohydrate Metabolism (Part D)
Edited by WILLIS A. WOOD

VOLUME 90. Carbohydrate Metabolism (Part E)
Edited by WILLIS A. WOOD

VOLUME 91. Enzyme Structure (Part I)
Edited by C. H. W. HIRS AND SERGE N. TIMASHEFF

VOLUME 92. Immunochemical Techniques (Part E: Monoclonal Antibodies and General Immunoassay Methods)
Edited by JOHN J. LANGONE AND HELEN VAN VUNAKIS

VOLUME 105. Oxygen Radicals in Biological Systems
Edited by LESTER PACKER

VOLUME 106. Posttranslational Modifications (Part A)
Edited by FINN WOLD AND KIVIE MOLDAVE

VOLUME 107. Posttranslational Modifications (Part B)
Edited by FINN WOLD AND KIVIE MOLDAVE

VOLUME 108. Immunochemical Techniques (Part G: Separation and Characterization of Lymphoid Cells)
Edited by GIOVANNI DI SABATO, JOHN J. LANGONE, AND HELEN VAN VUNAKIS

VOLUME 109. Hormone Action (Part I: Peptide Hormones)
Edited by LUTZ BIRNBAUMER AND BERT W. O'MALLEY

VOLUME 110. Steroids and Isoprenoids (Part A)
Edited by JOHN H. LAW AND HANS C. RILLING

VOLUME 111. Steroids and Isoprenoids (Part B)
Edited by JOHN H. LAW AND HANS C. RILLING

VOLUME 112. Drug and Enzyme Targeting (Part A)
Edited by KENNETH J. WIDDER AND RALPH GREEN

VOLUME 113. Glutamate, Glutamine, Glutathione, and Related Compounds
Edited by ALTON MEISTER

VOLUME 114. Diffraction Methods for Biological Macromolecules (Part A)
Edited by HAROLD W. WYCKOFF, C. H. W. HIRS, AND SERGE N. TIMASHEFF

VOLUME 115. Diffraction Methods for Biological Macromolecules (Part B)
Edited by HAROLD W. WYCKOFF, C. H. W. HIRS, AND SERGE N. TIMASHEFF

VOLUME 116. Immunochemical Techniques (Part H: Effectors and Mediators of Lymphoid Cell Functions)
Edited by GIOVANNI DI SABATO, JOHN J. LANGONE, AND HELEN VAN VUNAKIS

VOLUME 117. Enzyme Structure (Part J)
Edited by C. H. W. HIRS AND SERGE N. TIMASHEFF

VOLUME 118. Plant Molecular Biology
Edited by ARTHUR WEISSBACH AND HERBERT WEISSBACH

VOLUME 119. Interferons (Part C)
Edited by SIDNEY PESTKA

VOLUME 120. Cumulative Subject Index Volumes 81–94, 96–101

VOLUME 121. Immunochemical Techniques (Part I: Hybridoma Technology and Monoclonal Antibodies)
Edited by JOHN J. LANGONE AND HELEN VAN VUNAKIS

VOLUME 122. Vitamins and Coenzymes (Part G)
Edited by FRANK CHYTIL AND DONALD B. MCCORMICK

VOLUME 123. Vitamins and Coenzymes (Part H)
Edited by FRANK CHYTIL AND DONALD B. MCCORMICK

VOLUME 124. Hormone Action (Part J: Neuroendocrine Peptides)
Edited by P. MICHAEL CONN

VOLUME 125. Biomembranes (Part M: Transport in Bacteria, Mitochondria, and Chloroplasts: General Approaches and Transport Systems)
Edited by SIDNEY FLEISCHER AND BECCA FLEISCHER

VOLUME 126. Biomembranes (Part N: Transport in Bacteria, Mitochondria, and Chloroplasts: Protonmotive Force)
Edited by SIDNEY FLEISCHER AND BECCA FLEISCHER

VOLUME 127. Biomembranes (Part O: Protons and Water: Structure and Translocation)
Edited by LESTER PACKER

VOLUME 128. Plasma Lipoproteins (Part A: Preparation, Structure, and Molecular Biology)
Edited by JERE P. SEGREST AND JOHN J. ALBERS

VOLUME 129. Plasma Lipoproteins (Part B: Characterization, Cell Biology, and Metabolism)
Edited by JOHN J. ALBERS AND JERE P. SEGREST

VOLUME 130. Enzyme Structure (Part K)
Edited by C. H. W. HIRS AND SERGE N. TIMASHEFF

VOLUME 131. Enzyme Structure (Part L)
Edited by C. H. W. HIRS AND SERGE N. TIMASHEFF

VOLUME 132. Immunochemical Techniques (Part J: Phagocytosis and Cell-Mediated Cytotoxicity)
Edited by GIOVANNI DI SABATO AND JOHANNES EVERSE

VOLUME 133. Bioluminescence and Chemiluminescence (Part B)
Edited by MARLENE DELUCA AND WILLIAM D. MCELROY

VOLUME 134. Structural and Contractile Proteins (Part C: The Contractile Apparatus and the Cytoskeleton)
Edited by RICHARD B. VALLEE

VOLUME 135. Immobilized Enzymes and Cells (Part B)
Edited by KLAUS MOSBACH

VOLUME 136. Immobilized Enzymes and Cells (Part C)
Edited by KLAUS MOSBACH

VOLUME 137. Immobilized Enzymes and Cells (Part D)
Edited by KLAUS MOSBACH

VOLUME 138. Complex Carbohydrates (Part E)
Edited by VICTOR GINSBURG

VOLUME 139. Cellular Regulators (Part A: Calcium- and Calmodulin-Binding Proteins)
Edited by ANTHONY R. MEANS AND P. MICHAEL CONN

VOLUME 140. Cumulative Subject Index Volumes 102–119, 121–134

VOLUME 141. Cellular Regulators (Part B: Calcium and Lipids)
Edited by P. MICHAEL CONN AND ANTHONY R. MEANS

VOLUME 142. Metabolism of Aromatic Amino Acids and Amines
Edited by SEYMOUR KAUFMAN

VOLUME 143. Sulfur and Sulfur Amino Acids
Edited by WILLIAM B. JAKOBY AND OWEN GRIFFITH

VOLUME 144. Structural and Contractile Proteins (Part D: Extracellular Matrix)
Edited by LEON W. CUNNINGHAM

VOLUME 145. Structural and Contractile Proteins (Part E: Extracellular Matrix)
Edited by LEON W. CUNNINGHAM

VOLUME 146. Peptide Growth Factors (Part A)
Edited by DAVID BARNES AND DAVID A. SIRBASKU

VOLUME 147. Peptide Growth Factors (Part B)
Edited by DAVID BARNES AND DAVID A. SIRBASKU

VOLUME 148. Plant Cell Membranes
Edited by LESTER PACKER AND ROLAND DOUCE

VOLUME 149. Drug and Enzyme Targeting (Part B)
Edited by RALPH GREEN AND KENNETH J. WIDDER

VOLUME 150. Immunochemical Techniques (Part K: *In Vitro* Models of B and T Cell Functions and Lymphoid Cell Receptors)
Edited by GIOVANNI DI SABATO

VOLUME 151. Molecular Genetics of Mammalian Cells
Edited by MICHAEL M. GOTTESMAN

VOLUME 152. Guide to Molecular Cloning Techniques
Edited by SHELBY L. BERGER AND ALAN R. KIMMEL

VOLUME 153. Recombinant DNA (Part D)
Edited by RAY WU AND LAWRENCE GROSSMAN

VOLUME 154. Recombinant DNA (Part E)
Edited by RAY WU AND LAWRENCE GROSSMAN

VOLUME 155. Recombinant DNA (Part F)
Edited by RAY WU

VOLUME 156. Biomembranes (Part P: ATP-Driven Pumps and Related Transport: The Na,K-Pump)
Edited by SIDNEY FLEISCHER AND BECCA FLEISCHER

VOLUME 157. Biomembranes (Part Q: ATP-Driven Pumps and Related Transport: Calcium, Proton, and Potassium Pumps)
Edited by SIDNEY FLEISCHER AND BECCA FLEISCHER

VOLUME 158. Metalloproteins (Part A)
Edited by JAMES F. RIORDAN AND BERT L. VALLEE

VOLUME 159. Initiation and Termination of Cyclic Nucleotide Action
Edited by JACKIE D. CORBIN AND ROGER A. JOHNSON

VOLUME 160. Biomass (Part A: Cellulose and Hemicellulose)
Edited by WILLIS A. WOOD AND SCOTT T. KELLOGG

VOLUME 161. Biomass (Part B: Lignin, Pectin, and Chitin)
Edited by WILLIS A. WOOD AND SCOTT T. KELLOGG

VOLUME 162. Immunochemical Techniques (Part L: Chemotaxis and Inflammation)
Edited by GIOVANNI DI SABATO

VOLUME 163. Immunochemical Techniques (Part M: Chemotaxis and Inflammation)
Edited by GIOVANNI DI SABATO

VOLUME 164. Ribosomes
Edited by HARRY F. NOLLER, JR., AND KIVIE MOLDAVE

VOLUME 165. Microbial Toxins: Tools for Enzymology
Edited by SIDNEY HARSHMAN

VOLUME 166. Branched-Chain Amino Acids
Edited by ROBERT HARRIS AND JOHN R. SOKATCH

VOLUME 167. Cyanobacteria
Edited by LESTER PACKER AND ALEXANDER N. GLAZER

VOLUME 181. RNA Processing (Part B: Specific Methods)
Edited by JAMES E. DAHLBERG AND JOHN N. ABELSON

VOLUME 182. Guide to Protein Purification
Edited by MURRAY P. DEUTSCHER

VOLUME 183. Molecular Evolution: Computer Analysis of Protein and Nucleic Acid Sequences
Edited by RUSSELL F. DOOLITTLE

VOLUME 184. Avidin–Biotin Technology
Edited by MEIR WILCHEK AND EDWARD A. BAYER

VOLUME 185. Gene Expression Technology
Edited by DAVID V. GOEDDEL

VOLUME 186. Oxygen Radicals in Biological Systems (Part B: Oxygen Radicals and Antioxidants)
Edited by LESTER PACKER AND ALEXANDER N. GLAZER

VOLUME 187. Arachidonate Related Lipid Mediators
Edited by ROBERT C. MURPHY AND FRANK A. FITZPATRICK

VOLUME 188. Hydrocarbons and Methylotrophy
Edited by MARY E. LIDSTROM

VOLUME 189. Retinoids (Part A: Molecular and Metabolic Aspects)
Edited by LESTER PACKER

VOLUME 190. Retinoids (Part B: Cell Differentiation and Clinical Applications)
Edited by LESTER PACKER

VOLUME 191. Biomembranes (Part V: Cellular and Subcellular Transport: Epithelial Cells)
Edited by SIDNEY FLEISCHER AND BECCA FLEISCHER

VOLUME 192. Biomembranes (Part W: Cellular and Subcellular Transport: Epithelial Cells)
Edited by SIDNEY FLEISCHER AND BECCA FLEISCHER

VOLUME 193. Mass Spectrometry
Edited by JAMES A. MCCLOSKEY

Section I

Phospholipase Assays, Kinetics, and Substrates

[1] Assay Strategies and Methods for Phospholipases

By LAURE J. REYNOLDS, WILLIAM N. WASHBURN, RAYMOND A. DEEMS, and EDWARD A. DENNIS

This chapter reviews the numerous methods currently used to measure phospholipase (PL) activity. We first present a brief overview of some general issues that must be considered when choosing an assay and then briefly describe and compare the specific assays. No details of individual methods will be given; these can be found elsewhere in this volume, in the references cited, or in earlier reviews.[1-4] This discussion focuses heavily on phospholipase A_2 (PLA_2) since it is the best characterized phospholipase.[5] The concepts are applicable, however, to other phospholipases. Some additional considerations for conducting assays in the presence of inhibitors are included at the end.

General Considerations

There are three main issues to consider when selecting a phospholipase assay. First, an appropriate substrate must be chosen from the large number of phospholipids available. Second, since phospholipids form many aggregated structures in water, one must decide which physical form to employ in the assay. And finally, an appropriate detection method must be found that is compatible with both the substrate and its physical form.

Phospholipid Substrates

The general classes of phospholipids and the stereospecific numbering (*sn*) of the glycerol backbone are shown in Fig. 1. Diacyl phospholipids are biosynthetically derived from L-glycerol 3-phosphate and contain two fatty acids esterified to the *sn*-1 and *sn*-2 hydroxyl groups. The *sn*-1 fatty acid is predominantly saturated while unsaturated fatty acids are found

[1] H. van den Bosch and A. J. Aarsman, *Agents Actions* **9,** 382 (1979).

[2] M. Waite, "Handbook of Lipid Research, Volume 5: The Phospholipases," Plenum, New York, 1987.

[3] H. van den Bosch, *in* "Phospholipids" (J. N. Hawthorne and G. B. Ansell, eds.), p. 313. Elsevier Biomedical Press, New York, 1982.

[4] A. A. Farooqui, W. A. Taylor, C. E. Pendley, J. W. Cox, and L. A. Horrocks, *J. Lipid Res.* **25,** 1555 (1984).

[5] E. A. Dennis, *in* "The Enzymes" (P. Boyer, ed.), 3rd Ed., Vol. 16, p. 307. Academic Press, New York, 1983.

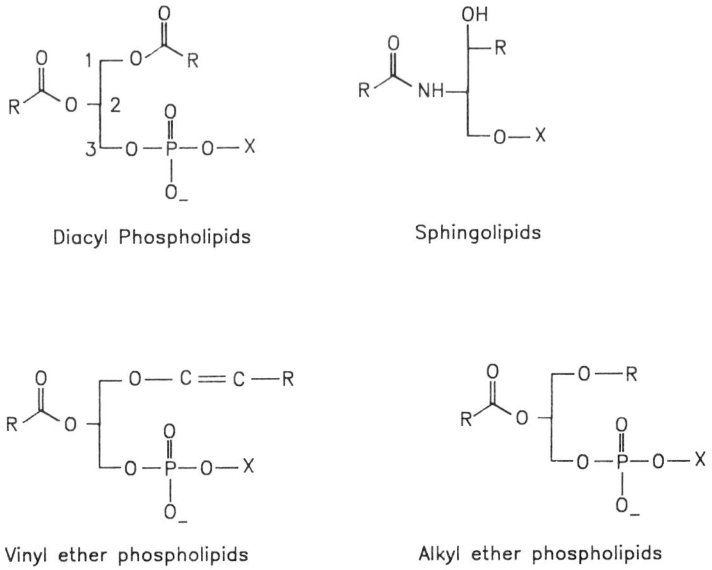

Diacyl Phospholipids

Sphingolipids

Vinyl ether phospholipids

Alkyl ether phospholipids

FIG. 1. Naturally occurring phospholipid and sphingolipid substrates for phospholipases. R represents an alkyl chain, and X represents a polar head group.

predominantly in the *sn*-2 position. Vinyl ether and alkyl ether phospholipids[6] are derived biosynthetically from dihydroxyacetone phosphate and long-chain fatty alcohols. In the vinyl ether phospholipids, also known as plasmalogens, the *sn*-1 alkyl chain is attached to the glycerol backbone via a vinyl ether linkage. The alkyl ether phospholipids have an ether linkage in the *sn*-1 position. Platelet-activating factor (PAF) is the most prominent member of this phospholipid class; it contains an acetate group in the *sn*-2 position. Sphingolipids derive from serine and palmitic acid rather than L-glycerol 3-phosphate. They resemble phospholipids structurally and functionally but contain a hydroxyalkyl chain in place of the *sn*-1 acyl group and a fatty acid amide at what would be the *sn*-2 position. Phospholipids containing a single fatty acid chain are called lysophospholipids and are metabolic products of phospholipid hydrolysis.

The major classes are further subdivided according to head group. In phospholipids, the head group is composed of a phosphate esterified to the glycerol backbone and to one of the following polar moieties: choline, ethanolamine, serine, glycerol, or inositol. The sphingolipids commonly contain phosphorylcholine, glucose, galactose, or other carbohydrates as

[6] F. Paltauf and A. Hermetter, this volume [12].

a head group, attached directly to what would be the *sn*-3 hydroxyl group. The head group strongly influences the physical properties of the phospholipid. Phosphatidylcholine (PC) and phosphatidylethanolamine (PE) are zwitterionic at neutral pH, whereas phosphatidic acid (PA), phosphatidylserine (PS), phosphatidylglycerol (PG), and phosphatidylinositol (PI) are anionic. Phospholipids with PC as a head group will readily form mixed micelles and vesicles; PE is much harder to solubilize. Ca^{2+}, a cofactor for many phospholipases, decreases the solubility of anionic phospholipids and PE.

The specificity of an enzyme often narrows the choice of phospholipids that can be employed in an assay. Most phospholipases specifically hydrolyze only one of the four phospholipid ester bonds. Satisfying the ester bond specificity of the enzyme is usually straightforward, whereas deciding which fatty acid and head group to use is not so easy. While a particular head group or fatty acid may be preferred, most phospholipases will hydrolyze other substrates at slower rates. This lack of specificity makes selecting an appropriate substrate very difficult. It also complicates the comparison of rates obtained with different assay methods or substrates. This lack of specificity does have one beneficial effect; it has made possible the development of assays based on synthetic substrates which contain fluorophores or chromophores.

There are two ancillary concerns that should be addressed when assaying phospholipases. First, phospholipids should be stored and handled with care since they are prone to oxidation, especially those with extensive unsaturation such as arachidonic acid. Phospholipids with a PE or PS head group appear to be more labile than those with PC.[7] The decomposition of substrate can be guarded against by storage in the cold and dark, under nitrogen, and in radical quenching solvents such as toluene and ethanol. While phospholipids containing saturated fatty acids are more stable, they are also often less soluble. At normal assay temperatures, saturated phospholipids may also undergo thermotropic phase transitions which can affect enzyme activity.

Second, all cells contain the basic repertoire of lipolytic enzymes required to maintain phospholipid metabolism. Because crude enzyme preparations contain many of these enzymes, similar products can often be produced by several different routes. For example, radiolabeled fatty acid can be liberated from the *sn*-2 position of a phospholipid by a phospholipase A_2, by the combined action of a phospholipase A_1 and a lysophospholipase, or by the action of a phospholipase C and a lipase. This ambiguity

[7] G. Rouser, G. Kritchevsky, A. Yamamoto, G. Simon, C. Galli, and A. J. Bauman, this series, Vol. 14, p. 291.

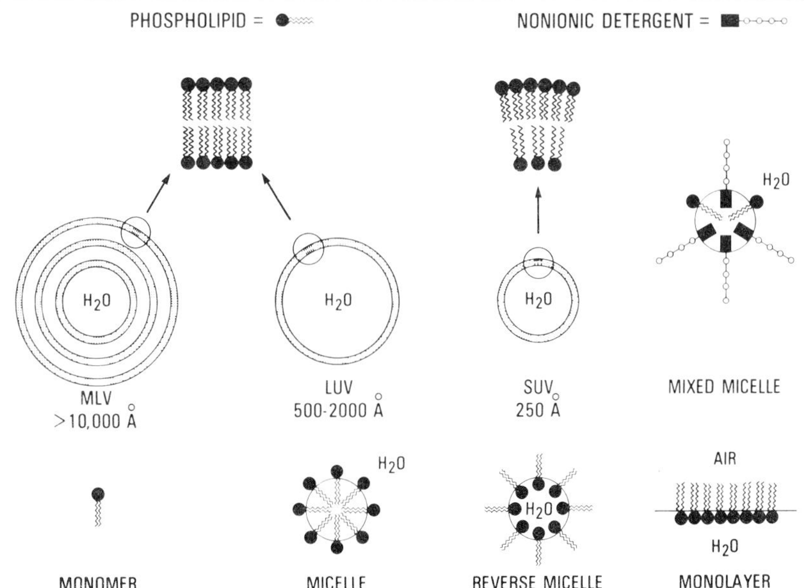

FIG. 2. Phospholipid structures commonly used in phospholipase assays. Multilamellar vesicles (MLV) and large unilamellar vesicles (LUV) have such large diameters that the outer surface is essentially flat at the molecular level. For small unilamellar vesicles (SUV), however, the outer surface is highly curved. In the presence of detergent, the phospholipids are solubilized into mixed micelles, as illustrated for the nonionic detergent Triton X-100. Synthetic phospholipids with short fatty acid chains and lysophospholipids are monomers at concentrations below their CMC and micelles at concentrations above. In certain mixed solvent systems, phospholipids form reverse micelles or microemulsions with a small amount of water in the center. Phospholipids also form monolayers at air–water interfaces. (Reprinted with permission from Ref. 5.)

precludes one from readily determining the positional specificity of an enzyme in such preparations.

Substrate Physical Forms

Phospholipase assays are beset by several unique problems. The most important stem from the insolubility of the substrate in water. Because of this insolubility, phospholipids exist in many different aggregated forms in water (Fig. 2). Studies of the extracellular phospholipases have shown that the rate of hydrolysis depends not only on the particular substrate used but also on the physical state of that phospholipid. For example, some of these enzymes hydrolyze aggregated substrates faster than monomeric ones, with the monomeric rates usually being quite low.[5] Each physical form has its advantages and disadvantages, and there is no one best form. Their characteristics must be matched to the requirements of the assay and the overall experiment.

Synthetic short-chain phospholipids and lysophospholipids exist either as monomers or micelles. At concentrations below their critical micelle concentration (CMC) they are monomeric, while above their CMC they aggregate into micelles. The CMC of a particular phospholipid decreases with increasing fatty acid chain length. For example, dihexanoyl-PC has a CMC of about 10 m*M*[8] whereas diheptanoyl-PC has a CMC of about 2 m*M*.[9,10] The CMC values of the corresponding lysophospholipids are much higher. Thus, at assay concentrations, a significant portion of the lysophospholipid is usually monomeric, and this must be taken into account when determining the actual micelle concentration.[11] Because of their long fatty acid chains, natural phospholipids have negligible monomer concentrations, estimated to be 10^{-10} *M* for dipalmitoyl-PC.[12]

Phospholipids with long fatty acid chains will not readily form micelles, but they can be solubilized into mixed micelles with detergents.[13] The most commonly used detergent for this purpose has been the uncharged Triton X-100.[14] Deoxycholate has also been used; however, the anionic nature of this detergent makes the micelle highly sensitive to Ca^{2+}, pH, and ionic strength. This complex system is difficult to characterize kinetically.[15] Long-chain phospholipids also form vesicles and liposomes. The vesicles can be either small unilamellar, large unilamellar, or multilamellar. These structures have the bilayer organization of natural membranes and can serve as membrane models. Phospholipid packing and conformation in these structures are highly dependent on the particular phospholipid used and on its thermotropic phase transition.

Vesicles and mixed micelles are the two phospholipid forms most commonly employed in phospholipase assays. They each bring distinctly different advantages and disadvantages to an assay. While vesicles do serve as a membrane model, their phospholipid packing and phase transition characteristics are a disadvantage since these two factors have been

[8] A. Plückthun and E. A. Dennis, *J. Phys. Chem.* **85,** 678 (1981).

[9] R. J. M. Tausk, J. Karmiggelt, C. Oudshoorn, and J. T. G. Overbeek, *Biophys. Chem.* **1,** 175 (1974).

[10] J. H. van Eijk, H. M. Verheij, R. Dijkman, and G. H. de Haas, *Eur. J. Biochem.* **132,** 183 (1983).

[11] R. E. Stafford, T. Fanni, and E. A. Dennis, *Biochemistry* **28,** 5113 (1989).

[12] R. Smith and C. Tanford, *J. Mol. Biol.* **67,** 75 (1972).

[13] D. Lichtenberg, R. J. Robson, and E. A. Dennis, *Biochim. Biophys. Acta* **737,** 285 (1983).

[14] E. A. Dennis, *Arch. Biochem. Biophys.* **158,** 485 (1973).

[15] F. F. Davidson and E. A. Dennis, submitted for publication (1991).

shown to dramatically affect enzymatic activity.[16–18] This presents a problem when comparing the hydrolysis rates for different phospholipids. It is difficult to determine if an apparent preference of an enzyme for one phospholipid over another represents the true specificity of the enzyme or is simply due to the different phospholipid packing or phase transition temperatures of the two substrates. These factors can also be a problem when performing assays containing other surface active agents, such as inhibitors. In mixed micelles, the presence of excess detergent prevents the micelle structure from being greatly affected by changes in phospholipid or by the introduction of inhibitors.[19] This offers an important advantage for specificity and inhibition studies. Vesicles and mixed micelles also differ in their ability to exchange phospholipids. On an assay time scale, the rate of exchange of phospholipids and products into or out of vesicles is negligible, while exchange between mixed micelles occurs readily.[20] The choice between the vesicle or mixed micelle systems for a phospholipase assay depends on the purpose of the assay and on the preference of the particular enzyme for one form or the other.

In addition to the physical forms listed above, purified phospholipids can also be incorporated into reverse micelles and monolayers. Reverse micelles are formed in mixed solvent systems such as ether-water and have been used for a phospholipase assay,[21] but the system is not well defined and its use is uncommon. Phospholipid monolayers will form at an air–water interface. In this system, phospholipase activity depends on surface pressure and hydrolysis can be monitored using a variety of specialized methods.[22]

The search for an assay system which more closely resembles the natural state has led some researchers away from purified phospholipid systems to the use of radiolabeled, autoclaved *Escherichia coli*.[23] This substrate form has proved useful, especially for the purification of mammalian nonpancreatic phospholipases A_2.[24,25] However, this substrate has limited use for kinetic and inhibitor studies, since the composition of the

[16] C. R. Kensil and E. A. Dennis, *J. Biol. Chem.* **254,** 5843 (1979).

[17] C. R. Kensil and E. A. Dennis, *Lipids* **20,** 80 (1985).

[18] J. A. F. Op den Kamp, M. T. Kauerz, and L. L. M. van Deenen, *Biochim. Biophys. Acta* **406,** 169 (1975).

[19] A. A. Ribeiro and E. A. Dennis, *Chem. Phys. Lipids* **14,** 193 (1975).

[20] J. W. Nichols, *Biochemistry* **27,** 3925 (1988).

[21] R. L. Misiorowski and M. A. Wells, *Biochemistry* **13,** 4921 (1974).

[22] S. Ransac, H. Moreau, R. Rivière, and R. Verger, this volume [4].

[23] P. Elsbach and J. Weiss, this volume [2].

[24] R. M. Kramer, C. Hession, B. Johansen, G. Hayes, P. McGray, E. P. Chow, R. Tizard, and R. B. Pepinsky, *J. Biol. Chem.* **264,** 5768 (1989).

[25] R. M. Kramer and R. B. Pepinsky, this volume [35].

membrane is a complex mixture of phospholipids and the concentration of the substrate in the assay is extremely low.[26]

Detection Methods

The most important decision to make in choosing a phospholipase assay is the detection method. A number of methods have been developed (Table I). The choice between these detection methods depends partly on the purpose of a particular experiment. For example, some assays can be used on purified enzymes but are incompatible with crude systems, some methods provide a continuous assay and generate a time course while others do not, and some methods are amenable to automation while others are not. However, the most important consideration in the choice of a detection method is the sensitivity required for a particular enzyme. The required sensitivity depends on the quantity of enzyme available and on its specific activity. This point is especially important for the assay of mammalian nonpancreatic phospholipases A_2, which are found in lower quantities and are, in general, less active than their counterparts from pancreas or venom.

The sensitivity of an assay is influenced by a number of factors, the most important of which is the detection limit of a particular method. The detection limit is determined by the physical properties of the assay method itself such as the extinction coefficient of a chromophore or the physical limitations of the instrumentation. Table I lists the estimated detection limits for the various assay methods. When comparing assays, it is important to recognize that the detection limit alone does not necessarily reflect the sensitivity of an assay for a particular enzyme. For example, based on the detection limits, one would expect the thio assay to be more sensitive than the pH-stat assay. In our experience with cobra venom PLA_2,[27] however, the thio assay with didecanoylthio-PC (substrate **1**, Fig. 3) has approximately the same sensitivity as the pH-stat assay with dipalmitoyl-PC. Both assays are routinely run with the same amounts of enzyme. The apparent disparity can be accounted for by the lower V_{max} of the cobra enzyme with the thio substrate and by a difference in assay temperature and volumes. If another phospholipase were to hydrolyze both substrates equally well, the thio assay would indeed be much more sensitive. This is a more serious problem with many of the fluorescent assays, where a method with an intrinsically low detection limit loses much of its sensitivity because it employs a poor substrate. Thus, in

[26] F. F. Davidson, E. A. Dennis, M. Powell, and J. R. Glenney, *J. Biol. Chem.* **262,** 1698 (1987).

[27] L. J. Reynolds and E. A. Dennis, this volume [33].

TABLE I
PHOSPHOLIPASE ASSAYS

Method	Detection equipment	Detection limit[a]	Continuous monitoring[b]	Natural substrate[c]
Titrametric	pH stat	20 nmol	+	+
Acidimetric	pH meter	100 pmol	+	+
Radiometric	Scintillation counter			
^3H		1 fmol	−	−
^{14}C		1 pmol	−	−
E. coli		—	−	−
NMR	NMR spectrometer			
^1H		1 μmol	−	+
^{13}C		1 μmol	−	−
^{31}P		1 μmol	−	+
Monolayer	Monolayer trough	—	+	+
Polarographic	Polarograph	40 nmol	+	+
Spectrophotometric	Spectrophotometer			
Thio		1 nmol	+	−
SIBLINKS		200 pmol	+	−
Dye release		200 pmol	+	−
Indicator dye		10 nmol	+	+
Phosphate		1 nmol	−	+
Fatty acid soaps		25 nmol	−	+
CoA-coupled		5 nmol	−	+
Plasmalogen coupled		200 pmol	+	+
Turbidometric		—	+	+
Hemolytic		—	−	+
Fluorometric	Fluorometer	1 pmol	±	−
ESR	ESR spectrometer	1 nmol	+	−

[a] These limits are approximate and are given to indicate the general sensitivity of the assays for comparison purposes only. For continuous assays, the detection limits given are per minute. No adjustments have been made for differences in assay volumes.

[b] All assays can be run to a fixed time point; (+) indicates assays which can also be continuously monitored.

[c] (+) indicates assays which can generally use natural substrates; (−) indicates substrates that require special synthesis.

addition to the detection limit, the sensitivity of an assay depends on the turnover rate of the particular enzyme, the rate of the background reaction, and, in some cases, on the assay time and volume. Further comments on the sensitivity of the individual assays can be found below.

Specific Assays

This section provides a brief description of some of the detection methods available for phospholipase assays. Included are general advantages and disadvantages of each method as well as information on the sensitivity of the assay, whether it is continuous or not, and the occurrence of any important side reactions.

Titrametric

The titrametric (pH-stat) assay has been one of the two most frequently used phospholipase assays and is the workhorse in studies of the extracellular enzymes.[28-30] The assay is relatively straightforward and requires relatively inexpensive equipment. In this assay, the pH is held constant by titrating the liberated fatty acid. Enzymatic activity is followed by observing the amount of base consumed. This is a continuous assay that can detect the hydrolysis of any substrate, natural or synthetic. The detection limit is 20–100 nmol/min, which is not sensitive enough to detect mammalian intracellular phospholipases before purification. This assay does not detect a reaction product directly, but rather the proton released upon fatty acid ionization. Thus, one must be aware of any factors which could contribute to an alteration of the free hydrogen ion concentration. The extent of ionization depends on the pK_a of the fatty acid and the pH of its immediate environment. Assays cannot be performed at acidic pH values because the fatty acid will not be sufficiently ionized. This method has also been used to assay phospholipases C[31] and D.[32]

Acidimetric

The acidimetric assay also detects the protons released from the liberated fatty acid.[33] It does so by measuring the change in pH with time. This is also a continuous assay that can detect the hydrolysis of any substrate. The assay has a potential sensitivity of 100 pmol/min at pH 8.0. However, there are several serious drawbacks to this assay. The first is that the pH of the assay varies continuously during the assay. The second is that the change in pH produced by the release of a given number of protons critically depends on the presence of any buffers, which includes the substrate and enzyme. The third is that the sensitivity is determined by the starting pH, and, thus, it also varies during the time course.

[28] E. A. Dennis, *J. Lipid Res.* **14,** 152 (1973).

[29] R. A. Deems and E. A. Dennis, this series, Vol. 71 [81].

[30] W. Nieuwenhuizen, H. Kunze, and G. H. de Haas, this series, Vol. 32 [15].

[31] R. F. A. Zwaal, B. Roelofsen, P. Comfurius, and L. L. M. van Deenen, *Biochim. Biophys. Acta* **233,** 474 (1971).

[32] T. T. Allgyer and M. A. Wells, *Biochemistry* **18,** 5348 (1979).

[33] J. I. Salach, P. Turini, R. Seng, J. Hauber, and T. P. Singer, *J. Biol. Chem.* **246,** 331 (1971).

Radiometric

The radioactive assay is the most sensitive and most widely used phospholipase assay. It requires the use of synthetic, radiolabeled phospholipids, many of which are available commercially and can be quite expensive. The sensitivity of the assay depends on the specific radioactivity of the labeled substrate. Typical specific activities for commercially available phospholipids are 50–100 Ci/mmol for ^3H and 50–100 mCi/mmol for ^{14}C, which correspond to detection limits of about 1 fmol and 1 pmol, respectively. ^{32}P-Labeled phospholipids are generally not commercially available, and there is seldom any need for the added sensitivity that the use of ^{32}P would yield. The radioactive assay follows hydrolysis by directly measuring the liberation of one of the hydrolysis products. The assay is discontinuous and requires the separation of the radioactive substrate from the labeled products by thin-layer chromatography (TLC), high-performance liquid chromatography (HPLC), solvent extraction, or centrifugation as discussed below. These separations are laborious and time consuming and are a major disadvantage of the assay.

TLC Assay. Thin-layer chromatography is by far the most commonly used separation technique.[1,34,35] Various modifications of the Bligh and Dyer technique,[36] which uses a chloroform, methanol, and acetic acid extraction, have been used to extract the phospholipids from the aqueous assay mixture. In assays of phospholipases C and D, unreacted starting material and the diacylglycerol or diacylglycerol phosphate products are extracted into the organic phase while the head group product remains in the aqueous phase. If the substrate is radiolabeled in the head group, then the extent of reaction can be determined by counting an aliquot of the aqueous phase. In phospholipase A assays, all of the reactants and products extract into the organic phase. The organic phase is then dried. The lipids are redissolved in a small volume of chloroform and methanol, spotted on a silica gel plate, and eluted. The various spots are then scraped into scintillation fluid and counted. An important advantage of the TLC assay is that the reactants and products can be identified and quantitated. From this information, accurate ratios of products and reactants are obtained. These ratios can then be used to calculate the actual number of moles of substrate hydrolyzed per unit time. If all of the compounds are extracted equally, any losses in the extraction will affect all components equally. Therefore, even if the absolute amounts of the components vary,

[34] M. D. Lister, R. A. Deems, Y. Watanabe, R. J. Ulevitch, and E. A. Dennis, *J. Biol. Chem.* **263,** 7506 (1988).

[35] Y. Y. Zhang, R. A. Deems, and E. A. Dennis, this volume [43].

[36] E. G. Bligh and W. J. Dyer, *Can. J. Biochem. Physiol.* **37,** 911 (1959).

the ratio of the components, as well as the rate calculated from it, will be unaffected.

Dole Assay. A second radioactive assay is based on the Dole extraction.[37] In this assay,[1,35,38] the lipids are extracted using 2-propanol, heptane, and sulfuric acid. The free fatty acid extracts into the heptane layer while the majority of the phospholipid and lysophospholipid remains in the aqueous layer. An aliquot of the heptane layer is counted. A small amount of phospholipid and lysophospholipid is also extracted into the heptane layer and can interfere with the fatty acid counts. For this reason, silicic acid is added to the extraction mixture to remove the remaining phospholipid and lysophospholipid from the heptane phase. Some assays require a silicic acid column to remove the phospholipid.[1] While this technique is faster and more convenient than the TLC assay, it does suffer one serious drawback. The counts in the heptane layer are not identified; therefore, any phospholipid or lysophospholipid that is extracted cannot be distinguished from fatty acid. Also, a loss of fatty acid into the aqueous layer cannot be detected. The efficiencies of the extractions must be reanalyzed when any change in the assay is made. This is often the case in inhibition studies since many inhibitors are surface active and can alter the extractions.

When using the radioactive assay, we prefer the TLC separation assay over the Dole extraction assay for accuracy in kinetic experiments, although we often employ the Dole extraction when monitoring column fractions during enzyme purification.

Escherichia coli Assay. The *E. coli* assay is another frequently used radioactive assay.[23,25] In this assay, *E. coli* membranes containing radiolabeled phospholipids are incubated with the enzyme. At the end of the assay, the membranes are centrifuged out of the assay mixture, leaving the products in solution. An aliquot of the supernatant is counted. This procedure is much faster than either of the methods described above. However, disadvantages are that the phospholipid composition is controlled by the *E. coli* and cannot be altered, the amount of phospholipid substrate present in an assay is extremely low,[26] and, again, the radioactivity in the supernatant is not identified.

Nuclear Magnetic Resonance

While nuclear magnetic resonance (NMR) can be used to follow phospholipase hydrolysis,[39] the very low sensitivity of this method prevents it from being employed as an assay in most instances. Of the various nuclei

[37] V. P. Dole, *J. Clin. Invest.* **35,** 150 (1956).

[38] H. van den Bosch, J. G. N. de Jong, and A. J. Aarsman, this volume [34].

[39] M. F. Roberts, this volume [3].

that can be followed, ^{31}P and ^{13}C have been most commonly used. In most cases, there are no appropriate proton resonances that can be easily followed. Natural abundance ^{31}P can be used where the ^{31}P signals of the product and reactants are sufficiently separated.[40] Substrate concentrations above 10 mM are generally required to obtain a sufficient signal, and 10% hydrolysis is required to have enough product to accurately quantitate. The concentration of product must be 1 mM or greater, or roughly 0.3 μmol per assay. Because of the natural abundance of ^{31}P, natural substrates can be used and synthesis of substrates is not required. ^{13}C has a low sensitivity and natural abundance; thus, its levels in the substrate must generally be enhanced.[41] Even then, its detection limit is not quite as good as for ^{31}P.

Monolayer Assay

In the monolayer assay, phospholipid monolayers are formed in a monolayer trough. Phospholipase activity can be followed by measuring the decrease in the surface area of phospholipid which is required to maintain a constant surface pressure.[22] The assays are extremely sensitive and can utilize natural substrates. Because very low concentrations of substrate are used, it is difficult to study enzymes near or above the K_m values.[14] This assay requires specialized equipment not generally available in most laboratories.

Polarographic Assay

A continuous polarographic assay for phospholipase A_2 has been developed using a coupled reaction with soybean lipoxygenase.[42] The technique uses unsaturated phospholipids as substrates and monitors oxygen uptake upon oxidation of the fatty acid products by lipoxygenase. This assay requires that either linoleic or arachidonic acid occupy the *sn*-2 phospholipid position, but egg yolk PC, which contains these two fatty acids, can be used as a substrate. The sensitivity of the assay is comparable to that of the pH-stat assay, and rates of approximately 40 nmol O_2 uptake/min have been observed.

Spectrophotometric Assays

Spectrophotometric assays, in general, are desirable because they are rapid, convenient, often continuous, and can be adapted to assay large numbers of samples. There have been many attempts to develop a good

[40] A. Plückthun and E. A. Dennis, *Biochemistry* **21,** 1750 (1982).

[41] S. P. Bhamidipati and J. A. Hamilton, *Biochemistry* **28,** 6667 (1989).

[42] P. H. Gale and R. W. Egan, *Anal. Biochem.* **104,** 489 (1980).

spectrophotometric assay for PLA_2 using a variety of approaches.[4] The best spectrophotometric assays to date utilize synthetic phospholipid substrates that generate a chromophore upon hydrolysis. Other spectrophotometric assays with more limited applicability have been developed which follow the pH change during the reaction, quantitate the amount of products formed using coupled reactions, follow the decrease in turbidity of a phospholipid solution upon hydrolysis, or measure heme release from red blood cells. The latter two phenomena are difficult to quantitate in terms of the amount of phospholipid hydrolyzed, so their detection limits have not been included in Table I.

Thio Assay. Phospholipases are capable of cleaving thio ester bonds in addition to oxy ester bonds. This property has been utilized to develop continuous spectrophotometric assays for PLA_2,[43,44] PLA_1,[45] phospholipase C (PLC),[46] lysophospholipase,[47] and lipase.[47] A number of thiophospholipids have been synthesized as substrates, including stereospecific didecanoylthio-PC, substrate **1** (see Fig. 3)[43,48,49] and its PE analog. *rac*-Dihexanoylthio-PC, a monomeric substrate, is also used in an assay.[44] A lengthy synthesis is required for the preparation of these substrates. The assay detects the free thiol which is formed on hydrolysis of the thio ester bond using either 5,5′-dithiobis(2-nitrobenzoic acid) or 4,4′-dithiobispyridine. The extinction coefficients of these thiol reagents vary with pH.[43,50] The assay is incompatible with free thiols or any other substance that would significantly reduce the diaryl disulfide bond in the thiol reagent and, as a consequence, is generally limited to the measurement of pure phospholipases. The thio assay is more convenient and more reproducible than the pH-stat assay and has a detection limit of about 1 nmol/min. The K_m for the thio ester substrates is lower than that observed for normal oxy esters.[1] Substrate **1** binds approximately 5-fold more tightly than natural PC, thereby reducing the amount of substrate required per assay.[48,51]

SIBLINKS Assay. Another approach to developing a spectrophotometric assay is to utilize a synthetic substrate which releases a chromo-

[43] L. Yu and E. A. Dennis, this volume [5].

[44] J. J. Volwerk, A. G. R. Dedieu, H. M. Verheij, R. Dijkman, and G. H. de Haas, *Recl. Trav. Chim. Pays-Bas* **98,** 214 (1979).

[45] G. L. Kucera, C. Miller, P. J. Sisson, R. W. Wilcox, Z. Wiemer, and M. Waite, *J. Biol. Chem.* **263,** 12964 (1988).

[46] J. W. Cox, W. R. Snyder, and L. A. Horrocks, *Chem. Phys. Lipids,* **25,** 369 (1979).

[47] A. J. Aarsman, L. L. M. van Deenen, and H. van den Bosch, *Bioorg. Chem.* **5,** 241 (1976).

[48] H. S. Hendrickson and E. A. Dennis, *J. Biol. Chem.* **259,** 5734 (1984).

[49] L. Yu, R. A. Deems, J. Hajdu, and E. A. Dennis, *J. Biol. Chem.* **265,** 2657 (1990).

[50] D. R. Grassetti and J. F. Murray, *Arch. Biochem. Biophys.* **119,** 41 (1967).

[51] R. A. Deems, B. R. Eaton, and E. A. Dennis, *J. Biol. Chem.* **250,** 9013 (1975).

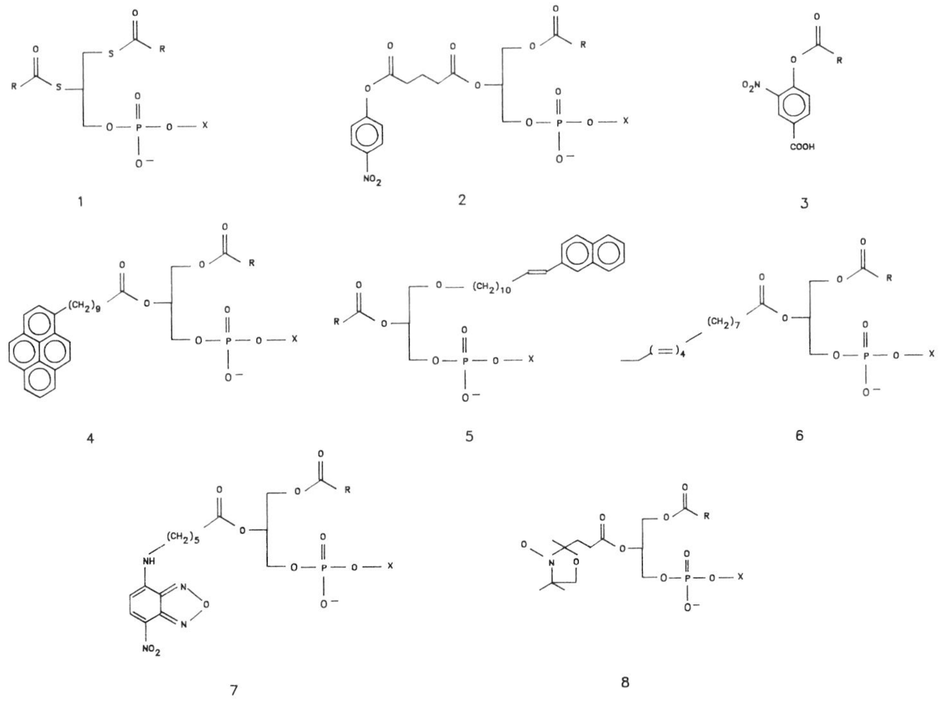

FIG. 3. Synthetic phospholipid substrates **(1–8)**. R represents an alkyl chain, and X represents a polar head group, usually choline.

phoric dye on enzymatic hydrolysis. In general, phospholipids which contain a dye linked directly to the *sn*-2 carbonyl are not hydrolyzed by phospholipases A_2. The SIBLINKS substrate **(2)** contains *p*-nitrophenol linked to the *sn*-2 position of a phospholipid substrate *via* a glutaric acid moiety. This compound *is* a substrate.[52] Hydrolysis of the *sn*-2 ester generates a free acid which then cyclizes and eliminates *p*-nitrophenoxide. The *p*-nitrophenoxide is easily detected spectrophotometrically. The reaction also generates a reactive anhydride product which diffuses away from the enzyme and is quenched by water. This substrate is easily prepared from commercially available compounds and provides a continuous, thiol-compatible assay with a detection limit of about 200 pmol/min. As with the thiol assay, the SIBLINKS substrate binds more tightly than normal PC substrates so less substrate is required to saturate the enzyme. The maximal velocity of cobra venom PLA_2 on this substrate is less than

[52] W. N. Washburn and E. A. Dennis, *J. Am. Chem. Soc.* **112,** 2040 (1990).

that observed for dipalmitoyl-PC, but K_{cat}/K_m is similar.[52] This assay is incompatible with strong nucleophiles such as hydroxylamine and with enzyme preparations containing general esterase activity.

Dye Release Assays. Two other assays have been developed which rely on the release of a chromophoric dye from the substrate upon enzymatic hydrolysis. 3-Nitro-4-(octanoyloxy)benzoic acid **(3)**[53,54] will act as a substrate for phospholipase A_2 even though the compound bears little resemblance to a phospholipid. Hydrolysis by phospholipase A_2 yields a *p*-nitrophenol product which is easily detected. The substrate is monomeric, and most phospholipases act poorly on monomeric substrates. Assays cannot be performed at substrate concentrations above 100 μM since irreversible acylation of the enzyme occurs. This substrate is synthesized in one step from commercially available material. The detection of a *p*-nitrophenol product is also the basis of a phospholipase C assay which uses *p*-nitrophenylphosphorylcholine as a substrate.[55] The K_m for this substrate is very high (0.2 *M*), and the assay requires large amounts of sorbitol or glycerol in order to observe adequate hydrolysis rates.

Indicator Dye Assay. The indicator dye assay utilizes a colorimetric pH indicator to monitor the change in pH due to the liberation of free fatty acids.[56] The choice of dye is critical to this assay since hydrophobic dyes inhibit some phospholipases[56,57] and the pK_a of the dye may change if it is incorporated into a micellar substrate.[56] With phenol red as an indicator, the assay is approximately as sensitive as the pH-stat assay. The assay is continuous, uses natural substrates, and can be useful for rapid screening of large numbers of samples. One disadvantage of the assay is that the indicator dyes have a high absorbance, and the method requires monitoring small differences between these large numbers.

Phosphate Assay. When radiolabeled phospholipids are unavailable, the TLC assay can be used in combination with a phosphate assay to determine the levels of phospholipid and lysophospholipid in an assay mix. After extraction of the phospholipid from the silica, the phosphate concentration can be determined using a number of different methods.[58,59] This method of assaying phospholipases is discontinuous and quite laborious, but the overall sensitivity of the assay is good. With the Bartlett

[53] W. Cho, M. A. Markowitz, and F. J. Kezdy, *J. Am. Chem. Soc.* **110,** 5166 (1988).

[54] W. Cho and F. J. Kézdy, this volume [6].

[55] S. Kurioka and M. Matsuda, *Anal. Biochem.* **75,** 281 (1976).

[56] A. L. de Araujo and F. Radvanyi, *Toxicon* **25,** 1181 (1987).

[57] R. E. Barden, P. L. Darke, R. A. Deems, and E. A. Dennis, *Biochemistry* **19,** 1621 (1980).

[58] G. R. Bartlett, *J. Biol. Chem.* **234,** 466 (1959).

[59] P. S. Chen, T. Y. Toribara, and H. Warner, *Anal. Chem.* **28,** 1756 (1956).

technique, as little as 1 nmol of phosphate can be determined.[60] PLC can be assayed without a TLC separation by quantitating the amount of phosphate in the aqueous phase after a Bligh and Dyer extraction.

Assay of Fatty Acid Soaps. The concentration of free fatty acids in an assay can be determined using a discontinuous colorimetric assay based on the formation of Co^{2+} soaps.[1,61,62] Fatty acids are extracted from the assay mix using the Dole extraction procedure then incubated with an aqueous cobalt nitrate solution. The cobalt-fatty acid soaps are extracted into an organic phase, and the quantity of cobalt present is determined by a colorimetric reaction with 1-nitroso-2-naphthol. The detection limit for this assay is approximately 25 nmol fatty acid.[62] The sensitivity of the procedure can be improved by using radioactive ^{60}Co or ^{63}Ni.[63,64] The presence of phospholipids in the Dole extract can interfere with the assay.[65]

CoA-Coupled Assay. The amount of fatty acid product in an assay can also be quantitated in a spectrophotometric assay by coupling the reaction to coenzyme A (CoA) metabolism.[66,67] Fatty acids are incorporated into acyl-CoA using acyl-CoA synthetase then oxidized to 2,3-*trans*-enoyl-CoA with acyl-CoA oxidase. The latter step also generates peroxide as a product which, in the presence of peroxidase, leads to the formation of a chromophore by oxidative coupling of 4-aminoantipyrine to either 3-methyl-*N*-ethyl-*N*-(β-hydroxyethyl)aniline or 2,4,6-tribromo-3-hydroxybenzoic acid. Commercial reagent kits are available for this assay. The procedure is discontinuous and has been adapted for automation using clinical samples.[67] This method can detect down to 5 nmol of fatty acid.[66]

Plasmalogen-Coupled Assay. Phospholipase A_2 activity toward plasmalogen substrates can be followed spectrophotometrically in another coupled assay.[68,69] Hydrolysis of plasmalogens by PLA_2 yields a lysoplasmalogen product. This product is in turn hydrolyzed by lysoplasmalogenase to glycerophosphocholine and a free aldehyde. The free aldehyde is reduced by alcohol dehydrogenase with concomitant oxidation of NADH. The reaction is followed spectrophotometrically by watching the disap-

[60] R. Asmis and E. A. Dennis, unpublished observations (1990).
[61] M. Novak, *J. Lipid Res.* **6,** 431 (1965).
[62] M. K. Bhat and T. V. Gowda, *Toxicon* **27,** 861 (1989).
[63] R. J. Ho and H. C. Meng, *Anal. Biochem.* **31,** 426 (1969).
[64] R. J. Ho, *Anal. Biochem.* **36,** 105 (1970).
[65] J. W. DePierre, *Anal. Biochem.* **83,** 82 (1977).
[66] J. Kasurinen and T. Vanha-Perttula, *Anal. Biochem.* **164,** 96 (1987).
[67] G. E. Hoffmann and U. Neumann, *Klin. Wochenschr.* **67,** 106 (1989).
[68] M. S. Jurkowitz-Alexander, Y. Hirashima, and L. A. Horrocks, this volume [7].
[69] Y. Hirashima, M. S. Jurkowitz-Alexander, A. A. Farooqui, and L. A. Horrocks, *Anal. Biochem.* **176,** 180 (1989).

pearance of NADH. This assay is rapid and continuous and can detect to 200 pmol/min/ml. Purification of the lysoplasmalogenase is required.

Turbidometric Assay. Phospholipase A_2 will hydrolyze the phospholipids in an egg yolk suspension, resulting in a decrease in the turbidity of the solution. The egg yolk clearing can be followed spectrophotometrically and used to assay PLA_2 activity.[70] The assay is continuous, rapid, simple, and inexpensive, but it is not very sensitive. The assay is also not quantitative and provides relative activities rather than moles of lipid hydrolyzed. The turbidity of the solution is very dependent on the particular conditions used because pH, certain metal ions, and other agents can affect the solubility of the substrate. This assay has also been performed using a multibilayer liposome substrate.[71]

Hemolytic Assay. A discontinuous hemolytic assay has been developed which is based on the phospholipase A_2-induced hemolysis of guinea pig erythrocytes.[72] The hemolysis is mediated by cardiotoxin and lipoproteins. As with the turbidometric assay, the assay provides information on relative enzyme activity rather than moles of lipid hydrolyzed. The assay is approximately as sensitive as the pH-stat assay but less time consuming, and it can be useful for rapid screening of large numbers of samples. Utilization of the hemolytic assay is restricted to samples with high levels of phospholipase activity and to situations where additional assay components will not affect the hemolytic process.

Fluorometric Assays

Fluorescence assays are second to the radiometric assays in sensitivity and have been used to assay even the low levels of phospholipase activities found in intracellular preparations.[73,74] The assays utilize synthetic phospholipid substrates which are labeled on one of the fatty acid chains with a fluorophore. Fluorescence assays, in general, have been discontinuous and, like the radioactive assays, rely on phase separation, HPLC, or TLC to separate substrate and products. A few continuous fluorescence assays have also been developed. In a continuous assay, the fluorescence of the substrate must be distinguishable from that of the product. This may or may not require a partitioning of the fluorescent product in a separate phase from the substrate. The conditions to achieve this vary for different

[70] G. V. Marinetti, *Biochim. Biophys. Acta* **98,** 554 (1965).

[71] C. Vigo, G. P. Lewis, and P. J. Piper, *Biochem. Pharmacol.* **29,** 623 (1980).

[72] C.-W. Vogel, A. Plückthun, H. J. Muller-Eberhard, and E. A. Dennis, *Anal. Biochem.* **118,** 262 (1981).

[73] H. S. Hendrickson, E. K. Hendrickson, and T. J. Rustard, *J. Lipid Res.* **28,** 864 (1987).

[74] F. Radvanyi, L. Jordan, F. Russo-Marie, and C. Bon, *Anal. Biochem.* **177,** 103 (1989).

substrates and may require either the addition or exclusion of albumin and other lipid-binding proteins which can pull labeled fatty acids out of the vesicles into solution. One should be aware that the addition of any inhibitors or proteins to the assay may affect the partitioning of the products.

Several fluorescent substrates have been developed which contain a pyrene fluorophore on the *sn*-2 fatty acid.[74–78] Many of these substrates can be used to measure both phospholipase A_1 and A_2 activity. At present, a fluorescence assay utilizing substrate **4** with albumin present provides the most sensitive continuous assay.[74] Assays based on pyrene utilize the high fluorescence of monomeric pyrene at 398 nm to obtain sensitivities as low as 1 pmol/ml of product. All of the pyrene-containing substrates reach half-maximum velocity around 1 μM. They all have poor turnover numbers with the exception of substrate **4,** whose specific activity with cobra venom PLA_2 is about 300 μmol/min/mg. For the pyrene-based assays, the critical issues are suppression of background fluorescence of the starting phospholipid and prevention of quenching of the hydrolysis products due to eximer formation. A two-step synthesis is required to prepare any of these substrates from the appropriate lysophospholipid and ω-pyreno-substituted carboxylic acid. Not all of the latter acids are commercially available.

Three other non-pyrene-based fluorescence assays have been reported. An exceedingly sensitive (1 pmol/min) discontinuous assay requiring HPLC separation of the products utilizes the vinylnaphthalene fluorophore of substrate **5.**[73] The ether linkage at the *sn*-1 position makes this substrate resistant to phospholipase A_1 and lysophospholipase activities. The vinylnaphthalene group is less perturbing than the pyrene fluorophore. Consequently, **5** is a good substrate with activities on the cobra and pancreatic enzymes comparable to that observed for natural phospholipids. The products can also be separated by TLC and quantitated by fluorescence scanning of the TLC plate.[79,80] With the TLC procedure, activities as low as 10 pmol/min can be measured. At present, this substrate is not commercially available and requires a lengthy synthesis. The substrate has also been synthesized with dansyl and dabsyl groups in place of the vinylnaphthalene.[79,80] Another fluorescence assay utilizes the change in the hyperpolar-

[75] T. Thuren, J. A. Virtanen, P. Vainio, and P. K. J. Kinnunen, *Chem. Phys. Lipids* **33,** 283 (1983).

[76] T. Thuren, J. A. Virtanen, P. J. Somerharju, and P. K. J. Kinnunen, *Anal. Biochem.* **170,** 248 (1988).

[77] T. Thuren, J. A. Virtanen, M. Lalla, and P. K. J. Kinnunen, *Clin. Chem.* **31,** 714 (1985).

[78] H. S. Hendrickson and P. N. Rauk, *Anal. Biochem.* **116,** 553 (1981).

[79] H. S. Hendrickson, this volume [8].

[80] H. S. Hendrickson, K. J. Klotz, and E. K. Hendrickson, *Anal. Biochem.* **185,** 80 (1990).

ization of the parinaric moiety contained in substrate **6** as a function of hydrolysis.[81] This substrate is easily prepared from parinaric acid and the appropriate lysophospholipid in a two-step synthesis. When run in the presence of albumin, the assay is continuous and capable of detecting 300 pmol/min. The sensitivity is comparable to some spectrophotometric assays. The general utility of the assay may be limited by the necessity to avoid photobleaching and air oxidation of the substrate. The remaining fluorescence assay is discontinuous and uses a phospholipid labeled with NBD (7-nitrobenzo-2-oxa-1,3-diazol-4-yl) **(7)** as a substrate.[82] This substrate is available commercially, but the assay is rather insensitive.

Electron Spin Resonance Assay

A continuous electron spin resonance (ESR) assay has been reported which utilizes a spin-labeled phospholipid substrate.[83] The procedure monitors the change in signal of 4-doxylpentanoic acid upon hydrolysis from the substrate 1-palmitoyl-2-(4-doxylpentanoyl)glycerophosphocholine **(8)**, and rates as low as 1 nmol/min can be detected. Very small volumes are required for this sensitive assay, and samples as low as 1 μl can potentially be accommodated. The substrate binds well to bee venom phospholipase A_2, but the specific activity is only 50 μmol/min/mg.

Assays in Presence of Inhibitors

The choice of assay is especially important in inhibitor studies where the particular assay conditions can profoundly affect the perceived inhibition.[84] The addition of an inhibitor can affect the physical state of the phospholipid substrate making it more or less susceptible to attack by the enzyme. An extreme example of this is the inhibition of phospholipase A_2 by lipocortin via a substrate-depletion mechanism.[26,85,86] These effects of inhibitors on the nature of the substrate can be minimized by using higher substrate concentrations or by using a more potent inhibitor whose concentration in the assay is much less than that of the substrate. Another approach for hydrophobic inhibitors is to solubilize the substrate and inhibitor into mixed micelles. With a large excess of detergent, the effect

[81] C. Wolf, L. Sagaert, and G. Bereziat, *Biochem. Biophys. Res. Commun.* **99,** 275 (1981).

[82] L. A. Wittenauer, K. Shirai, R. L. Jackson, and J. D. Johnson, *Biochem. Biophys. Res. Commun.* **118,** 894 (1984).

[83] C. Lai, J. Zhang, and J. Joseph, *Anal. Biochem.* **172,** 397 (1988).

[84] E. A. Dennis, *Bio/Technology* **5,** 1294 (1987).

[85] F. F. Davidson and E. A. Dennis, *Biochem. Pharmacol.* **38,** 3645 (1989).

[86] F. F. Davidson, M. D. Lister, and E. A. Dennis, *J. Biol. Chem.* **265,** 5602 (1990).

of the inhibitor on the physical state of the micelle is minimized. However, if the concentration of inhibitor in the interface increases significantly, the mole fraction of substrate phospholipid decreases, resulting in lower enzymatic activity. This surface dilution of substrate is a problem only with weak inhibitors and should not be confused with enzyme inhibition.[49]

An alternative to incorporating an inhibitor into mixed micelles is to incorporate it into vesicles with a phosphatidylmethanol substrate.[87,88] These vesicles have been shown to bind phospholipase A_2 irreversibly. Thus, this method has been successful in distinguishing agents which interfere with the binding of the enzyme to the lipid interface from those which affect catalysis.

The kinetic treatment of phospholipases acting on lipid/water interfaces is very complex, and standard kinetic treatments do not apply.[5,51] Thus, most inhibition constants for these enzymes are reported as IC_{50} or apparent K_i values. Under these conditions, the apparent effectiveness of an inhibitor is very dependent on the particular assay chosen. An inhibitor may *appear* more potent in a radioactive vesicle assay with 100 μM substrate than in a mixed micelle assay with 5 mM substrate and 20 mM detergent. The apparent inhibition also depends on the K_m of the particular substrate; an inhibitor may appear to be more effective when assayed against a substrate with a high K_m.

Inhibitors which cause an irreversible inactivation of phospholipases are generally less sensitive to the particular assay employed than are reversible inhibitors. Irreversible inhibitors such as *p*-bromophenacyl bromide[89] and manoalogue[90] are usually preincubated with the enzyme and then diluted into the assay after a specified time to measure residual enzyme activity. Since the inhibition occurs during the preincubation step, the assay itself has little effect on the amount of inactivation observed. And, as long as the dilution factor is high, the inhibitor has minimal effect on the assay since its concentration in the assay will be quite low.

Summary

Of the general considerations discussed, the two issues which are most important in choosing an assay are (1) what sensitivity is required to assay a particular enzyme and (2) whether the assay must be continuous. One can

[87] M. K. Jain and M. H. Gelb, this volume [10].

[88] M. K. Jain, W. Yuan, and M. H. Gelb, *Biochemistry* **28,** 4135 (1989).

[89] M. F. Roberts, R. A. Deems, T. C. Mincey, and E. A. Dennis, *J. Biol. Chem.* **252,** 2405 (1977).

[90] L. J. Reynolds, B. P. Morgan, G. A. Hite, E. D. Mihelich, and E. A. Dennis, *J. Am. Chem. Soc.* **110,** 5172 (1988).

narrow the options further by considering substrate availability, enzyme specificity, assay convenience, or the presence of incompatible side reactions. In addition, the specific preference of a particular phospholipase for polar head group, micellar versus vesicular substrates, and anionic versus nonionic detergents may further restrict the options.

Of the many assays described in this chapter, several have limited applicability or serious drawbacks and are not commonly employed. The most commonly used phospholipase assays are the radioactive TLC assay and the pH-stat assay. The TLC assay is probably the most accurate, sensitive assay available. These aspects often outweigh the disadvantages of being discontinuous, tedious, and expensive. The radioactive *E. coli* assay has become popular recently as an alternative to the TLC assay for the purification of the mammalian nonpancreatic phospholipases. The assay is less time consuming and less expensive than the TLC assay, but it is not appropriate when careful kinetics are required. Where less sensitivity is needed, or when a continuous assay is necessary, the pH-stat assay is often employed. With purified enzymes, when free thiol groups are not present, a spectrophotometric thiol assay can be used. This assay is approximately as sensitive as the pH-stat assay but is more convenient and more reproducible, although the substrate is not available commercially. Despite the many assay choices available, the search continues for a convenient, generally applicable assay that is both sensitive and continuous. The spectrophotometric SIBLINKS assay and some of the fluorescent assays show promise of filling this need.

Acknowledgments

Support for preparation of the manuscript was provided by National Institutes of Health Grants GM 20501 and HD 26171 and National Science Foundation Grant DMB 88-17392.

[2] Utilization of Labeled *Escherichia coli* as Phospholipase Substrate

By PETER ELSBACH and JERROLD WEISS

Introduction

The inherent difficulties in the kinetic analysis of the hydrolysis by phospholipolytic enzymes of their hydrophobic substrates in an aqueous environment has led to the use of numerous methods of presentation of different phospholipids that vary greatly in their physicochemical properties, depending on fatty acyl and polar head group composition. How well these various assays, in which the phospholipids are presented in different physical forms, for example, as monomers, vesicles, micelles in the presence or absence of detergents, or as monolayers, reflect the action of phospholipases on their natural substrates in the biological environment is not clear.

Escherichia coli marked with ^{14}C-labeled fatty acid have proved to be convenient for the assay of deacylating phospholipases under relatively more physiologic conditions. They have been used in two ways: (1) as intact bacteria and (2) after autoclaving.[1-4] As is the case for phospholipids in the membranes of most unperturbed cells, the phospholipids of intact *E. coli* are refractory to the action of both endogenous and added deacylating phospholipases. Consequently, it has been possible to study the ability of various agents to activate selectively or indiscriminately phospholipolysis by a broad range of secretory and cellular phospholipases.[5-7] The availability of *E. coli* mutants lacking one or both of the major phospholipases A[8-12] allows exclusion of the bacterial enzymes and assessment of the

[1] P. Elsbach, J. Weiss, R. C. Franson, S. Beckerdite-Quagliata, A. Schneider, and L. Harris, *J. Biol. Chem.* **254,** 11000 (1979).

[2] P. Patriarca, S. Beckerdite, P. Pettis, and P. Elsbach, *Biochim. Biophys. Acta* **280,** 45 (1972).

[3] R. Franson, P. Patriarca, and P. Elsbach, *J. Lipid Res.* **100,** 380 (1974).

[4] P. Elsbach and J. Weiss, *in* "Bacteria–Host Cell Interaction," (M. A. Horwitz, ed.), p. 47. Alan R. Liss, New York, 1988.

[5] J. Weiss, S. Beckerdite-Quagliata, and P. Elsbach, *J. Biol. Chem.* **254,** 11010 (1979).

[6] S. Forst, J. Weiss, and P. Elsbach, *J. Biol. Chem.* **257,** 14055 (1982).

[7] P. Elsbach and J. Weiss, *Biochim. Biophys. Acta* **947,** 29 (1988).

[8] M. Ohki, O. Doi, and S. Nojima, *J. Bacteriol.* **110,** 864 (1972).

[9] O. Doi and S. Nojima, *J. Biochem.* (*Tokyo*) **74,** 667 (1973).

[10] M. Abe, J. Okamoto, O. Doi, and S. Nojima, *J. Bacteriol.* **119,** 543 (1974).

[11] J. Weiss and P. Elsbach, *Biochim. Biophys. Acta* **466,** 23 (1977).

METHODS IN ENZYMOLOGY, VOL. 197

role of added phospholipases in the phospholipolysis observed in a given setting.[13]

Alternatively, autoclaving of *E. coli* tagged with ^{14}C-labeled fatty acid inactivates all bacterial phospholipases and exposes the phospholipids in the *E. coli* membranes to all deacylating phospholipases tested so far. The use of autoclaved labeled *E. coli* as substrate has now been widely adopted. Advantages are the ease of preparation and the high degree of sensitivity of the assay (Fig. 1); moreover, the retention of the typical rod shape and hence of the gross envelope structure of the *E. coli* after autoclaving suggests that the presentation of the bacterial phospholipids is more similar to that of phospholipids in natural membranes than is the case for phospholipids in most of the other commonly used assays.

Other advantages include the following: (1) The labeling of *E. coli* with ^{14}C-labeled fatty acid during several generations produces uniform labeling of the *E. coli* phospholipids so that the distribution of the label fairly reflects the bacterial phospholipid composition.[14,15] In *E. coli* as well as most other Enterobacteriaceae, phosphatidylethanolamine is the most abundant phospholipid (~70%); phosphatidylglycerol and cardiolipin (diphosphatidylglycerol) represent, respectively, approximately 20 and 8% of the total bacterial lipid. Only trace amounts of other lipids have been recognized in *E. coli*. (2) Where examined the rate of hydrolysis of the phosphatidylethanolamine and phosphatidylglycerol is similar.[2,16] (3) The rate of hydrolysis of the phospholipids of autoclaved *E. coli* is the same as of phospholipids extracted and dispersed in the aqueous assay mixture,[6] suggesting that nonlipid bacterial components do not alter the access to the substrate. (4) The phospholipid and fatty acid composition of lipids extracted from autoclaved and untreated *E. coli* are the same, indicating that the exposure of the bacteria to the autoclave conditions (120° and 2.7 kg/cm^2 for 15 min) does not appreciably change the bacterial phospholipids. This must owe in part to the fact that *E. coli* do not contain (readily oxidizable) polyunsaturated fatty acids. (5) The incorporation of saturated fatty acids (palmitic acid) nearly exclusively into the *sn*-1 position (>90%) and of unsaturated fatty acid (oleic or *cis*-vaccenic acid) into the *sn*-2

[12] P. de Geus, I. van Die, H. Bergmans, J. Tommassen, and G. H. de Haas, *Mol. Gen. Genet.* **190,** 150 (1983).

[13] P. Elsbach, J. Weiss, G. Wright, and H. Verheij, *in* "Cell Activation and Signal Initiation: Phospholipase Control of Inositol Phosphate, PAF and Eicosanoid Production," (E. A. Dennis, T. Hunter, and M. Berridge, eds.), p. 323. Alan R. Liss, New York, 1989.

[14] N. Wurster, P. Elsbach, J. Rand, and E. J. Simon, *Biochim. Biophys. Acta* **248,** 282 (1971).

[15] P. Patriarca, S. Beckerdite, and P. Elsbach, *Biochim. Biophys. Acta* **260,** 593 (1972).

[16] R. Franson, S. Beckerdite, P. Wang, M. Waite, and P. Elsbach, *Biochim. Biophys. Acta* **296,** 365 (1973).

position (>95%) of the *E. coli* phospholipids[17] provides an easy means of distinguishing positional preference of deacylating activities tested by analysis of labeled products of hydrolysis. (6) Since many deacylating phospholipases are more readily detected when phosphatidylethanolamine is the substrate than with phosphatidylcholine, the autoclaved *E. coli* assay is useful for initial screening of phospholipase activities.

Procedures

Preparation of Escherichia coli with ^{14}C-Fatty Acid-Labeled Phospholipids

Fatty acid incorporation by a range of wild-type and mutant *E. coli* strains is brisk during exponential growth. Its rate parallels the division time and hence depends mainly on the composition of the growth medium.

Fresh bacterial cultures grown overnight in triethanolamine buffered (pH 7.7–7.9) minimal salts medium,[18] or nutrient broth, are diluted 1 : 20 in fresh medium and subcultured at 37° for 3 hr (roughly six generations) in the presence of 1 mCi/ml of the appropriate fatty acid, usually either [^{14}C]oleic acid or [^{14}C]palmitic acid. The labeled fatty acid in organic solvent is added to the empty culture flask, and the solvent is removed by evaporation under a stream of nitrogen, after which the bacterial suspension is added. We have omitted adding albumin to complex and disperse the fatty acid adherent to the vessel wall, because incorporation by the bacteria is at least as good in the absence of albumin. At the end of incubation the bacteria are sedimented by centrifugation at 3000 *g* for 12 min at room temperature, resuspended in fresh growth medium, and reincubated for 30 min at 37° to chase adherent unincorporated radiolabeled fatty acid into ester positions. Finally, the labeled bacteria are washed once with 1% (w/v) bovine serum albumin (commercial Cohn fraction V, e.g., from United States Biochemical Corp., Cleveland, OH) to remove unincorporated radiolabeled precursors.

More than 50% of the added precursor is recovered in the *E. coli* phospholipids. The percentage of total bacterial radioactive lipid in free fatty acid should not exceed 5%. The lower this background radioactivity in the free fatty acid fraction, the greater the sensitivity of the measurement of subsequent hydrolysis. Bacterial concentrations are determined by measuring the OD_{550}, followed by resuspension to the desired concentration

[17] J. Weiss, R. Franson, K. Schmeidler, and P. Elsbach, *Biochim. Biophys. Acta* **436,** 154 (1976).

[18] N. Wurster, P. Elsbach, E. J. Simon, P. Pettis, and S. Lebow, *J. Clin. Invest.* **50,** 1091 (1971).

in sterile physiologic (0.9%) saline. The labeled bacteria can be used either as live organisms to examine the effect of agents that may be able to activate bacterial and/or added phospholipases[5,7,11] or after autoclaving as a substrate for the quantitative assay of phospholipase activity.

Measurement of Phospholipid Hydrolysis

In Intact Bacteria. Since *E. coli* retain their viability and metabolic integrity under a broad range of environmental conditions, their use for the study of phospholipase action on the envelope phospholipids of initially intact *E. coli* requires no prescription of the incubation conditions nor of the composition of the incubation medium. In our own investigations we have aimed generally at providing a reasonably physiologic setting with respect to electrolytes and pH.

Assay of Phospholipase A Activity with Autoclaved Bacteria as Substrate. The reaction mixture (total volume 250 μl) is composed of 2.5×10^8 autoclaved *E. coli* labeled with ^{14}C-fatty acid [a mix of labeled and unlabeled bacteria to provide ~10,000 counts per minute (cpm) of radioactivity] containing approximately 5 nmol of phospholipid, 40 m*M* Tris-HCl (pH 7.5) or appropriate other buffers for determination of pH dependence or for assay of enzymes with known different pH optima, and 10 m*M* $CaCl_2$. For termination of the reaction, because in the live as well as in the autoclaved *E. coli* the substrate is particulate and because product free fatty acids and lysophospholipids are quantitatively trapped in the extracellular medium by added bovine serum albumin (BSA),[2] the reaction is effectively arrested by addition of ice-cold 0.5% BSA (w/v) and prompt centrifugation of the reaction mixture in an Eppendorf microfuge at 10,000 *g* for 2 min at room temperature. The sedimented bacteria contain the undegraded phospholipids, and the albumin-complexed products of hydrolysis are in the supernatant. A measured sample of the supernatant is counted by liquid scintillation for quantitation of hydrolysis.

If the identification of hydrolysis products is desired, reaction mixtures are extracted by the Bligh and Dyer procedure,[19] and the lipid extracts are analyzed by thin-layer chromatography.[3] Activity is expressed in arbitrary units: one unit equals 1% hydrolysis/hr, representing approximately 10^{-6} μmol/min (IU).

Sensitivity of Phospholipase Assays Using Autoclaved Escherichia coli

Our use of autoclaved *E. coli* has been largely restricted to assays of phospholipases A_2. Phospholipid hydrolysis is approximately linear with increasing time and protein concentration until 20–25% of the substrate

[19] E. G. Bligh and W. J. Dyer, *Can. J. Biochem. Physiol.* **37,** 911 (1959).

has been hydrolyzed. The activity of a wide range of pure phospholipases A_2 is readily detected at protein concentrations of approximately 10^{-11} M, at a substrate concentration of about 2×10^{-5} M (Fig. 1). The sensitivity of the assay is further exemplified by the detection of phospholipase A_2 activity in samples of extracts of rabbit granulocytes, representing $1–2 \times 10^5$ cells.[1] The granulocyte phospholipase A_2 has been purified to homogeneity.[20] Based on recovery data we can therefore estimate the enzyme content of this number of cells at 100–200 pg (representing ~0.001% of total granulocyte protein). This demonstrates that trace amounts of the typical low-abundance cellular phospholipases A_2 can be detected readily in small tissue samples. The use of autoclaved phospholipid-labeled *E. coli* for assay of other phospholipases (C and D) should also be explored.

Other Uses of Escherichia coli Labeled with ^{14}C-Fatty Acid

Identification of Structural Determinants of Functional Diversity among Highly Conserved Phospholipases A_2. In contrast to the closely similar enzymatic activity of a wide range of phospholipases A_2 toward autoclaved *E. coli*, these enzymes differ markedly in their ability to act on the phospholipids of intact *E. coli* mutants, lacking activatable endogenous phospholipases A, treated with membrane-perturbing peptides or proteins[5,13,21] (Fig. 1). Such differences are most pronounced when phospholipase A-deficient *E. coli* are treated with a purified antibacterial protein of polymorphonuclear leukocytes (PMN), the bactericidal/permeability-increasing protein (BPI).[7] The combination of structural comparison of enzymes active or inactive against BPI-treated *E. coli*, chemical modification of phospholipase A_2,[6] and genetic manipulation of inactive enzymes resulting in conversion to active enzymes[22] has allowed initial identification of properties of variable regions of phospholipases A_2 that are determinants of BPI responsiveness.[7] Thus, the availability of phospholipase A-less *E. coli* mutants and the ability of certain membrane-damaging agents to trigger selectively the action of added phospholipases A_2 provide a convenient means of exploring structural determinants of functional differences among these conserved enzymes.

Role of Endogenous Phospholipase(s) in Biological Events. By genetic methods isogenic strains of *E. coli* have been generated that differ only in

[20] G. Wright, C. E. Ooi, J. Weiss, and P. Elsbach, *J. Biol. Chem.* **265,** 6675 (1990).

[21] P. Elsbach, J. Weiss, and S. Forst, *in* "Lipids and Membranes: Past, Present and Future" (J. A. F. Op den Kamp, B. Roelofsen, and K. W. A. Wirtz, eds), p. 259. Elsevier, Amsterdam, 1986.

[22] G. Wright, J. Weiss, C. van den Bergh, H. Verheij, and P. Elsbach, *Clin. Res.* **37,** 444A (1989).

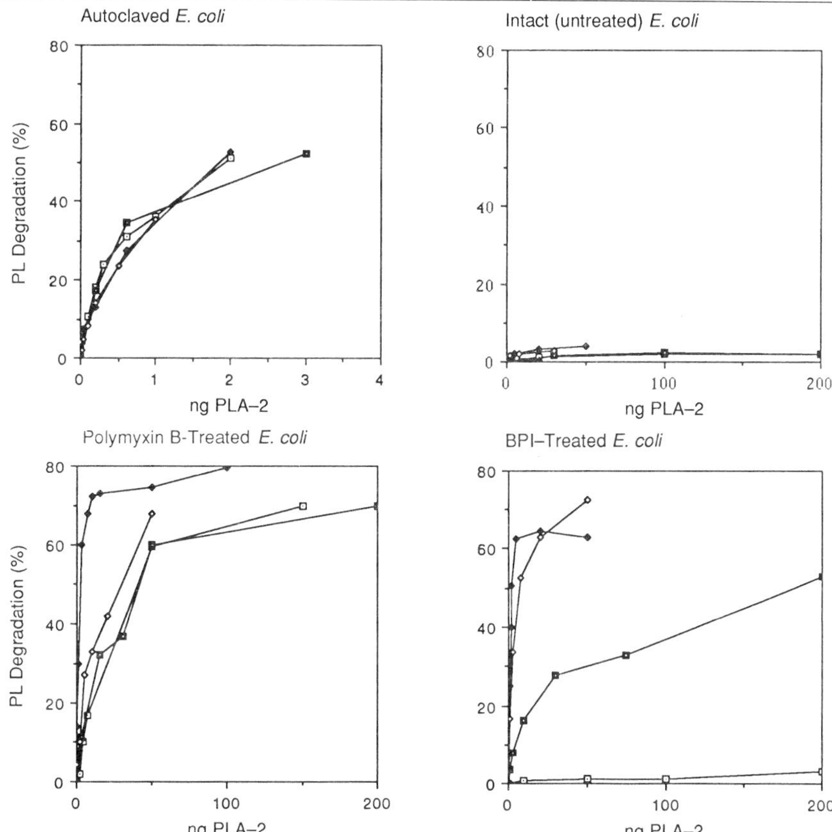

FIG. 1. Action of different phospholipases A_2 on variously treated *E. coli*. Incubations (see also description in the text) were carried out with 2.5×10^8 autoclaved *E. coli* or with 10^7 intact *E. coli* 1303 (*pldA*$^-$), either untreated or treated with polymyxin B (0.1 μg) or BPI (1–2 μg), at 37° for either 15 min (autoclaved *E. coli*) or 60 min (intact *E. coli*). ⊡, Pig pancreas; ◆, agkistrodon halys blomhoffii venom phospholipase A_2 (from Ref. 25); ■, agkistrodon piscivorus piscivorus venom phospholipase A_2; ◇, ascritic fluid phospholipase A_2 (from Ref. 26).

their content of the *pldA* gene that encodes an outer membrane phospholipase A of *E. coli*. Of three strains, one lacks the *pldA* gene and its product, the wild-type strain contains from 200 to 500 phospholipase A molecules/cell, and the third strain contains a multicopy plasmid with the inserted *pldA* gene, raising the phospholipase content 20- to 200-fold.[12] Comparison of the response of the three strains to agents that trigger bacterial phospholipolysis has shown the participation of the *pldA* gene product in the antibacterial action of intact PMN, purified BPI, bacteriophage infection, and bacteriocin release.

Role of Intramembrane Ca^{2+} in Activation of Ca^{2+}-Dependent Phospholipases A. Mobilization of Ca^{2+} to (cellular) sites of action by Ca^{2+}-dependent phospholipases is generally considered to be an essential step in activation. Maximal activity of phospholipases A_2 toward autoclaved *E. coli* requires at least 1 m*M* Ca^{2+}, and our standard assay mixtures contain 10 m*M* Ca^{2+}. The purified *pldA* gene product, which is the major envelope phospholipase A of *E. coli*, has the same Ca^{2+} dependence in its action toward micellar lipid. In contrast, both the endogenous bacterial and the added deacylases maximally hydrolyze the phospholipids of polymyxin B- or BPI-treated *E. coli* at ambient Ca^{2+} concentrations of no higher than 0.03 m*M*.[23] In fact, hydrolysis is less if 5–10 m*M* Ca^{2+} is added, presumably because the added divalent cations reduce binding of polymyxin B and BPI.[4,23] The cationic, membrane-inserting polymyxin B and BPI compete for anionic sites on the outer membrane lipopolysaccharides that are normally occupied by Ca^{2+} and Mg^{2+} and thereby displace bound Ca^{2+} near where the phospholipids are exposed to the phospholipases.[13,23] This bacterial membrane Ca^{2+} pool can be reversibly decreased or increased by incubating *E. coli* in Ca^{2+}-depleted or Ca^{2+}-repleted media (containing Mg^{2+}), generating reversible alterations in the sensitivity of polymyxin B or BPI-treated *E. coli* to Ca^{2+}-dependent phospholipases A (but without effect on Ca^{2+}-independent phospholipases).[13]

Selection of One Substrate versus Another in Phospholipase Assays

The ease of preparation of autoclaved *E. coli* labeled with ^{14}C-labeled fatty acid to serve as substrate in phospholipase (A_2) assays and the high sensitivity of the assay, allowing detection of pmolar concentrations of enzyme, have led to the widespread use of this substrate. It has to be recognized, however, that different phospholipases, including the highly conserved phospholipases A_2, have different preferences for the wide range of phospholipids, presented in disparate ways, employed by the many investigators studying phospholipases. Hence, the use of any single substrate is likely to result in the detection of some, but the exclusion or the underestimation of other phospholipolytic enzymes. This is evident, for example, in our experience that genetically altered phospholipases A_2 may have similar specific activity in the autoclaved *E. coli* assay but may exhibit reduced specific activity when other substrates are used (unpublished observations with H. Verheij and G. de Haas of the University of Utrecht). Thus, conclusions based on results obtained with a given substrate must be tempered by the fact that the choice of any substrate

[23] P. Elsbach, J. Weiss, and L. Kao, *J. Biol. Chem.* **260,** 1618 (1985).

imposes certain limitations, and it points to the need to compare the properties of crude as well as pure enzyme preparations vis-à-vis more than one set of substrate conditions. For a detailed analysis of the strengths and limitations of the many assays now in use, readers are referred to the recent comprehensive treatise by Waite.[24]

Acknowledgments

These investigations are supported by U.S. Public Health Service Grants 5R37DK 05472 and AI 18571.

[24] M. Waite, "The Phospholipases." Plenum, New York, 1987.
[25] S. Forst, J. Weiss, P. Blackburn, B. Frangione, F. Goni, and P. Elsbach, *Biochemistry* **25,** 4309 (1986).
[26] S. Forst, J. Weiss, and P. Elsbach, *Biochemistry* **25,** 8381 (1986).

[3] Nuclear Magnetic Resonance Spectroscopy to Follow Phospholipase Kinetics and Products

By MARY F. ROBERTS

Introduction

Nuclear magnetic resonance (NMR) spectroscopy has proved to be an excellent technique for describing phospholipid structures used as substrates for phospholipases. High-resolution ^{1}H, ^{13}C, and ^{31}P NMR studies have been used to characterize lipid structure and dynamics in micelles and small unilamellar vesicles.[1-5] The same aggregates are often used as substrates for phospholipases.[6-9] Multilamellar vesicles, while less frequently used as substrates, have been extensively studied by ^{2}H NMR.

[1] A. G. Lee, N. J. M. Birdsall, Y. K. Levine, and J. C. Metcalfe, *Biochim. Biophys. Acta* **255,** 43 (1972).
[2] I. C. P. Smith, *Can. J. Biochem.* **57,** 1 (1979).
[3] A. A. Ribeiro and E. A. Dennis, *Biochemistry* **14,** 3746 (1975).
[4] R. A. Burns, Jr., and M. F. Roberts, *Biochemistry* **19,** 3100 (1980).
[5] P. L. Yeagle, *in* "Phosphorus NMR in Biology" (C. T. Burt, ed.), p. 95. CRC Press, Boca Raton, Florida, 1978.
[6] E. A. Dennis, *Arch. Biochem. Biophys.* **158,** 485 (1973).
[7] M. Y. El-Sayed, C. D. DeBose, L. A. Coury, and M. F. Roberts, *Biochim. Biophys. Acta* **837,** 325 (1985).
[8] C. A. Kensil and E. A. Dennis, *J. Biol. Chem.* **254,** 5843 (1979).
[9] N. E. Gabriel, N. V. Agman, and M. F. Roberts, *Biochemistry* **26,** 7409 (1987).

TABLE I
PROPERTIES OF NUCLEI RELEVANT TO USING NMR TO ASSAY PHOSPHOLIPASE HYDROLYSIS OF PHOSPHOLIPIDS

Nucleus	Spin	NMR frequency[a] (MHz)	Natural abundance (%)	Relative sensitivity	Relative detectability at natural abundance
^{1}H	$\frac{1}{2}$	300.08	99.98	100	100
^{2}H	1	46.07	0.0156	0.96	0.00015
^{13}C	$\frac{1}{2}$	75.45	1.1	1.6	0.017
^{15}N	$\frac{1}{2}$	30.42	0.37	0.1	0.0004
^{31}P	$\frac{1}{2}$	121.48	100	0.066	0.066

[a] In a field of 7.0 tesla.

Analyses of line shapes and quadrupolar splitting have aided in understanding the motions of lipids in solution.[10]

NMR spectroscopic methods can also be used to monitor hydrolysis of lipids in different aggregates by phospholipases. Other techniques such as pH-stat and fluorescence assays provide information on the extent and rate of the reaction. The NMR methodology provides additional information on substrate and product structure and dynamics. It is less useful for accurate initial rate data, since in general at least 10% hydrolysis is needed for accuracy. Sensitivity is also considerably less and depends on the nucleus observed (Table I). In general 5–50 m*M* substrate solutions are necessary, with greater than 1 m*M* product a rough estimate for the minimum detected. Lipid systems must also be highly buffered, since hydrolysis reactions liberate protons into the solution. The major advantage of NMR methods is that changes in the substrate physical state as a function of enzyme action can be monitored. Different nuclei are sensitive to different parameters, and the choice of nucleus observed depends on the nature of the aggregate system and the phospholipase reaction to be studied. For example, ^{13}C NMR using enriched phosphatidylcholine (PC) is excellent for assaying the ester bond specificity of a phospholipase; ^{31}P NMR is a better choice for enzyme action on mixed phospholipids in micellar systems or for phospholipase C where one of the products (the one with the phosphorus atom) is water soluble. The usefulness of NMR methods for assaying phospholipases is best shown with several examples which illustrate the type of kinetic and structural information obtainable.

[10] J. Seelig, *Q. Rev. Biophys.* **13,** 19 (1980).

^{1}H NMR Assays

^{1}H NMR spectra of phospholipids are easily obtained but show little chemical shift difference between resonances for substrate phospholipids and most products (lysophospholipid, diglyceride, phosphatidic acid). Only fatty acids show an appreciable chemical shift difference, most noticeably in the α-CH_2 position, although there is a smaller shift difference for the fatty acyl β-CH_2 protons and substrate PC (and lyso-PC) β-CH_2 protons. Because of this, ^{1}H NMR spectroscopy is not very useful for generating accurate rates of product formation except for phospholipase A_2 or A_1 activity. In general, the linewidths for lipids in bilayer vesicles are broader than for lipids in detergent-mixed micelles or pure short-chain PC micelles. Therefore, using this nucleus to assay phospholipases is further restricted to using monomeric or short-chain phospholipid micelles or detergent-mixed micelles (as long as the appropriate CH_2 groups are not obscured by detergent resonances) as substrates. For 5 m*M* substrate, spectra can easily be accumulated every 5–10 min.

As shown in Fig. 1A, when phospholipase A_2 is added to diheptanoyl-PC micelles, the only dramatic spectral change is at the α-CH_2 resonance, which moves upfield about 0.25–0.3 ppm. Integrating these two regions and looking at the ratio of the fatty acyl peak to that of total PC substrate (which also includes the *sn*-1 α-CH_2 of the lyso-PC product) and fatty acid allows one to follow the rate of lipid hydrolysis. Greater than 10 mol % hydrolysis must occur for reasonable accuracy in monitoring the reaction. For phospholipase C quite different behavior is detected by ^{1}H NMR spectroscopy (see Fig. 1B, where 5 m*M* of diheptanoyl-PC was incubated with phospholipase C from *Bacillus cereus*). The phospholipase C reaction product diheptanoylglycerol is not very water soluble[11]; its solubility in diheptanoyl-PC micelles is limited and it eventually phase separates in the D_2O solutions. More importantly, when even a few percent diglyceride are generated, the micellar particles grow dramatically in size (K. M. Eum and M. F. Roberts, unpublished results). The practical implication of this is that substrate resonances decrease in intensity, and following the decrease in integrated intensities of the $N(CH_3)_3$, $(CH_2)_n$, and CH_3 resonances reflects enzyme activity, although not directly. Eventually, a sharp $N(CH_3)_3$ resonance for phosphocholine is detected slightly downfield (0.02 ppm) of the diheptanoyl-PC choline peak.

Hydrolysis of the short-chain PC by phospholipase D can also be monitored by ^{1}H NMR. This enzyme yields another water-soluble product, choline, and phosphatidic acid. The choline $N(CH_3)_3$ group is shifted

[11] M. Y. El-Sayed and M. F. Roberts, *Biochim. Biophys. Acta* **831,** 133 (1985).

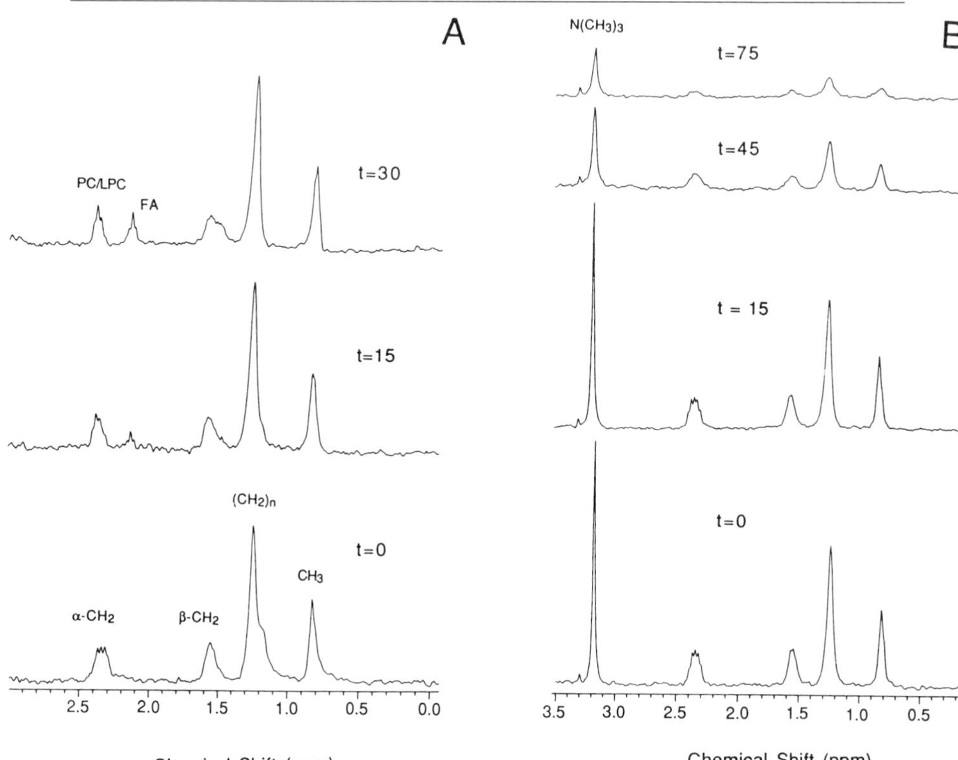

FIG. 1. ^{1}H NMR (300 MHz) spectra of 5 m*M* diheptanoyl-PC in D_2O with 20 m*M* inorganic phosphate, uncorrected pH 7.0, as a function of time after the addition of (A) 10 ng phospholipase A_2 (*Naja naja naja*) and 1 m*M* Ca^{2+}, or (B) 40 ng phospholipase C (*Bacillus cereus*). Incubation times are indicated. The identity of the different proton resonances are also shown.

slightly upfield (0.04 ppm) of the PC resonance, and is easily detected with >10% hydrolysis. Phosphatidic acid fatty acyl resonances are not distinguishable from PC resonances in the spectra. Since choline, phosphocholine, and phosphatidylcholine $N(CH_3)_3$ groups all have slightly different chemical shifts, it is easy to distinguish phospholipase C from phospholipase D activity.

^{31}P NMR Assays

^{31}P NMR is a much more useful technique for following the activity of phospholipases.[9,12–14] Resonances for the phospholipids, lysophospholip-

[12] M. F. Roberts, M. Adamich, R. J. Robson, and E. A. Dennis, *Biochemistry* **18,** 3301 (1979).

ids, and phosphate mono- and diesters are detected; buffers, ions, etc., do not interfere unless they contain phosphate groups. The ^{31}P isotope is 100% abundant, and signals for millimolar concentrations of phosphates can be detected in 10 min on moderately high-field spectrometers (4–7 tesla). Coupling of ^{31}P to protons three bonds or more removed (e.g., $POCH_2$), while small, broadens resonances. In practice, spectra are acquired with ^1H decoupling (WALTZ decoupling schemes on modern spectrometers effectively decouple ^1H without sample heating). T_1 relaxation times for small vesicles and micelles are usually less than 2–3 sec; for water-soluble phosphates this value can be considerably longer (e.g., 13 sec for dibutyryl-PC[15]). Depending on magnetic field strength and the reaction monitored, corrections for the T_1 values may need to be applied if recycle times do not allow all ^{31}P to relax completely.

Nuclear Overhauser effects (NOEs) are relatively small for phospholipids in micelles or vesicles at moderate field strengths (e.g., at 109.5 MHz the NOE is ~1.4 for the phosphorus nucleus in short-chain PC micelles and 1.2 for long-chain PC in small unilamellar vesicles[15]). If the products are partitioned into the same aggregate as the substrate, they will have similar NOEs, and direct integrated intensities can be compared. If a product is free in solution (as is the phosphomonoester produced by the phospholipase C reaction), one needs to measure substrate and product NOE values and correct integrated intensities if they are different. To monitor product formation, starting at 10% reaction, substrate (micelles or small vesicles) should be at least 20 m*M* for kinetics, with time points acquired every 5–10 min. Almost any organic buffer can be used, typically 50 m*M* Tris-HCL.

^{31}P is particularly useful for analyzing phospholipase activity toward mixtures with different lipid head groups. In order to do this by other methods, an extra separation step [thin-layer chromatography (TLC), high-performance liquid chromatography (HPLC)] would be necessary. As with ^1H, the linewidth is influenced (albeit not directly or in a simple fashion) by particle size and mobility.[5,15] ^{31}P NMR spectra of small unilamellar vesicles at moderately high fields often shows separate phosphorus resonances for molecules on the inner and outer leaflets of the particles.[5] The difference in chemical shift can be enhanced with the addition of an impermeable paramagnetic shift reagent such as Pr^{3+}.[16] This offers the

[13] A. Pluckthun and E. A. Dennis, *Biochemistry* **21,** 1750 (1982).

[14] J. C. Jones and G. R. A. Hunt, *Biochim. Biophys. Acta* **820,** 48 (1985).

[15] R. A. Burns, Jr., R. E. Stark, D. A. Vidusek, and M. F. Roberts, *Biochemistry* **22,** 5084 (1983).

[16] W. C. Hutton, P. L. Yeagle, and R. B. Martin, *Chem. Phys. Lipids* **19,** 255 (1977).

opportunity of monitoring selective monolayer hydrolysis. For phospholipid aggregates, there is an optimum H_0 for detection.

Chemical shift anisotropy (CSA) becomes an important relaxation mechanism for phosphorus nuclei at higher fields.[17] The CSA contribution increases as the square of the magnetic field; hence, for vesicles, magnetic field strengths of 4.7 to 7.0 tesla are usually optimal. To minimize the CSA problem a substrate system for phospholipases should be of moderate size and have good mobility. The best system, namely, one with narrow, well-separated lines, is a micellar one, most commonly detergent-mixed micelles. In bile salt or Triton X-100 mixed micelles, good resolution of phospholipids, lysophospholipids, and phosphomonoesters exists. The shifts of these different species are listed in Table II. For mixed micelles it should be noted that the exact chemical shift depends somewhat on the nature of the detergent. Phosphatidylcholine in Triton X-100 is about 0.2–0.3 ppm upfield of the same species in cholate. Nonetheless, relative shift differences between substrate and product phosphate groups are maintained. This means that phospholipid substrates hydrolyzed by phospholipases A_1 (product: 2-lysophospholipid), A_2 (product: 1-lysophospholipid), C (product: phosphate monoester), and D (product: phosphatidic acid) can be monitored by ^{31}P NMR spectroscopy. This nucleus should be particularly useful in observing phosphatidylinositol (PI) hydrolysis by PI-specific phospholipase C enzymes. The products of that enzyme-catalyzed reaction, phosphomonoesters, have chemical shifts that are well separated from substrate (Table II, Ref. 18). If a cyclic phosphodiester intermediate is generated, as has been suggested to occur for this enzyme from bacterial sources, this should also be well resolved (~15 ppm downfield of normal phosphodiester linkages).[19]

Three published examples illustrating the use of this nucleus to assay phospholipases include (1) phospholipase A_2 preferential hydrolysis of phosphatidylethanolamine (PE) in PE/PC/Triton X-100 mixed micelles,[12,13] (2) short-chain PC hydrolysis by phospholipase A_2 and phospholipase C in sphingomyelin/diheptanoyl-PC small bilayer particles,[9] and (3) integrity of a vesicle bilayer when phospholipase A_2 has hydrolyzed most of the PC.[14] The first two of these stress the use of ^{31}P in monitoring hydrolysis of individual phospholipid species in lipid mixtures.

In PC/PE/Triton X-100 mixed micelles, ^{31}P resonances for the two phospholipids and their corresponding lyso-PC compounds are well re-

[17] C. T. Burt (ed.) "Phosphorus NMR in Biology." CRC Press, Boca Raton, Florida, 1987.

[18] T. R. Brown, R. A. Graham, B. S. Szwergold, W. J. Thoma, and R. A. Meyer, *Ann. N.Y. Acad. Sci.* **508,** 229 (1987).

[19] R. Taguchi, Y. Asahi, and H. Ikezawa, *Biochim. Biophys. Acta* **186,** 196 (1978).

TABLE II

^{31}P CHEMICAL SHIFTS OF PHOSPHOLIPIDS AND RELATED SPECIES[a]

Phospholipid	δ_P (ppm from external H_3PO_4)	Ref.
Phosphatidylcholine		
Micellar		
Cholate	−0.65	*b*
Triton X-100	−0.85	*c*
Monomer		
Dibutyryl-PC	−0.60	*c*
1-Lyso-PC		
Cholate	−0.34	
Triton X-100	−0.38	*b*
Phosphocholine	3.1	*d*
Phosphatidylethanolamine	0.00	*b*
Lyso-PE	0.44	
Phosphoethanolamine	3.4	*d*
Phosphatidylserine	−0.12	*b*
Lyso-PS	0.4	
Phosphatidic acid	3.8	*b*
Phosphatidylglycerol	0.43	*b*
Phosphatidylinositol	−0.40	
myo-Inositol 1,2-cyclic phosphate	16.5	*e*
Inositol monophosphate	4.4	
Sphingomyelin	0.00	*b*

[a] For monomeric species at pH 7.0 or species solubilized in detergent-mixed micelles (cholate unless otherwise noted).

[b] E. London and G. W. Feigenson, *J. Lipid Res.* **20,** 408 (1979).

[c] A. Pluckthun and E. A. Dennis, *Biochemistry* **21,** 1743 (1982).

[d] C. T. Burt, S. M. Cohen, and M. Barany, *Annu. Rev. Biophys. Bioeng.* **8,** 1 (1979).

[e] T. R. Brown, R. A. Graham, B. S. Szwergold, W. J. Thoma, and R. A. Meyer, *Ann. N.Y. Acad. Sci.* **508,** 229 (1987).

solved, and hydrolysis is easily followed (Fig. 2A). The preferential hydrolysis of PE in these mixtures, as opposed to its lower rate of hydrolysis when solubilized separately in Triton X-100 micelles, was used to suggest that cobra venom phospholipase A_2 has two distinct binding sites for phospholipids, one catalytic, the other an activator site.[12] Other binary phospholipid mixtures where one lipid is in an aggregate and the other is monomeric have also been examined by ^{31}P NMR to monitor the effects of phospholipase A_2 action. For example, the hydrolysis of the binary mixture PE (in Triton X-100 micelles) and dibutyryl-PC (which is monomeric) can be accurately followed over the course of the entire reaction (Fig. 2B), and in this case very accurate rates could be obtained.[13]

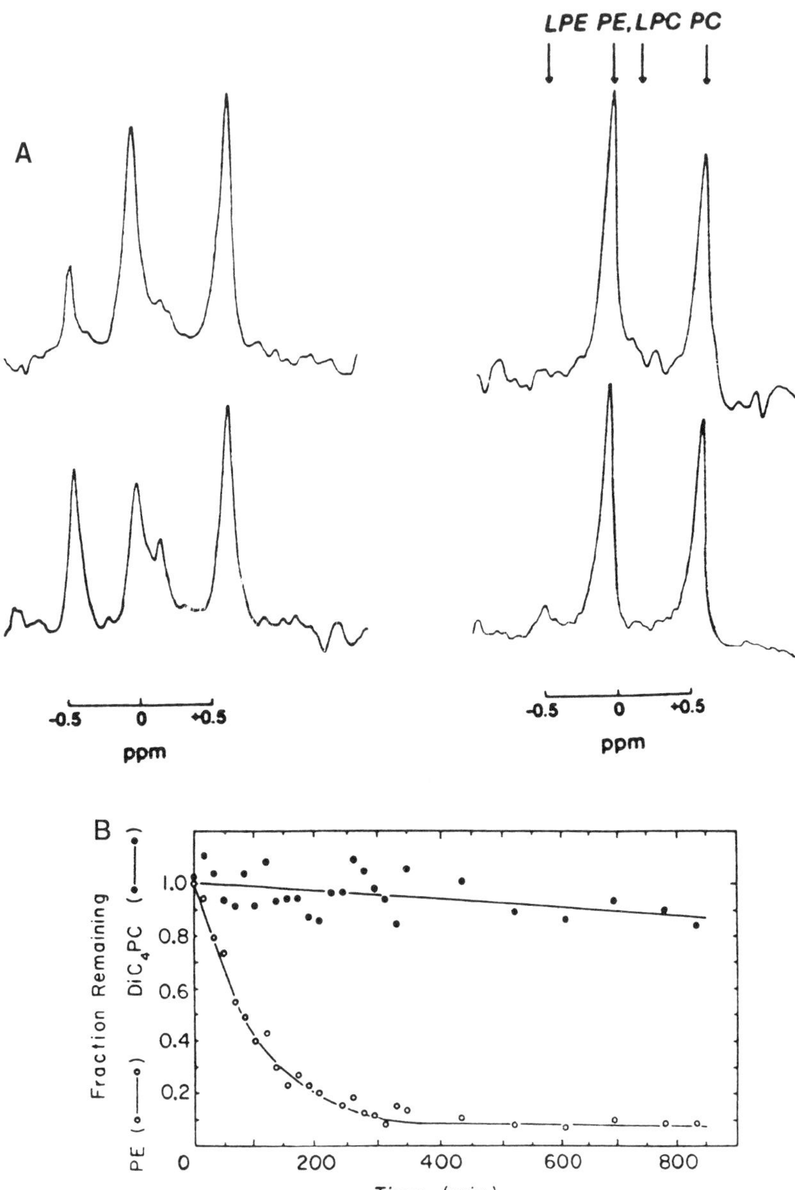

FIG. 2. (A) ^{31}P NMR spectra (40.5 MHz) of the hydrolysis of 2.6 mM egg PC and 3.0 mM egg PE in 48 mM Triton X-100, 50 mM Tris-HCl, 10 mM $CaCl_2$, pH 8.0, 40°, by phospholipase A_2 (50 μg). Peaks corresponding to lyso-PE (LPE), phosphatidylethanolamine (PE), lyso-PC (LPC), and phosphatidylcholine (PC) are indicated by arrows. [Reproduced with permission from M. F. Roberts, M. Adamich, R. J. Robson, and E. A. Dennis, *Biochemistry* **18,** 3301

Vesicular species have broader phosphorus resonances, and the resolution of different phospholipid head groups is more difficult. Sphingomyelin/diheptanoyl-PC (20 m*M* : 5 m*M*) bilayer particles, which are stable small particles below the T_m of the sphingomyelin, show distinct ^{31}P resonances for the two phospholipids separated by approximately 0.3 ppm; the upfield short-chain PC resonance is narrower than the resonance for the gel-state sphingomyelin. Only the short-chain PC is a substrate for phospholipases. When phospholipase A_2 hydrolyzes diheptanoyl-PC, one sees a decrease in the intensity of the ^{31}P resonance for that species and an increase in a new peak consistent with the lyso-PC product, 1-heptanoyl-PC; the sphingomyelin intensity remains constant, and a resonance for lyso-PC appears and increases in intensity.[9] Linewidths of the different lipid species do not change over the course of the reaction. Because of this, reaction rates can be quantified by monitoring peak heights in this system (Fig. 3A). In contrast, phospholipase C hydrolysis of the same bilayer lipid aggregate produces a sharp phosphocholine resonance while the substrate diheptanoyl-PC and matrix sphingomyelin broaden dramatically (Fig. 3B). This broadening occurs when approximately 20% of the diheptanoyl-PC has been hydrolyzed (determined by removing aliquots from the NMR sample and assaying for hydrolysis products by TLC). This means that the presence of 5 mol % short-chain diglyceride has destabilized the small bilayer particles, and large particles where the lipid phosphate groups are considerably broader have been produced.

^{13}C NMR Assays

^{13}C is only 1% naturally abundant. At 50–100 m*M* lipid concentration, excellent resolution of nearly all the carbons in the molecules is achieved, particularly carbons near the ester bonds to be hydrolyzed. Spectra are obtained with ^1H decoupling; for micellar material and water-soluble molecules with carbons with a directly bonded proton, the NOE is approximately 3. For the same nuclei on lipids solubilized in vesicles, the NOE decreases. With modern high-field spectrometers, although good signal-to-noise ratios can be obtained for 50–100 m*M* lipid in small aggregates,

(1979).] (B) Time course for hydrolysis of a binary mixture of 6.5 m*M* PE and 6.1 m*M* PC (dibutyryl-PC) in 96 m*M* Triton X-100 with 10 m*M* $CaCl_2$ by 0.11 μg phospholipase A_2 (*N. naja naja*) as followed by ^{31}P NMR at 40.5 MHz. The fraction of remaining substrate is calculated for PE from the integrated intensities of the PE and lyso-PE signals. For dibutyryl-PC, no lyso product was observed; therefore, the intensity of the PC signal relative to that at $t = 0$ is plotted. [Reproduced with permission from A. Pluckthun and E. A. Dennis, *Biochemistry* **21,** 1743 (1982).]

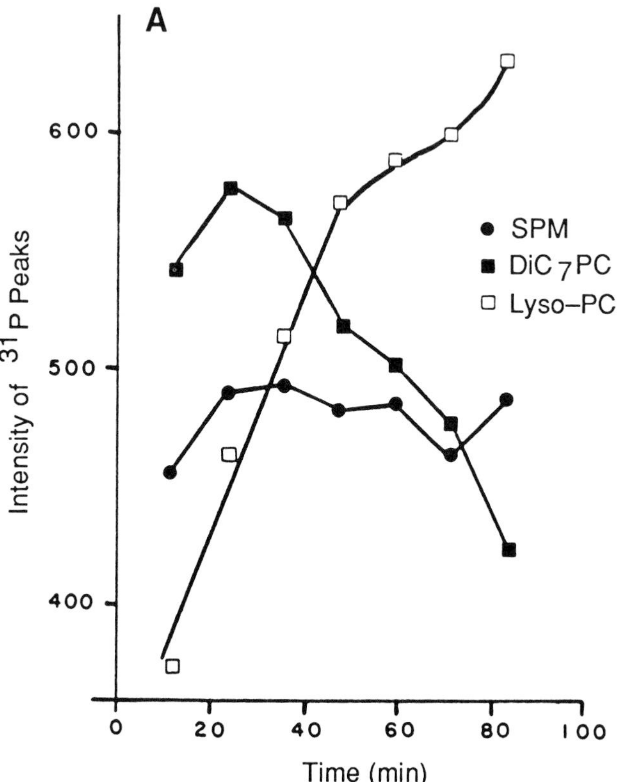

FIG. 3. (A) Time course for phospholipase A_2-catalyzed hydrolysis of sphingomyelin/diheptanoyl-PC (20 mM : 5 mM) as followed by ^{31}P NMR (109.3 MHz) spectroscopy at 25°. The concentration of enzyme (from *N. naja naja*) was 0.1 μg/ml. Since resonance linewidths remained constant over the time course of the incubation, peak height, rather than integrated area, is plotted as a function of time: (■) diheptanoyl-PC; (●) sphingomyelin; (□) 1-heptanoyl-PC. [Reproduced from N. E. Gabriel, N. V. Agman, and M. F. Roberts, *Biochemistry* **26,** 7409 (1987), with permission.] (B) ^{31}P NMR spectra of phospholipase C (*Bacillus cereus*) hydrolysis of the same particles as a function of time after the addition of enzyme (60 ng/ml). The separate resonances for sphingomyelin (SPM) and diheptanoyl-PC (diC$_7$PC) are identified. The arrow indicates the formation of the phosphate monoester (PME) phosphocholine.

at least 20% hydrolysis would be needed to reliably detect product in 0.5 to 1 hr. The substrate must be enriched in ^{13}C for greater sensitivity. With 99% enrichment, substrate concentrations should be at least in the 10 mM range.

The best choice of labeled phospholipid is the ester carbonyl carbon: the NOE is relatively small since there are no directly bonded protons,

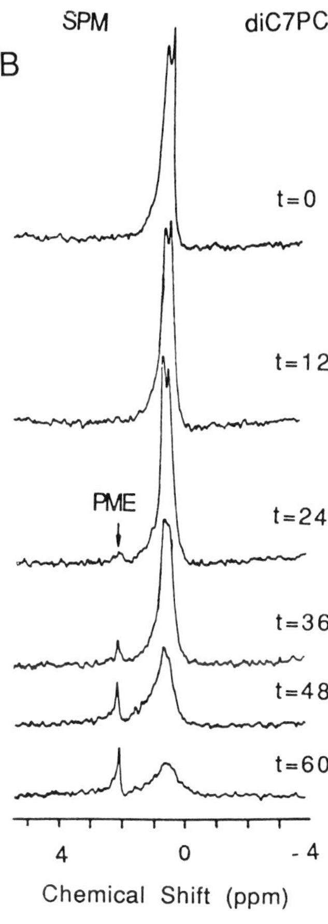

FIG. 3. (*continued*)

but this ensures that the linewidth is not so sensitive as CH_2 groups to particle size. With this probe, vesicles of even moderately large size can be observed for the effects of phospholipases. Phosphatidylcholines can be synthesized inexpensively with both carbonyls or the *sn*-2 carbonyl group selectively ^{13}C-labeled.[20,21] The carbonyl chemical shift has been shown to be sensitive to both intrinsic (chemical bonding and structure) and extrinsic factors (H-bonding and local environment).[22] This means

[20] J. T. Mason, A. V. Broccoli, and C.-H. Huang, *Anal. Biochem.* **113,** 96 (1981).

[21] C. M. Gupta, G. Radhakrishnan, and H. G. Khorana, *Proc. Natl. Acad. Sci. U.S.A.* **74,** 4315 (1977).

[22] J. A. Hamilton and D. P. Cistola, *Proc. Natl. Acad. Sci. U.S.A.* **83,** 82 (1986).

TABLE III
^{13}C CHEMICAL SHIFTS[a] OF CARBONYLS IN DIFFERENT PHOSPHATIDYLCHOLINE STRUCTURES

Phospholipid structure	δ (ppm)	Ref.
Monomer		
Dihexanoyl-PC	177.9; 177.2[e]	*b*
Lysohexanoyl-PC	177.0	
Micelle		
Diheptanoyl-PC	174.1	*b*
Dioctanoyl-PC	174.0	*b*
1-Hexanoyl-2-octanoyl-PC	174.3	
Octanoate	184.1	
1-Hexanoyl-2-octanoylglycerol	173.2	
Small unilamellar vesicle		
Egg PC	173.6 (i); 73.8 (o)[f]	*c*
Oleate (pH 4)	179.0	*c*
Lyso-PC	174.6 (i); 174.7 (o)	*c*
Diglyceride	173.3	*d*

[a] Referenced to external TMS.
[b] R. A. Burns and M. F. Roberts, *Biochemistry* **19,** 3100 (1980).
[c] S. P. Bhamidipati and J. A. Hamilton, *Biochemistry* **28,** 6667 (1989).
[d] S. P. Bhamidipati and J. A. Hamilton, *Biophys. J.* (1990).
[e] The two resonances represent *sn*-2 and *sn*-1 carbonyl groups.
[f] (i), Inner monolayer; (o), outer monolayer.

that the carbonyl resonances for phospholipids, lysophospholipids, fatty acid, and diglyceride are distinct (Table III). The carbonyl resonances for different head group lipids are not well resolved. Therefore, this nucleus is not a good choice for looking at mixed phospholipids, unless only one of them is enriched. Both chemical shifts and linewidths can provide information on the disposition of products in the system, namely, what is partitioned into the water, what is partitioned into the lipid aggregate, and any evidence for exchange. The reactions must be done in highly buffered solutions. Inorganic phosphate is a good choice if the phospholipase examined does not require any exogenous metal ions. If metal ions are necessary (e.g., Ca^{2+} for phospholipase A_2), an organic buffer such as Tris or MES at 500–100 m*M* can be used. Several examples follow.

Bhamidipati and Hamilton[23] have observed phospholipase A_2 hydrolysis of liquid crystalline dipalmitoyl-PC at 10–25 mol% in dihexadecyl-PC

[23] S. P. Bhamidipati and J. A. Hamilton, *Biochemistry* **28,** 6667 (1989).

small unilamellar vesicles prepared by sonication (Fig. 4). The ether-linked phospholipid is a nonhydrolyzable substrate analog, and it must be included in analysis of the rates if kinetic parameters (V_{max}, K_m) are desired. With substrate as a minor component of the vesicles, hydrolysis can occur without a large change in the overall particle. The ^{13}C NMR spectrum shows distinct resonances for inner and outer monolayer carbonyl resonances; the *sn*-1 and *sn*-2 carbonyls can be distinguished if the PC is synthesized with only one chain labeled. As seen in Fig. 4 only the PC on the outer leaflet was hydrolyzed under the conditions employed by Bhamidipati and Hamilton. From its chemical shift, the product fatty acid was shown to be partitioned almost exclusively in the bilayer and 50% ionized. Furthermore, it could be removed with addition of bovine serum albumin. This clearly indicates the chain specificity of the phospholipase and shows that vesicles with product are relatively stable.

Riedy *et al.*[24] have examined phospholipase hydrolysis of short-chain PC with the *sn*-2 carbonyl selectively ^{13}C-enriched. In pure micelles or in binary lipid bilayer systems (with gel state long-chain PC), the chemical shifts for the short-chain species are shifted approximately 0.5 ppm downfield from the long-chain carbonyl esters (Table III). This indicates a more hydrated environment around the carbonyl groups for these short-chain compounds compared to long-chain PC in bilayers.[25] Linewidths are relatively narrow ($\Delta\nu_{1/2}$ ~9–10 Hz), indicating significant mobility. In the small bilayer particles hydrolysis catalyzed by phospholipases is easily followed if the amount of enzyme added is such that at most 10% hydrolysis occurs in about 10 min. In the bilayer particles with 20 m*M* dihexadecyl-PC as the long-chain species, only the short-chain PC (5 m*M* 1-hexanoyl-2-[1-^{13}C]octanoyl-PC) is hydrolyzed. The enrichment is about 90% at the *sn*-2 chain and 10% at the *sn*-1 position; since both carbonyls are labeled above background, both products can be followed as a function of hydrolysis. Following either the decrease in intensity of PC (keeping in mind that only 90% of the integrated intensity of the resonance is from the *sn*-2 carbonyl) or the increase in labeled fatty acid (Fig. 5A) yields a rate consistent with that found by pH-stat assays.[9] The integrated intensity of the natural abundance $N(CH_3)_3$ resonance (which includes both PC components) serves as a control; as long as it is constant, there are no large particles being produced which might broaden carbonyls beyond detection. Products with short chains do not always stay associated with the bilayer. As the short-chain PC is hydrolyzed by phospholipase A_2, the 1-hexanoyl-

[24] G. Riedy, K.-M. Eum, and M. F. Roberts, submitted for publication.

[25] C. F. Schmidt, Y. Barenholz, C. Huang, and T. E. Thompson, *Biochemistry* **16,** 3948 (1977).

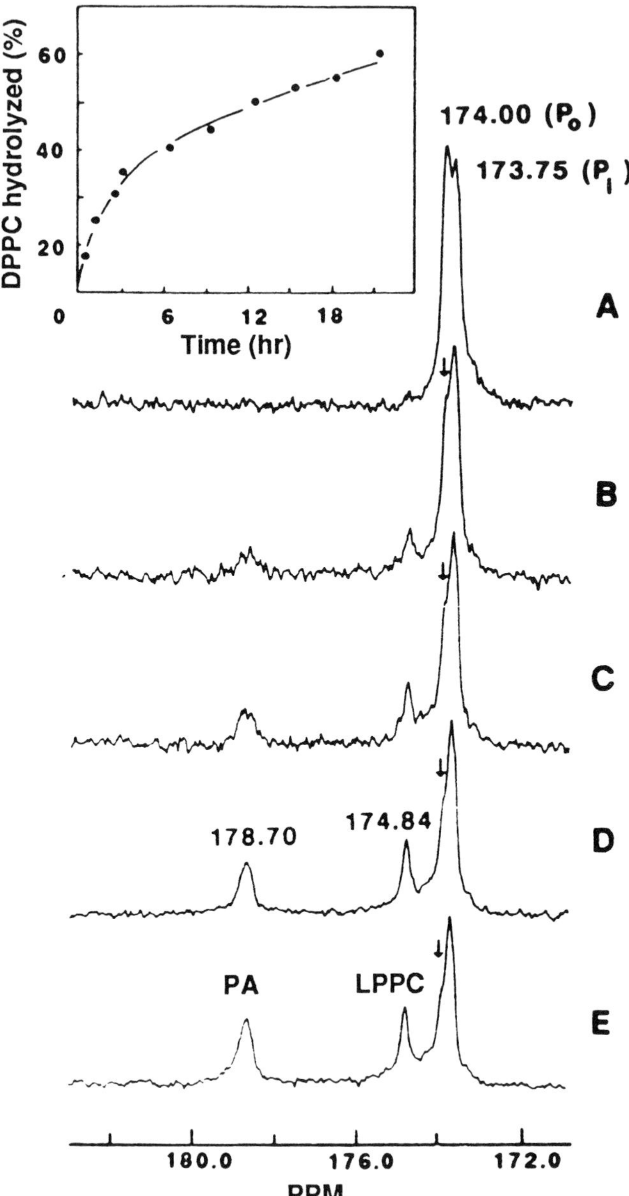

FIG. 4. Carbonyl region from 50.3-MHz ^{13}C NMR spectra of 10 mol % *sn*-1 and *sn*-2 [^{13}C]carbonyl-labeled dipalmitoyl-PC/90 mol % dihexadecyl-PC vesicles hydrolyzed by phospholipase A_2 (*Crotalus adamanteus*) at 46° as a function of time: (A) before the addition of enzyme and (B) 0.5 hr, (C) 2.5 hr, (D) 9.0 hr, and (E) 18.0 hr after the addition of enzyme.

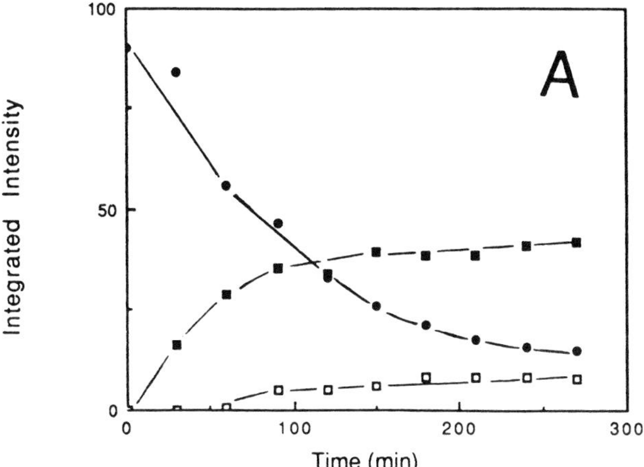

FIG. 5. (A) ^{13}C integrated intensities of different lipid resonances for 10 ng phospholipase A_2 (*N. naja naja*) hydrolysis of 5 m*M* 1-hexanoyl-2-[1-^{13}C]octanoyl-PC solubilized in bilayer particles with 20 m*M* dihexadecyl-PC : 1-hexanoyl-2-[1-^{13}C]octanoyl-PC (●); [1-^{13}C]octanoate (■); 1-hexanoyl-PC (□). The reaction was carried out at 25° in 50 m*M* Tris-HCl, pH 8, with 1 m*M* $CaCl_2$. (B) ^{13}C integrated intensities of 1-hexanoyl-2-[1-^{13}C]octanoyl-PC in the same particles as a function of time of incubation with 30 ng phospholipase C (*B. cereus*) : PC *sn*-2 carbonyl (●); natural abundance $N(CH_3)_3$ (O). (C) Linewidth (■) and chemical shift (□) of the PC carbonyl as a function of incubation time with phospholipase C. The large increase in linewidth above 100 min indicates that this region will not provide accurate rate data.

PC product (linewidth, $\Delta\nu_{1/2}$ ~4 Hz) is predominantly partitioned into the aqueous phase while the octanoate ($\Delta\nu_{1/2}$ ~9 Hz) is associated with the bilayer particles. This is information provided by the relative linewidths and chemical shifts of the products (Table III). Furthermore, the chemical shift of the fatty acid indicates it is completely ionized.

With phospholipase C added to the same binary lipid small bilayers, rates for hydrolysis of 5 m*M* 1-hexanoyl-2-[1-^{13}C]octanoyl-PC can also be determined (Fig. 5B). In this case a carbonyl resonance for the short-chain diglyceride is not detected, and all resonances begin to broaden (Fig. 5C; the linewidths and chemical shifts for the substrate are shown as a function of time) after about 25% of the short-chain PC has been hydrolyzed. The

---

P_o and P_i represent dipalmitoyl-PC carbonyl signals from the outer and the inner monolayers, respectively. Spectra (A)–(C) represent 12,000 transients with a 2.0-sec recycle delay and 2.0-Hz line broadening. Spectra (D) and (E) represent 2000 transients. The inset shows the percentage of dipalmitoyl-PC hydrolyzed (estimated from the loss in total dipalmitoyl-PC signal) as a function of time. [Reproduced from S. P. Bhamidipati and J. A. Hamilton, *Biochemistry* **28,** 6667 (1989), with permission.]

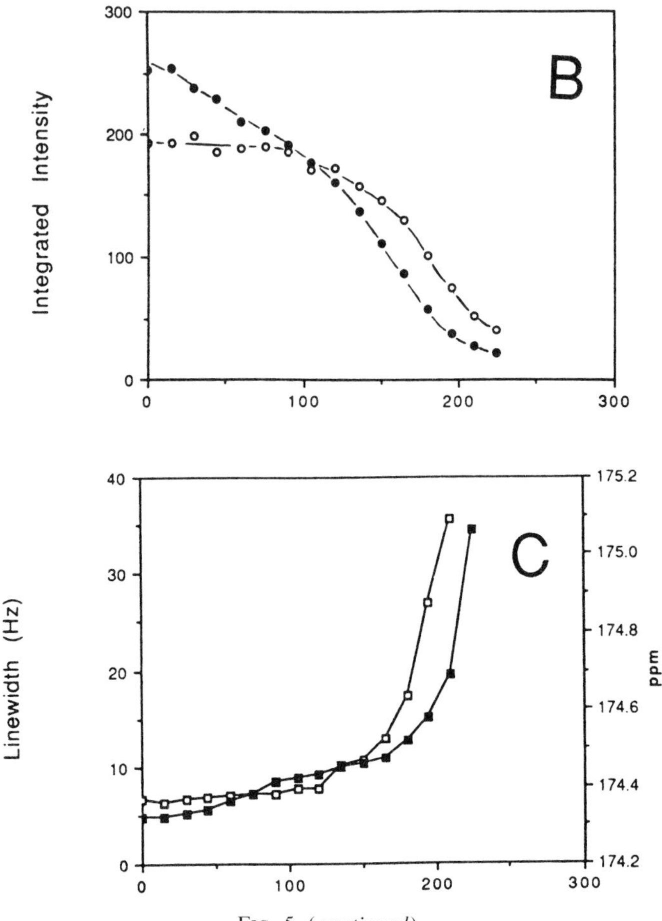

FIG. 5. (*continued*)

large linewidth and reduced intensity of the diglyceride carbonyl suggest that this species is either phase separating or exchanging between different environments. The chemical shift of the PC varies in parallel to the linewidth when the particles begin to grow. Thus, accurate rates can be determined only for the time course where such particle changes are minimal; again, a constant value for the integrated intensity of the natural abundance $N(CH_3)_3$ resonance serves as a good control (Fig. 5B). For the example shown, the phospholipase C (*Bacillus cereus*) specific activity estimated for less than 100 min was 940 μmol/min/mg. The extra information the ^{13}C NMR spectrum provides (in particle linewidth and chemical shift information) is that, after approximately 20% of the substrate is hydrolyzed (leading to 1 m*M* of 1-hexanoyl-2-[1-^{13}C]octanoyldiglyceride), the bilayer particles, which also contain 20 m*M* dihexadecyl-PC, grow to the point where much of the ^{13}C signal is no longer NMR visible. Phase separation into a lipid-rich multilamellar phase or fusion of the small bilayer particles to large multilamellar ones could be responsible for such behavior.

Other Nuclei

The examples so far have stressed the use of ^{1}H, ^{31}P, and ^{13}C NMR in assaying phospholipases. Other nuclei could also be exploited for assays, in particular ^{2}H and ^{15}N. In order to observe these nuclei in phospholipids, enriched substrates must be synthesized. ^{2}H is quadrupolar, and even with enrichment its sensitivity is not great. The ^{2}H spectrum for a C–D group will be an isotropic line when the lipid is solubilized in micelles and unilamellar vesicles.[10,26] When the C–D group is organized in a multilamellar particle, quadrupolar splitting (5–20 kHz, depending on the position of the C–D group in the phospholipid molecule) is observed.[26] Two types of ^{2}H experiments can be envisioned. Both focus not on enzyme rates, but on the distribution of substrates and products in different aggregate structures: (1) For observation of chain-deuterated (or specifically labeled) phospholipid in multilamellar vesicles as a function of time after the addition of phospholipases, products, as substrate is hydrolyzed, may cause formation of smaller vesicles which give rise to isotropic lines. (2) For observation of the ^{2}H signal for the substrate initially in an isotropic particle, if the action of a phospholipase alters this particle to a large aggregate, or if product phase separates into a lipid-rich phase, then quadrupolar coupling should be resolved. Again, this nucleus is not useful for routine assays but for changes in lipid substrate aggregate structure. Clearly, it would be good for assessing particles at the end of hydrolysis or after substantial hydrolysis has occurred.

^{15}N NMR observation of phospholipids would also require enrichment since the natural abundance of this isotope is only approximately 0.4%. The ^{15}N chemical shift range is large, and separation of PE, PC, and phosphatidylserine (PS) head groups should be quite easy. Water-soluble products should also have distinct chemical shifts; although values for lysophospholipids have not been tabulated, these are expected to be resolved from the corresponding substrates. At pH 7, two-dimensional (2D) indirect detection techniques (e.g., HMQC) cannot be used for detecting the ^{15}N of amino groups because of N–H exchange; hence, direct detection

[26] M. F. Brown, *J. Chem. Phys.* **77,** 1576 (1982).

methods are more suitable. ^{15}N can have a negative NOE, but it can also become positive. The exact value depends on the correlation time of the nucleus which is in turn related to the particle size and dynamics. T_1 values for water-soluble species can also be large so that, to obtain rate data, one needs to measure both the NOE and T_1 values for products and substrates and correct integrated intensities accordingly. ^{15}N might be particularly relevant for looking at preferential hydrolysis of lipids with different head groups. An example where it might be useful is in assaying the activity of phospholipases toward PE or PS substrates. For an amino nitrogen, the ^{15}N chemical shift will be very sensitive to pH. Differences in the PE chemical shift in different substrate particles may allow monitoring of environmental differences in particles which can be related to rates of hydrolysis. Shifts for PC species would be expected to be less sensitive to such parameters.

Conclusion

NMR spectroscopy is a useful technique for monitoring not only the kinetics of phospholipase action but the physical state of substrates and products produced. It is not a routine assay method and not so sensitive as others (~0.5 m*M* or more product must be produced in order to detect signals in a reasonable length of time), but it is excellent for providing information on product disposition and changes in the substrate physical state, information other assay systems cannot provide. NMR spectroscopy may prove to be very useful for mammalian phospholipases as long as a reasonable substrate system is chosen. With the right nucleus, buffers, detergents, etc., may be present, but these will not hinder observation of the desired resonances.

[4] Monolayer Techniques for Studying Phospholipase Kinetics

By S. RANSAC, H. MOREAU, C. RIVIÈRE, and R. VERGER

Introduction: Why Use Lipid Monolayers as Phospholipase Substrates?

There are at least five major reasons for using lipid monolayers as substrates for lipolytic enzymes (readers are referred to previous reviews for details[1-4]): (1) The monolayer technique is highly sensitive, and very little lipid is needed to obtain kinetic measurements.This advantage can often be decisive in the case of synthetic or rare lipids. Moreover, a new phospholipase A_2 has been discovered using the monolayer technique as an analytical tool.[5] (2) During the course of the reaction, it is possible to monitor one of several physicochemical parameters characteristic of the monolayer film: surface pressure, potential, radioactivity, etc. These variables often give unique information. (3) With this technique the lipid packing of a monomolecular film of substrate is maintained constant during the course of hydrolysis, and it is therefore possible to obtain accurate, presteady-state kinetics measurements with minimal perturbation caused by increasing amounts of reaction products. (4) Probably most importantly, it is possible with lipid monolayers to vary and modulate the "interfacial quality," which depends on the nature of the lipids forming the monolayer, the orientation and conformation of the molecules, the molecular and charge densities, the water structure, the viscosity, etc. One further advantage of the monolayer technique as compared to bulk methods is that, with the former, it is possible to transfer the film from one aqueous subphase to another. (5) Inhibition of phospholipase activity by water-insoluble substrate analog can be precisely estimated using a "zero-order" trough and mixed monomolecular films in the absence of any synthetic, nonphysiological detergent. The monolayer technique is therefore suitable for modeling *in vivo* situations.

[1] R. Verger and G. H. de Haas, *Annu. Rev. Biophys. Bioeng.* **5,** 77 (1976).
[2] R. Verger, this series, Vol. 64, p. 340.
[3] R. Verger and F. Pattus, *Chem. Phys. Lipids* **30,** 189 (1982).
[4] G. Piéroni, Y. Gargouri, L. Sarda, and R. Verger, *Adv. Colloid Interface Sci.* **32,** 341 (1990).
[5] R. Verger, F. Ferrato, C. M. Mansbach, and G. Piéroni, *Biochemistry* **21,** 6883 (1982).

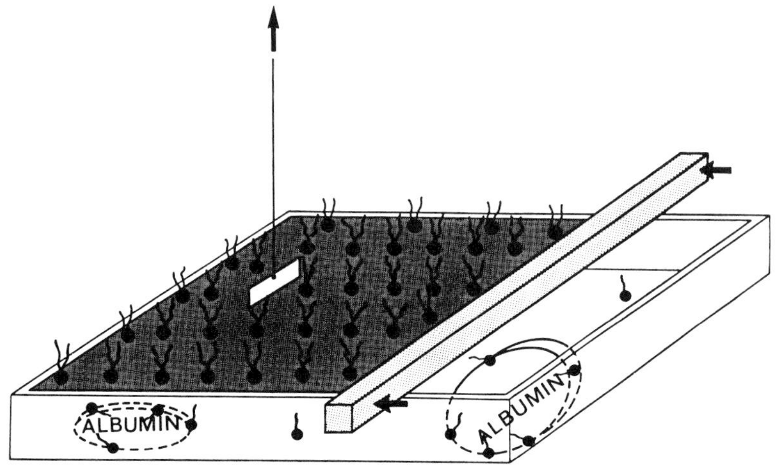

FIG. 1. Method for studying the hydrolysis of long-chain phospholipid monolayers with controlled surface density. A large excess of serum albumin has to be present in the aqueous subphase in order to solubilize the lipolytic products.

Pure Lipid Monolayers as Phospholipase Substrates

A new field of investigation was opened in 1935 when Hughes[6] used the monolayer technique for the first time to study enzymatic reactions. He observed that the rate of the phospholipase A-catalyzed hydrolysis of a lecithin film, measured in terms of the decrease in surface potential, diminished considerably when the number of lecithin molecules per square centimeter was increased. Since this early study, several laboratories have used the monolayer technique to monitor lipolytic activities, mainly with glycerides and phospholipids as substrates. These studies can be tentatively divided into four groups. Long-chain lipids were used and their surface density (number of molecules/cm^2) was not controlled, short-chain lipids were applied, again without controlling the surface density, or short-chain lipids were used at constant surface density. As shown in Fig. 1, C. Rothen (personal communication, 1980) has developed a fourth method at our laboratory for studying the hydrolysis of long-chain phospholipid monolayers by various phospholipases A_2 involving controlled surface density. A large excess of serum albumin has to be present in the aqueous subphase in order to solubilize the lipolytic products. This step was found not to be rate limiting. The linear kinetics obtained by these authors with

[6] A. Hughes, *Biochem. J.* **29**, 437 (1935).

natural long-chain phospholipids were quite similar to those previously described in the case of short-chain phospholipids using a zero-order trough with the barostat technique.[3]

A shortcoming of the monolayer technique is the denaturation of many enzymes that occurs at lipid–water interfaces. This is attributable to the interfacial free energy. Fortunately, this phenomenon occurs slowly with the highly resistant phospholipases.

Triggering of Monolayer Activity of Phospholipase A_2 by Electrical Field or "Vertical Compression"

Thuren *et al.*[7] showed that the action of phospholipase A_2 can be triggered by applying an electric field across a 1,2-didodecanoyl-*sn*-3-phosphoglycerol monolayer lying between an alkylated silicon surface and water. When the silicon wafer served as a cathode, rapid activation of porcine pancreatic phospholipase was observed and was found to depend on the magnitude of the applied potential. The degree of activation differed depending on whether pancreatic phospholipase A_2 or snake or bee venom enzymes were used. Maximally, a 7-fold activation of pancreatic phospholipase A_2 was observed when the applied potential was 75 V. The effective field over the lipid film was estimated to be approximately 25–175 mV, which is within the range of the membrane potentials found in cells. On the basis of these results, it was suggested that changes in membrane potential might be an important factor in the regulation of the action of intracellular phospholipases A_2 *in vivo*.

The same authors "vertically compressed" the phosphoglycerol monolayers by substituting an alkylated glass plate for air while maintaining a constant surface pressure.[8] Subsequently, the activities of phospholipases A_1 and A_2 toward the monolayers were measured both in the presence and in the absence of the support. While phospholipase A_1 activity was increased 4-fold by the support, the activity of phospholipase A_2 was reduced to 15% of the activity measured in the absence of the alkylated surface. These findings indicate that this "vertical compression" exerted on the monolayer is likely to have induced a conformational change in the phospholipid molecules, which in turn may have caused the above reciprocal changes in the activities of phospholipases A_1 and A_2.

[7] T. Thuren, A. P. Tulkki, J. A. Virtanen, and P. K. J. Kinnunen, *Biochemistry* **26,** 4907 (1987).

[8] T. Thuren, J. A. Virtanen, and P. K. J. Kinnunen, *Biochemistry* **26,** 5816 (1987).

Hydrolytic Action of Phospholipase A_2 in Monolayers in the Phase Transition Region

Grainger *et al.*[9,10] have characterized optically the phospholipase A_2 (*Naja naja*) during its action against a variety of phospholipid monolayers using fluorescence microscopy. By labeling the enzyme with a fluorescent marker (fluorescein), the hydrolysis of lipid monolayers in their liquid–solid phase transition region could be directly observed with the assistance of an epifluorescence microscope. Visual observation of hydrolysis of different phospholipid monolayers in the phase transition region in real-time could differentiate various mechanism of hydrolytic action against lipid solid phase domains. Lipid monolayers were spread over a buffer subphase at temperature necessary (30° for dipalmitoylphosphatidylcholine) to reach the monolayer liquid–solid phase transition at 22 mN/m for each respective lipid. Dipalmitoylphosphatidylcholine solid phase domains were specifically targeted by phospholipase A_2 and were observed to be hydrolyzed in a manner consistent with localized packing density differences, as illustrated in Fig. 2. Dipalmitoylphosphatidylethanolamine lipid domain hydrolysis showed no such preferential phospholipase A_2 response but did demonstrate a preference for solid/lipid interfaces. In all cases, after critical extents of monolayer hydrolysis in the phase transition region, highly stable, organized domains of enzyme of regular sizes and morphologies were consistently seen to form in the monolayers. Enzyme domain formation was entirely dependent upon hydrolytic activity in the monolayer phase transition region and was not witnessed otherwise.

Zero-Order Trough

Several types of troughs have been used to study enzyme kinetics. The simplest of these is made of Teflon and is rectangular in shape (Fig. 1), but it gives nonlinear kinetics.[11] To obtain rate constants, a semilogarithmic transformation of the data is required. This drawback was overcome by a new trough design (zero-order trough, Fig. 3), consisting of a substrate reservoir and a reaction compartment containing the enzyme solution.[12] The two compartments are connected to each other by a narrow surface canal. The kinetic recordings obtained with this trough are linear, unlike the nonlinear plots obtained with the usual one-compartment trough. The buffer is filtered and the pH is adjusted immediately prior to use. Before

[9] D. W. Grainger, A. Reichert, H. Ringsdorf, and C. Salesse, *FEBS Lett.* **252,** 73 (1989).

[10] D. W. Grainger, A. Reichert, H. Ringsdorf, and C. Salesse, *Biochim. Biophys. Acta* **1023,** 365 (1990).

[11] G. Zografi, R. Verger, and G. H. de Haas, *Chem. Phys. Lipids* **7,** 185 (1971).

[12] R. Verger and G. H. de Haas, *Chem. Phys. Lipids* **10,** 127 (1973).

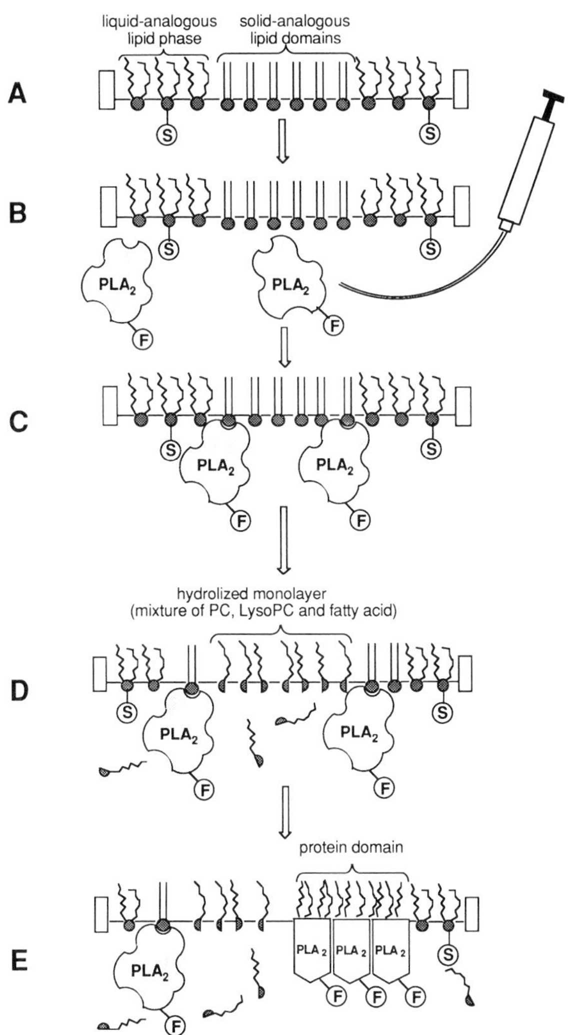

FIG. 2. Hydrolysis of a lipid monolayer by phospholipase A_2 (schematic): (A), monolayer in the phase transition region with solid analogous lipid domains in a fluid analogous matrix mixed with a sulforhodamine marker; (B), injection of the FITC-labeled phospholipase A_2; (C), specific recognition of the substrate lipids by the enzyme and preferential attack at boundaries between the solid analogous and the liquid analogous lipid phase; (D), hydrolysis of the solid analogous lipid domains and accumulation of the hydrolysis products in the monolayer; (E), aggregation of the enzyme to domains of regular morphology. [From D. W. Grainger, A. Reichert, H. Ringsdorf, and C. Salesse, *Biochim. Biophys. Acta* **1023,** 365 (1990).]

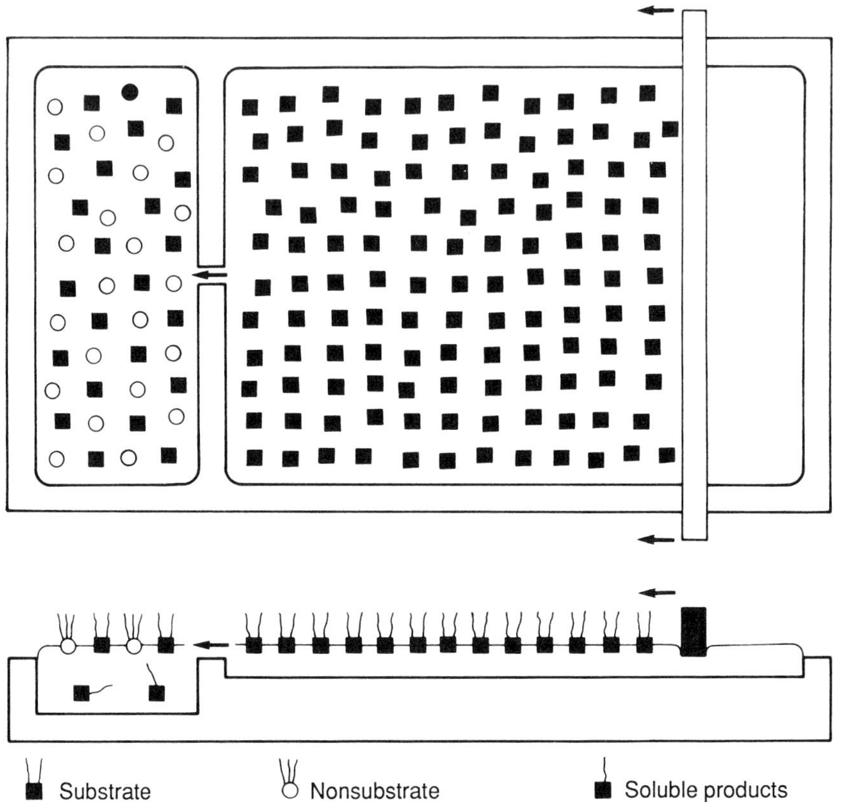

FIG. 3. Principle of the method for studying enzymatic lipolysis of mixed monomolecular films. [From G. Piéroni and R. Verger, *J. Biol. Chem.* **254,** 10090 (1979).]

each experiment, the trough is carefully cleaned with a paintbrush using absolute ethanol, rinsed several times with tap water until the Teflon surface no longer retains drops of water, and then rinsed twice with distilled water. Set-up and completion of a typical kinetic experiment requires about 30 min. Unlike previous experiments, no detergent solution is employed. The surface pressure is maintained constant automatically by the surface barostat method described elsewhere.[12] Fully automatized monolayer systems of this kind are now commercially available (KSV, Helsinki, Finland).

Film Recovery and Estimation of Bound Enzyme

The main difference between the monolayer and the bulk system lies in the ratios of interfacial area to volume, which differ by several orders of magnitude. In the monolayer system, this ratio is usually about 1 cm^{-1}, depending on the depth of the trough, whereas in the bulk system it can be as high as 10^5 cm^{-1}, depending on the amount of lipid used. Consequently, under bulk conditions, the adsorption of nearly all the enzyme occurs at the interface, whereas with a monolayer only one enzyme molecule out of a hundred may be at the interface.[11] Owing to this situation, a small but unknown amount of enzyme, responsible for the observed hydrolysis rate, is adsorbed on the monolayer. In order to circumvent this limitation, two different methods were proposed for recovering and measuring the quantity of enzymes absorbed at the interface.[13-15]

After performing velocity measurements, Momsen and Brockman[15] transferred the monolayer to a piece of hydrophobic paper, and the adsorbed enzyme was then assayed titrimetrically. The paper was pretreated by soaking for several hours in deionized water at pH 7 and then equilibrated overnight in air saturated with water vapor. When the mobile bar was about 4 cm from the end of the trough, the paper was lowered onto the surface and left in contact for about 15–20 sec. Leaving the paper on the surface up to 5 min did not affect the amount of lipase adsorbed. The paper was pulled gently over the edge of the trough, and any residual subphase droplets were shaken off. Subphase carryover at pressures of 10, 16, and 22 mN/m was found to be 20 ± 4 μl (n = 14) when measured gravimetrically. After correcting for the blank rate and subphase carryover, the amount (moles) of adsorbed enzyme was calculated from the net velocity and the specific enzyme activity.

In assays performed with radioactive enzymes,[13] the film was aspirated by inserting the end of a bent glass capillary into the liquid meniscus emerging above the ridge of the Teflon compartment walls, as depicted in Fig. 4C. The other end of the same capillary was dipped into a 5-ml counting vial connected to a vacuum pump. As aspiration proceeded, the film was compressed with a mobile barrier to facilitate quantitative recovery (0.5 ml of liquid with a film of 120 cm^2). The capillary was broken into pieces which were added to the vial before counting. As radioactive molecules dissolved in the subphase were unavoidably aspirated with the film constituents, the results had to be corrected by counting the radioactivity in the same volume of aspirated subphase. The difference between the two values, which actually expressed a certain excess of radioactivity existing at the surface, was attributed to the enzyme molecules bound to the film.[13]

The surface-bound enzyme includes not only those enzyme molecules directly involved in the catalysis but also an unknown amount of protein

[13] J. Rietsch, F. Pattus, P. Desnuelle, and R. Verger, *J. Biol. Chem.* **252,** 4313 (1977).
[14] S. G. Bhat and H. L. Brockman, *J. Biol. Chem.* **256,** 3017 (1981).
[15] W. E. Momsen and H. L. Brockman, *J. Biol. Chem.* **256,** 6913 (1981).

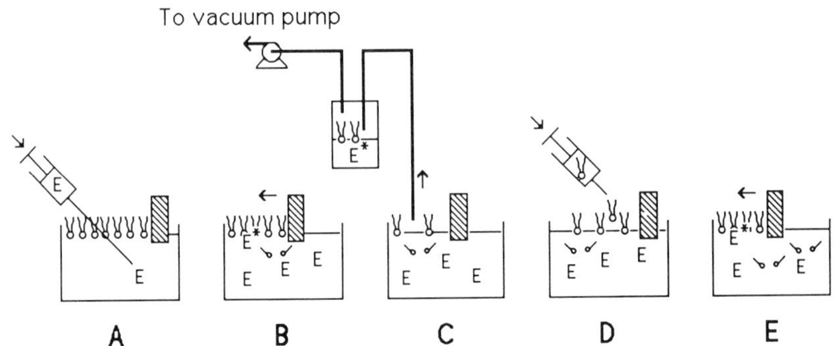

FIG. 4. Diagram of the method used for enzyme kinetic experiments after film respreading. (A) Enzyme injection. (B) First kinetic recording. (C) Film aspiration. (D) Film respreading. (E) Second kinetic recording.

present close to the monolayer. These enzyme molecules are not necessarily involved in the enzymatic hydrolysis of the film. Since it is possible with the monolayer technique to measure the enzyme velocity (expressed in μmol/cm^2/min) and the interfacial excess of enzyme (mg/cm^2), it is easy to obtain a value of enzymatic specific activity, which can be expressed as usual (in μmol/min/mg).

As illustrated in Fig. 4, Ransac *et al.*[16] devised a simple technique for estimating the amount of unlabeled enzyme bound to the monomolecular film. This technique involves the dual kinetic recording of enzyme velocity on two monomolecular films consecutively spread over the same enzyme solution. As usual, the enzyme was injected at a final concentration of 50 p*M* (Fig. 4A) into the subphase of the reaction compartment (100 cm^2) of a zero-order trough, and the first kinetic recording of enzyme velocity was performed during 10 min (Fig. 4B). The film was then collected (Fig. 4C) as described previously. The second substrate film was spread at the same final surface pressure over the remaining subphase containing the residual enzyme (Fig. 4D), and a second kinetic recording was performed (Fig. 4E). Such an experiment takes about 40 min. By comparing the enzymatic rates between the two kinetic experiments, it is possible to estimate the fraction of enzyme removed during the film aspiration. This technique turned out to be particularly well suited for determining the amount of native phospholipase A_2 bound to monolayers of substrate analogs.[16]

[16] S. Ransac, to be published.

Lag Periods in Lipolysis

A fairly common observation reported by many authors working on the kinetics of lipolytic enzymes is the occurrence of lag periods in the hydrolysis of emulsions, liposomes, micelles, and monolayers.[17–29] The zero-order trough gave, as might be expected, linear kinetics after injection of venom enzyme under the monolayer.[21] This was not the case, however, when pancreatic phospholipase A_2 was injected under a film of dinonanoyl-lecithin. In fact, the velocity, as given by the slope of the recorded curve, was found to increase with time and seemed to approach an asymptotic limit: the intercept between the asymptote and the time axis is the lag time τ. This behavior contrasts strongly with the kinetics obtained after injection of pure phospholipase A_2 from snake or bee venom under the same lecithin film. The exact reasons for the unusually long lag periods are still open to debate.[30] Bianco *et al.*[31] have observed, however, that changes in surface potential parallel the variation observed in the enzyme velocity with time. A plateau was found to occur corresponding to a decrease in potential between 20 and 70 mV. This decrease in surface potential depends linearly on the enzyme concentration in the subphase, and it probably reflects the enzyme adsorption at the interface.

Demel *et al.*[32] and Verheij *et al.*[33,34] measured the influence of the surface pressure on the lag time of several phospholipases A_2. The authors classified these enzymes unambiguously depending on their penetration power. With each enzyme, there exists a characteristic critical substrate

[17] H. L. Brockman, J. H. Law, and F. J. Kézdy, *J. Biol. Chem.* **248,** 4965 (1973).

[18] O. A. Roholt and M. Schlamowitz, *Arch. Biochem. Biophys.* **94,** 364 (1961).

[19] A. F. Rosenthal and M. Pousada, *Biochim. Biophys. Acta* **164,** 226 (1968).

[20] R. H. Quarles and R. M. C. Dawson, *Biochem. J.* **113,** 697 (1969).

[21] R. Verger, M. C. E. Mieras, and G. H. de Haas, *J. Biol. Chem.* **248,** 4023 (1973).

[22] M. K. Jain and R. C. Apitz-Castro, *J. Biol. Chem.* **253,** 7005 (1978).

[23] D. O. Tinker and J. Wei, *Can. J. Biochem.* **57,** 97 (1979).

[24] F. Pattus, A. J. Slotboom, and G. H. de Haas, *Biochemistry* **13,** 2691 (1979).

[25] G. C. Upreti, S. Rainier, and M. K. Jain, *J. Membr. Biol.* **55,** 97 (1980).

[26] G. C. Upreti and M. K. Jain, *J. Membr. Biol.* **55,** 113 (1980).

[27] B. Borgström, *Gastroenterology* **78,** 954 (1980).

[28] K. Hirasawa, R. F. Irvine, and R. M. C. Dawson, *Biochem. J.* **193,** 607 (1981).

[29] M. Menashe, D. Lichtenberg, C. Gutierrez, and R. L. Biltonen, *J. Biol. Chem.* **256,** 4541 (1981).

[30] M. K. Jain and O. G. Berg, *Biochim. Biophys. Acta* **1002,** 127 (1989).

[31] I. D. Bianco, G. D. Fidelio, and B. Maggio, *Biochem. J.* **258,** 95 (1989).

[32] R. A. Demel, W. S. M. Geurts van Kessel, R. F. A. Zwaal, B. Roelofsen, and L. L. M. van Deenen, *Biochim. Biophys. Acta* **406,** 97 (1975).

[33] H. M. Verheij, M. R. Egmond, and G. H. de Haas, *Biochemistry* **20,** 94 (1981).

[34] H. M. Verheij, M. C. Boffa, C. Rothen, M. C. Brijkaert, R. Verger, and G. H. de Haas, *Eur. J. Biochem.* **112,** 25 (1980).

packing density above which the enzyme cannot penetrate the film. This type of information can be of practical value when choosing a lipolytic enzyme for degrading the lipid moiety of biological membranes[32] or for predicting the anticoagulant properties of a given phospholipase A_2.[34]

Mixed Monolayers as Phospholipase Substrates

Most studies on lipolytic enzyme kinetics have been carried out *in vitro* with pure lipids as substrates. Actually, however, virtually all biological interfaces are composed of complex mixtures of lipids and proteins. The monolayer technique is ideally suited for studying the mode of action of lipolytic enzymes at interfaces using controlled mixtures of lipids. There exist two methods of forming mixed-lipid monolayers at the air–water interface: spreading a mixture of water-insoluble lipids from a volatile organic solvent or injecting a micellar detergent solution into the aqueous subphase covered with preformed insoluble lipid monolayers.

A new application of the zero-order trough was proposed by Piéroni and Verger[35] for studying the hydrolysis of mixed monomolecular films at constant surface density and constant lipid composition (Fig. 3). A Teflon barrier was disposed transversely over the small channel of the zero-order trough in order to block surface communication between the reservoir and the reaction compartment. The surface pressure was first determined by placing the platinum plate in the reaction compartment, where the mixed film was spread at the required pressure. Then surface pressure was measured after switching the platinum plate to the reservoir compartment where the pure substrate film was subsequently spread. The surface pressure of the reservoir was equalized to that of the reaction compartment by moving the mobile barrier. The subphase of the reaction compartment, composed of a standard buffer (10 m*M* Tris/acetate, pH 6.0, 0.1 *M* NaCl, 21 m*M* $CaCl_2$, 1 m*M* EDTA, thermostatted at 25° ± 0.5°), was stirred at 250 rpm with two magnetic bars. The barrier between the two compartments was then removed in order to allow surfaces to communicate. The pressure change during these operations did not exceed 0.25 dyne/cm. The enzyme was then injected into the reaction compartment and the kinetics recorded as described.[12]

The authors[36] studied the hydrolysis of mixed monomolecular films of phosphatidylcholine/triacylglycerol by pancreatic phospholipase A_2. The quantity of enzyme adsorbed on the interface was concomitantly determined with ^3H-amidinated phospholipase. At phospholipid packing levels

[35] G. Piéroni and R. Verger, *J. Biol. Chem.* **254,** 10090 (1979).

[36] G. Piéroni and R. Verger, *Eur. J. Biochem.* **132,** 639 (1983).

above the critical penetration pressure, triacylglycerol considerably enhances phosphatidylcholine hydrolysis. On the other hand, the activity of pancreatic phospholipase A_2 on a mixed film is inhibited by the action of pancreatic lipase.

Using the same methodology, Alsina *et al.*[37] studied the enzymatic lipolysis by pancreatic phospholipase A_2 and by phospholipase A_2 of *Vipera berus* on monomolecular films of mixture of natural lipids: cholesterol-egg lecithin and triolein-egg lecithin, by adding serum albumin to the aqueous subphase in order to imitate the physiological conditions. Later on Alsina *et al.*[38] reported a complementary study on the lipolysis of didecanoyl phosphatidylcholine/triolein mixed monolayers by phospholipase A_2.

Grainger *et al.*[39] have characterized physically and have studied the enzymatic hydrolysis of mixed monolayers of a natural phospholipid substrate and a polymerizable phospholipid analog. Enzyme hydrolysis showed large differences in the ability of the enzyme to selectively hydrolyze the natural phosphatidylcholine component from the monomeric as opposed to the polymeric mixtures. The results clearly show a strong influence of molecular environment on phospholipase A_2 activity, even if differences in the physical state of mixed monolayers are not detectable with isotherm and isobar measurements.

Inhibition of Phospholipases Acting on Mixed Substrate/Inhibitor Monomolecular Films

Many drugs and phospholipid analogs have been reported to act as phospholipase "inhibitors." A priori, these compounds can be said to interfere with phospholipase A_2 (PLA_2) activity by interacting either directly with the enzyme or indirectly by affecting the "interfacial quality" of the substrate. A number of "membrane-active" compounds have marked effects. For instance, low concentrations of alcohols and detergents can stimulate PLA_2-catalyzed hydrolysis of lipid bilayers, whereas higher concentrations are inhibitory. The effects of these compounds are related somehow to their amphiphilic nature, and it is believed that they may act as spacer molecules facilitating the penetration of the bilayer by the enzyme. An illustration has been provided by Vainio *et al.*,[40] who

[37] A. Alsina, O. Valls, G. Piéroni, R. Verger, and S. Garcia, *Colloid Polym. Sci.* **261,** 923 (1983).

[38] M. A. Alsina, M. L. Garcia, M. Espina, and O. Valls, *Colloid Polym. Sci.* **267,** 923 (1989).

[39] D. W. Grainger, A. Reichert, H. Ringsdorf, C. Salesse, D. E. Davies, and J. B. Lloyd, *Biochim. Biophys. Acta* **1022,** 146 (1990).

[40] P. Vainio, T. Thuren, K. Wichman, T. Luukkainen, and P. J. K. Kinnunen, *Biochim. Biophys. Acta* **814,** 405 (1985).

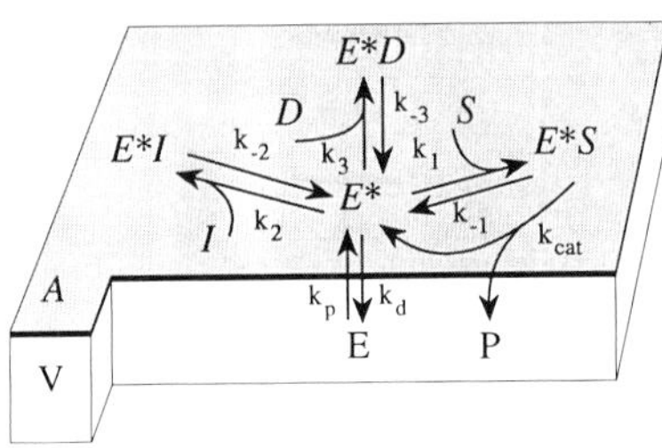

FIG. 5. Proposed model for competitive inhibition at interfaces. A, Total interfacial area; V, total volume; E, enzyme concentration; P, product concentration; S, interfacial concentration of substrate; I, interfacial concentration of inhibitor; D, interfacial concentration of detergent. [From S. Ransac, C. Rivière, J. M. Soulié, C. Gancet, R. Verger, and G. H. de Haas, *Biochim. Biophys. Acta* **1043,** 57 (1990).]

reported that human spermatozoa hydrolyzed only phosphatidylglycerol monolayers. Inhibition of the phospholipase activity by gossypol may contribute to the unknown contraceptive effects of this nonsteroid male antifertility agent.

It is now becoming clear from the abundant literature on lipolytic enzymes that any meaningful interpretation of inhibition data has to take into account the kinetics of enzyme action at the lipid–water interface. As shown in Fig. 5, Ransac *et al.*[41] have devised a kinetic model which is applicable to water-insoluble competitive inhibitors in the presence of detergent in order to quantitatively compare the results obtained at several laboratories. Furthermore, with the kinetic procedure developed, it was possible to make quantitative comparisons with the same inhibitor placed under various physicochemical situations, namely, micellar or monolayer states.

The addition of a potential inhibitor to the reaction medium can lead to paradoxical results. Usually, variable amounts of inhibitor are added, at a constant volumetric concentration of substrate and detergent. The specific area, the interfacial concentration of detergent, and the interfacial concentration of substrate are continuously modified accordingly. In order to minimize modifications of this kind, it was proposed to maintain constant the sum of inhibitor and substrate ($I + S$) when varying the inhibitory molar fraction [$\alpha = I/(I + S)$], that is, to progressively substitute a molecule of inhibitor for each molecule of substrate.[42] This is the minimal change that can be made at increasing inhibitor concentrations. Of course, with this method, the classic kinetic procedure based on the

[41] S. Ransac, C. Rivière, J. M. Soulié, C. Gancet, R. Verger, and G. H. de Haas, *Biochim. Biophys. Acta* **1043,** 57 (1990).

[42] G. H. de Haas, M. G. van Oort, R. D. Dijkman, and R. Verger, *Biochem. Soc. Trans.* **17,** 274 (1989).

Michaelis–Menten model is not valid because both inhibitor and substrate concentrations vary simultaneously and inversely. However, by measuring the inhibitory power (Z) as described by Ransac *et al.*[41] or $X_i(50)$ as used by Jain *et al.*,[43] it is possible to obtain a normalized estimation of the relative efficiency of various potential inhibitors.

Using the mixed-film technique and the zero-order trough described in Fig. 3, Ransac *et al.*[14] studied the hydrolysis of monomolecular films of L-dilauroylphosphatidylcholine or L-dilauroylphosphatidylglycol mixed with their corresponding phospholipid analogs bearing an amide bond instead of an ester bond at the 2-position of the glycerol backbone. These amidophospholipids, incorporated into mixed micelles, strongly and stereoselectively inhibited phospholipases A_2 with an increased inhibitory power in the case of the anionic derivatives.[44]

Kinetic experiments were performed at various molar fractions of inhibitor (α) present in mixed monomolecular films of substrate/inhibitor. Figure 6 shows the decrease in the relative phospholipase activities as a function of the inhibitor molar fraction at the optimum surface pressure with each substrate used. Both inhibitors of the L-phosphatidylcholine and L-phosphatidylglycol series exhibited strongly concave-shaped curves, whereas those obtained with inhibitors of the D series were rather close to the diagonal, which probably corresponds to a special case where $K_I^* = K_m^*$. The inhibitory powers (Z) were given by the slopes of the straight lines obtained by plotting R_v as a function of α (see insets). Inhibitors of the L series were found to be stronger inhibitors than those of the D series, and L-amido-C_{12}-Pglycol ($Z = 38$) was more potent than L-amido-C_{12}-PC ($Z = 25$). Figure 6A also shows that the lag times (τ) due to the presence of inhibitors of the L-phosphatidylcholine series increased as a function of the inhibitor molar fraction (α). This is in good agreement with the simulation curve obtained from the kinetic model presented in Fig. 5.

Enzymatic Activity of Phospholipase C (α Toxin from *Clostridium perfringens*) Using Phospholipid Monolayers as Substrate

Several research groups[32,45,46] have published kinetics studies on phospholipase C using surface radioactivity as an index to the splitting and the solubilization of the phosphorylcholine moiety in the water subphase. This method can give interesting qualitative information about the initial steps

[43] M. K. Jain, W. Yuan, and M. H. Gelb, *Biochemistry* **28,** 4135 (1989).

[44] G. H. de Haas, R. D. Dijkman, M. G. van Oort, and R. Verger, *Biochim. Biophys. Acta* **1043,** 75 (1990).

[45] A. D. Bangham and R. M. C. Dawson, *Biochim. Biophys. Acta* **59,** 103 (1962).

[46] I. R. Miller, and J. M. Ruysschaert, *J. Colloid Interface Sci.* **35,** 340 (1971).

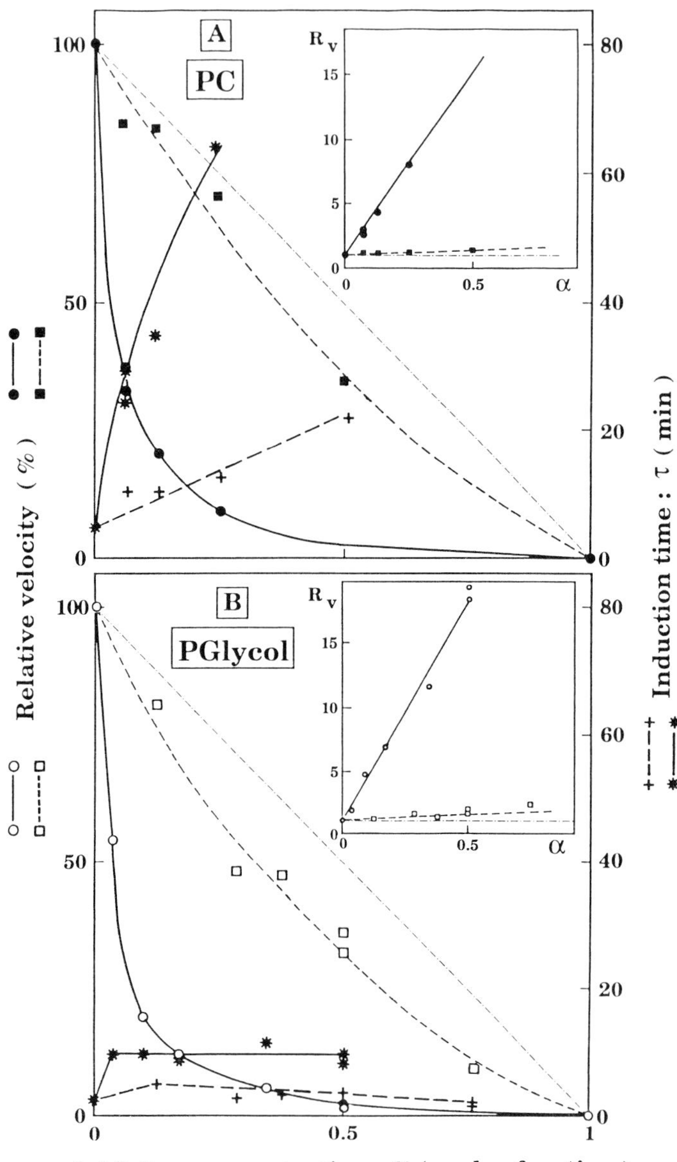

FIG. 6. Relative velocity, induction time (τ), and R_v (inserts) as a function of inhibitor molar fraction (α). Inhibitors of the D series (---). Inhibitors of the L series (——). (A) L-DiC_{12}-PC used as substrate, L-amido-C_{12}-PC (●) and D-amido-C_{12}-PC (■) used as inhibitors. Surface pressure was 10 mN/m. (B) L-DiC_{12}-Pglycol used as substrate, L-amido-C_{12}-Pglycol (○) and D-amido-C_{12}-Pglycol (□) used as inhibitors. Surface pressure was 30 mN/m.

during phospholipase C action. However, the main limitation of this technique is that insoluble diglyceride molecules accumulate with time, perturbing the initial phospholipid film. In other words, one of the products (diglycerides) of the reaction catalyzed by phospholipase C remaining in the interface can have a profound influence and progressively reduce the enzyme velocity.

In order to circumvent these product accumulation problems, Moreau *et al.*[47] have developed a new method based on the subsequent rapid lipase hydrolysis of the diglyceride generated by phospholipase C. In these assays, the authors used a large excess of pancreatic lipase saturated by colipase (ranging from 0.15 to 10 n*M* lipase and a concentration of 40 p*M* of phospholipase C). It is worth noting that the pancreatic lipase/colipase system is not able to hydrolyze a pure phosphatidylcholine film. These experiments were carried out under the following experimental conditions: L-dilauroylphosphatidylcholine film spread at a surface pressure of 14 mN/m over a subphase of 20 m*M* Tris-HCl buffer, pH 7.2, 0.15 *M* NaCl, 5 m*M* $CaCl_2$, and 0.1 m*M* ZnO_4.

Using the barostat technique,[1,12] it was checked that the enzyme kinetics were linear and that the velocity was directly dependent on the amount of phospholipase C added. This system can be used to study in detail the enzymatic kinetics of phospholipases C on lipid monolayers. This method is applicable to either medium-chain phospholipid monolayers or long-chain phospholipid monomolecular films providing that, in the latter case, bovine serum albumin (BSA) is introduced into the subphase to dissolve the fatty acid and monoglyceride formed. Bianco *et al.*[48] have recently used this new method to study the effects of sulfatide and gangliosides on phospholipase C activity.

Film Transfer Experiments Showing Role of Ca^{2+} and Zn^{2+} Ions in Phospholipase C Activity

When EDTA (0.1 m*M*) was introduced into the aqueous subphase of the reaction compartment, the phospholipase C activity measured under optimal conditions was immediately and completely abolished although, under these conditions, pancreatic lipase is known to be fully active.[13] Hydrolysis was not restored after addition of either Zn^{2+} (0.5 m*M* final concentration) or Ca^{2+} (5 m*M* final concentration) alone. The simultaneous presence of the two cations in the aqueous subphase was necessary for the activity of the phospholipase C to be recovered.

[47] H. Moreau, G. Piéroni, C. Jolivet-Reynaud, J. E. Alouf, and R. Verger, *Biochemistry* **27,** 2319 (1988).

[48] I. D. Bianco, G. D. Fidelio, and B. Maggio, *Biochim. Biophys. Acta,* in press (1990).

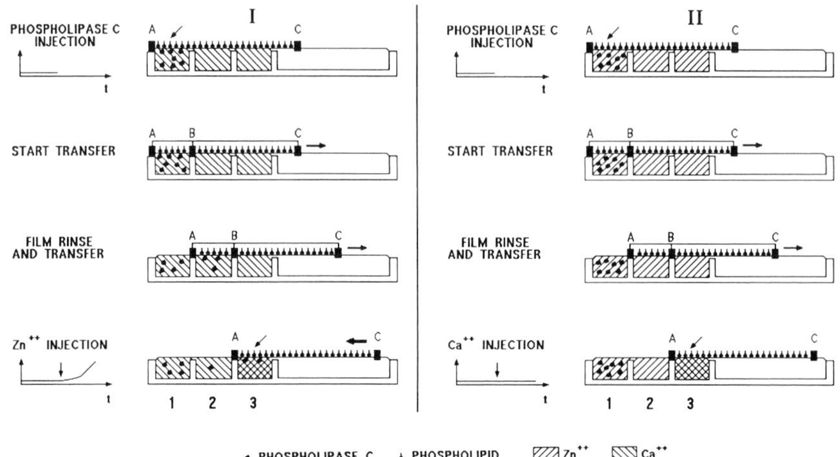

FIG. 7. Transfer experiments showing the role of Ca^{2+} and Zn^{2+} ions in phospholipase C activity. A, B, and C represent the three barriers; barrier C was mobile during the reaction recording. The graphs on the left-hand side of panels I and II are the kinetic curves recorded before (upper graph in I and II) and after (lower graph) the film transfer. Lipase saturated with colipase was present in all the reaction compartments. [From H. Moreau, G. Piéroni, C. Jolivet-Reynaud, J. E. Alouf, and R. Verger, *Biochemistry* **27,** 2319 (1988).]

To determine the respective roles of Ca^{2+} and Zn^{2+} ions on phospholipase C activity, transfer experiments were performed as described in Fig. 7. Lipase (1 n*M* final concentration) saturated with colipase was added to all three compartments containing 20 m*M* Tris-HCl buffer, pH 7.2, 0.1 m*M* EDTA, and 0.15 *M* NaCl. The portion of the L-dilauroylphosphatidylcholine film at a surface pressure of 14 mN/m and located over compartment 1, where phospholipase C was first injected (40 p*M* final concentration), was isolated by barrier B and transferred to the surface of compartment 3 with a film rinse over compartment 2. In experiment II, Zn^{2+} ions (0.5 m*M* final concentration) alone were first present in the three compartments, and, after film transfer, Ca^{2+} (5 m*M* final concentration) was injected into the aqueous subphase of compartment 3; no enzymatic activity was observed. As a control, a further injection of phospholipase C into the aqueous subphase of compartment 3 was found to initiate film hydrolysis at a rate comparable to that observed previously. In experiment I, Ca^{2+} ions (5 m*M* final concentration) were initially present in all three compartments. After film transfer, injection of Zn^{2+} (0.5 m*M* final concentration) into the aqueous subphase of compartment 3 initiated an immediate phospholipase C activity, amounting to 60% of the optimal value previously determined.

When only Zn^{2+} ions were present in the aqueous subphase, phospholipase C was not associated with the substrate film and consequently was not active. In the presence of Ca^{2+} ions only, phospholipase C was associated with the film, but its activity was dependent on the presence of Zn^{2+} ions. In conclusion, Ca^{2+} ions appear to be involved in the binding of the enzyme to the lipid interface, unlike Zn^{2+} ions, which are necessary for the expression of catalytic activity.

Acknowledgments

Our thanks are due to Dr. C. Rothen (Bern, Switzerland) and Professor G. H. de Haas (Utrecht University, the Netherlands) for personal communications prior to publication. S.R. acknowledges fellowship support from Groupement de Recherche de Lacq du Groupe Elf Aquitaine, France (Dr. C. Gancet and Dr. J. L. Seris). The authors are grateful to Dr. J. Blanc for revising the English and to M. T. Nicolas (Marseille) for typing the manuscript.

[5] Thio-Based Phospholipase Assay

By LIN YU and EDWARD A. DENNIS

Introduction

The thio assay possesses many characteristics that recommend it as a general assay for phospholipases. The most important are that it is a continuous, spectrophotometric assay which is very convenient, it directly detects one of the products liberated upon hydrolysis, it is one of the more sensitive assays, and it is also suitable for detailed kinetic studies.[1,2] The thio assay can be used for phospholipases A_1[3] and A_2[4] and with appropriate modification of the substrate would be applicable to other phospholipases.[5] However, owing to the lack of commercial availability of the thiophospholipid substrate and its complicated synthesis, the thio assay has not been used extensively. We[6] have recently modified the original[4] synthetic procedure so that gram quantities of chiral thiophospholipid can be readily

[1] H. S. Hendrickson and E. A. Dennis, *J. Biol. Chem.* **259,** 5734 (1984).
[2] H. S. Hendrickson and E. A. Dennis, *J. Biol. Chem.* **259,** 5740 (1984).
[3] G. L. Kucera, C. Miller, P. J. Sisson, R. W. Wilcox, Z. Wiemer, and M. Waite, *J. Biol. Chem.* **253,** 12964 (1988).
[4] H. S. Hendrickson, E. K. Hendrickson, and R. H. Dybuig, *J. Lipid Res.* **24,** 1532 (1983).
[5] J. W. Cox, W. R. Snyder, and L. A. Horrocks, *Chem. Phys. Lipids* **25,** 369 (1979).
[6] L. Yu, R. A. Deems, J. Hajdu, and E. A. Dennis, *J. Biol. Chem.* **265,** 2657 (1990).

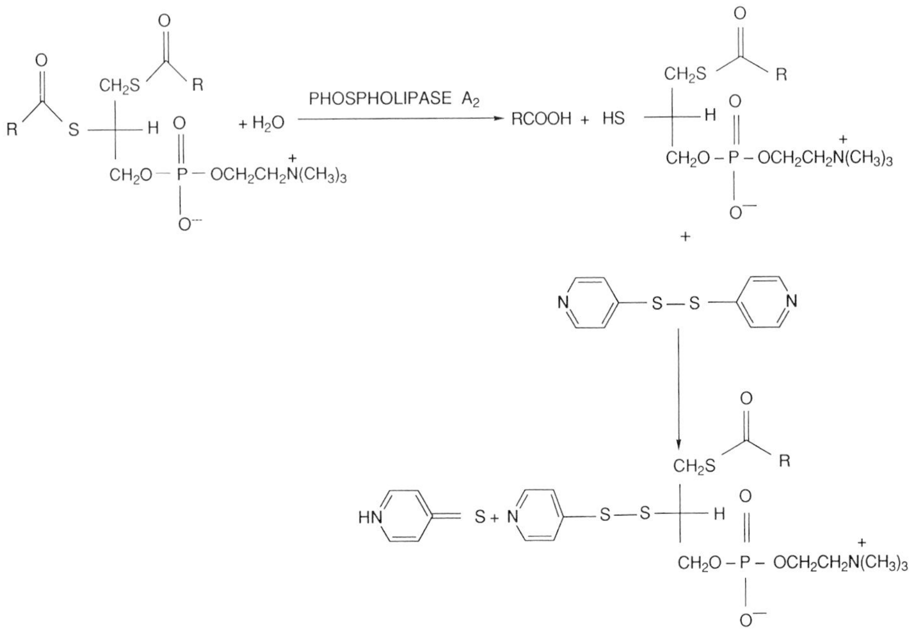

FIG. 1. Scheme for the spectrophotometric determination of phospholipase A_2 activity using thio ester analogs of phospholipids.

produced. In this chapter, we describe the thio assay for phospholipase A_2 and the synthesis of the thiophospholipid substrates.

Enzyme Assay

Principle. Phospholipase A_2 cleaves the *sn*-2 oxy ester of phospholipids; it will also hydrolyze an *sn*-2 thio ester. The fact that this hydrolysis releases a free thiol group has been utilized as the basis for the spectrophotometric assay, as shown in Fig. 1. The liberated thiol is allowed to react with a thiol-sensitive reagent, which is included in the assay, and the formation of the resulting chromophore is measured continuously by monitoring the increase in absorption associated with its production.

Two thiol reagents, 4,4′-dithiobispyridine (DTP) and 5,5′-dithiobis(2-nitrobenzoic acid) (DTNB), are routinely used to detect free thiol groups; both are commercially available from Aldrich (Milwaukee, WI). The choice of which thiol reagent to use depends on several factors: solubility in water, extinction coefficient, and effect on the enzyme. DTP has limited solubility in water; therefore, stock solutions must be prepared in organic solvents such as alcohol or acetone. These solvents are then introduced into the assay along with the reagent DTNB, on the other hand, is sufficiently soluble in water that the stock solutions are aqueous, and consequently no organic solvents are introduced into the assay. Since the sensitivity of the assay is limited by the extinction coefficient of the chromophore, use of reagent that has the higher extinction coefficient would produce the more sensitive assay. However, the extinction coefficient for both DTP and DTNB varies dramatically with pH (see Fig. 2). Thus, the assay pH is an important consideration when deciding which thiol reagent will yield the more sensitive assay. Both of these points become secondary considerations if the thiol reagent affects the activity of the enzyme, for phospholipase assays are complicated enough without the addition of extraneous modulators. We have found that DTNB activates the cobra venom phospholipase A_2 whereas DTP does not. Thus, we have chosen to use DTP in our studies of cobra venom phospholipase A_2,[1,2,6] but DTNB may be advantageous with other enzymes.

Reagents

Buffer A: 25 m*M* Tris-HCl (pH 8.5 at 30°), 0.1 *M* KCl, 10 m*M* $CaCl_2$

Buffer B: 25 m*M* Tris-HCl (pH 8.5 at 30°), 0.1 *M* KCl, 10 m*M* $CaCl_2$, 0.24 m*M* Triton X-100

Triton X-100: 10 m*M* Triton X-100 in buffer B (total Triton = 10.24 m*M*)

Thiophospholipid: about 10 m*M* in chloroform (determined by phosphorus analysis)

DTP: 50 m*M* 4,4′-dithiobispyridine in ethanol

Procedure. The standard assay for the cobra venom phospholipase A_2 employs mixed micelles of Triton X-100 and phospholipid, although monomers, micelles, and vesicles can also be used. The standard assay contains 0.5 m*M* L-dicaprylthiophosphatidylcholine, 4.24 m*M* Triton X-100, 10 m*M* Ca^{2+}, 0.1 *M* KCl, and 25 m*M* Tris-HCl in a total volume of 0.3 ml. The assay is carried out at 30° and pH 8.5. A cuvette, containing 0.3 ml of assay mixture and 5 μl of DTP (50 m*M* in ethanol), is placed in the compartment of a spectrophotometer thermostatted at 30° and balanced against a blank containing water. After equilibration of the substrate for 2 min at 30°, a stable baseline is recorded for 1 min. Phospholipase A_2 (5 μl containing 50 ng protein) is added to initiate the reaction. Phospholipase A_2 activity is determined by measuring the increase in absorption at 324 nm for 1 min. The reaction rate is calculated using the appropriate extinction coefficient (see below). Between each run, the cuvette is rinsed with a 5% solution of sodium hypochlorite (Clorox), 0.1 *M* HCl, deionized water, and acetone. This cleaning procedure is essential for achieving reproducible assays and accomplishes this by ensuring that the phospholipid and en-

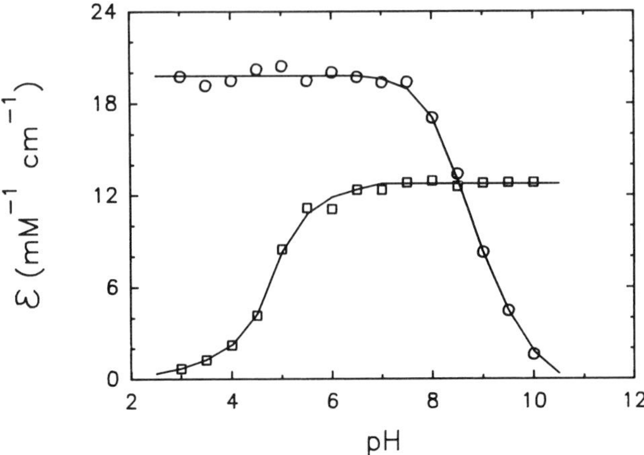

FIG. 2. Effect of pH on the extinction coefficient (ε) of 4-thiopyridine at 324 nm (○) (the chromophore liberated by the reaction of DTP with a free thiol) and 5-thio-2-nitrobenzoic acid at 412 nm (□) (the chromophore liberated by the reaction of DTNB with a free thiol). The extinction coefficient was determined in the presence of 100 m*M* KCl, 10 m*M* $CaCl_2$, and 25 m*M* of one of following buffers (at 30°): glycine, Tris, acetate, HEPES.

zyme (both of which show a propensity for binding to glass) are removed from the cuvette.

To obtain reliable activities, care must also be taken in preparing the substrate. The best results are obtained when the assay mixture is prepared fresh daily. The desired amount of thiophospholipid in chloroform is dried under a stream of nitrogen and, then, *in vacuo* for 10 min. If monomeric or pure micellar substrate is desired, an appropriate volume of buffer A is added. To prepare mixed micelles, the appropriate amount of the Triton X-100 stock solution is added, and the volume is brought to the final volume with buffer B. Buffer B contains 0.24 m*M* Triton X-100, roughly its critical micelle concentration (CMC), to ensure that the entire amount of Triton X-100 added via the stock is present as micelles. This is particularly important when dealing with very low substrate concentrations. The lipid mixture is sonicated in a bath sonicator until a clear solution is obtained. The phospholipid can also be dispersed by heating the solution to 40° and vortexing; however, this is more time consuming, and greater care must be used to ensure that all of the phospholipid is solubilized.

When the cobra venom phospholipase A_2 is being assayed, the enzyme is diluted to about 10 ng/μl with buffer B. Several of the extracellular phospholipases A_2 show a marked tendency to adhere to both glass and plastic. The presence of 0.24 m*M* Triton X-100 appears to prevent this from happening with the cobra venom enzyme.

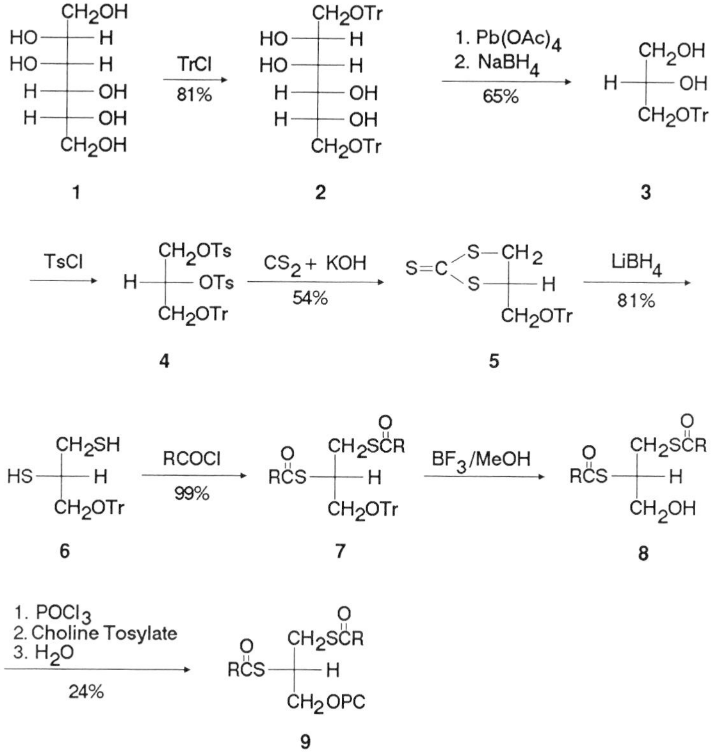

FIG. 3. Steps and yields in the synthesis of thiophospholipid substrates. Compounds **4** and **8** are not purified but are used directly in the steps that follow. The yield shown for compound **5** is based on compound **3**, and the yield shown for compound **9** is based on compound **7**. Tr, Trityl; Ts, tosyl.

The thio assay can be used to measure a wide range of phospholipase A_2 activity. This method can detect rates as low as 1 nmol/min and as high as 400 nmol/min with reproducibility within 5% across the entire range. However, deviations in buffer pH can dramatically affect the reproducibility of the assay. The extinction coefficients of both 4-thiopyridine (derived from DTP) and 2-nitro-5-thiobenzoic acid (derived from DTNB) are pH dependent. As shown in Fig. 2, the extinction coefficient of 4-thiopyridine decreases rapidly at pH values above 7 but is constant below pH 7. In contrast to 4-thiopyridine, 2-nitro-5-thiobenzoic acid has a constant extinction coefficient above pH 7 but a decreasing extinction coefficient as the pH drops below 7. Therefore, the buffer used in the assay must have sufficient buffering capacity to maintain the pH at the appropriate level, and the pH of the substrate solution should be checked before each experiment. If possible, it is better to use DTNB in basic solutions and DTP in acidic solutions. Clearly, this assay cannot be used in a system where free thiols have been added to stablize the enzyme or in crude cellular extracts that contain free thiol groups.

Preparation of Substrates

We first describe the complete synthesis of the chiral thiophosphorylcholine (thio-PC) as shown in Fig. 3. Then we describe the synthesis of racemic thio-PC, which is accomplished by replacing the first four steps of the chiral synthesis with a simple two-step procedure. The racemic thio-PC substrate is suitable for some applications, such as screening inhibitors. Finally, we describe a procedure for changing the head group of the chiral compound using the conversion of thio-PC to thiophosphorylethanolamine (thio-PE) as an example.

Synthesis of 1,2-Bis(acylthio)-1,2-dideoxy-sn-glycero-3-phosphorylcholine (Chiral Thio-PC)

1,6-Ditrityl-D-mannitol **(2).** D-Mannitol (20 g, 110 mmol) and trityl chloride (61 g, 220 mmol) are dissolved in 200 ml of dry pyridine.[7] After stirring at room temperature for 3 days, the pyridine solution is poured into 1 liter of ice water. After decanting the aqueous solution, a pasty product is obtained. The yellow pasty mixture is washed several times with ice water until it becomes a solid. The white solid is filtered and washed 3 times with water. The crude product is dissolved in a minimal volume of toluene (~100 ml) and purified by flash chromatography[8] on silica gel 60 (300 g), Merck 230–400 mesh (Aldrich, Milwaukee, WI) (column dimensions 5.0 × 30 cm). The column is eluted with a step gradient of 1 liter of toluene, 1 liter of toluene/diethyl ether (9 : 1), and the product is finally eluted in 1 liter of toluene/diethyl ether (4 : 1). The chromatography is monitored by thin-layer chromatography (TLC) in hexanes/acetone (7 : 3) (R_f value 0.3 for 1,6-ditritylmannitol). TLC is carried out on Analtech (Newark, DE) silica gel G-25 glass plates. Compounds are generally detected by spraying the plates with 2 *M* sulfuric acid followed by charring on a hot plate. Compounds containing the trityl moiety are visible after only gentle warming as bright yellow spots. A white solid is obtained (64.5 g, 88% yield) after drying under vacuum.

1-Trityl-sn-glycerol **(3)**. The procedure of Virtanen *et al.*[7] has been modified in our laboratory. The excess lead tetraacetate used in original

[7] J. A. Virtanen, J. R. Brotherus, O. Renkonen, and M. Kates, *Chem. Phys. Lipids* **27,** 185 (1980).

[8] W. O. C. Still, M. Kahn, and A. Mitra, *J. Org. Chem.* **43,** 2923 (1978).

method[7] oxidizes the product, 1-trityl-*sn*-glyceraldehyde, to 1-tritylglycolaldehyde. By decreasing the molar ratio of lead tetraacetate to compound **2**, the yield can be significantly increased. 1,6-Ditrityl-D-mannitol (58 g, 87.0 mmol) is dissolved in 350 ml of ethyl acetate and cooled in an ice bath. To this solution, lead tetraacetate (45 g, 101.4 mmol) is added in small portions with stirring. The suspension is stirred at 0° until the starting material disappears as judged by TLC in hexanes/diethyl ether (4 : 6) (R_f values 0.1 for 1,6-ditrityl-D-mannitol and higher for all oxidation products). The reaction usually takes about 1 hr. After the reaction is complete, excess lead tetraacetate is reduced by adding several drops of a saturated oxalic acid solution. The reaction mixture is filtered to remove the precipitated lead diacetate, and the precipitate is washed with 300 ml of diethyl ether. The filtrate and wash are combined and washed with water, 0.5 *M* sodium hydrogen carbonate until basic, and finally water. The solution is dried over $MgSO_4$, and the product is used in the next step without purification.

The crude aldehydes obtained from last step are reduced by sodium borohydride. Sodium borohydride (3.3 g, 87.4 mmol) is suspended in 250 ml of ethanol and added to the ether solution of aldehydes. The mixture is allowed to stir at room temperature and is monitored by TLC in hexanes/diethyl ether (4 : 6) (R_f value 0.1 for 1-trityl-*sn*-glycerol). The disappearance of aldehydes is checked by spraying with Schiff's reagent.[9] After the reaction is completed, the solution is washed twice with water and dried over anhydrous sodium sulfate. The solvent is removed on a rotary evaporator. The crude product is dissolved in 50 ml of hexanes/diethyl ether (1 : 1) and applied to a silica gel column (300 g silica gel; column dimensions 5.0 × 25 cm). The column is eluted with 1 liter of hexanes, 1 liter of hexanes/diethyl ether (1 : 1), and 1 liter of hexanes/diethyl ether (1 : 2). The column is monitored by TLC in hexanes/diethyl ether (4 : 6). A white solid is obtained (37.6 g, 65%).

2,3-Bis(tosyl)-1-trityl-sn-glycerol **(4)**. 1-Trityl-*sn*-glycerol (10 g, 30 mmol) is dissolved in 30 ml of dry pyridine.[4] After cooling to 5°, tosyl chloride (16.9 g, 90 mmol) is added to the solution in small portions with stirring over 10 min. The reaction is then stirred at room temperature for 3 hr. The completion of the reaction is verified by TLC in hexanes/diethyl ether (4 : 6) by the disappearance of 1-trityl-*sn*-glycerol. After the reaction is completed, a mixture of benzene–hexanes (3 : 1) (200 ml) is added. The solution is washed with ice water, 0.5 *M* HCl until acidic, water, 5% $NaHCO_3$, and water. The oganic solution is dried over anhydrous $MgSO_4$. The product shows one main spot, with a few minor spots, on TLC in hexanes/acetone (7 : 3) (R_f value 0.7 for 2,3-bis(tosyl)-1-trityl-*sn*-glycerol).

[9] M. Kates, "Techniques of Lipidology." North-Holland, Publ., Amsterdam, 1972.

After evaporating the solvents, a syrup is obtained (~18 g). The product is used in the next step without further purification.

3-Trityl-1,2-dideoxyl-1,2-(thiocarbonyldithio)-sn-glycerol **(5)**. Potassium hydroxide (13 g) is carefully dissolved in 100 ml of methanol.[4] After cooling the methanol solution in an ice bath, carbon disulfite (27 ml) is added dropwise with stirring. This solution is slowly added to an acetone solution (150 ml) of 2,3-bis(tosyl)-1-trityl-*sn*-glycerol (18 g, 28 mmol). After the reaction mixture is stirred at 40° for 24 hr, 500 ml of benzene/hexanes (2 : 1) is added. The organic solution is washed 4 times with water and dried over anhydrous $MgSO_4$. The solvents are removed *in vacuo* to give a yellow solid. The product is purified by crystallization. The crude product is dissolved in 45 ml of hot benzene; hexanes (~160 ml) are then added slowly until the solution becomes cloudy. This solution is allowed to stand at 4° and yields bright yellow crystals (6.1 g, 54% yield, mp 155°–157°).

3-Trityl-1,2-dideoxyl-1,2-dimercapto-sn-glycerol **(6)**. To a solution of 3-trityl-1,2-dideoxy-1,2-(thiocarbonyldithio)-*sn*-glycerol (6 g, 14.7 mmol) in tetrahydrofuran (60 ml) is added lithium borohydride (44 mmol).[6,10] The reaction suspension is refluxed with magnetic stirring, and methanol (6.6 ml) is added dropwise over a period of 15 min. After another 15 min, the yellow color is gone and the reaction is complete. The hot reaction mixture is cooled first to room temperature and then in an ice bath. Excess hydride is decomposed by cautious addition of water. The suspension is acidified with 6 *M* HCl and extracted twice with ether (total volume of ether 200 ml). The organic layer is washed with 5% $NaHCO_3$ and then water. The ether layer is dried over Na_2SO_4, and the solvent is removed *in vacuo*. The crude mercaptan is dissolved in 25 ml of acetone and applied to a silica gel column that has been packed in hexanes/acetone (50 : 1). Prior to packing the column, the silica gel is deoxygenated under vacuum and stored under argon. The column (250 g silica gel, column dimensions 5.0 × 25 cm) is eluted with 1.2 liters of hexanes/acetone (50 : 1). The chromatography is monitored by TLC developed with hexanes/acetone (7 : 1) (R_f value 0.65 for 3-trityl-1,2-dideoxy-1,2-dimercapto-*sn*-glycerol). The dimercaptan product is obtained as a clear oil (4.7 g, 87% yield).

3-Trityl-1,2-bis(acylthio)-1,2-dideoxy-sn-glycerol **(7)**. Different fatty acid chain lengths can be incorporated into the phospholipid analog at this step by using the corresponding acyl chloride. The following steps have been successfully applied to the synthesis of dibutyroyl-, dihexanoyl-, and dicaprylthiophosphatidylcholine. The synthesis of 1,2-bis(caprylthio)-1,2-dideoxy-*sn*-glycerolphosphorylcholine is used to illustrate the basic synthetic procedures.[6]

[10] K. Soai and A. Ookawa, *J. Org. Chem.* **51**, 4000 (1986).

3-Trityl-1,2-dideoxy-1,2-dimercapto-*sn*-glycerol (3.64 g, 9.9 mmol) is dissolved in 40 ml of hexanes and 6 ml of dry pyridine. Decanoyl chloride (5.72 g, 30.0 mmol) in 10 ml of hexanes is slowly added. The reaction mixture is stirred at room temperature for 7 hr. After the reaction is complete, 100 ml of benzene is added. The mixture is washed with water, 0.5 *M* NH_4OH in methanol/water (3 : 1), and methanol/water (1 : 1). The organic phase is dried over anhydrous $MgSO_4$ and evaporated *in vacuo*. A clear oil is obtained (6.59 g, 99% yield) which yields a single spot on TLC in hexanes/acetone (7 : 1) (R_f 0.75).

1,2-Bis(caprylthio)-1,2-dideoxy-sn-glycero-3-phosphorylcholine **(9)**. A 14% methanol solution of boron trifluoride (5 ml, 10.33 mmol) is added to a solution of 1-trityl-1,2-bis(caprylthio)-1,2-dideoxy-*sn*-glycerol **(7)** (6.59 g, 9.76 mmol) in 150 ml of $CHCl_3$.[6] The reaction is stirred at room temperature for 2 hr. After the reaction goes to completion, as judged by TLC in hexanes/ethyl acetate (8 : 1) (R_f value 0.35 for detritylated product), the reaction mixture is washed 3 times with ice water. The organic phase is dried over anhydrous Na_2SO_4 and evaporated *in vacuo*. The viscous liquid is usually used immediately in the next step without further purification, but it can be stored at $-70°$.

The crude detritylated compound **8** is dissolved in 130 ml of freshly distilled dry $CHCl_3$ containing 1.0 ml dry pyridine (12.5 mmol). To a solution of freshly distilled $POCl_3$ (1.8 g, 11.7 mmol) in 30 ml of dry $CHCl_3$ is added the solution of compound **8**. The reaction is stirred under argon at 50° for 1 hr. The disappearance of the starting material is followed by TLC in hexanes/acetone (2 : 1) (R_f 0.45). After the reaction is complete, the mixture is cooled to room temperature. Choline *p*-toluene sulfonate (3.64 g, 9.73 mmol) and 3.0 ml of dry pyridine are added. Choline *p*-toluene sulfonate is prepared as described by Brockerhoff and Ayengar[11] and dried over phosphorus pentoxide for 2 days prior to use. After stirring for 3 hr at room temperature, the reaction mixture is washed with two portions of water; methanol is added to break the emulsion. The chloroform solution is evaporated *in vacuo*. The crude product is dissolved in 50 ml chloroform/methanol/water (65 : 25 : 4) and mixed with 300 ml Rexyn I-300 (Fisher, Pittsburgh, PA) which has been prewashed and preequilibrated with the same solvent, the solution separated, and the Rexyn washed three times. [With short chain fatty acids, the Rexyn step is replaced by a silica gel column with chloroform/methanol/ammonia (50 : 100 : 5) as solvent.] The process is monitored by TLC in chloroform/methanol/water/concentrated ammonia (95 : 35 : 5.5 : 2) (R_f 0.64). The solution is evaporated *in vacuo* and lyophilized. The solid is dissolved in 30 ml of chloroform/methanol/

[11] H. Brockerhoff and N. K. N. Ayengar, *Lipids* **14,** 88 (1979).

water (65 : 25 : 4) and purified on a silica gel column (200 g) using the same solvent mixture. After lyophilization from dry benzene, a white solid (1.40 g, 24% yield from compound **7**) is obtained.

Synthesis of 1,2-Bis(acylthio)-1,2-dideoxy-rac-glycerol-3-phosphorylcholine (rac Thio-PC)

Trityl Glycidol. Trityl chloride (75 g, 270 mmol) and dry glycidol (35 g, 470 mmol) are dissolved in 500 ml of dry toluene containing 35 g of triethylamine.[12] The toluene solution is stirred at room temperature overnight. The reaction mixture is washed 3 times with water to remove most of the excess glycidol and triethylamine hydrochloride formed during the reaction. The solution is dried over $MgSO_4$. The product is purified by crystallization from ethanol and light petroleum. The yield is 80% [R_f value 0.85 for trityl glycidol in TLC in toluene/ether (85 : 15)].

3-Trityl-1,2-dideoxy-1,2-(thiocarbonyldithio)-rac-glycerol. The compound is synthesized by the same procedure as for compound **5**.[4] However, the conversion reaction requires only 6 hr at 50°. From 19.2 g of trityl glycidol, 12.5 g of product is obtained (yield 51%). From this point, the synthesis of chiral thio-PC is followed substituting this product for compound **5**.

Synthesis of 1,2-Bis(acylthio)-1,2-dideoxy-sn-glycero-3-phosphorylethanolamine (Chiral Thio-PE)

Chiral thio-PC is transformed into chiral thio-PE[1,6] by the action of phospholipase D according to the procedure of Comfurius and Zwaal.[13] The enzymatic synthesis of 1,2-bis(decanoylthio)-1,2-dideoxy-*sn*-glycero-3-phosphorylethanolamine is used as an example. Thiophospholipid analogs with other polar head groups such as phosphatidylserine and phosphatidylglycerol, can also be made by this method. Chiral thio-PC (150 mg 0.25 mmol) is dissolved in 3 ml of highly pure diethyl ether and 5 ml of acetate buffer (100 m*M*, pH 5.6) containing 100 m*M* $CaCl_2$. Ethanolamine hydrochloride (1.0 g) is added to the reaction mixture. The reaction is initiated by the addition of cabbage phospholipase D (10 units) and followed by TLC in chloroform/methanol/water (65 : 25 : 4) [R_f values 0.33 for PC, 0.6 for PE, 0.7 for phosphatidic acid (PA)]. The vial is sealed to prevent the evaporation of the ether, and the mixture is vigorously stirred at 45° for several hours. When the reaction is complete, a solution of EDTA (100 m*M*, 5 ml) is added to stop the reaction. The reaction mixture is

[12] C. M. Loch, *Chem. Phys. Lipids* **22,** 323 (1978).

[13] P. Comfurius and R. F. A. Zwaal, *Biochim. Biophys. Acta* **488,** 36 (1977).

extracted twice with chloroform. Methanol is added to break the emulsion formed during the extraction. The crude product is purified by flash chromatography on a silica gel column using chloroform/methanol/water (65 : 25 : 1) as the eluting solvents. This thio-PE is obtained in 91% yield (127 mg, 0.23 mmol) as a white powder.

Acknowledgments

Support for this work was provided by the National Institutes of Health (GM-20,501) and National Science Foundation (DMB 89-17392).

[6] Chromogenic Substrates and Assay of Phospholipases A_2

By WONHWA CHO and FERENC J. KÉZDY

Introduction

The high specificity of phospholipases A_2 (PLA_2) toward aggregated substrates renders the assay of these enzymes a particularly challenging analytical task. Sensitive assays require the use of aggregated substrates, such as micelles, mixed micelles, single bilayer vesicles, or monomolecular layers, where the activity of the enzyme depends critically on the physical state and the exact composition of the nonaqueous phase. For reproducible assays with such heterogeneous systems the experimental conditions must be strictly controlled since the presence of minor lipid impurities, the accumulation of small amounts of reaction products in the early phases of the reaction, or even slight changes in temperature or buffer composition may and often do elicit large changes in the rate of the enzymatic reaction. Homogeneous reaction kinetics, which are conducive to readily reproducible kinetics, could only be achieved with short-chain lecithins, such as dibutyryllecithin, toward which phospholipases A_2 display a rather low specificity, and the assays based on these substrates are not sensitive enough for most purposes. The sensitivity of the spectrophotometric assay described herein is between the two extremes, and the method, by virtue of its simplicity, is readily adaptable to a variety of purposes ranging from routine analyses to investigations of mechanistic details of the enzymatic reaction.

A variety of acyloxynitrobenzoic acids have been shown to be hydrolyzed by snake venom and pancreatic phospholipases A_2. The reaction is Ca^{2+}-dependent, and the pH dependency of the catalysis is also consistent with a reaction mechanism making full use of the catalytic apparatus

TABLE I
SPECIFICITY OF PHOSPHOLIPASES A_2 TOWARD COMPOUNDS **1** AND **2**[a]

Enzyme	k_{cat}/K_m ($\times 10^{-3}$ M^{-1} sec^{-1}) Compound **1**	Compound **2**
Agkistrodon piscivorus piscivorus D-49[2]	2.91 ± 0.05	18.9 ± 0.3
Crotalus atrox PLA_2[3]	7.01 ± 0.28	46.4 ± 1.3
Porcine pancreatic PLA_2	0.34 ± 0.03	2.3 ± 0.2

[a] Assays were conducted at 37° in 10 m*M* Tris-HCl buffer, pH 8.00, 0.1 *M* NaCl, 10 m*M* $CaCl_2$, 1.6% acetonitrile.

observed in the hydrolysis of lecithins.[1] All phospholipases A_2 we have tried reacted well with 4-nitro-3-(octanoyloxy)benzoic acid, but the specificity toward acyloxynitrobenzoic acids with the carboxylate in the meta or para position with respect to the scissile bond vary from enzyme to enzyme. Thus, for a novel enzyme some experimentation might be required in order to achieve optimal assay conditions. In Table I[2,3] the specificity constants, k_{cat}/K_m, are given for 4-nitro-3-(octanoyloxy)benzoic acid **(1)** and 3-nitro-4-(octanoyloxy)benzoic acid **(2)** with three representative phospholipases A_2.

We have observed that at substrate concentrations in excess of 100 μM monomeric phospholipases A_2 are slowly and irreversibly activated during the hydrolysis of 4-nitro-3-(octanoyloxy)benzoate.[4] Again, for a novel enzyme it should be ascertained experimentally that the reaction kinetics obey simple pseudo-first-order kinetics and that no irreversible activation complicates the assay. At $S_o < 1 \times 10^{-4}$ M the condition $S_o < K_m$ is satisfied, and with $S_o \gg E_o$ the enzymatic reaction is first-order with respect to the substrate, with $k_{exp} = k_{cat} E_o/K_m$. The pH dependence of k_{cat}/K_m is bell-shaped, with a plateau at pH 7 to 8. Thus, the exact pH of the reaction mixture is not critical in this pH interval. For compound **1** the pK_a of the phenol product is 7.1, and the apparent molar absorptivity of the product is slightly pH-dependent around pH 8; however, the calculation of the experimental first-order rate constant does not involve this quantity.

[1] W. Cho, M. A. Markowitz, and F. J. Kézdy, *J. Am. Chem. Soc.* **110,** 5166 (1988).
[2] J. M. Maraganore, G. Merutka, W. Cho, W. Welch, F. J. Kézdy, and R. L. Heinrikson, *J. Biol. Chem.* **259,** 13839 (1984).
[3] W. Hachimori, A. M. Wells, and D. J. Hanahan, *Biochemistry* **10,** 4084 (1971).
[4] W. Cho, A. G. Tomasselli, R. L. Heinrikson, and F. J. Kézdy, *J. Biol. Chem.* **263,** 11237 (1988).

Assay Method

Substrates. Acyloxynitrobenzoic acids are readily synthesized from commercially available acyl chlorides and hydroxynitrobenzoic acids.[1] The method is illustrated with the synthesis of 4-nitro-3-(octanoyloxy)benzoic acid. 3-Hydroxy-4-nitrobenzoic acid (183 mg, 1.0 mmol) and *N,N*-diisopropylethylamine (259 mg, 2.0 mmol) are dissolved in 2 ml of dry tetrahydrofuran and added dropwise to a stirred solution of octanoyl chloride (163 mg, 1.0 mmol) in 10 ml of dry tetrahydrofuran at 0°. The mixture is slowly warmed to room temperature and stirred overnight. The precipitate is filtered and washed with three 10-ml aliquots of ether. The combined filtrates are concentrated *in vacuo* and dissolved in 50 ml of ether/hexane, 2 : 1 (v/v). The organic layer is washed with two 50-ml aliquots of 10 m*M* HCl and with two 50-ml aliquots of water, dried with anhydrous Na_2SO_4, and evaporated *in vacuo*. The residue is fractionated by flash chromatography using Merck kieselgel 60 and hexane–ether–acetic acid, 20 : 10 : 1 (v/v/v), and the eluate is monitored by TLC [R_f 0.44 on a silica gel plate with hexane–ether–acetic acid, 20 : 10 : 1 (v/v/v)]. The appropriate fractions are pooled and evaporated *in vacuo*. The product is recrystallized from hexane–ethyl acetate, 10 : 1 (v/v), mp 142°–143°. The purity of the product is readily ascertained spectrophotometrically after alkaline hydrolysis of a weighed sample of the ester; for the phenolate ion of 3-hydroxy-4-nitrobenzoic acid the ε_{425} value is 4990 M^{-1} sec^{-1} and for that of 4-hydroxy-3-nitrobenzoic acid the ε_{410} value is 4550 M^{-1} sec^{-1}.

Reagents

Assay buffer: Tris-HCl, 10 mM, pH 8.00, containing 10 m*M* $CaCl_2$ and 0.1 M NaCl

Substrate stock solution: Compound **1**, 3.1 m*M* (0.96 mg/ml), in dry acetonitrile

Enzyme stock solution: Ideally in the range of 1×10^{-5} to 1×10^{-4} *M*, (0.1–2 mg/ml), dissolved in the assay buffer; however, any buffer which is mixed in a 1 : 60 ratio with the assay buffer does not appreciably change the pH of the latter. The enzyme is dissolved directly in the assay buffer and used without dilution as the assay medium. Enzymes of concentrations as low as 1 μg/ml can be assayed.

Assay Procedure. Three milliliters of the assay buffer is equilibrated at 37° in a thermostatted 1 cm path length cuvette in the sample compartment of the spectrophotometer. A 50-μl aliquot of the substrate stock solution is added from the flattened tip of a glass rod, and the solution is vigorously stirred. In monitoring the reaction at 410 nm a very slight buffer-catalyzed hydrolysis might be observed at this point, depending on the exact value of the pH. The enzymatic reaction is initiated by the addition of 50 μl of enzyme stock solution, and the increase in absorbance at 410 nm is recorded. Complete hydrolysis should produce a total change of about 0.2 absorbance units (AU). Ideally, the reaction should be allowed to continue for at least four half-lives.

Analysis of Results. The experimental first-order rate constant can be calculated from the absorbance versus time data using any of the commercially available nonlinear least-squares fit programs. The program MULTI,[5] written in BASIC, has been found convenient to use with data collected in a digital form by an on-line microcomputer. If A_i and A_f are the absorbance readings at the beginning and the end of the enzymatic reaction, respectively, then the absorbance A at time t is given by

$$A = A_i + (A_f - A_i)(1 - e^{-k_{exp}t})$$

The first few, usually noisy, points are discarded, and a minimum of 20 A,t data pairs are analyzed for one experiment. The parameters to be optimized for A_i, A_f, and k_{exp}. If pure enzyme is available, the method is first calibrated with a known enzyme concentration, and the value of k_{cat}/K_m is calculated from k_{exp}. Then, with this value, any further assay yields the enzyme concentration in the reaction mixture by $E_o = k_{exp}/(k_{cat}/K_m)$ and the concentration of the enzyme stock solution by $E_s = E_o \times 0.05/3.10$. Alternatively, k_{exp} can be determined graphically from the slope of a of $\log(A_f - A)$ versus time.

With very low enzyme concentrations, the initial rate is calculated from the initial rate of absorbance change, $V_a = \Delta A/\Delta t$, corrected for the nonenzymatic, buffer-catalyzed rate. Under these conditions, the exact pH and the exact substrate concentration, S_o, must be known in order to convert the data to a rate constant according to the equation

$$\frac{k_{cat}E_o}{K_m} = V_a\left(\frac{1 + 10^{(7.10 - \mathrm{pH})}}{4990\, S_o}\right)$$

Scope and Limitations. The assay described in this chapter provides a rapid, accurate, and convenient method for measuring phospholipase A_2 concentrations as low as 1×10^{-8} *M*. Since nitrophenyl esters are susceptible to hydrolysis by a variety of esterases and peptidases, it is recommended that, when working with crude extracts and partially purified enzymes, appropriate control experiments be carried out, where the 10 m*M* C^{2+} of the assay buffer is replaced by 0.1 m*M* EDTA, and that the

[5] K. Yamaoka, Y. Tanigawara, T. Nakagawa, and T. Uno, *J. Pharmacobio–Dyn.* **4,** 879 (1981).

k_{exp} of this control be subtracted from the value of k_{exp} of the enzymatic reaction. Finally, the adherence of the progress curve of the reaction to first-order kinetics should be verified for each novel enzyme in order to ascertain that $S_o \ll K_m$ and that no irreversible activation occurs under the chosen experimental conditions.

[7] Coupled Enzyme Assays for Phospholipase Activities with Plasmalogen Substrates

By M. S. Jurkowitz-Alexander, Y. Hirashima, and L. A. Horrocks

Introduction

Ether-linked glycerophospholipids of the plasmalogen type (1-alk-1′-enyl-*sn*-2-acylglycerophosphocholine or -ethanolamine) are present in virtually all mammalian cell membranes.[1] Among the enzymes known to function in plasmalogen metabolism in mammalian tissues are the hydrolytic enzymes lysoplasmalogenase[2–5] and plasmalogen-selective phospholipase A_2.[6,7]

Phospholipase A_2 (PLA_2) catalyzes the hydrolysis of plasmalogen at the *sn*-2 position, producing free fatty acid and lysoplasmalogen [Eq. (1)].

$$\text{Plasmalogen} + H_2O \xrightarrow{PLA_2} \text{lysoplasmalogen} + \text{fatty acid} \tag{1}$$

Lysoplasmalogenase catalyzes the hydrolysis of the alkenyl ether bond of lysoplasmalogen at the *sn*-1 position, forming free aldehyde and glycerophosphocholine [Eq. (2)]. Both of these enzymes have been purified recently. Plasmalogen-selective phospholipase A_2 was purified from the cytosol of sheep platelets[8] and canine myocardium,[6] and lysoplasmalogenase was purified from liver microsomes.[5,9]

[1] L. A. Horrocks, *in* "Ether Lipids: Chemistry and Biology" (F. Snyder, ed.), p. 177. Academic Press, New York, 1972.

[2] H. R. Warner and W. E. M. Lands, *J. Biol. Chem.* **236,** 2404 (1961).

[3] J. Gunawan and H. Debuch, *J. Neurochem.* **39,** 693 (1982).

[4] J. Gunawan and H. Debuch, *Hoppe-Seyler's Z. Physiol. Chem.* **362,** 445 (1981).

[5] M. Jurkowitz-Alexander, H. Ebata, J. S. Mills, E. J. Murphy, and L. A. Horrocks, *Biochim. Biophys. Acta* **1002,** 203 (1989).

[6] S. L. Hazen, R. J. Stuppy, and R. W. Gross, *J. Biol. Chem.* **265,** 10622 (1989).

[7] R. A. Wolf and R. W. Gross, *J. Biol. Chem.* **260,** 7295 (1985).

[8] L. A. Loeb and R. W. Gross, *J. Biol. Chem.* **260,** 10467 (1986).

[9] M. S. Jurkowitz-Alexander and L. A. Horrocks, this volume [46].

METHODS IN ENZYMOLOGY, VOL. 197

$$\text{Lysoplasmalogen} + H_2O \xrightarrow{\text{lysoplasmalogenase}} \text{glycerophosphocholine} + \text{aldehyde} \quad (2)$$

In order to detect and quantitate lysoplasmalogenase activity in native liver microsomes and in purified enzyme fractions, we developed a continuous spectrophotometric assay (assay I).[5] The activity of lysoplasmalogenase in native microsomes is 40 nmol/min/mg protein. The assay is sensitive to 0.3 nmol/min/mg protein. Assay II[10] was developed for analysis of plasmalogen-selective phospholipases A_2. In developing assay II, we used *Naja naja naja* venom phospholipase A_2 as the enzyme source because of its purity and high activity with the plasmalogen substrate.[11,12] The assay is sensitive to 0.3 nmol/min/mg and could probably be used with the phospholipase A_2 of canine myocardial cytosol (activity 1.5–4.6 nmol/min/mg).[6] Assays I and II are relatively specific for the respective enzymes. They do not require separation of the lipid and aqueous phases; thus, many samples can be assayed in a relatively short time.

In previous studies, lysoplasmalogenase was estimated by chemical assay of the alkenyl ether content which remained following the enzymatic reaction.[2,13] A more sensitive radioassay was used in the identification and characterization of the liver and brain enzyme.[3,4] Both assays are discontinuous and require extraction of the lipids prior to determination of product formation or substrate loss. Recently, a continuous assay was developed to measure the fatty aldehyde released from plasmalogen in brain tissue using alcohol dehydrogenase as a coupling enzyme.[14] We have adapted and extended this method in the development of assays I and II.

Theoretical Considerations in Coupled Enzyme Assays. In the lysoplasmalogenase reaction, formation of the product fatty aldehyde cannot be measured directly. Therefore, it is removed by the coupling enzyme alcohol dehydrogenase [Eq. (3)]. Every molecule of aldehyde that is formed results in the loss of a molecule of NADH, so absorption at 340 nm decreases. The coupling enzyme, alcohol dehydrogenase, is in this case the "indicator" enzyme.[15] The enzymes for reactions 1 and 2 in Eq. (3) are lysoplasmalogenase and alcohol dehydrogenase, respectively.

[10] Y. Hirashima, M. S. Jurkowitz-Alexander, A. A. Farooqui, and L. A. Horrocks, *Anal. Biochem.* **176,** 180 (1989).

[11] E. L. Gottfried and M. M. Rapport, *J. Biol. Chem.* **237,** 329 (1962).

[12] K. Waku and Y. Nakazawa, *J. Biochem.* (*Tokyo*) **72,** 149 (1972).

[13] R. V. Dorman, A. D. Toews, and L. A. Horrocks, *J. Lipid Res.* **18,** 115 (1977).

[14] N. M. Freeman and E. M. Carey, *Anal. Biochem.* **128,** 377 (1983).

[15] R. K. Scopes, *in* "Protein Purification" (C. R. Cantor, ed.), p. 228. Springer-Verlag, New York, 1986.

$$\text{Lysoplasmalogen} \xrightarrow{k_1} \text{aldehyde} \xrightarrow{k_2} \text{alcohol} \qquad (3)$$

$$\text{NADH} + \text{H}^+ \quad \text{NAD}^+$$

In coupled enzyme assays a steady state is set up in which the rate of action of the indicator enzyme is equal to that of the measured enzyme. In the reactions of assay system I [Eq. (3)] several assumptions are made.[16] K_1 is a zero-order rate constant, assured by maintaining a saturating concentration of lysoplasmalogen or by guaranteeing that only a small fraction of this substrate is converted to aldehyde during the period of measurement. Second, k_2 is a first-order rate constant, since the concentration of aldehyde will be much lower than its Michaelis constant. The second substrate of alcohol dehydrogenase, NADH, must be saturating. Both reactions in Eq. (3) are irreversible.

In assay I the amount of alcohol dehydrogenase is chosen to ensure that it is not rate limiting.[16] The K_m(tetradecanal) and V_{max} values for alcohol dehydrogenase, determined under the conditions employed to assay the activity of the primary enzyme, are 0.10 mM and 1.8 IU/min/mg, respectively. In order to calculate V_2, the units of alcohol dehydrogenase to add per milliliter to attain near steady-state concentrations of aldehyde in a given amount of time, we first set a maximal rate for NADH oxidation of 0.015 unit/ml, which is equivalent to a ΔA_{340} of 0.093 min^{-1}. This velocity is controlled by the units of lysoplasmalogenase added per assay. The time to attain 98% of the true rate (t^*) is set at 1 min after initiating the reaction. The rate constant k_2 is equal to V_2/K_m and has the units minutes^{-1}. By substituting into $V_2 = [\ln(1 - 0.98)K_m]/t^*$ [adapted from Eq. (6) of Ref. 16] we obtain $k_2 = 3.9$ min^{-1}, and therefore $V_2 = 3.9$ min^{-1} $\times$ K_m(tetradecanal) $= 0.39$ IU/ml.

In assay II for plasmalogen-specific phospholipase A_2 the same indicator enzyme, alcohol dehydrogenase, is used. But this time lysoplasmalogenase is included as a coupling enzyme. One molecule of NADH is oxidized for every molecule of plasmalogen hydrolyzed [Eq. (4)]. The enzymes for steps 1, 2, and 3 of Eq. (4) are phospholipase A_2, lysoplasmalogenase, and alcohol dehydrogenase, respectively. In assay II, where two coupling enzymes are used [Eq. (4)], the same assumptions are made for steps 1 and 2 as are made with the single coupled enzyme assay. Addition-

$$\text{Plasmalogen} \xrightarrow{k_1} \text{lysoplasmalogen} \xrightarrow{k_2} \text{aldehyde} \xrightarrow{k_3} \text{alcohol} \qquad (4)$$

$$\text{NADH} + \text{H}^+ \quad \text{NAD}^+$$

[16] W. R. McClure, *Biochemistry* **8,** 2782 (1969).

ally, the assumptions are made that k_3 is a first-order rate constant and that reaction 3 is irreversible.

From Eq. (6),[16] the number of units of coupling enzymes required to reach 98% of the true catalytic rate 1 min after the start of the reaction is calculated. In this assay, the maximum value of k_1 is set at 0.004 mM min^{-1}. K_m(lysoplasmalogen) = 0.006 mM[5]; therefore, k_2 = (IU of lysoplasmalogenase/ml)/0.006 mM. Similarly, k_3 = (IU of alcohol dehydrogenase/ml)/0.100 mM, as above. If t^* is chosen equal to 1 min, substituting into Eq. (6)[16] yields k_2 and k_3 values equal to 3.9 min^{-1}; therefore, the concentrations of lysoplasmalogenase and alcohol dehydrogenase are 0.024 and 0.39 IU/ml, respectively.

The other substrate for alcohol dehydrogenase, NADH, is present at saturating concentrations. NADH does not affect the activity of the primary enzymes measured in assay I or II.

Assay Procedure

Materials for Assays I and II. Bovine brain ethanolamine glycerophospholipids and bovine heart choline glycerophospholipids are obtained from Serdary Research Laboratories (London, Ontario, Canada). The detergent 1-*O*-*n*-octyl-β-D-glucopyranoside (octylglucoside) and the lipase, from *Rhizopus delemar*, are from Boehringer Mannheim (Mannheim, FRG). All other chemicals, including *Naja naja naja* phospholipase A_2, are of reagent grade from Sigma Chemical Co. (St. Louis, MO).

Preparation of Substrates for Assays I and II. Ethanolamine and choline plasmenyl substrates are purified from the commercially available glycerophospholipids from bovine brain and heart, respectively. The plasmalogens are isolated using the phospholipase A_1 activity of *Rhizopus delemar* lipase.[17] Ethanolamine and choline lysoplasmalogens are prepared by mild alkaline hydrolysis of the diradyl glycerophospholipids.[18] The choline glycerophospholipids are digested with base for 30 min. Plasmalogens and lysoplasmalogens are analyzed for purity using a two-dimensional TLC system[19] and are between 85 and 94% pure. The major contaminants are the mono- and diradyl 1-alkylglycerophospholipids which are not separable from the alkenyl compounds using this procedure. The stock solutions of plasmalogens and lysoplasmalogens are stored at 20 mM solutions in chloroform/methanol (2:1, v/v) at −20° until the day of experiment.

[17] Y. Hirashima, A. A. Farooqui, E. J. Murphy, and L. A. Horrocks, *Lipids* **25,** 344 (1990).
[18] J. Gunawan and H. Debuch, *J. Neurochem.* **44,** 370 (1985).
[19] G. Y. Sun and L. A. Horrocks, *J. Neurochem.* **18,** 1963 (1971).

Reagents and Solutions

Glycylglycine buffer: An 80 m*M* stock solution is prepared by dissolving 10.6 g of glycylglycine in about 900 ml of distilled water. The pH is adjusted to 7.0 (at 25°) by the addition of 1 *M* NaOH. The volume is adjusted to 1 liter with water. Store at −20°.

Reagent A (assays I and II): To 60 ml of a stock solution of 80 m*M* glycylglycine buffer is added 1.0 ml of a 20 mg/ml stock solution of fatty acid-free bovine serum albumin fraction V (0.35 mg/ml), 9 mg of dithiothreitol (1.0 m*M*), and 10.2 mg of NADH disodium salt (0.20 m*M*). The final concentrations in the reaction mixture are indicated in parentheses. The solution is maintained at 0° until approximately 20 min before the experiment, when the temperature is brought to 37°. Prepare daily.

Octylglucoside: Add 730 mg of octylglucoside (molecular weight 292) to 100 ml of water to obtain a 25 m*M* solution. Store at −20°.

Horse liver alcohol dehydrogenase for assays I and II: The nominal specific activity is 2 units/mg protein, measured with ethanol as substrate in 10 m*M* Tris-HCl buffer, pH 7.4 (Sigma). The specific activity and the apparent K_m with tetradecanal, determined using the standard reaction assay conditions for lysoplasmalogenase, are 1.8 IU/mg protein and 0.10 m*M*, respectively. An aqueous solution of 5 mg/ml is prepared daily and maintained at 0°.

Aqueous suspensions (1.5 m*M*) of plasmalogen and lysoplasmalogen substrates (assays I and II): Prepare daily. About 2 μmol of substrate is dried in the bottom of a test tube under nitrogen gas, after which 1.3 ml of water is added. Substrates are mixed by vortexing for 1 min. Lysoplasmalogens are sonicated for 1 to 2 sec, and plasmalogens for 6 to 10 sec. All sonications are carried out at a sonifier setting of 2 mV. Alternatively, plasmalogens (or lysoplasmalogen) may be prepared as mixed micelles with octylglucoside detergent. Dry 2 μmol plasmalogen under nitrogen gas, add 1.0 ml of 25 m*M* octylglucoside, and sonicate for 5 sec. Maintain substrates at 0°.

Tetradecanal: Prepare a 0.10 *M* stock solution in ethanol; store at −20°, tightly capped. To obtain a 2 m*M* aqueous solution, dry 0.1 ml of the 0.1 *M* stock solution under nitrogen gas. Add 5 ml of 25 m*M* octylglucoside. Vortex 1 min and sonicate 5 sec. Maintain tightly capped at 0°; prepare daily.

Lysoplasmalogenase: The preparation and kinetic parameters of the microsomal and purified liver lysoplasmalogenase[5] are given in [46], this volume.[9] For assay I, liver microsomes, solubilized microsomes, or purified fractions are used depending on the objective of experiment. For enzyme assay II, the DEAE- or HPHT-purified lysoplasmalogenase is used.

(a) Liver microsomes: Liver microsomal suspensions contain 10 mg protein/ml in 0.12 *M* sucrose, 70 m*M* glycylglycine buffer. The specific activity is 0.040 IU/mg protein.

(b) Solubilized microsomes: The "solubilized microsomes" are obtained in the 100,000 *g* supernatant of the octylglucoside-solubilized microsomes.[9] They contain approximately 3.5 mg protein/ml, with a specific activity of 0.05 IU/mg protein. They are suspended in 47 m*M* sucrose, 23 m*M* KCl, 30 m*M* glycylglycine buffer (pH 7), 0.40 m*M* dithiothreitol, and 50% (v/v) glycerol. Store at −20°.

(c) Purified fraction: The DEAE-purified lysoplasmalogenase is taken directly from the DEAE column, diluted 1 : 1 with glycerol, and stored at −20°. The suspending buffer contains approximately 8 m*M* octylglucoside, 0.8 *M* KCl, 0.5 m*M* dithiothreitol, 10 m*M* MOPS buffer, pH 7.0, and 50% (v/v) glycerol.

Enzyme Reaction for Assay I

The reactions are carried out at 37° in a total volume of 1.0 ml in quartz cuvettes. The absorbance is monitored at 340 nm with a Beckman DU-65 spectrophotometer equipped with a kinetics software package programmed to determine the rates of absorbance change per minute and to calculate linear regression curves for each line. Six cuvettes are run simultaneously. In the absence of a recording spectrophotometer, absorbance values are recorded before initiating the reactions with lysoplasmalogenase, and at various time points following this addition.

A typical experimental protocol with essential controls is shown in Fig. 1. Into six cuvettes the following additions are made: 0.90 ml of reagent A, 0.5 IU of alcohol dehydrogenase (0.5 IU/ml), and 0.04 ml of 1.5 m*M* choline or ethanolamine lysoplasmalogen (60 μM). The final concentrations in the reaction mixture are given in parentheses. The volumes are adjusted to 1.0 ml with glycylglycine buffer. Contents are mixed and preincubated for 2 min at 37°. The reaction is initiated by the addition of lysoplasmalogenase (0.001 and 0.010 IU/ml), and the contents are mixed rapidly by inverting the cuvettes. The recording of the absorbance readings is begun about 45 sec after the enzyme is added, and the rates are recorded for 2.5 min.

The rate of absorbance change is linear to approximately 15 nmol aldehyde reduced/ml (ΔA 0.09), at which point the rate declines. This may be due to inhibition of alcohol dehydrogenase by long chain alcohols.[14] The initial linear rate of absorbance change is obtained and converted to micromoles of NADH oxidized per minute per milliliter, using the extinction coefficient for NADH (6220 liters/mol × cm).

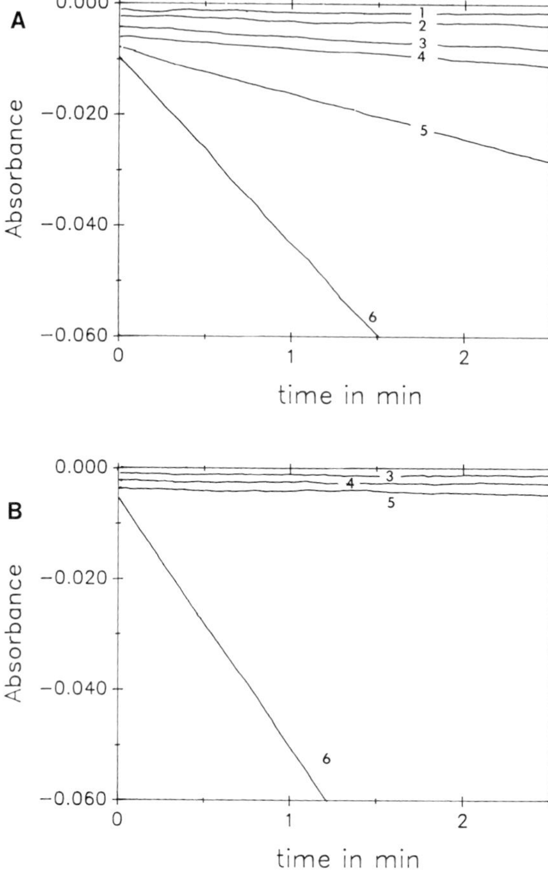

FIG. 1. Decrease in absorbance at 340 nm in spectrophotometric assay I for lysoplasmalogenase. The reaction mixture (1.0 ml final volume) included 70 m*M* glycylglycine buffer, 0.2 m*M* NADH, 0.35 mg/ml bovine serum albumin, 0.5 IU/ml alcohol dehydrogenase, and 60 μM choline lysoplasmalogen, according to the protocol below. The assay mixture was preincubated for 2 min before the addition of 100 μg solubilized microsomes (specific activity 0.052 IU/mg protein) to cuvettes 3–6 (A) or 0.56 μg of HPHT-purified enzyme (specific activity 11.0 IU/mg protein) to cuvettes 3–6 (B).

Reagent	Cuvette 1	2	3	4	5	6
Reagent A	+	+	+	+	+	+
Alcohol dehydrogenase	−	+	−	+	−	+
Choline lysoplasmalogen	+	+	−	−	+	+
Lysoplasmalogenase	−	−	+	+	+	+

$$\text{NADH } (\mu\text{mol}) \text{ oxidized/min/ml} = (\Delta A/\text{min})(1/6220 \text{ liter/mol})(1 \text{ liter}/1000 \text{ ml})(1 \times 10^6 \ \mu\text{mol/mol}) \quad (5)$$

External standards of aldehyde are necessary in order to test the competency of the coupling enzyme, alcohol dehydrogenase, under each new experimental condition. At the end of an incubation, between 1 and 10 nmol tetradecanal is added to cuvettes 2 and 4. The rate and extent of the resulting absorbance change are recorded.

The absorbance changes which resulted from a typical experiment with solubilized microsomal and HPHT-purified lysoplasmalogenase are shown in Figs. 1A and 1B, respectively. With both microsomes and purified enzymes the corrected rate for lysoplasmalogenase is obtained by Eq. (6).

$$\text{Lysoplasmalogenase activity} = (\text{rate } 6 - \text{rate } 4) - (\text{rate } 2 - \text{rate } 1) \quad (6)$$

With the purified enzyme the absorbance change is totally dependent on the presence of exogenously added alcohol dehydrogenase. (Compare rate 6 with rate 5, Fig. 1B.) With solubilized microsomes, however, the rate in the absence of alcohol dehydrogenase is about 25 to 33% of the rate in the presence of alcohol dehydrogenase. (Compare rate 6 with rate 5, Fig. 1A.) This is due to the presence of endogenous alcohol dehydrogenase activity in the solubilized microsomes. During chromatographic resolution of the liver microsomal proteins on the DEAE-cellulose column, alcohol dehydrogenase is separated from the lysoplasmalogenase.[20]

The rate of absorbance change is also dependent on the presence of exogenous lysoplasmalogen (rate 6 compared with rate 4). The solubilized microsomes and purified enzyme do not contain significant levels of lysoplasmalogen. Rate 2 minus rate 1 represents nonenzymatic release of aldehyde from the substrate; this difference is negligible in the experiment shown in Fig. 1. Rate 3 represents oxidation of NADH which is catalyzed by an enzyme(s) other than alkenyl hydrolase, for example, NADH oxidases. This rate is negligible with purified enzyme (Fig. 1B) but is significant with the native microsomes (Fig. 1A). Glycerol, used to stabilize the lysoplasmalogenase during storage at $-20°$, may contain small amounts of glyceraldehyde. Rate 4 minus rate 3 represents this aldehyde. From the data of Fig. 1, the activities of the HPHT-purified enzyme and of the solubilized microsomes are 11.0 and 0.052 μmol/min/mg protein, respectively.

General Considerations Concerning Method. The most common sources of error include the following: (1) Light scattering changes due to turbidity may be misinterpreted as NADH concentration changes. To

[20] M. S. Jurkowitz-Alexander, H. Ebata, J. S. Mills, Y. Hirashima, and L. A. Horrocks, *FASEB J.* **2,** A6226 (1988).

control for light scattering changes, (a) repeat incubations 1–6 (Fig. 1) in the absence of NADH and read at 340 nm. (b) Ensure that the components of the reaction mixture are as optically clear as possible, using only low concentrations of microsomes (<150 μg protein/ml) and plasmalogen. Use no more than 150 μM plasmalogen concentrations in reaction mixtures, unless adding as micelles in 25 mM octylglucoside. (2) The presence of contaminating enzymes may lead to side reactions which obscure the test. For example, NADH oxidase activity is present in native microsomal fractions. Inhibitors of these side reactions are needed, particularly for utilization of the assay with heart or brain microsomes.

The enzyme assay is specific for aldehydes. Thus, aldehydes produced from reactions other than that catalyzed by lysoplasmalogenase must be differentiated. In theory, free aldehydes produced from free radical-mediated reactions, or other reactions with other enzymes such as alkyl monooxygenase, could be detected by assay I. These reactions were not occurring in our studies. We verified that the reaction was a true lysoplasmalogenase reaction by an independent method employing lipid analyses of both the disappearance of substrate and the appearance of products by HPLC and TLC methods.[5] It is essential when using indirect measurements of an enzyme activity, such as in these coupled enzyme assays, to verify the reaction by a second method.

The most critical change we made in the Freeman and Carey assay[14] was the omission of Triton X-100 from our standard reaction mixture. This detergent inhibited lysoplasmalogenase by 100% at the final concentrations (0.05%) used by Freeman and Carey in their assays. When present at a concentration of 0.005%, Triton X-100 inhibited the enzyme by 50%.

Coupled Enzyme Assay II for Plasmalogen-Specific Phospholipase A_2 Activity[10]

Reagents and Solutions, in Addition to Those Described Above

Lysoplasmalogenase: DEAE–purified lysoplasmalogenase contains 1.6 IU/mg protein and is described under assay I. The purified lysoplasmalogenase contains no alcohol dehydrogenase, NADH oxidase, or plasmalogenase activities.[5]

Phospholipase A_2: Phospholipase A_2 is from *Naja naja naja* snake venom.[10] It contains phospholipase A_2 activity with 1,2-diacylglycerophosphocholine, 1-alkyl-2-acylglycerophosphocholine, and 1-alkenyl-2-acylglycerophosphocholine.[10] The specific activity is 4.9 IU/mg protein with the choline plasmalogen substrate. The value of K_m(plasmalogen) is greater than 0.30 mM.[10] The enzyme contains no plasmalogenase or lysoplasmalogenase activity. It is stored at $-20°$ as an aqueous suspension.

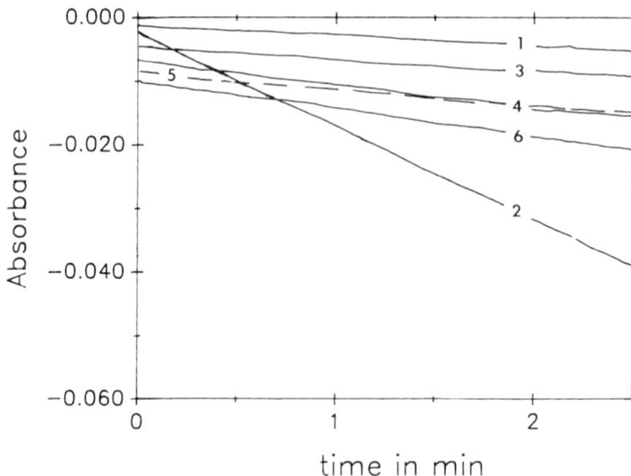

FIG. 2. Decrease in absorbance at 340 nm in spectrophotometric assay II for plasmalogen-specific phospholipase A_2. The reaction mixture (1.0 ml final volume) included choline plasmalogen (150 μM), DEAE-purified lysoplasmalogenase (0.024 IU/ml), bovine serum albumin (0.35 mg protein/ml), and horse liver alcohol dehydrogenase (0.5 IU/ml) as outlined below. The mixture was preincubated for 3 min, and the reaction was initiated by the addition of 240 ng of *Naja naja naja* phospholipase A_2 (specific activity 4.9 IU/mg protein).

Reagent	Cuvette 1	2	3	4	5	6
Reagent A	+	+	+	+	+	+
Alcohol dehydrogenase	−	+	−	+	−	+
Choline lysoplasmalogen	+	+	+	+	+	+
Lysoplasmalogenase	+	+	−	−	+	+
Phospholipase A_2	+	+	+	+	−	−

Enzyme Reaction for Assay II. The reaction assays are carried out in quartz cuvettes at 37° as described for assay I. A typical protocol is shown in Fig. 2. Into six cuvettes the following additions are made: 0.85 ml of reagent A, 0.05 ml of alcohol dehydrogenase (0.5 IU/ml), 0.05 ml of 3.0 mM choline plasmalogen (150 μM), 0.015 ml lysoplasmalogenase (0.024 units/ml). The volumes are adjusted to 1.0 ml with 80 mM glycylglycine buffer, pH 7.0. The final concentrations are given in parentheses. The cuvettes are inverted several times, and the absorbance changes at 340 nm are monitored for 3 min.

The phospholipase A_2 reaction is initiated by the addition of 0.0002 to 0.003 IU/ml of *Naja naja naja* phospholipase A_2. Enzymatic activity is monitored at 340 nm for 3 min. The linear rate between 1 and 2 min is used in calculating the rate of absorbance change per minute. The ΔA/minute values are converted to micromoles NADH per minute with Eq. (5).

The absorbance changes which result from the experiment outlined in Fig. 2 illustrate the controls necessary for a valid assay. In this experiment, 0.00116 IU/ml PLA_2 was added. The corrected rate is given by Eq. (7).

$$\text{Rate of phospholipase } A_2 = (\text{rate } 2 - \text{rate } 1) - [(\text{rate } 4 - \text{rate } 3) + (\text{rate } 6 - \text{rate } 5)] \quad (7)$$

The rate obtained by subtracting cuvette 1 from cuvette 2 is the provisional rate of phospholipase A_2 activity. Rate 4 minus rate 3 is low because the phospholipase A_2 is pure and contains no contaminating plasmalogenase or lysoplasmalogenase activities. Rate 6 minus rate 5 is also low because of the purity of lysoplasmalogenase, which contains no plasmalogenase or phospholipase A_2 activity.

Routinely, only 150 μM or less plasmalogen is used because higher concentrations produce turbidity in the reaction mixture and are therefore not appropriate for a spectrophotometric assay. However, this is below the K_m(plasmalogen) value for the phospholipase A_2. Therefore, the rates are assayed for only a short period of time so that k_1 [Eq. (4)] is zero-order with respect to substrate.

Tetradecanal standards (1–5 nmol) are added to cuvettes 1 through 6 to assay the efficacy of the coupling enzyme alcohol dehydrogenase under the various experimental conditions as explained for assay I. Lysoplasmalogen standards (1–5 nmol) are also used to test the efficacy of lysoplasmalogenase, as well as of alcohol dehydrogenase, under the various experimental conditions. The results obtained for *Naja naja naja* phospholipase A_2 activity with this coupled enzyme assay were verified using an independent method, in which the lipids were extracted, and product formation was quantitated by TLC and phosphorus analysis.[10]

The sources of error for assay II are the same as those described for assay I. In addition, the use of plasmalogen substrate concentrations below the K_m value requires careful measurement of the substrate in order to ensure the validity of the procedure.[10] Assay II should be appropriate for measuring phospholipase A_2 from mammalian sources, taking into consideration the problems encountered when crude subcellular fractions are used. Coupled enzyme assays I and II give the most reliable results if the enzymes to be assayed have high specific activities relative to side reactions (e.g., NADH oxidase). This occurs with partially purified enzymes or crude enzymes with high specific activities, such as liver microsomes.

In conclusion, we have described the procedures for two rapid, sensitive, and continuous methods for determination of lysoplasmalogenase and plasmalogen-selective phospholipase A_2.

[8] Phospholipase A_2 Assays with Fluorophore-Labeled Lipid Substrates

By H. STEWART HENDRICKSON

Introduction

Fluorophore-labeled lipid substrates have been used in two different ways to assay phospholipase A_2 (PLA_2). Continuous assays utilize changes in the fluorescence properties of the fluorophore on hydrolysis. Pyrene-labeled phospholipids have been used in such assays. The assays involve changes in excimer to monomer fluorescence on hydrolysis.[1-5] In discontinuous assays, the fluorophore is used to aid in detection of the substrate and/or products. Fluorophores such as the NBD (7-nitrobenzo-2-oxa-1,3-diazol-4-yl) group have been used to determine PLA_2 activity in intact cells[6] and on biomembrane phospholipids.[7] Trinitrophenylaminolauric acid-labeled phospholipids were used by Gatt *et al.*[8] to assay various lipases. These methods involve separation of the fluorescent product followed by measurement of its fluorescence. Simpler, sensitive methods involve chromatographic separation of fluorescent substrate and product by high-performance liquid chromatography (HPLC) or thin-layer chromatography (TLC) and direct determination of the ratio of substrate to product by fluorescence detection. Two such methods involving HPLC with fluorescence detection[9] and fluorescence scanning of TLC plates[10] are described here.

[1] H. S. Hendrickson and P. N. Rauk, *Anal. Biochem.* **116,** 553 (1981).

[2] T. Thuren, J. A. Virtanen, M. Lalla, and P. K. J. Kinnunen, *Clin. Chem.* **31,** 714 (1985).

[3] T. Thuren, J. A. Virtanen, M. Lalla, P. Banks, and P. K. J. Kinnunen, *Adv. Clin. Enzymol.* **5,** 149 (1987).

[4] T. Thuren, J. A. Virtanen, P. J. Somerharju, and P. K. J. Kinnunen, *Anal. Biochem.* **170,** 248 (1988).

[5] F. Radvanyi, J. Fordan, F. Russo-Marie, and C. Bon, *Anal. Biochem.* **177,** 103 (1989).

[6] A. Dagan and S. Yedgar, *Biochem. Int.* **15,** 801 (1987).

[7] N.-L. Saris and P. Somerharju, *Acta Chem. Scand.* **43,** 882 (1989).

[8] S. Gatt, Y. Barenholz, R. Goldberg, T. Dinur, G. Besley, Z. Leibovitz-Ben Gershon, J. Rosenthal, R. J. Desnick, E. A. Devine, B. Shafit-Zagardo, and F. Tsuruki, this series, Vol. 72, p. 351.

[9] H. S. Hendrickson, E. K. Hendrickson, and T. J. Rustad, *J. Lipid Res.* **28,** 864 (1987).

[10] H. S. Hendrickson, E. K. Hendrickson, and K. J. Kotz, *Anal. Biochem.* **185,** 80 (1990).

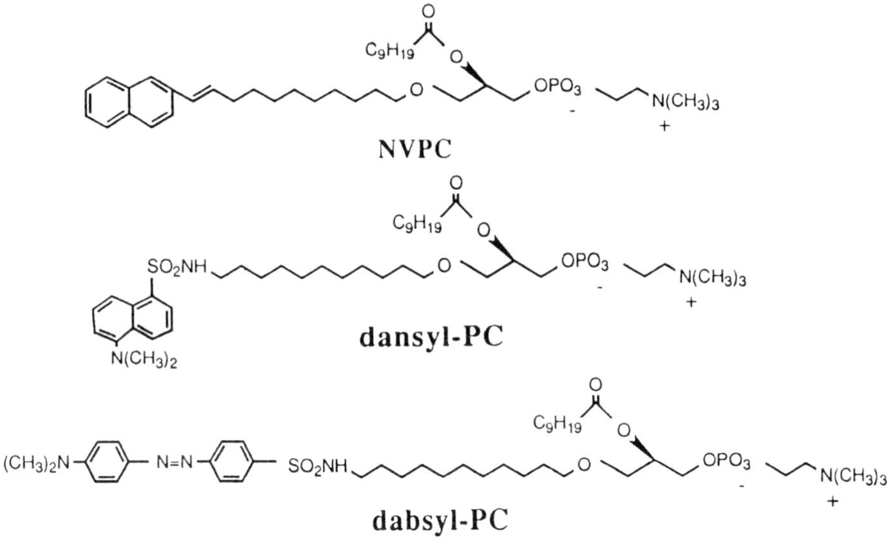

FIG. 1. Structures of phospholipid substrates.

Assay

Fluorescent phospholipid → fluorescent lysophospholipid + fatty acid

Principles. The fluorophore-labeled lipid substrates used (Fig. 1) have a fluorophore attached to the end of a hydrocarbon chain linked to the glycerol backbone by an ether linkage at the *sn*-1 position. Normal fatty acids are esterified at the *sn*-2 position. These substrates are resistant to phospholipase A_1 action, and the products are resistant to lysophospholipase action, making the assay quite specific for PLA_2. The phospholipid substrates and lysophospholipid products, both fluorescent, are separated by HPLC or TLC. The ratio of these two, determined by fluorescence detection, is used to calculate the amount of product released. Phospholipases C and D produce fluorescent hydrolysis products which are differentiated from that of PLA_2 by their chromatographic behavior.

Naphthylvinyl-PC {NVPC, 1-*O*-[12-(2-naphthyl)dodec-11-enyl]-2-*O*-decanoyl-*sn*-glycero-3-phosphocholine} is used in the HPLC assay.[9] It absorbs strongly at 250 nm and emits above 340 nm. The naphthylvinyl group is rather nonperturbing; it gives about the same activity as natural phosphatidylcholine with snake venom and pancreatic PLA_2s, and the platelet-activating factor (PAF) analog of NVPC is nearly as active as the natural agonist. This lipid is better suited for the HPLC assay since its low-wavelength emission is more subject to background interference on

TLC plates. Dansyl-PC [1-*O*-(*N*-dansyl-11-amino-1-undecyl)-2-*O*-decanoyl-*sn*-glycero-3-phosphocholine] and dabsyl-PC [1-*O*-(*N*-dabsyl-11-amino-1-undecyl)-2-*O*-decanoyl-*sn*-glycero-3-phosphocholine] are best suited for the TLC assay.[10] Dansyl-PC absorbs at 256 nm and gives a strong emission above 400 nm. Dabsyl-PC is not fluorescent, but it has an intense red color (absorbs at 540 nm) which can readily be visualized by the unaided eye. The dansyl and dabsyl groups are more perturbing; the lipids are slightly poorer substrates for snake venom PLA_2, and the PAF analog of dansyl-PC is a very poor agonist.

Synthesis of Substrates. NVPC is synthesized according to the procedure of Hendrickson *et al.*[9] Dansyl-PC is synthesized according to the procedure described by Schindler *et al.*[11] by dansylation of the amino lipid 1-*O*-(11-amino-1-undecyl)-2-*O*-decanoyl-*sn*-glycero-3-phosphocholine. Dabsyl-PC is synthesized by a similar procedure using dabsyl chloride instead of dansyl chloride to acylate the amino lipid.

Assay Procedure

Reagents

Buffer: 0.395 *M* NaCl, 66 m*M* Tris, 13.2 m*M* $CaCl_2$, pH 7
NVPC, 1 m*M* in $CHCl_3$
Dansyl-PC, 1 m*M* in $CHCl_3$
Dabsyl-PC, 1 m*M* in $CHCl_3$
Triton X-100, 10 m*M* in water
Quenching solvent: hexane/2-propanol/acetic acid (6 : 8 : 1.6, v/v)
HPLC solvent: hexane/2-propanol/water (6 : 8 : 1.6, v/v)
TLC solvent: $CHCl_3/CH_3OH$/concentrated ammonia/water (90 : 54 : 5.5 : 2, v/v)

Stock Substrate Solution. NVPC is used for assays involving HPLC. Dansyl-PC and dabsyl-PC are used for assays involving TLC. A measured amount (50 μl of a 1 m*M* stock solution) of substrate in chloroform is placed in a small test tube and dried under a stream of nitrogen and then under high vacuum for several minutes to remove the last traces of solvent. Triton X-100 (10 μl of 10 m*M*) and 190 μl of buffer are added, and the tube is repeatedly vortexed and sonicated in a bath sonicator until the lipid is completely dissolved and the solution is clear.

Enzyme Reaction. The stock substrate solution (40 μl) is mixed (vortex) with 60 μl of enzyme (about 2 ng of pure snake venom PLA_2) in a 6 × 50

[11] P. W. Schindler, R. Walter, and H. S. Hendrickson, *Anal. Biochem.* **174,** 477 (1988). Dansyl-PC is available from Molecular Probes, Eugene, OR.

mm glass tube and incubated at room temperature. The final concentrations are as follows: 0.15 *M* NaCl, 0.1 m*M* substrate, 0.2 m*M* Triton X-100, 5 m*M* $CaCl_2$, 25 m*M* Tris.

HPLC Analysis with NVPC. At various time intervals over a period of 30–60 min, 10-μl aliquots are removed from the reaction mixture, added to 90 μl of quenching solvent in a 0.5-ml microcentrifuge tube, and vortexed immediately. The tube is centrifuged at 15,600 *g* (microcentrifuge) for 2 min to remove any precipitated protein. Centrifugation is not necessary when only nanogram amounts of pure enzyme are used. Twenty microliters of the quenched, diluted sample is analyzed by HPLC on a 4.6 mm × 15 cm silica column [Waters Associates (Milford, MA), #85774] protected with a guard column. An autosampler (Spectra Physics, San Jose, CA) allows automated analysis of samples. A flow rate of 1 ml/min of HPLC solvent is maintained [Kratos (Ramsey, NJ) Spectroflow 400 pump], and the eluate is analyzed by a fluorescence detector (Kratos Spectroflow 980: excitation, 250 nm; emission, >320 nm; PMT, 775 volts; range, 0.1). The signal from the detector is analyzed using an integrator/plotter (Spectra Physics 4290). The integrator is programmed to calculate the fraction of the lyso-NVPC peak area (retention time 9.98 min) relative to the total peak areas for lyso-NVPC and NVPC (retention time 3.84 min). This value times the initial amount of NVPC present in the assay reaction (10 nmol) equals the amount of lyso-NVPC produced. A plot of nanomoles lyso-NVPC produced versus time (usually linear up to at least 5% hydrolysis) is used to determine the initial rate (activity, μmol/min). A control without enzyme shows no hydrolysis. Activities as low as 1 pmol/min in an assay volume of 100 μl can easily be measured.

TLC Analysis with Dansyl-PC. At various time intervals over a period of 30–60 min, 5-μl aliquots of the reaction mixture are removed and spotted directly onto TLC plates [HPTLC plates, 10 × 10 cm, Analtech (Newark, DE), #60077]. Substrate and products are separated by developing the plates in the TLC solvent. The plates are then scanned in a fluorescence densitometer [Shimadzu (Columbia, MD), CS-9000 with a fluorescence accessory]. Excitation is set at 256 nm, and emission is measured above 400 nm (filter #2). The amount of lysodansyl-PC produced is calculated from the fraction of lysodansyl-PC relative to total lysodansyl-PC and dansyl-PC in the same way as described for the HPLC analysis.

TLC Analysis with Dabsyl-PC. The procedure is similar to that with dansyl-PC with the following modifications. The assay requires at least 0.5 m*M* substrate and 1 m*M* detergent due to the less sensitive detection of the dabsyl group on TLC plates. After solvent development, the orange-red color tends to fade with time but can be enhanced (more red in color) and stabilized by spraying with a dilute solution of HCl. The plates are scanned in the dual-wavelength absorbance mode: sample absorbance at 540 nm and reference background at 610 nm.

Choice of Enzyme and Conditions

Snake venom PLA_2s are quite active in the presence of Triton X-100, but other PLA_2s (particularly pancreatic PLA_2) may be much less active with this detergent. PLA_2 in human synovial fluid is best assayed with NVPC by the HPLC method using hexadecylphosphorylcholine instead of Triton X-100.[12] Pancreatic PLA_2 is best assayed using sodium cholate as the detergent. The concentration of substrate and the ratio of substrate to detergent may be varied as desired. High salt concentrations (>1 *M* NaCl) may cause the quenched sample for HPLC analysis to phase separate and give artifactual peaks on HPLC.

Applications

The HPLC and TLC assays are sensitive enough to detect low levels of PLA_2 in crude physiological sources such as synovial fluid (see previous section). Detailed time-course analyses are feasible because of the small amount of sample required and speed of analysis. The TLC assay is well suited for rapid and efficient screening of PLA_2 inhibitors since many assays can simultaneously be analyzed on a single TLC plate. Qualitative results may be obtained by simple viewing of the plate under UV light for dansyl-PC or ordinary light for dabsyl-PC.

Derivatives of NVPC, dansyl-PC, and dabsyl-PC are potentially useful in the assay of other enzymes of lipid metabolism. Lysodansyl-PC and the PAF analog of dansyl-PC were used as substrates in TLC assays of lyso-PAF acyltransferase, lyso-PAF acetyltransferase, and PAF acetylhydrolase in polymorphonuclear leukocytes.[11] Since the probe remains attached to the glycerol backbone, simultaneous assay of these enzymes of the PAF cycle is possible.

Acknowledgments

I thank Peter Schindler (Pharma Forschung, Hoechst AG, Frankfurt) for fruitful discussions leading to the idea of this assay, particularly the design of dansyl-PC. Elizabeth Hendrickson synthesized the substrates; Margaret Benton, Laura Knoll, and Kim Kotz worked on the development of these assays. This work was supported by a grant (GM 33606) from the National Institutes of Health.

[12] H. S. Hendrickson and M. E. Benton, unpublished results (1988).

[9] Assays of Phospholipases on Short-Chain Phospholipids

By MARY F. ROBERTS

Introduction

Phospholipase kinetics against aggregated substrates can be divided conceptually into two parts: (1) binding of the enzyme to the aggregated substrate (interfacial phenomena) and (2) kinetic processing of the substrate. In most phospholipid aggregates (e.g., bilayers, mixed micelles) changes in the composition, components, or temperature can affect particle structure and/or enzymatic activity. Thus, determining whether an observed altered activity is the result of changed interfacial binding or kinetic processing can be extremely difficult.

A rational approach to analyzing the kinetics of surface-active enzymes is to use an interface formed of components that can be mixed in any proportions without drastically altering the interfacial characteristics. Such aggregates need to be studied carefully for any altered physical properties. The enzyme kinetics are then analyzed in light of the observed aggregate physical chemistry. Micellar species form ideal interfaces for such studies, and synthetic short-chain phosphatidylcholine (PC) molecules are particularly useful building blocks of complex but characterizable micellar particles. These PCs exist as monomers in aqueous solution and form micelles above a critical micelle concentration (CMC), which is related to the total number of carbons in both chains. The two fatty acyl chains can be the same chain length (forming symmetric PCs) or have different lengths (asymmetric PCs). Their detergent properties allow them to solubilize other lipophilic substances such as triglycerides and cholesterol and still form micelles. While other short-chain phospholipids (phosphatidylethanolamine, phosphatidylserine) have been used occasionally, they have not been extensively characterized, and thus are not so useful at present in monitoring phospholipase activity. As an assay system short-chain phospholipids offer the opportunity of comparing phospholipase activity toward the same molecule as a monomer, in micelles, and in bilayer aggregates, to examine how other low concentrations of amphiphiles affect enzyme activity, and to examine how chain length and other structural modifications of the substrate affect activity.

Properties of Short-Chain Lecithins

Synthesis

Short-chain PC species do not occur naturally, and hence they must be synthesized from glycerol phosphorylcholine or an appropriate lyso-PC and fatty acid. The symmetric PCs dibutyroyl-PC, dihexanoyl-PC, diheptanoyl-PC, and dioctanyol-PC have been synthesized by coupling activated acid to glycerophosphorylcholine.[1-3] Reagents used to activate the fatty acyl group are the same as have been used with long-chain PC syntheses. These diacyl-PC species are available commercially[4] and can usually be used without further purification. Dibutyroyl-PC does not form micelles up to 80 m*M*; dihexanoyl-PC has a high CMC (14 m*M*) but does aggregate to form small micelles.[5-7] Dioctanoyl-PC, with the lowest CMC, tends to phase separate as soon as micelles begin to form.[8,9] In order to suppress this phase separation, a variety of anions can be added (e.g., 0.1 *M* LiI, 0.5 *M* KSCN[8]). High anion concentrations, however, can affect phospholipase activity; for example, phospholipase C (*Bacillus cereus*) has been shown to be inhibited by high concentrations of anions.[10] This makes the 14-carbon species (di-C_7PC) the most useful compound because it has a critical micellar concentration in the millimolar range (1.4 m*M*), does not phase separate, and forms relatively small micelles that have been well studied.

Asymmetric short-chain lecithins have been synthesized by reacylation of the appropriate lyso-PC (formed by phospholipase A_2 hydrolysis of the diacyl-PC or purchased if commercially available), using either carbonyldi-

[1] P. P. M. Bonsen, G. J. Burbach-Westerhuis, G. H. de Haas, and L. L. M. van Deenen, *Chem. Phys. Lipids* **8,** 199 (1972).

[2] W. F. Boss, C. J. Kelley, and F. R. Landsberger, *Anal. Biochem.* **64,** 289 (1975).

[3] T. G. Warner and A. A. Benson, *J. Lipid Res.* **18,** 548 (1977).

[4] Short-chain diacyl-PCs can be obtained from Calbiochem (San Diego, CA), Sigma (St. Louis, MO), and Avanti Biochemicals (Birmingham, AL).

[5] R. J. M. Tausk, J. Karmiggelt, C. Oudshoorn, and J. T. G. Overbeek, *Biophys. Chem.* **1,** 175 (1974).

[6] R. J. M. Tausk, J. Van Esch, J. Karmiggelt, G. Voordouw, and J. T. G. Overbeek, *Biophys. Chem.* **1,** 184 (1974).

[7] R. A. Burns, Jr., M. F. Roberts, R. Dluhy, and R. Mendelsohn, *J. Am. Chem. Soc.* **104,** 430 (1982).

[8] R. J. M. Tausk, C. Oudshoorn, and J. T. G. Overbeek, *Biophys. Chem.* **2,** 53 (1974).

[9] G. M. Thurston, D. Blankschtein, M. R. Fisch, and G. B. Benedek, *J. Chem. Phys.* **84,** 4558 (1986).

[10] S. E. Aakre and C. Little, *Biochem. J.* **203,** 799 (1982).

imidazole as the fatty acid-activating agent[11] (for fatty acids longer than 5 carbons) or the fatty acid anhydride with 1 mol equivalent of 4-pyrrolidino-pyridine[12] and typically a 5-fold excess of fatty acid to lyso-PC. A wide variety of asymmetric PCs have been synthesized in excellent yield in this fashion.[13]

Micelle Structure

Detailed physical characterization of short-chain PCs has involved a variety of techniques. Small angle neutron scattering (SANS) external and internal contrast variation experiments have been used to determine the structure of monodisperse (aggregation number 19 ± 1 from 0.027 to 0.361 *M*) dihexanoyl-PC micelles.[14] The micelle is represented by a prolate ellipsoid with two uniform regions (Fig. 1A). The hydrocarbon chains form a close-packed spheroidal core with a minor axis equal to the fully extended fatty acyl chains (7.8 Å) and a major axis equal to 24 Å. Surrounding the hydrocarbon core are the polar head groups, which are distributed in a shell with a thickness of 10 Å in the direction of the minor axis and 6 Å along the major axis. The space between the hydrophilic head groups is considerably larger in a micelle structure (102 $Å^2$ per head group) than in a monolayer or bilayer structure (65 $Å^2$). The thickness of the hydrophilic shell region implies that the head group must have an orientation parallel to the surface of the hydrocarbon core. There is no preferential ordering near the center of the micelles: the terminal CH_3 groups are distributed throughout the hydrocarbon core with a slightly higher probability of being toward the center. This yields a highly disordered core structure, consistent with current views of micelle chain packing.

The addition of one or more CH_2 groups to each acyl chain produces micelles that grow significantly with increasing PC concentration. For diheptanoyl-PC, SANS studies[15] and a thermodynamic analysis have produced a picture of these micelles as spherocylinders that grow in the longitudinal direction (Fig. 1B). The growth is due to the free energy of insertion of a monomer in the straight section of the spherocylinder being lower than that in the end caps of the micelle. This free energy difference

[11] C. D. DeBose, R. A. Burns, Jr., J. M. Donovan, and M. F. Roberts, *Biochemistry* **24,** 1298 (1985).

[12] J. T. Mason, A. V. Broccoli, and C.-H. Huang, *Anal. Biochem.* **113,** 96 (1981).

[13] K. Lewis, J. Bian, A. Sweeney, and M. F. Roberts, *Biochemistry* **29,** in press (1990).

[14] T.-L. Lin, S.-H. Chen, N. E. Gabriel, and M. F. Roberts, *J. Am. Chem. Soc.* **108,** 3499 (1986).

[15] T.-L. Lin, S.-H. Chen, N. E. Gabriel, and M. F. Roberts, *J. Phys. Chem.* **91,** 406 (1987).

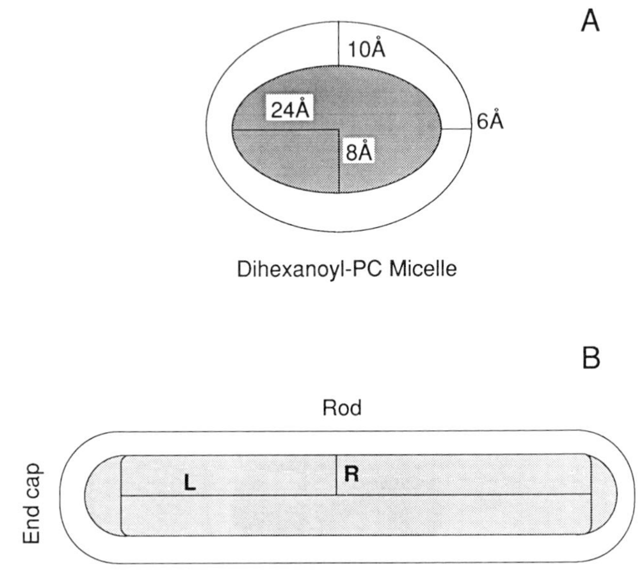

FIG. 1. (A) Micelle structure of dihexanoylphosphatidylcholine based on SANS data. The inner ellipsoid defines the hydrocarbon core, whereas the outer ellipsoid defines the head group and glycerol backbone area. (B) Schematic structure for diheptanoyl-PC micelles based on SANS data. L denotes the length of the hydrocarbon cylinder while R indicates the radius. The phosphocholine headgroups occupy the outer region around the hydrocarbon spherocylinder.

is attributed to the smaller surface area of the hydrocarbon core in contact with water for a monomer in the straight section compared to that in the end caps. All asymmetric PCs examined fit spherocylinder models with the same radial structure and different lengths.[16] Their growth characteristics (i.e., length of the rod) are similarly predicted by a thermodynamic model. This means that at any given concentration, not only the average particle dimensions, but the distribution of sizes can be computed. Therefore, we have a series of short-chain PC species with CMC values in the 0.1–15 mM range and sizes from 19 to 5000 monomers per micelle, depending on exact total number of fatty acyl carbons and the distribution between the two chains (Table I). If a phospholipase is suspected to be sensitive to differences in monomer and micelle, or to micelle size, then these short-chain PC molecules make an ideal assay system.

A short-chain PC molecule in a micelle has conformational, dynamic,

[16] T.-L. Lin, S.-H. Chen, and M. F. Roberts, *J. Am. Chem. Soc.* **109,** 2321 (1987).

TABLE I
PROPERTIES OF SHORT-CHAIN PHOSPHATIDYLCHOLINES

Short-chain PC[a]	CMC (mM)	N_0[b]	N_{15}[c]	Area/head group (Å^2)	Ref.
$(C_6)_2$-PC	14	19 ± 1	19	100	14
1-C_6-2-C_7-PC	4.6	25 ± 2	37	95; 76[d]	16
$(C_7)_2$-PC	1.5	27 ± 1	100	98; 80	15
1-C_6-2-C_8-PC	1.6	32 ± 5	120	93; 74	16
1-C_8-2-C_6-PC	1.4	27 ± 5	80	95; 77	16
1-C_{10}-C_4-PC	0.8	—	—	—	13
1-C_{12}-C_2-PC	0.6	—	—	—	13
1-C_7-C_8-PC	0.6	48 ± 10	4000	85; 70	16
1-C_8-C_7-PC	0.6	45 ± 10	2800	88; 72	16
$(C_8)_2$-PC	0.2	Phase separates			8

[a] $(C_x)_2$-PC, Diacylphosphatidylcholine; for 1-C_x-2-C_y-PC, x denotes the number of carbons in the fatty acid esterified to the *sn*-1 position, and y denotes the number of carbons in the fatty acid esterified to the *sn*-2 position.

[b] Number of monomers in the minimum size micelle.

[c] Number of monomers in the average size micelle at 15 mM PC.

[d] For spherocylinders the area for PC molecules in the end caps is larger than the area for a PC molecule in the cylindrical section.

and packing features analogous to those of naturally occurring PCs in bilayers.[7,14–18] The two fatty acids of PC molecules are nonequivalent in micelles as well as in bilayers, with the *sn*-2 ester group close to the interface and this chain slightly shorter than the *sn*-1 chain, although this difference is less than that in the bilayer structure. For asymmetric PC micelles with the same total number of carbons this conformational difference affects thermodynamic parameters such as the size and CMC. Chain dynamics, studied by nuclear magnetic resonance (NMR), indicate considerable segmental motion.

Chemically modified short-chain PCs behave much like the linear species. Substitution of an ether linkage for the acyl linkage yields short-chain PC molecules with lower CMC values, quantitatively consistent with the more hydrophobic nature of the ether moiety.[19] Differences in the CMC values of symmetric branched short-chain PCs are understandable in terms of the differential solubilities of the hydrophobic portion of the molecules. In this case the difference expected appears to be due to substitution of a methyl and methine group for two methylenes.[12] For these altered PCs,

[17] R. A. Burns, Jr., and M. F. Roberts, *Biochemistry* **19,** 3100 (1980).

[18] R. A. Burns, R. E. Stark, D. A. Vidusek, and M. F. Roberts, *Biochemistry* **22,** 5084 (1983).

[19] R. A. Burns, Jr., J. M. Friedmann, and M. F. Roberts, *Biochemistry* **20,** 5945 (1981).

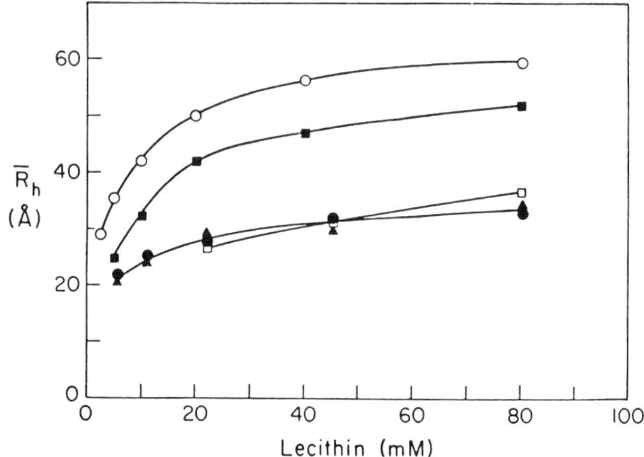

FIG. 2. Mean hydrodynamic radius, $\bar{R}_H$ (determined by quasi-elastic light scattering), of seven-carbon fatty acyl chain PC micelles as a function of PC concentration: (○) diheptanoyl-PC, (■) bis(5-methylhexanoyl)-PC, (●) bis(3-methylhexanoyl)-PC, (▲) bis(2-methylhexanoyl)-PC, and (□) bis(4,4-dimethylpentanoyl)-PC. [From C. D. DeBose, R. A. Burns, Jr., J. M. Donovan, and M. F. Roberts, *Biochemistry* **24,** 1298 (1985), with permission.]

micelle size reflects the location of the branch methyl (Fig. 2): when near the ester group the CMC is higher and micelle growth is small (like dihexanoyl-PC), and when the branch methyl is at the end of the chains, the molecule behaves much like diheptanoyl-PC in CMC and growth characteristics.

Kinetic Studies with Phospholipases

Detection of Interfacial Activation

Simple symmetric short-chain PCs were used early on to characterize the activity of water-soluble phospholipases.[20–22] Hydrolysis of short-chain PCs is usually followed by the pH-stat method, with a pH 8 end point and substrate concentrations in the millimolar region. In addition, 5 m*M* $CaCl_2$ is added for phospholipase A_2-catalyzed hydrolysis of phospholipids. Hydrolysis (1–5%) can easily be monitored continuously by this technique, and with this level of hydrolysis there is no change in the micelle size

[20] G. H. De Haas, P. P. M. Bonsen, W. A. Pieterson, and L. L. M. Van Deenen, *Biochim. Biophys. Acta* **239,** 252 (1971).

[21] M. A. Wells, *Biochemistry* **13,** 2248 (1974).

[22] C. Little, *Acta Chem. Scand. B* **31,** 267 (1977).

distribution. Whereas with long-chain PCs products can complicate the kinetic analysis, the products of phospholipase A_2 hydrolysis of these short-chain species is much simpler. The lyso-PCs produced have 10-fold higher CMC values than the diacyl species, and thus will usually partition into the aqueous phase as monomers. The fatty acids will also partition between water and the micelle. In this way, the micelle surface charge does not develop as rapidly as it does with long-chain PCs in either detergent micelles or bilayers. The pK_a of a short-chain fatty acid in the PC micelle is only slightly shifted up from the monomer value. This enables one to collect pH-stat kinetics over a wider pH range without worrying about the effectiveness of titrating the product fatty acid. For phospholipase C the short-chain diglyceride, while hydrophobic, is not as insoluble as the longer chain species. Solubilization of these diglycerides by the short-chain PC micelles is high, although eventually they phase separate.

Pure short-chain PCs have been extremely useful in comparing lipolytic enzyme activity toward monomeric and micellar substrates. Because a symmetric species such as dihexanoyl-PC has a high CMC (~14 m*M*), the enzyme can be incubated with both monomeric substrate (<10 m*M*) and micellar (>15 m*M*) species. For diheptanoyl-PC enzyme-catalyzed hydrolysis of substrate below 1 m*M* will monitor monomer, and greater than 1 m*M* substrate will represent micelle activity. Most phospholipases examined appear to show a unique kinetic feature under these conditions: they exhibit higher rates of hydrolysis when the substrate is packed in a micelle compared to when it is monomeric. This increase in V_{max} has been termed "interfacial activation." Thus, as the short-chain PC increases in concentration above its CMC, the enzyme activity increases. For phospholipase A_2 enzymes, the rate enhancement on micellization is often 20- to 100-fold.[23] An example for the enzyme from cobra venom is shown in Fig. 3A. For phospholipase C, on the other hand, this activation is only 2- to 3-fold (Fig. 3B).[24]

One can estimate kinetic parameters (K_m and V_{max}) for the monomeric and micellar substrates. For example, the data for phospholipase C-catalyzed hydrolysis of diheptanoyl-PC below 1 m*M* can be fit with a simple Michaelis–Menten treatment: V_{max} = 1340 μmol/min/mg, and K_m = 0.20 m*M*. As micelles form the rate of diheptanoyl-PC hydrolysis increases. If one assumes simple Michaelis–Menten behavior for micelles, a steady-state, and a phase separation model for micelle formation (i.e., the concen-

[23] M. F. Roberts and E. A. Dennis, *in* "Phosphatidylcholine Metabolism," (D. E. Vance, ed.), p. 121. CRC Press, Boca Raton, Florida, 1989.

[24] M. Y. El-Sayed, C. D. DeBose, L. A. Coury, and M. F. Roberts, *Biochim. Biophys. Acta* **837,** 325 (1985).

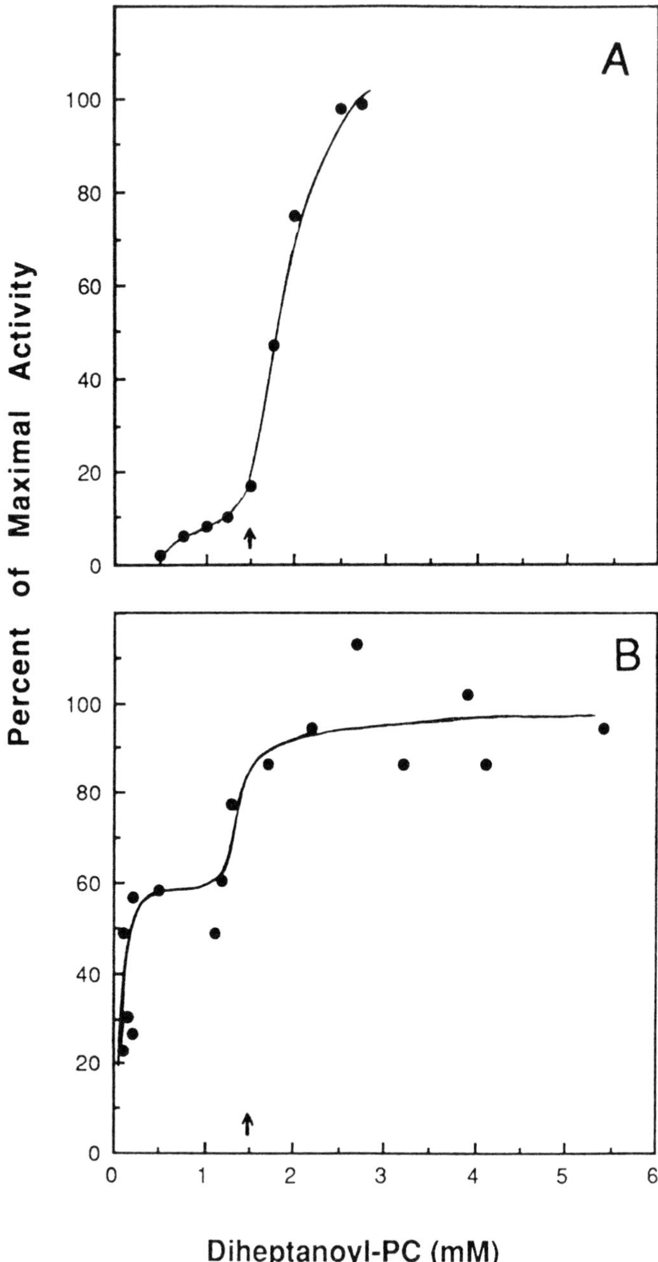

FIG. 3. Percent maximal activity of (A) cobra venom phospholipase A_2 [adapted from T. L. Hazlett, Ph.D. Dissertation, University of San Diego, La Jolla, California (1986), with permission] and (B) *Bacillus cereus* phospholipase C toward diheptanoyl-PC as a function of PC concentration [adapted from M. Y. El-Sayed and M. F. Roberts, *Biochim. Biophys. Acta* **831,** 133 (1985)]. The CMC of diheptanoyl-PC is indicated by an arrow.

tration of monomer is held constant at the CMC), a simple expression for analysis of observed velocities is derived:

$$V = \frac{V_{max}^{mic}[(S_o - CMC)/K_{sm}] + V_{max}^{mon}(CMC/K_s)}{1 + (S_o - CMC)/K_{sm} + CMC/K_s} \tag{1}$$

where S_o is the total PC concentration, K_{sm} and K_s are the Michaelis constants for micellar and monomeric species, and V_{max}^{mic} and V_{max}^{mon} are the maximum velocities for micelle and monomer species. For phospholipase C this treatment generates apparent micelle kinetic constants of $K_{sm} = 0.03$ mM and $V_{max}^{mic} = 2650$ μmol/min/mg. In this way the effect of substrate aggregation on enzyme activity can be quantified.[24] For several short-chain PC and lyso-PC species K_m decreases and V_{max} increases on micelle formation of the substrate.

Recently, van Oort *et al.* have used short-chain PCs (D and L isomers) to examine phospholipase A_1 and lysophospholipase activity of *Staphylococcus hyicus* lipase.[25] No interfacial activation was observed toward diheptanoyl-PC, although such behavior was observed for tributyrin. Phospholipase A_2 from *Naja melanoleuca* also exhibited high enzymatic activity below the CMC of diheptanoyl-PC and no further increase above the CMC.[26] For these enzymes the high activity below the CMC could be explained by the capacity of the enzyme to form lipid–protein aggregates well below the CMC. Reexamination of monomer/micelle preferences with other short-chain PCs (dihexanoyl-PC or 1-hexanoyl-2-heptanoyl-PC) with higher CMC values might shed more light on this observation.

Detergent Effects

Because they form micelles, short-chain PCs can be used with other detergents to form mixed micelles where the mole fraction of PC varies from 1.0 to 0. In other words, they form a perfect system for examining "surface dilution" and other detergent-related effects. With long-chain PCs, which prefer to pack as bilayers, an excess of detergent is often required to produce a mixed micelle population and to solubilize all bilayers.[27,28] This reduces the PC range from 0.3 to 0 mole fraction (e.g., with Triton X-100). A wide variety of detergents have been shown to act as competitive inhibitors of phospholipase C when present at high relative

[25] M. G. van Oort, A. M. T. J. Deveer, R. Dijkman, M. L. Tjeenk, H. M. Verheij, G. H. de Haas, E. Wenzig, and F. Gotz, *Biochemistry* **28,** 9278 (1989).

[26] J. Van Eijk, H. M. Verheij, R. Dijkman, and G. H. de Haas, *Eur. J. Biochem.* **132,** 183 (1983).

[27] E. A. Dennis, *Arch. Biochem. Biophys.* **158,** 485 (1974).

[28] A. A. Ribeiro and E. A. Dennis, *Biochim. Biophys. Acta* **332,** 26 (1974).

concentrations (i.e., when the ratio of detergent to phospholipid is ≥ 2).[29] At low mole fractions they could also affect enzyme activity, but this cannot be studied in long-chain PC mixed-micelle systems. To do this one needs a water-soluble PC such as diheptanoyl-PC. Experiments of this type with phospholipase C (*Bacillus cereus*) show that the presence of low mole fractions (below the K_i) of zwitterionic and nonionic detergents do not affect enzyme activity, while the presence of cationic or anionic detergents at the same low levels increases activity as much as 2-fold.[30] This rate enhancement ("detergent activation") occurs with both positively and negatively charged detergents and is an interfacial phenomenon, since similar concentrations of dihexanoyl-PC, which are monomeric, and charged detergent show no activation (Fig. 4).

Physical studies of various PC/detergent and detergent/diglyceride mixed micelles suggest product release of the diglyceride as the cause of the detergent activation. Interfacial activation of phospholipase C can then be explained as follows: lipids dispersed in aggregate structures are hydrolyzed more efficiently than those monomolecularly dispersed in solution, because diglyceride can partition from the enzyme into a micellar aggregate easier than into an aqueous environment.

Definition of Substrate-Binding Requirements

For enzymes with soluble, monomeric substrates, chemical modification of the substrate molecule can often be correlated with enzyme-binding requirements. For surface-active enzymes, such as the phospholipases, the added complication of altered aggregate structure must also be considered. The short-chain PC micellar system allows one to vary PC chemical structural features and assess the effect on phospholipase kinetics without having to worry about phase transitions, packing, or other effects. Thus far, short-chain PC modifications have primarily involved modifications of the fatty acyl chains: substituting ether linkages for the ester moiety,[19] branching the fatty acyl chains by the addition of methyl groups,[12] addition of phenylalkanoyl chains,[24] and production of PCs with the same total number of carbons but a varying asymmetric distribution in the *sn*-1 and *sn*-2 chains.[13] Because these modified lipids have CMC values, dynamics, and conformations similar to the standard linear symmetric compounds, they should mix ideally with the linear species and, hence, are also useful for competition assays.

Phospholipase A_2 (*Naja naja naja*) shows particularly interesting kinet-

[29] R. A. Burns, Jr., M. Y. El-Sayed, and M. F. Roberts, *Proc. Natl. Acad. Sci. U.S.A.* **79,** 4902 (1982).

[30] M. Y. El-Sayed and M. F. Roberts, *Biochim. Biophys. Acta* **831,** 133 (1985).

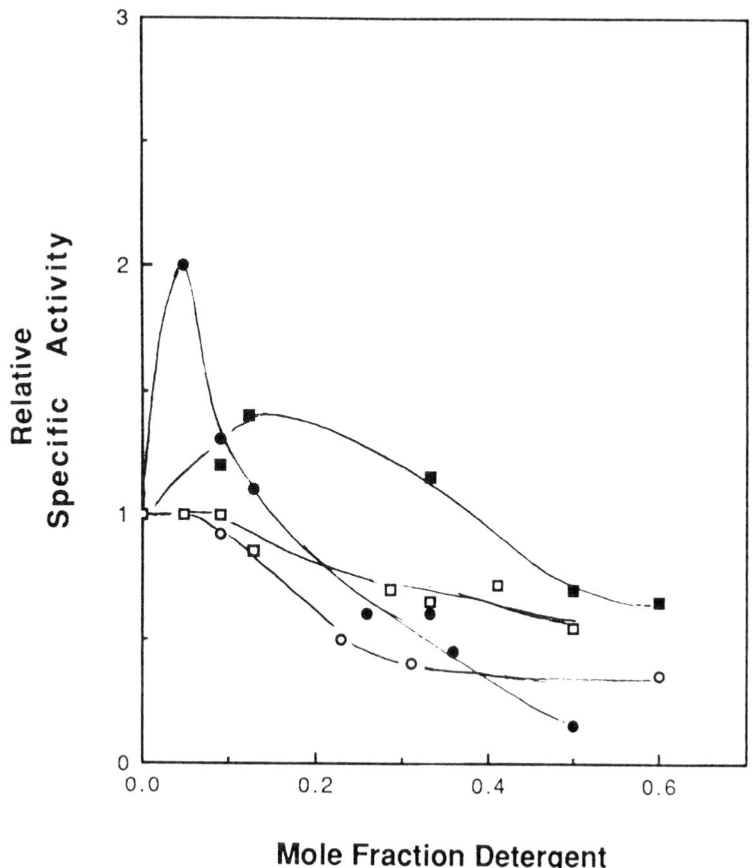

FIG. 4. Percent maximal activity of phospholipase C (*Bacillus cereus*) toward 10 m*M* diheptanoyl-PC as a function of the mole fraction of detergent: cetyltrimethylammonium bromide (●), sodium deoxycholate (■), Zwittergent 3-14 (○), and Triton X-100 (□). [Adapted from M. Y. El-Sayed and M. F. Roberts, *Biochim. Biophys. Acta* **831,** 133 (1985), with permission.]

ics with the modified PCs. Compared to diheptanoyl-PC, the PCs with both chains branched are poor substrates and good inhibitors of this enzyme.[12] When only one acyl chain bears a methyl group, the hybrid PC is easily hydrolyzed by phospholipase A_2. This occurs even when the added methyl group is adjacent to the *sn*-2 lipid ester bond that is to be cleaved. Because micellar structure and behavior and lipid conformation are not different in these systems, one can postulate that the inhibition of phospholipase A_2 by branched-chain PCs is due to a steric interaction of both bulky fatty acyl chains with the enzyme that prevents a catalytically efficient change in the protein. If, in fact, two substrate molecules are required to bind for catalysis, as has been suggested for the cobra venom enzyme, and both bind in the same vicinity, the active center may be too crowded to accommodate a second branched PC molecule after one branched chain PC is bound. The net result is that both fatty acyl chains are important in determining the rate of phospholipase A_2 hydrolysis of a lipid. The bis(methylhexanoyl)-PC molecules are poor substrates because of altered enzyme conformation or aggregation state, not because of altered phospholipid conformation or interfacial behavior.

The chain length specificity of phospholipase A_2 can also be probed with modified short-chain PC species. In most work investigators have looked at enzyme activity toward symmetric short-chain PCs. Dibutyroyl-PC is a very poor substrate; however, this could be attributed to the fact that it is a monomeric substrate, although activity may also be low because the enzyme requires a certain length *sn*-2 fatty acyl chain for efficient binding and catalysis. Fourteen-carbon asymmetric PCs can define this chain length requirement. In Fig. 5A, the activity of phospholipase A_2 toward 6 m*M* substrate is shown as a function of *sn*-2 chain length. There is a dramatic dependence of activity on the number of carbons in the *sn*-2 chain. For 7 or more carbons activity is high and shows no further dependence on chain length.[13]

The same short-chain PC modifications provide insight into the mechanism of phospholipase C action. In particular, they are useful in determining the structural features of a PC molecule important for binding to phospholipase C (*B. cereus*). Because dibutyroyl-PC and dihexanoyl-PC have high CMC values, one can compare monomer kinetic parameters for the two species. Activity is high toward dihexanoyl-PC monomers and poor toward dibutyroyl-PC,[24] suggesting that 6 or more carbons are necessary on a fatty acyl chain for effective enzyme binding and catalysis. Any increases in the length of both chains are irrelevant in the short-chain PC assay system. To see if there is any specificity for *sn*-1 or *sn*-2 chains phospholipase C can be assayed with asymmetric short-chain PCs. If the total number of carbons is maintained at 14, the short-chain PC will have CMC values around 1 m*M*. As can be seen in Fig. 5B, even the 1-dodecanoyl-2-acetyl-PC species is an excellent substrate for phospholipase C. Therefore, the chain requirement is a nonspecific need for a hydrophobic moiety to anchor the substrate to the enzyme, and whether the sufficiently long chain is at the *sn*-1 or *sn*-2 position appears irrelevant.[13]

Ether substitution dramatically affects phospholipase C activity. Short-chain ether-linked PCs are very poor substrates and not good inhibitors of the enzyme.[19] Therefore, a carbonyl moiety must be necessary for

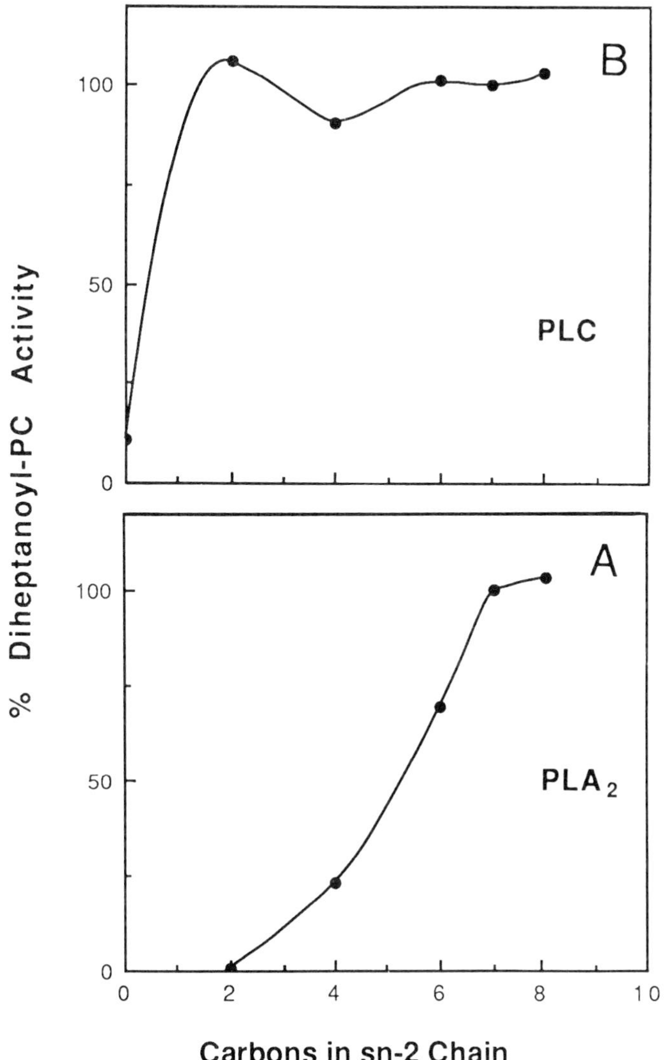

FIG. 5. Specific activity of (A) phospholipase A_2 (*Naja naja naja*) and (B) phospholipase C (*Bacillus cereus*) toward 6 m*M* asymmetric 14-carbon short-chain PCs as a function of the number of carbons in the *sn*-2 chain.

aligning the substrate PC at the phospholipase C active site. Consistent with this is the fact that methyl branching inhibits activity only when the CH_3 is adjacent to the carbonyl carbons. In a similar fashion, PCs with phenylalkanoyl chains become poor substrates only when the phenyl group is near the acyl linkage.[24] It appears that an ester at the *sn*-2 position is more effective, since the activity of the enzyme is higher toward 1-dodecanoyl-2-acetyl-PC than toward lysomyristoyl-PC. By using these modified short-chain PCs one finds that a certain amount of phospholipid chain hydrophobic binding is important for this water-soluble surface-active enzyme, but fatty acyl carbonyls (preferably unhindered) are more critical to productive binding and subsequent hydrolysis.

Short-Chain Phosphatidylcholines in Bilayer Matrix

A short-chain PC in a micelle is a good substrate for phospholipases (typically 1000–2000 μmol/min/mg for micelles), whereas long-chain PC molecules are poor substrates in bilayers (typically <50 μmol/min/mg) and good substrates in detergent mixed micelles. In order to understand these kinetic differences, one would like to take a short-chain PC molecule and put it into a bilayer structure. A kinetic analysis of this hybrid system could, in principle, help explain the decreased activity of many phospholipases toward bilayer substrate. If 20 mol % short-chain PC is cosolubilized with long-chain PC, bilayer structures form.[31] If the long-chain PC is below its gel-to-liquid crystalline phase transition (T_m), small (R_H ~90 Å) homogeneous particles are produced.[32] When the long-chain PC is liquid crystalline, very large lamellar structures are formed.[32] In these binary bilayer aggregates made with PC species, phospholipases have access to substrates with the same head group but different fatty acyl chain lengths. In Fig. 6, a partial phase diagram is shown for the diheptanoyl-PC/dipalmitoyl-PC system (25 m*M* total lipid) below the T_m of the long-chain component. Bilayer aggregates occur uniquely between 0.15 and 0.4 mole fraction diheptanoyl-PC. At lower mole fractions, multilamellar vesicles coexist with these small bilayers; above 0.5 mole fraction only mixed micelles (which are lytic to red blood cells) are detected. Therefore, for standard assays a 1:4 ratio of short-chain PC (5 m*M*) to long-chain PC (20 m*M*) appears reasonable, since only a single phase (small bilayer particles) is present.

The thermal and motional properties of both components in the small bilayer region have been investigated by differential scanning calorimetry (DSC), Fourier transform infrared spectroscopy (FT-IR), and

[31] N. E. Gabriel and M. F. Roberts, *Biochemistry* **23,** 4011 (1984).

[32] N. E. Gabriel and M. F. Roberts, *Biochemistry* **25,** 2812 (1986).

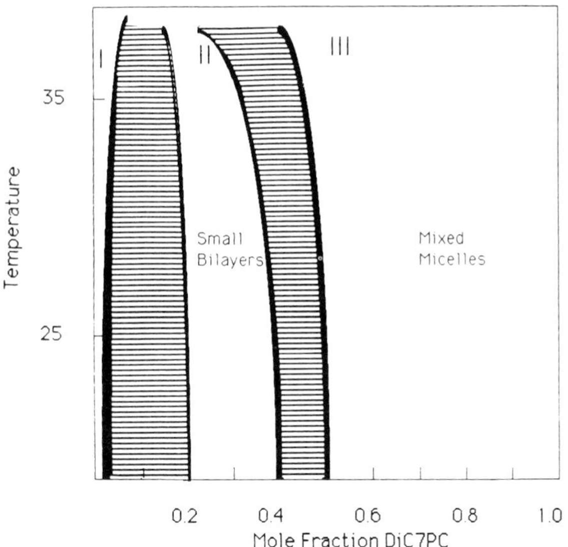

FIG. 6. Partial phase diagram of diheptanoyl-PC/dipalmitoyl-PC, 25 mM total phospholipid, below the T_m of the long-chain component (based on pyrene excimer/monomer data, erythrocyte hemolysis, and ^{2}H NMR data). The single-phase regions indicated correspond to (I) multilamellar vesicles, (II) stable small bilayer aggregates, and (III) mixed micelles. At mole fractions of diheptanoyl-PC in between these regions two types of aggregates will coexist.

NMR.[32–34] The chain conformation of the long-chain PC is similar to those of pure PC. Furthermore, the long-chain PC in these bilayer aggregates has thermal properties (e.g., T_m) similar to those of the pure PC, although the short-chain PC does decrease the size of the cooperative unit (ΔH is reduced). The short-chain PC does not interact appreciably with its gel-state neighbors. For diheptanoyl-PC/dipalmitoyl-PC mixtures, SANS suggests a bilayer thickness of 46 Å for the hydrocarbon section.[35] When the aggregates are incubated at a temperature equal to the T_m of the long-chain PC, a reversible size change occurs. The small particles fuse to very large particles which are multilamellar.[36]

Phospholipases prefer the short-chain PC in these bilayer aggregates, with hydrolysis rates about 2-fold less than in micelles.[33,37] The hydrolysis of the long-chain PC occurs at 1/100 to 1/10 the rate of the short-chain

[33] N. E. Gabriel and M. F. Roberts, *Biochemistry* **26,** 2432 (1987).

[34] M. F. Roberts and N. E. Gabriel, *Colloids Surf.* **30,** 113 (1988).

[35] T.-L. Lin, M.-Y. Tseng, S.-H. Chen and M. F. Roberts, unpublished results, 1990.

[36] K.-M. Eum, G. Riedy, K. H. Langley, and M. F. Roberts, *Biochemistry* **28,** 8206 (1989).

[37] N. E. Gabriel, N. V. Agman, and M. F. Roberts, *Biochemistry* **26,** 7409 (1987).

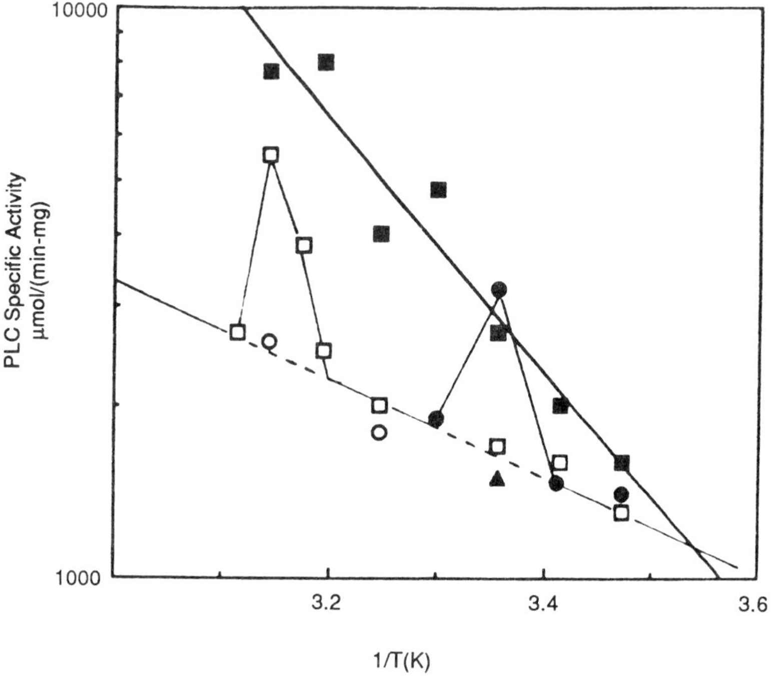

FIG. 7. Arrhenius plot for phospholipase C (*B. cereus*) activity toward diheptanoyl-PC (5 m*M*) in micelles (■) and in bilayer aggregates prepared with different long-chain PCs (20 m*M*): dipalmitoyl-PC (□), dimyristoyl-PC (●), egg PC (○), and dihexadecyl-PC (▲). Note the increase in activity toward diheptanoyl-PC in vesicles (to values almost as high as those for pure diheptanoyl-PC micelles) around the gel-to-liquid crystalline transition of the long-chain PC. [From M. F. Roberts and E. A. Dennis, in "Phosphatidylcholine Metabolism" (D. E. Vance, ed.), p. 121. CRC Press, Boca Raton, Florida, 1989.]

species for phospholipase C and phospholipase A_2, respectively. Both enzymes show an anomalous increase in specific activity around the T_m of the long-chain PC.[37] For phospholipase C, the enzyme-specific activity 5°–10° below and above the T_m is consistent with an activation energy of 4.5–6 kcal/mol (Fig. 7); this implies that differences in rates cannot be ascribed to substrate diheptanoyl-PC held in a fluid versus a gel matrix. Since the short-chain PC does not exhibit a phase transition, this must reflect fluctuations in head group area or vertical motions of the short-chain PC caused by the surrounding long-chain PC molecules.

A physical picture of the bilayer aggregates provides an explanation for the differences in enzyme activity toward the two PC substrates and an explanation for the interfacial activation phenomenon. The short-chain PC has greater mobility (particularly the phosphocholine head group as detected by the $Pr^{3+}/^{1}H$ NMR titrations) than the long-chain phospholipid in the particles and is "isolated" motionally and physically from the long-chain species. Thus, either head group conformational flexibility of the short-chain PC or the isolated behavior of the acyl chains in the long-chain matrix must be responsible for the preferential hydrolysis of the short-chain species. There is a considerable difference in binding affinity of the enzymes toward long- and short-chain PCs (1–2 orders of magnitude). Whether this is related to head group or chain differences is unclear at the moment. Both phospholipase A_2 and phospholipase C need to bind the phospholipid head group, glycerol backbone, and initial segment of the fatty acyl chain for optimal catalysis.

An interpretation of these kinetics with short-chain PC/long-chain PC systems is that while phospholipases can bind to the surface, in some cases with apparent K_m values comparable to values in mixed micelles, this interaction cannot lead to catalysis unless the enzyme can "isolate" or "uncouple" the substrate molecule from its neighbors.

Short-Chain Phosphatidylcholine Mixed Micelles

Short-chain PCs have been shown to solubilize a wide variety of amphiphiles, including cholesterol,[38] diglyceride,[30] and triglyceride.[39,40] The triglyceride-containing mixed micelles have been characterized by SANS, and a complete particle size distribution has been presented.[41] Such mixed micelles can be used to see how amphiphilic additives alter phospholipase activity. For example, both phospholipase A_2 and phospholipase C specific activities decrease with increasing cholesterol concentration in the short-chain PC micelle, although the changes are small (only 10–18% cholesterol can be solubilized).[37] A similar inhibition has also been noted in Triton X-100 mixed micelles.[42] With triglycerides which are only present at the particle surface to a minor extent, no decrease in phospholipase activity is observed.[39]

Summary

Short-chain phospholipids are extremely useful compounds for analysis of interfacial and substrate requirements of water-soluble phospholipases. A wide variety of PCs are easily prepared and can be characterized

[38] R. A. Burns, Jr., and M. F. Roberts, *Biochemistry* **20,** 7102 (1981).
[39] R. A. Burns, Jr., and M. F. Roberts, *J. Biol. Chem.* **256,** 2716 (1981).
[40] R. A. Burns, Jr., J. M. Donovan, and M. F. Roberts, *Biochemistry* **22,** 964 (1983).
[41] T.-L. Lin, S.-H. Chen, N. E. Gabriel, and M. F. Roberts, *J. Phys. Chem.* **94,** 855 (1990).
[42] M. F. Roberts, M. Adamich, R. J. Robson, and E. A. Dennis, *Biochemistry* **18,** 3301 (1979).

in detail to the point where at a given PC concentration, the micelle size distribution is known, and an area per head group can be estimated. Since they can be presented to enzymes as monomers, micelles, or in bilayer structures, short-chain phospholipids are ideal for examining kinetic preferences of lipolytic enzymes. In the future detailed studies of short-chain phospholipids with different head groups should facilitate kinetics with phospholipases which show a preference for anionic substrates.

[10] Phospholipase A_2-Catalyzed Hydrolysis of Vesicles: Uses of Interfacial Catalysis in the Scooting Mode

By MAHENDRA KUMAR JAIN and MICHAEL H. GELB

Introduction

Interfacial catalysis by phospholipase A_2 (PLA_2) is adequately described by Scheme I.[1] As a first step, the enzyme in the aqueous phase (E) binds to the substrate interface, and the bound enzyme (E*) undergoes the catalytic turnover in the interface according to the classic Michaelis–Menten formalism as shown in the box. Thus, the binding of the enzyme to the interface and the binding of a phospholipid substrate to the active site of the bound enzyme (E*) to produce the enzyme–substrate complex (E*S) are distinct steps. This is implicit in the proposal that the enzyme contains an interfacial binding surface that is topologically and functionally distinct from the catalytic site. This kinetic view of interfacial catalysis has been the working hypothesis in several laboratories.[1–5]

A critical experimental test of Scheme I has not been possible until recently.[1] In most studies, especially with micelles[3,4] and monolayers,[2] kinetic and equilibrium contributions of the E to E* step are not readily discernible. Even in bilayers, many of the anomalies seen in the kinetics of action of PLA_2 on insoluble lipid substrates are due to a significant, yet variable, contribution of the E to E* step to the steady-state enzymatic turnover.[1,2] This is illustrated in the following two examples (see Refs. 1

[1] M. K. Jain and O. G. Berg, *Biochim. Biophys. Acta* **1002,** 127 (1989).

[2] R. Verger and G. H. de Haas, *Annu. Rev. Biophys. Bioeng.* **5,** 77 (1976).

[3] H. M. Verheij, A. J. Slotboom, and G. H. de Haas, *Rev. Physiol. Biochem. Pharmacol.* **91,** 91 (1981).

[4] E. A. Dennis, *in* "The Enzymes" (P. D. Boyer, ed.), 3rd Ed., Vol. 16, p. 307. Academic Press, New York, 1983.

[5] See other contributions to this volume.

and 2 for others). First, a lag phase is seen in the reaction progress curve for the action of porcine pancreatic PLA_2 on vesicles of phosphatidylcholine.[1] This can be explained on the basis of the observation that the enzyme has a very weak affinity for vesicles of zwitterionic phospholipids,[6,7] but it binds several orders of magnitude more tightly to anionic vesicles.[8] Initially, with zwitterionic vesicles, most of the enzyme is in the aqueous phase and the reaction velocity is low. As the anionic fatty acid product accumulates in the interface, the reaction progress curve accelerates as more and more of the enzyme partitions into the interface where the lipolysis occurs.[8–10] The second example relates to the inhibition of PLA_2. Most inhibition studies reported in the literature are difficult to interpret because the effects of inhibitors on the E to E* step are not easy to characterize.[11,12] For example, many of the reported inhibitors of PLA_2 work not by binding tightly to the enzyme, but by promoting the desorption of enzyme from the interface.[12,13] Thus, in most of the assay procedures used for PLA_2 that have been reported, the fundamental features of the enzymology that occur within the interface are "blurred" by the reversible association of the enzyme with the interface, that is, the E to E* step (see Ref. 1 for a review and Ref. 13 for a resolution of these difficulties).

In order to minimize such kinetic complexities of interfacial catalysis, we have developed a methodology[14–17] for studying the action of PLA_2 on substrate vesicles in the "scooting mode,"[6] in which all of the enzyme is tightly bound to the interface. Here, after the initial binding of the enzyme to the interface, the E to E* step is no longer part of the catalytic turnover within the interface. In the scooting mode, the anomalous effects associated with the E to E* step, such as the latency period,[6,8] the effect of

[6] G. C. Upreti and M. K. Jain, *J. Membr. Biol.* **55,** 113 (1980).

[7] M. K. Jain, M. R. Egmond, H. M. Verheij, R. J. Apitz-Castro, R. Dijkman, and G. H. de Haas, *Biochim. Biophys. Acta* **688,** 341 (1982).

[8] R. J. Apitz-Castro, M. K. Jain, and G. H. de Haas, *Biochim. Biophys. Acta* **688,** 349 (1982).

[9] M. K. Jain and D. V. Jahagirdar, *Biochim. Biophys. Acta* **814,** 313 (1985).

[10] M. K. Jain, B. Yu, and A. Kozubek, *Biochim. Biophys. Acta* **980,** 23 (1989).

[11] M. K. Jain, M. Streb, J. Rogers, and G. H. de Haas, *Biochem. Pharmacol.* **33,** 2541 (1984).

[12] M. K. Jain and D. V. Jahagirdar, *Biochim. Biophys. Acta* **814,** 319 (1985).

[13] M. K. Jain, W. Yuan, and M. H. Gelb, *Biochemistry* **28,** 4135 (1989).

[14] M. K. Jain, J. Rogers, D. V. Jahagirdar, J. F. Marecek, and F. Ramirez, *Biochim. Biophys. Acta* **860,** 435 (1986).

[15] M. K. Jain, B. P. Maliwal, G. H. de Haas, and A. J. Slotboom, *Biochim. Biophys. Acta* **860,** 448 (1986).

[16] M. K. Jain, J. Rogers, J. F. Marecek, F. Ramirez, and H. Eibl, *Biochim. Biophys. Acta* **860,** 462 (1986).

[17] M. K. Jain, G. H. de Haas, J. F. Marecek, and F. Ramirez, *Biochim. Biophys. Acta* **860,** 475 (1986).

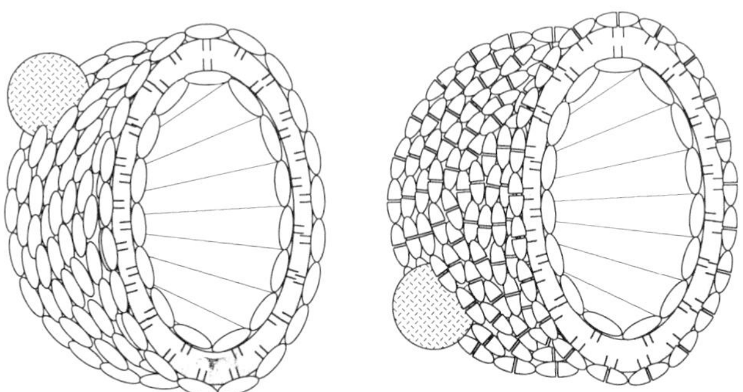

FIG. 1. Schematic representation of interfacial catalysis phospholipase by A_2 in the scooting mode. See text for details.

gel–fluid phase transitions,[6] and the activating or inhibiting effects of amphiphilic solutes,[1,12] are no longer observed.

We have found that PLA_2 hydrolyzes vesicles of anionic phospholipids, such as dimyristoylphosphatidylmethanol (DMPM), in the scooting mode. Here hydrolysis begins without any noticeable latency period (less than 5 sec), and the enzyme undergoes several thousand catalytic turnovers without leaving the interface. This interfacial catalysis in the scooting mode continues until all of the substrate in the outer layer of the enzyme-containing vesicles is hydrolyzed. Throughout the entire reaction progress, the vesicles retain their overall physical integrity. This is illustrated schematically in Fig. 1. The observation that the enzyme undergoes several thousand catalytic cycles while remaining attached to the same vesicle provides the strongest evidence for interfacial catalysis according to the kinetic Scheme I. A quantitative interpretation of the reaction in the

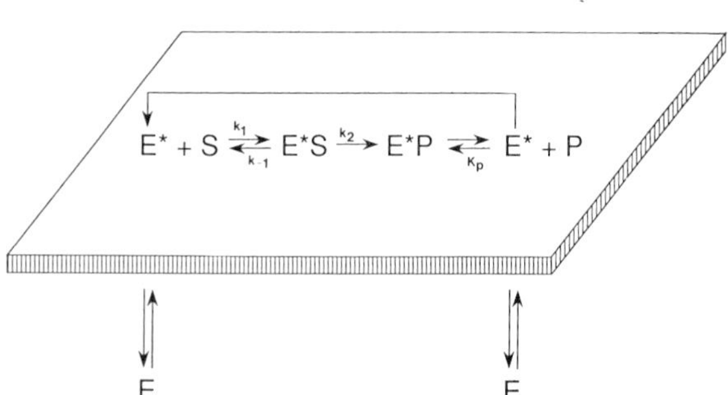

SCHEME I. Interfacial catalysis by phospholipase A_2.

scooting mode[1] requires an appreciation of the structure and molecular dynamics of the vesicles of anionic phospholipids.[18,19]

The substrate used for these studies are small unilamellar vesicles of DMPM containing no additives; however, the scooting phenomenon is observed with vesicles of a number of different phospholipids as long as they contain a critical mole fraction of anionic phospholipid (typically 5–10 mol %). PLA_2 binds essentially irreversibly to vesicles of DMPM (see below). Under the conditions used for such studies, neither the enzyme nor the substrate or products of hydrolysis exchange with other vesicles. In addition, complications arising from fusion of vesicles as well as transbilayer movement of phospholipids have been eliminated. In this chapter, we present a detailed description of the basic protocols for the study of PLA_2 in the scooting mode. In addition, we emphasize some applications of this unique technique in studies on lipolytic enzymes.

Kinetic Analysis of Phospholipase A_2 in Scooting Mode

Preparation of DMPM. DMPM can be easily prepared from commercially available dimyristoylphosphatidic acid as follows. Dimyristoylphosphatidic acid disodium salt (200 mg, 0.31 mmol, Avanti Polar Lipids, Birmingham, AL) is suspended in 16 ml of ether containing 1% (v/v) water in an Erlenmeyer flask with a stir bar. Sufficient HCl to protonate all of the phosphatidic acid sodium salt (0.15 ml of 6 *N* HCl) is added, and the mixture is stirred vigorously for several minutes until both layers become clear. Stirring is continued at room temperature, and freshly distilled diazomethane in ether is added dropwise until the characteristic yellow color of the diazomethane persists. An alcohol-free solution of diazomethane is prepared from Diazald (Aldrich, Milwaukee, WI) according to the manufacturer's directions. The mixture is stirred for an additional 30 min. The solution is then filtered through a pad of silica gel (2×3 cm) containing a 1 cm top layer of Celite in a sintered glass funnel. The silica pad is washed with an additional 50 ml of ether, and the combined solutions are transferred to a tared recovery flask and concentrated on a rotary evaporator at reduced pressure and at room temperature. The flask is attached to a vacuum pump for 30 min to remove traces of solvent.

To the oily residue of the phosphate triester (160 mg, 0.29 mmol) is added reagent-grade acetone (5 ml) and anhydrous LiBr (35 mg, 0.4 mmol, Fluka), and the mixture is refluxed with magnetic stirring for 6 hr. It is not necessary to provide moisture protection. The reaction flask is capped and

[18] M. K. Jain, "Introduction to Biological Membranes." Wiley, New York, 1989.
[19] H. Eibl, *Membr. Fluid. Biol.* **2,** 217 (1983).

is kept in a freezer at $-20°$ overnight. The solid is collected by suction filtration and washed with a few milliliters of cold acetone. The solid is dried *in vacuo* for 2 hr to give 130 mg of pure DMPM lithium salt. The material can be stored in a desiccator at $-20°$ indefinitely.

The material shows a single spot (R_f 0.5) by thin-layer chromatography on a silica plate with $CHCl_3$/methanol/acetic acid (65/15/2). DMPM can also be prepared from commercially available dimyristoylphosphatidylcholine using phospholipase D in the presence of methanol.[20] In this case, the product is contaminated with a small amount of dimyristoylphosphatidic acid which has little effect on the scooting assay. Most of the DMPM used in our earlier studies[14–17] was prepared from 1,2-dimyristoyl-*sn*-3-glycerol, but this starting material is less readily available.

Preparation of DMPM Vesicles. DMPM (20 mg of the Li salt) is weighed into a disposable glass culture tube (10 × 75 mm, Kimball Glass, Vineland, NJ). Two milliliters of distilled water is added, and the tube is capped with a rubber septum. For phospholipids containing saturated acyl chains, it is not necessary to keep the solutions under nitrogen. The suspension is briefly sonicated in a bath sonicator (Lab Supplies, Hicksville, NY, Model G112SPIT) for about 5 sec, and the cloudy solution is frozen (these frozen samples can be stored frozen and reused over several months). The frozen suspension is placed in the central cavitation zone of the optimally tuned sonicator bath. Besides the electronic adjustments in the power supply for the sonicator crystal as recommended by the supplier (generally done with the new bath), the sonicator bath is tuned just before use by adjusting the level of water until the energy is focused in a single cavitation zone in the center on the surface of the water. Sonication is allowed to proceed until an almost water-clear dispersion is obtained (typically 50–150 sec, depending on the tuning of the sonicator). The suspension can be kept at room temperature for more than 8 hr, and aliquots are diluted into the reaction solution for the kinetic studies.

Kinetic Studies. Hydrolysis of vesicles in the scooting mode is monitored with a pH stat (e.g., Radiometer ETS822 system) equipped with a high-speed mechanical stirrer (e.g., Radiometer TTA60) and a water-jacketed thermostatted vessel maintained at 21°. The corners of the stirrer paddle are rounded a bit with a file so that the stirring is still rapid (tested by injecting a dye solution into the reaction vessel and making sure that the color becomes dispersed within 1 or 2 sec) but not so violent so as to create turbulence, cavitation, or small bubble formation during the reaction progress. Nitrogen is continuously passed over the reaction solution (at a rate of approximately 50 ml/min) to prevent the absorption of carbon

[20] H. Eibl and S. Kovatchev, this series, Vol. 72, p. 632.

dioxide. Prior to a kinetic run, 4 ml of a solution of 0.6 m*M* $CaCl_2$ and 1 m*M* NaCl in distilled water is placed into the reaction vessel, and the pH is adjusted to 8.0 using the autoburet with 3 m*M* NaOH as titrant. The stock solution of vesicles in water (typically 0.5 to 1 mg DMPM) is added in one portion, and the pH is adjusted back to 8.0. After the baseline drift subsides (typically within a few minutes if there is no contamination from PLA_2), an aliquot of PLA_2 (typically 0.05–0.1 μg in 1–10 μl of water) is added in one portion. The reaction is maintained at pH 8.0 by continuous pH-stat titration with 3 m*M* NaOH. The reaction is allowed to proceed until the rate of consumption of titrant ceases (typically about 15–30 min).

Several additional factors are noted. If the solution of the enzyme is acidic or buffered, as is the case with most commercially available preparations, the initial part of the reaction progress curve will have contributions from the pH imbalance. Often, this can be manually subtracted by making a control run in which enzyme is added to vesicles in the absence of calcium. The amount of base used under these conditions corresponds to the amount of acid in the enzyme aliquot. Between kinetic runs, it is sufficient to wash the reaction vessel and pH electrode with distilled water since the enzyme sticks much tighter to the anionic vesicles than to the glass surfaces of the pH electrode or the vessel. In all pH-stat type assays, it is important that the pH electrode be working properly (low pH drift and a quick response time). Occasionally it is useful to clean the electrode surface with a mild detergent such as Micro (International Products Inc., Box 118, Trenton, NJ) after several enzymatic runs. Finally, on rare occasion, the observed reaction progress appears more rounded in shape compared to a purely exponential curve (Fig. 2 as discussed below). This is almost always due to polydispersion in the size of the vesicles, and the problem can be easily remedied by freezing and sonicating the stock solution of DMPM a second time. At the end of the enzymatic reaction, the consumption of titrant is very small (less than 3 nmol/min) under the conditions described above. This small (less than 2% of the initial rate) but finite consumption of base is largely due to slow fusion of vesicles and transbilayer movement of phospholipids, and it can be eliminated only at lower calcium concentrations, where other kinetic complications are observed.

Quantitative Analysis. A typical reaction progress curve for the hydrolysis of DMPM vesicles by PLA_2 is shown in Fig. 2. At 0.6 m*M* calcium, the reaction progress curve is first-order. Since the contribution of the E to E* step to the overall kinetics is eliminated,[1] this reaction progress curve is interpreted only in terms of the steps in the box in Scheme I. This is just the Michaelis–Menten formalism adopted for catalysis within an interface. The amount (moles) of titrant used at the end of the reaction

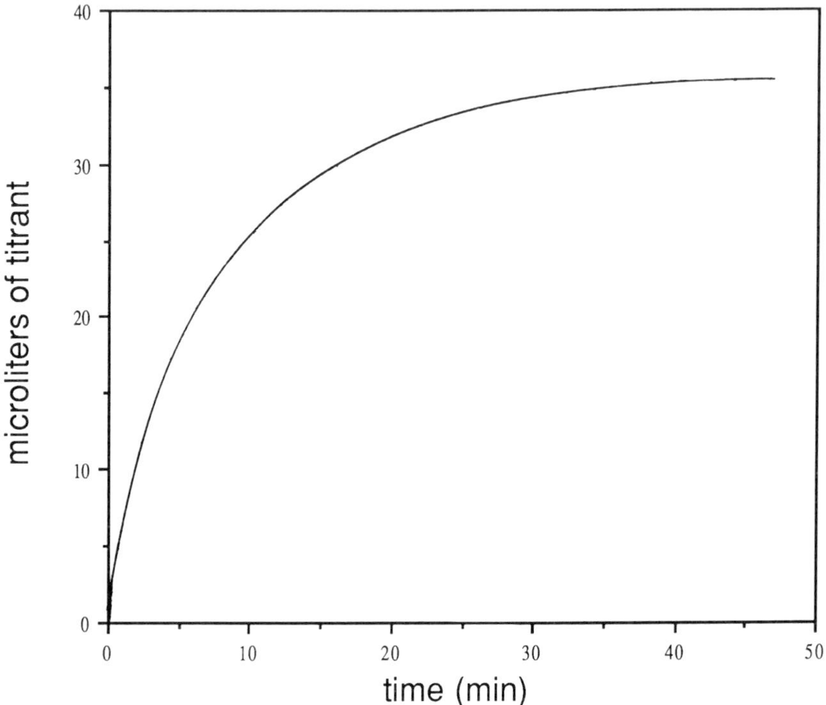

FIG. 2. Reaction progress curve for the hydrolysis of DMPM vesicles in the scooting mode. Conditions: 0.6 mg of DMPM in 4 ml of 0.6 m*M* $CaCl_2$, 1 m*M* NaCl, pH 8.0, 21°. The reaction was initiated by the addition of porcine pancreatic PLA_2 (0.1 μg in 2 μl of water). Based on the calibration of the titrant with myristic acid (see text for details), each microliter of titrant corresponds to 3.5 nmol of product.

(Fig. 2) is much less than the total moles of DMPM present in the reaction. This is because the ratio of vesicles to enzyme is large, and the enzyme does not hop to other vesicles. The partial extent of hydrolysis is not due to inactivation of the enzyme. At the end of the reaction, induction of vesicle fusion by raising the calcium concentration[14] or promotion of intervesicle exchange of enzyme by high salt[15] causes immediate resumption of hydrolysis. All other reasonable possibilities for the partial hydrolysis of the substrate were considered and have been ruled out.[14] These include severe product inhibition, the formation of multilamellar vesicles, and a decrease in the efficiency of pH titration.

As elaborated elsewhere[1] such curves (Fig. 2) are completely described by Eq. (1), where P_t is the amount of product produced (or titrant utilized) at time t, P_{max} is the extent of hydrolysis (the amount of product produced at the end of the reaction), and k_i is the first-order relaxation constant.

$$P_t = P_{max}[1 - \exp(-k_i t)] \quad (1)$$

This first-order reaction progress curve is typical of the Michaelis–Menten formalism under the condition that the apparent Michaelis constant, K_m^{app} is much larger than the concentration of substrate.[21,22] For interfacial catalysis, the concentration of substrate within the interface is expressed as its mole fraction. Since the maximum value of the substrate mole fraction in the vesicle is 1, the first-order nature of the reaction progress curve implies that $K_m^{app} \gg 1$. The interfacial K_m^{app} has its normal meaning[1,2,21–23] according to Eq. (2):

$$K_m^{app} = K_m(1 + 1/K_p) \quad (2)$$

where $K_m = (k_{-1} + k_2)/k_1$. According to Eq. (2), K_m^{app} differs from the classic K_m due to the inhibition by the reaction products. Here, the rate constants that make up K_m^{app} are given in Scheme I. K_p is the dissociation constant for the product, and, in the case of significant product inhibition (i.e., small K_p), the K_m^{app} will be magnified according to Eq. (2). The first-order relaxation constant, k_i, in Eq. (1) is given in Eq. (3)[1]:

$$N_T k_i = k_2/K_m^{app} \quad (3)$$

Here, N_T is the average number of DMPM molecules in the outer layer of the vesicles and has a value of about 13,000 for the example shown in Fig. 2 (however, see below). $N_T k_i$ is the turnover number for vesicles of DMPM, which has a value of about 2500–3000 min^{-1} under the assay conditions described above. Again, Eq. (3) is derived from an extension of the Michaelis–Menten formalism to interfacial catalysis.[1]

In the scooting mode, the enzymes bound to vesicles do not hop to other vesicles. In this case, the moles of product formed after the reaction comes to a halt, P_{max}, is simply equal to the moles of enzyme-containing vesicles multiplied by N_T. The value of P_{max} will be maximal when there is a large excess of vesicles over enzymes so that no vesicle will contain more than one enzyme. According to the Poisson distribution, the probability of having more than one enzyme per vesicle is only 0.5% if the vesicle to enzyme ratio is 6. This is the case for the concentrations of components described above in the assay procedure. As expected, in the presence of excess substrate vesicles, the value of P_{max} increases linearly with the amount of the enzyme in the reaction mixture.[14] Also, in the presence of a large excess of enzyme over vesicles, 63% of the total

[21] I. H. Segel, "Enzyme Kinetics." Wiley, New York, 1975.

[22] A. Fersht, "Enzyme Structure and Mechanism," 2nd Ed. Freeman, New York, 1985.

[23] A. Plückthun and E. A. Dennis, *J. Biol. Chem.* **260,** 11099 (1985).

substrate present in the reaction mixture can be hydrolyzed (this corresponds to the amount of substrate in the outer layer of small unilamellar vesicles). Such experiments demonstrate that only the substrate present in the outer monolayer of a vesicle is accessible for hydrolysis by the bound enzyme.[14] It should also be mentioned that Eq. (3) is valid only under the conditions of at most one enzyme per vesicle.

Values of k_i and P_{max} are easily obtained by fitting the reaction progress curve (Fig. 2) to Eq. (1). In order to convert the value of P_{max} in microliters of titrant used to moles of product formed, it is necessary to calibrate the titrant in the following way. A solution of DMPM vesicles in 0.6 mM $CaCl_2$, 1 mM NaCl is prepared as described above and adjusted to pH 8.0 in the pH stat. A solution of myristic acid in ethanol of known concentration (typically 5 μl of a 40 mM solution) is added, and the pH is brought back to 8.0. The amount of titrant used corresponds to the number of moles of myristic acid added. This procedure is more accurate than a simple titration of a strong acid, such as HCl, in water, since it takes into account the fact that the surface pK_a of myristic acid in the DMPM vesicle is not far below 8. Such an analysis has shown that the titration efficiency (moles of titrant used per mole of fatty acid added) is 0.90 for myristic acid in DMPM and 0.87 for myristic acid in DMPM vesicles at the end of the reaction progress curve when all of the substrate in the outer layer of the vesicles has been hydrolyzed. Thus, there is no significant change in the titration efficiency during the course of the enzymatic reaction.

Comments on Assay of Phospholipase A_2 in Scooting Mode

When interfacial catalysis by PLA_2 is examined in the scooting mode, many of the previously reported anomalous features of the kinetics are not observed. The hydrolysis of DMPM vesicles begins immediately after the addition of enzyme; the latency period in the reaction progress that is seen in other assays[8] does not appear with vesicles of DMPM. In this sense, this system with vesicles is also distinctly different than the monolayer system of Verger and de Haas,[2] where they assumed that the E to E* step is slow. On the other hand, as shown elsewhere by stopped-flow kinetic analysis, the rate of binding of the enzyme to the micelle or vesicle interface is rapid.[24] Therefore, we believe that the primary mechanism for the origin of the latency period is the much slower buildup of anionic product in the interface, which leads to a time-dependent modulation of the E to E* equilibrium. This is the major difference between our model[1] and the model proposed by Verger and de Haas.[2]

[24] M. K. Jain, J. Rogers, and G. H. de Haas, *Biochim. Biophys. Acta* **940,** 51 (1988).

The enzyme is still fully catalytically active at the end of the reaction progress,[14] and the cessation of the reaction is due to the consumption of all of the substrate in the outer layer of enzyme-containing vesicles. When the enzyme to vesicle ratio is large enough so that all of the vesicles contain one or more bound enzymes, 63% of the total DMPM present in small sonicated vesicles is hydrolyzed, which corresponds to the fraction of substrate in the outer layer of the vesicles. With large unilamellar vesicles, 50% of the total DMPM is hydrolyzed whereas only small amounts (typically about 8%) of the substrate is hydrolyzed with multilamellar vesicles. The vesicles remain intact, even after all of the substrate in the outer layer has been hydrolyzed.

Studies of vesicle fusion by light scattering and fluorescence resonance energy transfer have demonstrated that the rate of fusion of DMPM vesicles, under the above-described conditions, is much too slow to be of any significance except at the end of the first-order reaction progress curve (as discussed above).[14] Finally, the rates of transbilayer movement of phospholipids and of intervesicle exchange of enzyme, phospholipid substrate, and the products are also very slow.

Although detailed investigations of interfacial catalysis have been carried out with DMPM vesicles, similar results are seen with vesicles of other anionic phospholipids[16,25] or with codispersions of zwitterionic and anionic phospholipids.[26] It is particularly striking to note here that over 50 PLA_2s tested from different sources exhibit a first-order reaction progress curve of the type shown in Fig. 2. These include not only the enzymes from pancreas and their mutants, but also those from venoms of *Apis*, Elapidae, Viperidae, and Crotalidae, as well as bacteria like *Escherichia coli* and *Saccharomyces*.[27] Thus, the assay in the scooting mode with DMPM vesicles is by far the most generally useful procedure for studying PLA_2s reported to date. Other assay procedures for analyzing PLA_2s have been described, including the use of mixtures of phospholipids dispersed into detergents,[4,23,28] radiolabeled bacterial membrane fragments,[29] and vesicles of radiolabeled phospholipids.[30] These other assays are not generally applicable to all enzymes, and in many cases it is necessary to include a lipophilic additive to promote the binding of the enzyme to the interface.[23] All of these problems stem from the assay-dependent variation in the

[25] M. K. Jain and J. Rogers, *Biochim. Biophys. Acta* **1003**, 91 (1989).

[26] F. Ghomashchi, B. Yu, O. Berg, M. K. Jain, and M. H. Gelb, in preparation.

[27] M. K. Jain, G. N. Ranadive, B.-Z. Yu, and H. M. Verheij, in preparation.

[28] W. Niewenhuizen, H. Kunze, and G. H. de Haas, this series, Vol. 32B, p. 147.

[29] For example, see R. L. Jesse and R. C. Franson, *Biochim. Biophys. Acta* **575,** 467 (1979).

[30] For example, see J. Balsinde, E. Diez, A. Schüller, and F. Mollinedo, *J. Biol. Chem.* **263,** 1929 (1988).

amount of enzyme present in the interface (E to E* effects) during the steady-state catalytic turnover. These effects are not observed in the assay with DMPM since the E to E* step has been eliminated. Additional problems with micelle or mixed micelle assays due to substrate replenishment have been discussed.[1]

Finally it should be mentioned that although the assay with DMPM described in this chapter makes use of a pH stat to detect the production of product, it is also possible to adopt other methods of product analysis. For example, it is possible to use radiolabeled phospholipids followed by chromatographic separation of the products and analysis of the amount of released radioactive fatty acid by scintillation counting.[26] The important point is that with DMPM as a matrix, the anomalous kinetic effects arising from the E to E* step are avoided.

Uses of Scooting Assay

Interfacial catalysis in the scooting mode opens a new dimension in the study of PLA_2. The protocols described here can be used as such or adopted appropriately to address questions related to catalytic turnover, substrate specificity, interfacial rate constants, inhibition, and activation. Detailed studies on many of these topics are reported elsewhere. Because of space limitations, here we present a subset of the uses of the study of PLA_2 in the scooting mode.

Calibrating Solutions of Phospholipase A_2. Since the extent of hydrolysis, P_{max}, under the conditions of excess vesicles to enzyme is proportional to the moles of catalytically active enzyme, the concentration of functional enzyme active sites in a stock solution of enzyme can be determined directly. We have found this protocol to be particularly useful for ascertaining the purity and homogeneity of chemically modified[27] or semisynthetic forms of PLA_2 and genetically engineered enzyme mutants. The use of this assay with DMPM vesicles is the only known method for carrying out such a measurement. The usual methods of determining the concentration of PLA_2 are based on assays of the total protein content, regardless of whether all of the enzyme in the solution is catalytically active. To obtain the concentration of catalytically active enzyme from the extent of hydrolysis, P_{max}, one needs to know the value of N_T. We have found that preparations of DMPM, prepared as described above, reproducibly give a value of about 13,000 for N_T. However, the value of N_T may vary somewhat in different laboratories. The size of vesicles, prepared by sonication of DMPM, is known to be sensitive to a number of factors including the method of sonication, the presence of fatty acid heterogeneity in the synthesized substrate, and the presence of ions and other impurities. For this reason, it is recommended that the value of N_T be determined using a well-characterized PLA_2, such as the enzyme from porcine pancreas. This enzyme, when purified by the published procedure,[28] consistently has a specific activity of 1300–1500 μmol product/min/mg in an assay with egg yolk.[28] The concentration of porcine pancreatic PLA_2 in the stock solution is determined from the absorbance at 280 nm using the published extinction coefficient,[28] $E_{1\%}$, of 13.0. The value of N_T can then be obtained from a measurement of P_{max}, under the conditions of excess vesicles over enzyme, and knowing the amount of porcine pancreatic PLA_2 added to the assay. Once the value of N_T is determined in this manner, it can be used to calibrate solutions of other, less well-characterized PLA_2s.

Intrinsic Catalytic Activities of Phospholipases A_2. The value of $N_T k_i$ measured in the scooting assay is a characteristic property of a particular PLA_2 acting on a particular substrate vesicle. It has the same meaning as the k_{cat}/K_m value for an enzyme that operates in water on a particular substrate, except for the fact that rate constants have units which are appropriate for interfacial catalysis.[1] In the past, it has not been possible to obtain accurate values of the turnover numbers for PLA_2 since such calculations are based on maximal velocity measurements using different assays and require the assumptions that all of the enzyme in the solution is fully active and that the exchange of the enzyme, substrate, or products is not part of the catalytic turnover.

Detection of Phospholipase A_2 Impurities in Samples. The detection of trace amounts of PLA_2 in biological samples, for example, in preparations of melittin, can be accurately determined using the scooting assay. A second example that comes to mind concerns the activity of unusual or mutant forms of PLA_2. For example, a PLA_2 from *Agkistrodon piscivorus piscivorus* has been isolated that contains lysine in place of the catalytically important aspartate residue at position 49.[31] This enzyme was reported to have significant catalytic activity; however, this result was recently challenged by van den Bergh *et al.*,[32] who provided strong evidence that the lysine-49 enzyme is contaminated with a very active aspartate-49 PLA_2. This problem can be easily and conclusively resolved by analyzing the enzyme in the scooting assay. The moles of catalytically active enzyme in the stock solution can be determined from the values of P_{max} and N_T since every active molecule of enzyme will eventually hydrolyze all of the substrate in the outer layer of the vesicle to which it is bound. This is true even in the case that the particular PLA_2 under investigation has an

[31] J. M. Maraganore, G. Merutka, W. Cho, W. Welches, F. J. Kezdy, and R. L. Heinrikson, *J. Biol. Chem.* **259,** 13839 (1984).

[32] C. J. van den Bergh, A. J. Slotboom, H. M. Verheij, and G. H. de Haas, *J. Cell. Biochem.* **39,** 379 (1989).

extremely small interfacial turnover number, $N_T k_i$. If the absolute number of catalytically active enzyme molecules in the stock solution, as determined using the scooting assay, is much smaller than the total number of enzyme molecules, calculated using a standard protein assay (i.e., absorbance at 280 nm), one would conclude that an active PLA_2 impurity is present in a large amount of inactive PLA_2 protein. In all of these studies, it is important to check that the sample to be tested does not contain impurities which induce the fusion of vesicles.

Other Uses of Scooting Assay and Concluding Remarks

In addition to the above-mentioned analyses, the scooting assay has numerous other uses. For example, there has been considerable controversy in the literature as to whether a particular PLA_2 is catalytically active as a dimer or as a monomer.[4] This issue can be resolved using the scooting assay.[27] In the presence of excess vesicles over enzyme, the value of P_{max} will be twice as large for a given number of moles of monomeric enzyme compared to the same number of moles of a dimeric enzyme.

The scooting assay is also very useful in the discovery of compounds that function as true competitive inhibitors of PLA_2.[13] For example, many of the previously reported PLA_2 inhibitors function by promoting the desorption of enzyme from the interface (E to E* effects) rather than by binding directly to the bound enzyme. These compounds do not inhibit the action of PLA_2 in the scooting mode.[13] In the inhibition studies, the kinetics were determined at higher calcium concentrations (2.5 rather than 0.6 m*M*). Under these conditions the vesicles are much larger, and the shape of the reaction progress curve is different than that shown in Fig. 2 in that it contains a significant constant velocity at early times (zero-order phase). This is mainly a result of the fact that the buildup of product and the depletion of substrate take longer with larger vesicles, and the initial velocity is maintained for a longer period than in the small vesicles described here. The full details of the inhibition analysis at high calcium will be described.[33]

Perhaps the most important use of the scooting assay will be in the determination of the absolute substrate specificities and the interfacial rate constants of lipolytic enzymes. In previous studies of substrate specificities, relative velocities for the action of PLA_2s and other lipolytic enzymes on a variety of different substrate vesicles composed of pure phospholipid classes have been reported. In these studies, no effort has been made to normalize for the amount of enzyme bound to the interface. The use of

[33] M. K. Jain, B.-K. Yu, and O. G. Berg, in preparation.

the scooting assay, in which all of the enzyme is in the interface, provides an extremely powerful solution to this problem.[26]

At first glance, the scooting assay may appear somewhat unsettling and counterintuitive to traditional enzymologists. The irreversible nature of the interaction of the enzyme with the interface may not be an accurate representation of the action of the enzyme in a physiological environment, such as the inside of a cell. However, this is not a concern, since under physiological conditions all of the fundamental features of interfacial catalysis, whether they involve the selection of various substrates within the cell membrane or the inhibition of the enzyme, will occur within the interface. In other words, only events such as lipid exchange and other processes that occur in parallel to the events within the interface have been constrained in the scooting assay. The mechanism of action of the enzyme within the interface is not altered by this technique. The scooting assay provides an attractive *in vitro* method for the analysis of lipolytic enzymes that is free from the distortions caused by the relative affinities of the PLA_2 (E to E* effects) for a variety of vesicles composed of pure phospholipids.

Acknowledgments

This work was supported by Grants HL-36235 (M. H. G.) and GM-29703 (M. K. J.) from the National Institutes of Health and by a grant from Sterling Pharmaceutical Co. (M. K. J.). The authors would like to give special thanks to Farideh Ghomashchi for technical assistance.

[11] Assay of Phospholipases C and D in Presence of Other Lipid Hydrolases

By KARL Y. HOSTETLER, MICHAEL F. GARDNER, and KATHY A. ALDERN

Introduction

Phospholipases C and D were first identified in bacteria[1] and plants,[2] respectively, but they are now known to be present in mammalian cells. These activities are generally not difficult to measure, and a variety of methods are available. The methods are usually based on the release of a

[1] M. G. Macfarlane and B. C. J. Knight, *Biochem. J.* **35,** 884 (1941).
[2] D. J. Hanahan and I. L. Chaikoff, *J. Biol. Chem.* **169,** 669 (1947).

radioactive polar head group such as choline, ethanolamine, inositol, or their phosphates into the water phase of a lipid extraction.[3] Phospholipase C may also be assayed spectrophotometrically using thioester[4] or *p*-nitrophenyl-linked substrates.[5] Others rely on the measurement of the lipid-soluble product of the phospholipases, diacylglycerol (phospholipase C) or phosphatidic acid (phospholipase D), in the organic phase of a suitable lipid extraction.[6,7]

To establish the presence of a phospholipase C or D in a biological preparation it is important that both the lipid product and the water-soluble product of the reaction be identified. In some cases it may also be necessary to demonstrate that the water-soluble product did not arise from cleavage of a more complex water-soluble phosphodiester such as glycerophosphocholine, glycerophosphoethanolamine, or glycerophosphoinositol.

Clearly, it may be difficult to measure phospholipase C and/or phospholipase D in the presence of phospholipases A, lysophospholipases, and other phosphodiesterases. The purpose of this chapter is to present an approach for assessing phospholipase C and D activity in crude enzyme preparations containing other phospholipase, phosphodiesterase, and phosphatase activities. As an example, we focus on the hydrolysis of phosphatidylcholine at pH 4.4 by a crude lysosomal enzyme preparation[6] which contains phospholipase C, phospholipase A_1, lysophospholipase, acid phosphatase, and acid phosphodiesterases. Methods are described for analysis of both the lipid- and water-soluble products.

Isolation and Chromatography of Lipid Products of Phospholipase C and D Hydrolysis of Phosphatidylcholine

Experimental Procedure

Purified lysosomes are obtained from the liver of rats treated with Triton WR-1339 and a soluble protein fraction isolated as previously described.[8] The lysosomal soluble protein (100 μg) is incubated with 50 m*M* sodium acetate, (pH 4.4), 45 μM di[1-^{14}C]oleoylphosphatidylcholine

[3] H. van den Bosch, *in* "Phospholipids" (J. N. Hawthorne and G. B. Ansell, eds.), p. 313. Elsevier Biomedical, Amsterdam, 1982.

[4] A. A. Farooqui, W. A. Taylor, C. E. Pendley, J. W. Cox, and L. A. Horrocks, *J. Lipid Res.* **25,** 1555 (1984).

[5] S. Kurioka and M. Matsuda, *Anal. Biochem.* **75,** 281 (1976).

[6] Y. Matsuzawa and K. Y. Hostetler, *J. Biol. Chem.* **255,** 646 (1980).

[7] Y. Matsuzawa and K. Y. Hostetler, *J. Biol. Chem.* **255,** 5190 (1980).

[8] Y. Matsuzawa and K. Y. Hostetler, *J. Biol. Chem.* **254,** 5997 (1979).

([1-^{14}C]DOPC, specific activity 30 mCi/mmol) in a total volume of 0.200 ml for 0, 30, 60 and 180 min at 37° as previously described.[6,7] A total lipid extract is prepared by the method of Folch *et al.*[9] and chromatographed as described below.

Measurement of phosphatidic acid and/or diglyceride release from phosphatidylcholine by phospholipase C or D when lysophosphatidylcholine is present due to the action of phospholipase A can be accomplished by two-dimensional thin-layer chromatography. The lower phase of the lipid extract is concentrated to a small volume with a stream of nitrogen and applied to the origin of a 0.25 mm layer of silica gel H (20 × 20 cm) prepared with 50 m*M* magnesium acetate. (The method may also be adapted to utilize commercial silica gel G plates.) Fifty nanomoles each of the following reference lipids is applied to the origin: phosphatidylcholine, lysophosphatidylcholine, phosphatidic acid, lysophosphatidic acid, 1,2-diolein, 1-monoolein, and oleic acid. The plate is first developed with chloroform/methanol/concentrated ammonia/water (120/80/10/5, by volume). After air drying for 20 min at room temperature, the plate is reoriented and developed in the second dimension with chloroform/acetone/methanol/glacial acetic acid/water (100/40/30/20/12, by volume). After drying, the plate is developed with iodine vapors and photographed. The spots are marked and the iodine color allowed to fade. The lipid areas are scraped into liquid scintillation vials and counted.[6]

Results

The two-dimensional thin-layer chromatogram of standard lipid compounds is shown in Fig. 1. This system readily separates phosphatidic acid (produced by phospholipase D), lysophosphatidylcholine (produced by phospholipase A), and diglyceride (produced by phospholipase C). Diacylglycerol migrates in this system with monoacylglycerol but can be resolved in the one-dimensional system described below. Table I shows the time course of [1-^{14}C]DOPC conversion to products by a crude liver lysosomal soluble protein fraction. From this type of study it is readily apparent that lysosomes contain both phospholipase A and phospholipase C, as evidenced by the generation of lysophosphatidylcholine and diacylglycerol, respectively. Phospholipase D is absent as judged by the absence of phosphatidic acid. It is worth noting that this analytical method could also be used to measure phospholipase C and phospholipase D in the same sample by assessing the relative rates of release of diacylglycerol and phosphatidic acid.

[9] J. Folch, M. Lees, and G. H. Sloane Stanley, *J. Biol. Chem.* **193,** 497 (1957).

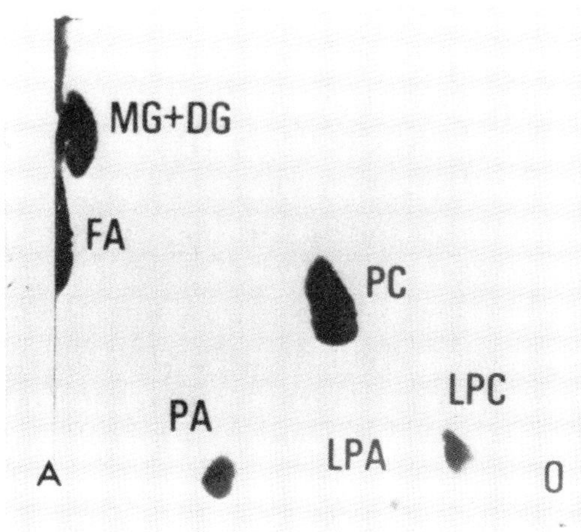

FIG. 1. Two-dimensional thin-layer chromatogram of reference lipids separated on silica gel G plates. Reference standards were applied to the origin of a 0.25 mm silica gel G plate prepared with 50 mM magnesium acetate. The plate was developed as noted in the text, air dried, developed with iodine vapors, and photographed. MG, Monoolein; DG, diolein; FA, oleic acid; PC, phosphatidylcholine; LPC, lysophosphatidylcholine; PA, phosphatidic acid; LPA, lysophosphatidic acid; O, origin. (Reproduced from Ref. 6 with permission.)

Measurement of Diglyceride, Monoglyceride, Fatty Acid, and Lysophosphatidylcholine Formation from Phosphatidylcholine

Experimental Procedures

If phospholipase D is absent, it is easier to measure phospholipase A and C activity using one-dimensional thin-layer chromatography to separate diacylglycerol, monoacylglycerol, and lysophosphatidylcholine.[7] Di[1-^{14}C]oleoylphosphatidylcholine (or other suitable glycerol- or fatty acid-radiolabeled phospholipid) is incubated with the protein fraction containing phospholipases A and C, and a lipid extract is prepared as noted above. Examples of incubation conditions for lysosomal phospholipase C measurement in the presence of phospholipase A may be found in Refs. 7 and 10. The lower phase is reduced to a small volume under a stream of nitrogen and the lipid extract applied to the origin of a silica gel H plate prepared with 10 mM magnesium acetate. Commercially prepared silica gel G plates may also be used. The plate is developed to a level of 7 cm above the origin with chloroform/methanol/water (65/35/5, by volume).

[10] K. Y. Hostetler and Y. Matsuzawa, *Biochem. Pharmacol.* **30,** 1121 (1981).

TABLE I
TIME COURSE OF PRODUCT FORMATION DURING HYDROLYSIS OF DI[1-^{14}C]OLEOYLPHOSPHATIDYLCHOLINE BY LYSOSOMAL SOLUBLE, DELIPIDATED PROTEIN[a]

Time (min)	PC	FA	MG + DG	LPC	PA	LPA
0	82.6	2.05	0.93	1.81	0.04	0.09
30	18.4	94.2	37.7	14.8	0.04	0.09
60	8.5	100.7	45.1	7.43	0.03	0.20
180	3.1	135.7	42.8	4.10	0.03	0.11

[a] Lysosomal protein was incubated with [1-^{14}C]DOPC and the lipid products as described in the text. PC, Phosphatidylcholine; FA, fatty acid; MG, monoglyceride; DG, diglyceride; LPC, lysophosphatidylcholine; PA, phosphatidic acid; LPA, lysophosphatidic acid. Values are given as nanomoles/mg protein. (Adapted from Ref. 6 with permission.)

The plate is allowed to dry for 20 min followed by development to the top of the plate (20 cm) with heptane/diethyl ether/formic acid (90/60/4, by volume). After drying, the plate may be scanned for radioactivity, or 0.5-cm portions may be scraped into liquid scintillation vials for counting.

Results

Figure 2 shows lipid standards developed on a one-dimensional plate, stained with iodine vapors and photographed. The position of free fatty acid, diacylglycerol, monoacylglycerol, phosphatidylcholine, and lysophosphatidylcholine is shown relative to the first and second solvent fronts (F1 and F2). This system can separate fatty acid, monoglyceride, diglyceride, and lysophosphatidylcholine, but it is not useful if phosphatidic acid (phospholipase D) is present. We have used this system to assess the inhibition of lysosomal phospholipases A and C by chloroquine and diethylaminoethoxyhexesterol[7] and other agents[10] and to detect phospholipase A and C in kidney,[11,12] heart,[13] and other tissues.[14] It is more useful than the two-dimensional system above because larger numbers of samples can be separated on one thin-layer plate (5 to 10 samples per plate). This system is generally used to analyze the metabolites of phosphatidylcholine. However, the method can also be adapted for other glycerophospholipid substrates by adjusting the solvent used in the first development so

[11] K. Y. Hostetler and L. B. Hall, *Proc. Natl. Acad. Sci. U.S.A.* **79,** 1663 (1982).
[12] K. Y. Hostetler and L. B. Hall, *Biochim. Biophys. Acta* **710,** 506 (1982).
[13] G. Nalbone and K. Y. Hostetler, *J. Lipid Res.* **26,** 104 (1985).
[14] K. Y. Hostetler and L. B. Hall, *Biochem. Biophys. Res. Commun.* **96,** 388 (1980).

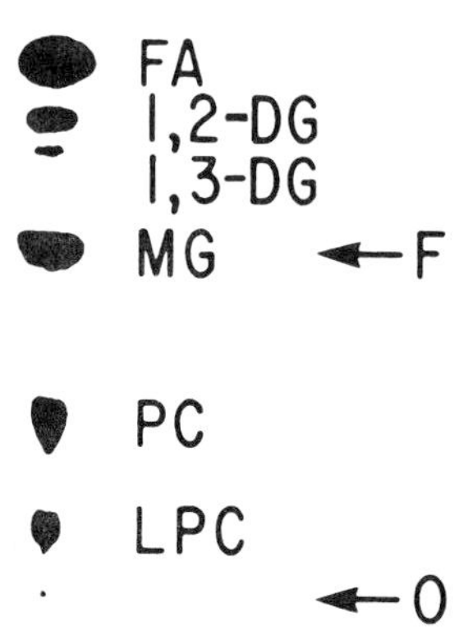

FIG. 2. Two-step one-dimensional thin-layer chromatographic separation of phosphatidylcholine hydrolysis products. Reference lipids were applied to the origin of a 0.25 mm layer of silica gel H prepared with 10 m*M* magnesium acetate and developed as noted in the text. FA, Oleic acid; 1,2-DG, 1,2-dioleolyglycerol; 1,3-DG, 1,3-dioleoylglycerol; MG, 2-monoolein; PC, phosphatidylcholine; LPC, lysophosphatidylcholine; O, origin; F1, first solvent front; F2, second solvent front. (Reproduced from Ref. 7 with permission.)

that the phosphoglyceride and its lysophospholipid are well separated. The second solvent need not be changed.

Isolation and Chromatography of Water-Soluble Products of Phosphatidylcholine Hydrolysis by Phospholipase C or D

Experimental Procedure

Phospholipase C from *Bacillus cereus* (10 μg) is incubated with 0.17 m*M* dipalmitoylphosphatidyl[*methyl*-^{3}H]choline ([^{3}H]DPPC), specific activity 9 μCi/μmol, 0.5 mg/ml Triton X-100, and 50 m*M* sodium acetate, pH 5.4, for 10 min at 37° in a shaking water bath. In other experiments, lysosomal soluble protein is incubated as described above except that the substrate is 0.2 m*M* [^{3}H]DPPC. Total lipid extracts are prepared by the method of Folch *et al.*[9] The upper phases are removed and washed once with 1 volume of Folch lower phase. One milliliter of the washed upper phase is diluted with 1 ml of buffer A and analyzed as noted below.

The radiolabeled water-soluble metabolites are separated using an HPLC system (Beckman, Fullerton, CA) equipped with a radioactivity flow detector (Radiomatic, Tampa, FL) using an adaptation of a method first described by Liscovitch *et al.*[15] An aliquot is injected into a prepacked Hibar Lichrosphere SI100 (10 μm) column (4.0 mm i.d. × 250 mm) at room temperature. The mobile phase consists of acetonitrile/water/ethanol/acetic acid/0.83 *M* sodium acetate, pH 3.6 (800/127/68/2/3, by volume) (buffer A), and buffer B consisting of the same components (400/400/68/53/79, by volume). The column is eluted with buffer A for 15 min at 2 ml/min followed by a linear gradient of 0 to 100% buffer B from 15 to 35 min and elution with buffer B to 50 min. Flo-Scint IV (Radiomatic) is used as the scintillator in the radioactivity flow detector. Standards of [*methyl*-^{14}C]choline, glycerophospho[*methyl*-^{14}C]choline, phospho[*methyl*-^{14}C]-choline. di[1-^{14}C]palmitoylphosphatidylcholine, di[1-^{14}C]oleoylphosphatidylcholine, and 1-[1-^{14}C]palmitoyllysophosphatidylcholine are also analyzed by the same method.

Results

High-performance liquid chromatographic (HPLC) analysis of reference standards of choline, glycerophosphocholine, and phosphocholine is shown in Fig. 3. As summarized in Table II, the retention times for these compounds were 22.2, 26.7, and 35.4 min, respectively. Our results are similar to the retention times published by Liscovitch *et al.*,[15] except that our retention time for phosphocholine is 35.4 versus 40 min; this is probably due to differences in flow rate (2.0 ml/min versus 2.7 ml/min) and column temperature (20° versus 45°).

It is important to note that intact phosphatidylcholine and its lyso compound have retention times similar to that of choline. As shown in Fig. 3C, lysophosphatidylcholine has a retention time of 21.7 versus 22.2 min for choline. Phosphatidylcholines also have retention times of 22–23 min (Table II). Thus, when studying the water-soluble metabolites of phosphatidylcholine, it is very important to carefully wash the aqueous phase of the lipid extract with lower phase in order to remove all traces of the lipid compounds so as to avoid interference with choline. Two washes are generally sufficient to remove the phosphatidylcholine and lysophosphatidylcholine from the upper phase of a Folch extraction.

Figure 4 shows the profile of labeled water-soluble metabolites after incubation with either phospholipase C from *B. cereus* or crude lysosomal soluble protein from rat liver. After a 10-min incubation with phospholipase C (Fig. 4A), a single peak of radioactivity was noted at 35.4 min,

[15] M. Liscovitch, A. Freese, J. K. Blusztajn, and R. J. Wurtman, *Anal. Biochem.* **151,** 182 (1985).

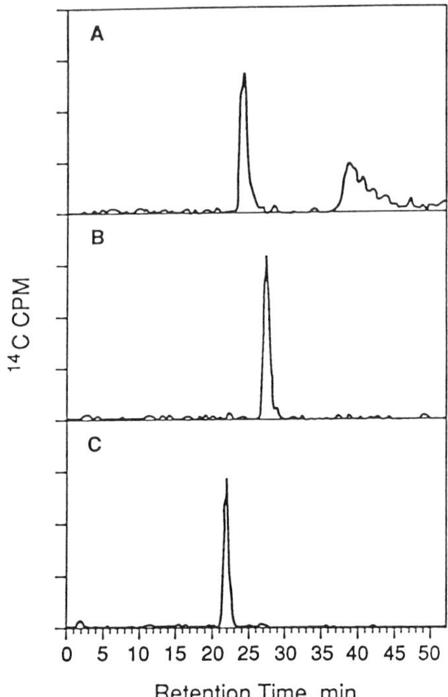

FIG. 3. HPLC of ^{14}C-labeled choline, glycerophosphocholine, phosphocholine, and lysophosphatidylcholine. Samples of the respective standards were analyzed by HPLC as described in the text. (A) Choline and phosphocholine; (B) glycerophosphocholine; (C) lysophosphatidylcholine.

indicating that phosphocholine is the sole product. Phospholipase D from cabbage produces a single peak of radioactive choline (data not shown). However, as shown in Fig. 4B, when crude lysosomal soluble protein was incubated with [^{3}H]DPPC for 1 hr, both phosphocholine and glycerophosphocholine were produced. After 24 hr, a large choline peak was also present, indicating conversion of the phosphocholine and glycerophosphocholine to choline by lysosomal acid phosphatase and/or phosphodiesterases (Fig. 4C).

The presence of three water-soluble metabolites of phosphatidylcholine in the lysosomal incubations emphasizes the importance of analyzing both the lipid-soluble products and the water-soluble products when measuring phospholipase C (or D). In the lysosome example shown in Table I and Fig. 4, the presence of phosphocholine can be attributed to the action of a lysosomal phospholipase C, whereas glycerophosphocholine is produced by the sequential action of phospholipase A_1 and lysophospholi-

TABLE II
RETENTION TIMES OF PHOSPHATIDYLCHOLINE AND METABOLITES[a]

Compound	Retention time
Choline	22.2 ± 0.4 (8)
GPC	26.7 ± 0.4 (8)
Phosphocholine	35.4 ± 1.1 (8)
DOPC	23.3 ± 1.3 (7)
DPPC	22.2 ± 0.5 (6)
Lyso-PC	21.7 ± 0.4 (3)

[a] Values are minutes ± S.D. Numbers in parentheses are the number of replicates. DOPC, Dioleoylphosphatidylcholine; DPPC, dipalmitoylphosphatidylcholine; lyso-PC, 1-palmitoylglycerophosphocholine; GPC, glycerophosphocholine.

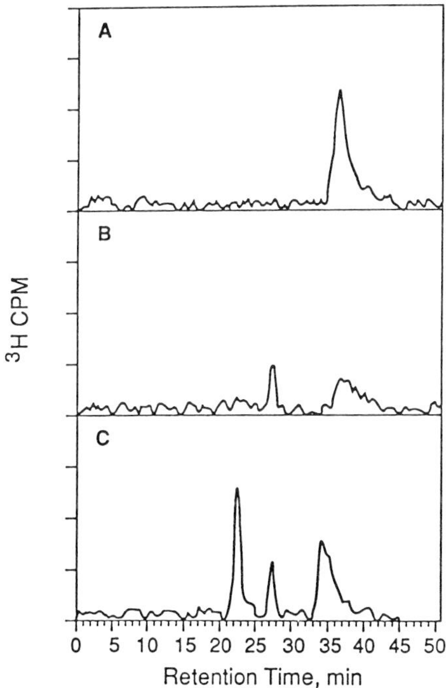

FIG. 4. HPLC analysis of the water-soluble products obtained on incubation of phospholipase C or lysosomal soluble protein with [^{3}H]DPPC. Phospholipase C (*B. cereus*) was incubated with [^{3}H]DPPC for 2 hr as described in the text and analyzed by HPLC (A). Lysosomal soluble protein was also incubated with [^{3}H]DPPC for 1 hr (B) or 24 hr (C).

pase which are all present in this protein fraction. The presence of choline is not due to the action of phospholipase D, but rather to further metabolism of phosphocholine and glycerophosphocholine by other lysosomal acid hydrolases. This analysis is confirmed by the fact that the lipid products (Table I) consist only of lysophosphatidylcholine and diglyceride. (Monoglyceride was also produced by the action of a lysosomal lipase.) The absence of phosphatidic acid in the lipid products shown in Table I further emphasizes the absence of phospholipase D in lysosomes.

Summary

The activity of a phospholipase C or phospholipase D may be assessed by measuring the radioactivity or phosphate released into the aqueous phase of a lipid extract. However, in crude enzyme fractions, this type of analysis may not be possible due to formation of water-soluble metabolites by other enzymatic reactions, as demonstrated here with a crude lysosomal enzyme fraction. In such instances, analysis of both water-soluble and lipid-soluble metabolites, at various times of incubation, may still provide clear identification of phospholipases C or D, even when a variety of lipases and other hydrolases are present.

[12] Preparation of Alkyl Ether and Vinyl Ether Substrates for Phospholipases

By FRITZ PALTAUF and ALBIN HERMETTER

Introduction

Glycerophospholipids containing an alkyl or 1′-alkenyl ether[1] group in position 1 of *sn*-glycerol are widely distributed in human and animal cell membranes and in some microorganisms.[2] Acyl ester and phospho ester bonds of ionic ether lipids are cleaved by the same phospholipases that hydrolyze the corresponding bonds in diacylglycerophospholipids, although the rates of reaction catalyzed by the same phospholipase may differ considerably depending on the phospholipid subclass (diacyl-, alkyl-

[1] 1-*O*-1′-Alkenylacylglycerophospholipids are commonly termed plasmalogens.

[2] See respective chapters in "Ether Lipids: Chemistry and Biology" (F. Snyder, ed.). Academic Press, New York and London, 1972.

acyl-, or 1′-alkenylacylglycerophospholipid) serving as the substrate.[3] Both the 1′-alkenyl and alkyl ether bonds resist phospholipases, and specific enzymes exist that are involved in their degradation.[4] Whereas the vinyl ether bond of plasmalogens is rather labile in that it is readily cleaved by acid hydrolysis, the ether bond in alkylglycerolipids is stable under acidic and alkaline conditions. Therefore, synthetic alkyl ether analogs of phospholipids with one or two nonhydrolyzable ether bonds have been used successfully for the assay of phospholipases under conditions where the presence of other acyl ester hydrolases would interfere if 1,2-diacylglycerophospholipids were used as the substrates.[3] For example, 1,2-dialkylglycerophospholipids are potential substrates for phospholipases C or D, but cannot be hydrolyzed by phospholipases A and B or by lipases. 1-*O*-Alkylglycerolipids resist the action of phospholipases A_1 and of 1(3)-regiospecific lipases.

It is the intent of this chapter to describe procedures for the isolation from natural sources, the semisynthesis, and the total chemical synthesis of selected alkyl ether and 1′-alkenyl ether glycerophospholipids that can serve as substrates for phospholipases A_2, C, or D, or for phosphatidate phosphohydrolase (EC 3.1.3.4, phosphatidate phosphatase). The choice of procedures reflects the experience of the present authors and, because of space limitation, is restricted to the preparation of representative phospholipid classes and species. In addition, variations in standard procedures are reported that facilitate the synthesis of radioactively labeled phospholipids for use in phospholipase assays. These adaptations can also be used for syntheses involving other expensive substrates. Laboratory directions for the synthesis of phospholipids containing ether bonds have been provided previously,[5,6] and the pertinent literature has been reviewed.[7]

Semisynthetic Procedures for Preparation of 1-*O*-Alkylacyl- and 1-*O*-1′-Alkenylacylglycerophospholipids

Synthesis starting from natural ether lipid substrates results in products which are heterogeneous with respect to the alkyl or 1′-alkenyl chain composition, but it is presently the only practical route for the preparation

[3] F. Paltauf, *in* "Ether Lipids: Biochemical and Biomedical Aspects" (H. K. Mangold and F. Paltauf, eds.), p. 211. Academic Press, New York, 1983.

[4] F. Snyder, T.-C. Lee, and R. L. Wykle, *in* "The Enzymes of Biological Membranes" (A. N. Martonosi, ed.), Vol. 2, p. 1. Plenum, New York, 1985.

[5] A. F. Rosenthal, this series, Vol. 35, p. 429.

[6] A. Hermetter and F. Paltauf, *in* "Ether Lipids: Biochemical and Biomedical Aspects" (H. K. Mangold and F. Paltauf, eds.), p. 390. Academic Press, New York, 1983.

[7] F. Paltauf, *in* "Ether Lipids: Biochemical and Biomedical Aspects" (H. K. Mangold and F. Paltauf, eds.), p. 49. Academic Press, New York, 1983.

of vinyl ether glycerophospholipids (plasmalogens). Total chemical syntheses of plasmalogens have been reported by several groups[7] but to our knowledge have not been adopted by the "users" of this phospholipid subclass.

A convenient source from which to isolate choline or ethanolamine plasmalogens are beef heart choline or ethanolamine phospholipids, which contain approximately 50% plasmalogens. The diacyl subclass can be selectively removed either by very mild alkaline hydrolysis[8] or by treatment with a regiospecific lipase, such as *Rhizopus* lipase,[9] which attacks only the acyl ester bond in the *sn*-1 position of diacylglycerophospholipids and converts them to lysophospholipids. The plasmalogens thus obtained are accompanied by small quantities (~5%) of the alkylacyl subclasses. By mild alkaline hydrolysis of choline glycerophospholipids, the diacyl subclass is completely hydrolyzed, whereas plasmalogens as well as alkylacylglycerophosphocholines are converted to the respective lyso derivatives, which can be separated into molecular species by reversed-phase high-performance liquid chromatography[10] (HPLC). The 1′-alkenyl groups in beef heart choline plasmalogen are mainly $C_{16:0}$ (62%), $C_{18:0}$ (11%), and $C_{18:1}$ (7%), with the remaining 20% containing $C_{17:0}$, $C_{15:0}$, and $C_{14:0}$.[11] Acylation of the lysoplasmalogens yields choline plasmalogens with a defined acyl chain composition.

Catalytic hydrogenation of the vinyl ether double bond converts plasmalogens or lysoplasmalogens to alkyacylglycerophosphocholines and alkylglycerophosphocholines, respectively. Acylation of the latter, for example, with acetic acid or fatty acid anhydrides, provides a straightforward method for synthesizing platelet-activating factor or alkylacylglycerophosphocholines of defined acyl chain composition. Modification of the head group, for instance, replacement of choline with ethanolamine, is achieved by phospholipase D-catalyzed transphosphatidylation.[12,13] By this procedure primary aliphatic alcohols or polyols with a chain length not exceeding 6 carbon atoms can be introduced.

General Procedures and Reagents

Anhydrous solvents required for syntheses are prepared by chromatography of commercially available solvents on basic alumina (0.063–0.2 mm, activity I from Merck, Darmstadt, FRG). Medium-pressure liquid

[8] O. Renkonen, *Acta. Chem. Scand.* **17,** 634 (1963).
[9] F. Paltauf, *Lipids* **13,** 165 (1978).
[10] M. H. Creer and R. W. Gross, *J. Chromatogr.* **338,** 61 (1985).
[11] H. H. O. Schmid and T. Takahashi, *Biochim. Biophys. Acta* **164,** 141 (1968).
[12] P. Comfurius and R. F. A. Zwaal, *Biochim. Biophys. Acta* **488,** 36 (1977).
[13] H. Eibl and S. Kovatchev, this series, Vol. 72, p. 632.

chromatography[14] (MPLC) is carried out on columns packed with silica gel 60 (0.04–0.06 mm; Merck). Purity of products is checked by thin-layer chromatography (TLC) on plates precoated with 0.2 mm silica gel 60 using the following developing solvents: solvent A, light petroleum (bp 40°–60°)–diethyl ether–acetic acid (80 : 20 : 2, by volume); solvent B, chloroform–methanol–25% ammonia (65 : 35 : 5, by volume); solvent C, chloroform–acetone–methanol–acetic acid–water (50 : 20 : 10 : 10 : 5, by volume); solvent D, chloroform–methanol–water (65 : 25 : 4, by volume).

Lipids are detected on TLC plates by exposure to iodine vapor, by charring after spraying with 50% H_2SO_4, by spraying with molybdic acid reagent[15] (for phospholipids), by heating after spraying with ninhydrin (0.2% in methanol; for phospholipids containing amino groups), or by spraying with 2,4-dinitrophenylhydrazine (0.4% in 2 *N* HCl; for plasmalogens) or with Dragendorff's reagent[16] (for choline-containing phospholipids). Phospholipids are stored either in the dry state at −20° or, if they contain polyunsaturated fatty acids, in toluene–ethanol (2 : 3, v/v) under argon at −20°. The vinyl ether linkage is sensitive to acid-catalyzed hydrolysis; therefore, all operations with plasmalogens should be carried out at neutral or slightly alkaline pH.

Isolation of Phospholipids from Beef Heart

Beef hearts obtained immediately after slaughter are freed of pericardial fat, and the tissue is homogenized in the presence of chloroform–methanol (2 : 1, v/v; 10 liters solvent/kg tissue).[17] After standing for 10 hr at 4° the mixture is filtered, and the residue is reextracted with chloroform–methanol (2 : 1, v/v; 3 liters/kg tissue). The extracts are combined and kept in the cold for 1–2 hr for the separation of water and organic phases. The upper water phase is removed, and the lower phase is washed with 0.2 volumes of water. After evaporation of the organic phase to dryness, the residue (~45 g) is dissolved in a 6-fold (v/w) volume of chloroform. From this solution phospholipids are precipitated by the addition of a 10-fold volume of acetone and incubation for 3 hr at 4°. The solvent is decanted and the precipitate isolated by centrifugation at 4°. Then total phospholipids (30 g) are dissolved in 150 ml chloroform, and the resulting solution is filtered through a thin (0.5 cm) layer of silica gel and applied to a column (5.5 × 45 cm) for MPLC separation. The

[14] H. Loibner and G. Seidl, *Chromatographia* **12,** 600 (1979).
[15] J. C. Dittmer and R. L. Lester, *J. Lipid Res.* **5,** 126 (1964).
[16] H. Wagner, L. Hörhammer, and P. Wolff, *Biochem. Z.* **334,** 175 (1961).
[17] Chloroform and methanol are toxic; therefore, all operations with these solvents must be carried out under a hood or in an efficiently ventilated laboratory.

phospholipids are eluted in 500-ml fractions using the following solvents: 4 liters chloroform–methanol (9 : 1, v/v), 4 liters chloroform–methanol (7 : 3, v/v), and 26 liters chloroform–methanol (1 : 1, v/v). Cardiolipin (3.3 g, fractions 7–16), ethanolamine glycerophospholipids (10.7 g, fractions 27–36), and choline glycerophospholipids (7.1 g, fractions 43–66) are obtained. Fractions are analyzed by TLC with solvent D.

Treatment of Choline and Ethanolamine Glycerophospholipids with Lipase from Rhizopus arrhizus

Identical procedures[9] can be followed for the treatment of either choline or ethanolamine glycerophospholipids with lipase. The phospholipid (100 mg) is dispersed in 20 ml of buffer [40 m*M* Tris-HCl, pH 7.6, 0.1 *M* NaCl, 3 mg/ml sodium deoxycholate, 4.5 mg/ml bovine serum albumin (Calbiochem, San Diego, CA), 5 m*M* $CaCl_2$] by shaking for 10 min, followed by sonication for 20 sec at 0°–5°. Then lipase (1 mg, 600 IU) from *Rhizopus arrhizus* (Boehringer, Mannheim, FRG) is added, and the mixture is incubated with stirring at 25° for 2 hr. After incubation the lipids are extracted with 100 ml chloroform–methanol (2 : 1, v/v), and the lower organic phase is recovered and evaporated under reduced pressure. Lipids can be isolated by silica gel chromatography on small columns according to the procedure described for the isolation of 1′-alkenyloleoylglycerophosphocholine (see below) or by preparative TLC on silica gel H using solvent B. The yield of chromatographically pure products is 80%, based on the original amount of plasmalogens present in choline and ethanolamine phospholipids. The plasmalogens are accompanied by alkylacylglycerophospholipids (~5 mol %).

1-O-1′-Alkenylglycerophosphocholines

To a solution of 5 g choline glycerophospholipids in 200 ml chloroform, a methanolic solution of NaOH (200 ml of 0.1 *N* NaOH) is added. After standing for 2 hr at 25° the mixture is chilled in an ice bath and neutralized with 200 ml of 0.1 *N* acetic acid in water. After addition of 200 ml chloroform the phases are separated, and the upper phase is reextracted with 400 ml chloroform–methanol (2 : 1, v/v). The lower phases are combined, washed with 0.2 volumes of methanol–water (1 : 1, v/v), and then evaporated under vacuum. The residue (4.6 g) is purified by MPLC on a 3 × 30 cm column using chloroform–methanol (1 : 1, v/v) as the solvent. With collection of 200-ml fractions, 1 g of pure lysoplasmalogen (63% based on choline plasmalogens) is obtained from fractions 16–30. TLC with solvent B shows a single spot, R_f 0.15. The product, 1-*O*-1′-alkenylglycerophosphocholine, contains approximately 5% of the 1-*O*-alkyl analog.

Chromatographic Separation of 1-O-1′-Alkenylglycerophosphocholines

Alkenylglycerophosphocholine obtained from beef heart phospholipids (see above) can be resolved into molecular species by reversed-phase HPLC on Ultrasphere ODS.[10] Using isocratic elution with methanol–water–acetonitrile (57 : 23 : 20, by volume) containing 20 m*M* choline chloride, the main species, 1-*O*-1′-hexadecenylglycerophosphocholine, elutes at 40 min, using a flow rate of 2 ml/min. Micromole quantities can be separated on a 4.6 mm × 25 cm stainless steel column, with UV absorbance at a wavelength of 203 nm as the method of detection.

1-O-1′-Alkenyl-2-oleoyl-sn-glycero-3-phosphocholine

Acylation is carried out by a combination of methods described by Gupta *et al.*[18] and Selinger and Lapidot.[19] To a solution of 1-*O*-1′-alkenylglycerophosphocholine (300 mg, 0.624 mmol) and oleic acid (352 mg, 1.25 mmol) in 8 ml chloroform, dicyclohexylcarbodiimide (579 mg, 2.81 mmol) and dimethylaminopyridine (305 mg, 2.5 mmol) are added. After stirring for 17 hr at 40° the mixture is cooled, and 52 ml chloroform plus 30 ml methanol are added. The solution is washed with 18 ml methanol–water (1 : 1, v/v) followed by removal of the solvent under vacuum. Crude 1′-alkenylacylglycerophosphocholine (500 mg) is purified by MPLC on a 2 × 22 cm column filled with silica gel 60. Lipids dissolved in 20 ml chloroform–methanol–25% ammonia (80 : 20 : 0.5, by volume) are applied to the column, and elution is performed with 80 ml of the same solvent mixture, followed by 80 ml chloroform–methanol–25% ammonia (70 : 30 : 0.5, by volume), 150 ml chloroform–methanol–ammonia (60 : 40 : 0.5, by volume), and 200 ml chloroform–methanol–ammonia (50 : 50 : 0.5, by volume). Fractions of 15 ml are collected with the product (350 mg, 76% yield) eluting in fractions 22–30.

Preparation of 1-O-1′-Alkenyl-2-oleoyl-sn-glycero-3-phosphoethanolamine by Phospholipase D

A convenient source of phospholipase D is white cabbage, from which the enzyme can be prepared by extraction with water.[13] All operations are performed at 0°–5°. Leaves of white cabbage (1.5 kg) are cut in small pieces and homogenized for 5 min in the presence of 250 ml distilled water. The homogenate is filtered by suction, and the turbid filtrate is centrifuged at 25,000 *g* for 20 min. The pellet is discarded, and the clear supernatant

[18] C. M. Gupta, R. Radhakrishnan, and H. G. Khorana, *Proc. Natl. Acad. Sci. U.S.A.* **74,** 4315 (1977).

[19] Z. Selinger and Y. Lapidot, *J. Lipid Res.* **7,** 174 (1966).

is adjusted to a protein concentration of 3 mg/ml, which can directly be used for transphosphatidylation. The enzyme solution can be stored at $-25°$ for more than 2 years without loss of phospholipase D activity.

Transphosphatidylation can be carried out on any scale between milligrams and hundreds of grams. The following procedure has been used for the preparation of radioactively labeled 1-*O*-1′-alkenyl-2-[1′-^{14}C]oleoyl-*sn*-glycero-3-phosphoethanolamine. 1-*O*-1′-Alkenyl-2-oleoyl-*sn*-glycero-3-phosphocholine (4.7 mg) is dissolved in 1 ml diethyl ether. To this solution 3 ml buffer[20] containing ethanolamine (20%) and crude phospholipase D (6 mg protein) are added. The mixture is vigorously stirred with a magnetic stirrer for 10 hr at 40°. Diethyl ether is removed from the reaction mixture by evaporation at reduced pressure, and the residue is extracted 3 times with 2-ml portions of chloroform–methanol (2 : 1, v/v). The solution is washed twice with 1 ml methanol–water (1 : 1, v/v). The lower phase is separated and evaporated to dryness. The lipid is dissolved in chloroform–methanol (9 : 1, v/v), and the product is isolated by preparative TLC on two 10 × 10 cm plates coated with 0.2 mm silica gel H. Plates are developed with solvent D, and the product is visualized by spraying the plates with water. The corresponding band at R_f 0.5 is scraped off, and the product is eluted from the wet silica gel with chloroform–methanol (1 : 4, v/v). Unreacted choline plasmalogen migrates to an R_f of 0.25 and can be recovered. The yield of pure 1′-alkenylacylglycerophosphoethanolamine is 3 mg (64%), which shows a single spot on TLC using solvents B or D.

1-O-1′-Alkenyl-2-oleoyl-sn-glycero-3-phosphate

If the reaction described in the previous section is carried out in the absence of ethanolamine, 1-*O*-1′-alkenyl-2-acyl-*sn*-glycero-3-phosphate is formed in quantitative yield. Thus, a solution of 200 mg of 1′-alkenylacylglycerophosphocholine dissolved in 200 ml diethyl ether stirred for 10 hr in the presence of 20 ml buffer and 600 mg crude phospholipase D gives 170 mg (97% yield) of the phosphatidic acid analog. This product is chromatographically pure when checked on TLC with solvent system B (R_f 0.1).

Preparation of 1-O-Alkyl-sn-glycero-3-phosphocholine by Catalytic Hydrogenation of 1-O-1′-Alkenyl-sn-glycero-3-phosphocholine

Platinum dioxide (30 mg) is added to a solution of alkenylglycerophosphocholine (100 mg) in 25 ml dry tetrahydrofuran in a flask equipped with inlet and outlet tubes and a magnetic stirrer. The flask is flushed with

[20] The buffer contains 0.2 *M* sodium acetate, 0.1 *M* calcium chloride, 20% ethanolamine, adjusted to pH 5.6 with HCl.

argon, then filled with hydrogen, and the outlet tube is closed. After 3 hr of stirring under hydrogen (35 psi) the hydrogenation is complete. The platinum catalyst is removed by filtration over a thin layer of Hyflo,[21] and the solution is evaporated to dryness. The product is obtained in pure form with quantitative yield. On TLC it migrates to the same R_f (0.15 in solvent B) as the lysoplasmalogen, but it does not show a color reaction when sprayed with dinitrophenylhydrazine. Individual molecular species of alkylglycerophosphocholine can be separated by reversed-phase HPLC following the protocol described above for the separation of the 1′-alkenylglycerophosphocholine species.

1-O-Alkyl-2-acetyl-sn-glycero-3-phosphocholine (Platelet-Activating Factor)

Dry alkylglycerophosphocholine (500 mg, ~1 mmol) is suspended in a mixture of 7 ml benzene and 5 ml acetonitrile, to which 0.4 ml (~4 mmol) of acetic anhydride is added, and the mixture is warmed until a clear solution is obtained.[22] After cooling to room temperature, 150 mg (1.2 mmol) of dimethylaminopyridine is added, and the suspension is kept at room temperature for 4–5 hr. The solvents are removed in a rotary evaporator at 40°, and the yellowish residue is dissolved in a small volume of chloroform and purified on preparative TLC plates coated with 0.5-mm layers of silica gel H, using methanol–water (2 : 1, v/v) as the solvent. The fraction containing the alkylacetylglycerophosphocholine is scraped off, and the product is eluted from the silica gel with methanol. After distilling off the solvent at reduced pressure, 10 ml of acetone is added to the colorless residue. The solution is kept at −10° for a few hours with occasional stirring, at which time the colorless precipitate is collected and dried in a vacuum desiccator over calcium chloride. The saturated alkylacetylglycerophosphocholine is obtained in an amount of 0.35 g (64% yield). The alkyl moieties consist of approximately 62% $C_{16:0}$ and 18% $C_{18:0}$, with smaller quantities of $C_{14:0}$, $C_{15:0}$, and $C_{17:0}$. Platelet-activating factor with a defined alkyl chain can be obtained by an analogous procedure starting from synthetic alkylglycerophosphocholines (see below).

1-O-Alkyl-2-arachidonoyl-sn-glycero-3-phosphocholine

Alkylglycerophosphocholine (7.7 mg, 15.3 μmol), arachidonic acid (9.2 mg, 30 μmol), dicyclohexylcarbodiimide (19.2 mg, 73 μmol), and 4-dimethylaminopyridine (9.6 mg, 78 μmol) are dissolved in 300 μl anhydrous chloroform and kept overnight at 40° under anhydrous conditions in an

[21] Reduced platinum catalyst will ignite when drying on filter paper in the presence of oxygen.
[22] T. Muramatsu, N. Totani, and H. K. Mangold, *Chem. Phys. Lipids* **29**, 121 (1981).

atmosphere of argon.[23] Then 1 ml chloroform is added, and the reaction mixture is applied to a small column (a Pasteur pipette plugged with glass wool and filled with 350 mg silica gel to a height of 4 cm). Lipids are eluted from the column with 4 ml chloroform, followed by mixtures of chloroform–methanol in the ratios 9 : 1 (3 ml), 8 : 3 (2 ml), 7 : 3 (1 ml), 6 : 4 (1 ml), and 1 : 1 (10 ml). Fractions of 1 ml are collected, and the product (11 mg, 91% yield) is recovered from fractions 12–17. It is pure as judged by TLC using solvents B, C, and D. In an analogous manner and with identical yields, the eicosapentaenoyl and docosahexaenoyl derivatives as well as the choline plasmalogen analogs have been prepared.

Chemical Synthesis of Alkyl Ether Substrates

The chemical synthesis of 1-*O*-alkylglycerols has been described in a previous volume of this series.[24] The procedure described below follows essentially the same route, but, owing to some modifications,[25] the reaction times are considerably shorter, the reaction conditions are milder, and the workup of products has been simplified. Educts for the synthesis of chiral ether phospholipids are the commercially available 1,2- or 2,3-isopropylidene-*sn*-glycerols; racemic products are obtained starting from racemic isopropylidene glycerol. Introduction of the triphenylmethyl group in position 3 of alkylglycerols, followed by acylation in the 2-position gives 1(3)-*O*-alkyl-2-acyl-3(1)-triphenylmethylglycerols, which on removal of the protecting group are converted to 1(3)-*O*-alkyl-2-acylglycerols. Removal of the triphenylmethyl group is best achieved with boron trifluoride/methanol at low temperature.[26] The product is used for the following phosphorylation step without purification in order to avoid acyl migration. Coupling alkylacylglycerophosphate with the appropriate (protected) head group alcohol, using triisopropylbenzenesulfonyl chloride as the condensing agent,[27] gives the final alkylacylglycerophospholipid (after deprotection, if necessary). Radioactively labeled fatty acids can be introduced by deacylation–reacylation of the respective choline or N-protected ethanolamine derivatives. Deacylation–reacylation is also recommended for the introduction of other expensive (e.g., fluorescently labeled) or oxygen-

[23] Owing to the oxygen sensitivity of polyunsaturated fatty acids all operations should be carried out under an atmosphere of inert gas, preferentially argon. An antioxidant such as di-*tert*-butyl-*p*-cresole (0.1 mg/ml solvent) can be added to the reaction mixture.

[24] G. A. Thompson, Jr., and V. M. Kapoulas, this series, Vol. 14, p. 668.

[25] R. Franzmair, personal communication (1989).

[26] A. Hermetter and F. Paltauf, *Chem. Phys. Lipids* **29,** 191 (1981).

[27] R. Aneja, J. S. Chadha, and A. P. Davies, *Tetrahedron Lett.* **48,** 4183 (1969).

sensitive polyunsaturated fatty acids in order to reduce losses of expensive material and/or to avoid oxidative damage of the products.

Phospholipase D-catalyzed transphosphatidylation of alkylacylglycerophosphocholines gives easy access to other phospholipid classes, and N-methylation of alkylacylglycerophosphoethanolamines[28] with [^{3}H]- or [^{14}C]methyl iodide allows the synthesis of alkylacylglycerophosphocholines radioactively labeled in the polar head group. Choline glycerophospholipids radioactively labeled in the head group have also been prepared by demethylation to the corresponding *N*-dimethylethanolamine phospholipid with sodium benzenethiolate, followed by quaternization with [^{14}C]methyl iodide.[29] The "unnatural" *sn*-1 isomers of alkylacylglycerophospholipids are accessible by treatment of the respective racemic phospholipids with phospholipase A_2,[30] which leaves the *sn*-1 isomers intact. The *sn*-3 lysophospholipids formed can then be acylated to give *sn*-3 alkylacylglycerophospholipids.

Alkylmethane Sulfonates

To a suspension of 1-*O*-octadecanol (100 g, 0.37 mol) in 500 ml dichloromethane, 44.9 g (0.444 mol) triethylamine is added at room temperature.[25] Then 50.8 g (0.443 mol) methane sulfochloride in 100 ml dichloromethane is added dropwise over a period of 50 min while cooling with water and stirring. After stirring for an additional 1 hr at room temperature, the dichloromethane is removed by distillation under reduced pressure, and 700 ml ethanol–water (1 : 1, v/v) is added to the residue. The product is precipitated in crystalline form and is collected by filtration after 30 min of stirring at room temperature.

If an unsaturated alkylmethane sulfonate is prepared, the product will not crystallize after the addition of ethanol–water. In this case the oily residue is collected by centrifugation, washed with water, and then dried by dissolving it in benzene followed by evaporation of the solvent under reduced pressure. The alkylmethane sulfonates are obtained in 99% yield and can be used for subsequent reactions without further purification.

1-O-Hexadecyl-sn-glycerol

The procedure described here is a modification[25] of the method described by Baumann and Mangold.[31] The isomeric purity of the commercially available chiral educt for the synthesis of 1-*O*-hexadecyl-*sn*-glycerol,

[28] K. M. Patel, J. D. Morrisett, and J. T. Sparrow, *Lipids* **14,** 596 (1979).

[29] W. Stoffel, this series, Vol. 35, p. 533.

[30] F. Paltauf, *Chem. Phys. Lipids* **17,** 148 (1976).

[31] W. J. Baumann and H. K. Mangold, *J. Org. Chem.* **29,** 3055 (1964).

2,3-isopropylidene-*sn*-glycerol, should be checked by measuring its optical rotation ($[\alpha_D]$ − 14.3°).[32]

To a solution of 2,3-isopropylideneglycerol (10 g, 75.7 mmol) in 85 ml dry dimethyl sulfoxide, powdered potassium hydroxide (7.3 g, 130.1 mmol) is added. The suspension is stirred for 1 hr at 50° with protection from atmospheric moisture. Then hexadecylmethane sulfonate (15.9 g, 49.6 mmol) is added, and stirring at 50° is continued. After 3 hr the reaction mixture is cooled to room temperature, 170 ml water is added, and the product is extracted with 3 portions (100 ml) of light petroleum. The solution is washed 3 times with water, then dried over anhydrous sodium sulfate, and the solvent is removed under reduced pressure. On TLC the product shows an R_f of 0.8 with light petroleum–diethyl ether (7 : 3, v/v) as the developing solvent. Small quantities of impurities (mainly dihexadecyl ether) do not interfere with the subsequent reaction and need not be removed.

The crude 1-*O*-hexadecyl-2,3-*O*-isopropylidene glycerol (21.5 g) is hydrolyzed by stirring for 3 hr at room temperature in a solution of 80 ml of 10% aqueous hydrochloric acid in 500 ml methanol. After the addition of 500 ml water the product is extracted with 2 portions (200 ml each) of diethyl ether. The combined ether extract is washed consecutively with 3 portions of water (100 ml each) and dried over anhydrous sodium sulfate. The solvent is evaporated, and the product is isolated after recrystallization from 80 ml light petroleum. The yield of pure product is 15.3 g (80% based on hexadecylmethane sulfonate), mp 65.6°. It shows a single spot on TLC (R_f 0.15) with light petroleum–diethyl ether (3 : 7, v/v) as the developing solvent.

1-O-Hexadecyl-3-O-triphenylmethyl-sn-glycerol

A mixture of 15.3 g 1-*O*-hexadecyl-*sn*-glycerol (48.3 mmol) and 27 g triphenylmethyl chloride (97 mmol) in 200 ml anhydrous pyridine is stirred for 24 hr at 50° under anhydrous conditions.[33] The solution is poured into ice-cold water (400 ml), and the mixture is extracted with 3 portions (200 ml each) of light petroleum. Undissolved triphenylmethanol is removed by filtration. The filtrate is washed 3 times with water (100 ml) and dried over anhydrous sodium sulfate. The solvent is evaporated, and light petroleum, bp 40°–60° (150 ml), is added to the residue. On standing overnight at 4° additional triphenylmethanol is precipitated. The solid is filtered off, and the filtrate is evaporated to dryness. TLC analysis (solvent A) of the resulting product (30.6 g) shows that the alkylglycerol has completely

[32] G. Hirth, H. Saroka, W. Bannwarth, and R. Barner, *Helv. Chim. Acta* **66,** 1210 (1983).
[33] G. K. Chacko and D. J. Hanahan, *Biochim. Biophys. Acta* **164,** 252 (1968).

reacted. The crude 1-*O*-hexadecyl-3-*O*-triphenylmethyl-*sn*-glycerol is directly used for the next reaction step.

1-O-Hexadecyl-2-oleoyl-3-O-triphenylmethyl-sn-glycerol

Oleic acid (15.5 g, 55 mmol) and carbonyldiimidazole (9.8 g, 60 mmol) in 100 ml water-free tetrahydrofuran are stirred at room temperature for 1 hr. The freshly prepared solution of oleoylimidazolide is added to 1-*O*-hexadecyl-3-*O*-triphenylmethyl-*sn*-glycerol (15.3 g, 27.4 mmol). Tetrahydrofuran is removed by evaporation under reduced pressure, and the residue is dissolved in 100 ml dimethyl sulfoxide. After addition of the catalyst, which is prepared by dissolving metallic sodium (1.35 mg, 58.7 mg atoms) in 100 ml dimethyl sulfoxide, the reaction mixture is kept at room temperature under anhydrous conditions and slightly shaken at intervals of about 10 min. The reaction mixture is then cooled in an ice bath and rapidly neutralized with 160 ml of 0.2 *N* acetic acid in water. The product is extracted with 3 portions (20 ml) of chloroform–methanol (2 : 1, v/v). The extracts are combined, washed with 0.2 volumes methanol–water (1 : 1) containing 1.5% concentrated ammonia, then with 0.2 volumes methanol–water (1 : 1), and dried over anhydrous sodium sulfate. After evaporation of the solvent under reduced pressure, the crude product (23 g) is purified by MPLC. The sample, dissolved in 100 ml light petroleum (bp 40°–60°), is applied to a silica gel column (3 × 30 cm)[34] and eluted with 300 ml light petroleum and 1 liter of light petroleum–diethyl ether (9 : 1, v/v); 100-ml fractions are collected. Pure alkylacyltriphenylmethylglycerol is obtained from fractions 4–7. The yield is 20.5 g (91% based on alkyltriphenylmethylglycerol). The product gives a single spot (R_f 0.6) on TLC developed with light petroleum–diethyl ether (9 : 1, v/v).

1-O-Hexadecyl-2-oleoyl-sn-glycerol

1-*O*-Hexadecyl-2-oleoyl-3-*O*-trityl-*sn*-glycerol (19 g, 23 mmol) is dissolved in 170 ml anhydrous dichloromethane containing 80 ml of a 14% (w/v) solution of boron trifluoride in methanol.[26] The mixture is stirred at 0° for 60 min and then extracted 3 times with ice-cold water. The organic phase is dried over anhydrous sodium sulfate, and the solvent is removed by evaporation under reduced pressure. An oily residue is obtained which is used without further purification for the preparation of the corresponding phosphatidic acid analog.

[34] The column is preconditioned with light petroleum saturated with aqueous concentrated ammonia to deactivate the solid phase. On activated silicic acid partial hydrolysis of the triphenylmethoxy group would occur.

1-O-Alkyl-2-oleoyl-sn-glycero-3-phosphate

The following procedure is a modification[30] of the method described by Berecoechea *et al.*[35] A solution of 11.3 g (20 mmol) alkyloleoylglycerol and anhydrous pyridine (3.2 g, 40 mmol) in 60 ml anhydrous tetrahydrofuran is added dropwise with stirring to freshly distilled phosphorus oxychloride (3.3 g, 22 mmol) dissolved in 20 ml tetrahydrofuran. Stirring is continued for 3 hr at 0°. Then 100 ml of 10% sodium bicarbonate is added at once, and the mixture is stirred for 15 min at 0°. The solution is then poured on ice water, acidified with HCl, and extracted with diethyl ether. After washing the ether phase with water and drying it over anhydrous sodium sulfate, the solvent is removed under reduced pressure, and 11.5 g of the product is obtained. TLC with solvent B shows the product at R_f 0.1 and slight impurities, at R_f 0.7, corresponding to dipyrophosphatidic acid.[35] The product is used for the subsequent reaction without purification.

1-O-Hexadecyl-2-oleoyl-sn-glycero-3-phosphocholine

1-*O*-Hexadecyl-2-oleoyl-*sn*-glycero-3-phosphate (1.65 g, 2.5 mmol) and choline toluene sulfonate[36] (1.4 g, 5 mmol) prepared according to Brockerhoff and Ayengar[37] are dried over phosphorus pentoxide under vacuum for 12 hr. The mixture is dissolved in 45 ml anhydrous pyridine, and 2,4,6-triisopropylbenzene sulfonyl chloride (1.82 g, 6 mmol) is added. The solution, protected from atmospheric moisture, is stirred for 1 hr at 70° and then for 4 hr at room temperature. After the addition of 2 ml water, the solvents are removed under vacuum on a rotary evaporator, and pyridine is removed by repeated evaporation in the presence of toluene. The residue is extracted with diethyl ether, and the solutions are decanted from undissolved material (mainly triisopropylbenzenesulfonic acid) and evaporated to dryness.

The solid material is dissolved in 50 ml chloroform–methanol (2 : 1), and the mixture is successively extracted with 10-ml portions of 3% aqueous Na_2CO_3, methanol–water (1 : 1), 2% HCl, and finally methanol–water (1 : 1). The organic phase is dried over sodium sulfate and evaporated under reduced pressure. The crude product (2 g) dissolved in 20 ml chloroform is applied to a 2 × 30 cm column for MPLC purification. The following solvents are used for chromatography: 200 ml chloroform, 50 ml each

[35] J. Berecoechea, M. Faure, and J. Anatol, *Bull. Soc. Chim. Biol.* **50,** 1561 (1968).

[36] To a 40% solution of choline hydroxide in water an equimolar amount of toluene sulfonic acid is added, and the solvents are removed under reduced pressure. The choline toluene sulfonate is recrystallized from acetone.

[37] H. Brockerhoff and N. K. N. Ayengar, *Lipids* **14,** 88 (1979).

of chloroform–methanol 9 : 1, 8 : 2, 7 : 3, and 6 : 4, followed by 500 ml chloroform–methanol 4 : 6. Fractions of 15 ml are collected, and the product (1.3 g, 70% yield) is obtained in fractions 28–42. On TLC with solvents B, C, and D it shows a single spot corresponding to authentic phosphatidylcholine from egg yolk.

1-O-Hexadecyl-2-oleoyl-sn-glycero-3-phosphoethanolamine

A mixture of *N*-triphenylmethylethanolamine (360 mg, 1.2 mmol; prepared[38] according to Aneja *et al.*[27]), triisopropylbenzenesulfonyl chloride (450 mg, 1.5 mmol), 1-*O*-hexadecyl-2-oleoyl-*sn*-glycero-3-phosphate (400 mg, 0.6 mmol) in 12 ml anhydrous pyridine, and 5 ml dry chloroform is stirred at room temperature for 3 hr. Then 2.5 ml of water is added, the mixture is evaporated under vacuum, and the residue is extracted 3 times with diethyl ether. After evaporation of the combined diethyl ether extracts, the crude product is purified by MPLC on a 3 × 35 cm column using the following solvents for elution: 200 ml chloroform, 200 ml chloroform–methanol (95 : 5), and 400 ml chloroform–methanol (9 : 1); 50-ml fractions are collected. The product (450 mg, 80% yield) is obtained from fractions 14 and 15. TLC shows a single spot, R_f 0.6, with chloroform–methanol (9 : 1) as the solvent.

For removal of the triphenylmethyl protecting group, trifluoroacetic acid (15 ml) is added to an ice-cold solution of hexadecyloleoylglycerophospho(*N*-triphenylmethyl)ethanolamine (450 mg) in 15 ml dichloromethane under protection from atmospheric moisture. The reaction mixture is left at 0° for 5 min and then poured rapidly into a vigorously stirred solution (50 ml) of 6% ammonia in water. The aqueous phase is separated and extracted twice with chloroform (100 ml). The combined organic phases are washed with water and evaporated to dryness. The crude product is purified by MPLC on a silica gel column using a chloroform–methanol gradient for elution. The product (270 mg, 80%) is eluted at a chloroform–methanol ratio of 7 : 3. TLC with solvent B shows a single spot, R_f 0.6.

Chemical Synthesis of Diether Substrates

The occurrence of 1,2-di-*O*-alkylglycerophosphocholines in nature has been reported, but obviously they constitute only a minute proportion of the choline glycerophospholipids isolated, for example, from bovine

[38] *N*-Triphenylethanolamine is prepared by reacting triphenylmethyl bromide and ethanolamine in chloroform in the presence of triethylamine.

heart.[39] Because they do not contain an acyl ester linkage, 1,2-dialkylglycerophospholipids cannot be hydrolyzed by phospholipase A or B, by lipases, or by other acyl ester hydrolases. However, they can serve as substrates for phospholipase C or D.[3] In addition, for studies on mechanisms and kinetics of phospholipases they can be used as inert matrices in which the appropriate cleavable diacylglycerophospholipid substrates are embedded at varying concentrations but in a constant total surface presented to the enzyme.

Racemic di-*O*-alkylglycerols with two identical alkyl chains are preferentially synthesized from tetrahydropyranylglycerol.[40] The enantiomers containing saturated alkyl chains are accessible via 1- or 3-*O*-benzylglycerols[41] which are commercially available. Di-*O*-alkylglycerols with two different alkyl chains (saturated or unsaturated) can be prepared from 1(3)-alkyl-3(1)-trityl-*sn*-glycerols[42] (see above) by reaction with alkylmethane sulfonates. By the same route optically active dialkylglycerols with identical, or different, saturated or unsaturated alkyl chains can be prepared. The subsequent steps of phosphorylation and attachment of head group alcohols are the same as with alkylacyl substrates (see preceding sections).

1-O-Hexadecyl-2-O-octadec-9'-enyl-sn-glycerol

1-*O*-Hexadecyl-3-*O*-trityl-*sn*-glycerol (6 g, 10.6 mmol; see above) and powdered potassium hydroxide (6 g) in 100 ml xylene are refluxed for 1 hr in a 500-ml flask fitted with a water-separation head, reflux condenser, dropping funnel, and magnetic stirrer. Then octadecenylmethane sulfonate (4.5 g, 13 mmol; see above) in 50 ml xylene is added dropwise during 15 min, and refluxing is continued for 6 hr. After cooling 100 ml water is added, the xylene phase is removed, and the water phase is extracted twice with 100 ml diethyl ether. The combined organic phases are dried over anhydrous sodium sulfate and evaporated under reduced pressure. The residue (8.6 g) is dissolved in 25 ml light petroleum and purified by MPLC on a 3×25 cm column. The following solvents were used for elution: light petroleum (200 ml), light petroleum–diethyl ether, 9 : 1 (100 ml), and light petroleum–diethyl ether, 8 : 2 (300 ml). Fractions (10 ml) are collected, and the product is recovered from fractions 8–28. The yield of pure product is 8 g (90%). The protecting triphenylmethyl group is removed by treating the product with boron trifluoride–methanol following the procedure described for the alkylacyl analog (see above). Pure 1-*O*-hexa-

[39] E. L. Pugh, M. Kates, and D. J. Hanahan, *J. Lipid Res.* **18,** 710 (1977).

[40] F. Paltauf and F. Spener, *Chem. Phys. Lipids* **2,** 168 (1968).

[41] M. Kates, B. Palameta, and L. S. Yengoyan, *Biochemistry* **4,** 1595 (1965).

[42] W. J. Baumann and H. K. Mangold, *J. Org. Chem.* **31,** 498 (1966).

decyl-2-octadec-9′-enyl-*sn*-glycerol (5 g) is obtained in 90% yield after MPLC purification on a 2 × 35 cm column, eluted with a light petroleum–diethyl ether gradient. It shows a single spot, R_f 0.2, on TLC with solvent A.

1-O-Hexadecyl-2-O-octadec-9′-enyl-sn-glycero-3-phosphocholine and -ethanolamine

Di-*O*-alkylglycerophospholipids are prepared from di-*O*-alkylglycerols by procedures described for the alkylacyl analogs[43] (see above). On TLC using solvents B, C, or D they migrate to the same R_f as the analogous alkylacyl- and diacylglycerophospholipids.

Acknowledgments

The authors acknowledge the excellent technical assistance of H. Stütz and thank R. Franzmair for stimulating discussions. The original work described in this chapter has been supported by the Fonds zur Förderung der wissenschaftlichen Forschung in Österreich (Project 5746B).

[43] Because dialkylglycerophospholipids are stable in dilute alkali or acid, workup procedures are more straightforward than with 1′-alkenylacyl or alkylacyl analogs. Acidic or basic byproducts can readily be removed by extraction from a chloroform–methanol phase, with dilute sodium hydroxide or hydrochloric acid, respectively.

[13] Inositol Phospholipids and Phosphates for Investigation of Intact Cell Phospholipase C Substrates and Products

By MICHAEL R. HANLEY, DAVID R. POYNER, and PHILLIP T. HAWKINS

Introduction

The receptor-coupled activation of inositol lipid-specific phospholipase C is conventionally analyzed in intact cells, as there has been comparatively limited success in developing cell-free or reconstituted systems. Thus, much of the information on the physiological substrates and resulting enzymatic products has been inferred from the kinetics of changes in the levels of labeled lipids and the production of labeled water-soluble products. The early work on this second messenger pathway used tissue slices or acutely isolated cells, but, where feasible, the recent emphasis has been placed on established cell lines, which have advantages in homogeneity, reproducibility, and accessibility to scaling-up, for the intact cell investigation of phospholipase C. In this chapter, illustrations are drawn from the use of a specific cell line, the NG115-401L neuronal line,[1] but the methodologies and results are representative of many cell populations and tissue preparations.

Analysis of ^{32}P-Labeled Phospholipids by High-Performance Thin-Layer Chromatography (HPTLC)

One of the simplest, but most powerful, techniques is the use of ortho-[^{32}P]phosphate to label cellular ATP pools, which are in rapid equilibrium with the monoester phosphates of the polyphosphoinositides.[2] Cells are labeled with ortho[^{32}P]phosphate (10 μCi/10^5 cells, 60 min at 37°; either monolayers or suspensions can be used) in a modified reduced-phosphate physiological saline to facilitate incorporation of labeled phosphate into lipid (low-phosphate Krebs–Ringer bicarbonate: 100 m*M* NaCl, 4.4 m*M* KCl, 2.37 m*M* $CaCl_2$, 1.11 m*M* $MgSO_4$, 22.3 μM KH_2PO_4, 23.4 m*M* $NaHCO_3$, pH 7.4, gassed with 95% O_2/5% CO_2). Cells are washed 3 times with physiological saline to remove excess radioactivity, and then cell stimulants are added.

At the end of the incubation, cells are quenched and extracted with 3.75 volumes of chloroform/methanol/HCl (40 : 80 : 1, v/v). If cells are analyzed as monolayers, they must be grown on glass coverslips and can be transferred directly to the extraction medium, whereas cell suspensions can be incubated in 0.25–1 ml in appropriate plastic vials [e.g., Beckman (Fullerton, CA) Bio-Vials]. The lipids are separated from the labeled water-soluble products by the addition of 1.25 volumes each of chloroform and 0.1 *M* HCl. The mixture can be split into two phases either by standing at room temperature or by a brief centrifugation in a benchtop centrifuge. The lower, organic phase contains the labeled lipids, and aliquots are taken for direct counting, to assess later yields, and for drying under a nitrogen stream. The dried sample is redissolved in 0.1 ml chloroform/methanol/water (75 : 25 : 2, v/v), and an aliquot is taken for determination of the total radioactivity.

The remaining redissolved sample is applied to a strip on a 10 × 10 cm HPTLC plate [Merck (Darmstadt, FRG) 5631 Kieselgel-60], which has been oxalate impregnated by prechromatography in 1% potassium oxalate, followed by activation at 110° for 10 min.[3] The HPTLC plates are devel-

[1] T. R. Jackson, T. J. Hallam, C. P. Downes, and M. R. Hanley, *EMBO J.* **6,** 49 (1987).

[2] J. A. Creba, C. P. Downes, P. T. Hawkins, G. Brewster, R. H. Michell, and C. J. Kirk, *Biochem. J.* **212,** 733 (1983).

[3] C. P. Downes, M. D. Dibner, and M. R. Hanley, *Biochem. J.* **214,** 865 (1983).

oped in chloroform/acetone/methanol/acetic acid/water (40 : 15 : 13 : 12 : 8, v/v), or another suitable solvent system,[4] for approximately 60 min, dried, and then exposed to X-ray film (Kodak X-OMat RP, Fuji RX) for 10–16 hr, depending on the radioactivity. The mobilities of standard unlabeled inositol lipids are determined by running a separate lane with 10 μg each of inositol phospholipids (Sigma, St. Louis, MO), which can be visualized after development of the autoradiogram by one of several standard techniques, such as iodine vapor staining.[5] If results need to be quantified, the resulting autoradiogram can be used as a template overlay, and the identified bands scraped and counted by Cerenkov counting in water. Figure 1 shows an HPTLC chromatogram of this type.

For the analysis of inositol phospholipids, HPTLC analysis of ^{32}P-labeled lipids cannot be used, without other information, for direct chemical estimation of the levels of these lipids, but it can be used to determine relative alterations on activation of phospholipase C. In Fig. 2, phospholipase C has been stimulated by the peptide bradykinin, and one observes rapid (<2 sec) and dramatic (up to 80–90% decline) changes in the presumptive major phospholipase C substrate, phosphatidylinositol 4,5-bisphosphate, with smaller changes in the levels of phosphatidylinositol 4-phosphate and phosphatidylinositol. From these results, it would be unclear whether phospholipase C specificity was restricted to phosphatidylinositol- 4,5-bisphosphate or whether it extended to hydrolyis of the other inositol lipids. For investigations into the precise substrate requirements in intact cells for phospholipase C, it is more appropriate to analyze products rather than substrates (see below).

Analysis of [^{3}H]Inositol-Labeled Phospholipids as Deacylation Products on High-Performance Liquid Chromatography (HPLC)

By using [^{3}H]inositol radiolabeling, the quantitative analysis of inositol lipid changes can be made more precise and can give accurate comparisons between different inositol lipids, which label to different specific activities with ^{32}P. The lipids are subsequently analyzed by chemical removal of the fatty acid chains and then resolved as glycerophosphoinositol phosphate species on high-performance liquid anion-exchange chromatography.[6]

[4] O.-B. Tysnes, G. M. Aarbakke, A. J. M. Verhoeven, and H. Holmsen, *Thromb. Res.* **40,** 329 (1985).

[5] S. I. Patterson, T. R. Jackson, M. Dreher, and M. R. Hanley, *in* "Neuropeptides: A Methodology" (G. Fink and A. J. Harmar, eds.), p. 245. Wiley, Chichester, New York, 1989.

[6] L. Stephens, P. T. Hawkins, and C. P.Downes, *Biochem. J.* **259,** 267 (1989).

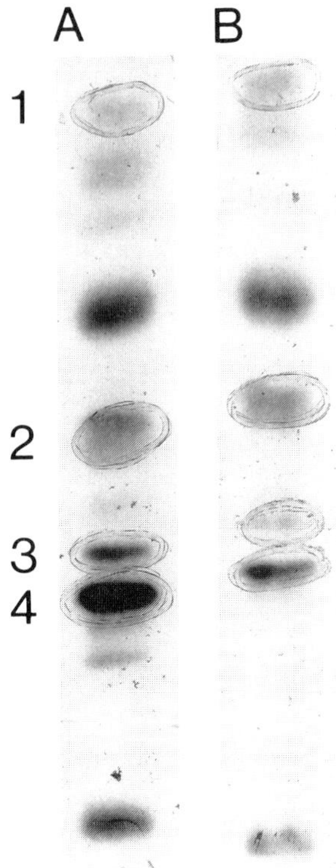

FIG. 1. Representative autoradiograph of ^{32}P-labeled phospholipids separated by high-performance thin-layer chromatography (HPTLC). Samples from ^{32}P-labeled NG115-401L neuroblastoma × glioma cells were prepared and chromatographed as described in the text. The plate was exposed for autoradiography for 20 hr. The positions of cochromatographed unlabeled phospholipid standards are numbered as follows: 1, phosphatidic acid; 2, phosphatidylinositol; 3, phosphatidylinositol 4-phosphate; 4, phosphatidylinositol 4,5-bisphosphate. (Lane A) Basal pattern of labeled phospholipids. (Lane B) Pattern of labeled phospholipids following bradykinin stimulation (maximal dose) for 15 sec. The autoradiograph has been marked with a soft pencil to imprint the migration positions of labeled lipids on the underlying chromatogram. These samples can then be scraped from the plate for Cerenkov counting.

Cells are labeled with [^3H]inositol (1–10 μCi/ml) for up to 3 days for actively growing monolayers. If cells are labeled to isotopic steady state, then the specific activity of the labeled lipids will not change over short activation periods, which permits interpretations about changes in absolute levels of the lipid species and gives an accurate indication of their true proportions relative to each other.

Cells are quenched with 2 ml ice-cold 20% trichloroacetic acid, and the

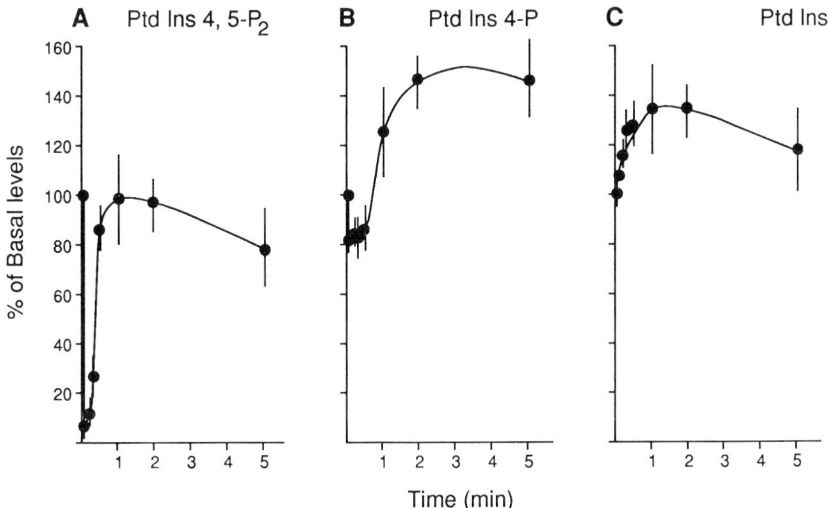

FIG. 2. Time courses of changes in radiochemical levels of ^{32}P-labeled inositol phospholipids in NG115-401L neuroblastoma × glioma cells following stimulation with a maximal dose of bradykinin. Results have been taken from HPTLC chromatograms, such as that shown in Fig. 1, using a thin-layer linear analyzer to determine levels of radioactivity. The resulting data have been normalized to percentages of unstimulated basal levels for each lipid and are plotted as means ± S.E.M. for four independent experiments. (A) Phosphatidylinositol 4,5-bisphosphate; (B) phosphatidylinositol 4-phosphate; (C) phosphatidylinositol.

lipid-containing precipitate is centrifuged and washed with 1 ml of 5% trichloroacetic acid/2 m*M* EDTA. The washed pellet is extracted by mixing with 1 ml of chloroform/methanol/10 m*M* EDTA, 0.1 m*M* inositol, 0.1 *M* HCl (5 : 9 : 4, v/v/v). Two phases are obtained by adding 0.37 ml each of chloroform and 10 m*M* EDTA, 0.1 m*M* inositol, 0.1 *M* HCl, and the upper phase is removed after a brief centrifugation. The lower phase and interfacial material are washed with 1.33 ml of synthetic upper phase, and the resulting upper phase is again discarded. The presence of EDTA during the lipid extraction is essential to remove divalent cations (e.g., Mg^{2+}, Ca^{2+}) which would interfere with the subsequent deacylation of polyphosphoinositides. The lower phase and particulate material are dried *in vacuo* and taken up in 2 ml methylamine deacylation reagent [57.66 ml methylamine in water (25–30%), 61.57 ml methanol, 15.41 ml *n*-butanol], which is heated at 53° for 15 min. The samples are dried in a vacuum centrifuge, redissolved in 2 ml water and 2.4 ml *n*-butanol/light petroleum (bp 40°–60°)/ethyl formate (20 : 40 : 1, v/v), and mixed well. The phases are split with a brief centrifugation, and the upper phase is removed. The

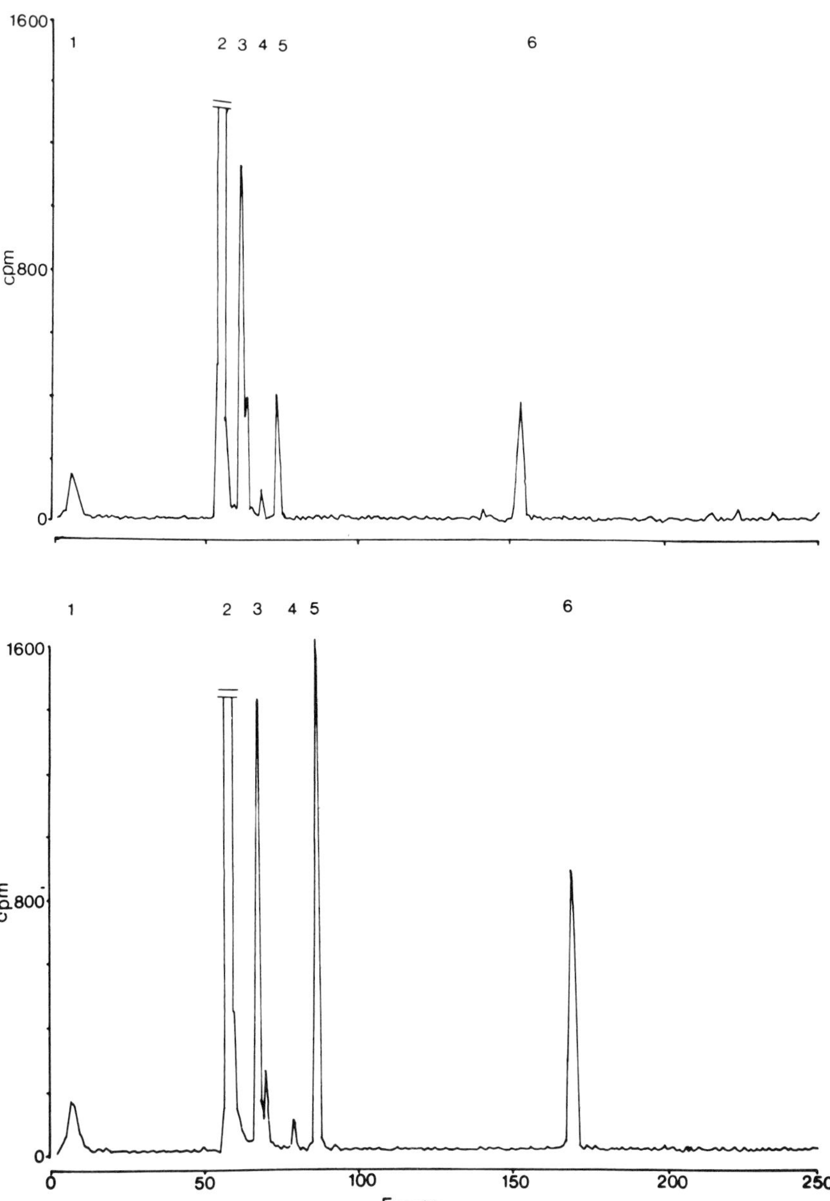

FIG. 3. Representative HPLC analysis of [^{3}H]inositol-labeled deacylated phospholipids. Samples from [^{3}H]inositol-labeled (20 μCi/ml, 24 hr) NG115-401L neuroblastoma × glioma cells were prepared and chromatographed as described in the text. The upper trace shows the pattern of radiolabeled products following stimulation with bradykinin (maximal dose)

lower phase is washed with 2.4 ml of the organic solvent mixture, and the lower phase is retained for analysis. An aliquot is taken for counting to assess yields, and the remainder can be stored at $-20°$ or, alternatively, freeze-dried and resuspended in 1 ml HPLC-grade water.

Samples are fractionated on HPLC using a Partisphere SAX column [Whatman (Clifton, NJ), 5 μm, 12.5 cm] with a solvent flow rate of 1 ml/min. The stepwise gradient elution uses HPLC-grade water (reservoir A) and 1.25 *M* ammonium phosphate/phosphoric acid, pH 3.8 (reservoir B) as follows: A alone, 5 min; 0–12% B, 40 min; 12–20% B, 7 min; 20–100% B, 12 min; maintain at 100% B for 12 min. The preparation of the reservoir B buffer merits special comment. Ammonium phosphate at high concentrations exhibits solubility difficulties at low pH; thus, 1 liter of buffer is prepared by dissolving the ammonium phosphate in 900 ml HPLC-grade water and then adding approximately 100 ml 85% phosphoric acid to adjust the pH. After completion of the HPLC run, the column is washed by a rapid linear gradient (5 min) back to the starting conditions, followed by a 20-min elution with water. Samples are collected every 20 sec, mixed with 1 ml methanol/water (1 : 1, v/v) to avoid precipitation of the high salt, and 4 ml of high-salt compatible scintillant (Quickszint, Zinsser) is added for counting. A representative chromatographic run is shown in Fig. 3. Internal standards labeled with ^{32}P can be prepared by deacylation of suitably labeled phospholipids fractionated on HPTLC as above. The changes observed upon phospholipase C activation using [^{3}H]inositol labeling can be markedly less than those observed with ^{32}P-labeling, which suggests that lipid phosphomonoesterases may contribute to the larger apparent reductions from resting levels in ^{32}P-labeled inositol lipids.

It is important to note that the higher resolving power of the HPLC fractionation of deacylated inositol lipids has been used to identify and examine the regulation of a novel inositol phosphospholipid, phosphatidylinositol 3-phosphate.[6,7] Although levels of phosphatidylinositol 3-phosphate decline in some cells with a time course paralleling activation of

[7] M. Whitman, C. P. Downes, M. Keeler, T. Keller, and L. Cantley, *Nature (London)* **332,** 644 (1988).

for 15 sec, which is to be compared to the basal pattern of radiolabeled products (control) shown in the lower trace. The numbered standards are as follows: 1, inositol; 2, glycerol inositol; 3, glycerophosphoinositol; 4, glycerophosphoinositol 3-phosphate; 5, glycerophosphoinositol 4-phosphate; 6, glycerophosphoinositol 4,5-bisphosphate.

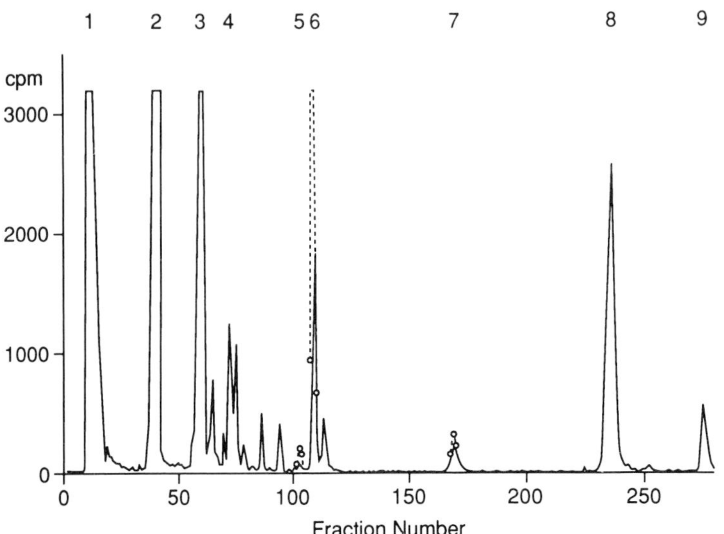

FIG. 4. Representative high-performance liquid chromatography (HPLC) analysis of [^{3}H]inositol-labeled acid-soluble metabolites. Samples from [^{3}H]inositol-labeled (20 μCi/ml, 24 hr) NG115-401L neuroblastoma × glioma cells were prepared and chromatographed as described in the text. The solid line indicates the basal levels of [^{3}H]inositol-labeled compounds, and the dotted line illustrates the stimulated changes in the inositol tris- and tetrakisphosphate fractions following bradykinin stimulation (15 sec, maximal dose). The positions of standards are as follows: 1, inositol; 2, glycerophosphoinositol; 3, inositol monophosphates; 4, inositol bisphosphates; 5, inositol 1,3,4-trisphosphate; 6, inositol 1,4,5-trisphosphate; 7, inositol tetrakisphosphates; 8, inositol pentakisphosphates; 9, inositol hexakisphosphate (phytic acid).

phospholipase C,[6] this lipid has not proved to be a substrate for any purified inositol–lipid phospholipase C[8]; however, it is a substrate for a highly specific phosphatidylinositol 3-phosphate phosphomonoesterase.[9] This result calls attention to the need to analyze the inositol products of phospholipase C hydrolysis in order to address substrate specificity.

Analysis of [^{3}H]Inositol-Labeled Inositol Phosphates by High-Performance Liquid Chromatography

The most frequently used technique for analyzing quantitatively the exact products of phospholipase C attack on inositol lipids is fractionation

[8] L. A. Serunian, M. J. Haber, T. Fukui, J. W. Kim, S. G. Rhee, J. M. Lowenstein, and L. C. Cantley, *J. Biol.Chem.* **264,** 17809 (1989).

[9] D. Lips and P. W. Majerus, *J. Biol. Chem.* **264,** 19911 (1989).

of the labeled water-soluble [^{3}H]inositol metabolites by HPLC. Cells are labeled with [^{3}H]inositol to isotopic steady state and quenched with 20% trichloroacetic acid as before. However, for analysis of the labeled water-soluble products, the extract is centrifuged and the supernatant is taken for analysis. The pellet is reextracted with 5% trichloroacetic acid/2 m*M* EDTA which is combined with the previous supernatant. The acid is removed from the combined extracts by addition of 9 ml of 1,1,2-trichlorotrifluoroethane (Freon)/tri-*n*-octylamine (1 : 1, v/v) and vigorous mixing.[10] The upper aqueous phase is removed and can be stored at $-20°$.

Samples are fractionated by HPLC on a Partisil SAX column (Jones Chromatography, 10 μm, 25 cm) using a flow rate of 1.25 ml/min. The stepwise gradient elution uses HPLC-grade water (reservoir A) and 3.5 *M* ammonium formate/phosphoric acid, pH 3.7 (reservoir B): A alone, 5 min; 0–20% B, 10 min; 20–50% B, 45 min; 50–100% B, 20 min; 100% B, 6 min. The preparation of reservoir buffer B should follow the same guidelines as noted in the preceding section for the preparation of ammonium phosphate buffer. The column is washed by a rapid return to starting conditions (5 min, linear gradient), followed by a 20-min rinse with reservoir A (water). Samples are collected every 20 sec, diluted with 1 ml methanol/water (1 : 1, v/v), and counted by the addition of 4 ml high-salt compatible scintillant. A representative analysis is shown in Fig. 4.

As with deacylated lipids, inositol phosphates may be provisionally assigned using radiolabeled internal standards, many of which are commercially available. However, the formal problems of inositol phosphate identification are now potentially quite elaborate, given the recognized complexity of inositol phosphate species.[11] Indeed, the predicted major product of phosphatidylinositol 4,5-bisphosphate breakdown, inositol 1,4,5-trisphosphate, can occur in an HPLC fraction with as many as five other inositol trisphosphate species.[12] However, on activation of intracellular phospholipase C, the only species of inositol trisphosphate in the "inositol 1,4,5-trisphosphate" HPLC fraction which exhibits increases in levels is authentic inositol 1,4,5-trisphosphate. This conclusion has been based in part on conversion of the labeled inositol polyphosphate to characteristic noncyclic polyols by sequential periodate oxidation, borohydride reduction, and dephosphorylation, which has been described in detail.[13] The polyol characteristic of D-inositol 1,4,5-trisphosphate is D-iditol.[13]

[10] P. T. Hawkins, L. R. Stephens, and C. P. Downes, *Biochem. J.* **238,** 507 (1986).

[11] S. B. Shears, *Biochem. J.* **260,** 313 (1989).

[12] L. R. Stephens, P. T. Hawkins, and C. P. Downes, *Biochem. J.* **262,** 727 (1989).

[13] L. R. Stephens, P. T. Hawkins, N. G. Carter, S. Chahwala, A. J. Morris, A. D. Whetton, and C. P. Downes, *Biochem. J.* **249,** 271 (1988).

Radioreceptor Assay for Inositol 1,4,5-Trisphosphate

The complications in the analysis of radiolabeled inositol lipids and phosphates, as well as the need to obtain absolute chemical levels of the biologically active second messenger product, inositol 1,4,5-trisphosphate, has led to the introduction of novel measurement techniques. The ability of extracted unlabeled inositol 1,4,5-trisphosphate to compete for [^{3}H]inositol 1,4,5-trisphosphate in binding to its intracellular receptor has been developed into a simple radioreceptor assay.[14,15]

[14] D. S. Bredt, R. J. Mourmey, and S. H. Snyder, *Biochem. Biophys. Res. Commun.* **159,** 976 (1989).

[15] R. A. J. Challis, E. R. Chilvers, A. L. Willcocks, and S. R. Nahorski, *Biochem. J.* **265,** 421 (1990).

[14] Chromatographic Analysis of Phospholipase Reaction Products

By MERLE L. BLANK and FRED SNYDER

Introduction

This chapter describes the current methodology available for investigating lipid products formed during phospholipase catalysis of phospholipids in both analytical and biochemical studies. Products expected from hydrolysis of phospholipids and lysophospholipids by phospholipases and lysophospholipases are indicated in Table I. However, the presence of multiple lipases in a cell or crude enzyme preparation can result in the production of other hydrolytic products via subsequent hydrolysis of the primary products. The analysis of hydrolytic products from phospholipase reactions are simplified when the experiments are conducted with labeled substrates of known structure and specific radioactivities. However, interpretation of phospholipase activities requires qualitative and quantitative analyses of catabolic products. Some form of chromatography is normally used for separation of the enzymatic products, with thin-layer chromatography (TLC) being the most widely utilized system. The variety of solvent systems used for the TLC separation of products from a single phospholipase activity are often as diverse as the number of laboratories conducting the research. Because of the nearly insurmountable task of listing all of the specific TLC solvent systems that have been used for separation of

TABLE I
HYDROLYTIC PRODUCTS PRODUCED FROM GLYCEROPHOSPHATIDES BY PHOSPHOLIPASE AND LYSOPHOSPHOLIPASE ENZYMES

Phospholipase	Products formed with substrates: 1,2-Diacylphospholipids	Products formed with substrates: 1-Ether-2-acylphospholipids[a]
A_1	Fatty acid + 1-lyso-2-acylphospholipid	No reaction
A_2	Fatty acid + 1-acyl-2-lysophospholipid	Fatty acid + 1-ether-2-lysophospholipid
C	1,2-Diacylglycerol + phospho base	1-Ether-2-acylglycerol + phospho base
D	1,2-Diacylglycerophosphate + base group	1-Ether-2-acylglycerophosphate + base group
Lysophospholipase	**Products formed with substrates: 1-Acyl-2-lyso- or 1-Lyso-2-acylphospholipids**	**Products formed with substrates: 1-Ether-2-lysophospholipids[a]**
A_1	Fatty acid + glycerophospho base	No reacion
A_2	Fatty acid + glycerophospho base	No reaction
D	No reaction	1-Ether-2-lysoglycerophosphate + base group

[a] The term ether represents either an alkyl or an 1-alkenyl group.

phospholipase reaction products, we attempt to discuss only those (albeit without some of the minor modifications) that have had the most general usage. In using this approach it is possible that we may inadvertently omit TLC systems that might be superior for specific applications other than those described. When applicable to the analyses of phospholipase reaction mixtures, chromatographic methods such as high-performance liquid chromatography (HPLC) and gas–liquid chromatography (GLC) are also discussed.

Analysis of Products from Phospholipases A_1 and A_2

Although phospholipase A_1 hydrolyzes acyl groups from the *sn*-1 position of diradylglycerophosphatides and phospholipase A_2 the acyl groups from the *sn*-2 position of diradylglycerophosphatides, the chromatographic analysis of products (free fatty acids and lysophospholipids) derived from these two phospholipases is essentially the same. Differentiation of the A_1 and A_2 phospholipase enzymes as well as the possible involvement of a subsequent lysophospholipase activity is usually accomplished by selection of phospholipid substrates that have radiolabeled acyl groups at either the *sn*-1 or *sn*-2 position. Some of the phospholipids containing radiolabeled *sn*-2 acyl groups are available from commercial sources, and others can be prepared from *Escherichia coli* labeled lipids[1] or by using rat liver microsomes as an enzyme source to acylate appropriate lysophospholipid acceptor molecules.[2] Various modifications of the latter procedure have been published, including one describing conditions to optimize the yield and specific radioactivity of the final product.[3] By using the specifically labeled phospholipid substrates, together with careful analysis of the hydrolytic products, researchers have often been able to discern the type(s) of phospholipase activity being studied (see Ref. 4, for example). Addition of unlabeled lysophosphatidylcholine to assays for phospholipase A_1 and A_2 activity has been used to minimize interference by lysophospholipases.[5,6]

Under ideal conditions, only free fatty acids and lysophospholipids should be produced by phospholipase A_1 and A_2 reactions, and these products should be released in stoichiometric amounts. There have been various solvent systems reported for the TLC separation of free fatty acids from lysophospholipids and the unhydrolyzed substrate; however, a solvent commonly used for this separation on silica gel G plates is hexane/diethyl ether/glacial acetic acid (80 : 20 : 1, v/v). Free fatty acids have an R_f value of about 0.25, and phospholipids remain at the origin in this solvent system.[7,8] The phospholipid substrate can be separated from the lysophospholipid product by TLC (usually on layers of silica gel H) of the samples in a more polar solvent system such as chloroform/methanol/glacial acetic acid/water (50 : 25 : 8 : 4, v/v) or chloroform/methanol/ammonium hydroxide (60 : 35 : 8, v/v).[8] In the acidic polar solvent system, free fatty acids and other neutral lipids migrate together at (or very near) the solvent front, but in the system containing ammonium hydroxide, free fatty acids have an R_f value only slightly higher than that of phosphatidylethanolamine. Neutral, polar solvent systems with a chloroform/methanol ratio of about 2 : 1 (v/v) and a water content of 4 to 6% are also used for the TLC separation of phospholipids. Increasing or decreasing the water content of the TLC solvent systems normally results in a corre-

[1] P. Elsbach and J. Weiss, this volume [2].

[2] A. F. Robertson and W. E. M. Lands, *Biochemistry* **1,** 804 (1962).

[3] A. M. Connor, P. D. Brimble, and P. C. Choy, *Prep. Biochem.* **11,** 91 (1981).

[4] L. Kaplan-Harris, J. Weiss, C. Mooney, S. Beckerdite-Quagliata, and P. Elsbach, *J. Lipid Res.* **21,** 617 (1980).

[5] S. W. Tam, R. Y. K. Man, and P. C. Choy, *Can. J. Biochem. Cell Biol.* **62,** 1269 (1984).

[6] Y.-Z. Cao, K. O, P. C. Choy, and A. C. Chan, *Biochem. J.* **247,** 135 (1987).

[7] M. L. Blank and F. Snyder, *in* "Analysis of Lipids and Lipoproteins" (E. G. Perkins, ed.), Chap. 4. The American Oil Chemists' Society, Champaign, Illinois, 1975.

[8] F. Snyder, *J. Chromatogr.* **82,** 7 (1973).

sponding change in R_f values of the phospholipids. For an example of modifications that can be made in these TLC solvent systems to obtain the desired separation of phosphatidylcholine, phosphatidylethanolamine, phosphatidylinositol, and their lyso derivatives, readers are referred to a recent paper by Balsinde *et al.*[9]

If it has been established that determination of released fatty acids alone will give an accurate measurement of the phospholipase activity being investigated,[10] then the nonchromatographic extraction method of Dole,[11] as modified for interfering phospholipids,[12,13] offers an attractive alternative to the more time-consuming analysis of free fatty acids by TLC. However, eicosanoid metabolites that may be derived from arachidonic acid (in some systems) would be more easily detected by using chromatographic methods. Most eicosanoid metabolites produced by either cyclooxygenase or lipoxygenase activities are more polar than free fatty acids.

Analysis of Products from Phospholipase C

Diradylglycerols and phospho bases are the two products produced by phospholipase C hydrolysis. The diradylglycerols are soluble in the chloroform phase of a Bligh and Dyer extraction,[14] whereas the phospho base portion remains in the upper methanol–water layer. Phospholipid substrates with a radiolabeled atom located in the diradylglycerol and/or the phospho base portion of the molecule are usually used in assays for phospholipase C activity. Location of the radiolabel in the substrate determines whether the analysis of products will involve chromatographic separation of diradylglycerols and/or the water-soluble phospho base moieties. The lipids that have received the most attention as substrates for phospholipase C are phosphatidylcholine and the inositol-containing phospholipids.

Water-soluble inositol phosphates (IP, IP_2, and IP_3) produced by phospholipase C hydrolysis of inositol-labeled phospholipids are usually separated by open-column ion-exchange chromatography.[15] However, high-

[9] J. Balsinde, E. Diez, A. Schuller, and F. Mollinedo, *J. Biol. Chem.* **263,** 1929 (1988).

[10] H. van den Bosch, A. J. Aarsman, and L. L. M. Van Deenen, *Biochim. Biophys. Acta* **348,** 197 (1974).

[11] V. P. Dole, *J. Clin. Invest.* **35,** 150 (1956).

[12] S. A. Ibrahim, *Biochim. Biophys. Acta* **137,** 413 (1967).

[13] K. M. M. Shakir, *Anal. Biochem.* **114,** 64 (1981).

[14] E. G. Bligh and W. J. Dyer, *Can. J. Biochem. Physiol.* **37,** 911 (1959).

[15] M. J. Berridge, R. M. C. Dawson, C. P. Downes, J. P. Heslop, and R. F. Irvine, *Biochem. J.* **212,** 473 (1983).

performance liquid chromatography has also been utilized for these separations.[16–18] The water-soluble products from phospholipase C (and D) hydrolysis of choline-labeled phospholipids can be analyzed by TLC on plates coated with silica gel G. Most solvent systems used for the separations are modifications of the solvent mixture described by Yavin.[19] One of the more popular solvent systems for TLC separation of water-soluble choline metabolites appears to be methanol/0.9% NaCl/NH_4OH (10 : 10 : 1, v/v).[20] A solvent system of methanol/0.5% NaCl/acetic acid (20 : 20 : 1, v/v) has also been used with silica gel H plates for these separations.[21] Silica gel G-coated TLC plates and a solvent mixture of methanol/0.5% NaCl/NH_4OH (10 : 10 : 1, v/v) has been described for the resolution of water-soluble ethanolamine products.[22]

Diradylglycerols are the other products produced by phospholipase C hydrolysis of diradylglycerol phosphatides. However, it should be noted that in the presence of ATP and an active diradylglycerol kinase the diradylglycerols may be converted to phosphatidic acid and, conversely, can arise by dephosphorylation of phosphatidic acid that was originally produced by a phospholipase D reaction. As with the separation of phospholipids, there have been numerous TLC solvent systems reported for the analysis of labeled diradylglycerols. One of the commonly used TLC solvent mixtures for resolving labeled diradylglycerols from free fatty acids, monoacylglycerols, triacylglycerols, and phospholipids consists of hexane (or petroleum ether)/diethyl ether/1% glacial acetic acid with ratios of hexane to diethyl ether ranging from 4 : 1 to 3 : 2 (v/v). The R_f value for diradylglycerols increases as the amount of diethyl ether is increased in the solvent systems. If sufficient amounts of radioactivity are available, the diradylglycerols can be converted to benzoate derivatives which can be further resolved as subclasses (1-alkenylacyl, alkylacyl, and diacyl types) by either TLC[23] or HPLC.[24] Isolated subclasses of the diradylglycerobenzoates can then be separated into molecular species by re-

[16] J. L. Meek and F. Nicoletti, *J. Chromatogr.* **351,** 303 (1986).

[17] D. Portilla, J. Morrissey, and A. R. Morrison, *J. Clin. Invest.* **81,** 1896 (1988).

[18] D. Portilla, M. Mordhurst, W. Bertrand, and A. R. Morrison, *Biochem. Biophys. Res. Commun.* **153,** 454 (1988).

[19] E. Yavin, *J. Biol. Chem.* **251,** 1392 (1976).

[20] D. E. Vance, E. M. Trip, and H. B. Paddon, *J. Biol. Chem.* **255,** 1064 (1980).

[21] R. A. Wolf and R. W. Gross, *J. Biol. Chem.* **260,** 7295 (1985).

[22] Z. Kiss and W. B. Anderson, *J. Biol. Chem.* **264,** 1483 (1989).

[23] M. L. Blank, M. Robinson, and F. Snyder, *in* "Platelet-Activating Factor and Related Lipid Mediators" (F. Snyder, ed.), Chap. 2. Plenum, New York, 1987.

[24] M. L. Blank, E. A. Cress, and F. Snyder, *J. Chromatogr.* **392,** 421 (1987).

versed-phase HPLC.[25] Methodology to accomplish the same type of analyses with other derivatives of the diradylglycerols, acetate derivatives, for example,[26] also exists.

It is often important to determine the amount of diradylglycerols produced as a result of stimulation of phospholipase C activity in cells. One sensitive method to obtain this information is through conversion of the unlabeled diracylglycerol to labeled phosphatidic acid using diglyceride kinase and [γ-^{32}P]ATP with a known specific radioactivity. The labeled phosphatidic acid formed is then separated by TLC and the radioactivity determined as a measure of the amount of diradylglycerols originally present.[27,28] After conversion of the [^{32}P]phosphatidic acid to the dimethyl derivative with diazomethane, it can be separated into various molecular species by both TLC and HPLC.[29,30] Amounts of unlabeled diradylglycerols have also been quantitated by photodensitometry of charred TLC plates[31] and by GLC of *tert*-butyldimethylsilyl derivatives of the isolated diradylglycerols.[32] The latter procedure yields information about the molecular species of diradylglycerols present. Benzoate derivatives of diradylglycerols have also been utilized to determine mass levels of alkylacyl- and diacylglycerols in human neutrophils by normal-phase HPLC.[33]

Analysis of Products from Phospholipase D

Phosphatidic acid and a free base group are the two products expected from phospholipase D hydrolysis of diradylglycerophospho base lipids. However, recent publications have indicated that some biological systems contain very active phosphohydrolases that produce diradylglycerols from the phosphatidic acid originally formed by phospholipase D hydrolysis of the intact phospholipid (see Refs. 33–39 for examples). This means that detailed analyses are needed to estimate the relative indirect contribution that phospholipase D makes to the formation of diradylglycerols compared

[25] M. L. Blank, M. Robinson, V. Fitzgerald, and F. Snyder, *J. Chromatogr.* **298,** 473 (1984).

[26] Y. Nakagawa and L. A. Horrocks, *J. Lipid Res.* **24,** 1268 (1983).

[27] D. A. Kennerly, C. W. Parker, and T. J. Sullivan, *Anal. Biochem.* **98,** 123 (1979).

[28] J. Preiss, C. R. Loomis, W. R. Bishop, R. Stein, J. E. Niedel, and R. M. Bell, *J. Biol. Chem.* **261,** 8597 (1986).

[29] D. A. Kennerly, *J. Chromatogr.* **363,** 462 (1986).

[30] D. A. Kennerly, *J. Biol. Chem.* **262,** 16305 (1987).

[31] S. Rittenhouse-Simmons, *J. Clin. Invest.* **63,** 580 (1979).

[32] M. S. Pessin and D. M. Raben, *J. Biol. Chem.* **264,** 8729 (1989).

[33] D. E. Agwu, L. C. McPhail, M. C. Chabot, L. W. Daniel, R. L. Wykle, and C. E. McCall, *J. Biol. Chem.* **264,** 1405 (1989).

to the amount produced directly by phospholipase C hydrolysis.[33-39] The most direct method to detect the presence of phospholipase D activity is to include ethanol in the assay system[34]; phosphatidylethanol formed via the base exchange properties of phospholipase D[40,41] indicates the extent to which this enzymatic activity is present. Phosphatidylethanol, phosphatidic acid, and diradylglycerols can then be separated from each other and other classes of lipids by TLC.[34]

As mentioned previously it is generally prudent to use substrates labeled in both the diradylglycerol and the phosphobase portions of the molecule in assays for phospholipase D activity. Thus, both the water-soluble and chloroform-soluble products can then be analyzed by the chromatographic methods discussed in the previous section on phospholipase C. The presence of labeled phosphatidic acid is usually indicated by radioactive material that migrates above phosphatidylethanolamine on TLC silica gel layers developed in chloroform/methanol/glacial acetic acid/water (50 : 25 : 8 : 4, v/v, or 9 : 1 : 1 : 0, v/v) and below phosphatidylcholine in a solvent system of chloroform/methanol/ammonium hydroxide (60 : 35 : 8, v/v).[8] The amount of radioactivity associated with phosphatidic acid should be nearly equal in both the acidic and basic solvent systems. Two alternative approaches are as follows: (1) isolating the radioactive material from TLC plates developed in the acidic solvent system and rechromatographing it in the basic solvent system or (2) utilizing two-dimensional TLC development with an acidic solvent in the first dimension and a basic solvent system in the second direction (90° from the first). As with any TLC system, unlabeled standards should be cochromatographed with the labeled materials because of day-to-day and laboratory-to-laboratory variations in TLC R_f values.

Analysis of Products from Lysophospholipases

As the term lysophospholipase suggests, these enzymes utilize only lysophospholipids as substrates. The most well-known lysophospholi-

[34] J-K. Pai, M. I. Siegel, R. W. Egan, and M. M. Billah, *Biochem. Biophys. Res. Commun.* **150,** 355 (1988).

[35] M. C. Cabot, C. J. Welsh, H-T. Cao, and H. Chabbott, *FEBS Lett.* **233,** 153 (1988).

[36] S. B. Bocckino, P. F. Blackmore, P. B. Wilson, and J. H. Exton, *J. Biol. Chem.* **262,** 15309 (1987).

[37] T. W. Martin, *Biochim. Biophys. Acta* **962,** 282 (1988).

[38] M. M. Billah, S. Eckel, T. J. Mullmann, R. W. Egan, and M. I. Siegel, *J. Biol. Chem.* **264,** 17069 (1989).

[39] A. P. Truett, R. Snyderman, and J. J. Murray, *Biochem. J.* **260,** 909 (1989).

[40] S. Yang, S. Freer, and A. Benson, *J. Biol. Chem.* **242,** 477 (1967).

[41] R. Dawson, *Biochem. J.* **102,** 205 (1967).

pases are those that hydrolyze acyl groups from the *sn*-1 and/or *sn*-2 positions of acyllysoglycerophosphatides (see Section V of this volume). In this instance the products of hydrolysis are free fatty acids and the water-soluble glycerophosphobase portions of the substrate. Free fatty acids can be analyzed by TLC as previously described for phospholipases A_1 and A_2. The glycerophosphobases can be separated from other labeled water-soluble metabolites by the same TLC methods described for the analysis of water-soluble products in the section on phospholipase C of this chapter.

Lysophospholipase D is unique in that it hydrolyzes the base group solely from ether-linked lysophosphatides. The alkyl- or 1-alkenyllysoglycerophosphatides produce a free base and the corresponding radyllysoglycerophosphate.[42] Alkyllysoglycerophosphatidylethanolamine,[43] alkenyllysoglycerophosphatidylethanolamine,[43] and alkyllysoglycerophosphatidylcholine[43-46] have been shown to be hydrolyzed by this enzyme, whereas the corresponding acyllyso analogs of these lipids are not substrates for lysophospholipase D.[45] Because active phosphohydrolases are also present in the tissue enzyme preparations, the major product formed in the absence of inhibitors is alkyl- or 1-alkenylglycerols. The phosphohydrolases can be inhibited by including sodium fluoride[43] or sodium vanadate[46] in the assays, so that the major product is the ether-linked analog of lysophosphatidic acid. TLC systems used for the resolution of the products formed by lysophospholipase D (and phosphohydrolase) are described in detail in the original references.[43-46]

Acknowledgments

This work was supported by the Office of Energy Research, U.S. Department of Energy (Contract No. DE-AC05-760R00033), the American Cancer Society (Grant BE-26U), the National Cancer Institute (Grant CA-41642-05), and the National Heart, Lung, and Blood Institute (Grant HL-27109-10).

[42] R. L. Wykle and J. C. Strum, this volume [58].

[43] R. L. Wykle and J. M. Schremmer, *J. Biol. Chem.* **249,** 1742 (1974).

[44] R. L. Wykle, W. F. Kraemer, and J. M. Schremmer, *Arch. Biochem. Biophys.* **184,** 149 (1977).

[45] R. L. Wykle, W. F. Kraemer, and J. M. Schremmer, *Biochim. Biophys. Acta* **619,** 58 (1980).

[46] T. Kawasaki and F. Snyder, *Biochim. Biophys. Acta* **920,** 85 (1987).

[15] Assays for Measuring Arachidonic Acid Release from Phospholipids

By FLOYD H. CHILTON

Introduction

Arachidonic acid and its oxygenated metabolites [prostaglandins, thromboxanes, leukotrienes, hydroxyeicosatetraenoic acids (HETEs)] have been shown to play a pivotal role in regulating a number of important inflammatory responses. The major stores of arachidonate in mammalian cells are esterified at the *sn*-2 position of glycerophospholipids.[1] During cell activation, phospholipids are acted on by different phospholipases which release arachidonic acid. This can be accomplished by phospholipase A_2, which directly cleaves arachidonic acid from the *sn*-2 position of the molecule.[1] Alternative enzymes such as phospholipase C or phospholipase D remove the phospho base or the base moiety, respectively, to generate metabolites (diglycerides and phosphatidic acid) that are subsequently acted on by lipases to mobilize arachidonate.[1-3] Once arachidonic acid is cleaved, it is metabolized via the lipoxygenase or cyclooxygenase pathway to produce hydroxylated products or is rapidly reacylated into complex lipids.[4]

Until recently, the phosphoglyceride sources of arachidonic acid for these enzymatic systems were typically identified by studies in which cells were prelabeled with radioactive arachidonic acid immediately before stimulation. Although these experiments have provided important information on precursor pools of arachidonate, they have often been based on assumptions that were not strictly valid. It is crucial in this type of experiment to take into account the sizes of different endogenous pools of arachidonate and the rates of incorporation of exogenously provided arachidonic acid into these pools. It is frequently assumed that exogenously provided arachidonic acid has reached a state of equilibrium among all phosphoglyceride molecular species after some relatively short period of time. In many cases, it is also assumed that the loss of labeled arachidonate from a phospholipid correlates with that phospholipid being a source

[1] R. F. Irvine, *Biochem. J.* **204,** 3 (1982).

[2] L. W. Daniel, M. Waite, and R. L. Wykle, *J. Biol. Chem.* **261,** 1928 (1986).

[3] J-K Pai, M. I. Siegel, R. W. Egan, and M. M. Billah, *J. Biol. Chem.* **263,** 12472 (1988).

[4] P. Needleman, J. Turk, B. A. Jakschik, A. R. Morrison, and J. B. Lefkowith, *Annu. Rev. Biochem.* **55,** 59 (1986).

of arachidonate for eicosanoids. In studies where cells have been labeled for short periods of time, many of the aforementioned problems have led to major discrepancies between studies using the data obtained by measuring labeled arachidonate and those using data obtained by determining the mass of arachidonate, for example, by gas chromatography–mass spectrometry (GC/MS).

Several methods to label cells with arachidonic acid and to quantify arachidonate release from phospholipids in the human neutrophil have been outlined in this chapter. Many of the assumptions and problems often encountered when using these methods have been emphasized. These include the effects of arachidonic acid concentration and of incubation time on the ability to obtain equilibrium labeling conditions, the loss of labeled arachidonate from phospholipids, and the formation of radiolabeled eicosanoids. Understanding the limitations of methods which use labeled arachidonate may provide insight into why there may be discrepancies in the literature between studies that measure the absolute mass of arachidonate and those which measure radiolabeled arachidonate to determine sources of arachidonate utilized for eicosanoid biosynthesis.

Experimental Procedures

Materials

[5,6,8,9,11,12,14,15-^{3}H]Arachidonic acid (83 Ci/mmol) was purchased from Du Pont–New England Nuclear (Boston, MA). All solvents were HPLC grade from Fisher (Fairlawn, NJ). Lipid standards used in thin-layer chromatography (TLC) were purchased from Avanti Polar Lipids (Birmingham, AL). Essentially fatty acid-free human serum albumin (HSA) was purchased from Sigma (St. Louis, MO). Hanks' balance salt solution (HBSS) was purchased from GIBCO (Grand Island, NY). *N*-(*tert*-Butyldimethylsilyl)-*N*-methyltrifluoroacetamide for making the *tert*-butyldimethylsilyl derivative was purchased from Aldrich (Milwaukee, WI). Octadeuterioarachidonic acid was obtained from Biomol Research Laboratories, Inc. (Plymouth Meeting, PA).

Methods

Preparation of Cells. Human neutrophils are prepared from venous blood obtained from healthy human donors immediately before the neutrophil isolation as previously described,[5] using the Percoll–plasma tech-

[5] C. Haglet, L. A. Guthrie, M. M. Lopaniac, D. B. Johnston, and P. M. Henson, *Am. J. Pathol.* **119,** 101 (1985).

nique. The cells are incubated with labeled arachidonic acid as described below.

Separation of Arachidonate-Containing Glycerophospholipids. In all experiments, lipids are extracted by the method of Bligh and Dyer.[6] Glycerolipid classes are separated by normal-phase HPLC using a modified method of Patton *et al.*[7] In this system, cellular glycerolipids are placed on an Ultrasphere-Si column (4.6 × 250 mm; Rainin Instrument Co., Woburn, MA) and eluted with 2-propanol/25 m*M* phosphate buffer (pH 7.0)/hexane/ethanol/acetic acid (245.0 : 15.0 : 183.5.5 : 50.0 : 0.3, v/v) at 1.0 ml/min for 5 min. After 5 min, the solvent is changed to 2-propanol/25 m*M* phosphate buffer (pH 7.0)/hexane/ethanol/acetic acid (245.0 : 25.0 : 183.5 : 40.0 : 0.3, v/v) over a 10-min period. The retention time of the glycerolipid classes in this system is very sensitive to the composition of phosphate buffer in the running solvent. Our method differs from the original method[7] in the fact that we utilize a gradient with phosphate buffer from 3.0 to 5.0%. This was found to be necessary to resolve early eluting glycerolipid classes such as the neutral lipids, phosphatidylethanolamine (PE), and phosphatidic acid.

Using a reversed-phase high-performance liquid chromatography (HPLC) system described by Patton *et al.*,[7] we have been able to resolve the major arachidonate-containing phosphoglycerides. This system has a major advantage over others which use diglyceride derivatives of phospholipids in that it requires much less sample preparation prior to chromatography. On the other hand, a disadvantage is that it does not allow for baseline resolution of all major phosphoglyceride molecular species. However, it provides for reasonable resolution of arachidonate-containing species. Prior to reversed-phase HPLC, glycerolipid classes are separated by normal-phase HPLC as described above. The various molecular species of choline-, ethanolamine-, and inositol-linked phosphoglycerides are then separated by reversed-phase HPLC using an Ultrasphere ODS column (4.6 × 250 mm, Rainin) eluted with methanol/water/acetonitrile (905 : 70 : 25, v/v) and 20 m*M* choline chloride at 2 ml/min at 40°. The running solvent is prepared by mixing choline chloride dissolved in water (filtered with a 0.45-μm filter, Millipore, Bedford, MA) with acetonitrile and methanol.

In some cases, purified choline and ethanolamine phospholipids are further separated into 1-acyl, 1-alkyl, and 1-alk-1-enyl subclasses as described by Nakagawa *et al.*[8] Briefly, the phospholipids are hydrolyzed to

[6] E. G. Bligh and W. J. Dyer, *Can. J. Biochem. Physiol.* **37,** 911 (1959).

[7] G. M. Patton, J. M. Fasulo, and S. J. Robins, *J. Lipid Res.* **29,** 190 (1982).

[8] Y. Nakagawa, Y. Ishima, and K. Waku, *Biochim. Biophys. Acta* **712,** 667 (1982).

diradylacylglycerols with phospholipase C (from *Bacillus cereus*) in 100 m*M* Tris-HCl buffer, pH 7.4. The diradylacylglycerols are then extracted and converted to 1,2-diradyl-3-acetylglycerols with acetic anhydride and pyridine for 3 hr at 37°. The 1,2-diradyl-3-acetylglycerols are then separated into 1-acyl, 1-alkyl, and 1-alk-1-enyl subclasses by TLC on layers of silica gel G (Analtech, Alexandria, VA) developed in benzene/hexane/ethyl ether (50 : 45 : 4, v/v). Each of the separate subclasses is then extracted from the silica gel.

Specific Activity of Arachidonate in Glycerophospholipids. Glycerophospholipid classes, subclasses, and molecular species are separated by the above procedures. The amount of label in each of the phosphoglycerides is determined by liquid scintillation counting (automatic external standard for quench correction). The arachidonate content of the fractions is characterized by fast atom bombardment–mass spectrometry. The mole quantity of unlabeled arachidonic acid is determined by hydrolyzing the phosphoglycerides with 2 *N* KOH in ethanol/water (75 : 25, v/v) for 15 min at 60° to liberate the free arachidonic acid. Octadeuterioarachidonic acid (500 ng) as an internal standard is added to the reaction mixture. After 15 min, additional water is added to a final volume of 3.0 ml, and the pH is adjusted to 3.0 with 6 *N* HCl. The free arachidonic acid is extracted with hexane/ether (1 : 1). The arachidonic acid is then converted to the *tert*-butyldimethylsilyl ester.[9] The *tert*-butyldimethylsilyl arachidonate is analyzed by GC/MS using selected ion recording techniques to monitor the M − 57 ion at m/z 361 to that of *tert*-butyldimethylsilyl octadeuterioarachidonate added as an internal standard (m/z 369) as described earlier.[10] The analysis is carried out on a Hewlett-Packard selective mass detection system (HP 5790, Palo Alto, CA) employing a cross-linked methylsilicone capillary column heated from 200° to 270° at a rate of 10°/min. Specific activities of each of the arachidonate-containing species are determined from these two measurements (radioactivity and mass) and expressed as nanocuries/nanomole.

Cellular Assays

Incorporation of Labeled Arachidonic Acid into Neutrophil Lipids: Effect of Arachidonic Acid Concentration. In labeling a cell such as the neutrophil with arachidonic acid, we have found that there are at least two important factors which must be taken into consideration. The first factor is the concentration of arachidonic acid presented to the cell. In theory,

[9] K. L. Clay, R. C. Murphy, J. L. Andres, J. Lynch, and P. M. Henson, *Biochem. Biophys. Res. Commun.* **121,** 815 (1984).

[10] F. H. Chilton and R. C. Murphy, *J. Biol. Chem.* **261,** 7771 (1986).

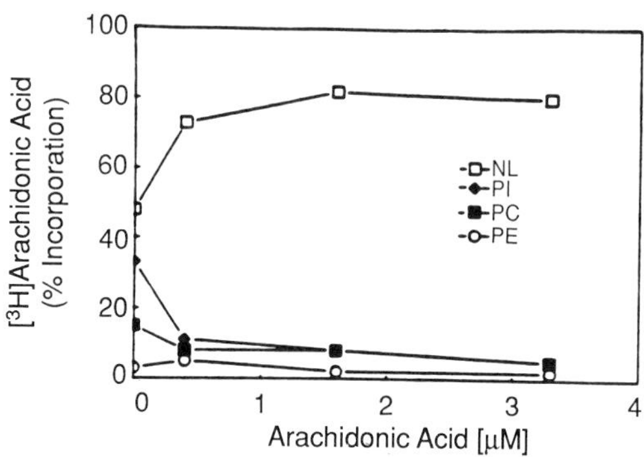

FIG. 1. Incorporation of [^{3}H]arachidonate into neutral lipid (NL), glycerophosphatidylinositol (PI), glycerophosphatidylcholine (PC), and glycerophosphatidylethanolamine (PE) classes of glycerophospholipids after a 5-min incubation with 0 to 3.3 μM arachidonate. The data are representative of three separate experiments. [From F. H. Chilton and R. C. Murphy, *Biochem. Biophys. Res. Commun.* **145,** 1126 (1987).]

arachidonic acid may have specific as well as nonspecific properties as a fatty acid when presented to the cell. For example, high-affinity enzymatic systems may recognize arachidonic acid as the unique eicosanoid precursor for incorporation into glycerophospholipids, whereas the recognition of arachidonic acid as a fatty substrate by low-affinity, high-capacity enzymes may be involved in the storage of this as a fatty substance. We have performed studies to examine the effect of adding various amounts of arachidonic acid (complexed to HSA) to the neutrophil for 5 min and found that there are significant differences in the levels to which glycerolipids incorporate exogenous arachidonic acid, depending on the concentration provided to the cell.[11] For example, increasing the concentration of arachidonic acid resulted in a smaller percentage of this fatty acid incorporated into the phospholipids with a concomitant increase in the incorporation into the neutral lipids (Fig. 1). Further examination of the neutral lipid fraction by TLC revealed that the major portion of the radioactivity migrated with triacylglycerol.

In addition to the differences in glycerolipid classes, the distribution of labeled arachidonate in diacyl, alkylacyl, and alk-1-enylacyl molecular species differed profoundly based on the initial concentration of arachidonate provided to the cell. Shown in Fig. 2 is the distribution of [^{3}H]-arachidonate in phosphoglyceride molecular species of phosphatidylcholine (PC) at various concentrations of arachidonic acid. At the lower concentrations of arachidonic acid, the majority of [^{3}H]arachidonic acid

[11] F. H. Chilton and R. C. Murphy, *Biochem. Biophys. Res. Commun.* **145,** 1126 (1987).

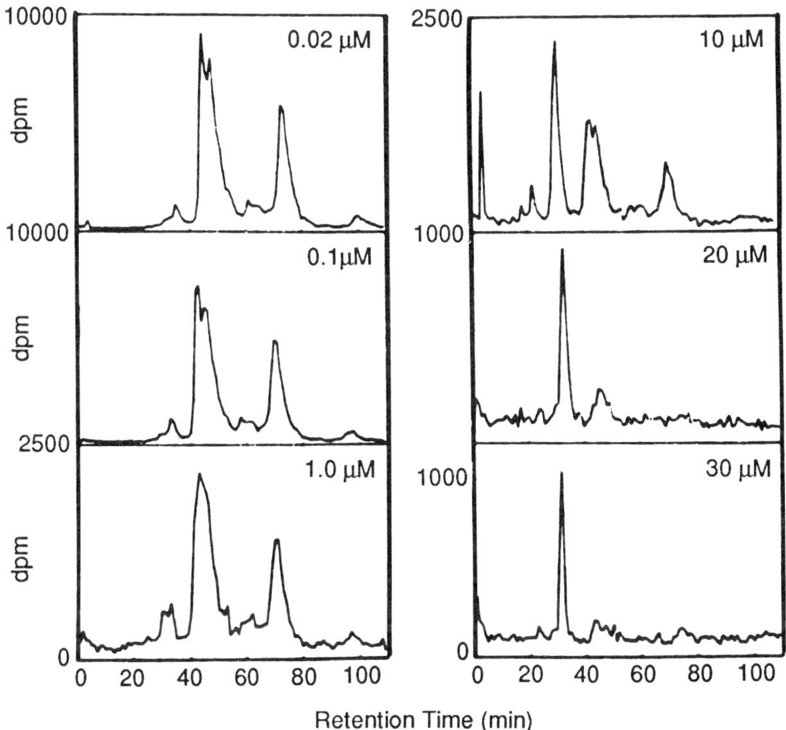

FIG. 2. Reversed-phase HPLC separation of arachidonate-containing molecular species of glycerophosphatidylcholines from human neutrophils incubated with 0.02 to 30 μM [^{3}H]arachidonate. The major labeled species from 0.02 to 1 μM arachidonate correspond to 16 : 0 acyl (45 min), 18 : 1 acyl (48 min), and 18 : 0 acyl (72 min) at *sn*-1. A different molecular species (34 min) became predominant when arachidonic acid was exposed to neutrophils at concentrations from 10 to 30 μM. The molecular species (a) which contains arachidonate has not been structurally characterized. [From F. H. Chilton and R. C. Murphy, *Biochem. Biophys. Res. Commun.* **145,** 1126 (1987).]

was incorporated into 1-acyl[2-^{3}H]arachidonoylglycero-PC containing 16:0, 18:0, and 18:1 at the *sn*-1 position.[12] As mentioned above, cells incubated with higher concentrations ($>10\ \mu M$) of arachidonic acid incorporated a much smaller percentage of exogenously added arachidonic acid into phosphatidylcholine. However, of the label incorporated into phosphatidylcholine, a major portion was found in a single molecular species of phosphatidylcholine (Fig. 2). This unique molecular species of phosphatidylcholine was identified as 1,2-diarachidonoyl-*sn*-glycero-3-

[12] F. H. Chilton and T. Connell, *J. Biol. Chem.* **263,** 5260 (1988).

phosphocholine based on the elution time on reversed-phase HPLC and on the mass spectrum utilizing fast atom bombardment–mass spectrometry.

These data emphasize the importance of identifying arachidonate-containing phospholipids when prelabeling cells with different concentrations of arachidonic acid for the study of sources of precursor arachidonic acid use in oxidative metabolism. Quite often in labeling studies, the rationale for using a high specific activity tracer such as [^{3}H]arachidonic acid or a low specific activity one such as [^{14}C]arachidonic acid is not clarified. The data suggest that the phosphoglycerides labeled using higher concentrations of [^{14}C]arachidonic acid may be quite different from those labeled with low concentrations of [^{3}H]arachidonic acid due to the disparity in specific activities. Without information regarding the distribution of arachidonic acid into different pools under different labeling conditions, it may be difficult to determine the contribution of a particular glycerolipid class or molecular species to eicosanoid production.

Incorporation of Labeled Arachidonic Acid into Neutrophil Lipids: Effect of Incubation Time. A second factor which must be taken into account when labeling a cell such as the neutrophil is the time that the cells are exposed to the arachidonic acid. We have found that different phospholipid molecular species are labeled depending on the time of exposure to arachidonic acid.[10] In some cases (such as the neutrophil), it may not be possible to reach conditions where the specific activity of all phospholipids are uniform considering the short life span of the cell. In the experiments presented below, isolated neutrophils (5.0×10^7, 0.95 ml) were incubated at 37° with a 50 μl of an aqueous solution containing 0.07 μM of [^{3}H]arachidonic acid (6 μCi) complexed with HSA. The reaction was terminated by the addition of 5.0 ml of ice-cold Hanks' balanced salt solution (HBSS) containing HSA (0.25 mg/ml). The cells were removed from the supernatant fluid by centrifugation (225 *g*, 8 min), washed 2 times with HBSS containing HSA (0.25 mg/ml) to rid the cells of exogenous arachidonic acid, and then suspended in 1.0 ml HBSS. In some experiments, the cells were incubated for an additional 120 min. The cells were then centrifuged (225 *g*, 8 min), and all incubations were terminated by adding methanol to the cell pellet.

In the resting neutrophil, the predominant pools of endogenous (unlabeled) arachidonate are found in ethanolamine-, choline-, and inositol-containing glycerolipids (Fig. 3). Phosphatidylserine (PS) and neutral lipids contain less than 3% of the arachidonate found in the human neutrophil. There were no significant changes in the mole quantity of arachidonate in any of the glycerolipid classes after incubation of the cells for 120 min with or without labeled arachidonic acid at the concentrations indicated above.

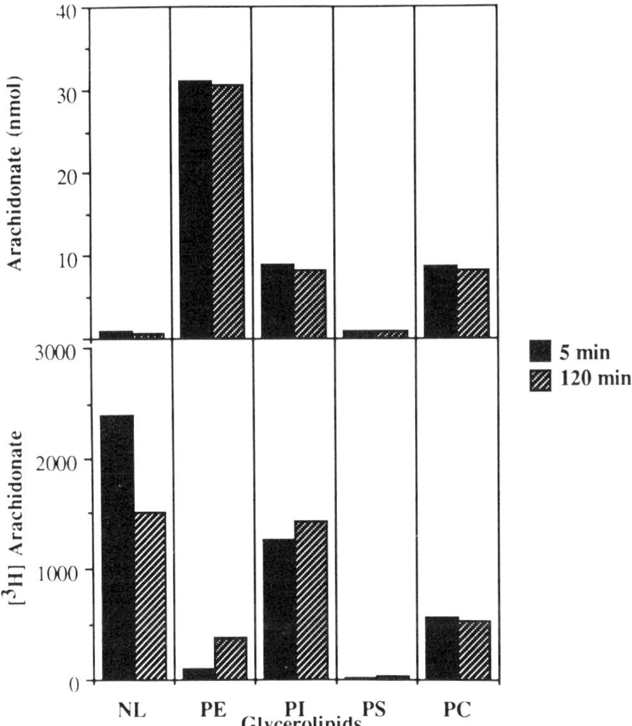

FIG. 3. Distribution of arachidonate in glycerolipid classes. Neutrophils were pulse-labeled for 5 min (solid) and analyzed, or pulse-labeled then incubated for an additional 120 min (hatched) and analyzed as described. The mole quantity of unlabeled arachidonate and the amount of labeled arachidonate were determined. The data are representative of three separate experiments.

When the cells were given exogenous [^{3}H]arachidonic acid, over 60% of the label was taken up by the cell within the first 5-min pulse. Of this, 45% had been incorporated into phospholipids (Fig. 3), primarily phosphatidylinositol (PI) (30%), PC (13%), and PE (2%). The remaining cell-associated label eluted on HPLC with neutral lipids (55%), and the bulk of this was identified as labeled triglycerides. When cells were pulse-labeled for 5 min followed by an additional 120-min incubation, PE gained a significant amount of label. The radioactivity in PE increased concomitantly with a decrease of label in the neutral lipid fraction. In addition, there was a small increase in PI during the 120-min incubation.

Taken together, the label and mass data indicated that PE contained the greatest quantity of endogenous arachidonate but incorporated exogenous labeled arachidonate at the slowest rates. In contrast, labeled neutral lipids (triglycerides) and PI contain the bulk of the labeled arachidonate at both 5 and 120 min, whereas by mass analysis, they were found to contain small pools of endogenous arachidonate. This discrepancy in the pools that contain the most endogenous arachidonate and those which incorporate most rapidly the labeled arachidonic acid resulted in large differences in the radiospecific activity of the phosphoglyceride classes of the neutrophil. The large nonuniformity in the specific activity among all major glycerophospholipid classes could still be observed even after 120 min of neutrophil incubation. This may reflect a difference in the turnover of arachidonate among the major phosphoglyceride classes.

In addition to examining the ratio of labeled to unlabeled arachidonate in phospholipid classes, we have also studied this ratio in individual phospholipid molecular species of each phospholipid class. Neutrophils were labeled for 5 and 120 min and phospholipid molecular species of each class were isolated as described above. Figure 4 shows the HPLC profile of $[^3H]$arachidonic acid in the different molecular species of phosphatidylcholine following the 5-min pulse (Fig. 4A) and the pulse with 120 min incubation (Fig. 4B). After the initial pulse, the majority of $[^3H]$arachidonate was incorporated into 1-acyl$[2\text{-}^3H]$arachidonoylglycero-PC containing 16 : 0, 18 : 0, and 18 : 1 at the *sn*-1 position. After the additional 120-min incubation, a major portion of the arachidonate was redistributed into 1-alkyl-$[2\text{-}^3H]$arachidonoylglycero-PC containing 16 : 0, 18 : 0, and 18 : 1 at the *sn*-1 position.

Similarly, $[^3H]$arachidonic acid was rapidly incorporated into 1,2-diacyl molecular species of ethanolamine-containing phospholipids (data not shown). However, in contrast to choline-linked phospholipids, ethanolamine-linked phosphoglycerides redistribute a large proportion of labeled arachidonate into the 1-alk-1-enyl fraction after 120 min and very little into the 1-alkyl-linked subclass. Finally, labeled arachidonic acid was distributed into 1-acyl-linked molecular species of PI after the 5-min pulse (data not shown). Unlike choline- and ethanolamine-linked phosphoglycerides, neutrophils incubated for an additional 120 min showed no change in the distribution of $[^3H]$arachidonate among PI molecular species.

Table I shows the radiospecific activities of all major arachidonate-containing phospholipids of the neutrophil after the 5-min pulse and 120-min incubation. The radiospecific activity of $[^3H]$arachidonate in all diacyl molecular species was high relative to that of alkylacyl and alk-1-enylacyl molecular species in both choline- and ethanolamine-linked phospholipids after 5 min. The specific activities of all the arachidonate-containing molecular species become more uniform as label appeared in ether and plasmalogen pools in choline- and ethanolamine-linked phosphoglycerides, respec-

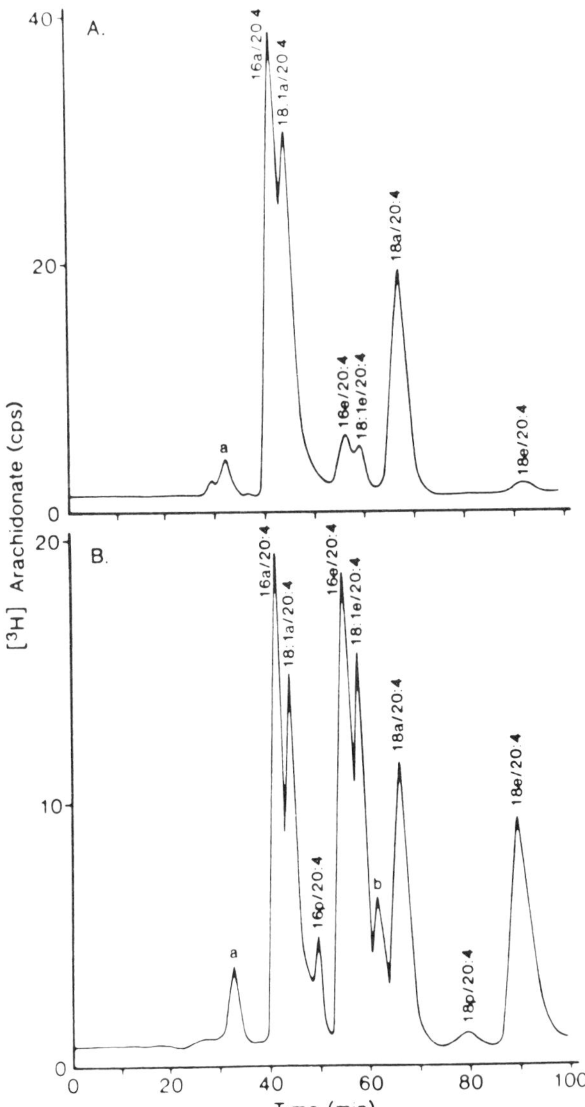

FIG. 4. Distribution of labeled arachidonate in choline-linked molecular species. Neutrophils were labeled as described in the text. The purified 1-radyl[2-^{3}H]arachidonoylglycero-PC was separated into various molecular species by HPLC system II. (A) Distribution of label in the various molecular species from cells pulse-labeled for 5 min. (B) Distribution of label from cells that were pulse-labeled followed by an additional 120-min incubation. The newly synthesized 1-radyl[^{3}H]arachidonoylglycero-PC contained 16 : 0, 18 : 0, and 18 : 1 fatty acid chains at the *sn*-1 position with *a* representing an acyl linkage at the *sn*-1 position, *e* representing an ether linkage at the *sn*-1 position, and *p* representing a plasmalogen linkage at the *sn*-1 position. Arachidonic acid (20 : 4) esterified at *sn*-2 is indicated following the terminal number. The data are representative of five separate experiments. [From F. H. Chilton and R. C. Murphy, *J. Biol. Chem.* **261,** 7771 (1986).]

TABLE I
SPECIFIC ACTIVITIES OF MAJOR ARACHIDONATE-CONTAINING GLYCEROLIPID MOLECULAR SPECIES[a]

Assay	Phosphatidylcholine					
	16:0a[b]	18:0a	18:1a	16:0b	18:0b	18:1b
5 min	279	179	442	22.5	18.0	23.9
120 min	113	88.8	126	58.1	59.4	70.4
	Phosphatidylethanolamine					
	16:0a[b]	18:0a	16:0b	18:0b	16:0c	18:0c
5 min	9.2	12.4	2.5	2.5	2.5	1.2
120 min	36.0	25.9	14.2	11.6	12.2	12.1
	Phosphatidylinositol, 18:0a[b]					
5 min		148.7				
120 min		143.8				

[a] Neutrophils were pulse-labeled for 5 min and analyzed (5 min), or incubated for an additional 120 min (120 min) as described. Incubations were terminated by extracting lipids from the cell pellet. 1-Radyl[2-^{3}H]arachidonoylglycero-PC, glycero-PE, and glycero-PI were separated from other glycerolipids by HPLC. Individual molecular species of 1-radyl[2-^{3}H]arachidonoylglycero-PC were isolated by HPLC. The amount of label in each PC molecular species was determined. Then, the molar quantity of arachidonic acid was determined as described. Specific activities of each individual molecular species were determined from the two measurements. The data are representative of three separate experiments.

[b] a, Acyl linkage at the *sn*-1 position; b, ether linkage at the *sn*-1 position; c, plasmalogen linkage at the *sn*-1 position.

tively, during the 120-min incubation. In contrast, the specific activity of the major inositol phospholipid molecular species 1-stearoyl-2-arachidonoylglycero-PI remained relatively constant throughout the 120-min incubation.

It is important to note that the largest pools of endogenous arachidonate in the neutrophil resides in 1-alk-1-enyl-2-arachidonoylglycero-PE. In our experiments, the specific activity of this pool of arachidonate, although moving toward equilibrium, is still 10 to 12-fold less than that in 1-acyl-2-arachidonoylglycero-PC and glycero-PI even after 120 min of incubation. Moreover, the specific activity of molecular species within any given phospholipid class may differ 2 to 3-fold after 120 min. These data indicated that it may not be possible to label the human neutrophil to anywhere near equilibrium in the short life span of the cell using classic labeling techniques.

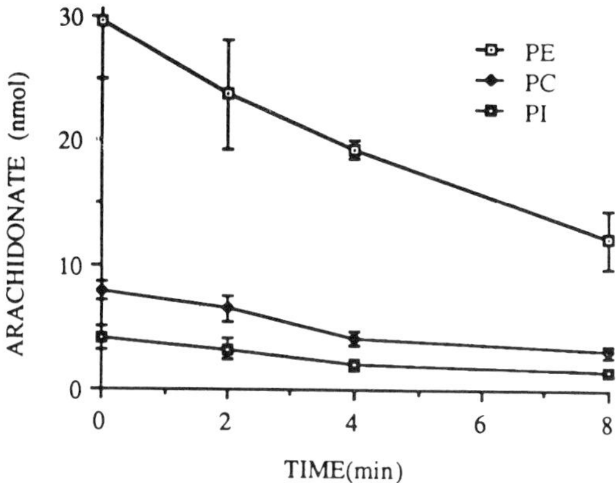

FIG. 5. Quantity of arachidonate in glycerolipid classes before and after treatment with ionophore A23187. Neutrophils (40×10^6) were incubated with ionophore A23187 for various periods of time. The cells were then removed from the medium by centrifugation at 4°. The nanomole quantity of arachidonate in each glycerolipid class was determined by GC/MS as described. The data are averages of four experiments (error bars show S.E.). The nanomole quantity of arachidonate found in all molecular species of phosphatidylethanolamine, phosphatidylcholine, and phosphatidylinositol are as shown. [From F. H. Chilton and T. Connell, *J. Biol. Chem.* **263,** 5260 (1988).]

Release of Arachidonate from Cellular Phospholipids of Neutrophils. In addition to examining the uptake of exogenous arachidonic acid, assays were designed to determine which of the arachidonate-containing glycerolipids were broken down most rapidly during cell activation.[12,13] In these experiments, we were particularly interested in comparing data obtained with radiolabeled arachidonate with data obtained using GC/MS to follow endogenous pools of arachidonate. Here, neutrophils were pulse-labeled as described above for 5 min. Subsequently, the cells were washed 2 times with HBSS containing HSA (0.25 mg/ml). The cells were then incubated at 37° for an additional 15 min to assure that all free [^{3}H]arachidonic acid had been esterified into complex lipids. Neutrophils labeled in this manner were incubated at 37° with ionophore A23187 or no stimulus for up to 8 min. Cell-associated phosphoglyceride classes were separated by normal-phase HPLC, and the mole quantity of arachidonate in each class was determined by GC/MS.

The distribution of endogenous arachidonate in each phosphoglyceride as a function of time after stimulation with the ionophore is shown in Fig. 5. Arachidonate was lost from all major arachidonate-containing phosphoglyceride classes during neutrophil activation with the ionophore. Ethanol-

[13] F. H. Chilton, *Biochem. J.* **258,** 327 (1989).

TABLE II
DISTRIBUTION OF RADIOLABELED ARACHIDONIC ACID IN NEUTROPHIL GLYCEROLIPIDS[a]

Glycerolipid	Treatment	% of Total disintegrations per minute	Change
Phosphatidylcholine	Control	25.6 ± 2.7	
	A23187	11.1 ± 0.5	−14.5
Phosphatidylinositol	Control	18.3 ± 0.4	
	A23187	6.3 ± 1.5	−12.0
Phosphatidylethanolamine	Control	8.4 ± 0.8	
	A23187	5.8 ± 1.7	−2.6
Neutral lipids	Control	46.5 ± 1.4	
	A23187	74.7 ± 3.4	+28.2

[a] Neutrophils, labeled as described, were incubated for 8 min with ionophore A23187 or no stimulus (control). Cells were removed from the medium by centrifugation. The lipids were extracted from the pellet and the glycerolipid classes separated by normal-phase HPLC. The amount of radioactivity in each phospholipid species is expressed as a percentage of the total radioactivity recovered in all glycerolipid classes. The mean value from three experiments are presented (±S.E.). The percent change was determined by the difference between control and A23187 challenge values.

amine-linked phosphoglycerides were the major source of arachidonate during neutrophil activation; in fact, on a mole basis, 3–4 times more arachidonate was lost from ethanolamine phosphoglycerides than from choline or inositol phosphoglycerides. Data acquired in these experiments are expressed as the mole quantity of arachidonate lost from each phosphoglyceride after the incubation; depletion of arachidonate from a given phosphoglyceride such as phosphatidylinositol may have arisen from the direct liberation of arachidonate via phospholipase A_2 or from the conversion of the phosphoglyceride to another product such as diacylglycerol via phospholipase C.

In the same cell extract that was used for the measurement of mass loss above, the loss of [^{3}H]arachidonate previously incorporated into cellular phosphoglycerides was determined. Ionophore A23187 induced the release of [^{3}H]arachidonate from all phosphoglycerides (Table II). However, the bulk of labeled arachidonate was lost from PC and PI, with PE contributing only 9% of the labeled arachidonate released during cell activation. This experiment pointed out that there were again major discrepancies when comparing absolute mass of and the amount of radiolabeled arachidonate lost from phospholipids. For example, labeling data from these experiments suggest that labeled arachidonate is lost from choline- and inositol-linked phosphoglycerides when it is clear from the mass data that ethanolamine-linked phosphoglycerides provide the bulk of arachidonate.

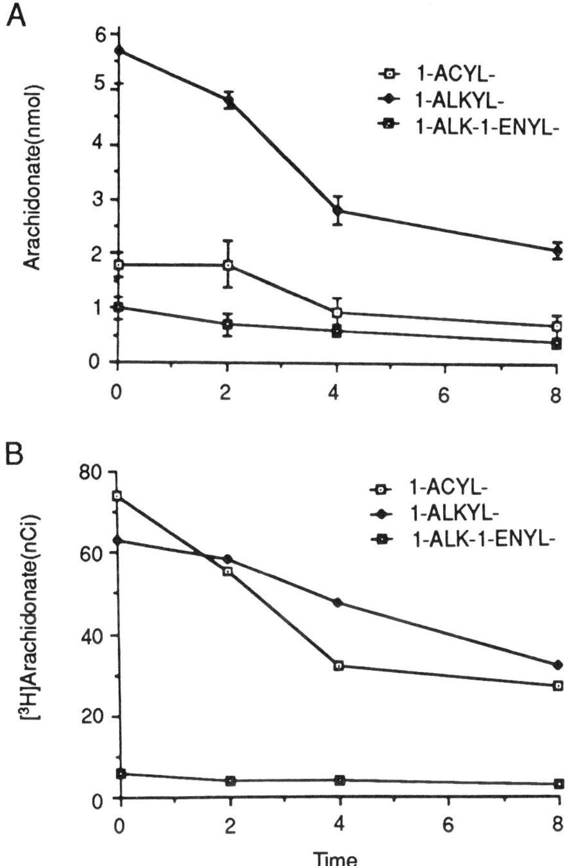

FIG. 6. Time course of the loss of mass and labeled arachidonate from choline-linked glycerolipid subclasses on challenge by ionophore A23187. Neutrophils labeled as described were incubated for various periods of time with ionophore A23187. Choline-linked glycerolipids were further separated as diglyceride acetates into 1-acyl, 1-alkyl-, and 1-alk-1′-enyl subclasses by TLC. The nanomole quantity of arachidonate in each subclass was determined by GC/MS and is presented in (A). The data in (A) are averages of three experiments (±S.E.). The quantity of radiolabeled arachidonate found in each subclass after ionophore stimulation is shown in (B). These data are representative of three separate experiments. The quantities (nmol or nCi) of arachidonate found in 1-acyl-, 1-alkyl, and 1-alk-1′-enyl subclasses are as shown. [From F. H. Chilton and T. Connell, *J. Biol. Chem.* **263,** 5260 (1988).]

The quantity of endogenous and labeled arachidonate was also determined in choline- and ethanolamine-linked subclasses. Figure 6A demonstrates that in the resting neutrophil, most of the arachidonate in choline-linked phospholipids is found at the *sn*-2 position of molecules which contain a 1-alkyl moiety at the *sn*-1 position. The remaining PC stores of

TABLE III
CHANGES IN QUANTITIES OF ARACHIDONATE IN GLYCEROLIPID CLASSES AND PRODUCTS[a]

Arachidonate in phospholipids	Change (nmol)	Leukotrienes and free arachidonic acid	Change (nmol)
Phosphatidylethanolamine	−7.6 ± 0.6	Leukotriene B_4	+1.8 ± 0.1
Phosphatidylcholine	−3.7 ± 0.2	20-Hydroxyleukotriene B_4	+1.1 ± 0.1
Phosphatidylinositol	−1.9 ± 0.1	Cellular AA	+1.6 ± 0.1
Phosphatidic acid	+1.3 ± 0.4	Supernatant AA	+0.2 ± 0.1

[a] Neutrophils were incubated for 8 min with A23187 or no stimulus. Cells were removed from the medium by centrifugation. The amount of free arachidonic acid (AA) in both the cell pellet and supernatant was determined by GC/MS. The nanomole quantity of arachidonate in each glycerolipid class was determined by GC/MS as described. The quantities of LTB_4 and 20-hydroxy-LTB_4 were determined following purification by reversed-phase HPLC using UV optical densities at 270 nm. Means of the differences in the nanomole quantity of arachidonate in each lipid following ionophore stimulation from four experiments are presented (±S.E.).

arachidonate are found in 1-acyl-linked phospholipids. During cell activation, arachidonate is lost from all three PC subclasses. However, loss of arachidonate from the 1-alkyl-linked fraction represents the major portion of arachidonate removed from choline-linked phospholipids. Labeled arachidonate was examined in PC subclasses in the same cell extracts used for mass analysis (Fig. 6B). Under the labeling conditions used in these experiments, [^{3}H]arachidonate is distributed in the resting cell in primarily 1-acyl-2-arachidonoylglycero-PC. On cell activation labeled arachidonate is lost from primarily 1-acyl-linked molecular species, with a smaller portion lost from the 1-alkyl-linked pool. This again points out a major discrepancy at the phospholipid subclass level between the results obtained using mass data and radiolabeling data. Similar discrepancies were observed for ethanolamine-linked subclasses (data not shown).

Metabolism of Released Arachidonic Acid by Neutrophil. Both leukotriene B_4 (LTB_4) and 20-hydroxyleukotriene B_4 were produced and released from the neutrophil during A23187 stimulation. In order to determine what percentage of released arachidonate formed leukotrienes, separate experiments were performed to determine the distribution of arachidonate among various products. As shown in Table III, ionophore A23187 induced large increases in the mole quantity of free fatty acid and leukotrienes found in the cell and supernatant fluid, respectively. Furthermore, there was an increase in the mole quantity of arachidonate in phosphatidic acid within stimulated cells. In addition to the aforementioned products, a small portion (10–20%) of arachidonate could be found as esterified glycerolipids in the cell supernatant fluid. The total amount of products accounted for 60–70% of the arachidonate lost from phosphoglycerides. An interesting point from this study was that leukotrienes accounted for only a small percentage of the total arachidonate lost from the cell. The fact that only a small fraction of arachidonate was metabolized to eicosanoids made it difficult to determine which of the phospholipids contributed the arachidonate that was converted to leukotrienes. Therefore, the loss of arachidonate from a phospholipid does not necessarily implicate that phospholipid as a source of substrate arachidonate for eicosanoids.

Closing Remarks

For several years, labeled arachidonic acid has been an important tool utilized to identify the source(s) of arachidonate for eicosanoid biosynthesis.[1] Although data from these labeling experiments have often provided important information, many times they have been in stark disagreement with data obtained using mass analysis. In this chapter, we have discussed some of the problems which can arise when using cells in which the phospholipids have been prelabeled with arachidonic acid. The major problem is the nonuniform uptake of exogenous arachidonic acid into phospholipid molecular species. In the neutrophil, the uptake of labeled arachidonic acid and its distribution into complex lipids are greatly influenced by (1) the amount of arachidonic acid provided to the cell and (2) the time of exposure to arachidonic acid. We have not been able to overcome the latter problem because the short life span of the neutrophil does not allow the radioactive arachidonate to equilibrate with the endogenous stores. The experiments presented here point out that even after 120 min of incubation, there can be differences as great as 10-fold between the radiospecific activity of arachidonate in major phospholipid molecular species.

Large differences in radiospecific activities between the phospholipids make it difficult if not impossible to extrapolate the data obtained from the losses of radioactivity to the mass of endogenous arachidonate. For example, PE contributes only a small proportion of labeled arachidonate lost during cell activation. In contrast, 3 to 4 times as much arachidonate (by mass) is lost from PE than from PC or PI during cell activation. In this case, the labeling data provide information which is totally inconsistent with the mass data. There are several reasons for the discrepancies between the labeling and mass data. As pointed out above, PE molecular species are labeled with arachidonate to a much lower specific activity than PC and PI. Consequently, a small loss of radioactivity from a large, low specific activity pool such as PE represents a large loss of endogenous arachidonate relative to that released from other high specific activity phospholipids such as PI. A second factor which can lead to discrepancies between mass and labeling studies is rapid changes in the specific activity of phospholipids during cell activation. For example, in the neutrophil, the specific activity of 1-acyl-2-arachidonoylglycero-PC dropped rapidly during cell activation (data not shown). Thus, the kinetic of labeled arachidonate lost after cell activation may not accurately reflect that of arachidonate mass.

Another assumption often made when performing labeling or mass studies is that loss of arachidonate from a phospholipid can be equated to that phospholipid being a source of arachidonate for eicosanoids. In experiments presented here, leukotrienes produced during cell activation represented only a small fraction of the total arachidonate lost from all phospholipids. Consequently, the loss of arachidonate from a phosphoglyceride molecular species does not necessarily implicate that phosphoglyceride molecular species as a source of arachidonate for eicosanoids. In the aforementioned experiment, ionophore A23187 was used as a stimulus; there may be cases where more physiological stimuli may result in a higher percentage of arachidonate lost from phospholipids found in eicosanoids. In preliminary experiments, we have found this to be the case in antigen-stimulated mast cells. Under these conditions the loss of arachidonate from a phospholipid may correlate more closely with its being a source for eicosanoids.

In spite of the many potential problems of using labeled arachidonate in cellular experiments, there are many instances where it can be a useful tool, particularly in cells where equilibrium labeling can be achieved. We have determined that [^{3}H]arachidonic acid ($<0.1\ \mu M$) will reach a constant radiospecific activity in all phospholipid molecular species of the mast cell after 24–36 hr in culture. When this is accomplished, the data obtained from the labeled arachidonate will reflect more accurately the behavior of endogenous arachidonate. Certainly in almost all cases, it is much easier to measure label by scintillation counting than to do mass analysis using GC/MS. For this reason, labeled fatty acids and arachidonate in particular will continue to be important tools used in experiments designed to examine the turnover of fatty acids in complex lipids.

Acknowledgments

Work discussed in this chapter was supported by National Institutes of Health Grants AI 24985 and AI 26771.

[16] Phosphoinositide-Specific Phospholipase C Activation in Brain Cortical Membranes

By MICHAEL A. WALLACE, ENRIQUE CLARO, HELEN R. CARTER, and JOHN N. FAIN

Introduction

One major pathway for neurotransmitter signaling involves phosphoinositide-specific phospholipase C (PLC). The enzyme catalyzes the breakdown of the phosphoinositides into two products: diacylglycerol and inositol phosphate(s). The former product is important for the regulation of protein kinase C and as an intermediate in overall lipid metabolism. The latter products have roles in the regulation of intracellular Ca^{2+} levels, most prominently, inositol 1,4,5-trisphosphate, which causes release of sequestered calcium from the endoplasmic reticulum. A variety of neurotransmitters have been shown to stimulate PLC activity in brain. These include, among others, norepinephrine, histamine, serotonin, and acetylcholine.[1,2] It is thought that these neurotransmitters activate PLC through guanine nucleotide-binding protein intermediates, analogous (or in some cases perhaps identical) to the G proteins which modulate adenylyl cyclase (adenylate cyclase) activity.[3] Effort is now being directed to the goal of elucidating the molecular events which occur when a neurotransmitter binds to its receptor, activates a G protein, and influences phospholipase C activity.

Here we describe an assay recently developed for carbachol-activated, guanine nucleotide-dependent phospholipase C stimulation using crude membrane preparations from rat brain cortex.[4] The substrate phosphoinositides can be supplied by direct addition of the radiolabeled phospholipids to the assay. Alternatively, phosphoinositides can be labeled with [^{3}H]inositol by incubation of cortical tissue slices in a physiological buffer with the radiolabel prior to membrane preparation.[5] The procedure is generally applicable to most brain regions, and it has been successfully used with rabbit brain, as well.

There are at least five isozymes of phosphoinositide-specific phospholi-

[1] S. K. Fisher and B. W. Agranoff, *J. Neurochem.* **48,** 999 (1987).
[2] D-M. Chuang, *Annu. Rev. Pharmacol. Toxicol.* **29,** 71 (1989).
[3] J. N. Fain, *Biochim. Biophys. Acta* **1053,** 81 (1990).
[4] E. Claro, M. A. Wallace, H.-M. Lee, and J. N. Fain, *J. Biol. Chem.* **264,** 18288 (1989).
[5] E. Claro, A. Garcia, and F. Picatoste, *Biochem. J.* **261,** 29 (1989).

pase C.[6] Brain is specifically enriched with the PLC-β form of the enzyme, most of which can be found in the cytosol or attached loosely to the cell membrane.[7] In addition, however, we have demonstrated that about 8% of the total PLC activity is tightly, perhaps integrally, bound to the cell membrane. Thus, our efforts have been concentrated at examining this tightly bound form of the enzyme as the most likely candidate for being part of a signal transduction path. We have recently purified to homogeneity the major PLC isozyme found tightly bound to rabbit brain cortex membranes, and we have identified the isozyme immunologically as being PLC-β.[7] Polyclonal antibodies directed specifically against this isozyme block carbachol plus guanine nucleotide stimulation of phosphatidylinositol bisphosphate (PIP_2) breakdown in these cortical membranes. Thus, this tightly bound form of PLC-β is regulated by muscarinic cholinergic receptors.

Preparation of Agonist-Responsive Rat Brain Cortical Membranes

Male Sprague-Dawley rats (75–150 g) are sacrificed by decapitation, and the brain is quickly removed and placed on ice. The meninges are carefully removed by blotting with paper towels, and the cortex is dissected free of the striatum and hippocampus. The cortex is then cross-chopped into 300 × 300 μm miniprisms using a McIlwain tissue chopper. The tissue is transferred to cold (4°) Krebs–Henseleit buffer, pH 7.4, equilibrated with 95% O_2/5% CO_2 (v/v). The buffer contains 116 m*M* NaCl, 4.7 m*M* KCl, 1.2 m*M* $MgSO_4$, 1.2 m*M* KH_2PO_4, 25 m*M* $NaHCO_3$, and 11 m*M* glucose. The miniprisms are then incubated for 15–30 min at 37° on a rotary water bath with vigorous shaking before preparation of membranes. This incubation period can be extended up to 6 hr and the endogenous membrane phosphoinositides labeled with [^{3}H]inositol prior to cell homogenization. In this regard it should be noted that omitting Ca^{2+} from the buffer enhances labeling with [^{3}H]inositol.

Preparation of membranes from the miniprisms is identical whether or not they have been prelabeled. All steps are carried out at 4°. The miniprisms are washed 3 times in 20 m*M* Tris-HCl buffer at pH 7.0 containing 1 m*M* EGTA, and they are then homogenized in 10 ml of this buffer per gram tissue with a Teflon/glass Dounce homogenizer using 10 strokes by hand. The homogenate is centrifuged at 3300 *g* for 15 min. The supernatant is discarded, and the pellet is homogenized in the same volume of buffer by 5 strokes by hand and incubated with vigorous shaking for 15 min at 4°, then homogenized again. The homogenate is centrifuged at 3300 *g* for

[6] S. G. Rhee, P-G. Suh, S-H. Ryu, and S. Y. Lee, *Science* **244,** 546 (1989).

[7] H. R. Carter, M. A. Wallace, and J. N. Fain, *Biochim. Biophys. Acta* **1054,** 119 (1990).

15 min, the supernatant is discarded, and the pellet is resuspended in the same volume of buffer and centrifuged again. After discarding the supernatant, the pellet is resuspended in buffer to a concentration of 2.5 mg protein/ml. One-milliliter aliquots are then dispensed to microcentrifuge tubes, which are spun at 10,000 *g* for 5 min. The supernatant is aspirated, and the pellets are stored frozen at −80°. There is no noticeable loss of carbachol responsiveness in these membranes for at least 1 month, although we have never systematically assessed the effects of longer storage.

We have found that the membranes contain a variety of enzymatic activities besides PLC which are involved in phosphoinositide and inositol phosphate metabolism. Phosphatidylinositol (PI) kinase, phosphatidylinositol 4-phosphate (PI4P) kinase, polyphosphoinositide phosphomonoesterase, inositol 1,4,5- and 1,4-phosphatases are present. Thus, the preparation is useful for examining the roles these enzymes play in modifying the PLC signaling path. On the other hand, these activities can confound the interpretation of some types of experiments. It has therefore been a boon to find that under some particular assay conditions PLC-β will degrade exogenously added [^{3}H]PI to diglyceride and inositol 1-phosphate in a neurotransmitter- and guanine nucleotide-dependent manner. This gives a straightforward method of looking at agonist receptor–G protein interactions with PLC in which the products of the interaction are not influenced by other enzymes.

Another interesting finding has been that the membranes can be washed with 2 *M* KCl and still retain full agonist responsiveness even though all peripheral membrane protein is removed by this procedure.[8] Clearly, such peripheral proteins are not required for agonist activation of PLC-β. Further, we can find no evidence for agonist- or guanine nucleotide-dependent release of PLC or of PLC-modulating factors from the membranes.[8] Thus, our current models for activation of PLC-β by agonists include only proteins firmly attached to the membrane.

Assay of PLC-β in Cortical Membranes Using Exogenous Substrates

Assay of PLC-β using [^{3}H]PIP$_2$ as substrate is performed in 100 μl final volume with 100 μg of membrane protein. The final concentrations of assay buffer components, which have been optimized for examining agonist and G protein effects, are 8 m*M* Tris–maleate, 1 m*M* deoxycholate, 2 m*M* ATP, 6 m*M* $MgCl_2$, 8 m*M* LiCl, 3 m*M* EGTA, and 0.27 m*M* $CaCl_2$ (which is sufficient to yield a final free Ca^{2+} concentration of 0.1 μM according

[8] H. R. Carter, M. A. Wallace, and J. N. Fain, *Biochim. Biophys. Acta* **1054,** 129 (1990).

to the calculation of Harafugi and Ogawa[9]). The pH is adjusted to 6.8 with KOH. The membrane protein pellets are homogenized and resuspended for assay in Tris–maleate containing $MgCl_2$, LiCl, and ATP. EGTA plus $CaCl_2$ are added separately. Labeled and unlabeled PIP_2 are mixed and evaporated to dryness under N_2. They are resuspended at a 10× concentration in 10 mM deoxycholate and sonicated with a Fisher (Fairlawn, NJ) Probe Dismembrator at setting 35 with 15-sec bursts. The final PIP_2 concentration in the assay is 30 μM at a specific activity of approximately 10^7 dpm (disintegrations per minute)/μmol.

The incubation period is usually 10 min at 37°. This is probably slightly beyond the point at which the agonist-stimulated PLC activity is linear, but agonist-stimulated versus basal activities are maximally distinct. When PI is used as substrate the agonist-stimulated activity is linear for at least 30 min. Assays are terminated by the addition of 1.2 ml chloroform/methanol (1 : 2, v/v). Chloroform (0.5 ml) and 0.25 M HCl (0.5 ml) are then added, and the mixture is thoroughly vortexed and then centrifuged to separate aqueous and organic phases. Aliquots of the upper phase can be directly counted in appropriate scintillation fluid to quantitate [^{3}H]inositol phosphate products. (One milliliter of upper phase can be efficiently counted with 4 ml of Ecolume from ICN Biomedicals Inc., Costa Mesa, CA.) Alternatively, the various inositol phosphates can be separated by Dowex ion-exchange chromatography[10] or by high-performance liquid chromatography (HPLC).[11] Analysis of the phospholipids in the organic phase can be done using thin-layer chromatographic (TLC) procedures.[12]

Our standard assay of PI uses a slightly different buffer in that the final Ca^{2+} concentration is 0.3 μM and the ATP is usually deleted. Maximal agonist effects are found when 100 μM exogenous PI is used as substrate, whereas when PIP_2 is used maximal agonist effects are not much changed over a PIP_2 concentration range of 0.3–100 μM. The magnesium concentration supporting optimal agonist effects with PIP_2 as substrate is 6 mM in the presence of ATP and 3 mM EGTA. With PI as substrate in the presence of 3 mM EGTA but no ATP, the optimal magnesium concentration falls to about 1 mM, probably owing to the fact that the MgATP needed to support PIP kinase activity is irrelevant under these conditions.

Of particular interest is the seeming requirement for addition of deoxycholate in order to see optimal effects of agonists and guanine nucleotides. The first demonstration of guanine nucleotide plus hormone activation of

[9] H. Harafuji and Y. Ogawa, *J. Biochem. (Tokyo)* **87,** 1305 (1980).

[10] C. P. Downs and R. H. Michell, *Biochem. J.* **198,** 133 (1981).

[11] R. A. Pittner and J. N. Fain, *Biochem. J.* **257,** 455 (1989).

[12] M. A. Wallace and J. N. Fain, this series, Vol. 109, p. 469.

exogenous PIP_2 breakdown was performed using membranes from blowfly salivary gland.[13] No requirement for deoxycholate was seen in such membranes. However, mammalian membranes have not generally been proved to respond in the same way. Further, although useful for membranes prepared from various different neural tissues, the assay method we describe here has not been successful with membranes derived from liver or fat cells. Thus, the reason for the deoxycholate requirement is not well understood. Many other detergents uncouple G protein and neurotransmitter receptor responses at very low doses, well below the concentrations at which they release membrane proteins (and specifically PLC-β). Deoxycholate, cholate, and octylglucoside tend to uncouple the agonist responses at the same concentrations at which they solubilize membrane proteins.[8] Therefore, these compounds appear to be the most useful for looking at neurotransmitter receptor–G protein–PLC interactions. Used at 1 mM (about one-fourth the critical micellar concentration), deoxycholate somehow primes the membrane responses whether exogenous [^{3}H]phosphoinositides or prelabeled membranes are used in the assays. Basal PLC activity is only slightly affected by this concentration of deoxycholate. Varying the protein content of the assay does not correspondingly vary the optimal concentration of deoxycholate needed to see maximal agonist effects. The bile salt effects appear specific since they induce a change in the membrane organization such as to allow PLC–G protein–receptor interactions which are otherwise inhibited or lost as a consequence of the homogenization and preparation of membranes from the cortex miniprisms.

Addition of KCL or NaCl to the assay as described has only inhibitory effects on the carbachol plus GTPγS-stimulated activity (pronounced inhibitions are found above 100 mM salt), with no effect on basal activity. Thus, these membranes are dissimilar to synaptosomal preparations where depolarization activates PLC. Inositol trisphosphate levels can be partially maintained by use of 6 mM 2,3-bisphosphoglycerate. Degradation of inositol 1,4,5-trisphosphate is only partially inhibited at this concentration, but higher levels inhibit agonist stimulation of PLC.

An important component of the buffer to be considered is Ca^{2+}. Maintaining a low (~100 nM) concentration of Ca^{2+} is optimal for assay of agonist or guanine nucleotide effects. Above 3 μM free Ca^{2+} PLC activity is very high and cannot be further stimulated. Conversely, the basal activity of PLC is very low below 10 nM free Ca^{2+}. Thus, the critical role intracellular Ca^{2+} plays in regulating PLC activity can be appreciated.

[13] J. N. Fain and I. Litosch, this series, Vol. 141, p. 255.

Agonist effects on PLC activity can be greatly modified by the level of Ca^{2+}.

The inclusion of ATP in the assays has several effects. In its absence, direct degradation of exogenously added PI to form inositol 1-phosphate is found, and this breakdown is stimulated by carbachol in a guanine nucleotide-dependent manner.[4] In the presence of ATP, most of the inositol phosphates generated from exogenously added PI are made subsequent to its conversion to PI4P or PIP_2. Inclusion of ATP in assays of exogenous PIP or PIP_2 breakdown does not affect the results in any major way, but it apparently serves to help maintain the level of polyphosphoinositides. The purified PLC-β has about a 5-fold greater affinity for PIP or PIP_2 over PI. The membrane preparation shows the same preference for PIP and PIP_2. Indeed, it has generally been found that under most conditions agonist stimulation of polyphosphoinositide degradation is much more readily apparent than is the breakdown of PI. However, the use of PI as the exogenously added substrate is often desirable in experiments to clearly differentiate ligand effects on PLC from effects on kinases and phosphatases.

Results

Results of studies on neurotransmitter action with the rat cortical membrane system can now be summarized. First, carbachol stimulation of PLC-β is seen only in the presence of GTP or a nonhydrolyzable GTP analog such as GppNHp or GTPγS. Other purine or pyrimidine nucleotides will not substitute. The responses mediated by GTPγS are effectively inhibited by GDP or GDPβS. Thus, carbachol is apparently working through a G protein to activate PLC-β. The GTPγS requirement for carbachol stimulation of PLC-β activity has an EC_{50} of about 0.1 μM (Fig. 1). When added in the absence of carbachol, GTPγS can by itself stimulate the breakdown of substrate but not to the same extent as seen with carbachol. GTPγS is much more potent and efficacious in this system than are GppNHp or GTP.

We find that G proteins apparently mediate inhibitory inputs into the phospholipase C signaling system as well. Dopamine inhibits the activation by carbachol of PLC-β in rat cortical membranes.[14] Pharmacological characterization reveals that the effect of dopamine occurs through stimulation of a D_1-type receptor. Most interestingly, the effect of dopamine is to shift to higher concentrations the EC_{50} of GTPγS for supporting carbachol stimulation (Fig. 1). The effects of dopamine on basal PLC-β activity are

[14] M. A. Wallace and E. Claro, *Neurosci. Lett.* **110,** 155 (1990).

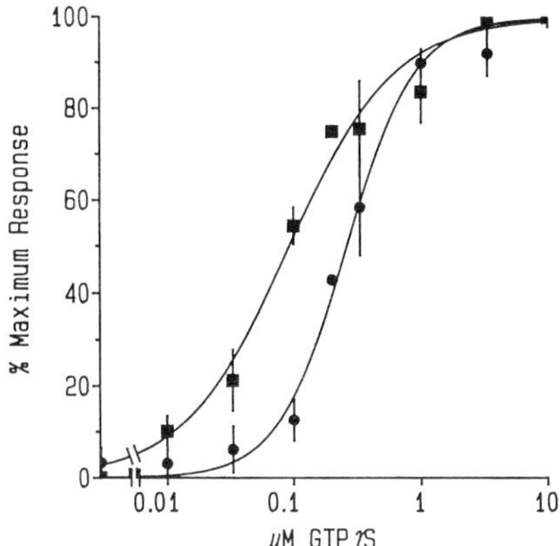

FIG. 1. Effect of dopamine on the response of phospholipase C to GTPγS in the presence of carbachol. Membranes were incubated in the presence of 1 m*M* carbachol without (squares) or with (circles) 200 μ*M* dopamine at the indicated concentrations of GTPγS. Results are normalized to the maximal stimulation of PLC activity obtained in the presence of 10 μ*M* GTPγS plus 1 m*M* carbachol. [Reprinted with permission from M. A. Wallace and E. Claro, *Neurosci. Lett.* **110,** 155 (1990).]

not striking, but stimulation of PI breakdown by carbachol in the presence of 0.1 μ*M* GTPγS is completely inhibited by dopamine with an IC_{50} of 14 μ*M*. At GTPγS concentrations above 0.3 μ*M*, however, no effect of dopamine on the stimulation of PLC-β by carbachol can be found. Thus, there is a subtle balance of positive and negative effectors working on PLC-β in these membranes which may be modified as a function of guanine nucleotide concentration.

A commonly used approach to studying G protein effects on adenylyl cyclase or on phospholipase C has been to use NaF plus $AlCl_3$ to directly activate the G protein being studied. Supposedly, AlF_4^- can complex with GDP bound to a G protein to mimic a γ-phosphate as would be found if GTP were the ligand.[15]

In rat brain cortical membranes a dual effect of NaF is found, however.[16] First, there is a stimulation of PLC activity, and, second, there is an inhibition of PI kinase. Addition of $AlCl_3$ with NaF potentiates the former effect but has no influence on the latter. The results therefore suggest that a G protein controls the activity of PLC-β but not that of PI kinase in cortical membranes. Caution must be exercised, though, in interpreting results obtained with a cation such as Al^{3+} when assaying

[15] J. Bigay, P. Deterre, C. Pfister, and M. Chabra, *FEBS Lett.* **191,** 181 (1985).
[16] E. Claro, M. A. Wallace, and J. N. Fain, *Biochem. J.* **268,** 733 (1990).

phospholipases. Aluminum has been shown to have multiple effects on membrane structure commensurate with its avid binding to both negatively charged and neutral phospholipids. Thus, aluminum stimulates degradation of PI but inhibits degradation of PIP_2 in assays of PLC.[17]

Along the same lines we have reported that mastoparan, an amphiphilic α-helix-forming peptide, can also greatly stimulate or inhibit the activity of purified PLC-β dependent on its interaction with substrate.[18] Future studies of the neurotransmitter receptor–G protein–PLC signaling path must take into consideration the membrane perturbant potential of the various purified protein components used to reconstitute the response and of the various ligands used to modify that response. An example is the widely used antipsychotic drug fluphenazine. We find that fluphenazine stimulates the breakdown of exogenous PIP_2 but inhibits the breakdown of PI (M. A. Wallace, unpublished data, 1990). This indicates that its actions are probably due to interaction with substrates.

Our current model for G protein interactions with PLC-β is based on simple analogy with the adenylyl cyclase system.[3] Carbachol would promote a GTP-liganded G protein α subunit to directly interact with PLC-β in the membrane so as to stimulate the otherwise dormant activity. Dopamine inhibition of carbachol stimulation could occur through competition of inhibitory and stimulatory α subunits as they bind to PLC-β. Alternatively, stoichiometric consideration of stimulatory α subunits and $\beta\gamma$ subunits from inhibitory G proteins may be involved. Whatever the mechanism, the membrane assays described here have allowed us to measure both positive and negative modulators of PLC-β interacting at a common postsynaptic membrane site. Thus, this assay system holds great promise for answering questions concerning receptor and G protein cross talk in neurotransmission.

[17] L. J. McDonald and M. D. Mamrack, *Biochem. Biophys. Res. Commun.* **155,** 203 (1988).
[18] M. A. Wallace and H. R. Carter, *Biochim. Biophys. Acta* **1006,** 311 (1989).

[17] Quantitative Analysis of Water-Soluble Products of Cell-Associated Phospholipase C- and Phospholipase D-Catalyzed Hydrolysis of Phosphatidylcholine

By DONALD A. KENNERLY

Introduction

Increasing evidence points to activation of cellular phospholipase C (PLC) and/or phospholipase D (PLD) on endogenous phosphatidylcholine (PC) in receptor-stimulated cells (reviewed by Exton[1]). Whereas PLC catalyzes the direct formation of 1,2-diacylglycerol, PLD may importantly contribute to the formation of this second messenger by an "Indirect Pathway" involving phosphatidic acid phosphohydrolase-catalyzed conversion of phosphatidic acid to diacylglycerol. Although studies have utilized [^{3}H]choline to prelabel cellular PC in order to examine the liberation of labeled choline and/or phosphorylcholine as a result of cell activation, this approach involves a number of problems inherent to the use of radioisotopes, particularly in cells which cannot be labeled to isotopic equilibrium. To overcome this limitation, we have significantly extended the mass-based methods initially developed by others[2,3] for quantitating choline.[4] Although other methods can provide similar sensitivity, they require high-performance liquid chromatography (HPLC) with postcolumn enzyme-catalyzed detection,[3] a technique frequently not in the repertoire of cell biologists and biochemists. Further, we have improved methods of differential extraction of cell-associated choline and phosphorylcholine and developed a new assay of the latter. An overview of the experimental approach is illustrated in Fig. 1.

Extraction of Cellular Choline

Principle. Choline present in the methanol–water upper phase of a Bligh and Dyer[5] extract forms a lipid-soluble ion pair with heptanone-soluble sodium tetraphenylboron.

[1] J. H. Exton, *J. Biol. Chem.* **265,** 1 (1990).
[2] W. D. Reid, D. R. Haubrich, and G. Krishna, *Anal. Biochem.* **42,** 390 (1971).
[3] N. M. Barnes, B. Costal, A. F. Fell, and R. J. Naylor, *J. Pharm. Pharmacol.* **39,** 727 (1987).
[4] J. J. Murray, T. T. Dinh, and D. A. Kennerly, *Biochem. J.* **270,** 63 (1990).
[5] E. Bligh and W. Dyer, *Can. J. Biochem. Physiol.* **37,** 911 (1959).

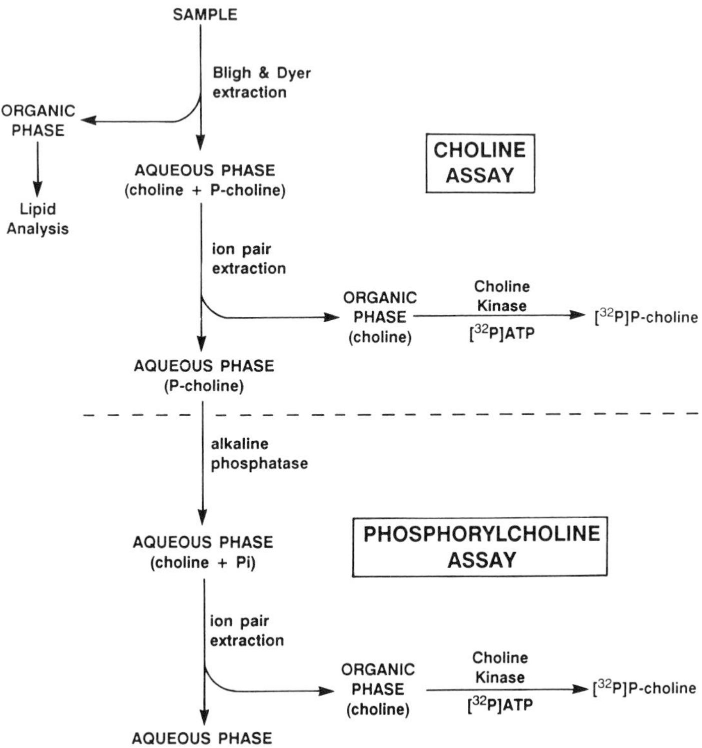

FIG. 1. Extraction and quantitation of cellular choline and phosphorylcholine (P-choline).

Reagents

Sodium tetraphenylboron, 5 mg/ml in 4-heptanone
Sodium phosphate buffer, 12 m*M* (pH 7.0)
HCl, 0.1 *N*
[^{3}H]Choline chloride, 4.5 μCi/ml in ethanol (specific activity 87 Ci/mmol)

Procedure. Unless otherwise indicated, all procedures were performed at room temperature. Lipids and water-soluble cellular components are extracted from a pellet containing 10^5–10^7 cells by adding 0.5 ml of methanol, 0.25 ml of chloroform, and 0.2 ml of water. After removing insoluble materials at 200 *g* for 10 min, the supernatant is transferred to a tube containing 45 nCi of [^{3}H]choline, and phase separation is accomplished by adding 0.25 ml of chloroform and 0.25 ml of water. A 0.85-ml sample of the 0.925-ml choline-containing upper methanol–water phase is removed for further analysis of choline, and the lipid-containing lower phase can be dried and for analysis of cellular lipids. After adding 0.65 ml of 12 m*M* sodium phosphate (pH 7.0) and mixing briefly, cellular choline is extracted into the upper heptanone phase generated by adding 1.0 ml of sodium tetraphenylboron in heptanone. After vigorous mixing and phase separation by centrifugation, 0.9 ml of the heptanone phase is removed and added to 0.3 ml of 0.1 *N* aqueous HCl (to disrupt the ion pair and extract free choline into the acidic aqueous lower phase). Following brief centrifugation in a Microfuge, the heptanone layer is removed by vacuum aspiration, 20 μl of the choline-containing aqueous phase is placed in a scintillation vial (to assess [^{3}H]choline recovery) and 0.25 ml of the aqueous phase is dried either by using a Speed Vac or by heating in an oven at 65° overnight.

Remarks. Although choline and phosphorylcholine are freely soluble in water, many investigators may find it useful to be able to examine both lipid- and water-soluble products of endogenous PLC and/or PLD activation. As a result, methods were developed to permit efficient differential extraction of choline and phosphorylcholine from the methanol/water upper phase of a Bligh and Dyer cell extract (neither choline nor phosphorylcholine is found to any significant degree in the lower chloroform-containing phase). We exploit the ability of the tetraphenylborate ion to selectively complex with choline[6] (but not with phosphorylcholine[4]) to render the choline–tetraphenylboron complex soluble in heptanone in preference to water/methanol. Aqueous sodium phosphate is added to the Bligh and Dyer upper phase to maintain a pH of approximately 7.0 and to reduce the concentration of methanol in order to increase the efficiency of extraction of the choline–tetraphenylborate complex into heptanone (80 ± 3 compared to 67 ± 5% when no additional water is added[4]). Purposefully incomplete sampling (69% of the theoretical maximum) is employed to improve the reproducibility of the multiple extractions required in the assay (average standard error of the mean equals 3.5%).

Although adding NaCl in a Bligh and Dyer extraction promotes phase separation and disrupts protein–lipid interactions, increasing the ionic strength interrupts the choline–tetraphenylboron ion pair and reduces the effectiveness of choline extraction into tetraphenylboron-containing heptanone. Because 25 m*M* NaCl causes a 10% reduction and 100 m*M* a 25% reduction in choline recovery,[4] the addition of unnecessary salt is discouraged.

Extraction of Cellular Phosphorylcholine

To obtain cellular phosphorylcholine, the 1.5-ml choline-depleted methanol/water phase described above is subjected to a second identical extraction with sodium tetraphenylboron in heptanone to further reduce

[6] F. Fonnum, *Biochem. J.* **113,** 291 (1969).

residual choline (to <5% of that initially present). A 1.0-ml sample of this doubly extracted methanol–water extract is dried either in a Speed Vac or by heating at 65° overnight.

Assay of Cellular Choline

Principle. As illustrated in Fig. 1, cellular choline (obtained by differential extraction as described above) is quantitated by radioenzymatic phosphorylation using [γ-^{32}P]ATP and the labeled phosphorylcholine product separated from [^{32}P]ATP by ion-exchange chromatography.

Reagents

- Reaction buffer 1: 1.25 m*M* [γ-^{32}P]ATP [200 dpm (disintegrations per minute)/pmol], 50 m*M* sodium phosphate (pH 7.0), 1 mg/ml dithiothreitol, 0.5 U/ml choline kinase (0.6 U/mg protein; from Sigma, St. Louis, MO), 15 m*M* $MgCl_2$
- Phosphorylcholine, 2.5 mg/ml
- Barium acetate, 0.3 *M*
- Ammonium acetate, 75 m*M* (pH 10)
- Dowex AG1-X8, formate form (200–400 mesh; Sigma, St. Louis, MO), equilibrated in ammonium acetate; a column containing 300 μl of resin is made in a 1-ml pipettor tip
- HCl, 2 *N*

Procedure. To the dried sample containing choline, 50 μl of reaction buffer 1 is added. After mixing, the reaction is allowed to proceed for 30 min at room temperature and is halted by adding 20 μl of barium acetate, 80 μl of unlabeled phosphorylcholine, and 100 μl of water. The insoluble barium salt of ATP is removed by centrifugation in a Microfuge and residual labeled ATP removed by subsequent ion-exchange chromatography. Specifically, a 200-μl aliquot of the supernatant is applied to a 300-μl Dowex anion-exchange column and [^{32}P]phosphorylcholine eluted with 900 μl of 75 m*M* ammonium acetate (pH 10) while ATP is bound to the resin. The eluate is collected into a 5-ml scintillation vial containing 225 μl of 2 *N* HCl (bringing the mixture to near neutral pH), and radioactivity is quantitated after adding 3.3 ml of scintillation fluid.

Remarks. Choline is efficiently radioactively phosphorylated using choline kinase (yields of 90% are typical). The formation of labeled phosphorylcholine is linearly related to the quantity of choline present in the cell extract from 10 pmol to 10 nmol, with a useful range of 5 pmol–30 nmol.[4] Although recovery was quite uniform from experiment to experiment, the use of [^{3}H]choline allowed exact correction for choline recovery for each tube. The sensitivity of the assay (5 pmol) is limited not by the ability to achieve sufficient radioactive specific activity, but rather by the presence of ^{32}P-labeled products eluting even in the absence of added choline kinase. Because increasing the amount of ion-exchange resin and changing a variety of chromatographic parameters were unsuccessful at reducing the elution of radioactive compounds (0.0026% of the [^{32}P]ATP initially added), it is suspected that the [^{32}P]ATP contained minor quantities of labeled contaminants failing to bind to the resin. Increased sensitivity could likely be achieved by further microtization of the assay from 50 μl to 10–20 μl. Physiologically relevant amino alcohols (serine and ethanolamine) do not interfere with the extraction and/or quantitation of choline even when 100 nmol is added.[4]

Assay of Cellular Phosphorylcholine

Principle. After extracting cellular choline from other methanol/water soluble cell constituents (described above), residual phosphorylcholine is obtained and quantitated by enzymatically converting it to choline *in vitro* followed by extraction and radioenzymatic assay as described above (illustrated in Fig. 1).

Reagents for Conversion of Phosphorylcholine to Choline

- Reaction buffer 2: 20 m*M* glycine (pH 10.4), 1 m*M* $ZnCl_2$, 1 m*M* $MgCl_2$
- Alkaline phosphatase, bovine intestinal mucosa (Sigma P-6899; 1000 U/ml in 3 *M* NaCl; 1000 U/mg protein)
- Sodium phosphate, 6 m*M* (pH 7.0)
- Sodium tetraphenylboron, 5 mg/ml in 4-heptanone

Procedure for Conversion of Phosphorylcholine to Choline. A 1.0-ml sample of the choline-depleted water–methanol phase of the Bligh and Dyer cell extract (dried using a Speed Vac or by heating) is dried, subsequently solubilized in 90 μl of reaction buffer 2, and then treated with 10 μl of alkaline phosphatase. After mixing, the reaction is allowed to proceed for 30 min at 37° and is terminated by heating to 105° for 5 min. The choline produced by alkaline phosphatase-mediated hydrolysis of phosphorylcholine is extracted into the organic phase by adding the following reagents: 1.4 ml of 6 m*M* sodium phosphate (pH 7.0) and 1 ml sodium tetraphenylboron in heptanone as described above for cell-derived choline.

Procedure for Quantitation of Choline Derived from Phosphorylcholine. The choline formed by alkaline phosphatase-catalyzed hydrolysis of cellular phosphorylcholine is quantitated exactly as described above for cell-derived choline, with the exception that recovery of [^{3}H]choline is not assessed.

Remarks. After phosphorylcholine extraction, alkaline phosphatase-catalyzed hydrolysis, and extraction of the enzymatically produced choline, the recovery of choline originally present as phosphorylcholine is 30 ± 4% of the total (corresponding to 65% of the maximum theoretical recovery of 46% imposed by purposefully incomplete sampling at several steps). The formation of [^{32}P]phosphorylcholine from cell-derived phosphorylcholine is linear from 30 pmol to 10 nmol of added phosphorylcholine and has a useful range of 10 pmol–30 nmol.[4] The assay for phosphorylcholine is less sensitive than that for choline due to (1) higher background caused by the introduction of small amounts of choline by alkaline phosphatase and (2) reduced recovery of phosphorylcholine-derived choline compared to cellular choline. The presence of cellular choline had little impact on the quantitation of phosphorylcholine since only 5% of the choline originally present remained after differential extraction. Further, since cellular levels of choline are often an order of magnitude less than those of phosphorylcholine (13.1 ± 1.2 pmol of choline and 182 ± 19 pmol of phosphorylcholine per 10^6 mast cells[4]), the modest residual presence of the former is not a serious limitation. No crossover of phosphorylcholine into the choline assay is observed.

Of note is the finding that the observed conversion of [^{3}H]phosphorylcholine to [^{3}H]choline occurred to a greater extent than did conversion of phosphorylcholine mass to choline mass. Since no isotope effect was detected with regard to extraction of either choline or phosphorylcholine, this effect may be due to preferential cleavage of [^{3}H]phosphorylcholine compared to unlabeled phosphorylcholine by alkaline phosphatase.

Concluding Comments

These methods provide a sensitive and relatively simple assay of choline and phosphorylcholine, the water-soluble products of PLD and PLC on PC. Principal methodologic improvements over existing methods include (1) increased sensitivity, speed, and ease in the assay of choline, (2) easier and more effective conversion of phosphorylcholine to choline, and (3) the ability to assay both choline and phosphorylcholine from a single cellular extract by separating choline and phosphorylcholine using simple differential extraction in the presence of an ion-pairing reagent. Using this approach, we have demonstrated receptor-dependent increases in cellular choline, but not of phosphorylcholine, in mast cells, suggesting the activation of a cell-associated PLD toward PC.[7] As increasing attention is focused on the importance of cellular PLC and PLD activity toward PC in the generation of lipid second messengers, the ability to quantitate the cell-

[7] T. T. Dinh and D. A. Kennerly, manuscript in review (1990).

associated water-soluble products of these reactions using the methods described in this chapter will likely prove increasingly valuable.

Acknowledgments

The author wishes to thank Ms. Tammy Dinh for superb technical assistance in the development of these techniques. This work was supported by a grant from the National Institutes of Health (1-RO1-AI22277). Dr. Kennerly was also supported by a Burroughs Wellcome/American Academy of Allergy and Immunology Developing Investigator Award in Immunopharmacology of Allergic Diseases.

Section II

Phospholipase Structure–Function Techniques

[18] Dissection and Sequence Analysis of Phospholipases A_2

By ROBERT L. HEINRIKSON

Introduction

Our understanding of the function of phospholipases A_2 (PLA_2) is dependent, in large measure, on a sophisticated level of structural information that has been generated for these enzymes. Indeed, although mechanistic details have yet to be deciphered, the PLA_2 rank among the best characterized enzymes in terms of structure. Complete primary structures have been elucidated for dozens of PLA_2, and partial sequences of many more have been reported in the literature. Readers are referred to two recent reviews which present tabulations of more than 30 primary structures for PLA_2 from a variety of venom and pancreatic sources.[1,2] Many of these sequences may be found in this chapter among the 46 structures listed in Fig. 1, a compilation which includes several sequences that have been determined more recently by the application of both protein and gene sequencing methodologies.

In addition, a detailed picture of the three-dimensional architecture of the PLA_2 has emerged from X-ray crystallographic analysis. Tertiary structures have been determined for bovine pancreatic PLA_2,[3] the enzyme dimer from rattlesnake venom,[4,5] and the variant from water moccasin with a lysine in place of the aspartyl residue at position 49, the so-called K-49 PLA_2.[6] As is clear from the comparisons listed in Fig. 1, all of the enzymes from snake venom and pancreatic sources show strong sequence homology. As would be expected, these similarities at the level of amino acid sequence are even more pronounced in comparisons of the three-dimensional structures of the three enzymes mentioned above which fold in essentially identical patterns. Furthermore, it has been satisfying to

[1] E. A. Dennis, *in* "The Enzymes" (P. D. Boyer, ed.), 3rd Ed., Vol. 16, p. 307. Academic Press, New York, 1983.

[2] H. M. Verheij, A. J. Slotboom, and G. H. de Haas, *Rev. Physiol. Biochem. Pharmacol.* **91,** 91 (1981).

[3] B. W. Dijkstra, J. Drenth, K. H. Kalk, and P. J. Vandermaalen, *J. Mol. Biol.* **124,** 53 (1978).

[4] C. Keith, D. S. Feldman, S. Deganello, J. Glick, K. B. Ward, E. O. Jones, and P. B. Sigler, *J. Biol. Chem.* **256,** 8602 (1981).

[5] R. Renetseder, S. Brunie, B. W. Dijksra, J. Drenth, and P. B. Sigler, *J. Biol. Chem.* **260,** 11627 (1985).

[6] D. K. Holland, L. L. Clancy, S. W. Muchmore, T. J. Rydel, H. M. Einspahr, B. C. Finzel, R. L. Heinrikson, and K. D. Watenpaugh, *J. Biol. Chem.* in press, (1990).

Source	Sequence 1 – 60
C adamanteus	SLVQFETLIM-KVAKRSGLLWYSAYGCYCGWGGHGRPQDATDRCCFVHDCCYG---KATN
C atrox	SLVQFETLIM-KIAGRSGLLWYSAYGCYCGWGGHGLPQDATDRCCFVHDCCYG---KATD
App dimer	DLMQFETLIM-KIAKRSGMFWYSAYGCYCGWGGQGRPQDATDRCCFVHDCCYG---KVTG
App D-49	NLFQFEKLIK-KMTGKSGMLWYSAYGCYCGWGGQGRPKDATDRCCFVHDCCYG---KVTG
App K-49	SVLELGKMIL-QETGKNAITSYGSYGCNCGWGHRGQPKDATDRCCFVHKCCYK---KLTD
Ammodytoxin A	SLLEFGMMIL-GETGKNPLTSYSFYGCYCGVGGKGTPKDATDRCCFVHDCCYG---NLPD
Mamushi	HLLQFRKMIK-KMTGKEPVISYAFYGCYCGSGGRGKPKDATDRCCFVHDCCYE---KVTG
Habu snake X	HLLQFRKMIK-KMTGKEPIVSYAFYGCYCGKGGRGKPKDATDRCCFVHDCCYE---KVTG
Crotoxin basic	HLLQFNKMIK-FETRKNAIPFYAFYGCYCGWGGQGRPKDATDRCCFVHDCCYG---KLAK
Halys acidic	SLIQFETLIM-KVAKKSGMFWYSNYGCYCGWGGQGRPQDATDRCCFVHDCCYG---KVTG
Halys neurotox	NLLQFNKMIK-EETGKNAIPFYAFYGCYCGGGGQGKPKDGTDRCCFVHDCCYG---RLVN
Habu snake	GLWQFENMII-KVVKKSGILSYSAYGCYCGWGGRGKPKDATDRCCFVHDCCYG---KVTG
Himehabu	HLMQFETLIM-KIAGRSGVWWYGSYGCYCGAGGQGRPQDPSDRCCFVHDCCYG---KVTG
Western sand vi	NLFQFAKMIN-GKLGAFSVWNYISYGCYCGWGGQGTPKDATDRCCFVHDCCYG---RVRG
Caudoxin	NLIQFGNMIS-AMTGKSSL-AYASYGCYCGWGGKGQPKDDTDRCCFVHDCCYG---KADK
Rhinoceros adder	DLTQFGNMIN--KMGQ-SVFDYIYYGCYCGWGGQGKPRDATDRCCFVHDCCYG---KMGT
Gabon adder	DLTQFGNMIN--KMGQ-SVFDYIYYGCYCGWGGKGKPIDATDRCCFVHDCCYG---KMGT
Human Platelet	NLVNFHRMIK-LTTGKEAALSYGFYGCHCGVGGRGSPKDATDRCCVTHDCCYK-RLEKRG
Bluering sea kr	NLAQFALVIKCADKGKRPRWHYMDYGCYCGPGGSGTPVDELDRCCKTHDQCYAQAEKK-G
Broadband blueI	NLVQFSNLIQCNVKGSRASYHYADYGCYCGAGGSGTPVDELDRCCKIHDNCYGEAEKM-G
Broadband b III	NLVQFTYLIQCANSGKRASYHYADYGCYCGAGGSGTPVDELDRCCKIHDNCYGEAEKM-G
Beaked sea snak	NLVQFSYVITCANHNRRSSLDYADYGCYCGAGGSGTPVDELDRCCKIHDDCYGEAEKQ-G
Notechis 5	NLVQFSYLIQCANHGRRPTRHYMDYGCYCGWGGSGTPVDELDRCCKIHDDCYSDAEKK-G
Notexin	NLVQFSYLIQCANHGKRPTWHYMDYGCYCGAGGSGTPVDELDRCCKIHDDCYDEAGKK-G
Mulga snake Pa11	NLIQFGNMIQCANKGSRPSLDYADYGCYCGWGGSGTPVDELDRCCQVHDNCYEQAGKK-G
Mulga snake Pa13	NILQFRKMIQCANKGSRAAWHYLDYGCYCGPGGRGTPVDELDRCCKIHDDCYIEAGKD-G
Taipoxin alpha	NLLQFGFMIRCANRRSRPvWHYMDYGCYCGKGGSGTPVDDLDRCCQVHDECYGEAVRRFG
Nigexine	NLYQFKNMIHCTVP-SRPWWHFADYGCYCGRGGKGTPIDDLDRCCQVHDNCYEKAGKM-G
Mozambique III	NLYQFKNMIHCTVP-SRPWWHFADYGCYCGRGGKGTPVDDLDRCCQVHDNCYEKAGKM-G
Mozambique I	NLYQFKNMIHCTVP-SRPWWHFADYGCYCGRGGKGTAVDDLDRCCQVHDNCYGEAEKL-G
Forest cobra I	NLYQFKNMIHCTVP-NRPWWHFANYGCYCGRGGKGTPVDDLDRCCQIHDKCYDEAEKISG
Forest cobraIII	NLYQFKNMIHCTVP-NRSWWHFANYGCYCGRGGSGTPVDDLDRCCQIHDNCYGEAEKISG
Monocled c III	NLYQFKNMIQCTVP-NRSWWDFADYGCYCGRGGSGTPVDDLDRCCQVHDNCYDEAEKISR
Chinese cobra	NLYQFKNMIQCTVP-SRSWWDFADYGCYCGRGGSGTPVDDLDRCCQVHDNCYNEAEKISG
Ringhals	NLYQFKNMIKCTVP-SRSWWHFANYGCYCGRGGSGTPVDDLDRCCQTHDNCYSDAEKISG
Shieldnose snak	NLYQFKNMIQCTVP-NRSWWHFADYGCFCGYGGSGTPVDELDRCCQTHDNCYSEAEKLSG
B multicinctus	NLYQFKNMIVCAG--TRPWIGYVNYGCYCGAGGSGTPVDELDRCCYVHDNCYGEAEKIPG
β-Bungarotx A3	NLINFMEMIRYTIPCEKTWGEYTDYGCYCGAGGSGRPIDALDRCCYVHDNCYGDDAAIRD
β-Bungarotx A1	NLINFMEMIRYTIPCEKTWGEYADYGCYCGAGGSGRPIDALDRCCYVHDNCYGDAEKKHK
Dog pancreas	AVWQFRNMIKCTIPESDPLKDYNDYGCYCGLGGSGTPVDELDKCCQTHDHCYSEAKKLDS
Human pancreas	AVWQFRKMIKCVIPGSDPFLEYNNYGCYCGLGGSGTPVDELDKCCQTHDNCYDQAKKLDS
Rat pancreas	AVWQFRNMIKCTIPGSDPLREYNNYGCYCGLGGSGTPVDDLDRCCQTHDHCYNQAKKLES
Cow pancreas	ALWQFNGMIKCKIPSSEPLLDFNNYGCYCGLGGSGTPVDDLDRCCQTHDNCYKQAKKLDS
Pig pancreas	ALWQFRSMIKCAIPGSHPLMDFNNYGCYCGLGGSGTPVDELDRCCETHDNCYRDAKNLDS
Pig variant	ALWQFRSMIKCTIPGSDPLLDFNNYGCYCGLGGSGTPVDELDRCCETHDNCYRDAKNLDS
Horse pancreas	AVWQFRSMIQCTIPNSKPYLEFNDYGCYCGLGGSGTPVDELDACCQVHDNCYTQAKELSS

FIG. 1. Comparisons of amino acid sequences of select phospholipases A_2 from snake venom and mammalian sources. The alignment was made with the aid of the CLUSTAL program [D. G. Higgens and P. M. Sharp, *Gene* **73,** 237 (1988)]. Key to references: (*a*) R. L. Heinrikson, E. T. Krueger, and P. S. Keim, *J. Biol. Chem.* **252,** 4913 (1977); (*b*) A. Randolph and R. L. Heinrikson, *J. Biol. Chem.* **257,** 2155 (1982); (*c*) W. Welches, I. M. Reardon, and R. L. Heinrikson, submitted for publication; (*d*) J. M. Maraganore and R. L. Heinrikson, *J. Biol. Chem.* **261,** 4797 (1986); (*e*) A. Ritonja and F. Gubensek, *Biochim. Biophys. Acta* **828,** 306 (1985); (*f*) S. Forst, J. Weiss, P. Blackburn, B. Frangione, F. Goni, and P. Elsbach, *Biochemistry* **25,** 4309 (1986); (*g*) R. M. Kini, S. I. Kawabata, and S. Iwanaga, *Toxicon* **24,** 1117 (1986); (*h*) S. D. Aird, I. I. Kaiser, R. V. Lewis, and W. G. Kruggel, *Arch. Biochem. Biophys.* **249,** 296 (1986); (*i*) Y. C. Chen, J. M. Maraganore, I. Reardon, and R. L.

Sequence 61–134	Ref.
-----NPKTVSYTYSEENGEIVCGG-DDPCGTQICECDKAAAICFRDNIPSYDNK-YWLFPPKDC-RQEPEPC	a
-----NPKTVSYTYSEENGEIICGG-DDPCGTQICECDKAAAICFRDNIPSYDNK-YWLFPPKDC-REEPEPC	b
-----DPKLDSYTYSVENGDVVCGG-NNPCKKEICECDRAAAICFRDNKVTYDNK-YWRFPPQNC-KEESEPC	c
-----NPKMDIYTYSVENGNIVCGG-TNPCKKQICECDRAAAICFRDNLKTYDSKTYWKYPKKNC-KEESEPC	c, d
-----NHKTDRYSYSWKNKAIICEEKN-PCLKEMCECDKAVAICLRENLDTYNK-KYKAYFKLKC-KKP-DTC	d
-----SPKTDRYKYHRENGAIVCGKGT-SCENRICECDRAAAICFRKNLKTYNY-IYRNYPDFLC-KKESEKC	e
-----KPKWDDYTYSWKNGDIVCGGDD-PCKKEICECDRAAAICFRDNLKTYKK-RYMAYPDILC-SSKSEKC	f
-----DPKWSYYTYSLENGDIVCGGDP-YCTKVKCECDKKAAICFRDNLKTYKN-RYMTFPDIFC-TDPTEGC	g
-----NTKWDIYRYSLKSGYITCGKGT-WCEEQICECDRVAAECLRRSLSTYKY-GYMFYPDSRC-RGPSETC	h
-----DPKMDVYSFSEENGDIVCGG-DDPCKKEICECDRAAAICFRDNLTLYNDKKYWAFGAKNCPQEESEPC	i
-----NTKSDIYSYSLKEGYITCGK-GTNCEEQICECDRVAAECFRRNLDTYNNGYMFYR-DSKC-TETSEEC	j
-----NPKLGKYTYSWQ-GNIVCGG-DDPCDKEVCECDRAAAICFRDNLDTYDRNKYWRYPASNC-QEDSEPC	k
-----NTKDEFYTYTEEEGAISCGG-NDPCLKEVCECDLAAAICFRDNLNTYDSKKYWMFPAKNCLESE-EPC	l
-----NPKLAIYSYSFKKGNIVCG-KNNGCLRDICECDRVAANCFHQNKNTYN-KNYKFLSSSRC-RQTSEQC	m
-----SPKMILYSYKFHNGNIVCG-DKNACKKKVCECDRVAAICFAASKHSYN-KNLWRYPSSKCT-GTAEKC	n
-----DTKWTSYKYEFQDGDIICG-DKDPQKKELCECDRVAAICFANSRNTYN-SKYFGYSSSKCT-ET-EQC	o
-----DTKWTSYNYEIQNGGIDC--DEDPQKKELCECDRVAAICFANNRNTYN-SNYFGHSSSKCT-GT-EQC	p
-----GTLFLSYKFSNSGSRITCAK-QDSCRSQLCECDKAAATCFARNKTTYN-KKYQYYSNKHC-RGSTPRC	q
CY-----PKLTMYSYYCGGDGPYCNS-KTECQRFVCDCDVRAADCFAR--YPYNNKNYNINTSKRCK-------	r
CY-----PKWTLYTYESCTDTSPCDE-KTGCQGFVCACDLEAAKCFAR--SPYNNKNYNIDTSKRCK-------	s
CY-----PKLTMYNYYCGTQSPTCDD-KTGCQRYVCACDLEAAKCFAR--SPYNNKNYNIDTSKRCK-------	s
CY-----PKMLMYDYYCGSNGPYCRNVKKKCNRKVCDCDVAAAECFAR--NAYNNANYNIDTKKRCK-------	t
CS-----PKMSAYDYYCGENGPYCRNIKKKCLRFVCDCDVEAAFCFAK--APYNNANWNIDTKKRCQ-------	u
CF-----PKMSAYDYYCGENGPYCRNIKKKCLRFVCDCDVEAAFCFAK--APYNNANWNIDTKKRCQ-------	v
CF-----PKLTLYSWKCTGNVPTCNS-KPGCKSFVCACDAAAAKCFAK--APYKKENYNIDTKKRCK-------	w
CY-----PKLTWYSWDCTGDAPTCNP-KSKCKDFVCACDAAAAKCFAK--APYNKANWNIDTKTRCK-------	w
CA-----PYWTLYSWKCYGKAPTCNT-KTRCQRFVCRCDAKAAECFAR--SPYQNSNWNINTKARCR-------	x
CW-----PYFTLYKYKCSKGTLTCNGRNGKCAAAVCNCDLVAANCFAG--APYINANYNIDFKKRCQ-------	y
CW-----PYFTLYKYKCSQGKLTCSGGNSKCGAAVCNCDLVAANCFAG--ARYIDANYNINFKKRCQ-------	z
CW-----PYLTLYKYECSQGKLTCSGGNNKCEAAVCNCDLVAANCFAG--APYIDANYNVNLKERCQ-------	z
CW-----PYIKTYTYESCQGTLT-CKDGGKCAASVCDCDRVAANCFAR--ATYNDKNYNIDFNARCQ-------	aa
CW-----PYIKTYTYDSCQGTLTSCGAANNCAASVCDCDRVAANCFAR--APYIDKNYNIDFNARCQ-------	ab
CW-----PYFKTYSYECSQGTLTCKNGNNACAAAVCDCDRLAAICFAG--APYNNNNYNIDLEARCQ-------	ac
CW-----PYFKTYSYECSQGTLTCKGGNNCAAAAVCDCDRLAAICFAG--APYNDNDYNINLKARCQE------	ad
CR-----PYFKTYSYDCTKGKLTCKEGNNECAAFVCKCDRLAAICFAG--AHYNDNNNYIDLARHCQ-------	ae
CK-----PYIKTYSYDCSQGKLTCSGNDDKCAAFVCNCDRVAAICFAG--APYIDDNYNVDLNERCQ-------	af
C-----NPKTKTYSYTCTKPNLTCTDAAGTCARIVCDCDRTAAICFAA--APYNINNFMISSSTHCQ-------	ag
C-----NPKTSQYSYKLTKRTIICYGAAGTCARVVCDCDRTAALCFGQ--SDYIEGHKNIDTARFCQ-------	ah
C-----NPKTSQYSYKLTKRTIICYGAAGTCGRIVCDCDRTAALCFGQ--SDYIEGHKNIDTARFCQ-------	ai
CKFLLDNPYTKIYSYSCSGSEITCSSKNKDCQAFICNCDRSAAICFSK--APYNKEHKNLDTKKYC--------	aj
CKFLLDNPYTHTYSYSCSGSAITCSSKNKECEAFICNCDRNAAICFSK--APYNKAHKNLDTKKYCQS------	ak
CKFLIDNPYTNTYSYKCSGNVITCSDKNNDCESFICNCDRQAAICFSK--VPYNKEYKDLDTKKHC--------	aj
CKVLVDNPYTNNYSYSCSNNEITCSSENNACEAFICNCDRNAAICFSK--VPYNKEHKNLDKKK-C--------	al
CKFLVDNPYTESYSYSCSNTEITCNSKNNACEAFICNCDRNAAICFSK--APYNKEHKNLDTKKYC--------	am
CKFLVDNPYTNSYSYSCSNTEITCNSKNNACEAFICNCDRNAAICFSK--APYNKEHKNLDTKKYC--------	an
CRFLVDNPYTESYKFSCSGTEVTCSDKNNACEAFICNCDRNAAICFSK--APYNPENKNLDSKRKCA-------	ao

Heinrikson, *Toxicon* **25,** 401 (1987); (*j*) K. Kondo, J.-K. Zhang, K. Xu, and H. Kagamiyama, *J. Biochem.* **105,** 196 (1989); (*k*) S. Tanaka, N. Mohri, H. Kihara, and M. Ohno, *J. Biochem. (Tokyo)* **99,** 281 (1986); (*l*) F. J. Joubert and T. Haylett, *Hoppe-Seyler's Z. Physiol. Chem.* **362,** 997 (1981); (*m*) I. Mancheva, T. Kleinschmidt, B. Aleksiev, and G. Braunitzer, *Hoppe-Seyler's Z. Physiol. Chem.* **365,** 885 (1984); (*n*) C. C. Viljoen, D. P. Botes, and H. Kruger, *Toxicon* **20,** 715 (1982); (*o*) F. J. Joubert, G. S. Townshend, and D. P. Botes, *Hoppe-Seyler's Z. Physiol. Chem.* **364,** 1717 (1983); (*p*) D. P. Botes and C. C. Viljoen, *J. Biol. Chem.* **249,** 3827 (1974); (*q*) R. M. Kramer, C. Hession, B. Johansen, G. Hayes, P. McGray, E. P. Chow, R. Tizard, and R. B. Pepinsky, *J. Biol. Chem.* **264,** 5768 (1989); (*r*) G. Guignery-Frelat, F. Ducancel, A. Menez, and J. C. Boulain, *Nucleic Acids Res.* **15,** 5892 (1987); (*s*) S. Nishida, H. S. Kim, and N. Tamiya, *Biochem. J.* **207,** 589 (1982); (*t*) P. Lind and D. Eaker, *Toxicon*

note that the structures of the rare cellular PLA_2 involved in regulation and implicated in inflammation also fit the same structural format.[7] This similarity, evident from the sequence of human platelet PLA_2 in Fig. 1,[8] carries with it the inference of a tertiary structure that will correspond closely to those of PLA_2 already determined. The gene-derived protein sequence of the PLA_2 from bee venom[9] has corrected errors in the structure generated by conventional protein sequencing[10,11] and, thus, provides concrete evidence that this structurally distinct PLA_2 is, as well, related to the others by a divergent evolutionary process.[12]

The present chapter concerns general approaches to the fragmentation and protein sequence analysis of the PLA_2. At first glance, one might wonder why such a topic was selected for inclusion in this volume, and it is, therefore, appropriate to consider the rationale behind this decision. Perhaps the most obvious reason for treating this subject is that most of our information concerning PLA_2 primary structure has been derived from

[7] S. Forst, J. Weiss, P. Elsbach, J. M. Maraganore, I. Reardon, and R. L. Heinrikson, *Biochemistry* **25,** 8381 (1986).

[8] R. M. Kramer, C. Hession, B. Johansen, G. Hayes, P. McGray, E. P. Chow, R. Tizard, and R. B. Pepinsky, *J. Biol. Chem.* **264,** 5768 (1989).

[9] K. Kuchler, M. Gmachl, M. J. Sippl, and G. Kreil, *Eur. J. Biochem.* **184,** 249 (1989).

[10] R. A. Shipolini, G. L. Callewaert, R. C. Cottrell, and C. A. Vernon, *Eur. J. Biochem.* **48,** 465 (1974).

[11] R. A. Shipolini, S. Doonan, and C. A. Vernon, *Eur. J. Biochem.* **48,** 477 (1974).

[12] J. M. Maraganore, R. A. Poorman, and R. L. Heinrikson, *J. Protein Chem.* **6,** 173 (1987).

19, 11 (1981); (*u*) J. Halpert and D. Eaker, *J. Biol. Chem.* **251,** 7343 (1976); (*v*) J. Halpert and D. Eaker, *J. Biol. Chem.* **250,** 6990 (1976); (*w*) S. Nishida, M. Terashima, and N. Tamiya, *Toxicon* **23,** 87 (1985); (*x*) P. Lind and D. Eaker, *Eur. J. Biochem.* **124,** 441 (1982); (*y*) S. Chwetzoff, S. Tsunasawa, F. Sakiyama, and A. Menez, *J. Biol. Chem.* **264,** 13289 (1989); (*z*) F. J. Joubert, *Biochim. Biophys. Acta* **493,** 216 (1977); (*aa*) F. J. Joubert, *Biochim. Biophys. Acta* **379,** 317 (1975); (*ab*) F. J. Joubert, *Biochim. Biophys. Acta* **379,** 329 (1975); (*ac*) F. J. Joubert and N. Taljaard, *Eur. J. Biochem.* **112,** 493 (1980); (*ad*) I. H. Tsai, S. H. Wu, and T. B. Lo, *Toxicon* **19,** 141 (1981); (*ae*) F. J. Joubert, *Eur. J. Biochem* **52,** 539 (1975); (*af*) F. J. Joubert, *J. Biol. Chem. Hoppe-Seyler* **368,** 1597 (1987); (*ag*) K. Kondo, H. Toda, and K. Narita, *J. Biochem. (Tokyo)* **89,** 37 (1981); (*ah*) K. Kondo, H. Toda, K. Narita, and C. Y. Lee, *J. Biochem. (Tokyo)* **91,** 1531 (1982); (*ai*) K. Kondo, K. Narita, and C. Y. Lee, *J. Biochem. (Tokyo)* **83,** 101 (1978); (*aj*) O. Ohara, M. Tamaki, E. Nakamura, Y. Tsuruta, Y. Fujii, M. Shin, H. Teraoka, and M. Okamoto, *J. Biochem. (Tokyo)* **99,** 733 (1986); (*ak*) R. Grataroli, R. Dijkman, C. E. Dutilh, F. Van der Ouderaa, G. H. de Haas, and C. Figarella, *Eur. J. Biochem.* **122,** 111 (1982); (*al*) E. A. M. Fleer, H. M. Verheij, and G. H. de Haas, *Eur. J. Biochem.* **82,** 261 (1978); (*am*) W. C. Puijk, H. M. Verheij, and G. H. de Haas, *Biochim. Biophys. Acta* **492,** 254 (1977); (*an*) W. C. Puijk, H. M. Verheij, P. Wietzes, and G. H. de Haas, *Biochim. Biophys. Acta* **580,** 411 (1979); (*ao*) H. M. Verheij, J. Westerman, B. Sternby, and G. H. de Haas, *Biochim. Biophys. Acta* **747,** 93 (1983).

protein sequencing. These enzymes are small, easily isolated in high yield and pure form from abundant sources such as pancreas and venoms, and are, therefore, amenable to sequence analysis. Moreover, the active forms of the PLA_2 are unblocked, and the existence of a free NH_2 terminus greatly facilitates generation of protein sequence information by automated or manual Edman degradation. In other words, any investigator is able to obtain sequence information on a new, or old, PLA_2 with relative ease.

Of course, most workers have little interest in determining the complete sequence of a PLA_2. In many instances, however, it is desirable to have a few guidelines as to how to proceed to fragment the protein and zero in on select peptides from particular parts of the molecule that may be of interest. For example, numerous studies have been carried out utilizing chemical derivatization as a means of exploring structure–function relationships. In order to document sites of modification it is necessary to isolate derivatized peptides and identify amino acid residues that have undergone reaction. Insofar as the modified amino acids are within 40 residues of the NH_2 terminus, such identifications can often be made simply by automated Edman degradation. In other cases, however, it may be necessary to break the molecule into peptides and to study the modified fragments. The other obvious application has to do with the design and analysis of PLA_2 molecules that have undergone site-directed mutagenesis. Because of the high homology among PLA_2 species and the vast storehouse of sequence data available for comparisons, patterns of conserved residues have emerged which help relate a new sequence to all of the others. These patterns are familiar to protein chemists and enzymologists working with these enzymes. It follows that by looking at so many different PLA_2 sequences, we have gained important insights as to residues or regions of these small molecules that are conserved and, therefore, possibly essential for function. In a sense, we are looking at Nature's mutagenesis, and from that analysis we can design more critical and meaningful site-directed mutagenesis in the laboratory. Now, as more and more mutagenized PLA_2 forms are being expressed in laboratories throughout the world, it is essential to prove that the desired changes have been effected by analysis of the recombinant protein. One wants to get that information in the most expeditious fashion possible, and this requires some effort in the cleavage and sequence analysis of the protein in question.

In the present chapter, therefore, I address some important ideas about PLA_2 that have come from protein chemistry, some of the structural hallmarks of these enzymes that the protein chemist looks for, and a general treatment of strategies that have proved useful in deriving sequence information. The treatment is not meant to be comprehensive in any sense but will, it is hoped, give the aspiring worker in this field a conceptual framework within which to approach problems of PLA_2 structure.

General Principles

The general level of understanding of protein sequence analysis has fallen into such a decline since the advent of molecular biology and gene sequencing that it is appropriate to spend a little time with some basic information that serves to guide our strategic approach. In all of what follows, we focus on natural PLA_2 from pancreas or venom sources or recombinant enzyme available in reasonable quantities.

Protein Purity

Since the PLA_2 are small, stable enzymes, they are usually easy to isolate, and purity has not been a critical factor in sequence analysis. There are some things to look out for, however. The pancreatic PLA_2 must be activated by trypsin from proPLA precursors, and this step opens the possibility of proteolytic fragmentation. Moreover, the porcine pancreatic enzyme has been shown to consist of a mixture of isozymes that are not usually separated in commercial sources. Snake venoms are interesting in that they usually contain multiple enzymes, sometimes with quite distinct structures and functions. In fact, it is the exception to find a venom source with only one PLA_2 present. One example is *Crotalus atrox*, the western diamondback rattlesnake, which has a single, dimeric PLA_2.[13] The acidic p*I* of this enzyme facilitates its purification simply by gel filtration and anion-exchange chromatography. The closely related eastern diamondback rattlesnake, *Crotalus adamanteus* has two dimeric PLA_2 which differ only in residue 117 which is Glu in α and Gln in β.[14] The venom of *Agkistrodon piscivorus* is more typical in having an acidic dimer and two basic monomeric PLA_2.[15–17] One of the latter is a "typical" PLA_2 in having an Asp at position 49, and the other has a Lys at this position,

[13] B. W. Shen, F. H. C. Tsao, J. H. Law, and F. J. Kezdy, *J. Am Chem. Soc.* **97,** 1205 (1975).

[14] R. L. Heinrikson, E. T. Krueger, and P. S. Keim, *J. Biol. Chem.* **252,** 4913 (1977).

[15] J. M. Maraganore, G. Merutka, W. Cho, W. Welches, F. J. Kezdy, and R. L. Heinrikson, *J. Biol. Chem.* **259,** 13839 (1984).

[16] J. M. Maraganore and R. L. Heinrikson, *J. Biol. Chem.* **261,** 4797 (1986).

[17] W. Welches, I. Reardon, and R. L. Heinrikson, manuscript submitted (1990).

as well as some changes in the calcium-binding loop.[15,18] These Lys-49 PLA_2 have been identified thus far only in crotalid venoms, including *Agkistrodon, Bothrops*,[15] and *Trimesurus*.[18] A Lys-49 PLA_2 is one of two major species in the copperhead snake venom.[19]

Another distinction that must be kept in mind when purifying PLA_2 from venom sources has to do with the possible presence in a single enzyme of both esterolytic and toxic activities. Notexin is a potent neurotoxin from the venom of the Australian tiger snake that has weak PLA_2 activity.[20] The venom of the Chinese water moccasin, *Agkistrodon halys palas*, contains an acidic and a basic PLA_2, together with a neutral neurotoxin with PLA_2 activity.[21] Venoms of more primitive elapid snakes have often been shown to consist of multiple PLA_2 forms, often with varying degrees of toxic activity. It is safe to say that much of the work reported on venom PLA_2 sequences has been done with major forms that are easily resolved from contaminants and other PLA_2 species. The rule, however, is that venoms will most likely present a variety of PLA_2, and this must be borne in mind when formulating purification protocols.

Some further generalizations may be useful here. First, the PLA_2 are unusually small and stable as compared to most proteins, and they withstand rather harsh conditions of pH and temperature. Their low molecular weight (14,000) makes them easily separable from the majority of contaminating proteins by simple gel-exclusion chromatography. Dimeric PLA_2 present a different problem in that they tend to move with proteases on such columns. However, if the columns are run in 5% (v/v) acetic acid, dimers dissociate and run with monomeric PLA_2 on gel filtration.[22] These acidic conditions, therefore, help to minimize proteolysis, although the seven disulfide bridges of PLA make these enzymes highly resistant to proteases. This first step usually provides enzyme in high yield and, if the source is snake venom, in a substantial state of purity. In fact, the only contaminants may be other PLA_2 species. Knowledge of the p*I* of the PLA_2 of interest will dictate use of an anion- or cation-exchange column procedure for the second step of purification. Examples of such protocols are many; readers are referred to that of resolution of multiple PLA_2 forms in *A. piscivorus* venom.[15-17] In this day of reversed-phase high-performance liquid chromatography (HPLC), it may be noteworthy to add that the small size and stability of these enzymes make them particularly

[18] K. Yoshizumi, S.-Y. Liu, T. Miyata, S. Aaita, M. Ohno, S. Iwanaga, and H. Kihara, *Toxicon* **28,** 43 (1990).

[19] R. L. Heinrikson, unpublished observations, 1984.

[20] J. Halpert and D. Eaker, *J. Biol. Chem.* **250,** 6990 (1975).

[21] Y.-C. Chen, J. M. Maraganore, I. Reardon, and R. L. Heinrikson, *Toxicon* **25,** 401 (1987).

[22] W. Welches, D. Felsher, W. Landshulz, and J. M. Maraganore, *Toxicon* **23,** 747 (1985).

well suited to such procedures; yields are high, and resolution is particularly fine. Interestingly, although PLA_2 act at a lipid interface, they do not show any unusual tendency to bind to these hydrophobic matrices.

Group I and Group II Phospholipases A_2

Years ago we classified PLA_2 into two groups based on select structural determinants.[14] This distinction was crude but remains today one of the criteria for defining any particular PLA_2. Group I enzymes were noted in elapid and hydrophid venoms and in mammalian pancreas. They are distinguished from Group II enzymes found in viperid and crotalid venoms by having a disulfide bridge connecting half-cystines 11 and 69 and by an insertion of three amino acids at residues 54 to 56, the so-called elapid loop.[14] Group II PLA_2 not only lack this bridge, but have a COOH-terminal extension of about six amino acids terminating in a half-cystine joined to Cys-50 near the catalytic site His-48. These structural distinctions may be easily discerned with reference to Fig. 1, which displays Group II PLA_2 followed by the Group I enzymes. To date, this classification has held true for all venom and pancreatic PLA_2 subsequently described. We had originally assumed that all mammalian PLA_2 would follow the group I motif seen in the pancreatic enzyme. To our surprise, however, the rare membrane PLA_2 from platelets[8] and that from ascites fluid[7] are Group II PLA_2, so the distinction is not one based on some evolutionary process peculiar to venomous animals. We still have not defined, however, a functional distinction that correlates with the structural differences between Group I and II PLA_2.

Compositional Analysis as Guide to Fragmentation

One of the basic pieces of information protein chemists use in formulating sequence strategies comes from amino acid analysis. This may seem obvious, but it is often forgotten in contemporary studies of protein structure linked to the new molecular biology. The most powerful tool protein sequencers have at their disposal is automated Edman degradation. Such instrumentation has been designed to generate long stretches of sequence from relatively large peptides. It is of interest, therefore, to quantitate the number of rare amino acids which can serve as sites for specific fragmentation, because the resulting peptides will, on average, be quite large.

Prominent among these amino acids which occur relatively infrequently in proteins and for which there exist highly selective cleavage procedures are methionine and arginine. Cyanogen bromide cleaves most bonds COOH terminal to Met residues in high yield, and PLA_2 usually contain one to three methionines, one or two near the NH_2 terminus and one often in the middle of the molecule near position 60 (see Fig. 1). Therefore, if Edman degradation of the intact molecule can proceed beyond 30 residues, the overlap between two or more cyanogen bromide fragments will usually be established. Isolation and analysis of the COOH-terminal fragment will provide placement of most of the approximately 120 amino acids in the chain. Specific cleavage at the COOH-terminal side of arginyl residues can be effected by enzymes like clostripain or by trypsin following citraconylation of the protein and blocking of lysyl side chains. The combination of these two procedures often suffices to provide peptides for the complete structural determination. Of course, reversed-phase HPLC constitutes a method of such high resolving capability that almost any enzyme may be used to digest the PLA_2 of interest, and this is important when one wishes to focus on a particular part of the molecule involved in, for example, chemical derivatization or mutagenesis.

Reduction and Alkylation of Disulfides

Before embarking on any sequencing strategy, it is important to disrupt the tight three-dimensional structure of the PLA. The native molecule is resistant to proteolysis, and the disulfide bridges should be broken and the half-cystines modified irreversibly in order to provide a molecule suitable for structural studies. Reduction of the disulfide bridges in a denaturing milieu such as guanidinium hydrochloride, followed by alkylation with iodoacetate, iodoacetamide, or 4-vinylpyridine, will provide molecules with negative, neutral, or positive substituents, respectively. General guidelines for such procedures may be found in several volumes in this series devoted to protein structure. It is important to bear in mind that alkylation will alter 14 of the 120 residues in the PLA_2, so charge considerations may be crucial to the success of subsequent steps, depending on the p*I* of the enzyme in question. Generally speaking, the S-alkylated PLA_2 derivatives are soluble at neutral pH conditions optimal for proteolysis.

Perhaps a more critical consideration has to do with which particular bonds one wishes to cleave. As mentioned above, Arg residues are often convenient sites for specific, limited hydrolysis, and one such residue is found almost invariably at position 43 (see Fig. 1). Note that this arginine is preceded by Asp-42 and followed by two half-cystines. Alkylation with iodoacetate results in carboxymethylcysteinyl residues at 44 and 45, and this coupled with the invariant Asp at 42 yields a derivative in which Arg-43 is surrounded by negative charge. Most proteolytic enzymes, trypsin included, abhor such an environment, and cleavage at Arg-43 in carboxymethylcysteinyl PLA_2 derivatives is difficult to achieve in reasonable yield.

An alternate choice of alkylating agent, then, might be iodoacetamide. Indeed, carboxamidomethylation yields neutral alkylcysteines at positions 44 and 45, and trypsin readily cleaves the Arg-43–carboxamidomethyl-Cys-44 bond. However, liberation of the free NH_2-terminal carboxamidomethylcysteine is followed by rapid cyclization and blocking of the peptide[19] in a manner analogous to that of NH_2-terminal glutamine cyclization to pyroglutamic acid. In both cases, the resulting peptides are refractive to Edman degradation. The best choice of reagent among those mentioned, therefore, would be 4-vinylpyridine if one wants to be able to cleave the Arg-43 bond in high yield. The resulting pyridylethylated PLA_2 is soluble, and the positively charged environment is amenable to tryptic cleavage. Moreover, pyridylethylcysteine is readily quantified both by amino acid analysis following acid hydrolysis and as the phenylthiohydantoin derivative liberated during the course of Edman degradation.

Of course, much of the foregoing is rooted in strong inferences concerning location of residues based on homology among many PLA_2. The focus given here on Arg-43 is justified by the fact that this residue is just upstream of the active site His-48 and provides a convenient entree into this region. Moreover, one can often proceed 40 to 50 residues into a PLA_2 sequence by automated Edman degradation of the intact protein, thus establishing an overlap through Arg-43. Fortunately, most PLA_2 do not have another Arg in sequence until about residue 90 (Fig. 1), so one can generate a large fragment beginning at residue 44 with which to provide a good amount of the total PLA_2 sequence.

Tour of the Molecule

With this background of information relative to the structure of PLA_2 in general, we next focus attention on specific regions of the molecule. We include comments as to how these regions may be opened up for chemical inspection and how they might be related to the tertiary structure and function of the enzyme.

Residues 1–11

Residues 1–11 form the NH_2-terminal amphiphilic α helix. Much work on this region of the molecule has been done by the group of de Haas *et al.* in Utrecht; the general conclusion is that adding or deleting a residue at the NH_2 terminus or changing the configuration of this residue from L to D destroys enzymatic activity.[23] Chemical derivatization in this helical segment may have quite different functional consequences, depending on

[23] A. J. Slotboom and G. H. de Haas, *Biochemistry* **14,** 5394 (1975).

the residue modified. Alkylation of Met-8 has been shown to lead to inactivation of the pancreatic enzyme,[24] whereas similar modification of the Met-10 in *C. atrox* venom PLA_2 actually causes a slight activation of the enzyme.[25,26] It is generally held, therefore, that the NH_2-terminal helix plays an important functional role. Indeed, crystallographic analysis shows that this region of the molecule forms a wall on one side of the catalytic site that could be involved in interactions with the substrate. There are no apparent catalytic components in this helix, but the distribution of amino acids is such as to give it an amphiphilic character. Highly conserved are the Gln, Phe, and Ile residues at positions 4, 5, and 9. In the Lys-49 PLA_2, residues 4 and 5 are Glu and Leu; this serves as an easily discernable sign that one has such an enzyme under study.

Residue 11 is half-cystine in Group I PLA_2 and something else, often Lys, in the Group II enzymes. In our studies[27] of autoacylation of the PLA_2, we have found that Group II PLA_2 undergo acylation of lysyl residues in the NH_2-terminal helix and that the resulting acylated PLA_2 dimerizes and is activated for hydrolysis. Group I PLA_2 are similarly activated by autoacylation that occurs elsewhere in the molecule[28]; the Cys-11 disulfide bond, missing in the Group II PLA_2 might influence the course of this autoderivatization. The NH_2-terminal peptide can often be isolated by cyanogen bromide cleavage due to occupancy of positions 8 or 10 by Met (see Fig. 1). In a study of the *C. atrox* PLA_2, we showed that removal of the NH_2-terminal decapeptide by cyanogen bromide produced an inactive COOH-terminal fragment that was not activated upon reconstitution with the decapeptide.[25]

Residues 12–24

The stretch of amino acids 12–24 seems to be fairly uninteresting and quite variable; certain classes of PLA_2 show deletions here of no more, however, than a single amino acid (Fig. 1). Crystallographic analysis indicates a short helix, followed by largely nondefined structure. It may be of interest that the disparate honeybee PLA_2 begins in this region and is

[24] F. M. van Wezel, A. J. Slotboom, and G. H. de Haas, *Biochim. Biophys. Acta* **452,** 101 (1976).

[25] A. Randolph and R. L. Heinrikson, *J. Biol. Chem.* **257,** 2155 (1982).

[26] R. L. Heinrikson *in* "Proteins in Biology and Medicine" (R. A. Bradshaw, L. Chih-chuan, R. L. Hill, T. Tien-chin, J. Tang, and T. Chen-lu, eds.), p. 131. Academic Press, New York, 1982.

[27] W. Cho, A. G. Tomasselli, R. L.Heinrikson, and F. J. Kezdy, *J. Biol. Chem.* **262,** 11237 (1988).

[28] A. G. Tomasselli, J. Hui, J. Fisher, H. Z. Neely, I. M. Reardon, E. Oriaku, F. J. Kezdy, and R. L. Heinrikson, *J. Biol. Chem.* **264,** 10041 (1989).

missing approximately 20 residues at the NH_2 terminus of conventional PLA_2.[9]

Residues 25–52

A good part of the discussion here is focused on residues 25–52 because this region encompasses the loop which binds the essential calcium and the catalytic site region containing His-48. Here we find a highly conserved distribution of amino acids (see Fig. 1):

-Tyr-Gly-Cys-Tyr-Cys-Gly-( )-Gly-Gly-( )-Gly-( )-Pro-( )-Asp-
25 30 35

( )-( )-Asp-Arg-Cys-Cys-( )-( )-His-Asp-Cys-Cys-Tyr-
40 42 43 44 45 48 49 50 51 52

Tyr-28, Gly-30, and Gly-32, as well as Asp-49, have been shown to be coordinated to calcium in the crystallographic analysis of the bovine PLA_2–calcium complex.[3] The Lys-49 PLA_2 do not bind calcium until substrate is bound, so the participation of the residue at 49 in the ternary enzyme–phospholipid–calcium complex remains conjectural. The Lys-49 PLA_2 also show variance in the calcium-binding loop not seen in the Asp-49 enzymes.[15,16] Much evidence exists to implicate His-48 in catalysis; any kind of derivatization of this residue leads to inactivation of the PLA_2.[29] Indeed, a histidyl residue is a common feature of esterolytic enzymes, and it is satisfying to see one here that is not only invariant among all PLA_2 but which is coordinated to Asp-99 in a prominent cavity of the enzyme.[3] Cys-50 is not present in Group I PLA_2, often being replaced by Asn or Asp; Tyr-52 is generally invariant. Crystallographic analysis has shown that the calcium-binding loop up to about residue 41 is in a disordered conformation, and that the rest of this region is part of one of the two long central α helices of the PLA_2 (the C-helix). The discussion above focused on Arg-43 as a means of generating peptides which correspond to this part of the catalytic network. The distribution of residues throughout, however, is a clear hallmark of the PLA_2.

It may be best at this point to end our discussion of the molecule without reference to numbering. After Tyr-52, the PLA_2 show a good deal of diversity in structure, for example, with respect to loops and deletions, while maintaining particular essential elements and patterns of structure consistent with their classification. Highly (::::) or absolutely (■) conserved amino acid residues are indicated in Fig. 1. It is clear that the PLA_2 begin to diverge considerably distal to Tyr-52. An exception is the region

Cys-( )-Cys-Asp-( )-( )-( )-Ala-( )-Cys-
99

[29] J. J. Volwerk, W. A. Pieterson, and G. H. de Haas, *Biochemistry* **13,** 1446 (1974).

where Asp-99 is coordinated to His-48 in a manner reminiscent of that seen for Asp-102 in the serine proteinases.

Two loops of structure that are characteristic of PLA_2 subtypes merit recognition, and they are clearly delineated in Fig. 1. The first is the three-residue segment immediately following residue 53, one we referred to as the "elapid loop."[14] In fact, it is a determinant for Group I PLA_2. More prominent, however, is the loop of structure that is peculiar to the pancreatic enzymes following Cys-61, an insertion that serves to distinguish these PLA_2 from other elapid and hydrophid Group I enzymes. The use of genetic engineering to remove this loop (residues 62–66) from the pig pancreatic PLA_2 gave a derivative with enhanced catalytic activity toward micellar lecithin substrates but with slightly decreased activity toward negatively charged substrates.[30] Crystallographic analysis showed no effect of the deletion on catalytic site residues. Interestingly, this loop of structure is very near Lys-56, a residue we have shown to undergo autocatalytic, substrate-level acylation during the course of hydrolysis by the pig pancreatic enzyme.[28] The resulting acylated PLA_2 is a dimer in solution and shows remarkably enhanced catalytic activity toward both natural and artificial substrates. It may be, therefore, that this region is involved in some way with protein–protein interactions which lead to dimerization and enhancement of substrate recognition.

As we proceed beyond the pancreatic loop region, we come to segments of the molecule that have been proposed to be involved in the activity of PLA_2 as toxins.[31] It is well established that many of the PLA_2 from venoms exhibit neuro- or myotoxicity, as well as anticoagulant activity. As we come out of the long central C-helix[3] we enter a region of β structure, sometimes referred to as the β wing, and then into the long central E-helix parallel to the C-helix. Arguments have been presented in support of the view that this region up to about residue 110 contains segments that impart particular toxic or other physiological activities to the PLA_2,[31] but this matter is still open to experimental verification.

As we come to the end of the molecules (Fig. 1), the dramatic structural element is the extension at the COOH terminus seen in Group II PLA_2. The significance of this stretch of amino acids remains to be established. Our early inference that the COOH-terminal Cys had to be disulfide-bonded to the other unique half-Cys of the Group II PLA_2 at position 50[14] has been borne out by crystallographic analyis.[4] Cys-50 is very close to the His-Asp pair at the catalytic site, and it is tempting to speculate that there must be some functional consequence of bringing a COOH-terminal

[30] O. P. Kuipers, M. M. G. M. Thunnissen, P. de Geus, B. W. Dijkstra, J. Drenth, H. M. Verheij, and G. H. de Haas, *Science* **244,** 82 (1989).

[31] R. M. Kini and H. J. Evans, *J. Biol. Chem.* **262,** 14402 (1987).

tail in close proximity to this region in the Group II PLA_2. Thus far, however, no functional distinction exists that would support such a view.

Conclusions

Our understanding of the complex relationship between the structure and the function of proteins remains primitive and rudimentary, despite enormous advances in the technology surrounding structural analysis. We still do not know how the information encoded in the amino acid sequence is translated into folding and function of the protein in question. Indeed, this area of research has been, and continues to be, one of the most challenging in contemporary biochemistry. As was mentioned above, we see Nature's mutagenesis in the dozens of PLA_2 sequences available for examination. Accordingly, we have gained insights as to what parts of the molecule are retained and, therefore, essential for function, be it with respect to calcium binding, esterolysis, protein–protein interaction, enzyme–substrate complexation, or toxicology. Nevertheless, it is clear that we have a long way to go in the process of unraveling the secrets of even so simple a class of enzymes as the PLA_2. It is hoped that the present chapter provides a rationale and chemical basis for probing structure–function relationships in the PLA_2 that will, in turn, help to answer some of these intriguing questions of more general significance.

[19] Cloning, Expression, and Purification of Porcine Pancreatic Phospholipase A_2 and Mutants

By H. M. VERHEIJ and G. H. DE HAAS

Introduction

Phospholipase A_2 (EC 3.1.1.4) attacks the acyl ester bond at position 2 of 3-*sn*-phosphoglycerides.[1] The *in vivo* importance of phospholipase A_2 (PLA_2) is reflected by the fact that this enzyme occurs ubiquitously in nature and that PLA_2 activity has been detected in a large number of cell types and cell organelles. In general their physiological role can be either

[1] L. L. M. van Deenen and G. H. de Haas, *Biochim. Biophys. Acta* **70,** 538 (1963).

a digestive or a regulatory one. The exracellular PLA_2s from mammalian pancreas and snake venom undoubtedly belong to the digestive enzymes whereas the cellular PLA_2s have been assigned regulatory functions mainly.[2]

The PLA_2s are relatively small proteins (14 kDa; about 120 amino acids) and display a very high stability to denaturing conditions, such as high temperature and low pH. The high stability probably is related to the large number (six or seven) of disulfide bridges. The primary structure has been elucidated for about 65 extracellular and 5 cellular PLA_2s. A comparison of these sequences[3] (see also this volume [18]) reveals a high degree of homology. In addition, crystallographic data show striking similarities in the three-dimensional structure of PLA_2s from bovine and porcine pancreas and enzyme from the venom of *Crotalus atrox*.[4] The sequence similarities and the structural resemblances suggest a general mode of action for all PLA_2s. For pancreatic PLA_2s a mechanism of action has been proposed[5] in which an Asp-His couple serves as a proton relay system, resembling that of the serine proteases. Contrary to these esterases, however, PLA_2 lacks a serine in the active center, and a water molecule is held responsible for the nucleophilic attack.

The cellular PLA_2s are membrane-associated enzymes which normally occur in a dormant form; their activation is a matter of debate (see [18] in this volume). The pancreatic PLA_2s are produced and stored in the pancreas in the form of an inactive precursor, and activation involves the removal by trypsin of an amino-terminal heptapeptide. The occurrence of this natural cleavage site has been the basis for the successful expression of porcine pancreatic PLA_2.

Cloning of Porcine Pancreatic Prophospholipase A_2

Cloning of PLA_2s from various sources has been achieved using synthetic genes,[6,7] genomic banks[8,9] and cDNA banks.[8,10–12] For the general

[2] M. Waite, *in* "Handbook of Lipid Research" (D. J. Hanahan, ed.), Vol. 5, p. 155. Plenum, New York, 1987.

[3] C. J. van den Bergh, A. J. Slotboom, H. M. Verheij, and G. H. de Haas, *J. Cell. Biochem.* **39,** 379 (1989).

[4] R. Renetseder, S. Brunie, B. W. Dijkstra, J. Drenth, and P. B. Sigler, *J. Biol. Chem.* **260,** 11627 (1985).

[5] H. M. Verheij, J. J. Volwerk, E. H. J. M. Jansen, W. C. Puijk, B. W. Dijkstra, J. Drenth, and G. H. de Haas, *Biochemistry* **19,** 743 (1980).

[6] J. P. Noel and M.-D. Tsai, *J. Cell. Biochem.* **40,** 309 (1989).

[7] T. Tanaka, S. Kimura, and Y. Ota, *Gene* **64,** 257 (1988).

[8] J. J. Seilhamer, T. L. Randall, Y. Miles, and L. K. Johnson, *DNA* **5,** 519 (1986).

[9] R. M. Kramer, C. Hession, B. Johansen, G. Hayes, P.McGray, E. P. Chow, R. Tizzard, and R. B. Pepinsky, *J. Biol. Chem.* **264,** 5768 (1989).

methodology and references, readers are referred to these publications. The banks were screened in most cases with the aid of oligonucleotides or genetic probes and occasionally in an expression library with antibodies against the PLA_2.[12] The cDNA banks of porcine pancreatic PLA_2 were obtained[8,10] starting from fresh pancreatic tissue that was frozen immediately in liquid nitrogen. From this tissue RNA and poly(A)$^+$ RNA were isolated by established procedures.[13,14] After cDNA synthesis, Seilhamer *et al.*[8] obtained about 1.6×10^5 independent λgt10 clones per microgram RNA. After synthesis of cDNA by the Gubler and Hoffman procedure,[15,16] de Geus *et al.*[10] obtained, per microgram RNA, about 1.5×10^3 independent clones carrying an insert in the *Pst*I site of pBR322. These banks were screened with the aid of degenerated synthetic oligonucleotides that were based on the known amino acid sequence. From the cDNA bank from porcine pancreas, a clone containing a 560-base pair cDNA insert was sequenced and was found to cover the complete coding region for the preproPLA_2.[10] The predicted amino acid sequence (Fig. 1) is in full agreement with the one known for porcine proPLA_2.[17] Expression experiments in eukaryotic cell lines confirmed the functional integrity of the signal peptide sequence, since the proPLA_2 was secreted into the culture medium.[18]

Expression of (Pro)phospholipase A_2

In general, the expression of cloned eukaryotic proteins, at levels that permit characterization of the recombinant protein with methods requiring substantial amounts of protein, is far from being straightforward. In particular, posttranslational modifications can cause serious complications. An example of such a modification is the formation of the disulfide bonds which happen to be abundant in all PLA_2s. Two alternative approaches

---

[10] P. de Geus, C. J. van den Bergh, O. P. Kuipers, H. M. Verheij, W. P. M. Hoekstra, and G. H. de Haas, *Nucleic Acids Res.* **15,** 3743 (1987).

[11] O. Ohara, M. Tamaki, E. Nakamura, Y. Tsuruta, Y. Fujii, M. Shin, H. Teraoka, and M. Okamoto, *J. Biochem.* (*Tokyo*) **99,** 733 (1986).

[12] K. Küchler, M. Gmachl, M. J. Sippl, and G. Kreil, *Eur. J. Biochem.* **184,** 249 (1989).

[13] J. M. Chirgwin, A. E. Przybyla, R. J. MacDonald, and J. Rutter, *Biochemistry* **18,** 5294 (1979).

[14] H. Aviv and P. Leder, *Proc. Natl. Acad. Sci. U.S.A.* **69,** 1408 (1972).

[15] U. Gubler and J. Hoffman, *Gene* **25,** 263 (1983).

[16] U. Gubler, this series, Vol. 152, p. 330.

[17] W. C. Puijk, H. M. Verheij, and G. H. de Haas, *Biochim. Biophys. Acta* **492,** 254 (1977).

[18] P. de Geus, O. P. Kuipers, M. van den Heuvel, H. M. Verheij, and G. H. de Haas, *Chim.* (Ottobre), p. 73 (1987).

```
                                                      50
———poly (G)———  GCT TTT GCT CAC CAA CCT GAC AGC AGG |ATG| AAA TTC
                                                    |Met| Lys Phe
                                                     -15
                                     100
CTC GTG TTG GCT GTT CTG CTC ACA GTG GGC GCT GCC▼CAG GAA GGC ATC AGC TCA AGG↓GCA
Leu Val Leu Ala Val Leu Leu Thr Val Gly Ala Ala Gln Glu Gly Ile Ser Ser Arg Ala
        -10                                 -1   I   II  III  IV   V   VI  VII  1
                             150
TTA TGG CAG TTT CGT AGC ATG ATT AAG TGC GCA ATC CCC GGC AGT CAC CCC TTG ATG GAT
Leu Trp Gln Phe Arg Ser Met Ile Lys Cys Ala Ile Pro Gly Ser His Pro Leu Met Asp
                                 10                                      20
                200
TTC AAC AAC TAT GGC TGC TAC TGT GGC CTA GGT GGA TCA GGG ACC CCT GTG GAT GAA CTG
Phe Asn Asn Tyr Gly Cys Tyr Cys Gly Leu Gly Gly Ser Gly Thr Pro Val Asp Glu Leu
                                 30                                      40
250                                                                 300
GAC AGG TGC TGC GAG ACA CAC GAC AAC TGC TAC AGA GAT GCC AAG AAC CTG GAC AGC TGT
Asp Arg Cys Cys Glu Thr His Asp Asn Cys Tyr Arg Asp Ala Lys Asn Leu Asp Ser Cys
                                 50                                      60
                                                    350
AAA TTC CTC GTG GAC AAT CCC TAC ACC GAA AGC TAC TCC TAC TCA TGT TCT AAC ACT GAG
Lys Phe Leu Val Asp Asn Pro Tyr Thr Glu Ser Tyr Ser Tyr Ser Cys Ser Asn Thr Glu
                                 70                                      80
                                        400
ATC ACC TGC AAC AGC AAA AAC AAT GCT TGT GAG GCC TTC ATC TGT AAC TGT GAC CGA AAT
Ile Thr Cys Asn Ser Lys Asn Asn Ala Cys Glu Ala Phe Ile Cys Asn Cys Asp Arg Asn
                                 90                                      100
                            450
GCT GCC ATT TGC TTC TCA AAG GCC CCA TAC AAC AAG GAG CAC AAG AAC CTG GAC ACC AAG
Ala Ala Ile Cys Phe Ser Lys Ala Pro Tyr Asn Lys Glu His Lys Asn Leu Asp Thr Lys
                                 110                                     120
            500
AAG TAC TGT TAG AGC TAA GTA TCA CCC
Lys Tyr Cys ***     ###
        124
```

FIG. 1. cDNA and deduced protein sequence of porcine pancreatic preproPLA_2. The triangle represents the cleavage site for signal peptidase. Roman numerals denote residues of the activation peptide of proPLA_2, which is cleaved by trypsin (arrow).

can be chosen to solve this problem. First, for expression eukaryotic cells can be used which are capable of *in vivo* disulfide bond formation. Second, a prokaryote like *Escherichia coli* can be used as expression host to produce (partly) incorrectly folded protein, and subsequent *in vitro* refolding and oxidation may then give folded protein with the correct disulfide bonds. Reduction and reoxidation has been shown to be an efficient process both for pancreatic (pro)PLA_2 and for snake venom PLA_2.[19,20] Another problem that is often encountered during heterologous expression deals with the instability of the protein of interest due to cellular proteases.

Expression in Escherichia coli

During our experiments aimed at the expression in *E. coli*, it became evident that porcine pancreatic proPLA_2 could not be expressed to any significant levels, either directly in the cytoplasm, or after processing into

[19] G. J. M. van Scharrenburg, G. H. de Haas, and A. J. Slotboom, *Hoppe-Seyler's Z. Physiol. Chem.* **361,** 571 (1980).

[20] S. Tanaka, Y. Takahashi, N. Mohri, H. Kihara, and M. Ohno, *J. Biochem. (Tokyo)* **96,** 1443 (1984).

the periplasmic space. For this reason we developed a strategy for the expression of PLA_2 as a 59-kDa fusion protein which precipitates inside the cell. The approach combines several published methods[10]: (1) purification of insoluble fusion proteins from the cytoplasm, (2) S-sulfonation and subsequent reoxidation and renaturation of recombinant proteins, and (3) site-specific cleavage of the fusion protein with hydroxylamine or trypsin. Hydroxylamine cleavage of the peptide bond beteen Asn and Gly residues, a sequence which is absent in porcine pancreatic PLA_2, was optimized by the insertion of three consecutive Asn-Gly sequences between proPLA_2 and the bacterial leader fragment of the fusion protein. For tryptic cleavage the "natural" cleavage site, namely, the Arg-Ala[1] bond, is used. For bovine pancreatic PLA_2 an expression system has been published which makes use of a preproPLA_2 construct giving rise to the precipitation of the desired protein inside the cell, and a protocol was given for small-scale purification.[6]

In our laboratory *E. coli* K12 strain AB1157 was used originally[10] as the host organism for the pEX-derived expression vector and for vector plasmid pCI857, carrying the temperature-sensitive λ repressor. At a later stage this strain was replaced by strain MC 4100,[21] a strongly growing strain with a chromosomal *lacZ* deletion. More recently the DNA fragment coding for the truncated Cro-*lacZ*-proPLA_2 fusion protein[10] has been introduced into vector pUEX,[22] which by itself carries the temperature-sensitive λ repressor. Although the protein yields are similar, the main advantage is the improved ease of manipulating one plasmid instead of two.

To produce fusion protein in shaking bottles, a fresh overnight culture, grown at 30° in the presence of antibiotics, is diluted 5-fold with warm (41°) medium. Induction is continued for 3 hr, after which the cells are harvested by centrifugation. In a fermentor (New Brunswick Microferm Fermentor Model MF-114, New Brunswick Scientific Co., New Brunswick, NJ) the overnight starter culture is used to inoculate a 10-fold volume of medium. Stirring is set at 400 rpm, and aeration is kept at 300 ml/min per liter medium. Growth is continued at 30° until the OD_{600} is 0.6 (about 2 hr) after which the temperature is shifted in about 20 min to 41°. The OD_{600} increases to about 3.5, and after 3 hr of induction cells are harvested by centrifugation. Cells are broken by sonication after lysozyme treatment, protein aggregates are isolated from cell lysates by a 30-min centrifugation step at 5000 *g*; and the protein pellet is washed once with 0.1% Triton X-100. Starting from a 10-liter culture, a typical yield is 200–400 mg

[21] M. J. Casadaban, *J. Mol. Biol.* **104,** 541 (1976).

[22] G. M. Bressan and K. K. Stanley, *Nucleic Acids Res.* **15,** 10056 (1987).

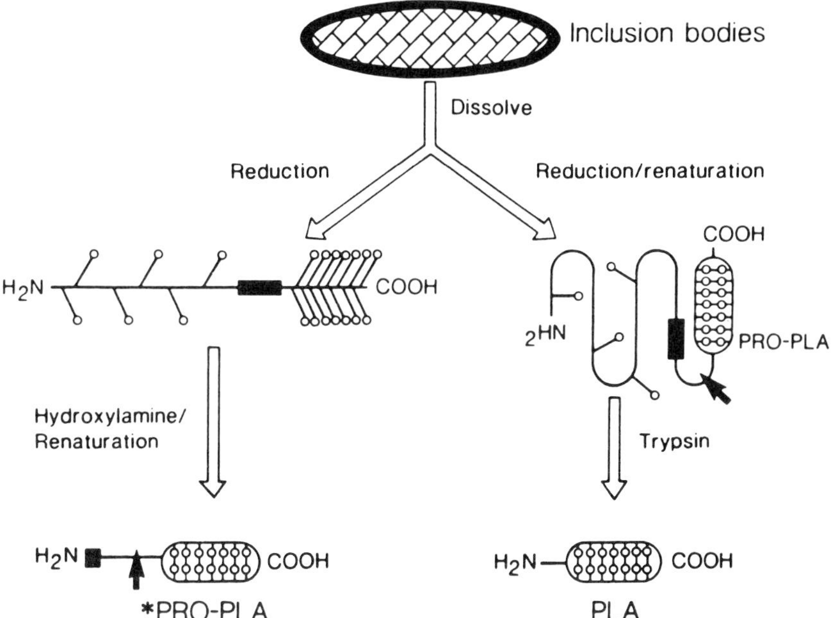

FIG. 2. Schematic representation of the refolding of PLA_2 and cleavage to proPLA_2 and active PLA_2, respectively. The black arrow represents the trypsin cleavage site, —○ a free SH group, —○—○— a disulfide bond, and ▭ the Asn-Gly linker.

fusion protein, being the major band on sodium dodecyl sulfate (SDS)–polyacrylamide gels.

The order of subsequent steps is outlined schematically in Fig. 2. The fusion protein is first sulfonated at 2–5 mg/ml, as described by Tannhauser and Sheraga.[23] The *S*-sulfo protein is precipitated by 5 volumes of a cold 1% (w/v) acetic acid solution. The precipitate is collected by centrifugation, and the pellet is washed several times with water. The pellet originating from a 1-liter culture is dissolved in 20 ml of 8 *M* urea, 100 m*M* sodium borate (pH 8.5), 20 m*M* EDTA. After dilution to 80 ml, reduced and oxidized glutathione are added to final concentrations of 2 and 1 m*M*, respectively. Renaturation usually reaches a maximal level within 24 hr at room temperature, as measured in an L-dioctanoyllecithin assay.[10] The (partly) renatured fusion protein is assayed for its content of active PLA_2 by adding a portion to the micellar assay reaction vessel, followed by 10 μg of trypsin. Within minutes the tryptic activation is complete as judged from the fact that the recorded hydroxide uptake becomes linear with time.

[23] T. W. Tannhauser and H. A. Scheraga, *Biochemistry* **24,** 7681 (1985).

After dialysis, preparative tryptic digestion of the renatured fusion protein is done in a buffer containing 20 m*M* Tris (pH 8) and 5 m*M* $CaCl_2$. Trypsin (5 μg/mg fusion protein) is added in the beginning and after 1 hr. Normally, after 2–3 hr of incubation at room temperature the PLA_2 activity reaches a maximum. The mixture is acidified to pH 4.5, dialyzed against 5 m*M* sodium acetate, pH 4.8, and centrifuged to remove insoluble peptides. Occasionally the pellet may contain residual PLA_2 activity. If so, the pellet is redissolved at pH 8, acidified to pH 4.8, and centrifuged. The combined supernatants are applied to a carboxymethyl-cellulose column (20 ml for 100 mg fusion protein) equilibrated at pH 4.8 with 5 m*M* sodium acetate. The PLA_2-containing fractions are further purified by two consecutive columns: a carboxymethyl-cellulose column at pH 6 and a diethylaminoethyl-cellulose column at pH 7.5–8.5, depending on the isoelectric point of the mutant PLA_2. In Fig. 3 the elution patterns for a *Y69F* mutant PLA_2 are depicted. Starting from a 10-liter culture, the yield of pure PLA_2 is 40–90 mg, reflecting a nearly quantitative conversion of 200–400 mg of the 59-kDa fusion protein.

ProPLA_2 is isolated from the fusion protein by cleavage with hydroxylamine. First, the inclusion bodies are solubilized under reducing conditions, at 5 mg/ml in 6 *M* guanidine, 50 m*M* Tris (pH 8), 1 m*M* EDTA and 5 m*M* 2-mercaptoethanol, for 1–3 hr under N_2. The mixture is then centrifuged to remove insoluble particles, and the supernatant is diluted 10-fold with cold 0.1% acetic acid. The precipitated fusion protein is collected by centrifugation, and the pellet is washed 3 times with dilute acid. The pellet is dissolved, at a protein concentration of 5 mg/ml, in 6 *M* guanidine, 2 *M* hydroxylamine, adjusted to pH 8.5 with LiOH. After a 3 hr incubation at 45°, the reaction is stopped by the addition of 0.1 volume acetic acid followed by dialysis against 1% acetic acid. Insoluble material is removed by centrifugation, and the soluble fraction is adjusted to pH 8 with NaOH. Additions are made to the following final concentrations: guanidine (1.5 *M*), Tris-HCl (50 m*M*, pH 8.0), Ca^{2+} (10 m*M*), reduced glutathione (2 mM), and oxidized glutathione (1 m*M*). Following renaturation and dialysis, proPLA_2 is purified as described above for the active enzyme. From a 10-liter culture about 15 mg of pure proPLA_2 is obtained.

Expression in Yeast

Whereas expression of PLA_2 in *E. coli* does not lead to the production of active (pro)PLA_2 because no or incorrect formation of disulfide bonds occurs in the cytoplasm, the yeast *Saccharomyces cerevisiae* is capable of both *in vivo* disulfide bridge formation and efficient secretion of heterologous proteins. Therefore, this yeast has been used as a host system for

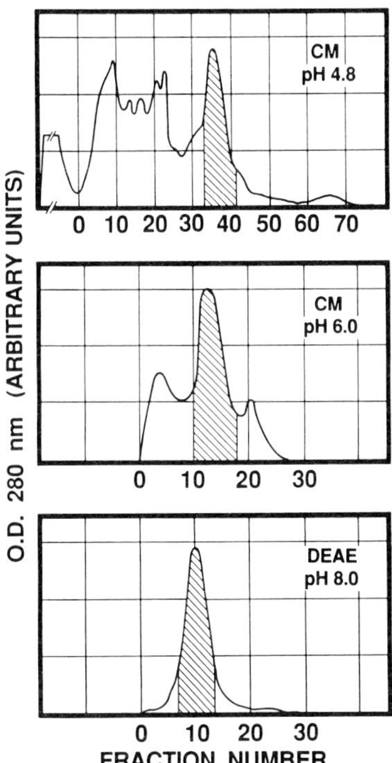

FIG. 3. Elution profiles of *Y69F* mutant porcine pancreatic PLA_2 after tryptic digestion of fusion protein. The recorded profiles were obtained starting from a 10-liter culture. Phospholipase A_2 activity is indicated by hatched areas. CM, pH 4.8, and CM, pH 6.0: carboxymethyl-cellulose columns run at pH 4.8 and 6.0, respectively; DEAE, pH 8.0: diethylaminoethyl-cellulose column run at pH 8.0. Proteins were eluted from the columns with linear gradients of NaCl of 0–0.5, 0–0.4, and 0–0.2 *M*, respectively. For details, see text.

the expression and secretion of several mammalian enzymes. Both bovine[7] and porcine[24–26] (pro)PLA_2 have been expressed and purified from this organism. In our laboratory constitutive expression of proPLA_2 in *S. cerevisiae* was obtained after fusing the proPLA_2 to the prepro sequence of the yeast α-mating factor.[24] On secretion, the fusion protein was cleaved by the *KEX2* protease, yielding a 140-amino acid long precursor form of the PLA_2. This protein, present in concentrations of about 600 μg/liter

[24] C. J. van den Bergh, A. C. A. P. A. Bekkers, P. de Geus, H. M. Verheij, and G. H. de Haas, *Eur. J. Biochem.* **140,** 241 (1987).

[25] C. J. van den Bergh, Ph.D. Thesis, University of Utrecht (1989).

[26] A. C. A. P. A. Bekkers, P. A. Franken, C. J. van den Bergh, J. Verbakel, H. M. Verheij, and G. H. de Haas, submitted for publication.

culture supernatant, was easily purified by ion-exchange chromatography. In later experiments[25] we have replaced the strong constitutive promoter of the yeast α-mating factor by the galactose-inducible *GAL7* promoter. The construct furthermore contains the invertase signal sequence preceding the α-mating factor prosequence–proPLA$_2$ fusion.[26] Cells harboring this construct target proPLA$_2$ into the culture medium to a level between 2.5 and 6 mg/liter. A similar level of production has been observed by Tanaka *et al.*[7] for the expression/secretion of bovine pancreatic proPLA$_2$ by yeast under control of the *PHO5* promoter. Whereas in our laboratory only proPLA$_2$ was excreted, Tanaka *et al.* obtained a mixture of precursor and active PLA$_2$. It is not clear whether this partial activation can be explained by the use of different expression constructs or whether it is strain dependent.

Purification of yeast proPLA$_2$ from fermentation supernatants of 10 liters or more can be accomplished with the use of ion-exchange chromatography,[24] thereby circumventing the use of Sephadex columns at an early stage of the purification method as was done by Tanaka *et al.*[7] After removal of the cells by centrifugation, the culture medium is acidified to pH 3.5, and 15 ml of preswollen SP-Sephadex C-25 is added per liter of medium. After gentle stirring for several hours at 4°, the beads are collected and washed first with 10 m*M* formate buffer, pH 3.5, and then with 10 m*M* acetate buffer, pH 5.0. In this step proPLA$_2$ binds nearly quantitatively to the ion-exchange resin, whereas most colored material remains unbound.

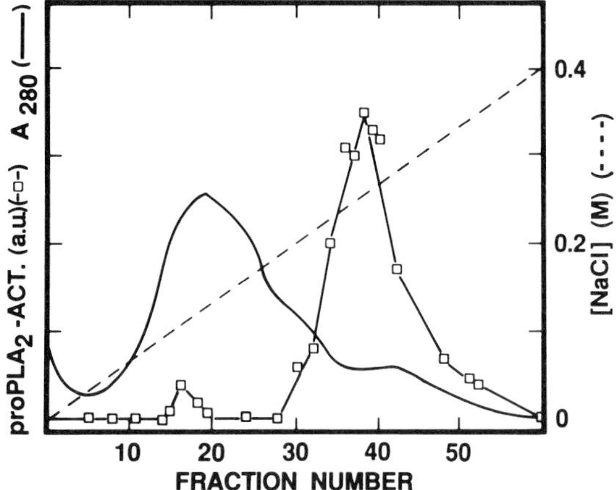

FIG. 4. Purification of yeast proPLA$_2$ by SP-Sephadex chromatography. The supernatant from 10 liters of culture broth of the strain harboring the *GAL7* construct encoding native porcine pancreatic PLA$_2$ (see text) was acidified to pH 3.5 and bound to 150 ml SP-Sephadex. After several wash steps the beads were loaded on top of a column containing 250 ml fresh SP-Sephadex. For details, see text.

The SP-Sephadex beads are then packed into a column on top of a fresh layer of SP-Sephadex C-25. The column is eluted with a salt gradient from 0 to 0.5 *M* NaCl. The proPLA_2 peak elutes after the majority of the strongly absorbing colored material which forms the large peak seen in Fig. 4. At this stage the proPLA_2 is the major band on SDS-polyacrylamide gel electrophoresis. After dialysis, the proPLA_2-containing fractions are further purified on a carboxymethyl-cellulose column at pH 6.0 in 5 m*M* acetate buffer. The resulting preparation is desalted and freed from traces of colored material on a Sephadex G-50 fine column in 10 m*M* acetic acid.

Yeast proPLA_2 has the same turnover number for monomeric substrates as native proPLA_2, although its activation peptide in the α-mating factor and in the *GAL7* constructs is 9 amino acids longer than in native proPLA_2. Apparently the extra amino acids do not hinder the entrance of substrate to the active site; they are probably fully exposed to the solvent and do not form contacts with any other part of the enzyme. This is also evident from the fact that yeast proPLA_2 is as rapidly cleaved by trypsin as is native proPLA_2. The resulting yeast PLA_2 is indistinguishable from authentic pancreatic PLA_2, showing that a protein with as many as seven disulfide bridges per 124 amino acid residues is correctly processed by the yeast secretory apparatus. Yeast, rather than *E. coli*, may therefore be the organism of choice to express PLA_2 mutants with impaired folding properties.

[20] Preparation of Antibodies to Phospholipases A_2

By MAKOTO MURAKAMI, KIYOSHI TAKAYAMA, MASATO UMEDA, ICHIRO KUDO, and KEIZO INOUE

Introduction

Phospholipases A_2 play important roles in several biological phenomena such as membrane phospholipid turnover, production of lipid mediators (eicosanoids, platelet-activating factor, etc.),[1] and the process of inflammation.[2,3] Phospholipases A_2 are classified into two groups in terms

[1] H. van den Bosch, *Biochim. Biophys. Acta* **604,** 191 (1980).

[2] P. Vadas and W. Pruzanski, *Lab. Invest.* **55,** 391 (1986).

[3] P. Vadas, S. Wasi, H. Z. Movat, and J. B. Hay, *Nature (London)* **293,** 583 (1981).

of primary structure.[4,5] The group I family includes the enzymes from Elapidae and Hydrophidae snake venom or mammalian pancreas, while the group II family includes the enzymes from Crotalidae and Viperadae snake venom. Recently, phospholipases A_2 belonging to group II have been detected in some mammalian nonpancreatic sources.[6] These enzymes have been shown to share common characteristics, such as a low molecular weight (~14,000), a high content of basic and hydrophobic amino acid residues, stability to acids, a high affinity for heparin, and more effective Ca^{2+}-dependent hydrolysis of phosphatidylethanolamine than phosphatidylcholine at alkaline pH.

To gain an understanding of the physiological functions, distribution, dynamics, and structural characteristics of phospholipases A_2, application of antibodies might provide useful information. Such antibodies would also be useful for identification and purification of the enzymes. Reports dealing with antibodies to phospholipases A_2 are listed in Table I.[7–9] In this chapter, we describe the preparation and possible application of antibodies to mammalian group II phospholipases A_2 obtained from both rat platelets[7,8] and human rheumatoid synovial fluid.[9]

Monoclonal Antibodies to Phospholipases A_2 from Rat Platelets[7] or Human Rheumatoid Synovial Fluid

Preparation of Antigen

Rat Platelet Phospholipase A_2. Blood is taken by cardiac puncture from ether-anesthetized Wistar rats (>400 g body weight, male) through a 20-gauge butterfly needle.[7] To minimize platelet activation, blood is collected into a plastic syringe containing 3.8% (w/v) sodium citrate in a ratio of 1 part to 9 parts blood. The blood is gently transferred to plastic tubes and centrifuged for 10 min at 270 *g* at room temperature to prepare platelet-rich plasma (PRP). After addition of a 1 : 5 (v/v) solution of 65 m*M* citric acid and 85 m*M* sodium citrate in 2% (w/v) glucose (ACD solution) to the PRP, platelets are separated from plasma by centrifugation at 1500

[4] R. L. Heinrikson, E. T. Krueger, and P. S. Keim, *J. Biol. Chem.* **252,** 4913 (1977).

[5] E. A. M. Fleer, H. M. Verheij, and G. H. de Haas, *Eur. J. Biochem.* **82,** 261 (1978).

[6] I. Kudo, H. W. Chang, S. Hara, M. Murakami, and K. Inoue, *Dermatologica* **179,** 72 (1989).

[7] M. Murakami, T. Kobayashi, M. Umeda, I. Kudo, and K. Inoue, *J. Biochem.* (*Tokyo*) **104,** 884 (1988).

[8] M. Murakami, I. Kudo, Y. Natori, and K. Inoue, *Biochim. Biophys. Acta* **1043,** 34 (1990).

[9] K. Takayama, I. Kudo, S. Hara, M. Murakami, M. Matsuta, T. Miyamoto, and K. Inoue, *Biochem. Biophys. Res. Commun.* **167,** 1309 (1990).

TABLE I
ANTIBODIES TO PHOSPHOLIPASES A_2

Enzyme source	Antibodies	Refs.
Pig, horse, ox, sheep pancreas	Polyclonal	*a*
Human pancreas	Polyclonal	*b, c*
Human seminal plasma	Polyclonal	*d*
Rat pancreas	Polyclonal	*e*
Swine intestine	Polyclonal	*f*
Porcine pancreas	Polyclonal	*g*
Naja naja venom	Polyclonal	*h*
Rat liver mitochondria	Monoclonal	*i*
Rat platelets	Monoclonal, polyclonal	*j, k*
Human rheumatoid synovial fluid	Monoclonal	*l*

[a] H. Meijer, M. J. M. Meddens, R. Dijkman, A. J. Slotboom, and G. H. de Haas, *J. Biol. Chem.* **253,** 8564 (1978).

[b] J. Nishijima, M. Okamoto, M. Ogawa, G. Kosaki, and T. Yamano, *J. Biochem.* (*Tokyo*) **94,** 137 (1983).

[c] B. Sternby and B. Akerstrom, *Biochim. Biophys. Acta* **128,** 788 (1985).

[d] M. Wurl and H. Kunze, *Biochim. Biophys. Acta* **834,** 411 (1985).

[e] M. Okamoto, T. Ono, H. Tojo, and T. Yamamoto, *Biochem. Biophys. Res. Commun.* **128,** 788 (1985).

[f] F. Senegas-Balas, D. Balas, R. Verger, A. de Caro, C. Figarella, F. Ferrato, P. Lechene, C. Bertrand, and A. Ribet, *Histochemistry* **81,** 581 (1984).

[g] D. Bar-Sagi, J. P. Susan, F. McCormick, and J. R. Feramisco, *J. Cell Biol.* **106,** 1649 (1988).

[h] J. Masliah, C. Kadiri, D. Pepin, T. Rybkine, J. Etienne, J. Chambaz, and G. Bereziat, *FEBS Lett.* **222,** 11 (1987).

[i] J. G. N. de Jong, H. Amerz, A. J. Aarsman, H. B. M. Lenting, and H. van den Bosch, *Eur. J. Biochem.* **164,** 129 (1987).

[j] Ref. 7.

[k] Ref. 8.

[l] Ref. 9.

g for 10 min at room temperature. The cells are then gently resuspended in 5 parts HEPES–Tyrode buffer [137 mM NaCl, 2.7 mM KCl, 10 mM HEPES, and 0.1% (w/v) glucose] and 1 part ACD, then washed once. The washed platelets are resuspended in HEPES–Tyrode buffer at a final concentration of approximately 2×10^9 cells/ml. The platelets are then activated with 2.5 units/ml thrombin (Sigma, St. Louis, MO) in the pressure of 2 mM $CaCl_2$ at 37° for 5 min. After platelet aggregation, the mixture is centrifuged at 3000 *g* for 5 min at 4° to obtain supernatant fluid. The supernatant fluid is used as an enzyme source after allowing it to stand at 4° for 5 hr and separating any fibrin clots by centrifugation (4000 *g*, 10 min). Starting from 70 rats, approximately 200 ml of supernatant fluid will be obtained.

Phospholipase A_2 in the fluid is purified by the sequential use of column chromatography on heparin-Sepharose (Pharmacia, Uppsala, Sweden) and TSK gel G2000SW (Tosoh, Tokyo, Japan) gel filtration [high-performance liquid chromatography (HPLC)].[10] Fractions containing phospholipase A_2 are pooled and used for immunization.

Human Rheumatoid Synovial Fluid Phospholipase A_2. Phospholipase A_2 is purified to near homogeneity from the synovial fluid[9] of patients with rheumatoid arthritis by the sequential use of column chromatographies on heparin-Sepharose, butyl-Toyopearl (Tosoh), and reversed-phase HPLC (Shiseido, Tokyo, Japan).[10]

Immunization by Intrasplenic Injection

Ten micrograms of purified phospholipase A_2 is absorbed on a nitrocellulose filter (1 × 1 cm). The filter is then minced, suspended in 250 μl of phosphate-buffered saline (PBS), pH 7.4 (137 m*M* NaCl, 2.7 m*M* KCl, 80 m*M* Na_2HPO_4, and 1.5 m*M* KH_2PO_4), and homogenized using a Polytron homogenizer (Kinematica GmbH, Luzern, Switzerland). The suspension is injected into a mouse by intrasplenic injection[11] as described below.

A BALB/c mouse (8- to 12-week-old female) is anesthetized by intraperitoneal injection of 1 mg of Nembutal (Abbott Laboratories, North Chicago, IL), and the antigen suspension is injected with a 25-gauge butterfly needle into the spleen, which is exposed after laparotomy. After the injection, the spleen is replaced in the peritoneal cavity and the peritoneal wall is sutured with platinum staples.

Production of Monoclonal Antibodies by Hybridomas

The fusion procedure is that described by Galfre *et al.*[12] and Kohler and Milstein *et al.*[13] with some modifications. Three days after the second intrasplenic injection (2-week interval), a spleen cell suspension is prepared by gently teasing the spleen while holding it with a pair of forceps in a petri dish. The cells are washed twice with Dulbecco's modified Eagle's medium (DMEM) (Nissui Pharmaceutical Co. Ltd., Tokyo, Japan) and resuspended in 10 ml of the same medium. Myeloma cells (P3X63-Ag.8.653) are collected at the logarithmic growth phase, washed twice, and resuspended in DMEM. Then 10^8 spleen cells and 10^7 myeloma cells are mixed and pelleted at 200 *g* for 5 min at room temperature. The

[10] S. Hara, H. W. Chang, K. Horigome, I. Kudo, and K. Inoue, this volume [36].

[11] M. Spitz, this series, Vol. 121, p. 33.

[12] G. S. Galfre, C. Howe, C. Milstein, G. W. Butcher, and J. C. Howard, *Nature (London)* **266,** 550 (1977).

[13] G. Kohler and C. Milstein, *Nature (London)* **256,** 495 (1975).

supernatant is carefully aspirated, removing as much fluid as possible, and the pellet is loosened by flicking the tube. One milliliter of a prewarmed (37°) 50% (w/v) polyethylene glycol (PEG) (molecular weight 4000; Merck, Darmstadt, FRG) solution in DMEM is added with gentle mixing to the cell slurry over a 1-min interval. The cells are incubated at 37° during the next minute, and then 10 ml of prewarmed (37°) DMEM is added drop by drop for 3 min to dilute the PEG. The cell suspension is pelleted by centrifugation (4°, 200 *g*, 5 min) and resuspended in 40 ml of HAT selecting medium consisting of DMEM supplemented with 2 m*M* glutamine (GIBCO Laboratories, Grand Island, NY), penicillin/streptomycin (100 U/ml and 100 μg/ml, respectively; Flow Laboratories, Irvine, Scotland), 1 m*M* nonessential amino acids (M. A. Bioproducts, Walkersville, MD), 20% fetal calf serum (GIBCO), 5% NCTC 109 (M. A. Bioproducts), 100 m*M* hypoxanthine (Flow), 0.4 m*M* aminopterin (Sigma), and 16 m*M* thymidine (Flow). The cell suspension is dispersed in 200-μl volumes into two 96-well tissue culture plates (Falcon, Lincoln Park, NJ) and incubated in a humid atmosphere of 5% CO_2–95% air (v/v). After 1 week, the cells are fed with 50 μl of HAT medium per well. Thereafter the cells are fed every 2–3 days according to the degree of cell growth. Supernatants of confluent wells are screened for the presence of specific antibodies by enzyme-linked immunosorbent assay (ELISA) (see below).

Positive wells are cloned by limiting dilution. Hybridoma cells are cultured in medium in the presence of BALB/c mouse thymocytes (4- to 6-week-old female) at 1×10^7 cells/ml as a feeder layer. After duplicate limiting dilution, the growing hybridoma clones are expanded and then injected intraperitoneally into BALB/c mice (10^7 cells/head) primed with pristane (0.5 ml/head; Aldrich Chemical Co., Milwaukee, WI). Ascites fluid is collected, and constituent antibodies are purified by 50% ammonium sulfate precipitation followed by affinity chromatography on protein A-Sepharose (Pharmacia) by the method of Ey *et al.*[14]

ELISA Procedure

The wells of microtiter plates (Immulon 2; Dynatech Laboratories, Alexandria, VA) are coated with 50 μl of purified phospholipase A_2 in PBS (1 μg/ml) for 2 hr. After blocking with 150 μl of 1% bovine serum albumin (BSA) (Sigma) in PBS (PBA solution) for 2 hr, the plates are treated sequentially with 50 μl of hybridoma supernatant for 2 hr and 50 μl of horseradish peroxidase (HRP)-conjugated anti-mouse immunoglobulin (Cappel Laboratories, Dowingtown, PA) in PBA (1 : 1000 dilution) for 2

[14] P. L. Ey, S. J. Prowse, and C. R. Jenkin, *Immunochemistry* **15,** 429 (1978).

hr, then stained with 100 μl of *o*-phenylenediamine (100 μg/ml in 0.1 *M* citrate–phosphate buffer, pH 5.0, containing 0.003% (v/v) H_2O_2; Wako, Osaka, Japan). The reaction is stopped by adding 50 μl of 4 *N* H_2SO_4. The optical density at 492 nm is determined by an ELISA reader. All of the above procedures are performed at room temperature.

Application of Monoclonal Antibodies

Western Blotting. Samples (such as tissue or cell homogenates) are resolved by sodium dodecyl sulfate (SDS)–polyacrylamide gel electrophoresis (SDS–PAGE) using a 15% (w/v) gel by the method of Laemmli[15] and transferred to a Millipore GVHP filter (Nihon Millipore Kogyo, K. K., Yonezawa, Japan) for immunoblot analysis [7.5 V/cm for 30 min and then 37.5 V/cm for 2 hr, in a buffer containing 25 m*M* Tris, 192 m*M* glycine, and 20% (v/v) methanol]. It should be emphasized that the electrophoresis must be carried out in the absence of 2-mercaptoethanol (2-ME), since treatment of the enzyme with 2-ME results in loss of reactivity with all of the antibodies so far established.[7–9] The filter is blocked with PBA solution for 2 hr and then incubated with biotinylated antibody (1 μg/ml) diluted with PBA for 2 hr. The biotinylated antibody is prepared by incubation of the antibody with biotin *N*-hydroxysuccinimide ester (Behring Diagnostics, La Jolla, CA) as described.[7] The filter is incubated for 1 hr in a solution containing 1 : 1000-diluted HRP-conjugated streptavidin (Zymed Laboratories, San Francisco, CA). After washing the filter with PBS, the color is developed by adding 5 mg of HRP color-developing reagent (2-chloronaphthol; Bio-Rad Laboratories, Richmond, CA) in 2.5 ml of methanol, 5 μl of 30% H_2O_2, and 10 ml of reaction buffer (20 m*M* Tris-HCl, pH 7.4, 0.5 *M* NaCl) and incubated for 5 min on a rocking platform. The reaction is stopped by rinsing the filter with distilled water. All the above procedures are carried out at room temperature.

Immunoaffinity Purification of Mammalian Group II Phospholipases A_2

Coupling of antibody to sepharose beads. Four milligrams of antibody is coupled to 2 ml of cyanogen bromide-activated Sepharose (Pharmacia) in 0.1 *M* citrate buffer (pH 6.8) for 30 min at room temperature. The efficiency of the coupling reaction is usually higher than 90% under these conditions. The beads are blocked by incubation with 1 *M* ethanolamine (pH 8.0) for 2 hr, packed into an Econo column (1 × 10 cm; Pharmacia),

[15] U. K. Laemmli, *Nature (London)* **227,** 680 (1970).

and then washed extensively with 10 m*M* Tris-HCl (pH 7.4) containing 0.15 *M* NaCl (TBS).

Purification of rat platelet phospholipase A_2. Washed rat platelets are sonicated 3 times for 30 sec each time. The lysate is incubated for 12 hr at 4° in 10 m*M* Tris-HCl (pH 7.4) containing 1 *M* KCl (>80% of enzyme activity is solubilized under these conditions) and centrifuged at 100,000 *g* for 1 hr, after which the supernatant is applied to an anti-rat platelet phospholipase A_2 antibody MD7.1-Sepharose affinity column. To avoid nonspecific adsorption, a precolumn of Sepharose CL-4B (1 × 10 cm; Pharmacia) is attached before the affinity column. The flow rate is below 10 ml/hr, and the effluent is monitored with a UV monitor (Pharmacia) at 280 nm. The column is washed extensively with TBS, and the bound enzyme is subsequently eluted with 50 m*M* glycine-HCl buffer (pH 2.3). The eluate at low pH gives a single protein band on SDS–PAGE, the molecular weight of which is estimated to be 14,000, corresponding to phospholipase A_2 (Fig. 1). The fractions are neutralized with 1 *N* NaOH immediately after the elution. The recovery of the enzyme activity is approximately 90%. From platelets of 70 rats in a single preparation, 1.3 mg of phospholipase A_2 is obtained.

This procedure can be applied to other immuno-cross-reactive phospholipases A_2. For example, rabbit platelet phospholipase A_2 has been purified by the same procedure as that for rat platelet enzyme using an anti-rat platelet phospholipase A_2 antibody MD7.1-Sepharose affinity column.[16]

Purification of human rheumatoid synovial fluid phospholipase A_2. Pooled synovial fluid from patients with rheumatoid arthritis is loaded onto an anti-human synovial fluid phospholipase A_2 antibody HP-1-Sepharose column, with a precolumn of Sepharose CL-4B attached before the antibody column. The flow rate is below 10 ml/hr. All of the phospholipase A_2 activity in synovial fluid is adsorbed by the antibody column. The column is washed extensively with 10 m*M* Tris-HCl (pH 7.4) containing 1 *M* NaCl and then subsequently eluted with 0.1 *M* glycine-HCl buffer (pH 2.3). The eluate at low pH contains one major protein band together with two minor ones on SDS–PAGE. Among them, only the major protein with a molecular weight of approximately 14,000 is reactive with the anti-human synovial phospholipase A_2 antibody when the eluate is examined by Western blotting. The fractions are neutralized with 1 *N* NaOH. The enzyme is purified approximately 1.6×10^5-fold in a single purification step. It should be noted that the recovery of total activity is 200–300%; thus, some inhibitory factor(s) might be removed during this purification step.

[16] H. Mizushima, I. Kudo, K. Horigome, M. Murakami, M. Hayakawa, D.-K. Kim, E. Kondo, M. Tomita, and K. Inoue, *J. Biochem.* (*Tokyo*) **105,** 520 (1989).

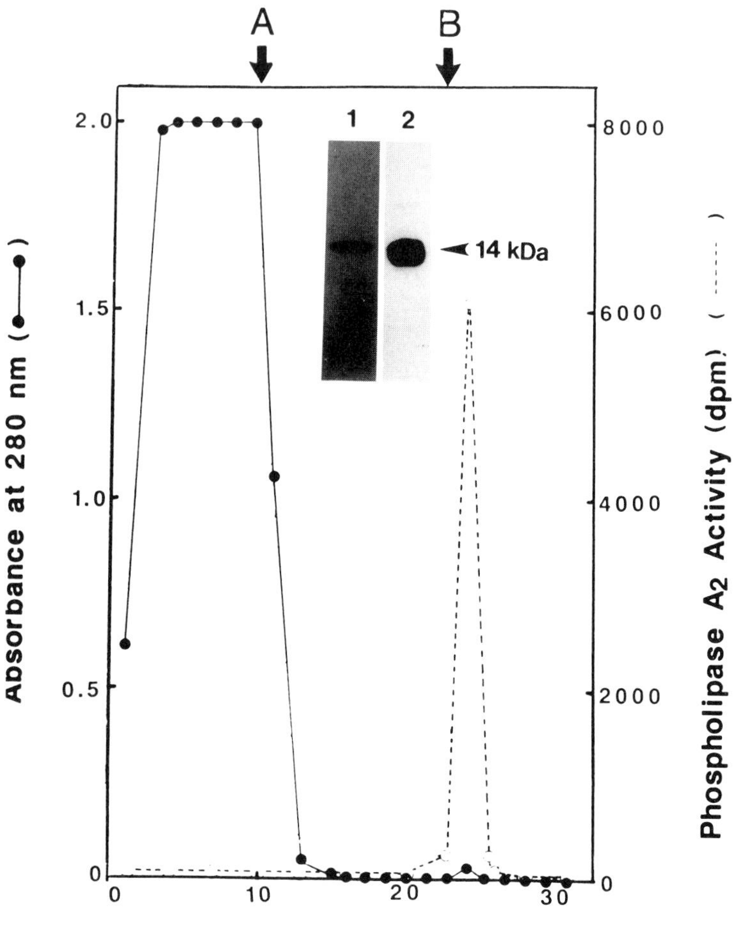

FIG. 1. Immunoaffinity purification of rat platelet phospholipase A_2. A rat platelet KCl extract was applied to an antibody MD7.1-conjugated Sepharose column. After washing with TBS (arrow A), the column was eluted with 50 m*M* glycine-HCl buffer, pH 2.3 (arrow B). Protein (●) and phospholipase A_2 activity measured by Dole's method [V. P. Dole and H. Meinertz, *J. Biol. Chem.* **235,** 2595 (1960)] using ^{14}C-labeled phosphatidylserine as a substrate (○) are shown. (Inset) Immunoaffinity-purified phospholipase A_2 (1 μg) was resolved by SDS–PAGE (lane 1; stained with Coomassie Brilliant blue) and subjected to Western blotting (lane 2; blotted with antibody MD7.1).

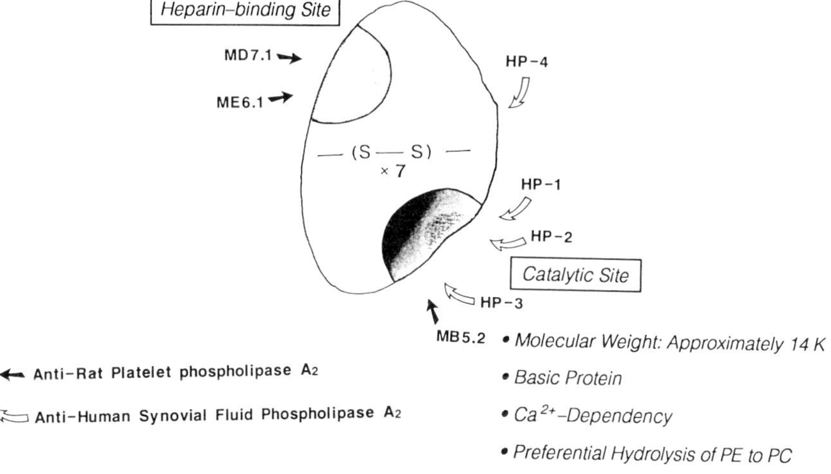

FIG. 2. Schematic structure of mammalian group II phospholipase A_2 and recognition sites for established monoclonal antibodies. PE, Phosphatidylethanolamine; PC, phosphatidylcholine.

Polyclonal Antibodies to Rat Platelet Phospholipase A_2

Preparation of Polyclonal Antibodies

Immunoaffinity-purified phospholipase A_2 is used as an immunogen for raising antibodies in Japanese albino rabbits (male, 2–3 kg body weight).[8] Intact or denatured (treated with 2-ME for 10 min at 100°) phospholipase A_2 is injected subcutaneously into the footpads according to the following schedule: 250 μg of phospholipase A_2 in Freund's complete adjuvant (Difco Laboratories, Detroit, MI) on day 0, then 150 μg in Freund's incomplete adjuvant (Difco) every 2 to 3 weeks for 2 months. Animals are bled 1 week after each booster injection. The titer of antiplatelet phospholipase A_2 antiserum is examined by ELISA. Each serum, with a dilution of up to 1 : 10,000, is applied to a protein A-Sepharose affinity column to obtain IgG fractions.

Application of Polyclonal Antibodies

Immunohistochemical Staining. Tissues from rats are fixed for 30 min at room temperature in 3.7% formaldehyde in PBS, dehydrated, and embedded in cytoparaffin. The tissue fragments (4–6 μm in thickness) are oriented under a sliding microtome. When staining single cells, the cells plated on glass slides are washed twice with PBS, air-dried, and fixed in

TABLE II
ESTABLISHED ANTIBODIES TO MAMMALIAN GROUP II PHOSPHOLIPASES A_2

	MD7.1	ME6.1	MB5.2	R377	R385	HP-1	HP-2	HP-3	HP-4
PLA_2[a] used as antigen	Rat platelet	Rat platelet	Rat platelet	Rat platelet	Rat platelet[b]	Human synovial	Human synovial	Human synovial	Human synovial
Characteristics	Monoclonal	Monoclonal	Monoclonal	Polyclonal	Polyclonal	Monoclonal	Monoclonal	Monoclonal	Monoclonal
Inhibition of enzymatic activity	No	No	Yes	Yes	Yes	Yes	Yes	Yes	No
Inhibition of binding to heparin	Yes	Yes	No	Yes	No	N.D.[c]	N.D.	N.D.	No
Binding to reduced antigen	No	No	No	No	Yes	No	No	No	No
Relative reactivity to PLA_2 from various sources									
Pancreas (I)[d]	–[e]	–	–	–	+	–	–	–	–
Snake venom (I, II)	–	–	–	–	+	–	–	–	–
Rat platelets (II)	+ + +	+ + +	+ +	+ + +	+ + +	–	–	+	–
Rabbit platelets (II)	+ +	–	+	–	+	+	–	–	–
Human synovial fluid (II)	+	N.D.	N.D.	+	+	+ + +	+ +	+ + +	+ + +

[a] PLA_2, Phospholipase A_2.
[b] Phospholipase A_2 treated with 2-mercaptoethanol.
[c] N.D., Not determined.
[d] (I) indicates group I enzyme, and (II) indicates group II enzyme.
[e] Purified antibodies (100 ng/ml) were tested for relative reactivity to various PLA_2 by ELISA as described (pp. 227–228). + + +, Optical density at 492 nm (OD_{492}) > 2.0; + +, 2.0 > OD_{492} > 1.0; +, 1.0 > OD_{492} > O; –, OD_{492} = 0.

3.7% formaldehyde in PBS. Then the samples are treated sequentially as follows: (1) 0.3% H_2O_2 in methanol for 40 min (to inactivate endogenous peroxidase), (2) 3% BSA in PBS for 30 min, (3) antiserum (1 : 500 dilution) for 1 hr, (4) biotinylated goat anti-rabbit immunoglobulins (1 : 1000 dilution; Seikagaku Kogyo, K.K., Tokyo, Japan) for 40 min, (5) HRP–streptavidin (1 : 1000 dilution; Seikagaku Kogyo) for 20 min. Finally, the samples are stained for 10 min with diaminobenzidine (10 mg in 37.5 ml of TBS containing 0.013% H_2O_2; Wako).

Determination of Phospholipase A_2 Content by Sandwich ELISA. The combination of polyclonal antibody R377 with monoclonal antibody MD7.1 gives a sensitive sandwich ELISA which can be applied to rat platelet-type phospholipase A_2 even in complex protein mixtures. Microtiter plates (Immulon 2) are coated with 1 μg/ml purified R377 for 2 hr. After washing, the wells are blocked with 1% PBA solution. After washing, samples or various concentrations of purified rat platelet phospholipase A_2 (as a standard) are added, and plates are further incubated for 1 hr. After washing, 1 μg/ml biotinylated monoclonal antibody MD7.1 is added, and the plates are incubated again for 1 hr, followed by washing and incubation with HRP–avidin (1 : 1000 dilution; Vector Laboratories, Burlingame, CA) for 1 hr. *o*-Phenylenediamine is added after washing, and the absorbance is measured at 492 nm. The sensitivity (35 fmol per well) of the ELISA is within the range of the phospholipase A_2 content detected at inflamed sites. One of the advantages of the method described here is that measurement is not affected by the presence of an endogenous inhibitor(s) of enzymatic activity.

Concluding Remarks

Antibodies which we have established are summarized in Table II and Fig. 2.[7–9] The antibodies are characterized in terms of their affinity to antigen, capacity to inhibit enzymatic activity or heparin-binding activity, and cross-reactivity with other phospholipases A_2. These antibodies are expected to be useful tools for the study of mammalian group II phospholipases A_2.

[21] Thermodynamics of Phospholipase A_2–Ligand Interactions

By RODNEY L. BILTONEN, BRIAN K. LATHROP, and JOHN D. BELL

Introduction

Phospholipases A_2 can be activated on interaction with aggregated phospholipid substrates.[1] This activation is time-dependent and can thus be studied in considerable detail.[2,3] Specific methods for simultaneous study of various time-dependent properties of the enzyme and lipid substrate during the time course of hydrolysis have been described in a separate chapter in this volume.[4] Such investigations provide important clues regarding the mechanisms of the activation process. An important step in the development and testing of hypotheses of activation is the identification and quantitative thermodynamic characterization of the various equilibria involved in the process. Important equilibrium processes involved in phospholipase A_2 catalysis include protein–protein, protein–calcium, and protein–lipid interactions. In this chapter, we describe the use of spectroscopic and calorimetric techniques to obtain thermodynamic information relating to the latter two types of interactions.

General Considerations

For the simple reaction $E + X \rightleftharpoons EX$ the equilibrium constant is given by

$$K = \frac{[EX]}{[E][X]} \tag{1}$$

If some property (e.g., fluorescence intensity) of the enzyme (E) or ligand (X) is proportional to the amount of complex (EX) formed, the fractional degree of association of E as a function of the *free* concentration of X is given by

$$f = \frac{K[X]}{1 + K[X]} = \frac{\Delta A}{\Delta A_{max}} \tag{2}$$

[1] H. M. Verheij, A. J. Slotboom, and G. H. de Haas, *Rev. Physiol. Biochem. Pharmacol.* **91,** 91 (1981).
[2] G. Romero, K. Thompson, and R. L. Biltonen, *J. Biol. Chem.* **262,** 13476 (1987).
[3] J. D. Bell and R. L. Biltonen, *J. Biol. Chem.* **264,** 12194 (1989).
[4] J. D. Bell and R. L. Biltonen, this volume [22].

where ΔA is the measured change in the observable at a given concentration of ligand and ΔA_{max} is the maximum change in the observable. ΔA_{max} and K can be estimated by nonlinear least-squares analysis of ΔA as a function of [X] or by linear regression analysis of the data in reciprocal form. The former method is preferred, but if linear regression of the reciprocal data is used, the data must be properly weighted.

In any experiment, ΔA_{max} is the total change in the observable between the protein in the *fully* liganded state and the unliganded state. In a calorimetric experiment, ΔA_{max} is the total heat change for the reaction (ΔQ_{max}), and the apparent enthalpy change per mole of enzyme is

$$\Delta H^\circ_{app} = \Delta Q_{max}/E_T \tag{3}$$

where E_T is the total *amount* of enzyme used in the calorimetric experiment.

The apparent Gibb's energy change for the association reaction is given by

$$\Delta G^\circ_{app} = -RT \ln K = \Delta H^\circ_{app} - T\Delta S^\circ_{app} \tag{4}$$

with ΔS°_{app} being the apparent entropy change for the reaction. If ΔH°_{app} is known, then ΔS°_{app} can be calculated. It must be noted that the values of ΔG°_{app} and ΔS°_{app} are dependent on the choice of reference state (i.e., concentration units of X used to calculate K). The authors prefer the use of a reference state of moles per liter or mole fraction. In any case, this reference condition should always be stated when reporting values of K, ΔG°_{app} and ΔS°_{app}.

Although calorimetric experiments directly provide estimates for ΔH°_{app}, ΔH°_{app} can also be estimated from the temperature dependence of K since

$$\Delta H^\circ_{app} = \frac{d(-R \ln K)}{d(1/T)} \tag{5}$$

It is important to note that since estimates of the apparent thermodynamic quantities obtained as just described are highly dependent on the assumed model, comparison of the ΔH°_{app} obtained calorimetrically and that obtained by van't Hoff analysis [Eq. (5)] is an excellent test of the model. Generally, the two estimates will agree *only* if the assumed model used to calculate K is correct.

It should be realized that the thermodynamic quantities derived as described above are apparent values. This is because it is generally unlikely that such an elementary representation of the interaction as given in Eq. (1) is correct; rather, the reaction may be coupled with other processes such as protein ionization. However, such coupling can be sorted out and in many cases used to advantage in experimental design. An example of this type of phenomenon is given in the discussion of calcium binding to porcine pancreatic phospholipase A_2 and is discussed in more general terms in a previous volume of this series.[5]

Calorimetry has been used to study a variety of protein–ligand interactions but has seen limited application to phospholipase A_2. All reported calorimetric studies of calcium and lipid binding to phospholipase A_2 have used LKB batch microcalorimeters equipped with twin gold cells,[6–8] and the procedure to be described applies specifically to such an instrument. Each of the cells consists of two compartments. On physical rotation of the calorimeter, the material in the two compartments of each cell is mixed, and the differential heat of mixing is measured. The reference cell contains buffer in both compartments, whereas the compartments of the sample cells contain the reactant solutions. The basic calorimetric experiment consists of mixing the two solutions which contain reactive components. The heat effect associated with the overall process is

$$Q_m = Q_r + Q_d + Q_p \qquad (6)$$

where Q_m, Q_r, Q_d, and Q_p are the measured heat, the heat of reaction, the heat of dilution of the components of the solutions, and the differential heat effect associated with the physical process of mixing. Q_p is determined by measuring Q_m when both compartments of the two cells contain buffer only. Q_d is determined by measuring $Q_m = Q_d + Q_p$ when one compartment of the sample cell contains protein or ligand in buffer and the other contains only the buffer. Q_d should be measured at every concentration of protein and ligand used in the experiment. Q_r is determined by measuring $Q_m = Q_r + Q_d + Q_p$ when one compartment of the sample cell contains protein in buffer and the other contains the ligand in buffer. The use of buffers is helpful if the protein–ligand interaction results in the release or adsorption of protons, which is likely to be the case with phospholipase A_2.

On mixing, the heat generated or absorbed produces a temperature difference between the two cells which is measured by the potential difference across two thermoelectric devices. This temperature difference is proportional to the differential rate of heat flow into or out of the sample

[5] R. L. Biltonen and N. Langerman, this series, Vol. 61, p. 287.

[6] G. R. Hedwig and R. L. Biltonen, *Biophys. Chem.* **19,** 1 (1984).

[7] P. S. de Araujo, M. Y. Rosseneu, J. M. H. Kremer, E. J. J. van Zoelen, and G. H. de Haas, *Biochemistry* **18,** 580 (1979).

[8] D. Lichtenberg, G. Romero, M. Menashe, and R. L. Biltonen, *J. Biol. Chem.* **261,** 5334 (1986).

compartment. The integrated signal over time is proportional to Q_m, which can be calculated using a calibration constant determined with an electrical heater within the cell.

The heat of reaction per mole of protein is given by

$$Q_{app} = \frac{Q_r}{E_T} = \frac{(Q_m - Q_d - Q_p)}{E_T} \quad (7)$$

Q_{app} includes the apparent heat of binding plus the heat associated with proton release or adsorption by the buffer. That is,

$$Q_{app} = Q + \Delta n \, \Delta H^\circ_b \quad (8)$$

where Q is the heat associated with the protein–ligand interaction. Δn is the number of protons released or absorbed on binding of ligand. ΔH°_b is the molar enthalpy change associated with the proton absorption by the buffer.

Alternatively, titration microcalorimeters could be used for binding experiments. Modern titration calorimeters are capable of resolving heats in the range of about 10–100 μcal. In a titration calorimeter, one reactant is present in the calorimeter cell and the other is injected from a syringe in a stepwise fashion. An advantage of a titration system is that multiple concentrations of one of the reactants can be added to a single sample of the other. Thus, the time and material required to obtain a complete binding curve are much less with a titration instrument than with the batch calorimeter. Freire and co-workers have published a calorimetric study of the binding of the B subunit of cholera toxin to lipid vesicles using such a titration calorimeter.[9] A detailed description of calorimetric instrumentation and its application to biological systems has been published in a previous volume in this series.[5,10]

Calcium Binding to Phospholipase A_2

Calorimetry

There has been only one calorimetric study of calcium binding to phospholipase A_2, and, thus, it must serve as the example for application of the technique.[6] The enthalpy of binding of calcium to phospholipase A_2 was determined from the heats of mixing of solutions of enzyme in buffer with solutions of $CaCl_2$ in buffer at the same pH and NaCl concentration.

[9] A. Schön and E. Freire, *Biochemistry* **28,** 5019 (1989).

[10] N. Langerman and R. L. Biltonen, this series, Vol. 61, p. 261.

Heats of dilution of the enzyme and $CaCl_2$ solutions were determined separately and subtracted from the measured heat of mixing to obtain the heat of reaction. The observed heats of reaction were in the range -0.3 to -1.5 mcal. To determine whether there is a concomitant proton uptake or release on calcium binding to phospholipase A_2 at pH 8.0, the experiments were carried out in both Tris-HCl and HEPES–NaOH buffer systems ($\Delta H°_b$ is different for each of the two buffers). In the mixing and dilution experiments, the calorimeter reference cell contained aliquots of buffer chosen to match closely the volumes in the reaction cell. The amount of solution added to each compartment of the calorimeter cell was determined gravimetrically. Prior to loading the calorimeter, the pH for both the enzyme and calcium solutions was adjusted to identical values, as necessary. After the experiment, the pH of the mixed solution was also determined to ascertain that the change in pH was never greater than 0.01 pH unit. Prior to mixing, the reactant solutions must be allowed to achieve thermal equilibrium with the calorimeter. The enzyme may adsorb to the walls of the gold reaction cell, and extreme care must be taken in the cleaning of the reaction cell at the end of each experiment. This adsorption is not peculiar to the gold cells but appears to be a property of phospholipase A_2. As discussed elsewhere, phospholipase A_2 has a tendency to adsorb also to glass surfaces.[3] The amount of protein adhering to the calorimetric cell surface in the calorimetric experiment described was small compared to the total amount of protein used in the experiment.

The observed enthalpy change of reaction of calcium with phospholipase A_2 in Tris buffer is shown as a function of the total calcium concentration in Fig. 1. A qualitatively similar response was obtained using HEPES buffer, but the values of ΔH were about 20% lower than with Tris buffer, indicating a net change in the degree of protonation of the enzyme on binding calcium. Since the heat of protonation of Tris and HEPES buffers are, respectively, -11.33 and -4.93 kcal/mol,[11,12] it follows that protons are released from the enzyme on binding calcium at pH 8.0. The ΔH data obtained in Tris buffer were analyzed assuming a 1 : 1 stoichiometry, yielding values for K (3507 M^{-1}) and ΔH (-4.99 kcal/mol).[6]

Spectroscopy

Calcium binding to phospholipase A_2 can also be monitored by measurement of any spectral parameter which changes on the interaction. Ultraviolet difference spectroscopy, circular dichroism spectroscopy, and

[11] I. Grenthe, H. Ots, and O. Ginstrup, *Acta Chem. Scand.* **24,** 1067 (1970).

[12] L. Beres and J. M. Sturtevant, *Biochemistry* **10,** 2120 (1971).

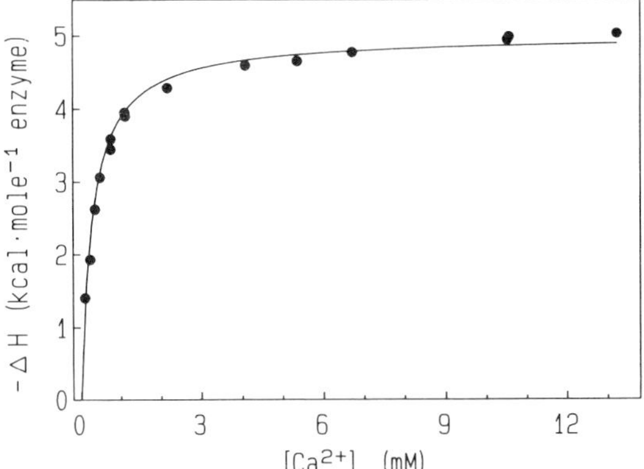

FIG. 1. Observed enthalpy of reaction of Ca^{2+} with pancreatic phospholipase A_2 as a function of the total calcium concentration: pH 8, 0.10 M NaCl, 25°; [enzyme] = 5.6×10^{-5} M; 50 mM Tris buffer. (Adapted with permission from G. R. Hedwig and R. L. Biltonen, *Biophys. Chem.* **19,** 1. Copyright 1984, Elsevier Science Publishers B.V.)

fluorescence spectroscopy have all been used.[1,6,13–15] The choice of observable dictates the protein concentration range that can be used (for ultraviolet absorption, [E] = 10^{-6} to 10^{-4} M; for fluorescence, [E] = 10^{-8} to 10^{-5} M, depending on the amino acid composition of the individual enzyme). Such binding assays are extremely convenient but not universally applicable because not all phospholipases A_2 exhibit spectral changes on calcium binding (e.g., porcine pancreatic enzyme fluorescence). In some cases, attachment of a fluorescent group such as 8-anilino-1-naphthalene sulfonate can be used to monitor calcium binding.[1,14]

One spectral technique useful in monitoring calcium binding to phospholipase A_2 is fluorescence. A typical Ca^{2+}-dependent change in the intrinsic fluorescence intensity of the monomer ASP-49 phospholipase A_2 from *Agkistrodon piscivorus piscivorus* to the binding of calcium is shown in Fig. 2. In this case, Ca^{2+} binding produces an enhancement of the fluorescence intensity and a red shift in the wavelength of the maximum intensity of about 3 nm. Least-squares analysis of the data of Fig. 2 yielded a value of K = 900 M^{-1}, which compares well with that obtained by Heinrikson and co-workers using equilibrium dialysis.[16]

[13] T. Miyake, S. Inoue, K. Ikeda, K. Teshima, Y. Samejima, and T. Omori-Satoh, *J. Biochem.* (*Tokyo*) **105,** 565 (1989).

[14] W. A. Pieterson, J. J. Volwerk, and G. H. de Haas, *Biochemistry* **13,** 1439 (1974).

[15] M. A. Wells, *Biochemistry* **12,** 1080 (1973).

[16] J. M. Maraganore, G. Merutka, W. Cho, W. Welches, F. J. Kézdy, and R. L. Heinrikson, *J. Biol. Chem.* **259,** 13839 (1984).

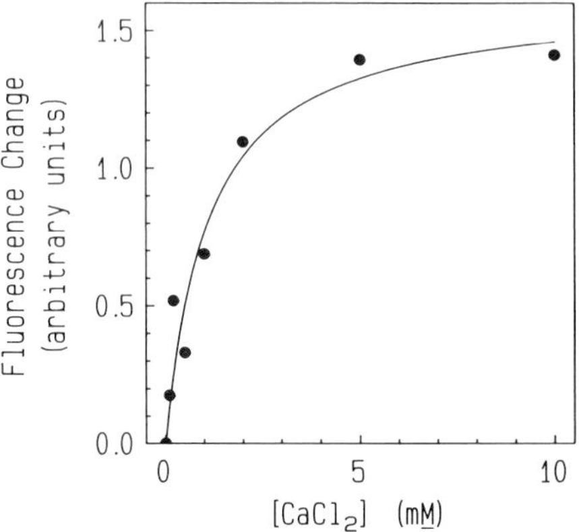

FIG. 2. Interaction of calcium ions with *A. piscivorus piscivorus* monomer phospholipase A_2 assessed by fluorescence spectroscopy. Enzyme (5.7×10^{-7} *M* final) was mixed with 100 m*M* KCl, 50 m*M* sodium borate at pH 8 and 28° at the indicated concentrations of $CaCl_2$. Since the presence of calcium appears to alter the amount of enzyme that adsorbs to the cuvette, the fluorescence intensity (excitation 280 nm, emission 346 nm) during the initial 10 sec after mixing (before appreciable adsorption to the cuvette) was averaged to obtain the values shown. The apparent association constant for calcium (determined by nonlinear least-squares fitting of the data) was 900 M^{-1}.

Miyake *et al.*[13] have also shown that the intrinsic tryptophan fluorescence emission intensity increases as a function of calcium concentration for phospholipase A_2 of *Trimeresurus flavoviridis* venom. They obtained binding constants for calcium at various pH and enzyme concentrations and thus obtained substantial information regarding the calcium-binding site and its relationship to the state of aggregation (monomer versus dimer) of the enzyme.[13]

ΔH for calcium binding to phospholipase A_2 could be obtained by determination of K at several temperatures and analyzing the data according to Eq. (8). To our knowledge, however, no such examination of the temperature dependence of K has been performed.

Other Techniques

Direct measurement of calcium binding to phospholipase A_2 has also been performed by equilibrium gel filtration and by equilibrium dialysis using ^{45}Ca.[13,15,16] The advantage of these techniques is that they measure binding directly, and the concentration of free calcium is known rather than calculated or assumed. The disadvantage is that such techniques require large amounts of protein to obtain a reasonable signal-to-background ratio due to the low affinity of the enzyme for calcium. When

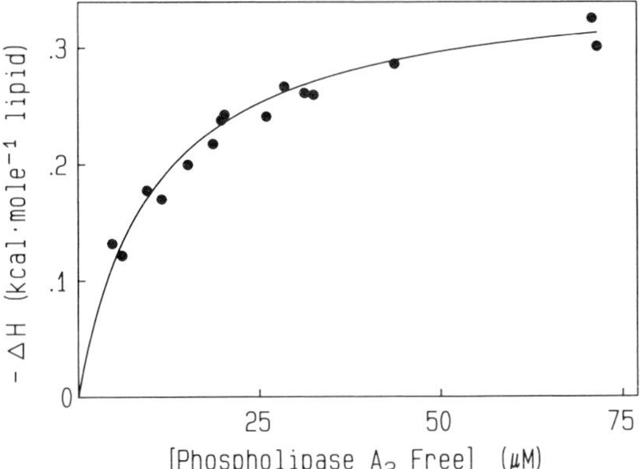

FIG. 3. Measured enthalpy changes for the interaction of pancreatic phospholipase A_2 and *n*-hexadecylphosphorylcholine micelles as a function of the phospholipase A_2 free concentration. Experiments were conducted at 25° at a lipid concentration of approximately 2 m*M* in 50 m*M* sodium acetate buffer (pH 6) and 100 m*M* NaCl. (Adapted with permission from P. S. Araujo, M. Y. Rossener, J. M. H. Kremer, E. J. J. van Zoelen, and G. H. de Haas, *Biochemistry* **18,** 580. Copyright 1979, American Chemical Society.)

applicable, however, these direct approaches are the method of choice. The use of these techniques is described elsewhere in this series.[17–19]

Binding of Phospholipase A_2 to Aggregated Lipid

Calorimetry

The basic procedure for the calorimetric assessment of phospholipase A_2 binding to lipids is identical to that described for the binding of calcium. de Araujo *et al.*, using an LKB batch microcalorimeter, measured the apparent heat of binding of porcine pancreatic phospholipase A_2 to *n*-hexadecylphosphorylcholine micelles.[7] The results of the experiment in which PLA_2 concentration was varied are shown in Fig. 3. In the analysis of this particular set of data it was assumed that each phospholipase A_2 interacted with n lipid molecules and that the free concentration of protein was given by

$$[E]_{free} = [E]_{total} - (Q/Q_{max})([lipid]/n) \tag{9}$$

where Q and Q_{max} are given in kcal/mol lipid, and

$$K = [E]_{bound}/\{[E]_{free}([lipid]/n - [E]_{bound})\} \tag{10}$$

[17] P. McPhie, this series, Vol. 22, p. 23.

[18] C. H. W. Hirs, this series, Vol. 47, p. 97.

[19] G. Amiconi and B. Giardina, this series, Vol. 76, p. 533.

TABLE I
THERMODYNAMICS OF BINDING OF PANCREATIC PHOSPHOLIPASE A_2 TO *n*-HEXADECYLPHOSPHORYLCHOLINE MICELLES[a]

Temperature (K)	*N* (Lipids/site)	K (M^{-1})[b]	$-\Delta H$[c]	$-\Delta H$[d]
288	47	1.5×10^5	0.27	12.7
298	39	0.9×10^5	0.36	14.0
308	26	0.8×10^5	0.52	13.5

[a] From Ref. 7.
[b] The reference state is mol/liter of sites and protein.
[c] kcal/mol lipid.
[d] kcal/mol protein.

Least-squares analysis of the data yielded estimates of the fitting parameters n, Q_{max}, and K. Estimates of these values at 15°, 25°, and 35° were obtained[7] and are tabulated in Table I.

An apparent temperature dependence of ΔH (per mole of lipid) and the stoichiometry appears to exist. However, the parameters n and Q_{max} are highly correlated in Eqs. (9) and (10). Since $\Delta H = n\ Q_{max}$, an apparent variation of n, Q_{max}, or both with temperature would give an apparent temperature dependence of ΔH. If both n and Q_{max} are allowed to vary independently in the fitting of data at each temperature, the large covariance of the parameters may result in artifactual interpretations of the temperature dependence. ΔH (per mole of protein) shows no temperature dependence, as shown in Table I. One must always be careful when interpreting binding data with models that contain correlated parameters.

Lichtenberg *et al.* have measured the apparent heat of binding of large dipalmitoylphosphatidylcholine vesicles to porcine pancreatic phospholipase A_2.[8] In that study the protein concentration was too large to obtain a binding isotherm from which an equilibrium constant could be deduced. However, from the data shown in Fig. 4, both ΔH and the stoichiometry of enzyme and lipid can be easily estimated. The obtained value of ΔH was -8 kcal/mol protein, and the stoichiometry ($n = 40$) was in good agreement with previous estimates.[7,20]

There are two potentially critical problems which should be noted in the assessment of experimental calorimetric data. The first is the likelihood that the binding thermodynamics of phospholipase A_2 will be highly depen-

[20] M. K. Jain, M. R. Egmond, H. M. Verheij, R. Apitz-Castro, R. Dijkman, and G. H. de Haas, *Biochim. Biophys. Acta* **688,** 341 (1982).

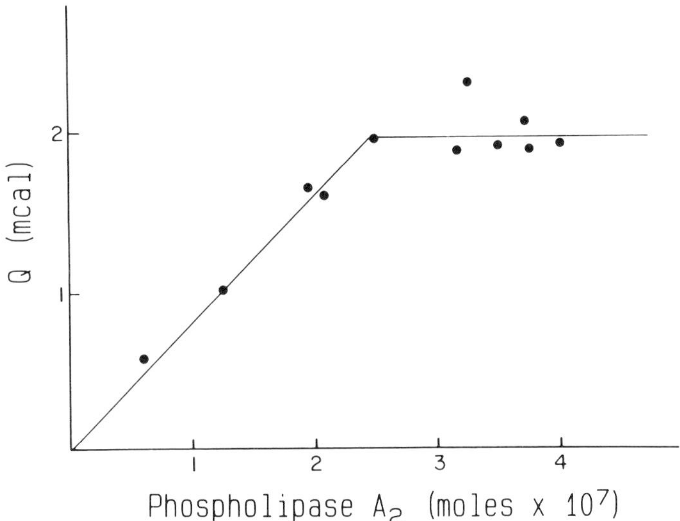

FIG. 4. Calorimetric titration of dipalmitoylphosphatidylcholine large unilamellar vesicles with pancreatic phospholipase A_2 (8.8 μmol, 2.9 mM final lipid concentration). Vesicles were mixed with the indicated amounts of pancreatic phospholipase A_2 in 50 mM KCl, pH 8, containing 1 mM EDTA. All experiments shown were carried out at 25°. The calculated ΔH of the reaction is -8 kcal/mol of enzyme. From the apparent stoichiometry of the binding reaction, each enzyme molecule apparently interacts with about 40 phospholipid molecules (20 if only the outer monolayer is taken into consideration). (Adapted with permission from D. Lichtenberg, G. Romero, M. Menashe, and R. L. Biltonen, *J. Biol. Chem.* **261,** 5334. Copyright 1986, The American Society of Biological Chemists, Inc.)

dent on the structure and composition of the lipid aggregate. For example, Pownell and co-workers have shown clearly that the binding of lipoproteins to phosphatidylcholine is strongly coupled to the main phase transition.[21] As shown in the next section, the apparent binding of phospholipase A_2 to small vesicles is very strongly coupled to a structural transition in the vesicle. Also, a number of studies have shown that fatty acids may greatly influence the binding thermodynamics.[20,22] Second, the rate of protein binding to the vesicle surface may become very slow at high protein to lipid ratios because the rate of association becomes limited, not by diffusion, but by reorganization of the protein on the lipid surface. In some of our experiments at high protein to lipid ratios, a slow secondary phase of interaction is noted. We suspect this is the result of such reorganization.

[21] H. J. Pownall, J. B. Massey, S. K. Kusserow, and A. M. Gotto, *Biochemistry* **18,** 574 (1979).

[22] M. K. Jain, B.-Z. Yu, and A. Kozubek, *Biochim. Biophys. Acta* **980,** 23 (1989).

Fluorescence Spectroscopy

In 1975, de Haas and co-workers demonstrated that the intrinsic fluorescence of the pancreatic phospholipase A_2 increases at least 2-fold on interaction with micelles of *n*-dodecylphosphorylcholine.[23] This fluorescence change, accompanied by a blue shift in the wavelength of maximum emission, was interpreted as indicating that the "microenvironment of the single tryptophan residue in the enzyme becomes less polar" on interaction of the phospholipase with the micelles.[23] Subsequently, the binding characteristics of the pancreatic and a number of snake venom enzymes to micelles and vesicles of various phospholipids have been determined using fluorescence spectroscopy.[1,20,24,25]

Care must be taken in designing and interpreting fluorescence data to assay phospholipase A_2. For example, one must be aware of time-dependent processes that may occur. Protein conformation changes, protein–protein interactions, and lipid reorganization are possible sources of time dependence. Also, some phospholipases A_2 such as the monomer aspartate-49 enzyme from *A. piscivorus piscivorus* binding tightly to the walls of the cuvette causing the measured fluorescence intensity to change with time.[3] Thus, one should always record the time course of protein fluorescence intensity for each sample. We have found that the time dependence of fluorescence due to binding of the protein to the spectroscopic cell is very reproducible. Therefore, for instantaneous binding, reliable estimates of the extent of binding can be obtained on mixing of the reactants in the spectroscopic cell. If the protein–lipid interaction is slow, it is best to allow the fluorescence intensity to become stable with time before assessing the extent of the fluorescence change due to the protein–lipid interactions.

It is important to establish that the process for which a binding constant is calculated is reversible. Reversibility can be tested by approaching the position of apparent equilibrium from different directions. If the apparent binding is temperature-dependent, for example, one can test reversibility by shifting the sample temperature from one at which the apparent binding is high to a temperature at which it is low and then returning to the original temperature. If the extent of binding at any temperature is independent of the direction of approach, then reversibility may be assumed.[24] Calcium concentration and pH are other variables that may be used to test revers-

[23] M. C. E. van Dam-Mieras, A. J. Slotboom, W. A. Pieterson, and G. H. de Haas, *Biochemistry* **14,** 5387 (1975).

[24] J. D. Bell and R. L. Biltonen, *J. Biol. Chem.* **264,** 225 (1989).

[25] J. Prigent-Dachary, M. C. Boffa, M. R. Boisseau, and J. Dufourcq, *J. Biol. Chem.* **255,** 7734 (1980).

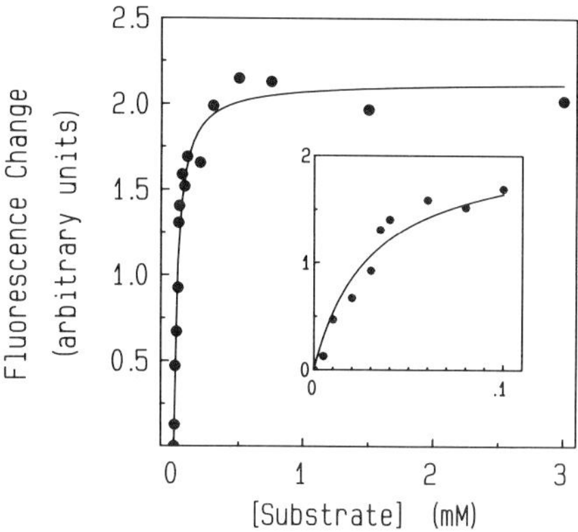

FIG. 5. Lipid concentration dependence of the fluorescence of the *A. piscivorus piscivorus* monomer phospholipase A_2 on interaction of small unilamellar vesicles of dihexadecylphosphatidylcholine. Enzyme (1.8×10^{-7} *M* final) was mixed with vesicles at the indicated lipid concentrations in 35 m*M* KCl, 10 m*M* $CaCl_2$, 10 m*M* sodium borate, and at least 1 m*M* NaN_3 (pH 8) at 25° in an SLM 8000C spectrofluorometer (excitation 280 nm, emission 340 nm). The fluorescence intensity was corrected for the light scattering and inner filter effects of the vesicles as described elsewhere in this volume.[4] The fluorescence in the absence of vesicles was then subtracted from the fluorescence in the presence thereof. *K* equals 3.0×10^4 M^{-1}.

ibility depending on the particular phospholipase A_2.[23,25] Of course, one must account for the effect of the variable on the intrinsic fluorescence of the protein in both the free and bound state to correctly assess binding.

Figure 5 details the fluorescence change exhibited by the monomer aspartate-49 phospholipase A_2 from the venom of *A. piscivorus piscivorus* on interaction with small, sonicated unilamellar vesicles of the nonhydrolyzable ether-linked phospholipid dihexadecylphosphatidylcholine. Analysis of the data in Fig. 5 yielded a value of 3×10^4 M^{-1} for the apparent association constant (expressed in terms of bulk lipid concentration). An apparent binding constant in terms of the number of binding sites may be calculated by taking into consideration the stoichiometry of binding.[1]

The apparent enthalpy change (ΔH) associated with the binding of phospholipase A_2 and phospholipid can be calculated from the temperature dependence of the equilibrium constant according to the van't Hoff equation [Eq. (5)]. The most direct method to obtain *K* as a function of temperature is to determine the substrate dependence of the fluorescence change at a variety of temperatures. This method requires a large number of experimental observations, and the large vesicle concentrations needed to obtain complete binding isotherms under conditions at which the binding is weak produce substantial optical artifacts due to sample turbidity. Alternatively, if one can establish that the maximum fluorescence change on

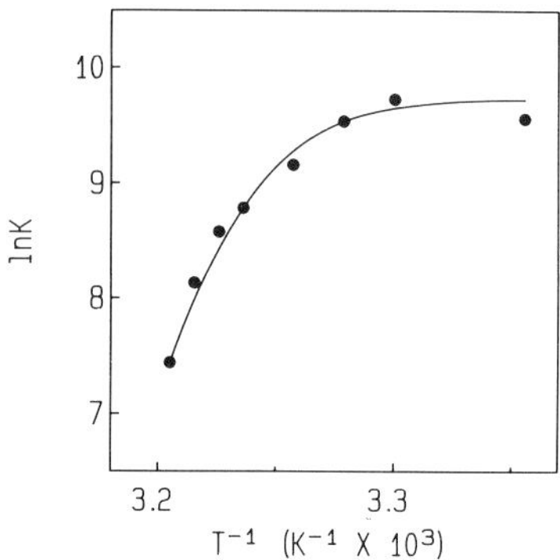

FIG. 6. van't Hoff plots of the interaction of *A. piscivorus piscivorus* monomer phospholipase A_2 with small unilamellar vesicles of dihexadecylphosphatidylcholine. The enzyme (2.9 $\times 10^{-7}$ *M* final) fluorescence was measured under the conditions described in Fig. 5. ΔF was calculated as the difference in the fluorescence intensity at apparent equilibrium before and after addition of vesicles at the indicated temperatures. *K* was calculated as described in the text using values for ΔF_{max} obtained from data such as shown in Fig. 5. The curve has no theoretical significance. [L] equals 0.1 m*M*, and ΔF_{max} equals 5.1 arbitrary units.

interaction of lipid and enzyme is relatively independent of temperature, the apparent association constant can be calculated using only one observation at a single lipid concentration at each temperature.[24] The association constant is given by the relationship $K = \Delta F/\{[L](\Delta F_{max} - \Delta F)\}$ where ΔF is the observed fluorescence change at a given temperature, ΔF_{max} is the maximum observed fluorescence change, and [L] is the concentration of phospholipid. Figure 6 shows a typical van't Hoff plot using equilibrium constants obtained in this manner for the monomer aspartate-49 enzyme from *A. piscivorus piscivorus* and small unilamellar vesicles of dihexadecylphosphatidylcholine. The data shown in Fig. 6 yield values for ΔH of the interaction from about 0 at 25° to about −100 kcal/mol at 38.5°. This large variation in apparent heat of interaction suggests that binding of the phospholipase A_2 to the bilayer is coupled to a thermotropic transition, most likely of the vesicle, since the lipid is known to undergo a thermotropic transition at about 29°.[26]

The presence of a fluorescence change on interaction of phospholipase A_2 with phospholipid depends on many factors including the structure of the particular phospholipase, the pH, the temperature, and the calcium concentration.[1,23–25] The fluorescence change may or may not reflect the binding step *per se* and could represent different processes in different enzymes. The absence of a fluorescence change on mixing of enzyme and

[26] D. Lichtenberg, M. Menashe, S. Donaldson, and R. L. Biltonen, *Lipids* **19,** 395 (1984).

lipid does not necesarily mean that binding has not occurred. Regardless of what specific process produces the fluorescence change, however, binding can be inferred from the phospholipid concentration dependence of the fluorescence change if that change is reversible.

Other Techniques

Several optical techniques such as light scattering and absorbance have also been used to assay the binding of the enzyme to phospholipid substrate.[7,23] As well, equilibrium[7,16] and nonequilibrium gel filtration[20] have been used, but these methods suffer from the requirement for large quantities of both enzyme and lipid. With all these techniques, many of the caveats and considerations described for the fluorescence spectroscopy are applicable and should be considered in interpreting the data.

Protein–Protein Interactions

Although this chapter does not review thermodynamic investigations of protein–protein interactions in detail, the following studies deserve mention. Monomer–dimer equilibria of certain phospholipase A_2 have been studied by spectroscopy,[27,28] ultracentrifugation,[28] and gel filtration.[13] Wells reported changes in the solvent perturbation ultraviolet difference spectra, circular dichroism spectra, and intrinsic tryptophan fluorescence as the monomer–dimer equilibrium of *Crotalus adamanteus* phospholipase A_2 was perturbed with urea.[27] Bukowski and Teller have provided a detailed description of the use of ultracentrifugation and intrinsic tryptophan fluorescence to characterize the monomer–dimer equilibrium.[28] The formation of enzyme dimers of the *Crotalus atrox* and *Naja naja* proteins was determined in the presence and absence of phospholipid and with calcium or strontium ions. The permutations of these conditions allowed investigation of the tendency of the enzyme to exist as a dimer when in free, substrate-bound, or active states. They also determined the pH dependence of the apparent dimerization constant and derived an elaborate scheme describing the dimerization mechanism for each enzyme.

Gel filtration can be used over a broad range of enzyme concentrations to determine apparent dimerization equilibrium constants. Miyake *et al.* investigated the monomer–dimer equilibrium of *T. flavoviridis* phospholipase A_2, using a 1×90 cm column of Sephadex G-100, as a function of pH and at protein concentrations in the range of 3×10^{-8} to about 1×10^{-5} M (concentration in the eluent).[13]

[27] M. A. Wells, *Biochemistry* **10,** 4078 (1971).

[28] T. Bukowski and D. C. Teller, *Biochemistry* **25,** 8024 (1986).

Summary

Future investigations into the role of the structure of phospholipid substrates and the interrelationships between substrate, calcium, and enzyme conformation in the activation process are clearly needed. Enzyme dimerization in the activation of phospholipase A_2 has been indicated, and a complex equilibrium between calcium, substrate, and monomer and dimer enzyme apparently exists.[2,29,30] The incorporation of proton binding further complicates the scheme, and one is quickly faced with obtaining a large number of equilibrium constants in order to describe the system explicitly. Nevertheless, similarly complex systems have been well characterized using thermodynamic approaches such as those described herein. An excellent example is the complex equilibrium involving the protonation of the histidine residues and the binding of a mononucleotide to ribonuclease A.[31,32] Achieving a complete thermodynamic description of that system allowed the investigators to make strong mechanistic statements about models for the catalytic mechanism of ribonuclease A. Since phospholipase A_2 is available for study at the same level of detail, one can anticipate a similar degree of quantitative detail regarding the important interactions of this enzyme to be forthcoming.

Acknowledgments

This work was made possible by Grants GM11838 and GM37658 from the National Institutes of Health, by Grant DMB8417175 from the National Science Foundation, and by Grant N00014-88-K-0326 from the Office of Naval Research.

[29] M. F. Roberts, R. A. Deems, and E. A. Dennis, *Proc. Natl. Acad. Sci. U.S.A.* **74,** 1950 (1977).

[30] W. Cho, A. G. Tomasselli, R. L. Heinrikson, and F. J. Kézdy, *J. Biol. Chem.* **263,** 11237 (1988).

[31] M. Flogel and R. L. Biltonen, *Biochemistry* **14,** 2603 (1975).

[32] M. Flogel and R. L. Biltonen, *Biochemistry* **14,** 2610 (1975).

[22] Activation of Phospholipase A_2 on Lipid Bilayers

By JOHN D. BELL and RODNEY L. BILTONEN

Introduction

The activation of phospholipase A_2 is a complex process that depends on the conformation of the enzyme[1-5] and the structure and dynamics of the lipid bilayer with which it interacts.[5-9] The physical nature of the lipid bilayer is of particular importance. Thus, one must choose carefully the experimental system to be used for studies of the mechanism of the activation process as well as studies of the mechanisms of action of putative activators or inhibitors of the enzyme. Ideally, the system would be simple, well-characterized physically, and amenable to investigations of time-dependent processes involved in enzyme activation. Substrates commonly used, such as sonicated vesicles of hen egg phosphatidylcholine or of saturated phosphatidylcholines, are frequently undesirable because at least some of the phospholipases A_2 are not very active toward vesicles in the liquid crystalline state, and hydrolysis time courses with sonicated vesicles in the gel state do not always exhibit a time-dependent activation. Also, sonicated vesicles of phosphatidylcholine are not stable in the gel state[10] and aggregate or fuse into structures that are not well-defined, thus complicating interpretation of the data.

One experimental system that offers several advantages for the study of the activation of soluble phospholipases A_2 is large unilamellar vesicles (LUV) of dipalmitoylphosphatidylcholine (DPPC): (1) Enzyme activation

[1] M. F. Roberts, R. A. Deems, and E. A. Dennis, *Proc. Natl. Acad. Sci. U.S.A.* **74,** 1950 (1977).

[2] D. O. Tinker and J. Wei, *Can. J. Biochem.* **57,** 97 (1979).

[3] G. Romero, K. Thompson, and R. L. Biltonen, *J. Biol. Chem.* **263,** 13476 (1988).

[4] W. Cho, A. G. Tomasselli, R. L. Heinrikson, and F. J. Kézdy, *J. Biol. Chem.* **263,** 11237 (1988).

[5] J. D. Bell and R. L. Biltonen, *J. Biol. Chem.* **264,** 12194 (1989).

[6] H. M. Verheij, A. J. Slotboom, and G. H. de Haas, *Rev. Physiol. Biochem. Pharmacol.* **91,** 91 (1981).

[7] M. Menashe, G. Romero, R. L. Biltonen, and D. Lichtenberg, *J. Biol. Chem.* **261,** 5328 (1986).

[8] N. Gheriani-Gruszka, S. Almog, R. L. Biltonen, and D. Lichtenberg, *J. Biol. Chem.* **263,** 11808 (1988).

[9] M. K. Jain, B.-Z. Yu, and A. Kozubek, *Biochim. Biophys. Acta* **908,** 23 (1989).

[10] M. Wong, F. H. Anthony, T. W. Tillack, and T. E. Thompson, *Biochemistry* **21,** 4126 (1982).

is slow and can be monitored by following the time course of vesicle hydrolysis.[3] (2) With certain phospholipases A_2, such as the monomer aspartate-49 enzyme from *Agkistrodon piscivorus piscivorus* (AppD49), changes in the intrinsic tryptophan fluorescence of the enzyme can be measured simultaneously with the hydrolysis reaction to correlate changes in the state of the enzyme during activation.[5] (3) Changes in the structure of the vesicles can be concurrently monitored using fluorescent probes to correlate the role of vesicle structure and dynamics in the activation process. (4) The vesicles can be well-characterized structurally and thermodynamically.[10,11] Although the methods described in this chapter focus on DPPC LUV and certain snake venom phosphlipases A_2, they are certainly adaptable to the study of other lipid hydrolases and substrates.

Preparation of Vesicles

The preparation of fused LUV of DPPC is based on procedures and principles previously described.[3,10] This method is applicable to DPPC and its ether analog but is not necessarily applicable to other lipids. The phospholipid, suspended in chloroform, is dried by evaporation. Final traces of the chloroform are removed with a lyophilizer. A solution of 50 m*M* KCl and at least 1 m*M* NaN_3 to prevent bacterial growth is added to the tube of dried lipid at a temperature of 45°–55°, a temperature greater than the gel–liquid crystalline transition temperature. The lipid sample should be maintained at this temperature for at least 1 hr and vortexed vigorously for several seconds about every 10 min. This procedure yields a suspension of multilamellar vesicles.[12]

The multilamellar vesicles are then dispersed by sonication at 45°–55°.[13] Sonication of the sample 3 or 4 times for 3 min each is generally sufficient to yield a dispersion of small unilamellar vesicles (diameter about 25 nm). The sample should appear transparent at the end of the sonication. Titanium grains from the sonicator probe and nondispersed lipid are removed from the sample by either ultracentrifugation[14] or a 5-min centrifugation at 14,000 rpm in a microfuge using 1.5-ml tubes. If ultracentrifugation is used, the sample must be maintained at no less than 45° during the separation. The upper two-thirds of the resulting supernatant is carefully removed and transferred to a glass tube and sealed with Teflon tape. The sample is allowed to stand at room temperature overnight and then

[11] J. Suurkuusk, B. R. Lentz, Y. Barenholz, R. L. Biltonen, and T. E. Thompson, *Biochemistry* **15,** 1393 (1976).

[12] A. D. Bangham, J. De Gier, and G. D. Greville, *Chem. Phys. Lipids* **1,** 225 (1967).

[13] C. Huang, *Biochemistry* **8,** 344 (1969).

[14] C. Huang and T. E. Thompson, this series, Vol. 32, p. 485.

transferred to storage at 4° for 3 weeks. The storage temperature is critical for reproducible vesicle fusion. Below 2°, the vesicles aggregate irreversibly. It should also be noted that the vesicles do not fuse properly in the presence of added $CaCl_2$.

At the end of the 3-week storage, the sample contains mostly large unilamellar vesicles with a diameter of about 90 to 100 nm.[10] The appearance of the sample should resemble dilute milk. Contaminating multilamellar vesicles that have formed during the fusion process must be removed to obtain a homogeneous sample. Two methods are useful for the separation. The sample may be applied to a column of Sepharose Cl-2B,[3] or it may be centrifuged for 5 min at 14,000 rpm in a microfuge as described above. The supernatant contains the LUV, and the vesicles appear to be stable for several months if stored at room temperature in the presence of NaN_3.[10] The purity of the LUV can be monitored by thin-layer chromotography[3] and by differential scanning calorimetry.[10,11] The calorimetric scan of DPPC LUV should show a single nearly symmetrical peak at 41.4° which is broader than that of multilamellar vesicles. The apparent T_m and width of the calorimetric scan will depend on the scan rate and calorimeter used.[11] Most importantly, a preparation of LUV containing *no* small unilamellar vesicles is characterized by the absence of a gel–liquid transition at about 37°.

Combined Enzyme Assay and Fluorescence Spectroscopy

Assay of the time course of vesicle hydrolysis by phospholipase A_2 using a pH stat has been described previously.[15] In order to make temporal correlations of changes in the intrinsic fluorescence intensity of the enzyme with changes in the catalytic activity during the time course, the pH stat instrument (Radiometer) has been used with a spectrofluorometer (SLM 8000C).[5] This is accomplished by inserting a small combination pH electrode (5 mm diameter) and the glass pipette tip from the pH stat buret in a two-hole rubber stopper as shown in Fig. 1. A hole is drilled in a lid for the fluorometer sample chamber to fit the stopper and to center the electrode and pipette over the sample cuvette. A stainless steel cannula is inserted through the rubber stopper for sample injection. The height of the rubber stopper is adjusted so that the electrode, pipette, and cannula tips are all submerged in the sample but do not interfere with the fluorometer light path. This is possible in the SLM 8000C fluorometer using a standard 1 cm^2 fluorometer cuvette with a stirring bar at the bottom and 2.5 ml of sample. Adjustments may be needed depending on the position of the

[15] W. Nieuwenhuizen, H. Kunze, and G. H. de Haas, this series, Vol. 32, p. 147.

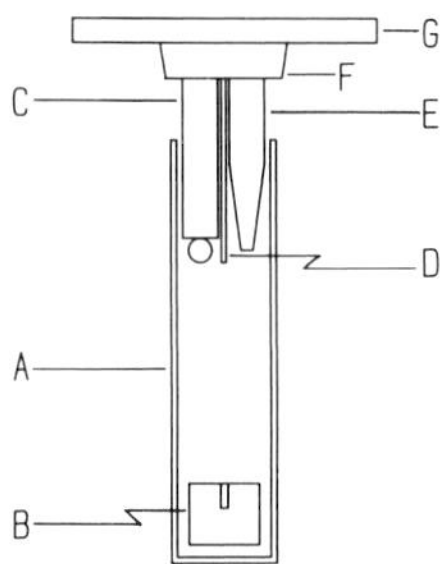

FIG. 1. Detail of the experimental setup for simultaneous measurement of enzyme fluorescence and hydrolysis. A, Standard fluorometer cuvette; B, magnetic stirring bar; C, combination pH electrode (from pH stat); D, cannula for enzyme injection; E, pipette tip (from pH stat buret); F, rubber stopper; G, fluorometer sample chamber lid.

optical windows in the sample chamber of other fluorometers. The rubber stopper assembly can be conveniently placed and removed as a unit between experiments as samples are exchanged. It is necessary for these experiments that the sample chamber of the fluorometer be temperature-controlled and equipped with a magnetic stirrer. Temperature is critical in studies involving DPPC (see below), and adequate stirring must be maintained for maximum resolution and sensitivity in monitoring the hydrolysis time course and to maintain sample homogeneity.

Prior to the experiment, the vesicles must be equilibrated in the appropriate solution; 10 m*M* $CaCl_2$, 35 m*M* KCl, and 1 m*M* NaN_3 has been commonly used.[5] After adjusting the solute concentration, the vesicles are slowly and repeatedly heated and cooled through the phase transition (41.5°) over a period of several hours to achieve equilibrium of all components across the vesicle bilayer. All solutions must be degassed to eliminate CO_2, which interferes with the pH stat assay by buffering the sample pH. Finally, the vesicles are temperature-equilibrated in the fluorometer sample chamber with the rubber stopper assembly in place and the magnetic stirrer turned on to ensure proper equilibration. The pH is then adjusted to 8.0 (or as desired) with the pH stat, and the experiment is initiated by injection of phospholipase A_2.

Figure 2 shows typical simultaneous fluorescence and hydrolysis time courses using the AppD49 phospholipase A_2. Note that the fluorescence intensity of the enzyme rapidly increases at the onset of rapid vesicle hydrolysis. The decrease in the intensity during the initial portion of the time course results from adsorption of the enzyme onto the quartz cuvette.[5] A number of methods to prevent this adsorption have been examined including treatment with silane, dextran, or polylysine. To date, only

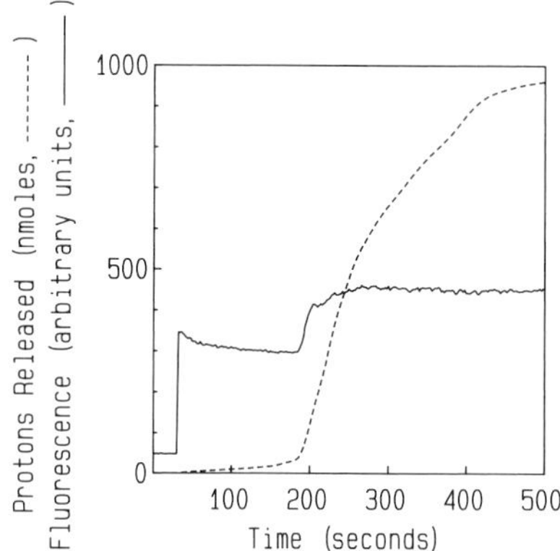

FIG. 2. Time courses of DPPC LUV hydrolysis and AppD49 phospholipase A_2 fluorescence monitored with the system described in Fig. 1. Vesicles (0.4 mM) in 2.5 ml of 35 mM KCl, 10 mM $CaCl_2$, and approximately 1 mM NaN_3 were equilibrated at 39° and adjusted to pH 8.0 using 10 mM NaOH as the titrant. Enzyme (20 μg/ml) was added at t = 30 sec, and fluorescence (excitation 280 nm, emission 340 nm) and vesicle hydrolysis (number of protons titrated with NaOH by the pH stat) were simultaneously recorded as described.

polylysine has been found to be effective in preventing the binding of enzyme to the cuvette. Unfortunately, polylysine appears to interfere with the hydrolysis reaction. More acidic phospholipases, such as the dimer enzymes from *Crotalus atrox* and *A. piscivorus piscivorus*, do not adsorb appreciably to the cuvette walls.

Changes in the physical properties of the vesicles during the hydrolysis time course can also be simultaneously monitored. Changes in vesicle light scattering suggest changes in vesicle size or refractive index during the time course and provide the data needed for correction of certain optical artifacts (see below). Light scattering can be monitored simultaneously by using a second emission monochromator mounted 90° from the excitation monochromator and 180° from the other emission monochromator. The wavelength of the second emission monochromator is set near the excitation wavelength. For example, the excitation wavelength for protein fluorescence is generally 280 nm, and one emission monochromator is set at 340 nm for the fluorescence and the second monochromator is set at 290 nm for measuring light scattering rather than at the excitation wavelength to avoid excessive intensity of the scattered light.

A fluorescent probe of membrane structure, trimethylammonium diphenylhexatriene (TMA-DPH), has been used to monitor other properties of the bilayer during the hydrolysis time course.[5] This probe is sensitive

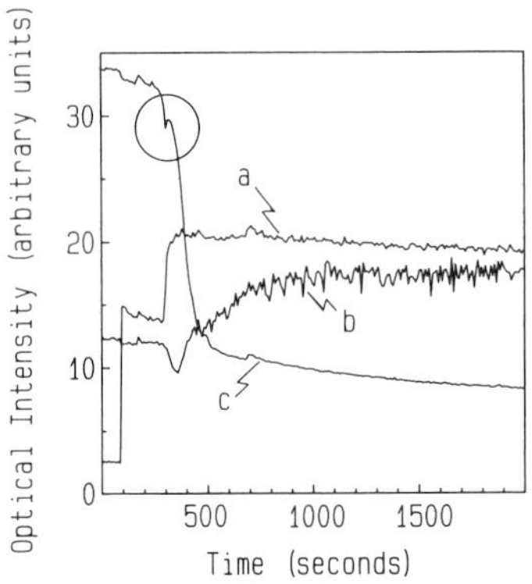

FIG. 3. Correlation of the time courses of phospholipase A_2 fluorescence, vesicle light scattering, and TMA-DPH fluorescence during the hydrolysis of DPPC LUV. Vesicles (0.4 m*M* DPPC) were equilibrated with 0.4 μM TMA-DPH in 2.5 ml of 35 m*M* KCl, 10 m*M* $CaCl_2$, 10 m*M* sodium borate (pH 8), and 1 m*M* NaN_3. AppD49 phospholipase A_2 (8 μg/ml) was added at t = 90 sec. Curve a, Enzyme fluorescence (excitation 280 nm, emission 340 nm); curve b, light scattering (excitation 280 nm, emission 290 nm); curve c, TMA-DPH fluorescence (excitation 360 nm, emission 430 nm).

to the fluidity of the phospholipid molecules within the bilayer[16] and to quenching by water. The TMA-DPH (2 m*M* in dimethylformamide) is diluted 5000-fold into the vesicle solution and equilibrated at least 1 hr at 45° in the dark. The concentrations of probe and vesicle are selected so that the ratio of lipid to probe in the bilayer will be at least 1000 : 1. Figure 3 shows a typical time course of simultaneous measurement of enzyme fluorescence (curve a), vesicle light scattering (curve b), and TMA-DPH fluorescence (curve c) during lipid hydrolysis. These three optical properties were obtained simultaneously by arranging the monochromators in the T format described above and by electronically switching the excitation wavelength between 280 nm (for the protein fluorescence and the light scattering) and 360 nm (for the TMA-DPH fluorescence). Likewise, one emission monochromator was switched between 340 nm (protein fluorescence) and 290 nm (light scattering). The other emission monochromator monitored the fluorescence at 430 nm (TMA-DPH). The switching of wavelengths is accomplished rapidly under computer control with the SLM 8000C instrument. The time required to complete the cycle is 9 sec.

As shown in Fig. 3, apparent changes in vesicle physical properties occur during the time course of hydrolysis. The light scattering decreases substantially after the increase in enzyme fluorescence and may reflect

[16] F. G. Prendergast, R. P. Haugland, and P. J. Callahan, *Biochemistry* **20,** 7333 (1981).

macroscopic changes in vesicle structure during the rapid phase of hydrolysis.[5] The TMA-DPH fluorescence decreases slowly during the slow phase of hydrolysis, begins to decrease more rapidly, increases transiently concurrent with the increase in enzyme fluorescence, and then decreases dramatically. The transient increase in TMA-DPH fluorescence (encircled in Fig. 3) is highly reproducible, though small, and is not an optical artifact. This change in TMA-DPH fluorescence concurrent with the increase in enzyme fluorescence has been interpreted to reflect a transition in bilayer structure arising from the accumulation of hydrolysis products that may be responsible for the rapid and sudden activation of the enzyme,[5] but it is not yet understood.

Fluorescent probes such as TMA-DPH may adsorb very tightly to the electrode and pipette tip from the pH stat. Consequently, it may not always be desirable to use the pH stat when a lipid probe is included in the reaction mixture. If the pH stat is not used, one must include a buffer in the reaction solution. Phosphate buffer precipitates with calcium and therefore cannot be used. Amine-containing buffers or sulfonic acid derivatives may be used, but they are highly sensitive to temperature. Borate buffer is not very temperature-sensitive and has been used successfully.[5]

Finally, proper experimental design with this system requires at least a phenomenological understanding of the temperature dependence of LUV hydrolysis by phospholipase A_2. The reaction is highly sensitive to the thermotropic phase transition of the lipid bilayer (41.4° for DPPC). Significant rates of hydrolysis are generally seen only when the temperature of the experiment is within 3° or 4° of the transition temperature. Significantly below the transition temperature, rates of hydrolysis are very slow and vesicle hydrolysis may require many hours. Significantly above the thermotropic phase transition, the maximum hydrolysis rates are high, but the lag times preceding rapid hydrolysis may become so long that sample evaporation interferes with the hydrolysis time course.

Data Analysis

The light scattered by the vesicles in solution poses two artifactual problems for quantitative analysis of the fluorescence data. This is especially true since the light scattering changes with time (Fig. 3). First, a small amount of light scattered by the vesicles creates a background at the emission wavelength. The intensity of this scattering is proportional to the concentration of vesicles and to the intensity of scattered light measured at a wavelength near the excitation wavelength. Figure 4A shows the relationship between scattered light measured at 290 nm and that obtained at 340 nm (excitation wavelength 280 nm in both cases). The calibration

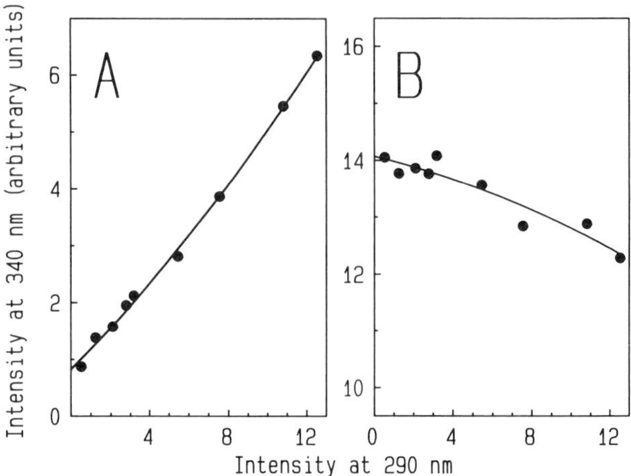

FIG. 4. Calibration curves for stray light and inner filter effects of light scattered by DPPC LUV. Various concentrations of LUV ranging from 0 to 0.4 m*M* were added to 2 ml of 35 m*M* KCl, 10 m*M* EDTA, 10 m*M* sodium borate (pH 8), and 1 m*M* NaN_3 at 37°. The intensity of light at 340 and 290 nm (excitation 280 nm) was recorded, tryptophan (1.5 μM final) was added, and the intensity at both wavelengths was again recorded. (A) Light intensity at 340 nm prior to the addition of tryptophan plotted as a function of the intensity at 290 nm. (B) Intensity of light at 340 nm prior to the addition of tryptophan subtracted from the intensity after addition of tryptophan, with the difference plotted as a function of the intensity at 290 nm. The calibration curves were fit by nonlinear regression to the function $Y = A + BX + CX^2$.

curve is calculated by nonlinear repression using the equation $Y = A + BX + CX^2$. One can then use this calibration curve and the measured light-scattering data obtained at 290 nm to calculate the background due to scattered light at 340 nm. This time-dependent background is subtracted from the fluorescence time course.

The presence of vesicles in the solution causes a second optical artifact. The turbidity of the solution, which is proportional to vesicle concentration, reduces the intensity of excitation light reaching the fluorophore and also reduces the amount of emitted light reaching the monochromator. This artifact can be quantified by adding various concentrations of vesicles to a solution of tryptophan. The background (presumably due to light scattering) is measured at both 290 and 340 nm (excitation 280 nm) prior to the addition of the tryptòphan. Fluorescence emission at 340 nm is then measured after addition of the amino acid. The background at 340 nm is subtracted from the fluorescence signal, and the difference is plotted as a function of the intensity of scattered light at 290 nm (Fig. 4B). A calibration curve is again obtained by nonlinear regression, and the fraction of the true fluorescence that is actually transmitted to the monochromator for a given amount of light scattering is inferred from the calibration curve. One can then correct the measured enzyme fluorescence for inner filter effects during the time course by first calculating the fractional transmittance at each time point using the calibration curve (Fig. 4B) and the time course of light scattering at 290 nm. The correction is made by subtracting the background light (as described above) and then dividing the observed fluorescence intensity at each time point by the corresponding fractional transmittance. In general, the amount of light scattered is relatively small with LUV at lipid concentrations below 0.2 m*M*. These corrections are only needed for quantitative analysis of the fluorescence changes with LUV at lipid concentrations of at least 0.4 m*M*.

Summary

So far, three phospholipases A_2 that display activation kinetics during the time course of hydrolysis of DPPC LUV have been found to undergo a fluorescence change coincident with the activation: the monomer (AppD49) and the dimer enzymes from *A. piscivorus piscivorus* and the dimer enzyme from *C. atrox*. The porcine pancreatic enzyme produces similar time courses of hydrolysis but does not display a concurrent fluorescence change.[17] It is assumed that other phospholipases A_2 will behave similarly in terms of the hydrolysis reaction. Which enzymes respond with a similar change in intrinsic fluorescence during the time course may well depend on the position of tryptophan residues and the amino acid sequence.

Even though a given phospholipase A_2 may not change its fluorescent properties on activation, the simultaneous monitoring of the hydrolysis reaction and the fluorescence of probes of the bilayer structure can be done with any phospholipase A_2. A variety of probes exist which are sensitive to slightly different membrane properties and could be used as described here for TMA-DPH. For example, 1,3-dipyrenylpropane is sensitive to the apparent microviscosity of the bilayer is terms of the ability of molecules to translationally diffuse in the membrane.[18] 6-Palmitoyl-2-{[2-(trimethylammonio)ethyl]methylamino}naphthalene chloride is sensitive to the ability of a molecule to rotate in the bilayer and displays large changes in its steady-state fluorescence as the anisotropy of the bilayer

[17] J. D. Bell and R. L. Biltonen, unpublished results (1989).

[18] R. L. Melorick, H. C. Haspel, M. Goldenberg, L. M. Greenbaum, and S. Weinstein, *Biophys. J.* **34,** 499 (1981).

changes.[19] 6-Propionyl-2-(dimethylamino)naphthalene is sensitive to the polarity and degree of hydration of its environment.[20] Finally, a compound titled NK-529 has recently been introduced that apparently monitors the lateral phase separation of fatty acids in the bilayer.[9]

The fact that activation of phospholipase A_2 can be monitored during the time course of hydrolysis of DPPC LUV makes this system an excellent choice for studying the mechanisms of activation and possible effects of various activators and inhibitors. The experimental system described here provides a way to determine whether such regulators exert their effects through alterations of the properties of the membrane and/or the enzyme. Importantly, this system allows one to seek temporal correlations of the various events in the process.

Acknowledgments

This work was made possible by funding from the National Institute of General Medical Science (Grants GM37658 and GM11838) and from the Office of Naval Research (N00014-88-K-0326).

[19] J. R. Lakowicz, D. R. Bevan, B. P. Maliwal, H. Cherek, and A. Balter, *Biochemistry* **22,** 5714 (1983).

[20] G. Weber and F. J. Farris, *Biochemistry* **18,** 3075 (1979).

[23] Phospholipase Stereospecificity at Phosphorus

By KAROL S. BRUZIK and MING-DAW TSAI

Introduction[1]

Two types of stereochemical information can be obtained from studies of enzymatic reactions utilizing substrates which are stereogenically[2] (chirally) labeled at phosphorus.[3–10] (1) The steric course of the nucleophilic

[1] This is paper 21 in the series "Phospholipids Chiral at Phosphorus." For Paper 20, see Ref. 16.

[2] K. Mislow and J. Siegel, *J. Am. Chem.Soc.* **106,** 3319 (1984).

[3] F. Eckstein, *Angew. Chem. Int. Ed. Engl.* **22,** 423 (1983); F. Eckstein, P. J. Romaniuk, and B. A. Connolly, this series, Vol. 87, p. 197.

[4] P. A. Frey, J. P. Richard, H.-T. Ho, R. S. Brody, R. D. Sammons, and K.-F. Sheu, this series, Vol. 87, p. 213; P. A. Frey, *Tetrahedron* **33,** 1541 (1984).

[5] J. A. Gerlt, J. A. Coderre, and S. Mehdi, *Adv. Enzymol. Relat. Areas Mol. Biol.* **55,** 291 (1983).

displacement at phosphorus may be elucidated through synthesis and configurational analysis of chirally labeled substrates, coupled with configurational analysis of the product. The results can provide information on the possible involvement of the intermediary covalent phosphoenzyme complex along the reaction pathway, thus enabling insight into details of the elementary steps of enzymatic catalysis.[3,6,7] Two types of chirally modified substrates can be used for such studies: those labeled with oxygen isotopes (^{17}O and/or ^{18}O) or phosphorothioates. (2) Analysis of the stereospecificity of the enzyme toward isomers of different configuration at phosphorus can provide information on the stereochemical constraint of substrate binding and the three-dimensional architecture of the active site.[3,9] This type of study uses phosphorothioates only.

This chapter describes application of such stereochemical approaches to study phospholipases and other phospholipid-metabolizing enzymes using "phospholipids chiral at phosphorus." Figure 1 shows the structures of these compounds (**1–9**) and explains their nomenclature. The enzymes mentioned and their abbreviations are as follows: phospholipase A_2 (PLA_2), phospholipase C (PLC), phospholipase D (PLD), phosphatidylinositide-specific PLC (PI-PLC), lecithin–cholesterol acyltransferase (LCAT), and phosphatidylserine synthase (PS synthase).

Synthetic Procedures

The synthetic steps for chiral thiophospholipids **1–6** are described below. The synthesis of [^{17}O,^{18}O]phospholipids **7–9** is omitted here since the procedures are quite lengthy and the potential application of these compounds to other enzymes is more limited.

DPPsE, DPPsC, DPPsI, and DPPsS (**1–4**).[11–18] The procedure is out-

[6] J. R. Knowles, *Annu. Rev. Biochem.* **49,** 877 (1980); S. L. Buchwald, D. E. Hansen, A. Hassett, and J. R. Knowles, this series Vol. 87, p. 279.

[7] G. Lowe, *Philos. Trans. R. Soc. London, Ser. B* **293,** 75 (1981); G. Lowe, *Acc. Chem. Res.* **16,** 244 (1983).

[8] M.-D. Tsai, this series, Vol. 87, 235.

[9] M. Cohn, *Acc. Chem. Res.* **15,** 326 (1982).

[10] W. J. Stec, *Acc. Chem. Res.* **16,** 411 (1983).

[11] K. Bruzik, R.-T. Jiang, and M.-D. Tsai, *Biochemistry* **22,** 2478 (1983).

[12] R.-T. Jiang, Y.-J. Shyy, and M.-D. Tsai, *Biochemistry* **23,** 1661 (1984).

[13] M. D. Mateucci and M. H. Caruthers, *J. Am. Chem. Soc.* **103,** 3185 (1981).

[14] K. S. Bruzik, G. Salamonczyk, and W. J. Stec, *J. Org. Chem.* **51,** 2368 (1986).

[15] G. Lin and M.-D. Tsai, *J. Am. Chem. Soc.* **111,** 3099 (1989).

[16] G. Lin, C. F. Bennett, and M.-D. Tsai, *Biochemistry* **29,** 2747 (1990).

[17] G. M. Salamonczyk and K. S. Bruzik, *Tetrahedron Lett.* **31,** 2015 (1990).

[18] W. Loffredo and M.-D. Tsai, *Bioorg. Chem.* **18,** 78 (1990).

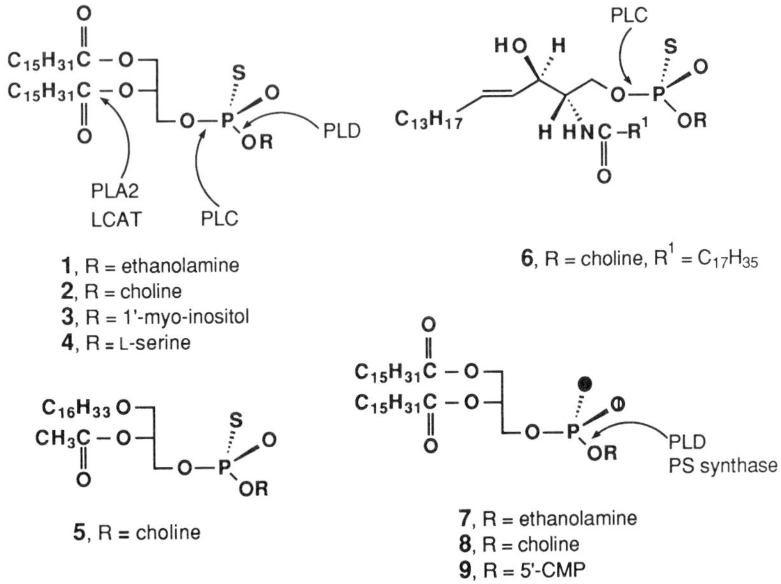

1, R = ethanolamine
2, R = choline
3, R = 1'-myo-inositol
4, R = L-serine

5, R = choline

6, R = choline, $R^1 = C_{17}H_{35}$

7, R = ethanolamine
8, R = choline
9, R = 5'-CMP

FIG. 1. Structures of chiral thiophospholipids (**1–6**) and [^{17}O,^{18}O]phospholipids (**7–9**). **1**, (R_p)-DPPsE; **2**, (R_p)-DPPsC; **3**, (S_p)-DPPsI; **4**, (R_p)-DPPsS; **5**, (R_p)-AGEPsC; **6**, (R_p)-SPsM; **7**, (R_p)-[^{17}O,^{18}O]DPPE; **8**, (R_p)-[^{17}O,^{18}O]DPPC; **9**, (S_p)-[^{17}O,^{18}O]CPD-DPG. It is a convention to use ⦶ and ● to denote ^{17}O and ^{18}O, respectively. In assigning the *R* and *S* configurations, it should be noted that the atomic number priority should be exhausted before considering atomic mass (i.e., OX has higher priority than ^{18}O) [R. S. Cahn, C. K. Ingold, and V. Prelog, *Angew. Chem., Int. Ed. Engl.* **5**, 385 (1966)] and that, even though all thiophospholipids shown have the same *relative* configuration, they have different *absolute* configurations due to different head group structures. For abbreviations of glycerophospholipids, the first two characters describe the structure of the glyceride moiety (e.g., DP, 1,2-dipalmitoyl; MP, 1-monopalmitoyl; DO, 1,2-dioleoyl). The third character, P/Ps, denotes phospho/thiophospho function. The fourth character denotes the head group structure (e.g., A, free acid; C, choline; E, ethanolamine; I, *myo*-inositol; S, L-serine). AGEPsC, 1-*O*-Hexadecyl-2-acetyl-3-thiophosphocholine; PAF, platelet-activating factor; CDP-DPG, cytidine 5′-diphospho-1,2-dipalmitoyl-*sn*-glycerol; SPsM, D-*erythro*-2,*N*-stearoylsphingosyl-1-thiophosphocholine.

lined in Fig. 2 (**10–15**). To the chloroform solution of 1,2-dipalmitoyl-*sn*-glycerol (**10,** 1 mmol, 5 ml) is added triethylamine (2 mmol) and *N,N*-diisopropylmethylphosphonamidic chloride (**11,** 1.2 mmol) at room temperature. After the completion of the reaction (5 min) the solvent and excess triethylamine are evaporated, and solid tetrazole (4 mmol) and the corresponding alcohol [1.1–3 mmol, *N*-tritylethanolamine or choline tosylate or (−)-2,3,4,5,6-pentabenzyl-*myo*-inositol or *N*-trityl-L-serine methoxymethyl ester] are added; all reactants are solubilized by adding acetonitrile–tetrahydrofuran (THF) (1 : 1, v/v). After 30 min solvents are

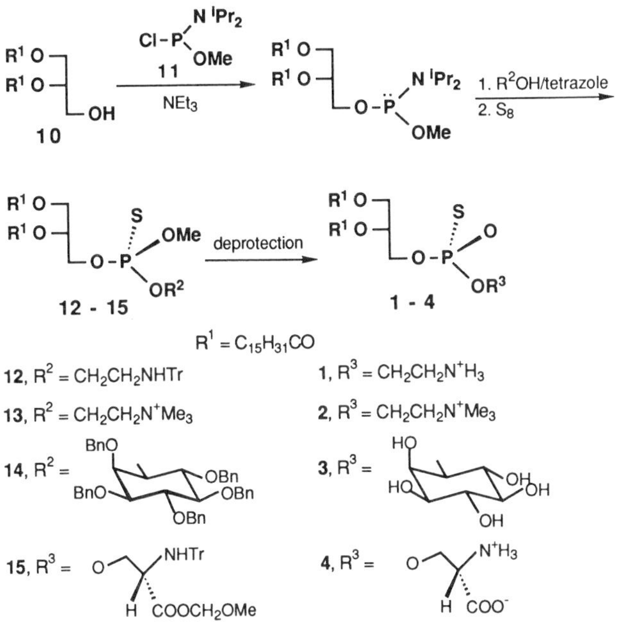

FIG. 2. Synthesis of phosphorothioate analogs of phospholipids **1–4** (via **10–15**). The products thus obtained should be mixtures of both diastereomers, but only one isomer is shown.

removed by evaporation and replaced with toluene. Elemental sulfur (10 mmol) is then added, and the reaction mixture is stirred for 12 hr at room temperature. The suspension is washed with triethylammonium bicarbonate buffer (1.5 *M*, pH 7.0), and the organic phase is concentrated and dried thoroughly under vacuum. Deprotection of the phosphate function in **12–14** is achieved by treatment of the triester with trimethylamine/toluene (1 : 1, v/v, 5 ml) at 50° for 20 hr. The deprotection of *N*-trityl-DPPsE is carried out with Zn^{2+}/acetic acid (200 mg/10 ml) at 45° during 9 hr. The pentabenzyl derivative of DPPsI is deprotected using ethanethiol/BF_3–etherate. In the case of DPPsS the deprotection of **15** is achieved by concentrated HCl in dry acetone followed by trimethylamine in toluene.

AGEPsC (**5**) *and SPsM* (**6**). Compounds **5** and **6** are obtained analogously as described above, except that the starting diacylglycerol **10** is replaced by 1-palmityl-2-palmitoyl-*sn*-glycerol for the synthesis of AGEPsC[19] (the 2-palmitoyl group is changed to a 2-acetyl group after the

[19] T. Rosario-Jansen, R.-T. Jiang, M.-D. Tsai, and D. J.Hanahan, *Biochemistry* **27,** 4619 (1988).

3-thiophosphocholine group has been introduced) and by D-*erythro*-3,*O*-*tert*-butyldiphenyl-2,*N*-stearoylsphingosine for the synthesis of SPsM.[20]

Separation of Diastereomers.[11] Separation of R_p and S_p isomers is achieved by stereospecific hydrolysis of the R_p isomer catalyzed by PLA_2 (see Results). (R_p + S_p)-DPPsC (1 g) is dissolved in chloroform/ether (60 ml, 1/5) and the solution treated with PLA_2 (bee venom, 1000 units) in buffer (3.5 ml, 10 m*M* Tris-Na, pH 7.2, 2 m*M* $CaCl_2$, 0.2 m*M* EDTA). The resulting emulsion is stirred at room temperature, and the reaction is monitored by ^{31}P nuclear magnetic resonance (NMR) (see below). When approximately 80% of (R_p)-DPPsC is hydrolyzed, the mixture is concentrated and the unreacted DPPsC separated from (R_p)-MPPsC (lyso-DPPsC) by chromatography on silica gel using chloroform/methanol/water (70 : 30 : 4, v/v). The fraction containing unreacted DPPsC is again subjected to hydrolysis by PLA_2 as described above, except the reaction is allowed to proceed until the formation of small quantities of (S_p)-MPPsC can be detected by ^{31}P NMR. Column chromatography on silica gel afforded pure (S_p)-DPPsC. The fraction containing pure (R_p)-MPPsC from the first column is subjected to reacylation using palmitoyl anhydride in the presence of 4-dimethylaminopyridine to give (R_p)-DPPsC. Alternatively, (R_p)-DPPsC can be obtained by the hydrolysis of (R_p + S_p)-DPPsC catalyzed by PLC from *Bacillus cereus,* which hydrolyzes (S_p)-DPPsC specifically.[11]

(R_p + S_p)-DPPsE is also separated into (R_p)- and (S_p)-DPPsE as described above. (R_p + S_p)-DPPsI is separated into (R_p)-DPPsI and (S_p)-MPPsI,[15,16] and (R_p + S_p)-DPPsS is separated into (R_p)-MPPsS and (S_p)-DPPsS by PLA_2.[18] Reacylation of (S_p)-MPPsI and (R_p)-MPPsS is not performed due to presence of other acylation sites in the molecules. Separation of (R_p + S_p)-DPPsI into (R_p)- and (S_p)-DPPsI is also achieved by column chromatography on silica gel at the stage of protected derivatives **14** using carbon tetrachloride/acetone (100 : 1, v/v) as the eluting solvent.[17]

Configurational Analysis

[^{17}O,^{18}O]DPPE.[21] Analysis of [^{17}O,^{18}O]DPPE is presented in Fig. 3. Diastereomers of [^{17}O,^{18}O]DPPE are first converted to [^{16}O,^{17}O,^{18}O]DPPA (**16**) and then to 1-[^{16}O,^{17}O,^{18}O]phosphopropane-1,2-diol (**17**). The configuration of **17** is determined by converting **17** to **18** followed by ^{31}P NMR analysis as described by Buchwald and Knowles.[22] The basis of the ^{31}P

[20] K. S. Bruzik, *J. Chem. Soc. Perkin Trans. 1,* p. 423 (1988).

[21] K. S. Bruzik and M.-D. Tsai, *J. Am. Soc. Chem.* **106,** 747 (1984).

[22] S. L. Buchwald and J. R. Knowles, *J. Am. Chem. Soc.* **102,** 6601 (1980).

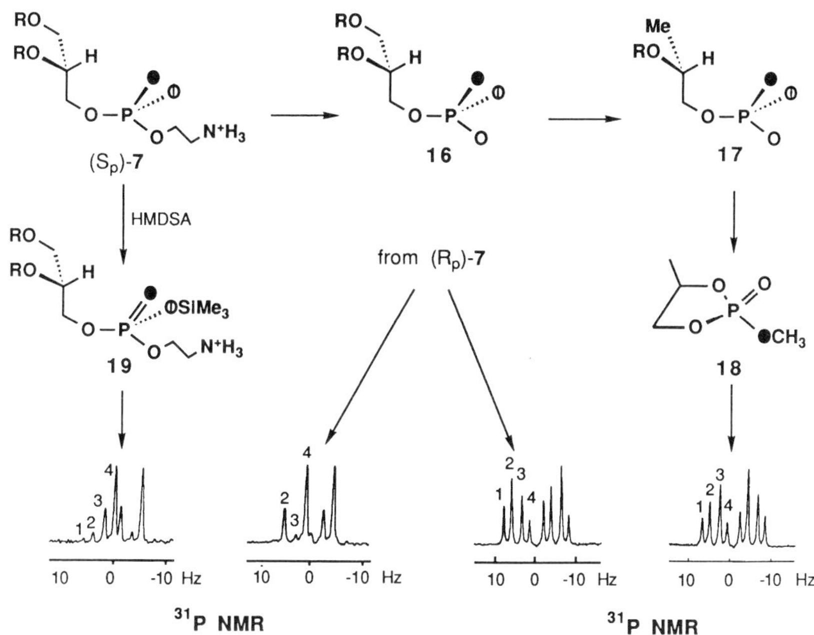

FIG. 3. Stereochemical analysis of P-chiral oxygen-labeled phospholipids. Although only single structures are shown for **18** and **19**, each consists of a mixture of two diastereomers arising from methylation in **18** and silylation in **19**. Each ^{31}P NMR spectrum consists of eight lines (four lines for each diastereomer of the triester derivative). Only one set is labeled (peaks 1–4) in each spectrum, which arise from different ^{18}O-labeled species: 1, unlabeled; 2, P—^{18}O; 3, P=^{18}O; 4, ^{18}O—P=^{18}O. The ratio of intensity of lines 2 and 3 determines the diastereomeric purity and configuration.

NMR method is explained in detail elsewhere.[8,23] Such a procedure has allowed elucidation of the absolute configuration of [^{17}O,^{18}O]DPPE. Now that the absolute configuration has been established, it is no longer necessary to go through the above steps. Instead, the distinction between diastereomers of [^{17}O,^{18}O]DPPE is made possible by silylation with hexamethyldisilazane (HMDSA) followed by ^{31}P NMR analysis of the resulting mixture of diastereomeric triesters (**19**), each having a distribution of various isotopomers (Fig. 3).[21,24] Diastereomers of singly labeled [^{18}O]DPPE can be analyzed by exactly the same method (see the spectrum in Fig. 4).

The configuration of other [^{17}O,^{18}O]phospholipids can be established

[23] M.-D. Tsai and K. S. Bruzik, *Biol. Magn. Reson.* **5,** 129 (1984).

[24] K. S. Bruzik and M.-D. Tsai, *J. Am. Chem. Soc.* **104,** 863 (1982).

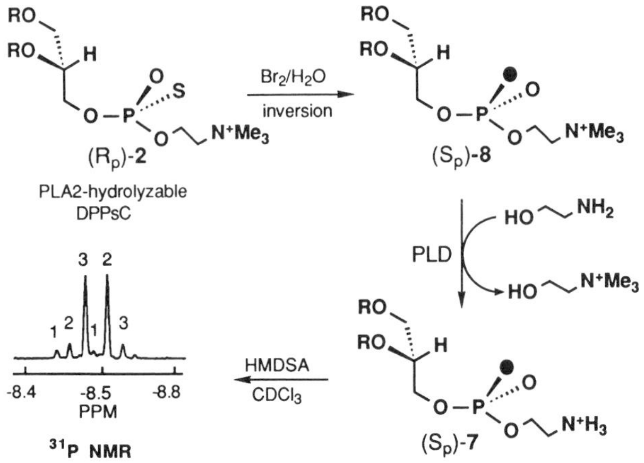

FIG. 4. Chemical correlation of the configuration of DPPsC to that of [^{16}O,^{18}O]DPPE. The method of configurational analysis of [^{16}O,^{18}O]DPPE is the same as that for [^{17}O,^{18}O]DPPE described in Fig. 3. The only difference is that peak 4 due to ^{18}O—P=^{18}O is missing, but the diastereomeric purity and configuration are determined only by the ratio peak 2/peak 3.

by chemical correlation with [^{17}O,^{18}O]DPPE based on the known steric course of PLD-catalyzed transphosphatidylation.[25]

Chiral Thiophospholipids. The configurations of chiral thiophospholipids are determined by chemical correlation with [^{18}O]DPPE as shown in Fig. 4. The PLA$_2$-hydrolyzable isomer of DPPsC (see Results) is desulfurized with bromine/$H_2{}^{18}O$, and the resulting [^{16}O,^{18}O]DPPC is subjected to the transphosphatidylation reaction in the presence of ethanolamine to give the corresponding [^{16}O,^{18}O]DPPE.[12] The configuration of this sample is determined as S_p by the silylation/^{31}P NMR method. Since the steric course of the desulfurization is inversion[26–28] and that of the transphosphatidylation reaction is retention[24] (see Results), the R_p configuration can be assigned to the PLA$_2$-hydrolyzable isomer of DPPsC. The configurations of the diastereomers of DPPsE are determined analogously.[12] The configurations of DPPsI,[15,16] DPPsS,[18] and AGEPsC[19] are determined on the basis of the stereospecific hydrolysis by PLA$_2$, whereas that of SPsM[20] is determined on the basis of the stereospecific hydrolysis by PLC. Now that the absolute configurations have been established, the diastereomers of chiral

[25] K. S. Bruzik and M.-D. Tsai, *Biochemistry* **23,** 1656 (1984).
[26] G. Lowe, G. Tansley, and P. M. Cullis, *J. Chem. Soc., Chem. Commun.*, p. 595 (1982).
[27] D. Sammons and P. A. Frey, *J. Biol. Chem.* **257,** 1138 (1982).
[28] B. A. Connolly, F. Eckstein, and H. H. Füldner, *J. Biol. Chem.* **257,** 3382 (1982).

TABLE I
^{31}P NMR CHEMICAL SHIFTS OF THIOPHOSPHOLIPIDS[a]

	DPPsC		DPPsE		DPPsS		DPPsI		AGEPsC		SPsM	
onditions	R_p	S_p	R_p	S_p	R_p	S_p	R_p	S_p	R_p	S_p	R_p	S_p
DCl$_3$	56.12	56.07	59.82	59.95	Unresolved		Unresolved		56.20	56.27	57.3	57.1
H$_3$OD	60.88	60.80	60.10	60.08	Unresolved							
$_2$O[b]	57.13	57.20			58.73	58.84	57.05	57.45			56.4	56.7
DCl$_3$/Et$_3$N					59.13	59.29						

[a] Data in ppm relative to 85% H_3PO_4. From Refs. 11, 12, 15, 18–20; also, DOPsC R_p 56.591, S_p 56.615 ppm ($CDCl_3$).
[b] Containing 5% Triton X-100, 0.1 *M* Tris, pH 8.0.

thiophospholipids can be differentiated directly by ^{31}P chemical shifts listed in Table I.

Results

Steric Course. The steric course of the reaction is illustrated by the transphosphatidylation catalyzed by PLD from cabbage.[24] As shown in Fig. 5, (R_p)-[^{16}O,^{18}O]DPPE was methylated, and the resulting (R_p)-[^{16}O,^{18}O]DPPC was used as a substrate for PLD (cabbage leaves) in the presence of ethanolamine. The resulting [^{16}O,^{18}O]DPPE was analyzed by ^{31}P NMR after silylation. Since the spectra of silyl derivatives obtained

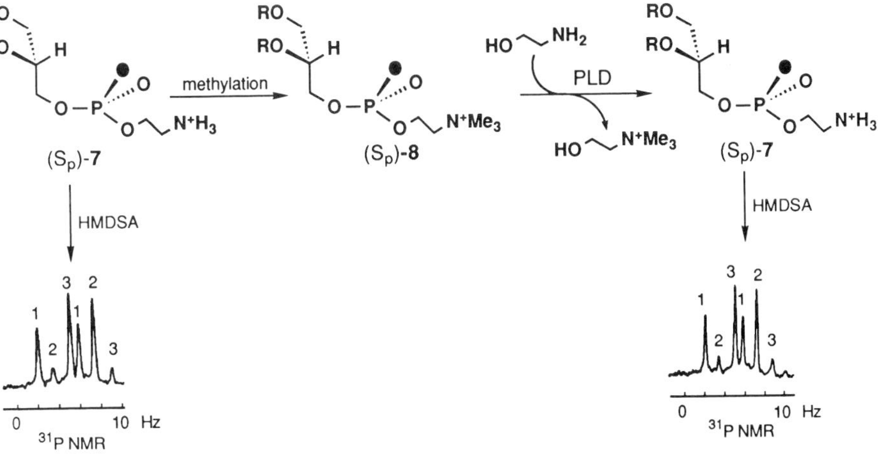

FIG. 5. Determination of the steric course of PLD-catalyzed transphosphatidylation.

TABLE II
STERIC COURSE OF PHOSPHOLIPID-METABOLIZING ENZYMES[a]

Enzyme	Source	Substrate	Product	Steric course	R
PLD	Cabbage	[^{17}O,^{18}O]DPPE	[^{16}O,^{17}O,^{18}O]DPPA	Retention	2
PLD	Cabbage	[^{18}O]DPPC	[^{18}O]DPPE	Retention	2
PS synthase	*Escherichia coli*	[^{17}O,^{18}O]CDP-DPG	[^{17}O,^{18}O]DPPS	Retention	[illegible]
PS synthase	Yeast	[^{17}O,^{18}O]CDP-DPG	[^{17}O,^{18}O]DPPS	Inversion	[illegible]
PI-PLC	*Bacillus cereus*	(R_p)-DPPsI	*exo*-cIPs[c]	Inversion	1
PI-PLC-I	Guinea pig	(R_p)-DPPsI	*exo*-cIPs[c]	Inversion	1
PI-PLC-II	Guinea pig	(R_p)-DPPsI	*exo*-cIPs[c]	Inversion	1

[a] When oxygen-labeled substrates were used, the experiments were usually performed for both and S_p isomers, whereas when thiophospholipid was used, only one isomer could be studied d to stereospecificity of the enzyme.

[b] C. R. H. Raetz, G. M. Carman, W. Dowhan, R.-T. Jiang, W. Waszkuc, W. Loffredo, and M.-Tsai, *Biochemistry* **26,** 4022 (1987).

[c] cIPs, *myo*-inositol 1,2-cyclic phosphorothioate.

from the substrate and from the product were nearly identical, the retention of configuration at phosphorus in the transphosphatidylation reaction was inferred. This suggests that the reaction proceeds by a two-step mechanism involving a phosphatidyl-enzyme intermediate. Results of the analysis of the steric course of reactions catalyzed by phospholipases and related enzymes are listed in Table II.

Stereospecificity. The results on stereospecificity are summarized in Table III. The enzymes which catalyze a P–O bond substitution or cleav-

TABLE III
STEREOSPECIFICITY OF PHOSPHOLIPID-METABOLIZING ENZYMES

Enzyme	Source	Substrate	Preferred isomer	Ref.
PLA_2	Bee venom	DPPsC, DPPsE	R_p	11, 12
PLA_2	*Naja naja*	DPPsC	R_p	11, 12
PLA_2	*Crotalus adamanteus*	DPPsC	R_p	11, 12
PLA_2	Porcine pancreas	DPPsC	R_p	11, 12
PLD	Cabbage	DPPsC	S_p	12
PLC	*Clostridium perfringens*	DPPsC, SPsM[a]	S_p	11, 12,
PLC	*B. cereus*	DPPsC, DPPsE, SPsM[a]	S_p	11, 12,
PI-PLC	*B. cereus*	DPPsI	R_p	15
PI-PLC-I	Guinea pig uterus	DPPsI	R_p	16
PI-PLC-II	Guinea pig uterus	DPPsI	R_p	16
LCAT	Human plasma	DPPsC, DOPsC	No preference	30
PAF receptor	Rabbit platelets	AGEPsC	S_p	19

[a] The configuration of SPsM was not determined by an independent method. Instead, the stereospificity of PLC was used to assign the configuration of SPsM.[20]

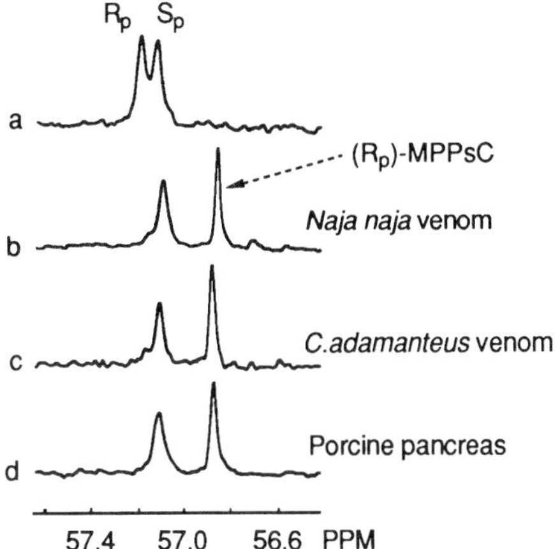

FIG. 6. ^{31}P NMR spectral analysis of the hydrolysis of (R_p + S_p)-DPPsC catalyzed by PLA_2 from various sources showing the stereospecific hydrolysis of (R_p)-DPPsC. (a) Starting (R_p + S_p)-DPPsC; (b–d) products of PLA_2 reactions. (From Bruzik *et al.*[11])

age (e.g., PLC, PLD, and PI-PLC) all exhibit a stereospecificity without surprise. The result of PLA_2 is more unexpected. In view of the rather broad specificity of this enzyme with respect to the structure of the phosphate head group, and the fact that the hydrolysis reaction occurs five bonds away from the negatively charged oxygen atoms of the phosphate group, the high stereospecificity of this enzyme with respect to configuration at phosphorus was not expected. However, ^{31}P NMR analysis in Fig. 6 indicates that PLA_2 is specific for the R_p isomer of DPPsC.

The result with PLA_2 has several impacts. (1) The stereospecificity of PLA_2 constitutes a basis for the enzymatic separation of various thiophospholipids as discussed above. (2) Since the stereospecificity is not expected to depend on the type of phospholipids for the same enzyme, PLA_2 and PLC can be used to determine the phosphorus configuration of other thiophospholipids. PLA_2 is particularly useful due to its broad substrate specificity, and it has been used to assign the configurations of DPPsI, DPPsS, and AGEPsC. (3) The result of PLA_2 was interpreted to suggest that the phosphate group functions as a "remote stereochemical control" in substrate binding via stereospecific coordination of Ca^{2+} to the *pro-S* oxygen of natural phospholipids.[29] Such an interpretation was further supported by detailed kinetic studies, particularly the observation that the ratio $V_{max}(R_p)/V_{max}(S_p)$ dramatically decreases on substitution of the hard metal Ca^{2+} by the soft metal Cd^{2+} (see data in Table IV). This result was

[29] T.-C. Tsai, J. Hart, R.-T. Jiang, K. Bruzik, and M.-D. Tsai, *Biochemistry* **24,** 3180 (1985).

TABLE IV
KINETIC PARAMETERS OF PLA_2 AND LCAT

Enzyme/metal ion	Substrate	K_m (mM)	V_{max} (mmol/min/mg)	Ref.
PLA_2/Ca^{2+}	DPPC	1.67	1850	29
	(R_p)-DPPsC	0.85	76	
	(S_p)-DPPsC	0.30	0.044	
PLA_2/Cd^{2+}	DPPC	6.4	17.6	29
	(R_p)-DPPsC	0.24	0.069	
	(S_p)-DPPsC	—	0.0044	
LCAT	DPPC	0.032	0.021	30
	(R_p)-DPPsC	0.064	0.023	
	(S_p)-DPPsC	0.07	0.02	
	DOPC	0.07	0.066	
	(R_p)-DOPsC	0.072	0.056	
	(S_p)-DOPsC	0.056	0.05	

also used to propose a substrate-binding model which is being tested by protein engineering techniques. (4) The same approach can be used to test whether the phosphate group of phospholipids is involved in a specific interaction in other systems. For example, in contrast to PLA_2, LCAT shows little variation in K_m and V_{max} on sulfur substitution and configuration change at phosphorus[30] (see Table IV). The results suggest that the interaction between the phospholipid substrate and LCAT does not involve stereospecific binding to the phosphate group. The approach has also been used in the interaction of AGEPC (PAF) with PAF receptors.[19]

An obvious question for the use of thiophospholipids is their reactivity relative to the natural substrate. As shown in Table IV, the active isomer of DPPsC has approximately 4% reactivity relative to DPPC. On the other hand, sulfur substitution has no effect in the case of LCAT. The effects in PLC and PLD have not been quantitatively determined, but it was estimated to be of the order of 1% for PLC and PI-PLC and much smaller for PLD.

Another question which has been raised occasionally is whether the observed stereospecificity could be due to differences in the physical properties of the two isomers, instead of differences in the binding to the active site. We have observed that some of the physical properties of the bilayers of chiral thiophospholipids, particularly the thermotropic property, are very sensitive to the configuration at phosphorus. This aspect is

[30] T. Rosario-Jansen, H. J. Pownall, J. P. Noel, and M.-D. Tsai, *Phosphorus Sulfur* **30,** 601 (1987); T. Rosario-Jansen, H. J. Pownall, R.-T. Jiang, and M.-D. Tsai, *Bioorg. Chem.* **18,** 179 (1990).

beyond the scope of this chapter. However, since all of the studies of phospholipases described in this chapter employed micelles or mixed micelles of substrates, the configurational effect on the physical properties should not be responsible for the observed stereospecificity.

Acknowledgments

The work in the laboratory of M.-D.T. was supported by a grant (GM30327) from the National Institutes of Health. The work in the laboratory of K.S.B. was supported by Grant CPBP.01.13.3.16 from the Polish Academy of Sciences.

[24] Phospholipase A_2: Microinjection and Cell Localization Techniques

By DAFNA BAR-SAGI

Introduction

Phospholipase A_2 (PLA_2) is a calcium-requiring esterase that catalyzes the hydrolysis of glycerophospholipids specifically at the *sn*-2 position to produce a fatty acid and a lysophospholipid.[1,2] As mentioned elsewhere in this volume, the activity of PLA_2 has been postulated to play an important regulatory role in several metabolic pathways. For example, PLA_2 catalyzes the release of arachidonic acid, the first and rate-limiting precursor in the biosynthesis of prostaglandins. In addition, the activity of the enzyme is part of the phosphoglyceride deacylation–reacylation cycle and as such mediates the rapid metabolic turnover of membrane phospholipids. Furthermore, there is increasing evidence in support of the participation of PLA_2 in the generation of receptor-mediated transmembrane signals.

This chapter is specifically concerned with two approaches, microinjection and cell localization, to analyze the biological properties of cellular PLA_2. As both approaches rely primarily on the availability of anti-PLA_2 antibodies, methods for obtaining suitable anti-PLA_2 antibodies are also included.

[1] H. van den Bosch, *Biochim. Biophys. Acta* **604,** 191 (1980).

[2] H. M. Verheij, A. J. Slotboom, and G. H. de Haas, *Rev. Physiol. Biochem. Pharmacol.* **91,** 91 (1981).

Preparation and Characterization of Anti-PLA_2 Antibodies

Immunization

As an immunogen, preparations of porcine pancreas PLA_2 purchased from Boehringer Mannheim Biochemicals (Indianapolis, IN) or Sigma Chemical Co. (St. Louis, MO) can be used. Although the pancreatic PLA_2 represents the secreted form of the enzyme, the rationale for using it for raising antibodies against the intracellular form is based on the evidence that the two forms are immunochemically related.[3] The enzyme solution (10 mg/ml) is provided as an ammonium sulfate suspension (3.2 mol/liter, pH 15.5) and is dialyzed against phosphate-buffered saline (PBS). Fifty percent of the dialyzed solution is supplemented with sodium dodecyl sulfate (SDS) to a final concentration of 1% and boiled for 2–3 min. The "SDS–denatured" PLA_2 solution and the "native" PLA_2 solution are each diluted to 1 mg/ml protein concentration with PBS, mixed at a ratio of 1 : 1 (w/w), and injected subcutaneously into rabbits according to the following schedule: 50 μg of PLA_2 in Freund's complete adjuvant on day 0, 100 μg of PLA_2 in Freund's incomplete adjuvant at 10-day intervals over a 2-month period, and 100 μg booster injection in Freund's incomplete adjuvant every 4–5 weeks. Animals are bled at 1-month intervals after the sixth injection, and the titer of the serum is determined by the enzyme-linked immunosorbent assay (using 50 ng of antigen per microtiter well). A high-titer serum is usually obtained within a 2 to 3-month period following the first injection.

Affinity Purification of Antibodies

Homogeneous porcine pancreatic PLA_2 (10 mg) is coupled to 1 ml of Affi-Gel 10 (Bio-Rad, Richmond, CA) by the procedure recommended by the manufacturer. The coupling efficiency exceeds 90%. The resin is poured into a polypropylene Econo-Column (Bio-Rad), and the column is equilibrated with 75 m*M* HEPES (pH 7.5). Crude antiserum is clarified by low-speed centrifugation and passed through the column 3 times, and the column is then washed extensively with 75 m*M* HEPES (pH 7.5). Specifically bound antibodies are desorbed with 12 0.5-ml washes of 50 m*M* glycine-HCl (pH 2.5), and the eluates are adjusted to neutral pH immediately by the addition of 0.5 *M* Tris (pH 7.5). Fractions are analyzed

[3] M. Okamoto, T. Ono, H. Tojo, and T. Yamano, *Biochem. Biophys. Res. Commun.* **128,** 788 (1985).

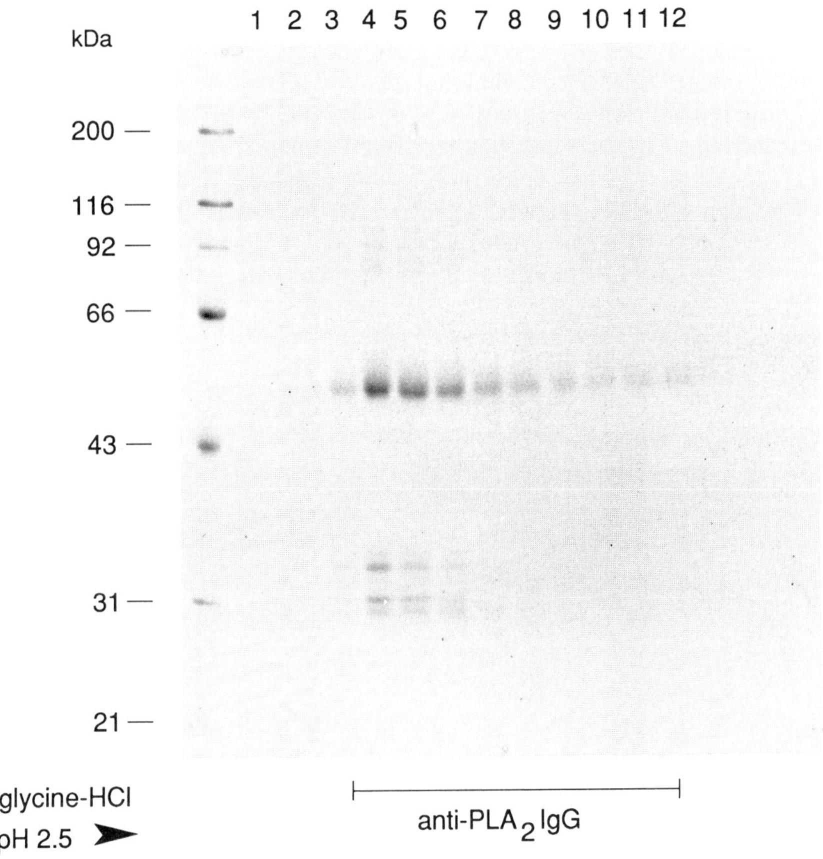

FIG. 1. Elution profile from a PLA_2 affinity column of rabbit anti-PLA_2 IgG. Ten-microliter aliquots of each 0.5-ml fraction were analyzed on the gel.

by SDS–polyacrylamide gel electrophoresis (SDS–PAGE) (Fig. 1). Combined eluates are dialyzed against PBS and concentrated 5- to 10-fold by centrifugal concentration (Centricon 10 microconcentrator, Amicon, Danvers, MA). After use, the column is rinsed extensively with 75 m*M* HEPES (pH 7.5) and stored in 75 m*M* HEPES containing 0.5% NaN_3. The resin can be reused at least 5 times without noticeable effects on the quality of antibodies purified.

Characterization of Antibodies

Two assays are used to establish the specificity of the antibody.

Immunoblotting. Cell extracts are fractionated by SDS–PAGE on a 12.5% gel. Proteins (100 μg/lane) are transferred to nitrocellulose sheets by the blotting procedure described by Towbin *et al.*[4] The nitrocellulose sheets are incubated in a blocking solution [3% bovine serum albumin (BSA) in PBS] overnight and then incubated for 2 hr at 20° with antibody solution (20–50 μg/ml in 1% BSA in PBS). The sheets are rinsed 3 times for 5–10 min each in PBS, then incubated in peroxidase-conjugated goat anti-rabbit IgG (Cappel Laboratories, Downington, PA) diluted 1 : 1500 in 1% BSA in PBS, for 1 hr at 20°. The sheets are again washed 3 times for 5–10 min each with PBS and then developed in 4-chloro-1-naphthol (3 mg/ml of 4-chloro-1-naphthol in methanol diluted with 5 volumes of PBS to which 0.01 volume of 30% hydrogen peroxide is added). This method is very informative since it establishes both the presence and the molecular weight of the antigen and any cross-reacting material. In cases where the abundance of the antigen is very low, the sensitivity of the detection method can be amplified by using biotinylated goat anti-rabbit antibodies followed by streptavidin–alkaline phosphatase.

Neutralizing Activity. The assay for neutralizing activity establishes the specificity of the antibody based on its ability to interfere with the enzymatic activity of the antigen. Cultured cells are labeled for 2 hr with 100 μCi/ml of carrier-free $^{32}PO_4$ in phosphate-free DMEM (Dulbecco's modified eagle's medium). At the end of the labeling period, cells are washed 5 times with phosphate-free DMEM and harvested in ice-cold hypotonic lysis buffer (20 m*M* Tris-HCl, pH 7.4, 10 m*M* NaCl, 0.1% 2-mercaptoethanol, 1% aprotinin). After incubation for 20 min on ice, cells are ruptured with 25–40 strokes (depending on cell type and density) of a Dounce homogenizer. Cell lysates are sedimented at low speed (500 *g* for 5 min at 4°) to remove nuclei, and the supernatant is centrifuged at 100,000 *g* for 30 min at 4°.

The resulting membrane pellet is resuspended in incubation buffer (50 m*M* Tris-HCl, pH 8.0, 120 m*M* NaCl) to a protein concentration of 0.2 mg/ml. Equal aliquots of the membrane suspension (100–200 μl) are preincubated for 2 hr at 4° with 1–15 μg of anti-PLA_2 IgG or control IgG. The reaction mixture is then supplemented with 5 m*M* $CaCl_2$ to trigger PLA_2 activity and incubated at 37° for various intervals. Reactions are terminated by the addition of an ice-cold mixture of methanol, chloroform, and HCl (2 : 1 : 0.02, v/v), and phospholipids are extracted as described previously.[5] Chromatographic separation of phospholipids is carried out on silica gel LK6D thin-layer chromatography (TLC) plates (Whatman, Inc, Clifton, NJ). To resolve the products of PLA_2 activity, namely, lyso-phosphatidylcholine (lyso-PC) and lysophosphatidylethanolamine (lyso-

[4] H. Towbin, T. Staehelin, and J. Gordon, *Proc. Natl. Acad. Sci. U.S.A.* **76,** 4350 (1979).

[5] D. Bar-Sagi, J. P. Suhan, F. McCormick, and J. R. Feramisco, *J. Cell Biol.* **106,** 1649 (1988).

PE), the solvent sysem of chloroform, methanol, ammonia, and water (45 : 30 : 3 : 5, v/v) is used. The ^{32}P-labeled phospholipids are visualized by autoradiography and identified by cochromatography with standards detected with iodine vapor. The TLC plates are exposed to film for approximately 12 hr. Quantification of ^{32}P incorporated into phospholipids is determined by scraping the labeled phospholipids off the plates and liquid scintillation counting. Total ^{32}P incorporation into lipids is approximately 50,000 counts per minute (cpm) under these conditions.

In this assay the activity of PLA$_2$ is time, temperature, and calcium dependent. Phosphatidylethanolamine is the preferred substrate for PLA$_2$ activity.[5] Inhibitory anti-PLA$_2$ antibodies result in a dose-dependent inhibition of the accumulation of lysophosphoglycerides, whereas control antibodies have no effect on PLA$_2$ activity (Fig. 2). Similar results are obtained when the activity of PLA$_2$ is monitored by the generation of free [^{3}H]arachidonic acid following labeling of cells with [^{3}H]arachidonic acid (1 μCi/ml) for 24 hr or by the hydrolysis of [^{14}C]oleate labeled *Escherichia coli* substrate.

Cell Injection

The microinjection technique is a valuable tool for the assay of the functional significance of cellular proteins. In principle, this technique allows one to modulate the level and/or activity of a particular protein by introducing defined amounts of the protein or antibodies directed against the protein into the cells. In view of the paucity of specific PLA$_2$ inhibitors, microinjection of inhibitory anti-PLA$_2$ antibodies makes this approach the method of choice to assay the role of phospholipase A$_2$ in various cellular processes. The technique of microneedle injection has been described in detail elsewhere.[6] In general, the procedure utilizes a glass capillary needle filled with the substance to be injected into the cell, a micromanipulator to place the needle into the cell, and a phase-contrast microscope to allow visualization of the injection process.

Cells

Cells for injection can be grown on either glass coverslips or on petri dishes provided that they adhere tightly to the substrate. It is difficult to predict the suitability of a given cell type for microinjection. As a rule, large or tall cells are easier to inject, whereas small or flat cells are difficult and sometimes even impossible to inject.

[6] A. Graessmann, M. Graessman, and C. Mueller, this series, Vol. **65,** p. 816.

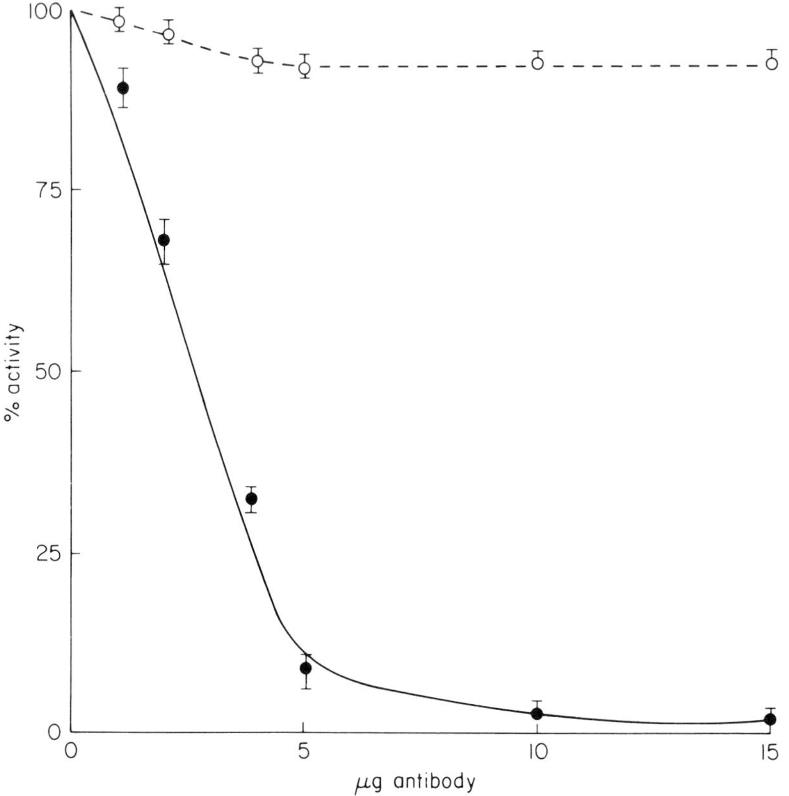

FIG. 2. Inhibition of the activity of membrane-associated PLA_2 as a function of antibody concentration. After preincubation for 2 hr at 4° with the anti-PLA_2 antibodies (filled circles) or nonimmune rabbit anti-rat antibodies (open circles), phospholipase A_2 activity was assayed as described in the text. Results are expressed as percentages of the initial activity of the enzyme.

Sample Preparation

The protein to be injected must be prepared in a buffer that will not have a deleterious effect on the cell. A variety of injection buffers have been used with success.[7] For example, a commonly used buffer for antibody injection is composed of 10 m*M* NaH_2PO_4, 70 m*M* KCl, pH 7.2. Protein concentrations for the injections should be adjusted according to the purpose of the experiments. Concentration ranges of 1–15 mg/ml can be used. The lower end of this range usually applies for the introduction

[7] K. Wang, J. R. Feramisco, and J. F. Ash, this series, Vol. **85,** p. 514.

of proteins, whereas the higher concentrations generally apply to the injection of antibodies.

Microinjection

Prior to injection, the protein solution is clarified by centrifugation for 5 min at 12,000 *g* at 4°. The solution is loaded into the needle by capillary action immediately prior to microinjection. Cells are removed from the incubator and are placed on the microscope stage. The area of injection can be marked by an ink circle using a marker objective. Cells are brought into focus, and the capillary is lowered until the tip is almost in focus. The cells are injected by further lowering the capillary tip and applying a gentle positive pressure. The injection process is marked by a slight swelling of the cell. The cytoplasm appears to lose contrast, whereas the nucleus gains contrast. These changes in appearance are transient, and within a few seconds injected cells should regain normal morphology. The volume injected per cell can be controlled by the amount of pressure applied on the syringe. It has been determined that the cytoplasmic injection of cells usually results in the introduction of 10^{-13} to 10^{-14} liter per cell.[6,8] Control injections should be done to determine the specificity of the effects monitored. In the case of microinjection of antibodies, control injections may include heat-denatured antibody, nonspecific antibody injected at the same concentration as the specific antibody, and buffer alone.

Analysis of Injected Cells

The analytical potential of microinjection experiments depends on two parameters: (1) the ability to positively identify injected cells and (2) the availability of a single cell assay that can be monitored visually. As for the first parameter, the immunofluorescence technique is ideal for the identification of injected cells. The staining procedures for immunofluorescence are detailed below (Immunocytochemical Localization of PLA_2). Following fixation and permeabilization of cells, injected rabbit anti-PLA_2 antibodies can be readily detected by staining with fluorescein-conjugated goat anti-rabbit IgG (Fig. 3). Antibody polypeptides injected at 5–10 mg/ml can be detected in cells up to 48 hr after injection.

A summary of the results obtained in microinjection experiments using anti-PLA_2 antibodies (Table I) provides examples for cellular responses that can be measured at the single cell level. In each of the studies described, the observed effects were dose dependent and were not detected in control injections.

[8] D. W. Stacey and V. G. Allfrey, *Cell* (*Cambridge, Mass.*) **9,** 729 (1976).

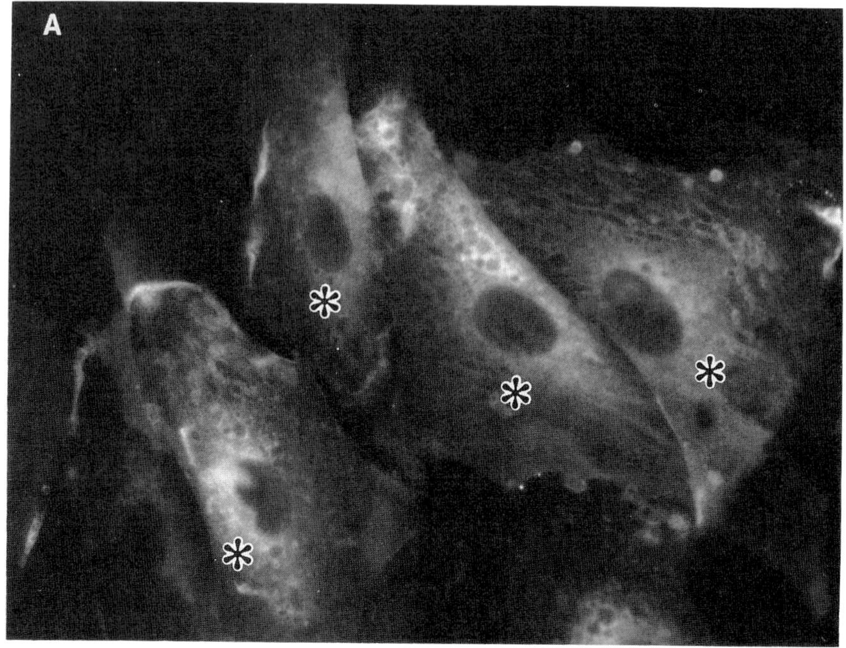
A

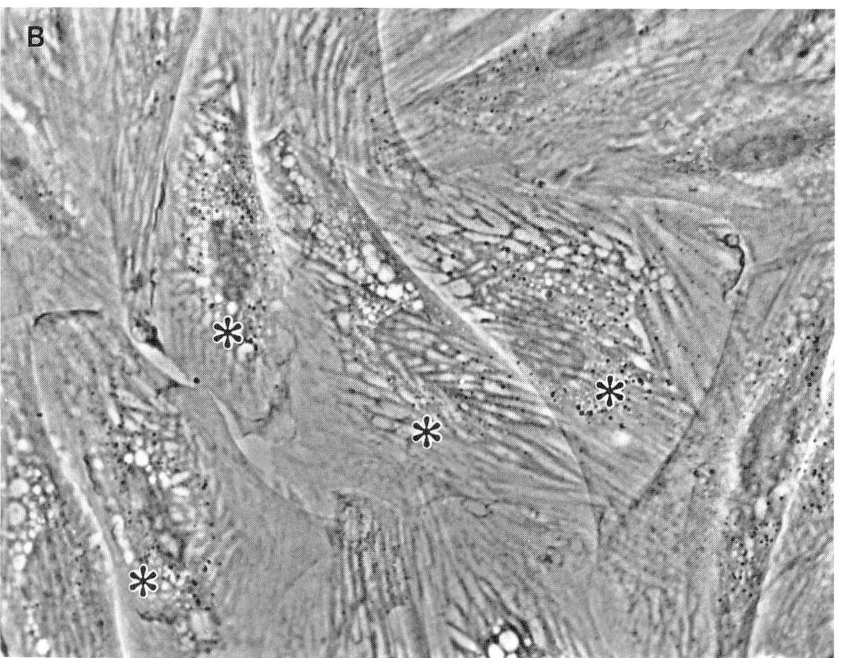
B

Immunocytochemical Localization of PLA_2

PLA_2 has been solubilized from several tissues by sonication,[1] indicating that the enzyme is not firmly bound to membrane structures. This makes it difficult to determine the cellular localization of the enzyme solely on the basis of biochemical fractionation. This section describes immunohistochemical techniques to determine the spatial aspects of PLA_2 distribution.

Preparation of Cells for Labeling

It is most convenient to grow cells on coverslips which can then be processed and mounted on slides for observation. In the case of cells that adhere poorly to glass, adhesion and cell growth can be improved by cleaning the coverslips with chromic acid or coating the surface with a positively charged polymer. Cell density can influence cell morphology and may affect the quality of the signal, especially in high-resolution fluorescence microscopy. Cell debris adsorb antibodies nonspecifically. For these reasons it is suggested that standard conditions for culture labeling be established. Prior to staining, it is necessary to fix the cells in order to preserve structure and then permeabilize them to allow labeling reagents to reach intracellular targets. A large number of fixation procedures have been described in the literature, with varying degrees of preservation of antigenic site and cellular morphology. Details of those procedures that were optimal for anti-PLA_2 antibody generated in our laboratory are described here.

Indirect Immunofluorescence

Cells plated on glass coverslips are washed twice with PBS and fixed at 20° for 30 min in 3.7% formaldehyde in PBS. Coverslips are rinsed twice with PBS, and cells are permeabilized by exposure to 0.2% Triton X-100 (Sigma) in PBS for 30 sec at 20°. The affinity-purified rabbit anti-PLA_2 antibody is used at a 1:50 dilution (20 μg/ml) in PBS containing

FIG. 3. Detection of microinjected anti-PLA_2 antibodies in rat embryo fibroblasts. Four cells in the field shown were injected with the antibody solution (5 mg/ml). At 5 hr postinjection, the cells were fixed, permeabilized, and stained with fluorescein-conjugated goat anti-rabbit IgG. (A) Fluorescence micrograph; (B) corresponding phase micrograph. Bar, 10 μm.

TABLE I
EFFECTS OF MICROINJECTION OF ANTI-PLA_2 ANTIBODIES

Cell type	Concentration of solution injected (mg/ml)	Assay	Results	Ref.[a]
ras-Transformed fibroblasts	5	Morphological reversion (cell flattening)	+	*1*
Normal fibroblasts (rat kidney or rat embryo cells)	10	Serum- or growth factor-induced stimulation of DNA synthesis[b]	No effect	*2*
Rat peritoneal mast cells	5	Ligand-induced exocytotic degranulation[c]	–	*3*
Endothelial cells	1	Thrombin-induced sustained increase in cytosolic calcium[d]	–	*4*

[a] Key to references: (*1*) D. Bar-Sagi, *in* "Cell Activation and Signal Initiation: Receptor and Phospholipase Control of Inositol Phosphate, PAF and Eicosanoid Production," p. 331. UCLA Symposium, Alan R. Liss, New York, 1989; (*2*) D. Bar-Sagi, unpublished observations (1988); (*3*) L. Graziadei and D. Bar-Sagi, in preparation (1990); (*4*) M. S. Goligorsky, D. N. Menton, A. Laszlo, and H. Lum, *J. Biol. Chem.* **264,** 16771 (1989).

[b] DNA synthesis is measured by [^{3}H]thymidine incorporation and emulsion autoradiography. [^{3}H]Thymidine (1 μCi/ml) is added to the medium within 1 hr after injection, and cells are further incubated for 12–24 hr. At the end of the incubation period the cells are fixed, and the dishes are coated with Nuclear Track Emulsion (NTB-2, Kodak) and processed for emulsion autoradiography.

[c] Degranulation is monitored by light microscopic visualization of the extrusion of granules [D. Bar-Sagi and B. Gomperts, *Oncogene* **3,** 463 (1988)].

[d] Concentration of cytosolic calcium in individual cells was monitored using the calcium indicator dye Fura-2.

0.5 mg/ml BSA. It should be noted, however, that the optimal concentration of the primary antibody varies widely, depending on titer and affinity, and therefore should be determined empirically. Incubation with the primary antibody is for 1 hr at 37° in a humidified atmosphere. Following two 10-min washes in PBS, cells are incubated with fluorescein-conjugated goat anti-rabbit antibody (Cappel) diluted 1 : 100 (10 μg/ml) in PBS containing 0.5 mg/ml BSA. Secondary antibody incubation is for 1 hr at 37° in a humidified atmosphere followed by two 10-min PBS washes and a distilled water rinse before mounting with Gelvatol (Monsanto Polymers and Petrochemicals Co., St. Louis, MO).[9] Reagents which slow down the photobleaching of fluorescein (*N*-propyl gallate or *p*-phenylenediamine) are generally added to the mounts. Before use, the secondary antibodies are

[9] J. R. Feramisco and S. H. Blose, *J. Cell Biol.* **86,** 608 (1980).

preadsorbed with methanol-fixed NRK (normal rat kidney) monolayers and clarified using a Beckman Airfuge to reduce nonspecific stickiness. Cells are examined with a Zeiss photomicroscope III equipped for epifluorescence. Phase micrographs are recorded using Kodak (Rochester, NY) Technical Pan film, and fluorescence micrographs are recorded on Kodak Tri-X film.

Controls are essential for interpreting immunofluorescence staining patterns. Some useful controls are the following: (1) omitting primary antibody from the staining procedure, (2) staining with nonimmune IgG, and (3) staining with specific IgG preadsorbed with antigen. All controls should give no specific staining pattern and be significantly dimmer than the experimental samples. It is particularly important to perform all controls in pilot experiments or whenever a new batch of antibodies or different cells are utilized.

Immunoelectron Microscopy

Analysis of the cellular distribution of PLA_2 at the electron microscopic level is extremely informative and should complement the analysis at the light microscopy level. A discussion of immunoelectron microscopy techniques, however, is beyond the scope of this chapter. In general, the preparation time for samples may take up to 1 week, and then the interpretation of results often calls for the assistance of an experienced electron microscopist. An important prerequisite for a meaningful ultrastructural analysis is a good preservation of cellular structures. However, good fixatives, while preserving fine structures, tend to alter or destroy antibody binding sites as well as to decrease accessibility of probes. Therefore, the effects of several different fixation procedures on the labeling pattern should be compared. Two staining methods can be employed: (1) indirect immunoperoxidase staining which utilizes peroxidase-conjugated secondary antibodies and (2) immunogold labeling which utilizes secondary antibodies attached to gold particles. The latter offers the possibility of using gold particles of different sizes (5–20 nm), thus allowing for the simultaneous labeling of two antigens in order to assess their spatial relationships.

Acknowledgments

We thank Madeline Wisnewski for secretarial assistance. This work was supported by National Institutes of Health Grant CA46370 and American Cancer Society Grant BC690.

[25] Analysis of Lipases by Radiation Inactivation

By E. S. KEMPNER, J. C. OSBORNE, JR., L. J. REYNOLDS, R. A. DEEMS, and E. A. DENNIS

Introduction

A particularly difficult problem in the measurement of the molecular weight or oligomeric state of phospholipases, lipases, or any protein that associates with a membrane is the membrane itself. Almost all methods that can ascertain the size of a macromolecule do so by measuring some property that is related, through size, to the molecular weight of the macromolecule. Therefore, if the molecule in its active form is associated with a membrane or bound to aggregated lipid, the size of the whole aggregate is determined, not the size of the individual macromolecule. Although sodium dodecyl sulfate-polyacrylamide gel electrophoresis (SDS–PAGE) can give the subunit molecular weight of a protein, the oligomeric state of the protein in its native environment (e.g., associated with or intercalated in a membrane) is not accessible to these techniques. This is a particularly troublesome problem because many membrane functions appear to depend on or are regulated by protein aggregation.

We present the technique of radiation inactivation which circumvents these problems. This technique can determine the size and aggregation state of a macromolecule even in a sea of phospholipid and can do so in intact cells. We summarize the theoretical aspects of this procedure, including its advantages and some precautions that must be taken to ensure the proper interpretation of the results. We also present data from studies of lipoprotein lipase as an example.

Radiation Inactivation

Methods used for the estimation of molecular weight can be divided into two general categories, those that depend on comparison with known standards, such as electrophoresis and chromatography, and those based on first principles, such as sedimentation equilibrium and light scattering, which depend only on the physical properties of the molecule of interest. In both of these categories the method of detection can be nonspecific, yielding molecular weights which correspond to all species in the sample, or specific, where the results correspond to discrete species. The technique of radiation inactivation is based on first principles and thus does not require comparison with known standards. Detection is usually based on enzyme activity, and results are thus specific for the active form of the desired molecule, even in crude or impure samples.

Protein molecular size determination using radiation inactivation is performed by exposing protein samples to increasing doses of ionizing radiation. The principal experimental requirement is that the irradiation be conducted on frozen samples; thus, the activity of interest must survive freezing and thawing. Lyophilized samples have been used, but these may lead to artifactually large molecular sizes.[1] Details of other radiation exposure parameters are available.[2-4] After irradiation, samples are assayed. Any functional or structural property of the protein can be measured, including enzymatic activity, ligand binding to receptors,[5] or even the remaining protein structure[6] or fluorescence.[7] The loss of activity or structure with increasing radiation dose is analyzed according to target theory to give a "target size" for the protein.[8] Reviews have appeared summarizing the radiation studies of enzymes,[9] receptors,[10] and transporters and pumps.[1] The technique has also been applied to complex systems involving regulatory units[11-13] and biological functions as complex as muscle contraction.[14] In the lipase field, radiation studies of both lipoprotein lipases[15,16] and phospholipases[17] have been reported.

[1] E. S. Kempner and S. Fleischer, this series, Vol. 172, p. 410.

[2] J. T. Harmon, T. B. Nielsen, and E. S. Kempner, this series, Vol. 117, p. 65.

[3] C. Y. Jung, *in* "Receptor Biochemistry and Methodology," (J. C. Venter and L. C. Harrison, eds.), Vol. 3, p. 193. Alan R. Lis, New York, 1984.

[4] G. Beauregard, A. Maret, R. Salvayre, and M. Potier, *Methods Biochem. Anal.* **32,** 313 (1987).

[5] T. L. Innerarity, E. S. Kemper, D. Y. Hui, and R. W. Mahley, *Proc. Natl. Acad. Sci. U.S.A.* **78,** 4378 (1981).

[6] G. Saccomani, G. Sachs, J. Cuppoletti, and C. Y. Jung, *J. Biol. Chem.* **256,** 7727 (1981).

[7] E. S. Kempner and J. H. Miller, *Science* **222,** 586 (1983).

[8] D. E. Lea, "Actions of Radiations on Living Cells," 2nd Ed. Cambridge Univ. Press, London and New York, 1955.

[9] E. S. Kempner, *in* "Advances in Enzymology," (A. Meister, ed.), Vol. 61, p. 107. Wiley, New York, 1988.

[10] W. Schlegel and E. S. Kempner, *in* "Investigation of Membrane-Located Receptors" (E. Reid, G. M. W. Cook, and D. J. Moore, eds.), p. 47. Plenum, New York, 1984.

[11] W. Schlegel, E. S. Kempner, and M. Rodbell, *J. Biol. Chem.* **254,** 5168 (1979).

[12] J. T. Harmon, C. R. Kahn, E. S. Kempner, and W. Schlegel, *J. Biol. Chem.* **255,** 3412 (1980).

[13] R. L. Kincaid, E. S. Kempner, V. C. Manganiello, J. C. Osborne, Jr., and M. Vaughn, *J. Biol. Chem.* **256,** 11351 (1981).

[14] R. Horowits, E. S. Kempner, M. E. Bisher, and R. Podolsky, *Nature (London)* **323,** 160 (1986).

[15] A. S. Garfinkel, E. S. Kempner, O. Ben-Zeev, J. Nikazy, S. J. James, and M. C. Schotz, *J. Lipid Res.* **24,** 775 (1983).

Radiation Physics and Target Theory

Ionizing radiation in the form of γ rays or high-energy electrons interact randomly throughout any sample placed in a radiation field. The probability of any particular molecule being hit by the radiation is directly related to the mass of that molecule. These interactions ("primary ionizations") are characterized by the transfer of large amounts of energy (~1500 kcal/mol) to the target molecule. Damage occurs from the interaction of radiation directly with these target molecules as well as indirectly by chemical reaction with radiation products produced elsewhere. Experimental conditions have been determined to maximize the direct actions of radiation with only a negligible contribution of the indirect mode. Under these conditions the radiation effects on macromolecules are independent of other molecules, such as the sea of lipids in a membrane.

When the radiation exposure is conducted at very low temperatures, the energy is dissipated throughout any polypeptide which is hit. Many different mechanisms of energy absorption are involved, some of which (such as the rupture of covalent bonds) result in irreversible changes in the protein. Because of the large amount of energy deposited in the molecule, irradiated proteins suffer gross structural damage and in general lose all biological activity. Molecules which escape radiation damage are unaltered in both structure and activity. Except in a few special cases, there are no molecules with partial activity, nor are there radiation fragments which retain activity.[18]

Target theory predicts that the decrease in function will be due to gross destruction of active molecules. In radiation studies involving enzyme activity (or ligand binding to receptors), radiation exposure results in reduced V_{max} (or B_{max}) values, with no change in the inherent K_m (or K_D) of the surviving molecules. Under appropriate experimental conditions, the amount of function which survives radiation exposure can be related to the mass of the active structure.[2] The measurement of function must be optimum, i.e., not rate-limited by assay conditions. Then the activity, A, determined in samples that have been exposed to a radiation dose, D (in rads), is given by

$$A = A_0 e^{-kD}$$

[16] T. Olivecrona, G. Bengtsson-Olivecrona, J. C. Osborne, Jr., and E. S. Kempner, *J. Biol. Chem.* **260,** 6888 (1985).

[17] P. Antaki, J. Langlais, P. Ross, P. Guerette, and K. D. Roberts, *Gamete Res.* **19,** 305 (1988).

[18] E. S. Kempner and J. H. Miller, *Biophys. J.* **55,** 159 (1989).

where A_0 is the activity in the unirradiated sample. The exponential term k is directly related to the mass of the active structure (in daltons) according to

$$k = 1.56 \times 10^{-12} \times S_t \times \text{mass}$$

where S_t is a factor dependent on the temperature at which the irradiation was conducted.[19,20] Thus,

$$\ln(A/A_0) = -1.56 \times 10^{-12} \times S_t \times D \times \text{mass}$$

The logarithm of the fraction of remaining activity after a dose D is all that is needed to determine the size of the biochemically active unit.

The target size determined by radiation inactivation is strictly a mass measurement: the chance of a molecule being hit by radiation is directly proportional to the mass of that molecule. Since lipid molecules are small in comparison to proteins, there is only a very small probability that a lipid molecule will be hit by radiation doses which destroy most protein molecules. Thus, the advantage of this technique for lipid-metabolizing enzymes is that protein size can be estimated in the presence of aggregated lipid substrates or cell membranes.

Interpretation of Target Sizes

Glycoproteins

Radiation damage to oligosaccharides does not spread to covalently attached proteins.[21] Thus, target sizes of glycoproteins apply only to the protein portion of these molecules.

Oligomeric Proteins

By radiation inactivation alone it is not known whether the target measured represents a single covalent structure or multiple subunits. This distinction can be made only by comparison with other experimental evidence, such as an SDS–PAGE-determined subunit size. For example, if a protein exists as a dimer, a direct hit on one subunit should destroy only that subunit and leave the other intact. If both subunits are required for activity, the surviving subunit will be inactive. Transfer of radiation energy from one subunit to another has been demonstrated where the

[19] E. S. Kempner and H. T. Haigler, *J. Biol. Chem.* **257,** 13297 (1982).

[20] E. S. Kempner, R. Wood, and R. Salovey, *J. Polym. Sci. Part B: Polym. Phys.* **24,** 2337 (1986).

[21] E. S. Kempner, J. H. Miller, and M. J. McCreery, *Anal. Biochem.* **156,** 140 (1986).

subunits are linked by a disulfide bond.[22] The possibility of energy transfer (the transfer of radiation-deposited energy between two noncovalently linked subunits) can be checked by subjecting the irradiated sample to SDS–PAGE and quantitating the amount of protein monomer remaining. If the biological activity decays as a dimer while the SDS–PAGE protein band decays as a monomer, the results indicate that the functional protein is a dimer. If, however, the protein band also decays as a dimer, it indicates that transfer of radiation energy from one subunit to another has occurred. Although these results would indicate that the protein is physically aggregated as a dimer, because of the energy transfer the results would not distinguish whether this association is required for biological activity or whether the aggregated subunits are behaving independently.

If biological activity decays as a monomer, the question of energy transfer does not apply. There is, however, a different possibility that should be explored for proteins with multiple subunits. On irradiation of an oligomeric protein, it is theoretically possible[23] for a surviving subunit to combine with other intact subunits and reform an oligomer (free subunits would not be observed in irradiated samples). This subunit exchange would result in the biological activity decaying as a monomer regardless of the number of subunits required for function. This equilibration phenomenon has been observed experimentally only in one case.[24]

Examples

Lipoprotein Lipase

Lipoprotein lipase (LpL) (in contrast with hepatic lipase,[25] which is a different enzyme) in rat heart and adipose tissues has been examined by radiation inactivation.[15] The loss of 99% of the enzymatic activity was shown to follow a simple exponential function of the radiation exposure (Fig. 1). Target analysis yielded a value of 126 kDa for the active unit. The cDNA for this enzyme in a variety of species has been reported (see

[22] H. T. Haigler, D. J. Woodbury, and E. S. Kempner, *Proc. Natl. Acad. Sci. U.S.A.* **82,** 5357 (1985).

[23] A. S. Verkman, K. Skorecki, and D. A. Ausiello, *Proc. Natl. Acad. Sci. U.S.A.* **81,** 150 (1984).

[24] T. D. Boyer and E. S. Kempner, unpublished observations (1989).

[25] M. C. Komaromy and M. C. Schotz, *Proc. Natl. Acad. Sci. U.S.A.* **84,** 1526 (1987).

[26] S. Enerback, H. Semb, G. Bengtsson-Olivecrona, P. Carlsson, M.-L. Hermanson, T. Olivecrona, and G. Bjursell, *Gene* **58,** 1 (1987).

[27] M. Senda, K. Oka, W. V. Brown, P. K. Qasba, and Y. Furuichi, *Proc. Natl. Acad. Sci. U.S.A.* **84,** 4369 (1987).

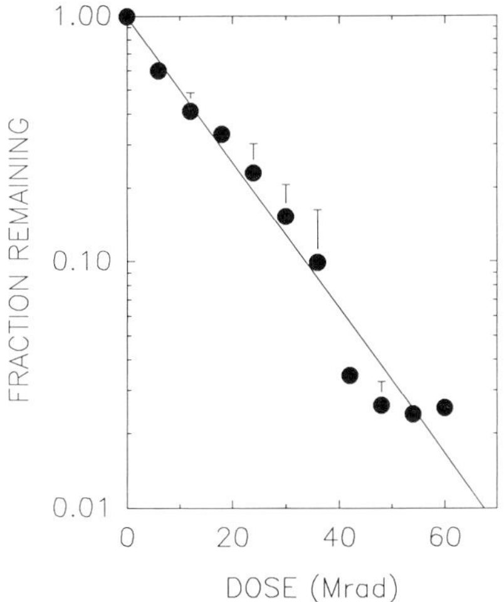

FIG. 1. Radiation inactivation of lipoprotein lipase activity in rat adipose tissue. A target size of 126 kDa was calculated. Data adapted from Ref. 15.

Table I).[26–29] The predicted amino acid sequences are very similar, yielding a subunit size of ~50,000 Da. Thus, the target value suggests that two subunits are involved in the expression of enzymatic activity.

Another radiation study of a LpL from bovine milk[16] reported a target size of 72 kDa associated with enzymatic activity. The same value was obtained when the LpL was irradiated under five different conditions (buffer with high salt and protein; with heparin; adsorbed to lipoprotein particles with and without activator; or in Triton–SDS). This target was interpreted as the dimer of a subunit whose protein size was 38,000 M_r.[30] Later studies of the cDNA[27] predicted a larger monomer size of 50 kDa for the native enzyme. This discrepancy suggests that there could have been a cleavage of the protein. If this putative truncated form was used in the radiation study, the interpretation of a 72 kDa dimer would be correct for the sample examined. Presumably, had an unmodified form of LpL been used, then a target size of 100–110 kDa would have been obtained.

[28] T. G. Kirchgessner, K. L. Svenson, A. J. Lusis, and M. C. Schotz, *J. Biol. Chem.* **262,** 8463 (1987).

[29] K. L. Wion, T. G. Kirchgessner, A. J. Lusis, M. C. Schotz, and R. M. Lawn, *Science* **235,** 1638 (1987).

[30] T. Olivecrona, G. Bengtsson-Olivecrona, and J. C. Osborne, Jr., *Eur. J. Biochem.* **124,** 629 (1982).

TABLE I
LIPOPROTEIN LIPASE MOLECULAR SIZES

Source	Subunit size (amino acid sequence predicted from cDNA) (kDa)	Target size (from radiation inactivation) (kDa)	Ref.
Rat cardiac		129 ± 8	15
Rat adipose		126 ± 24	15
Guinea pig adipocytes	51.0		26
Bovine milk	50.5	72 ± 8	16, 27
Mouse macrophages	50.3		28
Human	50.4	112 ± 12	29, 31

Other radiation studies on the human LpL yielded a target size of 112 kDa.[31] This result is also consistent with the concept that the expression of lipase function requires a dimer.

For lipoprotein lipase the results of radiation inactivation are consistent with the concept that an oligomeric species is required for enzymatic activity. Molecular sizes obtained were greater than the monomer in all cases and most consistent with a dimer being required for activity. Additional studies by active enzyme sedimentation equilibrium, sedimentation velocity, kinetic studies, and column chromatography[32,33] indicate that dissociation of the oligomer results in an irreversible loss of enzymatic activity. This loss was paralleled by a corresponding irreversible change in conformation, as evidenced by circular dichroic measurements.[32] Together with the radiation inactivation results, these data suggest that oligomer dissociation could play a role in the rapid changes in lipase activity which are observed *in vivo*. Secretion of an enzyme which self-destructs through oligomer dissociation would lessen the need for feedback regulation between the parenchymal cells (where lipase is synthesized) and nonparenchymal cells (where lipase functions as a hydrolase) in a given tissue. It should be noted that all lipases are also nonspecific hydrolases. Protection of cellular components, such as plasma membranes, could involve cofactor activation (lipoprotein particles contain specific activators of lipases) and a self-destruct mechanism.

[31] J. C. Osborne, Jr., and E. S. Kempner, unpublished observations (1984).

[32] J. C. Osborne, Jr., G. Bengtsson-Olivecrona, N. S. Lee, and T. Olivecrona, *Biochemistry* **24,** 5606 (1985).

[33] J. C. Osborne, Jr., and L. A. Zech, unpublished observations (1984).

Phospholipases

The phospholipases constitute another class of lipolytic enzymes for which questions of enzyme aggregation have been raised.[34] In particular, these questions apply to the extracellular phospholipase A_2, a small, water-soluble enzyme of molecular weight about 13,000 which exists in snake venoms, mammalian pancreas, and other mammalian tissues and cells. In aqueous solution, the pancreatic enzyme exists as a monomer whereas some snake venom enzymes, particularly those from rattlesnake, exist as tight dimers. Still other enzymes, such as the one from cobra venom, undergo concentration-dependent aggregation, existing as a monomer at low enzyme concentrations, as a dimer at high concentrations, and as higher order aggregates at even higher concentrations.[35] This variation in enzyme aggregation, plus the observation that phospholipid substrate induces aggregation of otherwise monomeric enzymes,[36–43] has raised the possibility that the active form of the enzyme is dimeric rather than monomeric.[35,44,45] However, none of the methods used to date have definitely demonstrated a functional requirement for dimers in the mechanism of action of phospholipase A_2. As discussed above, radiation inactivation has the potential to differentiate between functional monomers and functional dimers. Unfortunately, phospholipase A_2 is much smaller than most proteins and therefore must be subjected to higher doses of radiation than are normally used in such experiments.

Only one radiation inactivation study of phospholipase has been reported to date.[17] In this report, lyophilized human semen and spermatozoa were irradiated; surviving phospholipase A_2 activity yielded differing target sizes from 8 to over 60 kDa. The data were interpreted as evidence of

[34] E. A. Dennis, *in* "The Enzymes" (P. Boyer, ed.), 3rd ed., Vol. 16, p. 307. Academic Press, New York, 1983.

[35] R. A. Deems and E. A. Dennis, *J. Biol. Chem.* **250,** 9008 (1975).

[36] D. Lombardo and E. A. Dennis, *J. Biol. Chem.* **260,** 16114 (1985).

[37] T. L. Hazlett and E. A. Dennis, *Biochemistry* **24,** 6152 (1985).

[38] T. L. Hazlett and E. A. Dennis, *Biochim. Biophys. Acta* **958,** 172 (1988).

[39] T. L. Hazlett and E. A. Dennis, *Biochim. Biophys. Acta* **961,** 22 (1988).

[40] R. V. Lewis, M. F. Roberts, E. A. Dennis, and W. S. Allison, *Biochemistry* **16,** 5650 (1977).

[41] A. Plückthun and E. A. Dennis, *J. Biol. Chem.* **260,** 11099 (1985).

[42] J. D. R. Hille, M. R. Egmond, R. Dijkman, M. G. Van Oort, B. Jirgensons, and G. H. de Haas, *Biochemistry* **22,** 5347 (1983).

[43] J. D. R. Hille, M. R. Egmond, R. Dijkman, M. G. Van Oort, P. Sauve, and G. H. de Haas, *Biochemistry* **22,** 5353 (1983).

[44] M. F. Roberts, R. A. Deems, and E. A. Dennis, *J. Biol. Chem.* **252,** 6011 (1977).

[45] M. F. Roberts, R. A. Deems, and E. A. Dennis, *Proc. Natl. Acad. Sci. U.S.A.* 74, 1950 (1977).

the presence of multiple forms of phospholipase A_2 in these crude samples. It is not known whether the potential artifacts in target size introduced by lyophilization[1] apply in this system. No radiation studies have been reported on frozen or purified phospholipases to determine the oligomeric structure of the active protein; such studies are underway in our laboratories.

Acknowledgments

Support for this work was provided in part by the National Institutes of Health (GM-20,501) and the National Science Foundation (DMB 89-17392).

[26] Phosphorylation of Phospholipase C *in Vivo* and *in Vitro*

By JILL MEISENHELDER and TONY HUNTER

Introduction

In the 1980s, the role of phosphorylation in the transduction of mitogenic signals from receptors in the plasma membrane to the nucleus received much attention, particularly since many of the growth factor receptors themselves possess protein kinase activity. With increasing knowledge of the components of cell signaling pathways, an appreciation has developed for the way in which components of one signaling pathway interact directly with those of others, leading to an expansion of the original stimulus to affect a broad range of cellular functions.

Growth factor treatment of many cells results in an increased rate of phosphatidylinositol (PI) turnover. A key step in this turnover is the reaction whereby phosphatidylinositol 4,5-bisphosphate (PIP_2) is hydrolyzed to form diacylglycerol (DAG) and inositol 1,4,5-trisphosphate (IP_3). Both of these reaction products are second messengers, involved in the activation of protein kinase C and in control of intracellular Ca^{2+} fluxes, respectively. This hydrolysis is catalyzed by a family of enzymes called phospholipase C (PLC). At least 10 PLC isozymes have been identified so far; based on sequence homology these fall into four or five main subfamilies (α, β, γ, δ, ε).[1] In examining the link between receptor protein-tyrosine kinase activation and increased PI turnover, we and others have found

[1] S. G. Rhee, P.-G. Suh, S. H. Ryu, and S. Y. Lee, *Science* **244,** 546 (1989).

that treatment of appropriate cells with platelet-derived growth factor (PDGF), epidermal growth factor (EGF), or fibroblast growth factor (FGF) induces a rapid increase in tyrosine and serine phosphorylation of the PLC-γ isozyme.[2-5] Although it is not proven, it seems likely that these phosphorylations result in increased PLC activity. The following is a description of the methods we have used to study PLC phosphorylation.

Labeling of Cellular Proteins with Ortho[^{32}P]phosphate or ^{35}S-Labeled Amino Acids

Cell Culture

To study the phosphorylation of PLC following treatment of cells with various agonists of PI turnover, cellular proteins are labeled by incubating cells with ortho[^{32}P]phosphate or ^{35}S-labeled amino acids prior to their treatment and lysis. Tissue culture and labeling conditions vary depending on the cell line to be tested. Mouse fibroblasts (NIH3T3 and Swiss 3T3 cells), which are used to examine PLC's from resting versus growth factor-treated cells, are grown in 6-cm tissue culture dishes in Dulbecco's modified Eagle's medium (DMEM) including 10% (v/v) calf serum (CS) at 37°, 8% CO_2 until they are 95% confluent, at which point the medium is changed to DMEM with 0.5% CS. After 24–48 hr these "resting" cells (arrested in G_0) are ready for the addition of label. Cells of the human epidermoid carcinoma cell line A431, which is used to study the changes in PLC phosphorylation induced by EGF (these cells show increased PI turnover when treated with EGF unlike NIH3T3 cells and many other fibroblasts), are grown in DMEM with 10% fetal calf serum (FCS) until they are about 75% confluent (2–3 $\times$ 10^5 cells/6-cm dish); these cells are therefore still growing when the labeling is begun.

Labeling with Ortho[^{32}P]phosphate

To ensure maximum uptake of ortho[^{32}P]phosphate, cells are incubated in DMEM lacking phosphate which includes serum [either fetal calf serum (FCS) or CS as appropriate] that has been dialyzed against physiological

[2] J. Meisenhelder, P.-G. Suh, S. G. Rhee, and T. Hunter, *Cell (Cambridge, Mass.)* **57,** 1109 (1989).

[3] M. I. Wahl, S. Nishibe, P.-G. Suh, S. G. Rhee, and G. Carpenter, *Proc. Natl. Acad. Sci. U.S.A.* **86,** 1568 (1989).

[4] B. Margolis, S. G. Rhee, S. Felder, M. Mervic, R. Lyall, A. Levitzki, A. Ullrich, A. Zilberstein, and J. Schlessinger, *Cell (Cambridge, Mass.)* **57,** 1101 (1989).

[5] W. H. Burgess, C. A. Dionne, J. Kaplow, R. Mudd, R. Friesel, A. Zilberstein, J. Schlessinger, and M. Jaye, *Mol. Cell. Biol.* **10,** 4770 (1990).

saline (145 mM NaCl, 5 mM KCl, 0.5 mM $MgCl_2$, 1.8 mM $CaCl_2$) so that it, too, is phosphate-free. Cells are rinsed once in DMEM lacking phosphate, and then 2.5 ml labeling medium is added per 6-cm dish. Labeling medium contains 4% dialyzed serum for growing cells or 0.5% for resting cells; in the latter case, 15 mM HEPES–OH, pH 7.4, is also included to ensure maintenance of the pH of the labeling medium. Ortho-[^{32}P]phosphate (500–1200 mCi/ml H_3PO_4 in 0.02 N HCl, ICN Radiochemicals, Irvine, CA) is added to this medium to a final concentration of 2 mCi/ml for resting cells, which do not metabolize as rapidly, or 1 mCi/ml for growing cells. The volume of ortho[^{32}P]phosphate added is very small and does not affect the pH of the medium. In order to approach equilibrium labeling of phosphate in proteins, cells are incubated in labeling medium for 16–18 hr (overnight) at 37°, 8% CO_2 before treatment and lysis.

Labeling with ^{35}S-Labeled Amino Acids

To determine the stoichiometry of phosphorylation either by mobility shift detected by sodium dodecyl sulfate (SDS)–polyacrylamide gel electrophoresis or by equilibrium labeling methods (see below), it is necessary to label cellular proteins with ^{35}S-labeled amino acids (methionine, cysteine). For this, dishes of cells are set up as for ^{32}P labeling with respect to cell density, growth conditions, and serum concentrations. When ready, 6-cm dishes of cells are rinsed once in DMEM lacking methionine and cysteine and placed in 2.5 ml of such medium, which includes the appropriate concentrations of HEPES and dialyzed serum (the dialyzed sera are the same as those prepared for ^{32}P labeling since they lack free amino acids as well as phosphate). ^{35}S-Labeled amino acids (either Tran^{35}S label from ICN or Expre^{35}S^{35}S labeling mix from NEN, Wilmington, DE) are added to 100–150 μCi/ml final concentration; dishes are then incubated at 37°, 8% CO_2 overnight.

Radiation Hazards to Cells and People

It is worth noting at this point that several investigators have found that cells incubated with radioactivity (either ^{35}S or ^{32}P) may arrest in S phase, presumably due to radiation damage, and therefore even low density cells may not really be growing during these incubations.[6,7] Certain hematopoietic cell lines seem particularly susceptible to radiation damage; shorter incubation times (2–4 hr) as well as smaller amounts of label (0.5–1

[6] S. K. Hanks, unpublished observations, (1988).

[7] S. Shenoy, J.-K. Choi, S. Bagrodia, T. D. Copeland, J. L. Maller, and D. Shalloway, *Cell (Cambridge, Mass.)* **57,** 763 (1989).

mCi/ml ^{32}P) may be used to achieve more efficient labeling of proteins in these cells.

A word about radiation safety is also appropriate at this point. The quantities of ^{32}P used in these sorts of experiments are high, running into the tens of millicuries per single experiment. To avoid excessive exposure to these levels of radiation when adding ^{32}P to cells and when making cell lysates, radioactivity is handled behind 1-inch Plexiglas shields that have, as additional shielding against secondary radiation generated by the Plexiglas (Bremsstrahlung), 1/4-inch lead plates fixed to the outside of the shield. Dishes of cells are incubated in Plexiglas "houses" (boxes with one sliding removable side door that is only 3/4 the height of the box; this open slit in the otherwise closed box allows access to the CO_2/air mix in the incubator).

Although labeling with ^{35}S does not pose the same exposure hazard since the energy of the emitted β radiation is so much lower than that of ^{32}P, it is important to realize that a radioactive volatile compound is generated during such metabolic labeling, presumably a breakdown product of the labeled amino acids such as SO_2 or methyl mercaptan.[8] To avoid personal exposure and to limit the contamination of equipment, vials of ^{35}S-labeled amino acids are thawed in a special fume hood that has a charcoal filter in the back through which the hood is ventilated. Alternatively, a charcoal-filled syringe can be attached to a needle used to vent the vial; the activated charcoal readily adsorbs this compound. All incubations involving ^{35}S-labeled amino acids are carried out in a 37° incubator in which a charcoal filter (Florence Filter, San Diego, CA) is placed on the top shelf. The water used to humidify this incubator also becomes radioactive; it is kept in a glass pan in the bottom of the incubator and is therefore easily changed after every labeling.

Treatment, Cell Lysis, and Immunoprecipitation of Phospholipase C

Treating Cells

Resting cells are treated with growth factors or other agonists of PI turnover by simply adding these compounds directly to the cell labeling medium. The stock solutions of these factors are at high concentration, permitting the addition of 5 μl or less and thus not diluting the labeling medium relative to that on the untreated cells. After incubation at 37° for specific time intervals, the dishes of cells are moved to a cold room (4°), and the labeling medium is carefully removed with a Falcon disposable

[8] J. Meisenhelder and T. Hunter, *Nature (London)* **335,** 120 (1988).

pipette (#7524) and disposed of properly. The dishes are then rinsed twice with ice-cold Tris-buffered saline (137 m*M* NaCl, 5 m*M* KCl, 0.7 m*M* Na_2HPO_4, 25 m*M* Trizma base) to remove unincorporated extracellular radioactivity before the appropriate lysis buffer is added.

Making Cell Lysates

Many different cell lysis buffers have been formulated to solubilize cellular proteins. The choice of lysis buffer depends on the purpose of the experiment and on the antibody to be used: antibody–antigen recognition often depends on the antigen having a certain conformation. Some antibodies only recognize their antigen when it is fully denatured and the epitope is exposed, whereas for others recognition depends on the antigen retaining its native structure. Thus, while for some experiments SDS-boiled lysates (fully denatured) are appropriate, for experiments in which one wishes to study the association of a particular protein with other proteins this buffer is unsuitable since most noncovalent protein–protein interactions are lost at these high SDS concentrations.

In our studies of PLC we have used the monoclonal antibodies (MAbs) prepared in the laboratory of Rhee and co-workers[9]; these MAbs specifically recognize three of the different isoforms of PLC: β, γ, and δ (described fully in [48], this volume). These MAbs work equally well in all lysis buffers tested, allowing us to examine PLC and associated proteins under a variety of conditions. Six-centimeter dishes of either ^{32}P- or ^{35}S-labeled cells are routinely lysed in 1 ml cold RIPA buffer [10 m*M* sodium phosphate, pH 7.0; 0.15 *M* NaCl; 0.1% SDS; 1% Nonidet P-40; 1% sodium deoxycholate; 1% Trasylol (Mobay Corp., New York, NY), a protease inhibitor; 2 m*M* EDTA to inhibit postlysis kinase and phosphatase activities; 50 m*M* NaF, a serine phosphatase inhibitor; 100 μM Na_3VO_4 (orthovanadate) to inhibit tyrosine phosphatases], which solubilizes proteins in a seminative conformation. Cells are scraped off the dish using a disposable rubber policeman, and the lysate is allowed to sit in the dish at 4° for 10 min to ensure complete lysis and solubilization, after which the dish is scraped again and the lysate transferred to a 1.5-ml screw-cap Eppendorf tube (Sarstedt, Princeton, NJ). The lysate is then centrifuged 29,000 *g* at 4° for at least 20 min to clear it of all unsolubilized or aggregated material including nuclei; the supernatant is carefully transferred to a fresh plastic tube using a disposable plastic micropipette (RPI Corp., Mount Prospect, IL, #147500), avoiding any pelleted material. This is termed a "clarified" lysate.

[9] P.-G. Suh, S. H. Ryu, W. C. Choi, K.-Y. Lee, and S. G. Rhee, *J. Biol. Chem.* **263,** 14497 (1988).

To reduce immunoprecipitation background due to labeled RNA in ^{32}P-labeled lysates, 100–200 μg RNase A (Worthington Biochemical Corp., Freehold, NJ, RAF grade) is added per milliliter clarified lysate on ice. As for all subsequent additions, the solution is then vortexed for several seconds at high speed to ensure mixing. Although the number of proteins nonspecifically brought down in immunoprecipitates from RIPA lysates is reduced over that in similar precipitations from lysates made under gentler (no SDS or deoxycholate) conditions, even these few nonspecific proteins can confuse interpretation of results. To preclear the lysate of proteins that stick nonspecifically to *Staphylococcus aureus* (used later to recover the MAbs), simultaneous with the addition of RNase each milliliter of lysate is incubated on ice with 60–100 μl of a 10% suspension (w/v) of formalin-fixed *S. aureus* (Pansorbin from Calbiochem, La Jolla, CA, affectionately called "bugs") for 20–30 min. The bugs are then simply spun down at 4° in a Beckman J-6B centrifuge for 15 min at 3000 rpm and discarded, leaving the clarified, precleared, RNase-treated lysates (transferred to a new plastic tube) ready for the addition of antibodies.

Immunoprecipitation of Phospholipase C

In order to immunoprecipitate all of any particular PLC, 0.5 μg (total immunoglobulins) of specific monoclonal antibody mix is added per milliliter of cleared lysate. This concentration was initially chosen based on the calculation that there is roughly 500 pg protein in a typical fibroblast: if the PLC were an abundant protein (comprising, say, 0.1% of total cellular protein), there would then be 0.25 μg PLC in a 1-ml lysate of 5×10^5 cells (a confluent 6-cm dish). Since both PLC-γ and -β have molecular weights of about 150,000, as does IgG, the addition of 0.5 μg IgG to such a cell lysate puts MAb binding sites in (at least) 4-fold molar excess.

The mixes of MAb used to precipitate PLC-γ, -β, and -δ are as follows: for PLC-γ the mix is composed of MAbs B-6-4, B-20-3, D-7-3, E-8-4, E-9-4, and F-7-2; for PLC-β the mix contains K-32-3, K-82-3, and K-92-3; and for PLC-δ the mix contains R-29-1, R-32-1, R-39-2, S-11-2, and Z-78-5 (see Ref. 9). Use of such mixtures and concentrations of MAbs clears the lysate of PLC as shown by the failure of subsequent additions of MAb to precipitate any more PLC. Indeed, based on our (subsequent) calculations using ^{35}S-labeled lysates, there are roughly 42 ng PLC-γ per 6-cm dish of resting NIH3T3 cells; this then represents only 0.02% total cellular protein. Scanning densitometry of an autoradiogram of ^{35}S-labeled immunoprecipitates of all three forms of PLC from NIH3T3 cells indicates that PLC-β and -δ are, respectively, only 1/20 and 1/8 as abundant as PLC-γ.

After incubating the MAb mixture and lysate for 30 min on ice, 1.5 μg goat anti-mouse IgG (Cappel, Malvern, PA) is added for a further 30 min. Since some of the individual MAbs in the mixtures are IgG_1, which otherwise would not bind efficiently to the Pansorbin, this ensures that all the antibody is brought down in the immune complex. To complete formation of the immune complex, 40 μl (10% suspension) of bugs is then added for a further 30 min incubation on ice.

The lysate (and immune complexes) are layered by careful pipetting with a plastic micropipette onto a 1 ml RIPA/10% sucrose cushion in a plastic Falcon 2052 tube and spun in a Beckman J-6B centrifuge at 3000 rpm for 15 min at 4°. This sucrose cushion also helps to reduce the background of nonspecific proteins in the immune complex. The pellet of immune complexes is then washed 3 times by resuspending in 1 ml RIPA by vigorous vortexing and then repelleting by centrifugation at 3000 rpm for 8 min at 4°. After the third such wash, the pellet is transferred in 500 μl RIPA back to an Eppendorf tube, and the bugs are pelleted by centrifugation at room temperature for 2 min in a microfuge. After the supernatant is removed, the bugs are resuspended in Laemmli sample buffer (2% SDS, 10% glycerol, 50 m*M* Tris-Cl, pH 6.8, 20% 2-mercaptoethanol, with a dash of bromphenol blue). The samples are then incubated in a boiling water bath for 2 min to dissociate the immune complexes, the bugs are pelleted by spinning 2 min in a microfuge at 10,000 rpm, and the supernatants are applied to an SDS–polyacrylamide gel.

Gel Electrophoresis

We routinely run 1 mm thick, 14 cm long, either 7.5 or 10% SDS–polyacrylamide resolving gels using low bisacrylamide/polyacrylamide ratios as described in Anderson *et al.*[10] Electrophoresis usually takes 3.5 hr; the samples are run through the 4% stacking gel at 25 mA constant current and through the resolving gel at 35 mA until the dye front runs off the bottom. (A fan is used to cool the gel in the latter phase.)

Following electrophoresis, gels of ^{32}P samples are soaked for 15 min in water with 1 tablespoon of a mixed-bed ion-exchange resin (which helps to remove salts and unincorporated radioactivity from the gel) prior to being dried under vacuum onto Whatman 3MM paper. The paper is then marked with India ink that has been laced with ^{35}S-labeled amino acids (10 μCi/ml) so that the markings may be used later to align the autoradiograph with the gel to facilitate cutting out gel bands of interest. Autoradiography of ^{32}P gels is done overnight (8–16 hr) at 20° using Kodak XAR film. Bands readily seen after such exposures represent proteins which can, in general,

[10] C. W. Anderson, P. R. Baum, and R. F. Gesteland, *J. Virol.* **12,** 241 (1973).

be analyzed for their phosphoamino acid content or be used in tryptic mapping experiments. ^{32}P-Labeled PLC-γ immunoprecipitated from one 6-cm dish of resting NIH3T3 cells and run out in 1–2 lanes (0.8 cm wide) on a gel can be visualized in such an exposure.

Unless the labeled proteins are to be extracted for further analysis, gels on which ^{35}S-labeled samples have been run are impregnated with diphenyloxazole (PPO); use of this fluor amplifies the weak ^{35}S signal approximately 5-fold.[11] ^{35}S-Labeled PLC-γ immunoprecipitated from 0.5 ml NIH3T3 cell lysate (one-half of a 6-cm dish) run in one lane on a gel treated in this way may be visualized following an overnight exposure to presensitized XAR film at $-70°$.

Immunoprecipitation Controls

In order to be sure that a particular band seen on an autoradiograph is the PLC of interest, several different controls are used. When purified PLC is available 0.2–0.5 μg is run out on the gel in a lane next to the immunoprecipitated material, and the gel is stained with Coomassie blue immediately following electrophoresis to visualize the purified material. When purified antigen is not available we run parallel lanes of immunoprecipitates made from the same lysates using MAbs raised against unrelated proteins: comparison of the patterns of gel bands obtained then reveals which are specific to the individual MAbs and which are precipitated in all lanes and therefore nonspecific. Another way to prove the identity of an immunoprecipitated gel band is to do a "blocked" immunoprecipitation in which the lysate is split in half: to half of the lysate antibody is added in the manner described above, while to the other half is added antibody which has been preincubated with a saturating amount of purified PLC (1 μg) in 20 μl RIPA for 30 min on ice. The rest of the immunoprecipitation is done in similar fashion for both halves. The latter sample, the "blocked" immunoprecipitate, gives rise to all the nonspecific gel bands but no labeled bands specific to the antibodies. Figure 1 shows the patterns of gel bands obtained from both ^{32}P and ^{35}S-labeled NIH3T3 cells using MAbs directed against PLC-γ, -β, and -δ, as well as a blocked immunoprecipitate for PLC-γ.

To determine whether phosphorylation plays a role in the activation of PLC by a particular agonist of PI turnover, PLC immunoprecipitates from ^{32}P-labeled resting cells are compared with those from stimulated cells. Any increase in phosphorylation can be seen simply by comparing the intensity of the gel bands on an autoradiograph of PLC-γ immunoprecipi-

[11] M. K. Skinner and M. D. Griswold, *Biochem. J.* **209,** 281 (1983).

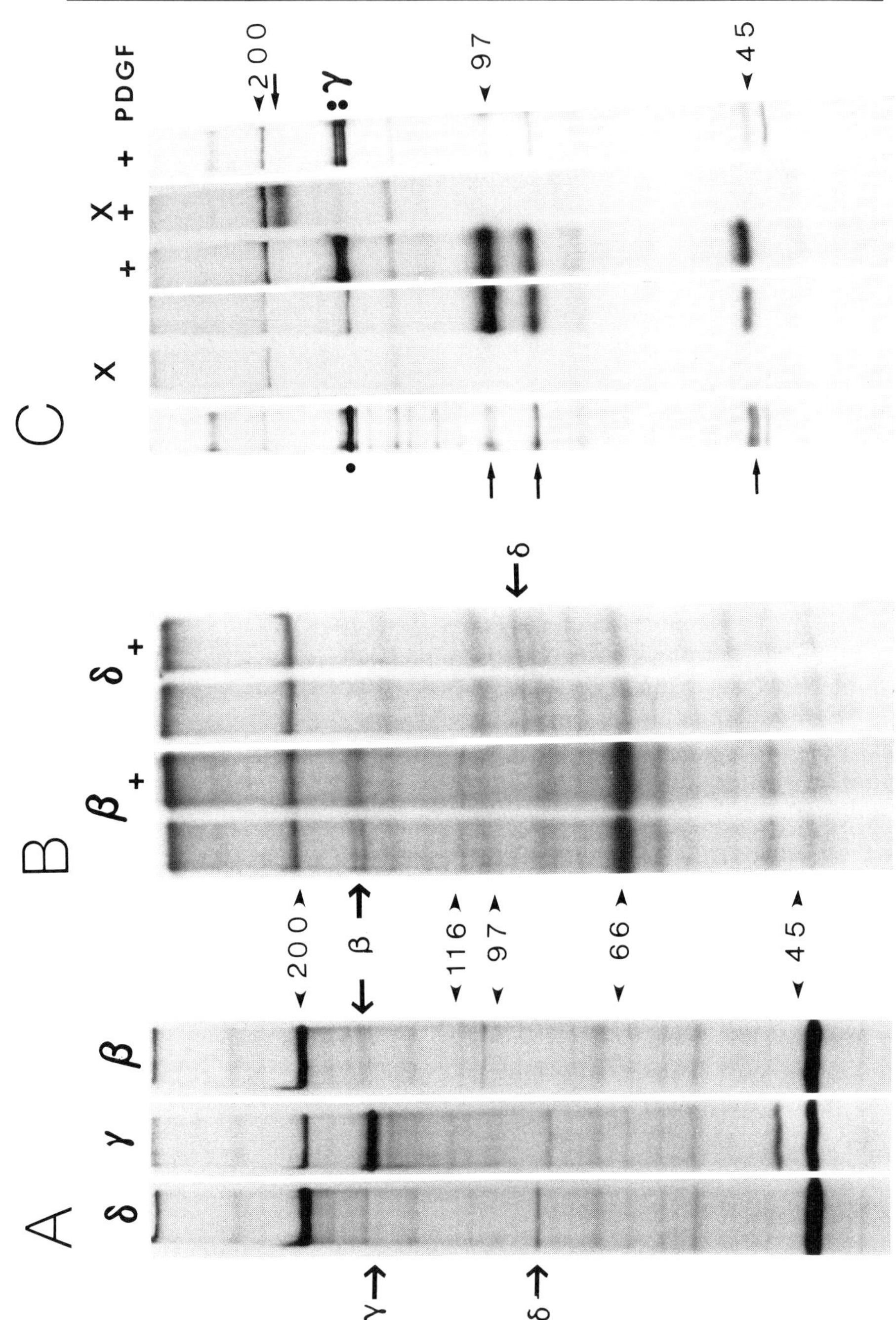
A
δ
γ
β
200
116
97
66
45
B
β +
δ +
C
X
+
X +
+ PDGF
200
γ
97
45

tated from equal numbers of cells which have or have not been treated with growth factor. In this manner we and others have shown that the level of phosphorylation of PLC-γ (but not of PLC-β or -δ) is increased in both mouse fibroblasts and human A431 cells when the cells are treated with PDGF or EGF, respectively.[2-4,12] An example of the stimulation of phosphorylation of PLC-γ by PDGF in NIH3T3 cells is shown in Fig. 1C.

Phosphatase Treatment of Immunoprecipitated Phospholipase C

Since phosphorylation of proteins commonly leads to reduced SDS–polyacrylamide gel mobility, one way of showing that phosphorylation of a protein occurs in response to stimulus is to analyze ^{35}S-labeled immunoprecipitates. In the case of PLC-γ, following growth factor treatment of cells, ^{35}S- as well as ^{32}P-labeled immunoprecipitated PLC-γ runs as a doublet on SDS–polyacrylamide gels rather than as the single (lower) band seen from untreated cells (see Fig. 1C). To prove that a shift in mobility is due to phosphorylation rather than to some other form of

[12] M. I. Wahl, N. E. Olashaw, S. Nishibe, S. G. Rhee, W. J. Pledger, and G. Carpenter, *Mol. Cell. Biol.* **9,** 2934 (1989).

FIG. 1. Immunoprecipitation of ^{35}S- and ^{32}P-labeled PLC-β-1, -δ, and -γ from resting or PDGF-treated NIH3T3 cells. (A) Resting cells were tagged with ^{35}S-labeled amino acids and lysed in RIPA as described in the text. Using mixtures of MAbs, individual PLC were immunoprecipitated from 1 ml lysate each (PLC-β-1 and PLC-δ, right and left lanes, respectively) or from 500 μl lysate (PLC-γ, middle lane) and run on a 10% polyacrylamide gel. The gel was impregnated with PPO, dried, and exposed to presensitized Kodak XAR film at $-70°$ for 53 hr. PLC are indicated by arrows; molecular weight standards ran as indicated. Note that PLC-β-1 runs as two bands at 150K, PLC-δ is a single band at 85K, and PLC-γ runs as a single band at 148K. (B) Resting cells labeled with ^{32}P were treated with PDGF for 5 min at 37° or left untreated as a control and lysed in RIPA as described. PLC-β-1 and PLC-δ were then immunoprecipitated from 1 ml lysate each and run on a 10% polyacrylamide gel that was dried and exposed to film for 48 hr. Comparison of the PLC immunoprecipitated from control or PDGF-treated (+) cells reveals no differences in the level of phosphorylation. (C) PLC-γ immunoprecipitated from 1 ml of lysates similar to those used for (B) was run on a 7.5% polyacrylamide gel, which was dried and exposed to film at 20° for 12 hr. Lanes marked X indicate immunoprecipitations wherein the MAb mix was preincubated with purified PLC-γ prior to addition to the lysate (a blocked immunoprecipitation); (+) indicates immunoprecipitations from lysates of cells treated with PDGF; S or P indicates ^{35}S- or ^{32}P-labeled proteins, respectively. Note that PLC-γ runs as a doublet when immunoprecipitated from PDGF-stimulated cells. It is associated with at least three other proteins (indicated by thin arrows, left-hand side) which are not coprecipitated in the blocked immunoprecipitations. In contrast, in blocked immunoprecipitations from PDGF-stimulated cells a new, approximately 180K protein is seen (putatively the PDGF receptor, thin arrow, right-hand side).

growth factor-induced modification, phosphatase treatment of immunoprecipitated PLC is carried out.

For example, in the case of PLC-γ, immune complexes of ^{32}P- and ^{35}S-labeled PLC-γ from either untreated or PDGF-treated cells are washed once in RIPA, twice in RIPA lacking SDS and deoxycholate, and 3 times in phosphatase buffer [20 mM MES–OH, pH 5.5, 1 mM $MgCl_2$, 0.1 mM dithiothreitol (DTT), 4 μg/ml leupeptin, 2% Trasylol, 4 μg/ml soybean trypsin inhibitor; the three protease inhibitors are included to neutralize any residual protease activity in the PAP preparation]. Immunoprecipitated material from two 6-cm dishes of cells is then divided into 3 equal parts, and each third is resuspended in 50 μl phosphatase buffer, phosphatase buffer containing 0.2 U potato acid phosphatase (PAP, Boehringer Mannheim Biochemicals, Indianapolis, IN, #108 197), or phosphatase buffer containing 0.2 U PAP plus 20 mM phosphate to competitively inhibit the phosphatase. (Alternatively, 1 mM Na_3VO_4 can be substituted for the phosphate as an inhibitor.) PAP is used rather than alkaline or calf intestinal phosphatases because we have found that PAP works more effectively as a general protein phosphatase. After the reactions are incubated for 1 hr at 37°, 1 ml RIPA is added, and the tubes are left for 20 min on ice. The immune complexes are then spun down in a microfuge, and the pellets are resuspended and boiled in 30 μl Laemmli sample buffer for gel analysis.

The ^{32}P-labeled immunoprecipitates serve as a control to demonstrate that the phosphatase is working. Greater than 90% of the ^{32}P label should be released from the protein. If the mobility of the ^{35}S-labeled PLC is increased by PAP treatment, one can conclude that phosphorylation causes the mobility shifts. For PLC-γ, 1 hr PAP treatment results in collapse of the doublet to a single, faster-migrating band with the same mobility as PLC-γ from untreated cells.[2] This collapse is not obtained when phosphate is included in the reaction mix, indicating that it is indeed due to the phosphatase activity and not to contaminating protease activity in the phosphatase preparation.

Stoichiometry of Phosphorylation of Phospholipase C

To quantify the extent of phosphorylation of a PLC more precisely, the stoichiometry of phosphorylation is determined in resting and stimulated cells. This is done by labeling cells to equilibrium with ^{32}P of known specific activity prior to stimulation and in parallel labeling cells to equilibrium with [^{35}S]methionine to determine the abundance of the protein. For a detailed account of the methods used to calculate the stoichiometry of phosphorylation, readers are referred to Refs. 13 and 14. Briefly, the procedure as applied to PLC-γ is as follows. Parallel 6-cm dishes of resting NIH3T3 cells are labeled for 48 hr with [^{35}S]methionine or ^{32}P. For both labels, each dish of cells is incubated in 2.5 ml of 80% DMEM lacking methionine, 20% DMEM, and 0.5% dialyzed CS. To this is added either 2 mCi ^{32}P or 125 μCi [^{35}S]methionine (1000 Ci/mmol, Amersham Corp., Arlington Heights, IL). After 24 hr the labeling medium is replaced with identical fresh radioactive medium for a further 24 hr.

Calculating Stoichiometry of Phospholipase C-γ Phosphorylation

Following treatments with growth factors, ^{35}S-labeled cells are rinsed twice in cold tris-buffered saline and lysed either in 1.0 *N* NaOH for protein determination or in RIPA (as described above) for immunoprecipitation of PLC-γ. The specific activity of total cellular protein is calculated as counts per minute (cpm) per microgram protein by doing protein determinations (Pierce BCA Protein Determination Kit, Pierce, Rockford, IL) on a fixed volume (30 μl) of the NaOH lysate in parallel with trichloroacetic acid (TCA) precipitation and scintillation counting of the protein from another 30-μl aliquot of the same lysate. The TCA precipitation is performed by adding 100 μl water to the 30-μl lysate followed by 30 μl of 100% TCA. After a 20-min incubation on ice the precipitate is collected by filtration onto a Whatman GFC glass fiber filter disk, which is then treated and counted in parallel with ^{35}S gel bands as below. The immunoprecipitated ^{35}S-labeled PLC-γ is run out on a gel, and the appropriate gel band is cut out and solubilized in the scintillation vial by incubation in 0.4 ml of 70% perchloric acid and 0.8 ml of 30% H_2O_2 for 5 hr at 60°, followed by addition of scintillation fluid (7 ml Ecolume, ICN) for counting. The radioactivity in the gel band is then used to calculate the number of moles of PLC-γ per dish, using the assumption that the specific activity of PLC-γ is the same as that of total cellular protein.

Immunoprecipitates from ^{32}P-labeled cells are used to calculate the number of moles of phosphate incorporated into PLC-γ. Assuming that after 48 hr of labeling the specific activity of phosphate in the cells is the same as that in the labeling medium, which contains 0.2 m*M* phosphate, the specific activity (cpm/mol) can be calculated by taking a sample of the medium (10 μl) for counting. The same dishes are treated, cells are lysed in RIPA, and the PLC-γ is immunoprecipitated as described above and run on a gel. After visualization with autoradiography, the [^{32}P]PLC-γ

[13] B. M. Sefton, this series, Vol. 200 (1991).

[14] B. M. Sefton, T. Patschinsky, C. Berdot, T. Hunter, and T. Elliot, *J. Virol.* **41,** 813 (1982).

gel bands are cut out and counted by simply putting them directly into scintillation fluid, and the number of moles of phosphate in PLC-γ per dish is calculated, based on the specific activity of the ^{32}P in the medium.

Both determinations are carried out in triplicate, and the averages for the ^{35}S and ^{32}P samples are used to determine the moles phosphate per mole PLC-γ. In a typical experiment we found that in resting, untreated NIH3T3 cells PLC-γ is phosphorylated to (at least) 0.1 mol phosphate per mole PLC-γ. Following a 5-min PDGF treatment of parallel dishes of cells the level of PLC-γ phosphorylation rose to 1.2 mol phosphate per mole PLC-γ, a 12-fold increase.[2] This estimate of stoichiometry is consistent with the observation that approximately 50% of the ^{35}S-labeled PLC-γ adopted a slower gel migration upon PDGF treatment (see Fig. 1C). However, the measurement of stoichiometry by this method makes a number of assumptions, and if enough material is available it would be better to use a direct microchemical phosphate determination method[15] or a two-dimensional gel-based method for determining phosphorylation stoichiometry.[16]

Phosphoamino Acid Analysis and Tryptic Phosphopeptide Mapping of Phospholipase C

To determine whether an observed increase in phosphorylation of a PLC is due to phosphorylations at new rather than existing sites, phosphoamino acid (PAA) analysis and tryptic phosphopeptide mapping studies can be carried out on ^{32}P-labeled immunoprecipitates from resting and growth factor-treated cells. These procedures are well described in Cooper *et al.*,[17] and an updated protocol for tryptic mapping may be found in Boyle *et al.*[18]

Briefly, the procedure as applied to PLC-γ is as follows. ^{32}P-Labeled immunoprecipitated PLC-γ is excised from a gel after visualization by autoradiography, and, after scraping off the paper backing with a razor blade, the gel bands are placed in a 1.5-ml Eppendorf tube and counted by Cerenkov counting. In general, a 6-cm dish of resting, untreated NIH3T3 cells yields approximately 400 cpm ^{32}P incorporated into PLC-γ. The recovery of PLC-γ from such a gel band is 60–70%. We usually remove 1/10 the sample for PAA analysis (or at least 50 cpm). Although a tryptic peptide map can be obtained from as little as 200 cpm in the final

[15] J. E. Buss and J. T. Stull, this series, Vol. 99, p. 7.
[16] J. A. Cooper, this series, Vol. 200 (1991).
[17] J. A. Cooper, B. M. Sefton, and T. Hunter, this series, Vol. 99, p. 387 (1983).
[18] W. J. Boyle, P. L. J. van der Geer, and T. Hunter, this series, Vol. 200 (1991).

sample, no further analysis, such as PAA determinations on individual peptides, is possible. It is often advisable to split a sample to run one part alone on one map and one part as a mix with another sample to compare migration of seemingly identical peptides. Also, one can get a result faster if more counts per minute are loaded! Owing to all these considerations, we most often analyze PLC-γ from at least two 6-cm dishes of cells per sample.

Extracting Protein from Gel Bands

To recover protein from the dried gel, bands are homogenized in 50 m*M* NH_4HCO_3, pH 7.2–7.4, in a 1.5-ml Eppendorf tube using a Kontes disposable pestle (Kontes, Vineland, NJ, #749520) inserted into a Black and Decker electric drill. Protein is eluted from the gel fragments by adding SDS and 2-mercaptoethanol, each to 1% final concentration, boiling the sample for 2 min, and then incubating with shaking at room temperature for at least 4 hr (and often overnight). The gel is spun down by centrifugation in a swing-out rotor at 3000 rpm for 5 min, and the supernatant is transferred to a National Scientific 1.5-ml Eppendorf tube (National Scientific Supply Co. Inc., San Rafael, CA, RN1700-GMT). The choice of tubes may be critical, as the proteolyzed protein sticks to certain brands, possibly owing to the mold release compound used in their manufacture.

A second elution is performed by adding enough NH_4HCO_3 so that the final (combined) volume of the eluate will be 1.3 ml. After a further (minimun) 2 hr (shaking) at room temperature, the gel is again spun down and the two supernatants are pooled. The eluate is then spun in a microfuge at 12,000 rpm for 5 min to pellet any gel and/or paper that was inadvertently included in the transfers. The clarified eluate is transferred to a fresh National Scientific tube and placed on ice. Twenty micrograms RNase A (Worthington, RAF grade) is added; this serves a dual function in that it acts as a carrier protein for PLC in the TCA precipitation and also digests any contaminating ^{32}P-labeled RNA, which will be rendered TCA-soluble. Following a 20-min incubation on ice with the RNAse, all the protein is precipitated by the addition of 250 μl of 100% TCA (final 16%). After 60 min of further incubation on ice, the precipitate is pelleted in a microfuge at 4°, washed once with 500 μl cold 100% ethanol to remove any residual TCA, and allowed to air dry.

At this point samples are split for PAA analysis or peptide mapping. The protein pellet is dissolved in 20 μl of 98% formic acid (BDH, Poole, UK) and the appropriate fraction is removed to another Eppendorf tube. Both parts are lyophilized to dryness using a SpeedVac concentrator (Savant Instruments, Hicksville, NY).

Phosphoamino Acid Analysis

For PAA analysis the (lyophilized) protein is dissolved in 30 μl constant boiling (6 *N*) HCl and hydrolyzed in the Eppendorf tube (with the top securely fixed down) in a 110° oven for 60 min; the HCl is then removed *in vacuo* using a SpeedVac concentrator with an NaOH trap. The sample is resuspended in 5 μl of a solution of 14 parts pH 1.9 buffer/1 part cold PAA standards (1 mg/ml each phosphoserine, phosphothreonine, and phosphotyrosine). After a brief spin in a microfuge to pellet any insoluble material the samples are applied, 0.3–0.5 μl at a time, and dried using a stream of cold air onto a 100-μm thin-layer cellulose plate (EM Science, Cherry Hill, NJ, #5716). Two-dimensional electrophoresis is performed as described[16]: the first dimension is run at pH 1.9 for 20 min at 1.5 kV, after which the plate is dried using a fan, and the second dimension is run at pH 3.5 for 16 min at 1.3 kV. When the plate is dry it is sprayed with ninhydrin (0.25% in acetone) to visualize the cold standards. The plates are then marked with radioactive ink and subjected to autoradiography with a fluorescence intensifying screen and presensitized Kodak XAR film at −70°. Typically, exposure times of 1–6 days are required. Another potentially more convenient protocol for determining PAA composition in which electrophoretically separated proteins are blotted onto Immobilon (Millipore, Bedford, MA) prior to acid hydrolysis has recently been described.[19]

Tryptic Phosphopeptide Mapping

For tryptic phosphopeptide mapping, the lyophilized sample is oxidized by incubation in 50 μl performic acid (formed by incubating a 1 : 9 mixture of 30% H_2O_2 and 98% formic acid for 30 min at room temperature) at 0° for 1 hr, followed by addition of 300 μl water, freezing, and lyophilization in a SpeedVac concentrator. The dried protein is exhaustively trypsinized by incubation in 50 m*M* NH_4HCO_3, pH 8.0, with 10 μg *N*-tosyl-L-phenylalanine chloromethyl ketone (TPCK)-trypsin at 37° for 18 hr (a further 10 μg of trypsin is added after the first 14 hr). Trypsinization is stopped by the addition of 300 μl water followed by lyophilization in a SpeedVac concentrator; residual NH_4HCO_3 is removed by 2 more lyophilizations using water. The sample is resuspended in 100 μl of pH 1.9 buffer, spun in a microfuge to remove any insoluble material, transferred to a new National Scientific Eppendorf tube, and lyophilized. This final sample is then resuspended in 5 μl of pH 1.9 buffer, spun in a microfuge again, and dried onto a thin-layer cellulose plate. Alternatively, Luo *et al.*[20] have

[19] M. P. Kamps and B. M. Sefton, *Anal. Biochem.* **176,** 22 (1989).
[20] K. Luo, T. R. Hurley, and B. M. Sefton, *Oncogene* **5,** 921 (1990).

recently reported a new method for proteolytic mapping of proteins transferred to Immobilon or nitrocellulose membranes that both increases recovery of peptides and reduces sample preparation time.

Tryptic phosphopeptides of PLC-γ are separated in two dimensions by running electrophoresis at pH 1.9 for 25 min at 1.0 kV, followed by ascending chromatography using "phosphochromo" buffer (75 volumes *n*-butanol : 50 volumes pyridine : 15 volumes acetic acid : 60 volumes water) until the buffer front reaches 1 inch from the top of the plate (about 7 hr). The plates are then subjected to autoradiography as described for PAA analysis. Typically, exposure times of 1–6 days are required.

The PAA composition of individual peptides can be determined by elution of radioactive spots from the map. For this purpose, cellulose from areas corresponding to peptides identified by alignment of the film with the plate is scraped from the plate into an "elution" tip as described.[18] The eluted ^{32}P-labeled peptide is dried and subjected to acid hydrolysis and PAA analysis as described above. This can be accomplished on a phosphopeptide taken from a single plate provided it gave a reasonably dark spot on an autoradiograph obtained from a 1-day exposure.

Using these methods we found that PLC-γ is rapidly phosphorylated on at least three new tyrosine sites following either PDGF treatment of NIH3T3 cells or EGF treatment of A431 cells.[2] Serine phosphorylation increases as well, but no novel serine sites are detected.[2] An alternative approach for analyzing tyrosine phosphorylation of PLC-γ is to imunoblot anti-PLC-γ immunoprecipitates with antiphosphotyrosine antibodies.[2,4] This method is convenient and rapid, and it has the virtue of avoiding labeling with ^{32}P. However, it cannot be used to assess serine phosphorylation or determine sites of tyrosine phosphorylation, and it relies on the specificity of the antiphosphotyrosine antibodies.

Phosphorylation of Phospholipase C by Purified Growth Factor Receptors

To test if a PLC can serve as a direct substrate for a growth factor receptor protein-tyrosine kinase or any other protein kinase, purified PLC is incubated with the purified protein kinase *in vitro* under conditions optimal for the particular protein kinase. For instance, we tested whether PLC-γ is a substrate for the EGF and PDGF receptors by incubating purified receptors with purified bovine brain PLC-γ and [γ-^{32}P]ATP under conditions favorable for receptor kinase activity.

The PDGF receptor protein-tyrosine kinase, purified from pig uterus (a gift from C.-H. Heldin, Uppsala), is activated by preincubation with PDGF. Routinely, 3 ng PDGF (AMGen, Thousand Oaks, CA, c-*sis*) is used to activate 36 ng PDGF receptor by coincubation for 10 min on ice in a 20-μl total volume of reaction buffer containing 10 mM HEPES–OH (pH 7.4), 6 mM $MnCl_2$, 50 μM Na_3VO_4, and 1 mM DTT. Mn^{2+} is a favored cofactor (over Mg^{2+}) for protein-tyrosine kinases in general; Na_3VO_4 is included as a tyrosine phosphatase inhibitor; DTT assures that the proteins remain in their normal reduced state. The A431 EGF receptor preparation (a gift of Gordon Gill, La Jolla, CA), purified by elution from an immunoaffinity column using EGF, is fully activated by preincubation of 12–36 ng EGF receptor in the Mn^{2+} buffer described for the PDGF receptor on ice for 10 min.

Following this incubation, to either receptor 0.5 μl (500 ng) purified bovine brain PLC-γ and 2 μl (20 μCi) [γ-^{32}P]ATP (3000 Ci/mmol, ICN) are added, and the kinase reaction is allowed to proceed for 10 min at 30°. To terminate the reaction, an equal volume (24 μl) of 2× Laemmli sample buffer is added, containing 100 μM Na_3VO_4 and 5 mM EDTA, so that the reaction can be run out directly on a polyacrylamide gel for analysis. Alternatively, 1 ml of RIPA is added, and the diluted reaction mix is put back on ice so that PLC-γ can be immunoprecipitated prior to gel separation. The latter immunopurification step is necessary to separate PLC-γ from the 150-kDa breakdown product of the EGF receptor which would otherwise contaminate the 148-kDa PLC-γ gel band and confuse further analysis.

To determine if the sites of *in vitro* phosphorylation are the same as those observed *in vivo*, ^{32}P-labeled PLC-γ bands are excised from the relevant lanes and subjected to tryptic peptide mapping as described above. Comparison of the tryptic phosphopeptide maps of receptor-phosphorylated PLC-γ with those of PLC-γ from the appropriate growth factor-treated cells reveals that the receptors can indeed phosphorylate PLC-γ *in vitro* on the same new tyrosine sites seen *in vivo*.[2]

Conclusions

The methods we have described in this chapter have been used to analyze the phosphorylation of PLC in response to various agonists of PI turnover. PDGF and EGF treatment of NIH3T3 and A431 cells, respectively, leads to a rapid increase in tyrosine and serine phosphorylation of PLC-γ, but not of PLC-β or PLC-δ. The increased phosphorylation occurs to a high stoichiometry, and three new sites of tyrosine phosphorylation are detected in PLC-γ, which can all be phosphorylated *in vitro* by the PDGF and EGF receptor protein-tyrosine kinases. Such treatments of these cells also increase PLC activity as measured by increased levels of specific inositol phosphates.[21,22] In contrast, when NIH3T3 cells are treated with EGF, which does not markedly stimulate PI turnover in these cells,[23] PLC-γ phosphorylation is not stimulated.[2] PLC-γ is also not phosphorylated on tyrosines in colony-stimulating factor (CSF-1)-treated 3T3 cells expressing the CSF-1 receptor protein-tyrosine kinase; this treatment also does not result in PLC activation as measured by Ca^{2+} mobilization.[24] Thus, a strong correlation exists between increased phosphorylation of PLC-γ and increased PI turnover. The kinetics of growth factor-induced phosphorylation are consistent with a causal role for these phosphorylations in the activation of PLC-γ. Thus, we conclude that the PDGF and EGF receptors phosphorylate PLC-γ directly, leading to increased intracellular PLC activity. Evidence that this is the case has recently been obtained by the expression of mutants of PLC-γ lacking tyrosine phosphorylation sites.[25]

Acknowledgments

The authors are indebted to James Durkin, Jennifer Price, and Ed Dennis for their generous supplies of patience.. The research described in this chapter was supported by a U.S. Public Health Service grant from the National Cancer Institute (CA39780) and by a grant from the American Business Foundation for Cancer Research.

NOTE ADDED IN PROOF. In this chapter, the molecule referred to as PLC-γ is, more specifically, now known as PLC-γ_1.

[21] M. I. Wahl and G. Carpenter, *J. Biol.Chem.* **265,** 7581 (1988).

[22] E. Nanberg and E. Rozengurt, *EMBO J.* **7,** 2741 (1988).

[23] M. Whitman and L. Cantley, *Biochim. Biophys. Acta* **948,** 327 (1988).

[24] J. R. Downing, B. L. Margolis, A. Zilberstein, R. A. Ashmun, A. Ullrich, C. J. Sherr, and J. Schlessinger, *EMBO J.* **8,** 3345 (1989).

[25] S. G. Rhee and J. Schlessinger, personal communications (1990).

Section III

Phospholipase A_1

[27] Detergent-Resistant Phospholipase A_1 from *Escherichia coli* Membranes

By YASUHITO NAKAGAWA, MORIO SETAKA, and SHOSHICHI NOJIMA

Introduction

Escherichia coli contains two kinds of phospholipase A_1. The first, termed detergent-resistant phospholipase A_1 (DR-phospholipase A_1), is bound to the outer membrane and retains its activity in the presence of detergent or in organic solvent. The other phospholipase A_1 is named detergent-sensitive phospholipase A_1 (DS-phospholipase A_1) and is present in the cytosol fraction. DR-phospholipase A_1 is a marker enzyme of the outer membrane of *E. coli*. This membrane-bound protein is solubilized from the membrane by treatment with sodium dodecyl sulfate (SDS) but not with Triton X-100. DR-phospholipase A_1 was the first of the membrane-bound phospholipases A_1 to be purified. It has been extensively purified to near homogeneity from B and K12 strains of *E. coli*[1,2] and partially characterized.

The gene region was identified and mapped at about 85 min on the chromosomes of a mutant of *E. coli* deficient in the synthesis of DR-phospholipase A_1 (*pldA* gene).[3] The structural gene coding for DR-phospholipase A_1 was cloned in K12 strain.[4] Plasmids containing the *pldA* gene were introduced into *E. coli,* and a strain which overproduced DR-phospholipase A_1 was obtained.[5] This strain was used for large-scale production of the enzyme. Analysis of the DNA sequence of the *pldA* gene provided a basis for predicting the structure of the enzyme.[6] The progress of studies on DR-phospholipase A_1 has been well reviewed by Waite[7] and van den Bosch.[8]

[1] C. J. Scandella and A. Kornberg, *Biochemistry* **10,** 4447 (1971).
[2] M. Nishijima, S. Nakaike, Y. Tamori, and S. Nojima, *Eur. J. Biochem.* **73,** 115 (1977).
[3] M. Abe, O. Doi, and S. Nojima, *J. Bacteriol.* **119,** 543 (1974).
[4] H. Homma, T. Kobayashi, Y. Ito, K. Kudo, K. Inoue, H. Ikeda, M. Sekiguchi, and S. Nojima, *J. Biochem.* (*Tokyo*) **94,** 2079 (1983).
[5] H. Homma, C. Chiba, K. Kobayashi, I. Kudo, K. Inoue, H. Ikeda, M. Sekiguchi, and S. Nojima, *J. Biochem.* (*Tokyo*) **96,** 1645 (1984).
[6] H. Homma, T. Kobayashi, N. Chiba, K. Karasawa, H. Mizushima, I. Kudo, K. Inoue, H. Ikeda, M. Sekiguchi, and S. Nojima, *J. Biochem.* (*Tokyo*) **96,** 1655 (1984).
[7] M. Waite, "The Phospholipases," p. 17. Plenum, New York, 1987.
[8] H. van den Bosch, *in* "Phospholipids" (A. Neuberger and L. L. M. van Deenen, eds.), p. 313. Elsevier, Amsterdam, 1981.

Assay Method

Principle

The activity of phospholipase A is generally determined by measuring fatty acid release from radioactive phospholipid. Thin-layer chromatography (TLC) has been widely used for the separation of released fatty acid from unhydrolyzed phospholipids and lysophospholipids. A more convenient and faster method than TLC for separation of fatty acid is the method developed by Dole and Meinertz, based on liquid–liquid partition.[9] Although it is difficult to establish suitable conditions for highly efficient fatty acid extraction and also for the complete separation of fatty acid from lysophospholipids, the method is useful to measure the phospholipase A activities of a large number of samples during the enzyme purification procedure.

Reagents

Tris-HCl buffer, 1 *M*, pH 8.0
Triton X-100, 1% (w/v)
$CaCl_2$, 100 m*M*
Dole's reagent (2-propanol/*n*-heptane/1 *N* H_2SO_4, 78 : 20 : 2, v/v)

Preparation of [¹⁴C]Acetate-Labeled Phospholipids

Labeled phospholipids such as phosphatidylethanolamine (PE), phosphatidylglycerol (PG), and diphosphatidylglycerol (DPG) can be conveniently prepared by biosynthesis using *E. coli* K12.[2] Cells are grown beyond the end of the log phase in 1 liter of Luria broth containing 1 mCi of [1-^{14}C]acetic acid sodium salt and spun down by centrifugation at 10,000 *g* for 30 min. Radioactive phospholipids are extracted according to the Bligh and Dyer method.[10] PE, PG, and DPG are separated by DEAE-cellulose column chromatography or preparative TLC.[2] The total amount of [^{14}C]PE is approximately 50 μmol, and the specific activity is 1700 counts per minute (cpm)/nmol. Radioactive PE is diluted with nonlabeled PE from *E. coli* to adjust the specific activity to 300 cpm/nmol. After evaporation under a stream of N_2, distilled water is added to the dry film of lipids to give a final concentration of 2 m*M*. The emulsion is dispersed with a sonicator in an ice-cold bath and used as a substrate.

[9] V. P. Dole and H. Meinertz, *J. Biol. Chem.* **235,** 2595 (1960).
[10] E. G. Bligh and W. J. Dyer, *Can. J. Biochem. Physiol.* **37,** 911 (1959).

Assay Procedure

A reaction mixture for standard assays contains 25 μl of [^{14}C]PE suspension, 25 μl of 1% Triton X-100, 25 μl of 100 m*M* $CaCl_2$, 13 μl of 1 *M* Tris-HCl buffer, 137 μl distilled water, and 25 μl enzyme solution. The mixture is incubated at 37° for 30 min, and the reaction is terminated by the addition of 0.5 ml chloroform, 1 ml methanol, and 0.15 ml water. After being stirred vigorously, the reaction mixture is separated into two phases by the addition of 0.5 ml chloroform and 0.6 ml water.[10] The lower phase (chloroform layer) is transferred to another tube, and the remaining upper phase is washed twice with 1 ml of chloroform. The chloroform fractions are combined and evaporated. The residue is dissolved in chloroform and applied to a silica gel G plate (Merck, Darmstadt, FRG), which is developed with petroleum ether/diethyl ether/acetic acid (60 : 40 : 1, v/v) for the separation of fatty acid and phospholipid. Lipids are stained with iodine, and the fatty acid spot is scraped off into a vial. Scintillation fluid {2.4 liters of toluene/Triton X-100 (4 : 1, v/v) containing 3 g of 2,5-diphenyloxazole, 0.1 g of 1,4-bis[2-(5-phenyloxazoyl)]benzene, and 50 ml water} is added for radioactivity counting.

For the liquid–liquid separation of liberated fatty acid from phospholipid, the reaction is terminated by addition of 3 ml of Dole's reagent.[7] The tubes are shaken vigorously with a vortex mixer for 15 sec. Distilled water (1.6 ml) and then heptane (1.8 ml) are added. Mixing of the heptane and water phases with vortex mixer for 10 seconds is essential for the quantitative extraction of fatty acid. The emulsion is clearly separated by centrifugation at 3,000 *g* for 10 min. A portion (1.5 ml) of the upper phase is removed to another tube containing 30 mg of silica gel (Wako gel, Wako Pure Chemical Industry Ltd., Osaka, Japan). After shaking for 20 sec, the silica gel is precipitated by centrifugation at 1000 *g* for 10 min. Heptane (0.8 ml) is removed for radioactivity counting in 7 ml of ACS II scintillation fluid (Amersham, Buckinghamshire, UK).

Purification Procedure[2]

Step 1: Preparation of Escherichia coli Homogenate. Cells of *E. coli* K12 grown in Penassay broth (Difco, Detroit, MI) medium at 37° until late logarithmic phase are harvested and frozen for storage. Frozen cells (200 g wet weight) are suspended in 420 ml of 50 m*M* Tris-HCl buffer (pH 8.0). The cells are disrupted in a French pressure cell at 400 kg/cm^2 (fraction I). Fraction I is centrifuged at 30,000 *g* for 60 min. The pellet is resuspended in 260 ml of 50 m*M* Tris-HCl buffer with a Teflon homogenizer (fraction II).

Step 2: Solubilization with Sodium Dodecyl Sulfate and 1-Butanol. The cell envelope is solubilized by the addition of powdered sodium dodecyl sulfate (SDS) and 100 m*M* EDTA (pH 7.4) to give final concentrations of 10 mg/ml and 1 m*M*, respectively. The solution is stirred for 30 min at 20° and then cooled to 0°. The solution is saturated with 1-butanol (0.15 volume), and the turbid suspension obtained is centrifuged at 30,000 *g* for 1 hr. The clear, yellowish supernatant (158 ml) is collected (fraction III).

Step 3: Acetate Extraction. Magnesium chloride (1 *M*, 0.4 ml) and sodium acetate (1 *M*, pH 5.2, 20 ml) are added to fraction III. The mixture is left to stand for 1 hr, then centrifuged at 8000 *g* for 15 min. The supernatant (160 ml) (fraction IV) is mixed with 300 ml of chilled acetone at −15° and left to stand at −15° for 30 min. DR-phospholipase A_1 is precipitated as a yellowish pellet by centrifugation at 8000 *g* for 15 min and stored at −20°.

Step 4: Acetone Fractionation. The pellets are resuspended in 85 ml of extraction buffer, which is an aqueous solution consisting of 10 mg/ml SDS, 1 m*M* EDTA, and 50 m*M* Tris-HCl buffer (pH 8.0) and saturated with 1-butanol. The suspension is centrifuged at 8000 *g* for 10 min. Magnesium chloride (1 *M*, 0.16 ml) and sodium acetate (1 *M*, pH 5.2, 8 ml) are added to the supernatant. A series of aliquots of chilled acetone, 15 ml (A-1), 9 ml (A-2), 10 ml (A-3), 10 ml (A-4), and finally 24 ml (A-5), are added carefully from a separatory funnel. After each addition of acetone, the mixture is stirred for 20 min at 0° and centrifuged at 8000 *g* for 5 min at 0°. The precipitates are suspended in 2 ml of 50 m*M* Tris-HCl buffer (pH 8.0). The precipitate from the A-3 fraction, which shows the highest specific activity (fraction V), is lyophilized. Fraction V is resuspended in 5 ml of 1-butanol and dispersed with a sonicator. The dispersion is incubated for 1 hr at 20°, and the supernatant, obtained by centrifugation at 5000 *g* for 5 min at 20°, is evaporated *in vacuo*.

Step 5: Sephadex G-100 Column Chromatography. The material is solubilized in 2 ml of 50 m*M* Tris-HCl buffer (pH 8.0) containing 0.5% SDS (w/v) with a sonicator. The suspension is centrifuged at 3000 *g* for 10 min at room temperature. The supernatant, containing 11 mg of protein, is passed through a Sephadex G-100 column (2.6 × 50 cm) equilibrated with 50 m*M* Tris-HCl buffer (pH 8.0) containing 0.5% SDS. Elution is carried out with the equilibration buffer at a flow rate of 18 ml/hr, and 5-ml fractions are collected. The enzyme activity appears in fractions 23–25, and these fractions are combined (fraction VI). Fraction VI is dialyzed twice against 2 liters of 50 m*M* Tris-HCl buffer (pH 8.0) containing 0.5% Triton X-100.

Step 6: DEAE-Cellulose Column Chromatography. The dialyzed sample (3.2 mg of protein) is applied to a DEAE-cellulose column (1.5 × 17

TABLE I
PURIFICATION OF DETERGENT-RESISTANT PHOSPHOLIPASE A_1 FROM *Escherichia coli*

Step (fraction)	Total protein (mg)	Total activity (units)[a]	Specific activity (units/mg)	Purification (-fold)	Recovery (%)
Homogenate (I)	32,200	90	2.8	1.0	100
Particles (II)	15,000	132	8.8	2.8	147
SDS/1-butanol solubilization (III)	1770	40	23	8.2	44
Acetate extraction (IV)	480	30	63	22.5	33
Acetone fractionation (V)	11	12	1090	389	13
Sephadex (G-100) (VI)	3	6	1553	554	6
DEAE-cellulose (VII)	0.6	3	4746	1695	3

[a] Micromoles of product (free fatty acid) formed per minute.

cm) equilibrated with 50 mM Tris-HCl buffer containing 0.5% Triton X-100. The column is eluted with a 200-ml linear gradient from 0 to 0.3 M KCl in the same buffer at 0°. Fractions 15 to 20, which show high activity, are combined and stored at −20° (fraction VII). The entire purification is summarized in Table I.

Comments

Escherichia coli bearing the *pldA* gene expresses DR-phospholipase A_1 and is a useful for the preparation of large amounts of the purified enzyme. The DR-phospholipase A_1 (about 4 mg) was purified to near homogeneity from 20 liters of culture of *E. coli* strain KL16-99 bearing a plasmid which contains the *pldA* gene (M. Setaka and S. Nojima, unpublished data). DeGeus *et al.* have also purified the phospholipase A_1 in high yield from an overproducing strain of *E. coli*.[11]

Properties

Purity. The protein obtained is homogeneous as judged by SDS–polyacrylamide gel electrophoresis, which reveals a single band containing all of the DR-phospholipase A_1 activity.

[11] P. DeGeus, N. H. Riegman, A. J. Horrevoets, H. M. Verheij, and G. H. de Haas, *Eur. J. Biochem.* **161,** 163 (1986).

Stability. The purified enzyme is stable for several months at $-70°$ in 50 m*M* Tris-HCl buffer (pH 8.0) containing 0.5% Triton X-100. The crude enzyme is quite stable to heat treatment and retains its activity even after incubation at 100° for 5 min. However, the purified enzyme is sensitive to heat treatment and irreversibly loses activity at 70°. Detergents as well as phospholipids above their critical micelle concentration (CMC) protect the enzyme against heat inactivation.[12]

Molecular Weight and Isoelectric Point. The determination of molecular weight was carried out by polyacrylamide gel electrophoresis under nondenaturing and denaturing conditions using the purified enzyme.[2] The molecular weight of DR-phospholipase A_1 was determined as 21,000 by SDS/polyacrylamide gel electrophoresis, whereas DR-phospholipase A_1 pretreated with SDS/urea/mercaptoethanol at 90° showed a molecular weight of 28,000. By Ferguson plot analysis, the molecular weight of DR-phospholipase A_1 was estimated as 28,000. The molecular weight determined by gel electrophoresis corresponds with that calculated from the DNA sequence for DR-phospholipase A_1 (30,809).[4] In polyacrylamide gel isoelectric focusing studies, the enzyme was recovered in a single peak with an isoelectric point of 5.

Substrate Specificity. The enzyme catalyzes the hydrolysis of the phospholipids of *E. coli,* in the following order of decreasing activity: PE, PG, DPG, and phosphatidylcholine (PC).[2] The chain length and degree of unsaturation of the acyl group of phospholipids have little effect on the activity of the enzyme. DR-phospholipase A_1 shows a wide range of positional specificity. The purified enzyme possesses both phospholipase A_1 and A_2 activities (phospholipase B activity). The enzyme also catalyzes the hydrolysis of 1-acyl- and 2-acyllysophospholipids (lysophospholipase L_1 and L_2 activities).

Inhibitors. The enzyme is stable on preincubation for 30 min at 37° with 0.5 m*M* dithiothreitol, reduced glutathione, iodoacetamide, *p*-chloromercuribenzoic acid, *N*-ethylmaleimide, or diisopropyl fluorophosphate. On the other hand, the enzyme activity is completely inhibited by 0.5 m*M* *p*-bromophenacyl bromide (M. Setaka and S. Nojima, unpublished data).

Divalent Cation. The enzyme requires 10 m*M* Ca^{2+} for maximal activity.[2] Cations such as Mg^{2+}, Ba^{2+}, and Sr^{2+} at concentrations of 5 m*M* increase the activity to about 50% of that with Ca^{2+}. The activity is completely inhibited by 5 m*M* Zn^{2+}, Fe^{2+}, Hg^{2+}, and EDTA.

Cloning and Sequence. The structural gene for DR-phospholipase A_1 (*pldA*) has been cloned in *E. coli*.[4] The strain of *E. coli* bearing the plasmid

[12] Y. Tamori, M. Nishijima, and S. Nojima, *J. Biochem.* (*Tokyo*) **86,** 1129 (1979).

R T L Q G W L L P V F M L P M A V Y A Q E A T V K E V H D A P A

R G S I I A N M L Q E H D N P F T L Y P Y D T N Y L I Y T Q T S

L N L E A I A S Y K W A E N A D L D E V L F Q L S S A F P L W D

I L G P N S V L G A S Y T Q L S W W Q L S N S E E S S P F D E T

Y E P Q L F L G F A T D Y R F A G W T L R T V E M G Y N H D S N

R S D P T S R S W N R L Y T R L M A E N G N W L V E V K P W Y V

G N T D D N P D I T L Y M G Y Y Q L K I G Y H L G Q A V L S A K

Q Y N W N T G Y G G A E L G L S Y P I T K H V R L Y T Q V Y S G

G E S L I D Y N F N Q T R V G V G V M L N D L F

FIG. 1. Amino acid sequence of DR-phospholipase A_1 of *E. coli* deduced from the DNA sequence.

with *pldA* gene produced DR-phospholipase A_1 in quantities 25–65 times that of the wild type.[5]

The nucleotide sequence of the *pldA* gene indicated that DR-phospholipase A_1 is first synthesized as a precursor form which has 20 extra amino acid residues as compared with mature DR-phospholipase A_1 (289 amino acids).[6] The extra peptide, Met-Arg-Thr-Leu-Gln-Gly-Trp-Leu-Leu-Pro-Val-Phe-Met-Leu-Pro-Met-Ala-Val-Tyr-Ala, has a sequence characteristic of signal peptides so far reported. DR-phospholipase A_1 contains 269 amino acid residues, with a calculated molecular weight of 30,809[6] (Fig. 1).

A characteristic feature of the amino acid sequence of DR-phospholipase A_1 is the absence of cysteine residues which form the disulfide bonds seen in pancreatic and venom phospholipases A_2. Another distinctive feature of DR-phospholipase A_1 as compared with other membrane-bound proteins is the hydrophobicity of this membrane-bound enzyme. Most of the amino acid sequence is rather hydrophilic, and there are no stretches of hydrophobic amino acids. No extensive sequence homologies are apparent between DR-phospholipase A_1 and other phospholipases A_2.

[28] Phospholipase A_1 Activity of Guinea Pig Pancreatic Lipase

By AMA GASSAMA-DIAGNE, JOSETTE FAUVEL, and HUGUES CHAP

Introduction

We previously reported the purification of two cationic lipases displaying a high phospholipase A_1 activity from guinea pig pancreas, which lacks the classic secretory phospholipase A_2.[1,2] Although other lipases have been shown to hydrolyze the *sn*-1 position of glycerophospholipids,[3–7] the cationic lipases were particularly useful in three main applications: determination of ether phospholipids,[8–11] studies on lipoprotein structure or metabolism,[12,13] and more recent investigations on the mechanism of lipocortin.[14]

Assay Methods

Owing to the broad substrate specificity of these enzymes, various assay methods have been used, including titrimetric determinations of

[1] J. Fauvel, M. J. Bonnefis, L. Sarda, H. Chap, J. P. Thouvenot, and L. Douste-Blazy, *Biochim. Biophys. Acta* **663,** 446 (1981).

[2] J. Fauvel, M. J. Bonnefis, H. Chap, J. P. Thouvenot, and L. Douste-Blazy, *Biochim. Biophys. Acta* **666,** 72 (1981).

[3] G. H. de Haas, L. Sarda, and J. Roger, *Biochim. Biophys. Acta* **106,** 638 (1965).

[4] A. J. Slotboom, G. H. de Haas, P. P. M. Bonsen, G. J. Burbach-Westerhuis, and L. L. M. van Deenen, *Chem. Phys. Lipids* **4,** 15 (1970).

[5] W. C. Vogel and E. L. Bierman, *Lipids* **5,** 385 (1970).

[6] C. Ehnholm and T. Kuusi, this series, Vol. 129, p. 716.

[7] G. L. Kucera, P. J. Sisson, M. J. Thomas, and M. Waite, *J. Biol. Chem.* **263,** 1920 (1988).

[8] A. El Tamer, M. Record, J. Fauvel, H. Chap, and L. Douste-Blazy, *Biochim. Biophys. Acta* **793,** 213 (1984).

[9] A. Diagne, J. Fauvel, M. Record, H. Chap, and L. Douste-Blazy, *Biochim. Biophys. Acta* **793,** 221 (1984).

[10] M. Record, A. El Tamer, H. Chap, and L. Douste-Blazy, Biochim. Biophys. Acta **778,** 449 (1984).

[11] A. El Tamer, M. Record, H. Chap, and L. Douste-Blazy, *Lipids* **20,** 699 (1985).

[12] B. P. Perret, F. Chollet, S. Durand, G. Simard, H. Chap, and L. Douste-Blazy, *Eur. J. Biochem.* **162,** 279 (1987).

[13] X. Collet, B. Perret, F. Chollet, F. Hullin, H. Chap, and L. Douste-Blazy, *Biochim. Biophys. Acta* **958,** 81 (1988).

[14] A. Gassama-Diagne, Thèse de Doctorat d'Etat ès Sciences, Toulouse, France, 1989.

lipase and phospholipase A_1 activities as well as radiochemical assays with phospholipids or neutral glycerolipids labeled with radioactive fatty acids. From our experience, lipase activity toward diacylglycerol, the preferred substrate of guinea pig pancreas cationic lipases, revealed to be the most convenient procedure to follow enzyme purification, but monoacylglycerol can also be used as well. To determine phospholipid hydrolysis, two assay procedures using 1-[^{3}H]palmitoyl-2-acyl-*sn*-glycero-3-phosphocholine (phospholipase A_1 assay) and 1-acyl-*sn*-glycero-3-phospho[*N-methyl*-^{3}H]choline (lysophospholipase assay) can be used.

Diacylglycerol and Monoacylglycerol Lipase Assays[15]

Materials

- Glycerol tri[9,10(*n*)-^{3}H]oleate, 540 mCi/mmol (The Radiochemical Center, Amersham, UK)
- Triolein, gum arabic, sodium deoxycholate, lipase from pig pancreas, grade VI-S (Sigma Chemical Co., St. Louis MO)
- Silica gel G 60, 0.25 mm thick plates (Merck, Darmstadt, FRG): miniplates used to follow the advancement of diacylglycerol accumulation during preparation of substrate are obtained by cutting small pieces (2 × 5 cm) with a glazier's diamond

Preparation of Di[^{3}H]oleoyl- and Mono[^{3}H]oleoyl-sn-glycerol. Glycerol tri[9,10(*n*)-^{3}H]oleate (72.4 μl, 362 mCi, toluene solution) is taken to dryness, mixed with 320 mg of triolein (362 μmol), and dispersed by sonication in 2 ml of 10% (w/v) gum arabic, which is dissolved just prior to use by slightly warming the buffer to around 37°, followed by cooling on ice. One volume of this emulsion is stirred with 14 volumes of 0.2 *M* Tris-HCl (pH 9.0) containing 9 mg/ml sodium deoxycholate. Then, 10 ml of this substrate preparation (10 m*M* triolein, final concentration) is incubated with 5 units of pig pancreatic lipase at 37° with gentle shaking. Every 30 min, a small aliquot is taken from the reaction mixture using a Pasteur capillary pipette, and it is directly deposited onto silica gel miniplates and developed with hexane/diethyl ether/formic acid (55 : 45 : 1, v/v, solvent A). This step is critical to improve the yield of di[^{3}H]oleoyl-*sn*-glycerol, which first accumulates and then disappears. Usually, incubation times between 1 and 2 hr give the best results, whereas maximum amounts of mono[^{3}H]oleoyl-*sn*-glycerol are obtained after 4 hr.

Incubation is blocked by adding 20 ml methanol, 20 ml chloroform, and 10 ml distilled water, followed by brief mixing with a vortex mixer at

[15] J. Fauvel, H. Chap, V. Roques, L. Sarda, and L. Douste-Blazy, *Biochim. Biophys. Acta* **792,** 65 (1984).

room temperature. Phase separation is obtained on centrifugation at 2000 *g*, and the lower phase is withdrawn. The upper phase is extracted again following addition of 10 ml chloroform. The combined chloroform extracts are taken to dryness under reduced pressure at 37° and dissolved in 3 ml of chloroform/methanol (1 : 1, v/v).

The total lipid extract is subjected to preparative thin-layer chromatography using solvent A and three silica gel plates, and lipids are immediately eluted with 250 ml of chloroform/methanol (1 : 1, v/v). After drying under reduced pressure, lipids are dissolved in chloroform/methanol (1 : 1, v/v) and kept at −20°. Usually, this procedure yields 13 μmol of di[^{3}H]oleoyl-*sn*-glycerol and 33 μmol of mono[^{3}H]oleoyl-*sn*-glycerol displaying specific radioactivities of 1500 and 750 disintegrations per minute (dpm)/nmol, respectively. Particular care must be taken in weighing triolein, since the specific radioactivity of products is calculated from the specific radioactivity of triolein subsequently hydrolyzed by lipase. Substrate concentrations are then determined from the radioactivity determination. On thin-layer chromatography with solvent A, diacylglycerol appears as a mixture containing around 75% 1(3),2-diacyl isomer and 25% 1,3-diacyl isomer. Using benzene/ethyl acetate/trimethyl borate (100 : 20 : 7.2, v/v) as a solvent, monoacylglycerol reveals the same degree of purity and contains around 85% 1(3)-monoacyl isomer and 15% 2-acyl isomer.

Procedure. From 10 determinations of either diacylglycerol or monoacylglycerol lipase, 10 μmol of either substrate is taken to dryness under a stream of nitrogen, followed by addition of 0.944 ml Tris-HCl (0.2 *M*, pH 9.0) containing 10% (w/v) gum arabic. Lipids are dispersed by sonication (twice for 15 sec, 0°) using an MSE sonicator at maximal output. Reaction mixtures contain 100 μl of substrate dispersion, the enzyme preparation, and enough Tris-HCl buffer to make the final volume 115 μl. Mixtures are incubated for 5 to 10 min at 37° while shaking, followed by addition of 1.95 ml of chloroform/methanol/heptane (1.21 : 1.41 : 1, v/v), and 0.63 ml of carbonate–borate buffer (pH 10.5). After centrifugation, the upper phase (1.3 ml) contains nonesterified fatty acids, which are extracted with a yield of 75%, as determined using the [^{3}H]oleic acid as a standard, whereas nonhydrolyzed glycerides remain in the lower phase. A 1-ml sample is taken from the upper phase and mixed with 10 ml Aqualuma (Kontron Analytic, Zurich, Switzerland); the radioactivity is determined using a liquid scintillation spectrometer equipped for quenching correction. Results are expressed as units of enzyme activity per milliliter of enzyme solution or per milligram of protein, 1 U corresponding to the release of 1 μmol of [^{3}H]oleic acid in 1 min.

Phospholipase A_1 Assay[1,16]

Materials

[9,10(*n*)-^{3}H]Palmitic acid, 500 mCi/mmol (The Radiochemical Center)
Coenzyme A, sodium deoxycholate (Sigma)
Silica gel G 60, 0.25 mm thick plates (Merck)

Preparation of 1-[^{3}H]Palmitoyl-2-acyl-sn-glycero-3-phosphocholine. The following procedure was developed by Waite and van Deenen[17] and van den Bosch *et al.*[18] Briefly, microsomes from the liver of a rat weighing 300–350 g are incubated with 1 mCi [^{3}H]palmitic acid (taken to dryness under nitrogen and dispersed by sonication) in 24 ml of 20 m*M* Tris-HCl (pH 7.35) containing 125 m*M* KCl, 45 m*M* ATP, 40 m*M* $MgCl_2$, and 1 m*M* coenzyme A. After a 60-min incubation at 37° with shaking, lipids are extracted according to Bligh and Dyer,[19] and phosphatidylcholine is purified by thin-layer chromatography, using chloroform/methanol/water/acetic acid (65 : 43 : 3 : 1, v/v, solvent B). The phospholipid concentration is determined by assay of phosphorus.[20] Thin-layer chromatography in solvent B reveals a radiopurity exceeding 95%, with 96% of [^{3}H]palmitic acid occupying the *sn*-1 position, as determined on hydrolysis with *Crotalus adamanteus* phospholipase A_2 (see below for incubation conditions).

Procedure. From 10 determinations of phospholipase A_1, 0.4 μmol of 1-[^{3}H]palmitoyl-2-acyl-*sn*-glycero-3-phosphocholine is taken to dryness under a stream of nitrogen and dispersed by sonication (2 times, 15 sec each) in 0.8 ml Tris-HCl (0.2 *M*, pH 8.5) containing 2.4 m*M* sodium deoxycholate. Reactions are started by mixing 80 μl of the substrate dispersion, enzyme solution, and enough Tris-HCl buffer without sodium deoxycholate to a final volume of 100 μl. Reactions are performed for 10 min at 37° while shaking, blocked with 0.3 ml chloroform/methanol (1 : 1, v/v), and followed by addition of 1 ml methanol, 1 ml chloroform, and 1 ml water. After phase separation via centrifugation, the lower chloroform phase is withdrawn, and a second lipid extraction is performed on addition of 0.5 ml chloroform. The pooled lipid extracts are taken to dryness and separated by thin-layer chromatography using solvent C (chloroform/methanol/water, 65 : 35 : 4, v/v). The various spots corresponding to lysophosphatidylcholine, phosphatidylcholine, and nonesterified fatty acids

[16] J. Fauvel, H. Chap, V. Roques, and L. Douste-Blazy, *Biochim. Biophys. Acta* **792,** 72 (1984).

[17] M. Waite and L. L. M. van Deenen, *Biochim. Biophys. Acta* **137,** 498 (1967).

[18] H. van den Bosch, L. M. G. van Golde, A. J. Slotboom, and L. L. M. van Deenen, *Biochim. Biophys. Acta* **152,** 694 (1968).

[19] E. D. Bligh and W. J. Dyer, *Can. J. Biochem. Physiol.* **37,** 911 (1959).

[20] C. J. F. Böttcher, C. M. van Gent, and C. Pries, *Anal. Clin. Acta* **24,** 203 (1961).

are identified by comparison to pure standards. These spots are scraped off, the pulver is directly added to 10 ml Aqualuma, and the radioactivity is determined. Phospholipase A_1 activity is calculated from the percentage of radioactivity present in nonesterified fatty acids, after subtraction of control values obtained in the absence of enzyme (usually less than 5%), taking into account the substrate concentration in the assay.

Comments. Although this is a rather lengthy procedure, owing to the chromatographic step, it was found to give the most reliable data. An alternative procedure can be adapted where fatty acids are selectively extracted according to Gatt and Barenholz.[21] In this case, the incubation conditions are the same, except that the reaction is interrupted by addition of 3 ml of heptane/2-propanol (1 : 1, v/v), 0.3 ml water, and 0.2 ml of 50 m*M* H_2SO_4. After a 3-min centrifugation at 1000 *g*, 0.8 ml of the upper phase (total volume 1.5 ml) is mixed with 25 mg silicic acid, and the radioactivity of 0.5 ml of the supernatant in 10 ml Aqualuma is determined. This procedure might appear more convenient for following the enzymatic protein during the course of purification. Also, 1-[^{3}H or ^{14}C]palmitoyl-2-oleoyl-*sn*-glycero-3-phosphocholine can be obtained from New England Nuclear Research–Du Pont (Dreieich, FRG), avoiding biosynthetic labeling of phosphatidylcholine substrate.

Lysophospholipase Assay[16]

Materials

- 1,2-Dipalmitoyl-*sn*-glycero-3-phospho[*N-methyl*-^{3}H]choline, 77 Ci/mmol (The Radiochemical Center)
- Phospholipase A_2 from *Crotalus adamanteus* venom, dipalmitoylphosphatidylcholine, sodium salt (Sigma)
- Trichloroacetic solution (10%, w/v, in water)
- Bovine serum albumin, Fraction V (Sigma), 1% solution (w/v) in water

Preparation of 1-Palmitoyl-sn-glycero-3-phospho[^{3}H]choline. Radioactive dipalmitoyl-*sn*-glycero-3-phospho[*N-methyl*-^{3}H]choline (10 μl, 10 μCi) is mixed with 10 μmol dipalmitoylphosphatidylcholine by pipetting from a stock solution (10 μmol/ml) in chloroform/methanol (1 : 1, v/v) and taken to dryness under nitrogen. Incubation is performed by vigorous stirring in capped tubes of 5 ml ethyl ether (dissolving the phospholipid substrate) with 1 ml of 0.1 *M* borax (pH 7.0) containing 5 m*M* $CaCl_2$ and 30 mg *C. adamanteus* venom (2 hr, room temperature). Lipids are ex-

[21] S. Gatt and Y. Barenholz, this series, Vol. 14, p. 167.

tracted with butanol and purified by column chromatography on Sepharose LH-20.[22]

Procedure. The incubation conditions are strictly identical to those described above for the phospholipase A_1 assay, except for the nature of the substrate and the concentration of sodium deoxycholate (7.5 m*M* in the dispersing buffer, i.e., 6 m*M* final concentration). After a 5- to 10-min incubation at 37°, 0.6 ml of trichloroacetic acid (10%, w/v) is added, followed by 0.3 ml bovine serum albumin solution (0.6%, w/v). Tubes are kept on ice for 10 min and centrifuged at 2000 *g* for 10 min (4°). This allows the precipitation of nonhydrolyzed lysophosphatidylcholine together with bovine serum albumin, whereas *sn*-glycero-3-phospho[^{3}H]choline produced by the enzyme remains soluble in supernatant. The radioactivity is then determined by mixing 0.6 ml of supernatant with 10 ml Aqualuma.

Comments. Because this procedure does not involve chromatographic separation, it can be used to follow enzyme activity during purification steps.

Purification of Enzyme[1]

Step 1: Pancreatic Extract. Guinea pigs (about 500 g) are sacrificed by cervical trauma, and each pancreatic gland is quickly removed and immersed in ice-cold 0.15 *M* NaCl. The various subsequent steps are then performed on ice or at 4°. It must be noted that guinea pig pancreas is more diffuse than that of other animal species, and care must be taken to remove the various parts of the gland. A typical purification can be performed starting from 30 pancreas (30 g), which are cleared from fat debris, sliced, and homogenized in 5 volumes of 0.1 *M* Tris-HCl (pH 8.5) using 10 strokes of a Potter–Elvehjem equipment. In order to promote disruption of zymogen granules, the homogenate is then treated in a Waring blendor at full speed for 2 min. The extract is frozen at −40° in an acetone bath containing CO_2 and kept at this temperature for 1 hr. After thawing, the extract is cleared by centrifugation at 900 *g* for 10 min, and the pellet is suspended in the same volume of Tris-HCl and centrifuged again. Pooled supernatants are then rid of cellular debris by centrifugation at 150,000 *g* for 60 min. This allows a recovery of about 70–80% of the lipase/phospholipase A_1 activity present in the original homogenate. Some improvement can be obtained on adding 1% (v/v) Triton X-100 to the homogenizing buffer, but this might complicate the following purification steps.

[22] A. J. Slotboom, G. H. de Haas, G. J. Burbach-Westerhuis, and L. L. M. van Deenen, *Chem. Phys. Lipids* **4,** 30 (1970).

Step 2: DEAE-Sepharose Chromatography. The enzyme extract is diluted twice with distilled water, and the pH is adjusted to 9.0 with 1 *M* NaOH. The diluted extract is applied to a column (5 × 15 cm) of DEAE-Sepharose (Pharmacia-LKB, Uppsala, Sweden) equilibrated with 50 m*M* Tris-HCl (pH 9.0). Cationic lipase with phospholipase A_1 activity (referred to as lipase I) is eluted in the first half of the void volume and can be used for further purification. Adsorbed proteins, which contain the classic lipase found in other animal species, can then be eluted using a linear NaCl gradient (0–400 m*M* in 50 m*M* Tris-HCl, pH 9.0). This step, however, can be omitted in this procedure.

Step 3: CM-Sepharose Chromatography. The pooled fractions containing lipase I are loaded onto a column of CM-Sepharose (2.6 × 20 cm, Pharmacia-LKB) equilibrated with 50 m*M* Tris-HCl (pH 9.0). After elution of nonadsorbed proteins, a linear NaCl gradient is applied (0–300 m*M* in 50 m*M* Tris-HCl, pH 9.0). Lipase/phospholipase A_1 activity is eluted as two poorly resolved peaks, which are pooled together, desalted onto a column of Sephadex G-25 (3.4 × 20 cm, equilibrated in 50 m*M* Tris-HCl, pH 9.0), and subjected to the same chromatographic step on CM-Sepharose. With a linear NaCl gradient of 0–150 m*M*, the protein profile displays three main peaks; the second and third peaks contain lipase/phospholipase A_1 activity. These are eluted at around 50 and 75 m*M* NaCl, and are referred to as lipase Ia and lipase Ib, respectively.

Step 4: Sephadex G-100 Gel Filtration. Each pooled fraction is concentrated to 2 ml using Centriflo membrane cones (CF 25, Amicon Co., Epernon, France) and applied to a Sephadex G-100 column (Pharmacia-LKB) equilibrated in 50 m*M* Tris-HCl (pH 9.0). Although the lipase Ia fraction displays a single protein peak, lipase Ib is further separated from lower molecular mass impurities.

Comments. Using nondenaturing polyacrylamide gel electrophoresis in an acidic buffer (pH 4.5, i.e., system III of Gabriel[23]), the two enzymes display one single band, indicating that they are virtually pure. Although the original procedure has not been repeated since our first publication,[1] several partially purified preparations have been obtained by a similar procedure employing DEAE–Sepharose chromatography. This procedure appears to be not only adequate for analytical studies on phospholipids,[8–11] but also necessary because the crude extract contains other lipolytic enzymes such as the phosphoinositide-specific phospholipase C. For instance, owing to its phospholipase B activity involving fatty acid migration in lysophospholipids (see below), the fraction "lipase I" has been successfully used to prepare *sn*-glycero-3-phosphoinositol from phosphatidylino-

[23] O. Gabriel, this series, Vol. 22, p. 565.

sitol, whereas only inositol 1-phosphate is obtained using the crude pancreatic extract.[24]

The whole purification procedure results in a 7% yield of phospholipase A activity.[1] This can certainly be improved by performing a single chromatography step on CM–Sepharose using the 0–150 m*M* NaCl gradient.

To follow the activity during the chromatographic steps, we would advise using the diacylglycerol lipase assay or the phospholipase A_1 assay which does not involve thin-layer chromatography. However, the latter will be preferable to detect phospholipase A_1 activity after each purification step.

Properties

Molecular Mass. Molecular mass has been estimated by gel filtration on Sephadex G-100 and found to be 37 and 42 kDa for lipases Ia and Ib, respectively. At the present time, we do not know whether the two enzymes represent products of the same gene differing in glycosylation or proteolytic processing. Neither of the enzymes appear to possess an inactive zymogen.[2]

Isoelectric Point. Analytical isoelectrofocusing reveals indistinguishable values of 9.3–9.4 for both proteins.

Stability. Despite a very low stability of dilute solutions even on freezing at $-20°$, the two enzymes were found to be very stable on storing at $-20°$ in 50 m*M* Tris-HCl (pH 8.5) containing 50% glycerol (v/v), as previously observed for phospholipase C from *Bacillus cereus*.[25] Indeed, a 100% recovery of diacylglycerol lipase and phospholipase A_1 activity could be measured after 9 years of storage under these conditions.

Substrate Specificity. The two enzymes did not reveal any significant differences in terms of specific activities toward various lipid substrates. When considering various neutral glycerides and glycerophospholipids, the following order of hydrolytic activity has been established using radiochemical assays with pure substrates under optimal conditions of measurement[15–16]: dioleoylglycerol > 1(3)-monooleoylglycerol > trioleoylglycerol > phosphatidylcholine = phosphatidylinositol > lysophosphatidylcholine (1-acyl-2-lyso-*sn*-glycero-3-phosphocholine) > phosphatidylethanolamine = phosphatidylglycerol. No activity could be measured toward cholesteryl oleate, methyl butyrate, or *p*-nitrophenyl acetate, indicating that the enzymes are different from carboxyl ester hydrolase.[26] The data differ

[24] H. Ribbes, M. Plantavid, P. J. Bennet, H. Chap, and L. Douste-Blazy, *Biochim. Biophys. Acta* **919,** 245 (1987).

[25] R. F. A. Zwaal and B. Roelofsen, this series, Vol. 32, p. 154.

[26] D. Lombardo, O. Guy, and C. Figarella, *Biochim. Biophys. Acta* **527,** 142 (1978).

somewhat from previous comparisons indicating identical lipase and phospholipase A_1 activities as determined by titrimetric assay using olive oil or egg yolk, respectively. The reasons for such a discrepancy are not yet understood.

Some apparent activity can be detected toward 2-acyl-*sn*-glycero-3-phosphocholine and 2-oleoyl-*sn*-glycerol, but this is certainly due to fatty acid migration to the external positions that are favored at acidic or alkaline pH. This explains why a total deacylation of glycerophospholipids can be observed (phospholipase B activity); however, it can be minimized by increasing the sodium deoxycholate concentration. As previously discussed, this probably reduces fatty acid migration from the *sn*-1 to the *sn*-2 position.[11] The absolute lack of activity against the 2-acyl ester bond, which is particularly evident with 1-*O*-alkyl-2-acyl-*sn*-glycero-3-phosphocholine,[2,8–11] is probably explained by the inability of the two enzymes to hydrolyze acyl ester bonds involving a secondary alcohol. However, the location of the phosphate group also affects enzyme activity; it is lowered by 6- to 15-fold when using either 1,3-dimyristoyl-*sn*-glycero-2-phosphocholine (β-lecithin) or 1-palmitoylglycol-2-phosphocholine.[16] Finally, the nearness of a hydroxyl group (or acyl ester group) greatly improves enzyme activity, as revealed by a 10-fold lower activity against 1-myristoylpropanediol-3-phosphocholine compared to 1-myristoyl-*sn*-glycero-3-phosphocholine. This is probably due to the inductive effect of the hydroxyl group, as suggested by Brockerhoff.[27]

Another striking feature which differentiates guinea pig pancreas phospholipases A_1 from classic phospholipases A_2 is the absolute lack of stereospecificity.

Activators and Inhibitors. Although the two enzymes display some low, but significant, activity toward phospholipid vesicles[1] or intact high-density lipoproteins,[12,13] they are poorly penetrating enzymes that require sodium deoxycholate for maximal activity. Depending on the phospholipid used, optimal concentrations of the detergent vary between 1 and 12 m*M*. No evidence has been found for an activation of these enzymes by colipase.

Among various agents tested, the two enzymes were found to be insensitive to *N*-ethylmaleimide, diisopropyl fluorophosphate, and *p*-bromophenacyl bromide. The latter reagent allows differentiation between classic phospholipases A_2 and guinea pig pancreatic phospholipases A_1, which also do not require calcium for activity. We recently took advantage of the latter property to confirm that inhibition of phospholipases by lipocortins is related to the ability of these proteins to interact with phos-

[27] H. Brockerhoff, *Biochim. Biophys. Acta* **159,** 295 (1968).

pholipid substrate,[14] as suggested by others.[28-31] Full activity of phospholipase A_1 can be measured against phospholipid dispersions in the presence of calcium or EGTA. This allowed us to observe that lipocortins inhibit phospholipase A_1 only in the presence of calcium, that is, under conditions where they are able to bind to phospholipid, fully supporting the so-called surface depletion model.[28]

Concluding Remarks

Cationic lipases with high concentrations of phospholipase A_1 from guinea pig pancreas represent unique enzymes in the animal kingdom, albeit a systematic search for their presence in the pancreas of other animal species has not yet been performed and is certainly hampered by the presence of pancreatic phospholipase A_2. Although these cationic lipases display some enzymatic properties similar to those of hepatic lipase,[6,7] they certainly play a role in (phospho)lipid digestion by the guinea pig. A study of ether phospholipid absorption in the guinea pig digestive tract led us to characterize and to purify from guinea pig enterocyte brush border membranes a new phospholipase B, which also represents a novel and unique enzyme.[31,32]

[28] F. F. Davidson, E. A. Dennis, and J. R. Glenney, Jr., *J. Biol. Chem.* **262,** 1698 (1987).
[29] A. J. Aarsman, G. Mynbeek, H. van den Bosch, B. Rothhut, B. Prieur, C. Comera, L. Jordan, and F. Russo-Marie, *FEBS Lett.* **219,** 176 (1987).
[30] K. Machoczek, M. Fischer, and H. D. Söling, *FEBS Lett.* **251,** 207 (1989).
[31] A. Diagne, S. Mitjavila, J. Fauvel, H. Chap, and L. Douste-Blazy, *Lipids* **22,** 33 (1987).
[32] A. Gassama-Diagne, J. Fauvel, and H. Chap, *J. Biol. Chem.* **264,** 9470 (1989).

[29] Purification of Rat Kidney Lysosomal Phospholipase A_1

By KARL Y. HOSTETLER and MICHAEL F. GARDNER

Introduction

Mammalian lysosomes contain acid hydrolases which degrade membrane lipids, lipoprotein lipids, and other materials. Lysosomes have been shown to contain several important lipolytic enzymes including phospholipase A_1,[1,2] phospholipase C,[3] lysophospholipase,[2] lipase,[4] and sphingomy-

[1] K. Y. Hostetler, P. J. Yazaki, and H. van den Bosch, *J. Biol. Chem.* **257,** 13367 (1982).
[2] M. Robinson and M. Waite, *J. Biol.Chem.* **258,** 14371 (1983).
[3] Y. Matsuzawa and K. Y. Hostetler, *J. Biol. Chem.* **255,** 5190 (1980).

elinase.[5] All these enzymes are glycoproteins and have acidic pH optima in the range of 4.0 to 5.5. The lysosomal phospholipases catalyze the initial steps in degradation of the glycerophospholipids to fatty acids, glycerol, and the various polar head groups. Phospholipases A and C have been implicated in an acquired phospholipid storage disease in humans and animals in which certain cationic amphiphilic pharmaceutical agents concentrate in the lysosomal interior where they inhibit the action of the phospholipases.[6,7] Purifications of lysosomal phospholipase A have been achieved from rat kidney[8] and rat liver.[1,2,9]

Purification of Rat Kidney Phospholipase A_1

Isolation and Solubilization of Lysosomal Phospholipase A_1. Eighteen to thirty rats of the Fischer 344 strain are fasted overnight, sacrificed by cervical fracture, and the kidneys are removed and placed in iced buffer A consisting of 0.25 *M* sucrose, 5 m*M* Tris (pH 7.4), and 2 m*M* EDTA at 4°. The tissue is excised, weighed, and rinsed in ice-cold buffer. Kidney cortex weighing 18–40 g is cut into small pieces, and a 20% homogenate in buffer A is prepared using a Potter–Elvehjem homogenizer. The homogenate is passed through four layers of cheesecloth and centrifuged at 160 *g* for 6 min. The pellet is resuspended in buffer A and recentrifuged at 160 *g* for 6 min; this washing is repeated 3 times, and the respective supernatants are combined with the original postnuclear supernatant. The nuclear pellet is discarded, and the combined supernatants are centrifuged at 20,000 *g* for 20 min. The pellet is taken up in 70–80 ml of buffer containing 10 m*M* sodium phosphate buffer, pH 7.2, and 62.5 m*M* NaCl and subjected to five cycles of freezing followed by thawing. The resulting suspension is centrifuged at 100,000 *g* for 60 min to sediment membranous material, which is discarded. The supernatant containing the soluble proteins is diluted with 0.25 volumes of cold glycerol, and this material is purified further as noted. Protein is measured by the method of Bradford[10] using rabbit γ-globulin as standard.

Hydroxyapatite/Concanavalin A-Sepharose. All operations are carried

[4] T. G. Warner, L. M. Dambach, J. H. Shin, and J. S. O'Brien, *J. Biol. Chem.* **256,** 2952 (1981).

[5] J. N. Kanfer, O. M. Young, D. Shapiro, and R. O. Brady, *J. Biol. Chem.* **241,** 1081 (1966).

[6] K. Y. Hostetler, *Fed. Proc., Fed. Am. Soc. Exp. Biol.* **43,** 2582 (1984).

[7] K. Y. Hostetler, *in* "Lipids and Membranes: Past Present and Future" (J. A. F. op den Kamp, B. Roelofsen, and K. W. A. Wirtz, eds.), p. 307. Elsevier, Amsterdam, 1986.

[8] K. Y. Hostetler, M. F. Gardner, and J. R. Giordano, *Biochemistry* **25,** 6456 (1986).

[9] B.-M. Löeffler and H. Kunze, *Biochim. Biophys Acta* **1003,** 225 (1989).

[10] M. M. Bradford, *Anal. Biochem.* **72,** 248 (1979).

out at 4° unless otherwise noted. The soluble protein in 8 m*M* sodium phosphate (pH 7.2)/50 m*M* NaCl containing 20% glycerol is applied to a column of hydroxyapatite (1.6 × 20 cm) connected in series to a column (1 × 4 cm) of concanavalin A-Sepharose. The sample (~60 ml) is applied to the hydroxyapatite at 20 ml/hr, and the columns are washed with an equal volume of the starting buffer solution. The columns are then disconnected, and the concanavalin A-Sepharose column is eluted with 90 ml of 0.5 *M* methyl α-mannoside at a flow rate of 20 ml/hr. The column is warmed to 20° and eluted with an additional 35 ml of methyl α-mannoside containing 5 m*M* EDTA. The methyl α-mannoside eluates are combined and taken to a small volume by using a 100-ml Amicon ultrafiltration cell with a YM10 membrane (Amicon Corporation, Lexington, MA).

Chromatofocusing. The sample is diluted to 50 ml with 25 m*M* Tris (pH 7.0) containing 20% glycerol and concentrated to a small volume as noted above. This process is repeated once, and the resulting sample (~4 ml) is applied to a 1 × 4 cm column of PB94 exchanger (Pharmacia, Piscataway, NJ) which has been equilibrated with 25 m*M* Tris (pH 7.0)/20% glycerol. The column is eluted at 14 ml/hr with 60 ml of Polybuffer 74 (pH 4.0) which is diluted 1:8 with distilled water and adjusted to a glycerol concentration of 20% (w/v). Fractions of 4 ml are collected and assayed for phospholipase A_1 activity.

Sephadex G-150. The active fractions from the chromatofocusing step are combined, concentrated rapidly to a small volume as noted above, and dilute with 0.15 *M* NaCl/5 m*M* Tris (pH 7.4)/20% glycerol. The process is repeated twice, and the sample (1.0 ml) is applied to a 1 × 30 cm column of Sephadex G-150 previously equilibrated with this buffer. The column is eluted, and the active fractions are combined. The combined fractions are adjusted to 50% glycerol and stored at −70° until use. The time required to purify phospholipase A_1 from the soluble protein fraction is approximately 60–72 hr.

Enzyme Assays. Phospholipase A_1 activity of the respective column fractions is measured as follows. Assay mixtures containing 50 m*M* sodium acetate (pH 4.4), 0.2 m*M* 1,2-di[1-^{14}C]oleoylphosphatidylcholine as sonicated vesicles, and protein are incubated for 20 min at 37°. The release of [1-^{14}C]oleic acid is measured using the Dole extraction[11] as modified by van den Bosch and Aarsman,[12] and aliquots of the upper heptane layer are counted by liquid scintillation spectrometry. The 1,2-di[1-^{14}C]oleoylphosphatidylcholine vesicles are prepared by sonication in dilute Tris buffer, pH 7.4, as previously described.[13] In some experiments 1,2-di[1-^{14}C]pal-

[11] V. P. Dole, *J. Clin. Invest.* **35,** 150 (1956).
[12] H. van den Bosch and A. J. Aarsman, *Agents Actions* **9,** 382 (1979).
[13] A. Pappu and K. Y. Hostetler, *Biochem. Pharmacol.* **33,** 1639 (1984).

TABLE I
PURIFICATION OF KIDNEY LYSOSOMAL PHOSPHOLIPASE A_1[a]

Step	Protein (mg)	Activity (nmol/mg/hr)	Total activity (nmol/hr)	Recovery (%)	Purification factor (-fold)
Homogenate	10,000	1.96	19,100	100	1.0
Lysosome fraction	2510	6.90	17,200	88	3.5
Soluble	680	4.58	2640	14	2.4
Hydroxyapatite/Con A	4.93	1200	6200	32	610
Chromatofocusing	1.63	2600	4270	22	1300
Sephadex G-150	0.154	15,000	1850	9.8	7600

[a] Data are averages of three separate preparations where the amount of homogenate protein has been set equal to 10,000 mg. The actual amount of protein in the three homogenates ranged from 10,300 to 27,800 mg. Phospholipase A_1 activity was measured as described in the text, using sonicated 0.2 mM [1-^{14}C]dioleoylphosphatidylcholine present as small unilamellar vesicles. Four different protein concentrations were used, and only data from the linear portion of the protein curve were used to calculate the rate. When the protein used was in the linear range, enzyme activity was linear with time to at least 20 min. The data in each column represent the means of three experiments. The means $\pm$ S.D. at the final purification step are as follows: protein, 0.154 $\pm$ 0.051; phospholipase A_1 activity, 15,000 $\pm$ 3500; total activity, 2200 $\pm$ 462; percent recovery, 9.8 $\pm$ 4.2; purification factor, 7600 $\pm$ 1800. Adapted from Ref. 8 with permission.

mitoylphosphatidylcholine with 0.5 mg/ml of Triton X-100 is used as substrate (mixed micelles). Lysophospholipase is assayed by measuring the release of [1-^{14}C]palmitic acid from [1-^{14}C]palmitoyllysophosphatidylcholine.[1]

Results

Table I shows the purification data for rat kidney phospholipase A_1.[8] The purification begins with the sedimentation from the homogenate of lysosomal phospholipase A_1 together with kidney lysosomes by centrifugation of the postnuclear supernatant at 20,000 g; the enzyme is obtained in an 88% yield and can be easily solubilized by freezing and thawing in hypotonic buffer. The yield of activity at this step appears to be only 14% but is actually much higher because proteins solubilized from kidney mitochondria strongly inhibit the enzyme (K. Y. Hostetler and J. R. Giordano, unpublished observations, 1987). In 8 mM phosphate buffer, kidney phospholipase A_1 does not bind to hydroxyapatite, and this column is connected in series to a concanavalin A-Sepharose column which binds the phospholipase A_1 eluting from the hydroxyapatite. The concanavalin

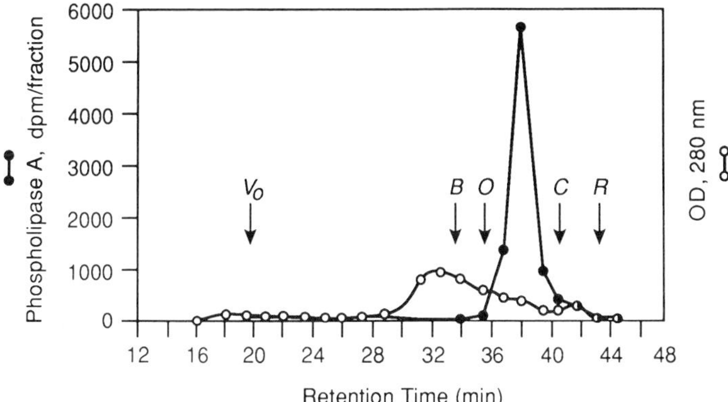

FIG. 1. Gel-permeation HPLC of purified kidney phospholipase A_1. The purified phospholipase A_1 was concentrated to a small volume in an ultrafiltration cell with a YM10 membrane followed by further concentration in a Centricon tube (both from Amicon). One hundred microliters was injected into a Superose 12 column and eluted with 0.15 *M* NaCl containing 5 m*M* Tris (pH 7.4) and 20% ethylene glycol at 0.4 ml/min. Fractions (0.5 ml) were assayed for phospholipase A_1 activity. V_0, Void volume; *B*, bovine serum albumin; O, ovalbumin; C, chymotrypsinogen; R, ribonuclease. (Adapted from Ref. 8 with permission.)

A column is then detached and the phospholipase A_1 activity eluted with 0.5 *M* methyl α-mannoside. This step is followed by chromatofocusing and gel-permeation chromatography using Sephadex G-150, resulting in a recovery of 9.8% of the phospholipase A_1 activity and a 7600-fold purification relative to the homogenate.

Sodium dodecyl sulfate-polyacrylamide gel electrophoresis (SDS–PAGE) under reducing conditions showed a major band at 34 kDa when silver stained, with a contaminant at 66 kDa and a minor band at 18 kDa. When the final protein fraction was subjected to Superose 12 high-performance liquid chromatography (HPLC),the 66- and 18-kDa proteins were found to be devoid of activity, as shown in Fig. 1. The phospholipase A_1 activity resides with the protein having an apparent molecular mass of 30 kDa.

Properties of Purified Rat Kidney Lysosomal Phospholipase A_1

Kidney phospholipase A_1 was not inhibited by divalent cations such as calcium, magnesium, and manganese nor by EDTA and was not affected by mercuric ion, *p*-bromophenacyl bromide, or diisopropyl fluorophosphate.[8] Purified kidney phospholipase A_1 was incubated with 2-[1-^{14}C]oleoylphosphatidylcholine and the reaction products analyzed by thin-layer chromatography. The major labeled product was lysophosphatidylcholine (96%), indicating that the enzyme is a phospholipase of the A_1 type. The V_{max} with dipalmitoylphosphatidylcholine/Triton X-100 mixed micelles was 53 μmol/mg/hr versus 48 μmol/mg/hr with dioleoylphosphatidylcholine small unilamellar vesicles as substrate; the apparent K_m values for the respective substrates were 0.57 and 1.1 mM, consistent with the fact that, in a mixed micelle, all of the phosphatidylcholine is available to the phospholipase versus only half available in the liposome vesicle. The purified enzyme preparation had weak lysophospholipase activity representing about 6% of the rate of hydrolysis observed with phosphatidylcholine. The apparent K_m for lysophosphatidylcholine was much lower, 0.04 mM.[8] Purified kidney lysosomal phospholipase A_1 is very stable when stored in 50% glycerol at $-70°$ and has been found to retain essentially full activity for at least 1 year.

Purification of Other Lysosomal Phospholipases

Recently, Löffler and Kunze reported a novel method for purifying lysosomal phospholipase A_1 from rat liver.[9] The method utilizes homogenization of the liver in a hypotonic buffer, acid precipitation of the solubilized phospholipase A, and combined dye binding using Yellow H-A and Red HE-3B agaroses followed by combined concanavalin A-Sepharose/Phenyl-Sepharose chromatography. The final fraction contained 88% of the total phospholipase A activity and contained two protein bands at 56 and 33 kDa. The authors concluded that the 56-kDa protein is the active enzyme. Others have reported lower molecular masses for the liver lysosomal phospholipase A_1 using different methodology.[1,2] No amino acid sequence information has been reported for the lysosomal phospholipases A_1.

[30] Purification and Substrate Specificity of Rat Hepatic Lipase

By MOSELEY WAITE, TOM Y. THUREN, REBECCA W. WILCOX, PATRICIA J. SISSON, and GREGORY L. KUCERA

Introduction

Hepatic lipase from rat liver is an acylhydrolase that cleaves the fatty acid from the 1(3)-position of a broad spectrum of glycerides.[1] In addition to the hydrolytic reaction, the enzyme can catalyze transacylation reactions in which the free hydroxyl of an acceptor lipid molecule becomes acylated rather than water, which is the acyl acceptor in hydrolysis.[2] As a consequence of the wide range of substrates and reactions catalyzed, the enzyme has been termed hepatic lipase, phospholipase, and monoacylglycerol acyltransferase. Although not yet verified, most likely the enzyme has an acyl intermediate covalently bound, and, since hepatic lipase is thought to be a serine hydrolase,[3] the purported acyl intermediate may be an acylserine residue. The cDNA obtained for the human[4] and rat[5] hepatic lipases predicts a molecular size for the peptide of 53,222 Da for the rat enzyme. However, hepatic lipase is a glycoprotein with perhaps more than two carbohydrate chains present,[6] which may account for the variation in the reported sizes of the enzyme, as determined by sodium dodecyl sulfate-polyacrylamide gel electrophoresis (SDS–PAGE). Two immunoreactive species were found in a cultured hepatoma line, a high mannose-containing species (55.4 kDa) and a secreted form with N-linked oligosaccharides (57.6 kDa).[6]

Evidence exists that hepatic lipase is synthesized in the parenchymal cells of the liver and following its release is bound by heparinlike receptors on the surface of liver endothelial cells.[7] However, other tissues involved in rapid uptake of cholesterol, such as the ovary, testes, and adrenals,

[1] M. Waite and P. Sisson, *J. Biol. Chem.* **249,** 6401 (1974).
[2] M. Waite and P. Sisson, *J. Biol. Chem.* **248,** 7985 (1973).
[3] O. Ben-Zeev, C. M. Ben-Avram, H. Wong, J. Nikazy, J. E. Shively, and M. C. Schotz, *Biochim. Biophys. Acta* **919,** 13 (1987).
[4] S. Datta, C.-L. Luo, W.-H. Li, P. van Tuinen, D. H. Ledbetter, M. A. Brown, S.-H. Chen, S.-W. Liu, and L. Chan, *J. Biol. Chem.* **263,** 1107 (1988).
[5] M. H. Doolittle, H. Wong, R. C. Davis, and M. C. Schotz, *J. Lipid Res.* **28,** 1326 (1987).
[6] L. A. Cisar and A. Bensadoun, *Biochim. Biophys. Acta* **927,** 305 (1987).
[7] P. K. J. Kinnunen, *in* "Lipases" (B. Borgstrom and H. L. Brockman, eds.), p. 307. Elsevier, Amsterdam, 1984.

have been shown to have a high content of bound hepatic lipase even though the cells of these organs do not appear to synthesize hepatic lipase.[8] It is thought that the enzyme synthesized and released by the liver can be bound specifically to these target cells. Various studies have implicated a role of hepatic lipase in reverse cholesterol transport, a process in which cholesterol in peripheral tissues is transported to cholesterol-consuming tissues such as liver and steroid hormone-producing tissues.[9] As proposed, catabolism of phospholipid on the coat of the cholesterol ester-rich HDL_2 favors uptake of cholesterol ester in the core of the high-density lipoprotein (HDL) with subsequent formation of the smaller HDL_3. On the other hand, hepatic lipase has been proposed to catabolize the triacylglycerol in chylomicron remnant lipoproteins and intermediate-density lipoproteins (IDL).[10] Since the enzyme has a broad substrate specificity and no clear requirement for an apolipoprotein, the physiologic function(s) of hepatic lipase remains unclear.

Materials

Heparin (porcine), Triton N-101 and X-100, and heparin-Sepharose (Type 1) were obtained from Sigma (St. Louis, MO), methyl-α-D-mannopyranoside from Calbiochem (San Diego, CA), DEAE-Sephacel from Pharmacia (Piscataway, NJ), and sodium pentobarbital from Barber Veterinary Supply. Radiolabeled lipid substrates either were purchased from New England Nuclear (Boston, MA) or Amersham (Arlington Heights, IL) or were synthesized as described earlier.[11] Thin-layer plates were prepared in the laboratory with silica gel H-60 (EM Science, Cherry Hill, NJ), although commercially available thin-layer chromatography (TLC) plates (Uniplate, Analtech, Inc. Alexandria, VA) are excellent for these assays. Thioester-containing substrates used in spectrophotometric assays were prepared as described by us previously.

Enzyme Assay and Purification

Routine assay of hepatic lipase is carried out with 1-oleoyl-[2-^{3}H]glycerol ([^{3}H]MO) for the steps in purification and to check enzyme stability. This has the advantage of eliminating the chromatographic separation of products since both hydrolysis and transacylation reactions release

[8] N. L. M. Persoon, W. C. Hulsmann, and H. Jansen, *Eur. J. Cell Biol.* **41,** 134 (1986).
[9] F. M. van't Hooft, T. van Gent, and A. van Tol, *Biochem. J.* **196,** 877 (1981).
[10] B. Landin, A. Nilsson, J.-S. Twu, and M. C. Schotz, *J. Lipid Res.* **25,** 559 (1984).
[11] G. L.Kucera, P. J. Sisson, M. J. Thomas, and M. Waite, *J. Biol. Chem.* **263,** 1920 (1988).

[^{3}H]glycerol that partitions into the methanol–water phase in the extraction procedure of Bligh and Dyer.[12] Since the remaining substrate partitions into the chloroform layer, enzyme activity is measured by scintillation counting of an aliquot of the upper phase.

Specifically, a cocktail of [^{3}H]MO is prepared by drying the chloroform solution of [^{3}H]MO under N_2 in a heavy walled tube. The necessary amount of water is added to make the final [^{3}H]MO emulsion in the cocktail 0.5 m*M*. A Tris-HCl buffer, pH 8.5, may be used for the cocktail preparation if a single pH value is to be used. The amount of [^{3}H]MO used should be sufficient for 50 nmol [10,000 disintegrations per minute (dpm)] per incubation. The tube is then sonicated for 3 min in a sonic bath (Heat Systems Ultrasonic, Farmingdale, NY). If the emulsion is not clear at this point, the sonication is continued. An aliquot of the sonication emulsion should be counted to determine that the [^{3}H]MO is totally suspended. An aliquot of 0.1 ml of the sonicated suspension is then pipetted into each reaction tube, brought to 0.49 ml with Tris-HCl buffer, pH 8.5 (0.1 *M* final concentration), followed by the addition of 10 μl of the enzyme sample to be assayed, and incubated for 10 min at 37°. (The enzyme volume may be adjusted if the preparation of enzyme is larger or smaller than that outlined here.) The reaction is terminated at the end of the 10-min incubation by the addition of 1.5 ml of a chloroform–methanol solution (1 : 2, v/v), mixed, 0.5 ml of water and 0.5 ml of chloroform are added, and the solution is thoroughly mixed a second time. Following the separation of the two phases in the reaction tube, a 0.5-ml aliquot of the upper (methanol–water) phase is added to a scintillation cocktail and the radioactivity determined.

Since these assays are for the localization of hepatic lipase in the chromatographic steps, the results can be expressed as disintegrations per minute of ^{3}H in the upper phase or percentage of hydrolysis calculated as follows:

$$\text{Hydrolysis (\%)} = \frac{\text{dpm in upper phase} \times 4}{\text{total dpm in reaction}} \times 100$$

A nonenzyme control is run with each series of assays run. The value for this control should be subtracted from the values obtained with enzyme. Because the substrate described here is not commercially available and must be synthesized,[13] other selective extraction assays that employ phospholipids may be used. If phospholipid is the chosen substrate, 1-[^{3}H]palmitoylphosphatidylethanolamine with an unsaturated fatty acid in the 2-position of the glycerol is preferred. Because of the physical properties of

[12] E. G. Bligh and W. J. Dyer, *Can. J. Biochem. Physiol.* **37,** 911 (1959).

[13] C. H. Miller, J. W. Parce, P. Sisson, and M. Waite, *Biochim. Biophys. Acta* **665,** 385 (1981).

this phospholipid substrate, the reaction mixture with phospholipid should contain Triton X-100.[11] Alternatively, chromatographic separation of products and substrate using either [^{3}H]MO or phospholipids may be used, although this approach is much more time-consuming.

Method of Enzyme Purification

The procedure employed is a modification of the methods of Jensen and Bensadoun[14] and Ehnholm and Kuusi[15] that involves perfusion of the liver followed by two chromatographic steps. The procedure described here has the advantage over those previously described in the time required and the yield and specific activity of the final product, based on the results from our laboratory. Two important points should be considered: first, the liver should be thoroughly washed by perfusion prior to the addition of heparin to remove as much contaminating protein as possible; and second, a minimum of time should be used in the preparation. The adoption of the procedure described here saves nearly 2 days in enzyme isolation and increases the specific activity of the final preparation 2- to 3-fold. Also, there are some indications that proteolysis may occur during preparation that may alter substrate specificity, although this has not been thoroughly documented.

For liver perfusion, male Sprague-Dawley rats (175–250 g) are anesthetized with 1.75 ml/kg sodium pentobarbital. The abdominal wall is opened and the portal vein cannulated. After beginning the flow of Krebs–Ringer bicarbonate perfusion buffer, pH 7.2, containing 30% glycerol (v/v), into the liver via the portal vein, the inferior vena cava is severed just below the liver to allow the efflux of perfusion buffer. Immediately after severing the vena cava, the chest cavity is opened and a second cannula is inserted through the right atrium of the heart and into the inferior vena cava. The inferior vena cava is ligated below the liver to allow for efficient circulation of the perfusion media through the liver. The liver is washed with 250 ml of Krebs–Ringer bicarbonate buffer, pH 7.2, that contains 30% glycerol (v/v). The Krebs–Ringer bicarbonate buffer is saturated with 95% O_2–5% CO_2 and maintained at 37°. This nonrecirculating perfusion system is used at a flow rate of 15 ml/min. After the initial washout, the liver is washed with 75 ml of Krebs–Ringer bicarbonate buffer, pH 7.2, 30% glycerol (v/v) that contains 10 units/ml heparin; the effluent is collected on 50 ml (bed volume) of heparin-Sepharose in an ice-water bath.

After the perfusate from eight rats is collected onto the heparin-Sepharose, the mixture is poured into a fritted disk funnel with a nominal

[14] G. L. Jensen and A. Bensadoun, *Anal. Biochem.* **113,** 246 (1981).

[15] C. Ehnholm and T. Kuusi, this series, Vol. 129 [43].

maximum pore size of 40–60 μm. The heparin-Sepharose is washed a minimum of 4 bed volumes of the following buffers that contain 10 m*M* potassium phosphate, pH 7.0: 0.3 *M* NaCl, 30% glycerol (v/v); 30% glycerol (v/v); 0.2% Triton X-100 (w/v), 10% glycerol (v/v); 30% glycerol (v/v). After the last wash, the heparin-Sepharose is packed into a column and an additional 2 bed volumes of the equilibrating buffer is passed over the heparin-Sepharose at the rate of 60 ml/hr. The hepatic lipase is then eluted with a linear gradient between 0.5 and 1.8 *M* NaCl in a total volume of 200 ml with 30% glycerol (v/v), 10 m*M* potassium phosphate, pH 7.0. Fractions of 5 ml are collected.

The fractions from the heparin-Sepharose with peak enzyme activity are pooled and dialyzed for 2–3 hr against 30 volumes of 50 m*M* Tris, pH 7.2, that contains 30% glycerol (v/v). When the reproducibility of enzyme elution from the heparin-Sepharose is established and related to the conductivity of the elution buffer, fraction pooling can be based on the conductivity and the protein peak. We routinely pool fractions between 0.4 and 0.9 *M* NaCl. After dialysis, the hepatic lipase sample is applied with a flow rate of 35 ml/hr to a 2.5 × 5 cm DEAE-Sephacel column equilibrated with 50 m*M* Tris, pH 7.2, 30% glycerol (v/v). After sample application, the column is washed with 4 bed volumes each of the equilibration buffer and the equilibration buffer minus the 30% glycerol (v/v). Purified hepatic lipase is eluted from the DEAE-Sephacel column with a single step of 80 m*M* NaCl, 0.2% (w/v) Triton N-101, 100 m*M* methyl-α-D-mannopyranoside, 50 m*M* Tris, pH 7.2. The active fractions usually elute within 20–30 ml following the addition of the NaCl solution.

The enzyme is stored in the elution buffer at −70° for a period of 1 month or more without significant loss of activity. It is possible to remove Triton N-101 from the preparation by passing the preparation through a 1–2 ml (bed volume) column of the heparin-Sepharose and eluting the enzyme with 2–3 ml of the elution buffer minus Triton N-101. The preparation minus Triton N-101 in 50% glycerol remains fairly stable but loses activity if stored for more than 2–3 weeks at −70°. The enzyme prepared in this manner has a single band on SDS–PAGE with a molecular weight of 57,000.

When assayed as described herein, the initial specific activity of the perfusate sampled prior to mixing with heparin-Sepharose was about 6 μmol/min/mg. The initially reported specific activity of the hepatic lipase following purification through DEAE-Sephacel was approximately 100.[11] However, with the modification of pumping the perfusate directly onto the heparin-Sepharose, the final specific activity was about 200, with a similar increase to the total activity recovered, about 30 μmol/min (150 μg of protein).

TABLE I
KINETICS OF LIPASE HYDROLYSIS OF PHOSPHOLIPIDS[a]

Substrate	V_{max} (μmol/min/mg)	K_s (mM)	K_m (mole fraction)
Phosphatidic acid	84	0.24	0.11
Phosphatidylethanolamine	56	0.19	0.10
Phosphatidylcholine	11	0.63	0.08
Phosphatidylserine	0.43	0.21	0.07

[a] K_s estimates the affinity for the bulk Triton–lipid substrate, whereas K_m estimates the affinity for the enzyme for the lipid substrate in the surface of the micelle and is therefore expressed as the surface mole fraction.

Substrate Specificity

The study of substrate specificity for any lipolytic enzyme is complicated by differences in the physical states of various lipids. For example, lipid substrates recognized by hepatic lipase can exist as oil droplets, micelles, bilayer vesicles, and hexagonal arrays. Attempts to study the relative activities on pure or mixed lipids must take this fact into account. Systems exist, however, that can minimize these differences by holding the lipids in a relatively constant milieu. Two methods available for this end are the monolayer system (described in [4] of this volume) and the matrix micelle system described by Hendrickson and Dennis.[16] Most simply stated, the substrate is dispersed in an inert matrix of Triton X-100 that is of a uniform state. Although not strictly accurate, this is a satisfactory assumption if the proper precautions are taken. The latter system has been used by us to study the substrate specificity for hepatic lipase and has worked successfully for most but not all substrates tested thus far. Unfortunately, triacylglycerol does not work well in this system, although both monoacylglycerol and diacylglycerol mixed micelles are satisfactory. Since the description of the preparation of the mixed micelles and the kinetic analyses are more than adequately described in the original works, readers should consult those papers.

By establishing case 1 and case 2 kinetics, we have determined the following kinetic values for the purified hepatic lipase hydrolysis of phospholipids (Table I). In this study the ratio of Triton X-100 to phospholipid was between 9 : 1 and 11 : 1.[11] As seen, there is a large difference in V_{max} values but not in the affinity constants. It must be noted, however, that the hepatic lipase prepared as described here has a higher V_{max} for phosphatidylserine (35–40 μmol/min/mg) than the value given in Table I. In that

[16] H. S. Hendrickson and E. A. Dennis, *J. Biol. Chem.* **259,** 5734 (1984).

TABLE II
EFFECT OF SUBSTRATE ESTER ON HYDROLYSIS

	V_{max}			
Substrate[a]	Experimental (μmol/min/mg)	Calculated[b]	K_s (m*M*)	K_m (mole fraction)
sn-3-OOPA	84.0	—	0.13	0.123
sn-3-TTPA	12.0	—	0.06	0.33
rac-TTPA	1.9	10	0.01	0.01
rac-TOPA	21.1	105	0.02	0.03
sn-1-TOPA	2.9	—	0.11	0.02

[a] *sn*-3-OOPA, 1-*O*-Oleoyl-2-*O*-oleoyl-*sn*-glycero-3-phosphate; *sn*-3-TTPA, 1-*S*-oleoyl-2-*S*-oleoyl-*sn*-glycero-3-phosphate; *rac*-TTPA, 1-*S*-oleoyl-2-*S*-oleoyl-*rac*-glycero-3-phosphate; *rac*-TOPA, 1-*S*-oleoyl-2-*O*-oleoyl-*rac*-glycero-3-phosphate; *sn*-1-TOPA, 3-*S*-oleoyl-2-*O*-oleoyl-*sn*-glycerol 1-phosphate.

[b] Calculated V_{max} values were determined by obtaining the effect of the *sn*-1 substrate on the observed hydrolysis velocity of the *sn*-3-substrate, a factor of 5.

study the enzyme was prepared as described by Jensen and Bensadoun.[14] Since some other properties of the enzyme have been noted to be different between the two methods of preparation, we conclude that enzyme modification may occur during preparation that alters substrate specificity.

Substitution of oxy by thioacyl esters in the substrate also had an effect on hydrolysis, when studied in the mixed micelle assay system (Table II).[17] The results show that the presence of two thio groups or the use of racemic substrate mixtures does alter the kinetics significantly. However, it is possible to establish a correction constant that can be used so that the results with thio ester substrates can be compared with those of radiolabeled substrates, at least for V_{max} values. It is clear that the affinity expressed as K_m of the hepatic lipase for the substrate with the phosphate in the 1-position either in racemic mixtures or as the pure compound is greater than that for the natural substrate with the phosphate in the 3-position (compare *sn*-3-OOPA versus *sn*-1-OOPA and *sn*-3-TTPA versus *rac*-TTPA). Also, the data show that the enzyme is not stereospecific.

Comparison of the three neutral lipids tested thus far (Table III) indicates that hepatic lipase prefers substrates without the polar head group. This is shown by the higher V_{max} and tighter binding constants. Indeed, the binding of the enzyme to both 1,3- and 1,2-dioleoylglycerols is so tight that affinity constants cannot be established using this kinetic analysis.

[17] G. L. Kucera, C. Miller, P. J. Sisson, R. W. Wilcox, Z. Wiemer, and M. Waite, *J. Biol. Chem.* **263,** 12964 (1988).

TABLE III
COMPARISON OF NEUTRAL LIPIDS AS SUBSTRATES

Substrate	V_{max} (μmol/min/mg)	K_s (mM)	K_m (mole fraction)
Monooleoylglycerol	145	11.0	0.0021
1,2-Dioleoylglycerol	162	—	—
1,3-Dioleoylglycerol	132	—	—

TABLE IV
EFFECT OF SUBSTRATE ON MICELLE SIZE

	Diameter of micelle (nm) based on mole percent dioleoylglycerol in Triton X-100	
Substrate	0.01%	0.1%
1,2-Dioleolyglycerol	7.8	10.2
1,3-Dioleoylglycerol	8.0	9.9

We have attempted to use the Triton mixed micelle assay system to determine the kinetic constants for triacylglycerol hydrolysis without success. However, we have found that the hepatic lipase prepared as described here has lower activity on triacylglycerol in gum arabic or Triton X-100 emulsions than that reported by others.[15] This may be due to differences in the enzyme preparation prepared as described here or to differences in the assay systems used by various laboratories, as pointed out by Ehnholm and Kuusi.[15]

These few examples show the utility of the mixed micelle assay system for the study of substrates. Alas, some drawbacks are to be noted. Unlike the *Naja naja* phospholipase A_2, the hepatic lipase has an appreciable affinity for Triton. For that reason the affinity constants (K_m and K_s) are only relative values and can be compared at comparable substrate to Triton ratios. Presumably, this affinity for Triton accounts for why the V_{max} calculated from the mixed micelle surface dilution study is lower than the observed specific activity of the enzyme as determined in the absence of Triton (145 versus an average of 200 μmol/min/mg without Triton). Also, the size and perhaps shape of the micelle change with a change in the substrate to Triton ratio, which potentially is another significant variable.[18] For example, Table IV shows that a 10-fold increase in the

[18] R. J. Robson and E. A. Dennis, *Biochim. Biophys. Acta* **508,** 513 (1978).

mole percentage of dioleoylglycerol increased the diameter of the micelle 20–25%, as determined by light scattering analysis (R. Wilcox, T. Thuren, and R. Hantgan, unpublished data, 1989). Despite these limitations, this kinetic analysis is extremely valuable in the determination of the substrate specificity of this and potentially other broad specificity (phospho)lipases.

[31] Human Postheparin Plasma Lipoprotein Lipase and Hepatic Triglyceride Lipase

By RICHARD L. JACKSON and LARRY R. MCLEAN

Introduction

Lipoprotein lipase (LpL; EC 3.1.1.34) and hepatic triglyceride lipase (H-TGL; triacylglycerol kinase, EC 3.1.1.3) are the major lipolytic enzymes responsible for the metabolism of lipoproteins in the circulation.[1,2] These enzymes catalyze the hydrolysis of the *sn*-1 ester bond of lipoprotein di- and triacylglycerols, phosphatidylcholines, and phosphatidylethanolamines. With lipoprotein substrates, the phospholipase A_1 activity is approximately 1% of that for triacylglycerols. The major lipoprotein substrates for LpL are chylomicrons and very low-density lipoproteins (VLDL), whereas lipoproteins of intermediate-density (IDL) and high-density (HDL), particularly the HDL_2 subfraction, are the preferred lipoprotein substrates for H-TGL (Fig. 1). These lipolytic enzymes are anchored to the plasma membrane of endothelial cells by electrostatic interactions with heparan sulfate proteoglycans. Intravenous heparin administration releases LpL and H-TGL from these sites, resulting in lipoprotein triglyceride hydrolysis, thus the term postheparin plasma lipolytic activity (PHLA). The major tissue sources of LpL are adipose tissue and muscle; H-TGL is synthesized in the periportal hepatocyte.[3]

The primary structures of human LpL and H-TGL have been deduced from the respective cDNAs.[4–7] LpL contains 448 amino acids, whereas

[1] R. L. Jackson, "The Enzymes," (P. D. Boyer, ed.), 3rd Ed., Vol. 16, p. 141. Academic Press, New York, 1983.

[2] R. H. Eckel, *N. Engl. J. Med.* **320,** 1060 (1989).

[3] A. J. M. Verhoeven and H. Jansen, *Biochim. Biophys. Acta* **1001,** 239 (1989).

[4] K. L. Wion, T. G. Kirchgessner, A. J. Lusis, M. C. Schotz, and R. M. Lawn, *Science* **235,** 1638 (1987).

[5] G. Stahnke, R. Sprengel, J. Augustin, and H. Will, *Differentiation* **35,** 45 (1987).

[6] S. Datta, C. C. Luo, W. H. Li, P. VanTuinen, D. H. Ledbetter, M. A. Brown, S. H. Chen, S. W. Liu, and L. Chan, *J. Biol. Chem.* **263,** 1107 (1988).

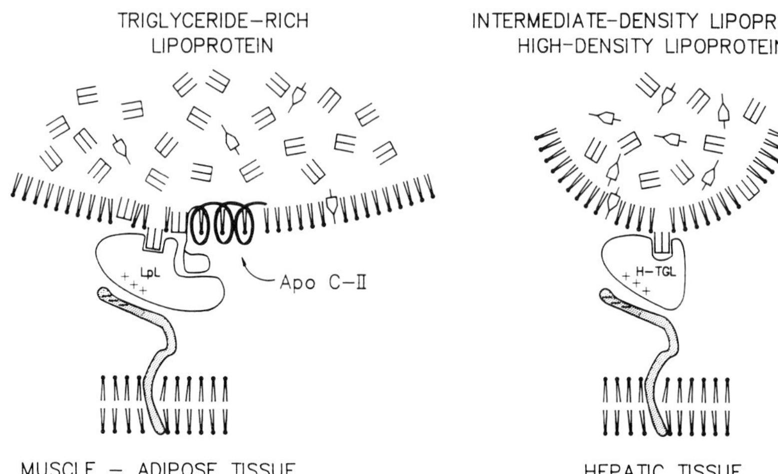

FIG. 1. Interaction of LpL and H-TGL with their lipoprotein substrates. The enzymes are shown interacting with heparan sulfate proteoglycans at the endothelial cell surface. Apolipoprotein C-II is shown residing in the monolayer surface of a triglyceride-rich lipoprotein (chylomicron or VLDL) and interacting with a specific domain of LpL. Triacylglycerol substrates reside primarily in the core of the lipoprotein, with a small percentage in the surface monolayer. The enzyme is shown interacting with a triacylglycerol substrate molecule.

H-TGL has 477 residues; the proteins are 47% homologous. The enzymes belong to a superfamily of lipases that also includes pancreatic lipase. Putative interfacial lipid-binding sites have been identified in both LpL and H-TGL. The sequence of the enzymes near the active site serine residue is homologous to that proposed for pancreatic lipase.[8]

Two major differences in the properties of LpL and H-TGL are as follows: (1) LpL requires a 77 amino acid protein cofactor, apolipoprotein C-II, for maximal activity, whereas H-TGL has no cofactor requirements, and (2) C-II, the enzymatic activity of LpL, is inhibited by high salt, whereas H-TGL is active in 1 *M* NaCl. The difference in salt sensitivity is the basis for many of the reported assays to measure PHLA. As shown in Fig. 1, the lipoprotein substrate consists of a neutral lipid core of triacylglycerols and cholesteryl esters and a surface monolayer of lipids and various apolipoproteins. The major lipid constituents of the lipoprotein

[7] G. A. Martin, S. J. Busch, G. D. Meredith, A. D. Cardin, D. T. Blankenship, S. J. T. Mao, A. E. Rechtin, C. W. Woods, M. M. Racke, M. P. Schaefer, M. C. Fitzgerald, D. M. Burke, M. A. Flanagan, and R. L. Jackson, *J. Biol. Chem.* **263,** 10907 (1988).

[8] A. Guidoni, F. Benkouka, J. De Caro, and M. Rovery, *Biochim. Biophys. Acta* **660,** 148 (1981).

monolayer are phosphatidylcholine, sphingomyelin, and unesterified cholesterol. The triacylglycerol substrate molecule is distributed between the surface and core of the lipoprotein particle. The concentration of triacylglycerol at the surface of a triglyceride-rich lipoprotein has been estimated at 2–4 mol %, relative to phospholipid.[9] Because of the uncertainty of the true substrate concentration, it is difficult to determine meaningful values for K_m and V_{max} with lipoprotein substrates.

The purpose of this chapter is to describe several of the methods used in our laboratory for the isolation and characterization of LpL and H-TGL from human postheparin plasma. In addition, we describe assays that have been used for assessing their enzymatic activities.

Isolation of Human Postheparin Plasma Lipases

Normal fasting (8–12 hr) subjects are given 100 units heparin per kg, and after 15 min blood is collected by venipuncture. After removing cells by low-speed centrifugation, the following are added (final concentrations) to the plasma: sodium azide (0.01%, v/v), EDTA (0.5 m*M*), glycerol (10%, v/v), and aprotinin (100 kallikrein inhibitory units/ml). The postheparin plasma is immediately mixed with an equal volume of ice-cold 10 m*M* potassium phosphate, pH 6.8, containing 0.45 *M* NaCl and 20% glycerol. Heparin-Sepharose CL-6B (Pharmacia) is added (1.0 ml packed gel per 10 ml of diluted plasma), and the mixture is incubated at 4° for 2 hr with gentle mixing on a rotating mixer; do not vortex the mixture.

For plasma volumes of less than 10 ml, heparin-Sepharose is poured into a column and eluted with a NaCl gradient as described below. For large volumes (>100 ml of post-heparin plasma), the gel can be conveniently collected on a sintered glass funnel and washed as previously described[10]; all eluants are added at 4°. After each addition, the gel is gently stirred and the liquid is removed by vacuum suction, care being taken to not let the gel go dry. The eluants added in order are as follows (volumes relative to gel): 10 volumes 5 m*M* potassium phosphate, pH 6.8, 0.3 *M* NaCl, and 1 m*M* EDTA; 10 volumes 5 m*M* potassium phosphate, pH 6.8, and 1 m*M* EDTA; 1 volume 5 m*M* potassium phosphate, pH 6.8, containing 0.2% Triton N-101 (Sigma, St. Louis, MO); 10 volumes 5 m*M* potassium phosphate, pH 6.8, and 1 mM EDTA; and 10 volumes of 5 mM potassium phosphate, pH 6.8, 0.3 M NaCl, and 1 mM EDTA. The advantage of washing the gel with detergent is that lipids are removed, facilitating the chromatography step.

[9] K. W. Miller and D. M. Small, *J. Biol. Chem.* **258,** 13772 (1983).

[10] R. L. Jackson, E. Ponce, L. R. McLean, and R. A. Demel, *Biochemistry* **25,** 1166 (1986).

The heparin-Sepharose is then poured into a column, and the protein is eluted with a NaCl gradient (0.3–2.5 M); the buffer is 10 m*M* Tris-HCl, pH 7.4, containing 1 m*M* EDTA and 10% glycerol. The volume of the gradient should not exceed 2 gel volumes. With these conditions, H-TGL elutes between 0.6 and 0.8 *M* NaCl and LpL between 1.2 and 1.6 *M*. As an alternative to elution with a salt gradient, the enzymes can be eluted batchwise, first with 0.8 *M* NaCl (H-TGL) and then with 2.5 *M* NaCl (LpL). Fractions should be assayed for lipolytic activity immediately and the appropriate fractions pooled; if no further purification is required, the enzymes should be stored in 50% glycerol at −20°. With these conditions of storage, LpL and H-TGL are stable for several months.

Typically, this simple one-step procedure results in purifications exceeding 10,000-fold. If purer preparations are required, H-TGL is further purified by chromatography on Phenyl–Sepharose. The heparin–Sepharose fractions described above are diluted with 5 volumes of 50 m*M* Tris-HCl, pH 8.6, 0.4 *M* NaCl, 1 m*M* EDTA and applied to a column of Phenyl–Sepharose CL-4B (Sigma). The volume of Phenyl–Sepharose is 1 ml per 100 ml of starting postheparin plasma. The column is then washed with 10 volumes of 50 m*M* Tris-HCl, pH 8.6, 0.4 *M* NaCl, and H-TGL is eluted with 25 m*M* sodium deoxycholate in 5 m*M* potassium phosphate, pH 6.8. To remove the detergent, fractions containing H-TGL activity are diluted with 10 volumes of 5 m*M* potassium phosphate, pH 6.8, and applied immediately to a column (same volume as the Phenyl–Sepharose) of heparin–Sepharose. After the sample enters the resin, the column is washed with 10 volumes of 5 m*M* potassium phosphate, pH 6.8, and then 10 volumes of 5 m*M* potassium phosphate, pH 6.8, containing 0.4 *M* NaCl. H-TGL is eluted with 50 m*M* Tris-HCl, pH 7.4, containing 1.0 *M* NaCl, 1 m*M* EDTA. The enzyme is stored in 50% (v/v) glycerol at −20°.

The lipolytic activities of postheparin plasma and purified LpL and H-TGL are routinely determined with a detergent-solubilized trioleoylglycerol emulsion. The substrate is prepared as follows: 200 mg of unlabeled trioleoylglycerol in chloroform and 200 μCi of tri[1-^{14}C]oleoylglycerol (~50 Ci/mol) are mixed and then evaporated to dryness in a 50-ml conical glass tube. Then 15 ml of 0.2% Triton N-101 (Sigma), 15 ml of 1 *M* Tris-HCl, pH 8.0, and 15 ml of 20% fatty acid-free bovine serum albumin (Sigma) are added. The mixture is emulsified by sonication (Branson 350 with microtip) for 4 min at 4°. The substrate can be aliquoted into 5-ml portions and stored at −20°. Once thawed the substrate should be vortexed and used immediately; it should not be refrozen.

For PHLA measurements, each assay is performed in triplicate tubes, namely, no-plasma blank, plus plasma, and plus plasma with 1 *M* NaCl.

All tubes (on ice) contain 75 μl substrate and 50 μl normal human plasma (as a source of apoC-II). For the plus-plasma tube (total PHLA), 10 to 50 μl postheparin plasma or the appropriate amount of purified enzyme is added. For the tubes with salt (H-TGL activity), 50 μl of 5 *M* NaCl is added. The final volume for all tubes is 250 μl. The tubes are then incubated at 37°. Typically, the amount of enzyme chosen for assay should give less than 10% hydrolysis in 10–30 min. The reaction is stopped by adding 3 ml of methanol/chloroform/heptane (140 : 125 : 100, v/v) and 1 ml of 140 m*M* potassium carbonate, 140 m*M* boric acid, pH 10.6. After vortexing (30 sec), the samples are centrifuged at 3000 rpm for 20 min (~2000 *g*) at room temperature; 1 ml of the top fraction (2.2 ml) is removed and radioactivity is determined. Enzyme activity is expressed as micromoles oleic acid released per hour per milliliter plasma or, for purified enzyme, per milligram protein. LpL activity is determined by subtracting H-TGL activity (plus 1 *M* NaCl) from total PHLA. The specific activity of pure LpL and H-TGL is approximately 30 mmol oleic acid released/hr/mg protein.

Synthetic Phospholipid Substrates

The primary advantage of synthetic phospholipid substrates in examining the phospholipase activity of LpL and H-TGL is the ease of manipulation of the physical characteristics of the substrates, which are generally well-defined. Two types of substrates are discussed: (1) phosphatidylcholines (PC) with relatively high critical micellar concentrations (CMC), which allow comparison of activities toward monomeric and micellar substrates,[11] and (2) detergent-solubilized PC in which the physical form of the substrate may be varied from bilayer structures to micelles by increasing the detergent to PC ratio.[12–14]

Short-Chain Phosphatidylcholines. Dihexanoyl-PC (di-C_6PC) and diheptanoyl-PC (di-C_7PC) have sufficiently high CMC values to allow comparison of the rates of phospholipase-catalyzed hydrolysis above and below the CMC of the substrate. The CMC is 9.3 m*M* for di-C_6PC and 1.0 m*M* for di-C_7PC. The reaction mixture contains 0.1 *M* Tris-HCl, pH 8.0, heparin (0.5 μg), and 3.5–60 m*M* PC in a total volume of 0.1 ml; the reaction is initiated by addition of enzyme. Heparin is added to stabilize the enzyme. Rates of enzyme catalysis are determined at 30° for 15–60

[11] M. Shinomiya and R. L. Jackson, *Biochem. Biophys. Res. Commun.* **113,** 811 (1983).

[12] M. Shinomiya, L. R. McLean, and R. L. Jackson, *J. Biol. Chem.* **258,** 14178 (1983).

[13] M. Shinomiya, R. L. Jackson, and L. R. McLean, *J. Biol. Chem.* **259,** 8724 (1984).

[14] M. Shinomiya and R. L. Jackson, *Biochim. Biophys. Acta* **794,** 177 (1984).

min at a concentration of enzyme which results in less than 20% hydrolysis. At the appropriate time points, the reaction mixture is applied directly to silica gel G thin-layer chromatography plates. The plates are developed in chloroform/methanol/water (65 : 35 : 5, v/v). The lipids are visualized with iodine vapor, the spots corresponding to lyso-PC and PC are scraped from the plate, and the lipid content is determined by the method of Bartlett[15] for phospholipid phosphorus.

Fluorescent Phosphatidylcholine.[16] LPL and H-TGL catalyze the hydrolysis of C_6-NBD-PC {1-acyl-2-[6-(7-nitro-2,1,3-benzoxadiazol-4-yl)amino]caproylphosphatidylcholine} in the *sn*-1 position, yielding the fluorescent product lyso-NBD-PC. Associated with catalysis of the C_6-NBD-PC is a 50-fold fluorescence enhancement with no shift in the emission maximum at 540 nm. The increase in fluorescence intensity allows continuous monitoring of enzyme catalysis. The rate of catalysis may be measured above and below the CMC (0.2 μM) of the lipid. The reaction mixture contains 10 mM Tris-HCl, pH 7.4, 0.1 M KCl, and 5×10^{-8} to 10^{-6} M C_6-NBD-PC (Avanti Biochemical Corp., Birmingham, AL) with or without apoC-II (2 μg) in a total volume of 1.0 ml. The reaction is initiated by the addition of enzyme and is monitored continuously with excitation at 470 nm and emission at 540 nm.

Phosphatidylcholine–Detergent Mixtures. Two methods are used to follow the LpL-catalyzed hydrolysis of PC in Triton: the pH-stat assay[17,18] and release of radioactive fatty acid products.[12–14] Substrates are prepared as follows: the phosphatidylcholine of interest (1.6 μmol) is mixed with PC radiolabeled in the fatty acyl chain in chloroform and dried first under N_2 and then in a lyophilizer. To the dry lipid is added 1.0 ml of 0.5 mM Bicine, pH 8.0, 0.15 M NaCl, 1 mM $CaCl_2$ (to bind released fatty acids) containing various amounts (up to 8 μmol) of Triton X-100. After vortexing, the lipid mixtures are incubated at 37° for 30 min. The assay mixtures are diluted to 0.16 mM PC. ApoC-II (0–200 μg/ml) is added for LpL assays. The substrate is split into two parts of 5 ml each. One part is used for measurement of hydrolysis by pH stat on a Radiometer automatic titrator. The temperature is maintained constant with a jacketed reaction vial and a recirculating water bath; the pH is maintained at 8.0 with 10 mM NaOH. The other part is incubated and 0.25-ml aliquots are taken at

[15] G. R. Bartlett, *J. Biol. Chem.* **234,** 466 (1959).

[16] L. A. Wittenauer, K. Shirai, R. L. Jackson, and J. D. Johnson, *Biochem. Biophys. Res. Commun.* **118,** 894 (1984).

[17] L. R. McLean, S. Best, A. Balasubramaniam, and R. L. Jackson, *Biochim. Biophys. Acta* **878,** 446 (1986).

[18] A. Balasubramaniam, A. Rechtin, L. R. McLean, and R. L. Jackson, *Biochem. Biophys. Res. Commun.* **137,** 1041 (1986).

intervals. The enzyme reactions are terminated in the aliquots by the addition of 3.25 ml of methanol/chloroform/heptane (100 : 88 : 70, v/v) and 1.0 ml of 0.14 *M* potassium borate buffer, pH 10.5. Released fatty acids are determined by liquid scintillation counting of the upper phase. Initial rates are determined for 10–15% hydrolysis in both assays.

[32] Phospholipase Activity of Milk Lipoprotein Lipase

By GUNILLA BENGTSSON-OLIVECRONA and THOMAS OLIVECRONA

Introduction

Lipoprotein lipase (LpL; EC 3.1.1.34) is the enzyme which hydrolyzes triglycerides carried in chylomicrons and very low-density lipoproteins (VLDL).[1,2] The released fatty acids are taken up by nearby cells for use in metabolic reactions or are recirculated in blood as albumin-bound free fatty acids. This reaction occurs at the endothelial surfaces of blood vessels, where the enzyme is bound by heparan sulfate proteoglycans.[3] The enzyme can be released from these sites by heparin, which has a higher affinity for the lipase than heparan sulfate has.[3]

The structure of milk LpL is known from cloning of its cDNA[4] and from direct protein sequencing.[5] The enzyme is structurally related to hepatic lipase and to pancreatic lipase.[6,7] An important corollary is that the three enzymes most likely have similar active sites.

LpL has a broad substrate specificity.[8] It has been shown to hydrolyze tri-, di-, and monoglycerides, phospholipids, and a variety of model sub-

[1] A. S. Garfinkel and M. C. Schotz, *in* "Plasma Lipoproteins" (A. M. Gotto, ed.), p. 335. Elsevier, Amsterdam, 1987.

[2] T. Olivecrona and G. Bengtsson-Olivecrona, *in* "Lipoprotein Lipase" (J. Borensztajn, ed.), p. 15. Evener, Chicago, 1987.

[3] T. Olivecrona and G. Bengtsson-Olivecrona, *in* "Heparin" (D. Lane and U. Lindahl, eds.), p. 335. Edward Arnold, London, 1989.

[4] M. Senda, K. Oka, W. V. Brown, P. K. Qasba, and Y. Furuichi, *Proc. Natl. Acad. Sci. U.S.A.* **84,** 4369 (1987).

[5] C.-Y. Yang, Z.-W. Gu, H.-X. Yang, M. F. Rohde, A. M. Gotto, Jr., and H. J. Pownall, *J. Biol. Chem.* **264,** 16822 (1989).

[6] S. Datta, C.-C. Lou, W.-H. Li, P. VanTuinen, D. H. Ledbetter, M. A. Brown, S.-H. Chen, S.-W. Liu, and L. Chan, *J. Biol. Chem.* **263,** 1107 (1988).

[7] F. S. Mickel, F. Weidenbach, B. Swarovsky, K. S. LaForge, and G. A. Scheele, *J. Biol. Chem.* **294,** 12895 (1989).

[8] T. Olivecrona and G. Bengtsson, *in* "Lipases" (B. Borgström and H. L. Brockman, eds.), p. 206. Elsevier, Amsterdam, 1984.

strates (e.g., *p*-nitrophenyl esters). In the action on glycerides it shows specificity for the *sn*-1,3 ester bonds, with a preference for the *sn*-1 ester bond.[9] It does not show any marked preference for different fatty acids, except that ester bonds involving long polyunsaturated fatty acids (e.g., arachidonic acid and eicosapentaenoic acids) appear to be cleaved more slowly than ester bonds involving saturated or monounsaturated C_{16} or C_{18} fatty acids.[10] This is true both for tri- and diglycerides and for phospholipids. Ester bonds involving phytanic acid are not cleaved, probably because the 3-methyl group causes steric hindrance.[11]

The action of LpL is stimulated by apolipoprotein C-II (apoC-II). This activation appears to occur through binding of the apolipoprotein to the enzyme, which changes its conformation and/or its orientation at the lipid–water interface. Readers are referred to recent reviews for a discussion of the mechanisms and the structural requirements of this activation.[2,8,12,13]

Another characteristic property of LpL is that it is inhibited by fatty acids.[14,15] It has been suggested that this is an important mechanism for feedback control of the enzyme by its main product. The major molecular mechanism is that the enzyme binds to fatty acid aggregates,[2,14] which results in sequestration of the enzyme away from substrate triglycerides. The inhibition is not overcome by activator protein but requires that the fatty acids be removed. *In vivo* the fatty acids are taken up by the underlying tissues and consumed in metabolic reactions. Albumin has a higher affinity for fatty acids than the lipase has, and it can be used to drive the reaction *in vitro*.

Scow and Egelrud[16] were the first to show that LpL hydrolyzes not only triglycerides but also phospholipids in lipoproteins (rat chylomicrons). They demonstrated that the action on phospholipids shows the same positional specificity as that on triglycerides, namely, the enzyme cleaves only the primary ester bond. Later studies have shown that when large triglyceride-rich lipoproteins are degraded by LpL, phospholipid and triglyceride hydrolysis initially balance each other.[17] As the particle

[9] N. Morley and A. Kuksis, *J. Biol. Chem.* **247,** 6389 (1972).

[10] B. Ekström, Å Nilsson, and B. Åkesson, *Eur. J. Clin. Invest.* **19,** 259 (1989).

[11] S. Laurell, *Biochim. Biophys. Acta* **152,** 80 (1968).

[12] R. L. Jackson, *in* "The Enzymes" (P. D. Boyer, ed.), 3rd Ed., Vol. 14, p. 141. Academic Press, New York, 1983.

[13] L. R. McLean, R. A. Demel, L. Socorro, M. Shinomiya, and R. L. Jackson, this series, Vol. 129, p. 738.

[14] G. Bengtsson and T. Olivecrona, *Eur. J. Biochem.* **106,** 557 (1980).

[15] I. Posner and J. DeSanctis, *Biochemistry* **26,** 3711 (1987).

[16] R. O. Scow and T. Egelrud, *Biochim. Biophys. Acta* **431,** 538 (1976).

[17] S. Eisenberg and T. Olivecrona, *J. Lipid Res.* **20,** 614 (1979).

shrinks, phospholipid hydrolysis can, however, no longer keep pace with triglyceride depletion, and excess surface is generated which is shed from the particle and ultimately forms high-density lipoproteins (HDL). One factor contributing to this sequence of events is that in chylomicrons, as secreted from the intestine, 10–20% of the phospholipid is phosphatidylethanolamine.[18] LpL hydrolyzes phosphatidylethanolamine more rapidly than phosphatidylcholine.

LpL acts mainly on chylomicrons and VLDL[12,19] and shows low activity against phospholipids in HDL.[19] This is a major difference from hepatic lipase, which appears to prefer HDL as a substrate over chylomicrons or VLDL.[12] Hepatic lipase and pancreatic lipase also have phospholipase activity. No rigorous comparison of the three enzymes has been made, but available evidence indicates that, for the same triglyceride activity, hepatic lipase has the highest phospholipase activity, and pancreatic lipase has the lowest. The enzymes also catalyze acyl transfer,[20] for example,

$$\text{Phospholipid} + \text{monoglyceride} \rightarrow \text{lysophospholipid} + \text{diglyceride}$$

Detailed studies of the phospholipase activity of LpL have been carried out by Stocks and Galton[21] and by Jackson and collaborators as part of their work on the mechanism by which apolipoprotein C-II stimulates LpL. These studies have been described in detail in an earlier volume in this series[13] and are not reviewed here. Our aim is to describe methods to prepare and handle milk LpL and its activator protein, as well as conditions which need to be considered in kinetic studies. We do not discuss different types of substrate presentation (monolayers, mixed emulsions, liposomes, micelles). This is covered by other chapters in this volume.

Purification

LpL has been purified from postheparin plasma and from animal tissues. A preferred starting material is bovine milk. Milk contains 1–2 mg active LpL per liter, and the enzyme can be purified by simple methods in yields approaching 50%. Milk does not contain significant amounts of other lipases or phospholipases. This has made LpL from bovine milk the most used model enzyme in LpL research.[2] In the lactating mammary gland LpL has an important role for uptake of lipids from plasma lipopro-

[18] Å. Nilsson, B. Landin, E. Jensen, and B. Åkesson, *Am. J. Physiol.* **252,** G817 (1987).

[19] S. Eisenberg, D. Schurr, H. Goldman, and T. Olivecrona, *Biochim. Biophys. Acta* **531,** 344 (1978).

[20] M. Waite and P. Sisson, *J. Biol. Chem.* **248,** 7985 (1973).

[21] J. Stocks and D. Galton, *Lipids* **15,** 186 (1980).

teins.[22] It is not clear, however, why there is so much LpL in milk. In earlier literature it was generally assumed that the lipase "leaked" into the milk because there was so much in the mammary gland. More recent studies have shown that LPL is secreted into milk as efficiently as other milk proteins are.[23] Hence, the possibility that the enzyme has a useful function in milk may have to be reconsidered. Some possibilities were discussed in a recent review.[23] Several methods to prepare LpL from milk have been described.[24–28] All are based on chromatography on heparin-Sepharose, with some different additional steps. We describe here a simple method suitable for preparation of LpL for kinetic studies.

Heparin-Agarose. Heparin-agarose is commercially available from Pharmacia LKB (Uppsala, Sweden), or it can also be prepared as described by Iverius.[29] Much material from skim milk adsorbs to the gel. To regenerate 100 ml of gel, it is suspended in 200 ml of 0.2 *M* Na_2CO_3 in 6 *M* urea, with gentle stirring for 1 hr. The slurry is filtered through a glass filter and washed with 200 ml of the same buffer and then with 1 liter of 2 *M* NaCl followed by at least 2 liters of water. The gel is stored in water with 0.5% (v/v) butanol and is stable for years.

Buffers. In earlier purification procedures[24–28] buffers of pH 7.4–8.5 have been used. We now use a 10 m*M* Bis–Tris/HCl buffer, pH 6.5 (Serva, Heidelberg, FRG). The lower pH enhances the binding of LpL to heparin.[30] Make a stock solution of 20 m*M* Bis–Tris/HCl and mix this with water or 5 *M* NaCl to obtain 10 m*M* Bis–Tris with the desired salt concentration.

Milk. There is considerable variation in the amount of LpL in milk from different cows.[23] It is therefore advantageous to test a group of cows and select a few with high LpL activity. For each preparation we use 20 liters of milk obtained by machine-milking. Collect the milk in two 10-liter plastic containers and chill in ice-water for 1–2 hr, until the temperature is below 10°. Chilling is important for three reasons: (1) LPL is not stable in milk at 37°, but loses activity with a half-time of about 4 hr[23]; (2) the physical properties of milk lipid droplets (cream) make them easier to remove by centrifugation of cold milk, because they form a mechanically

[22] R. O. Scow and S. S. Chernick, *in* "Lipoprotein Lipase" (J. Borensztajn, ed.), p. 149. Evener, Chicago, 1987.

[23] T. Olivecrona and G. Bengtsson-Olivecrona, *in* "Food Enzymology" (P. F. Fox, ed.), in press. Elsevier, Amsterdam, 1990.

[24] T. Egelrud and T. Olivecrona, *J. Biol. Chem.* **247,** 6212 (1972).

[25] P.-H. Iverius and A.-M. Östlund-Lindqvist, *J. Biol. Chem.* **251,** 7791 (1976).

[26] P.-H. Iverius and A.-M. Östlund-Lindqvist, this series, Vol. 129, p. 691.

[27] L. Soccorro, G. C. Green, and R. L. Jackson, *Prep. Biochem.* **15,** 133 (1985).

[28] I. Posner, C. S. Wang, and W. J. MacConathy, *Arch. Biochem. Biophys.* **226,** 306 (1983).

[29] P.-H. Iverius, *Biochem. J.* **124,** 677 (1971).

[30] G. Bengtsson-Olivecrona and T. Olivecrona, *Biochem. J.* **226,** 409 (1985).

stable fat cake; and (3) there is proteolytic activity in milk which degrades LpL to fragments.[31,32]

Procedure. Centrifuge the chilled milk for 15 min at 3500 *g*, 4°. A swing-out rotor makes the floating fat cake easier to handle. Holes are gently made in the fat cakes with a glass rod, and the skim milk (infranatant) is carefully decanted. The skim milk is filtered through glass wool to remove remaining fragments of the fat cake. The total volume is measured, and an aliquot is stored frozen for assay. Solid NaCl is added to the skim milk, 0.34 mol/liter. When the salt is fully dissolved, heparin-agarose is added (100 ml sedimented gel for 20 liters of milk). The mixture is gently stirred with a motor-operated stirring rod for 2 hr (4°). Avoid magnetic stirring since this may damage the agarose beads.

The heparin-agarose is collected on a large glass-filter funnel (Schott, Mainz, FRG, pore size 2) by vacuum filtration (10°). The filter is easily clogged if lipid remains in the skim milk. A sample of the filtered skim milk is stored frozen for assay. The agarose gel (still on the filter funnel) is washed with 2 liters of Bis–Tris with 0.5 *M* NaCl and then with 1 liter of Bis–Tris with 0.85 *M* NaCl to remove proteins which have lower affinity for heparin than LpL has. The gel is then suspended and poured into a short column (diameter ~80 mm). The wash is continued with Bis–Tris with 0.9 *M* NaCl until the absorbance at 280 nm of the effluent is below 0.03. This may take 1–2 hr at a flow rate of 2–3 ml/min. The salt concentration of this final wash must be adapted to the heparin-agarose preparation used, to remove the maximal amount of contaminating proteins but little or no LpL. LpL is then eluted with Bis–Tris with 1.5 *M* NaCl. Fractions of about 10 ml are collected, and peak fractions (from absorbance) are collected and pooled (usually 100–150 ml). They can be frozen and stored without significant loss of activity. An aliquot is stored frozen for assay, while the rest is dialyzed for 1 hr at 4° against 1 liter Bis–Tris with 0.5 *M* NaCl and then applied on a column of about 15 ml heparin-agarose (diameter 15 mm, flow rate 1 ml/min). The column is washed with Bis–Tris with 0.95 *M* NaCl until the absorbance is below 0.02. LpL is then eluted by a gradient of Bis–Tris with 0.95 to 2.0 *M* NaCl (125 plus 125 ml, flow rate 1 ml/min, fraction volume 5 ml). Fractions from the protein peak are assayed for LpL activity and total protein immediately or after storage.

The protein peak is usually asymmetrical. The specific activity is initially low but increases and approaches a constant value of about 600 U/mg (one unit is 1 μmol fatty acid released per minute at 25° and pH 8.5) in the later fractions. The low specific activity in the leading edge of the

[31] G. Bengtsson and T. Olivecrona, *Eur. J. Biochem.* **113,** 547 (1981).
[32] L. Socorro and R. Jackson, *J. Biol. Chem.* **260,** 6324 (1985).

peak is due to inactive forms of LpL and sometimes proteins other than LpL. With some milk samples antithrombin has been a major component preceding LpL.[2] In most samples of bovine milk the content of antithrombin is, however, low. LpL preparations always contain some proteolytic fragments.[31,32] They are held together by disulfide bonds, and the cleaved molecules migrate together with intact LpL on heparin chromatography. For kinetic experiments the entire LpL peak can usually be pooled and used. In situations when pure LpL protein is desired, one should only use the later part of the peak and consider additional steps of purification (see below). Select fractions from the specific activity. The absorption coefficient ($A^{1\%}$) for bovine LpL is 16.8 cm^{-1}.[33] The overall recovery of LpL activity is 25–40%.

Comments. In the first batch adsorption only 50–75% of the LPL activity binds to heparin-agarose. The incomplete binding is presumably due to presence in milk of substances which compete for binding. If the unbound fraction is adsorbed to a new batch of heparin-agarose a similar fraction of the LpL activity binds. In the second step binding to heparin-agarose is essentially complete. The amount of agarose gel should be kept low to reduce nonspecific binding. Variations and additions to this basic procedure have been proposed. Iverius and Östlund-Lindqvist used selective adsorption, size fractionation, and ammonium sulfate precipitation to further purify material from heparin-agarose.[25,26] Kinnunen included a nonionic detergent in the buffer to decrease nonspecific adsorption of protein to the gel.[34] Ben-Avram *et al.* used dextran sulfate-agarose instead of heparin-agarose.[35] Socorro *et al.*[27] advocate the use of phenylmethylsulfonyl fluoride (PMSF) to impede proteolytic cleavage of LpL during the preparation.

Storage. The purified lipase (protein concentration 0.2–0.6 mg/ml) can be stored at −70° or in liquid nitrogen for several months with minimal loss of lipase activity. The concentrated solution is also rather stable at 4° but is very unstable at higher temperatures, and it cannot be diluted in buffer without grave risk of loss of activity. It is advisable to include albumin (1–2 mg/ml), glycerol (20%, w/v), and/or detergents [e.g., 0.1 Triton X-100 or a combination of 1% Triton X-100 and 0.1% sodium dodecyl sulfate (SDS)]. Extensive dialysis (2–3 days) against 5 m*M* sodium deoxycholate, 10 m*M* Tris/HCl, pH 8.5, results in a rather stable solution of the enzyme.[36] Addition of 0.1 *M* sodium oleate or SDS to this further

[33] T. Olivecrona, G. Bengtsson, and J. C. Osborne, Jr., *Eur. J. Biochem.* **124,** 629 (1982).

[34] P. K. J. Kinnunen, *Med. Biol.* **55,** 187 (1977).

[35] C. M. Ben-Avram, O. Ben-Zeev, T. D. Lee, K. Haaga, J. E. Shively, J. Goers, M. E. Pedersen, J. R. Reeve, and M. C. Schotz, *Proc. Natl. Acad. Sci. U.S.A.* **83,** 4185 (1986).

[36] G. Bengtsson and T. Olivecrona, *Biochim. Biophys. Acta* **575,** 471 (1979).

improves stability, so that the enzyme becomes stable even at room temperature and at low concentrations (e.g., 0.1 μg/ml). This dramatic stabilization probably occurs through binding of anionic detergents to the enzyme.[36] The activity recovered when aliquots of LpL solutions are pipetted is often higher when the buffer contains detergents. This is due to better transfer of enzyme protein, as can be demonstrated with ^{125}I-labeled LpL. With dilute solutions of LpL without detergents a large fraction of the enzyme may stick to the pipette.

Assay of Lipase Activity. A number of convenient assays have been described which use fatty acid-labeled triglycerides as substrate. The released fatty acids are extracted and then quantitated by liquid scintillation counting. These assays have been discussed in a recent review by Nilsson-Ehle[37] and are not described here. They suffer from variation between individual preparations of the substrate emulsion (usually prepared by sonication). To calibrate such assays and to monitor the specific activity of lipase preparations over several years, we use the commercial lipid emulsion Intralipid as substrate.[24] Release of fatty acids is quantitated by titration.

Preparation of Apolipoprotein C-II

ApoC-II is usually prepared from human plasma. For this, VLDL are isolated by ultracentrifugation, delipidated, and the apolipoproteins separated by chromatography. This has been detailed by Jackson and Holdsworth in a previous volume of this series.[38] Apolipoproteins C are present on HDL as well as on VLDL. If a lipid emulsion is added to plasma, apolipoproteins C will transfer to the emulsion droplets, which can be recovered by centrifugation in a regular centrifuge. This step increases the yield of apolipoproteins C and obviates extensive ultracentrifugation. We have applied this method successfully to human, bovine,[39] and guinea pig plasma,[39a] both fresh and frozen.

Lipid Emulsion. We use Intralipid, which is a commercial emulsion of soybean triglycerides in egg yolk phosphatidylcholine used for parenteral nutrition (AB KABI Nutrition, Stockholm, Sweden). This emulsion is provided in two concentrations, 10 and 20%. Both contain the same amount of phospholipids, 12 g/liter, but differ in the amount of triglycerides, 100 or 200 g/liter. Some of the phospholipids are present as liposome-

[37] P. Nilsson-Ehle, *in* "Lipoprotein Lipase" (J. Borensztajn, ed.), p. 59. Evener, Chicago, 1987.

[38] R. L. Jackson and G. Holdsworth, this series, Vol. 128, p. 288.

[39] H. N. Astrup and G. Bengtsson, *Comp. Biochem. Physiol.* **72B,** 487 (1982).

[39a] Y. Andersson, L. Thelander, and G. Bengtsson-Olivecrona, *J. Biol. Chem.*, in press (1991).

like material. This amounts to about two-thirds of total phospholipids in the 10% preparation but only about one-third in the 20% preparation. Therefore, it is preferable to use the 20% preparation. Presumably, other lipid emulsions for parenteral nutrition can also be used.

Other reagents required are Tris-buffered saline (TBS: 0.15 *M* NaCl, 0.1 *M* Tris/HCl, pH 7.4), sucrose, and chloroform/methanol (1 : 1 and 2 : 1, v/v, cold).

Procedure. To separate the emulsion droplets from the excess phospholipid, sucrose is added to 10% (w/v) to increase the density of the Intralipid. The mixture is layered under 2 volumes of TBS and centrifuged at 4° for 90 min at 13,000 *g*. If possible, use a swing-out rotor. The fat cake is recovered and resuspended in TBS to the original volume. The washing procedure must be repeated twice to remove all excess phospholipid, but this is not necessary for semiquantitative preparative work. Then 5 volumes of bovine plasma is added per volume Intralipid, and the mixture is incubated in a water bath at 25° with gentle stirring for 30 min. Sucrose is then added to 10% (w/v), and the mixture is layered under 2 volumes of TBS and centrifuged for 90 min at 4° and 13,000 *g*. The fat cakes are resuspended in TBS and washed twice by centrifugation as above, but the final wash should be with water instead of TBS. The final fat cake should be recovered with as little liquid as possible and put into stainless steel centrifugation cups for delipidation. About 10 volume cold chloroform/methanol (2 : 1, v/v) is added. After stirring to a homogeneous sludge, another 20 volumes of chloroform/methanol (1 : 1, v/v) is added. After thorough stirring the mixture is centrifuged. The supernatant is carefully discarded, and the precipitated proteins are suspended in chloroform/methanol (2 : 1 and then 1 : 1, v/v) as described above. After another centrifugation the precipitated proteins are further washed with 10 volumes cold diethyl ether. After centrifugation the ether is carefully discarded and the protein left to dry at room temperature. Typically about 150 mg protein is obtained per liter of plasma.

The delipidated proteins are separated by chromatography. This has been described by Jackson and Holdsworth in an earlier volume in this series.[38] We dissolve the proteins in 6 *M* guanidinium chloride in 10 m*M* Tris/HCl, pH 8.2, which is added directly to the centrifuge cups. The proteins are first separated by size via gel filtration on Sephacryl S-200 (Pharmacia LKB). This results in three major peaks. The first, excluded peak is mainly apolipoprotein B. The second peak contains a variety of plasma proteins including albumin and apolipoproteins A1 and E. The third peak contains the apolipoproteins C. These are further separated by ion-exchange chromatography on DEAE–cellulose in 5 *M* urea. The main components are isoforms of apolipoprotein C-III. The apoC-II proteins are identified by their ability to activate LpL. For this any of the assay systems with radioactive triglyceride substrate[37] can be used. Activator is omitted from the assay mix, and a standard amount of LpL is added. The amount of enzyme should be chosen to give about 1% hydrolysis of the substrate in 30 min. Up to 5 μl of the column fractions can be added directly to a 200-μl assay. Larger volumes may result in inhibition of lipase by the guanidinium chloride or urea added.

Conditions for Incubations

Enzyme Stability. Enzyme stability is a major consideration for the design of incubation conditions. If one incubates LpL without heparin or lipid substrate the enzymes rapidly loses its activity, under some conditions, within minutes.[2] Binding to emulsion droplets stabilizes the enzyme. This is why it is possible to obtain linear release of fatty acids during assay of LpL activity with triglyceride emulsions. Other substrate systems, for example, liposomes, micelles, soluble substrates, may or may not stabilize the enzyme. Factors that improve the stability are as follows: heparin, presumably by forming enzyme–heparin complexes; apoC-II (or serum as a source of the apolipoprotein), probably by forming enzyme–activator complexes; and other proteins (e.g., albumin), probably by covering nonspecific binding sites on glassware.

Heparin. The diluted enzyme can usually be stabilized by heparin at concentrations of 1 μg/ml (corresponding to ~0.15 IU/ml and ~0.1 μM for most pharmaceutical preparations). The LpL–heparin complex is more stable and soluble than the enzyme alone[3] and is catalytically active. Therefore, heparin is included in most buffers used for extraction of LpL from tissue sources and for incubation of the enzyme in kinetic studies. There are reports that heparin may change the kinetic parameters for the enzyme under some conditions.[40]

pH. The pH dependency observed for LpL activity is often a bell-shaped curve with optimum around pH 8.5. The shape of the curve varies considerably, however, with the assay conditions. Closer study has revealed that the active site reaction itself increases with pH at least to pH 10.[41,42] The decrease at higher pH can be explained by decreased binding of the enzyme to the lipid–water interface and/or by decreased stability of the enzyme.[42]

Lipase activity is usually measured at alkaline pH because the higher

[40] I. Posner and J. Desanctis, *Arch. Biochem. Biophys.* **253,** 475 (1987).

[41] D. Rapp and T. Olivecrona, *Eur. J. Biochem.* **91,** 379 (1978).

[42] G. Bengtsson and T. Olivecrona, *Biochim. Biophys. Acta* **712,** 196 (1982).

rate makes the assay more sensitive. It should be pointed out, however, that the enzyme has high activity also at physiological pH (7.4). Other aspects of the reaction may be sensitive to pH, for example, lipid packing, lipid–protein interactions, and the rate at which partial glycerides/lysophospholipids are isomerized. Therefore, pH 7.4 is recommended for studies which explore physiological reactions, for example, lipoprotein interconversion.

Temperature. Already in 1968 Greten, Levy, and Fredrickson reported that LpL-catalyzed release of fatty acids was linear at room temperature but not at 37°.[43] This was because, under their assay conditions, the enzyme was not stable at 37°. This important observation has been overlooked in many subsequent studies. Unless there is a specific reason to use 37° (e.g., to study lipid transitions), it is recommended that incubations be carried at 25° or below.

Buffer Composition. LpL is sometimes defined as the salt-sensitive lipase. It is now clear, however, that the enzyme can exert full catalytic activity even in the presence of 1 *M* NaCl.[2,41,42] The effect of salt is not exerted at this level but on the physical state and the stability of the enzyme molecule. For most purposes NaCl concentrations of 0.05–0.15 *M* result in optimal activity and stability for kinetic studies.

Albumin. When triglycerides in emulsion droplets/lipoproteins are hydrolyzed by LpL, hydrophobic core molecules are converted to surface-active products which locate at the interface. In the absence of an acceptor, the fatty acids and monoglycerides form extensions of the surface film and separate bilayer structures.[44,45] Sequestration of the lipase in these structures is probably the main reason for the inhibition of lipase action.[2] In these systems it is necessary to include albumin to bind the fatty acids. The molar ratio of fatty acids to albumin should not exceed 5 at any time. Most incubation systems for triglyceride hydrolysis contain 30–60 mg/ml albumin. For most purposes it is not necessary to use fatty acid-free albumin preparations.

With liposomes and micelles the substrate molecules are surface-located, and hydrolysis does not result in overcrowding of the surface. Albumin is less important here. Its main effect will be to impede transesterification reactions. Furthermore, albumin at concentrations of a few milligrams per milliliter is useful to prevent adsorption of the lipase to glassware.

Some commercial albumin preparations contain traces of lipids and

[43] H. Greten, R. I. Levy, and D. S. Fredrickson, *Biochim. Biophys. Acta* **164,** 185 (1968).
[44] R. O. Scow and J. Blanchette-Mackie, *Prog. Lipid Res.* **24,** 197 (1985).
[45] D. P. Cistola, J. A. Hamilton, D. Jackson, and D. M. Small, *Biochemistry* **27,** 1881 (1988).

apolipoproteins.[19,46] This can have profound effects on the kinetics of LpL-catalyzed reactions and on the product profiles. We use bovine albumin, fraction V powder from Sigma Chemical Co (St. Louis, MO). For each new batch we check that the albumin, in high concentrations, does not cause inhibition of the lipase.

Activator. With triglyceride emulsions as substrate the activity of LpL is increased severalfold by apoC-II. The concentration of apoC-II needed varies, but 1 μg/ml (10^{-7} *M*) is usually enough for maximal or near-maximal activation. The relevant parameter is probably the two-dimensional concentration of apoC-II at the lipid–water interface.[47] Hence, more apoC-II is needed with higher concentrations of lipid substrate. ApoC-II tends to aggregate in solution. To prevent this we use 3 *M* guanidinium chloride in stock solutions (4.4 mg/ml, 0.5 m*M*) of the apolipoprotein. For assay purposes, whole serum or HDL can be used to provide activator.

The effect of C-II varies greatly with the substrate system. With most triglyceride emulsions the enzyme displays activity even without the activator. This basal activity is usually 5–20% but can approach 100% of optimal activity.[2,8] With phospholipid liposomes the basal activity is often low. This aspect must be checked for each substrate system studied and the optimal amount of activator determined.

Other Proteins. It has been reported that the activity of LpL is inhibited by a variety of proteins. The mechanism(s) of the inhibition has not been explored in detail. It is probably nonspecific since inhibition is seen with many different proteins.[2] In no case has a specific protein–protein interaction between the lipase and an inhibitory protein been demonstrated. The common denominator seems to be that the proteins bind to the surface of the substrate droplets.

Concluding Remarks

LpL should be a useful reagent to prepare 2-acyllysophospholipids. Because of its higher relative activity against phospholipids, hepatic lipase could be even better for this purpose. A drawback is that starting material for purification of this enzyme (usually postheparin plasma) is more difficult to obtain.

During its action on chylomicrons and VLDL, LpL hydrolyzes most of the triglycerides and some of the phospholipids. Other phospholipids are transferred to the HDL fraction. The balance between these two

[46] R. J. Deckelbaum, T. Olivecrona, and M. Fainaru, *J. Lipid Res.* **21,** 425 (1980).

[47] R. L. Jackson, S. Tajima, T. Yamamura, S. Yokoyama, and A. Yamamoto, *Biochim. Biophys. Acta* **875,** 211 (1986).

pathways may be an important determinant of plasma HDL levels. Little is known, however, of what factors govern the relative activity of LpL on phospholipids and triglycerides in emulsions and in lipoproteins. This is an interesting question for future studies.

Acknowledgments

Our studies on LpL are supported by Grant B13X-727 from the Swedish Medical Research Council.

Section IV

Phospholipase A_2

A. Phospholipase A_2
Articles 33 through 38

B. Platelet-Activating Factor Acetylhydrolase
Article 39

C. Lecithin–Cholesterol Acyltransferase
Article 40

[33] Cobra Venom Phospholipase A_2: *Naja naja naja*

By LAURE J. REYNOLDS and EDWARD A. DENNIS

Introduction

1,2-Diacyl-*sn*-glycero-3-phosphatide + $H_2O \rightarrow$ 1-acyl-*sn*-glycero-3-phosphatide + fatty acid

Phospholipase A_2 (PLA_2; EC 3.1.1.4) is a major component of most snake venoms and can be purified in large quantities from this source. Purification of the enzyme is complicated by the presence of multiple isozymes in many species. The venom from the Indian cobra (*Naja naja naja*) contains as many as 14 PLA_2 isozymes.[1,2] This chapter describes the purification of an acidic phospholipase A_2 from this source.[3] This purification is less harsh than a procedure previously published in this series[4] and yields a purer enzyme preparation.

Assay Method

Principle. Phospholipase A_2 catalyzes the hydrolysis of the fatty acid in the *sn*-2 position of phospholipids. The reaction is followed by titrating the fatty acid that is released using a pH-stat apparatus. The substrate for this reaction is egg phosphatidylcholine (PC), which is solubilized in Triton X-100 mixed micelles.

Reagents

Egg phosphatidylcholine (purified from egg yolks by the method of Singleton *et al.*[5]), 200 m*M* in $CHCl_3$
Triton X-100, 200 m*M*
$CaCl_2$, 200 m*M*
KOH, 5 m*M*

Procedure.[4,6] The standard assay mixture contains 5 m*M* egg phosphatidylcholine, 20 m*M* Triton X-100, and 10 m*M* $CaCl_2$. The appropriate volume of egg phosphatidylcholine solution is measured into a homogeni-

[1] M. K. Bhat and T. V. Gowda, *Toxicon* **27,** 861 (1989).
[2] J. Shiloah, C. Klibansky, and A. deVries, *Toxicon* **11,** 481 (1973).
[3] T. L. Hazlett and E. A. Dennis, *Toxicon* **23,** 457 (1985).
[4] R. A. Deems and E. A. Dennis, this series, Vol. 71 [81].
[5] W. S. Singleton, M. S. Gray, M. L. Brown, and J. L. White, *J. Am. Oil Chem. Soc.* **42,** 53 (1965).
[6] E. A. Dennis, *J. Lipid Res.* **14,** 152 (1973).

zation tube. The sample is dried first under a stream of nitrogen then under vacuum until all the chloroform has evaporated. The appropriate amounts of Triton X-100 and $CaCl_2$ are added, and the solution is brought to the final volume by addition of deionized water. The phospholipid is solubilized by heating to 40°–50° followed by vortexing.

The enzymatic reaction is followed using a Radiometer pH-stat apparatus (Westlake, OH) equipped with a 0.25-ml burette. Two milliliters of assay mix is brought to pH 8.0 and 40° under a stream of nitrogen. The reaction is initiated by addition of enzyme, and the fatty acids released are titrated with 5 m*M* KOH. Further details and comments on the assay procedure have been discussed previously in this series.[4]

Comments. The assay is also routinely performed with commercially available dipalmitoylphosphatidylcholine as a substrate. With this substrate, micelles are prepared in the same manner, except the solution is heated to 60° before vortexing. Assays with dipalmitoylphosphatidylcholine give comparable rates to the egg PC assays.

Purification Procedure

The purification scheme used here was first described by Hazlett and Dennis.[3] The PLA_2 is quite stable, and purification can be carried out at room temperature. Protein concentrations are determined by the method of Lowry *et al.*[7] using the correction factor for *Naja naja naja* phospholipase A_2 which was determined by Darke *et al.*[8] The Lowry assay overestimates the protein concentration for this enzyme when bovine serum albumin is used as a standard. Therefore, protein values obtained by this method must be multiplied by a factor of 0.66 to obtain the correct value. Although this correction is valid for the purified PLA_2 only and not necessarily for the crude venom, for consistency, the correction has been applied to all protein values reported in Table I.

Caution: When handling the lyophilized venom, care should be taken not to disperse or inhale the powder. Use of gloves and a mask is recommended at this stage. Non-PLA_2 column fractions which may contain toxins are treated with sodium hypochlorite or strong acid prior to disposal. After use, column packing materials, which may retain toxins, are usually disposed of and not reused.

Step 1. Deionized water (500 ml) is added to a flask containing approxi-

[7] O. H. Lowry, N. J. Rosenbrough, A. L. Farr, and R. J. Randall, *J. Biol. Chem.* **193,** 265 (1951).

[8] P. L. Darke, A. A. Jarvis, R. A. Deems, and E. A. Dennis, *Biochim. Biophys. Acta* **626,** 154 (1980).

TABLE I
PURIFICATION OF PHOSPHOLIPASE A_2 FROM COBRA VENOM[a]

Step	Total protein (mg)	Total activity (10^3 units)	Specific activity (units/mg)	Purification (-fold)
1. Soluble venom	5285	872	165	1.0
2. Affi-Gel Blue	374	552	1480	9.0
3. DE-11	220	416	1890	11.5
4. SP-Sephadex	163	356	2180	13.2

[a] Reprinted from Hazlett and Dennis[3] with permission.

mately 10 g (dry weight) of lyophilized venom [*Naja naja naja* (Pakistan), obtained from Miami Serpentarium Laboratories, Punta Gorda, FL]. Most of the venom is dissolved by swirling the mixture gently for several minutes. Much of the venom material remains insoluble and is removed by centrifugation of the mixture for 15 min at approximately 8000 *g*. The supernatant is diluted 1 : 1 with 50 m*M* ammonium acetate (pH 6.0). The pH is checked with pH paper and adjusted to pH 6.0 with acetic acid if necessary.

Step 2. A 200-ml volume of Affi-Gel Blue [Bio-Rad (Richmond, CA), 100–200 mesh, 75–150 μm] is suspended in approximately 100 ml of 20 m*M* ammonium carbonate (pH 10.5). This buffer is prepared by bringing a solution of ammonium carbonate to pH 10.5 with concentrated ammonium hydroxide, then diluting to the final volume with deionized water. The suspension is poured into a 5 cm diameter column and washed with the same buffer (~800 ml) at a rate of 2.5–3 ml/min to remove any free blue dye. The height of the packed column is 11.5 cm. The column is equilibrated with 50 m*M* ammonium acetate (pH 6.0) until the pH returns to about 6.0.

After equilibration of the column, the sample is loaded. The column effluent is monitored at 280 nm. A large protein peak elutes in the flow-through fraction. When loading is complete, the column is washed with 50 m*M* ammonium acetate (pH 6.0) until the absorbance returns to baseline. At this point, the second buffer, 50 m*M* ammonium bicarbonate (pH 8.0), is started. A small peak, which is sometimes barely observable, starts to elute after 250–300 ml of this buffer. When the absorbance again returns to baseline, the final buffer, 20 m*M* ammonium carbonate (pH 10.5), is started. This buffer elutes a large protein peak which contains 63% of the initial activity. The protein peak is pooled, neutralized to pH 7.5, and lyophilized, leaving a salt-free protein powder. We typically encounter two problems with the lyophilization. First, the sample usually foams when initially placed under vacuum and should be monitored carefully at this stage. This problem is minimized by having the sample well frozen before starting lyophilization. Second, during later stages of lyophilization the sample often thaws and must be diluted with water and then refrozen.

Step 3. The enzyme is next passed through a DEAE-cellulose column (Whatman DE-11 or DE-23, 4.2 × 30 cm). The resin is prepared as recommended by the manufacturer. Approximately 600 ml (settled volume) of prepared resin is stirred into 1 liter of 50 m*M* sodium phosphate (pH 7.5) containing 0.3 *M* NaCl. The pH of the slurry is readjusted to 7.5 with HCl. The excess buffer is decanted, and the column is packed with the remaining slurry. The column is equilibrated with 5 m*M* sodium phosphate (pH 7.5) at a rate of 2.8 ml/min. The lyophilized protein is dissolved in 200 ml of the same buffer and loaded onto the column. When loading is complete the column is washed with 5 m*M* sodium phosphate (pH 7.5). After the absorbance returns to baseline, a second buffer is started which is composed of 0.1 *M* NaCl in 5 m*M* sodium phosphate (pH 7.5). When the absorbance returns to the baseline, the third buffer (0.3 *M* NaCl in 5 m*M* sodium phosphate, pH 7.5) is started. The majority of the loaded protein is eluted by this high-salt buffer. The peak is pooled and dialyzed in Spectrapor 1 tubing (Spectrum Medical Industries, Los Angeles, CA) (6000–8000 MW cutoff) against 10 m*M* sodium phosphate (pH 6.0).

Step 4. SP-Sephadex C-25 (25 g) is swollen in 10 m*M* sodium phosphate (pH 6.0) and poured into a 2.4 cm diameter column to a final bed height of 35 cm. The column is equilibrated in the same buffer, at a flow rate of ml/min, before loading the dialyzed protein. The phospholipase A_2 passes through the column while a minor contaminant is retained. After loading, the column is washed with 10 m*M* sodium phosphate (pH 6.0) until baseline absorbance is again reached. The flow-through fractions that contain PLA_2 are pooled, separated into 40-ml aliquots, and stored at $-20°$. This protein preparation is stable for several years.

Properties

Purity. This procedure yields a PLA_2 preparation that displays a single protein band on analytical isoelectric focusing, sodium dodecyl sulfate–polyacrylamide gel electrophoresis (SDS–PAGE), and Ouchterlony double-diffusion precipitation tests.[3] Phospholipase A_2 purified by the previous method[4] contains a minor contaminant, not visible on SDS–polyacrylamide gels, which is removed by the third column in this procedure.

Stability. Cobra venom phospholipase A_2 contains seven disulfide bonds, which make the enzyme extremely stable. The enzyme retains activity after incubation in the absence of buffers for 12 hr at 60°.[9] Circular dichroism spectra show that the enzyme begins to denature at 80°, although this change is reversed upon cooling.[9] In fact, the enzyme can be heated at 100° for 10 min at pH 3–4 with only 5% loss in activity.[4] Cobra venom phospholipase A_2 is also fairly stable to chaotropic salts. The enzyme retains 60% of its activity when assayed in the presence of 6 *M* guanidine hydrochloride.[4] For complete denaturation, the protein must be subjected to 8 *M* guanidine hydrochloride at 40° in the presence of 10 m*M* dithiothreitol and 10 m*M* EDTA.[9] The enzyme is stable at pH 9.5[10] but is unstable at pH 10.5.[9]

Primary Structure. Cobra venom phospholipase A_2 is classified as a type 1 phospholipase based on its disulfide bond pattern.[9,11] It is composed of a single polypeptide chain of 119 residues. The amino acid sequence of the enzyme has been determined.[9] The molecular weight, calculated from the sequence, is 13,348. Analysis of the circular dichroism spectrum of the native protein indicates that it is composed of 42–50% α helix.[9]

Physical Properties. The extinction coefficient of the enzyme, $E_{278}^{0.1\%}$, is 2.2.[8] The protein is acidic, with an isoelectric point of 5.1.[10] The enzyme exhibits a broad activity optimum between pH 7 and 9.[10] Below 0.05 mg/ml cobra venom phospholipase A_2 exists as a monomer.[12] Above this concentration the enzyme undergoes a concentration-dependent aggregation and exists as a dimer or higher-order aggregate.

Cofactor. Cobra venom phospholipase A_2 displays an absolute requirement for the presence of Ca^{2+} for catalysis. The kinetically determined K_m for Ca^{2+} is 1 m*M*[13] whereas the K_D for Ca^{2+} has been estimated at 0.15 m*M*.[10]

Specificity. This phospholipase catalyzes the hydrolysis of the fatty acid ester bond at the *sn*-2 position of L-phospholipids. The enzyme will slowly catalyze the hydrolysis of monomerically dispersed phospholipids, but enzyme activity is much greater when hydrolyzing lipids that are aggregated at a lipid–water interface.[14] The enzyme is sensitive to the nature of the interface and displays greater activity toward detergent-mixed micelles than to phospholipid vesicles. The activity of the enzyme in mixed micelles is also sensitive to the phospholipid–detergent ratio. The highest activity is seen at high mole fractions of phospholipid. Increas-

[9] F. F. Davidson and E. A. Dennis, *Biochim. Biophys. Acta* **1037,** 7 (1990).

[10] M. F. Roberts, R. A. Deems, and E. A. Dennis, *J. Biol. Chem.* **252,** 6011 (1977).

[11] R. L. Heinrikson, E. T. Kreuger, and P. S. Keim, *J. Biol. Chem.* **252,** 4913 (1977).

[12] R. A. Deems and E. A. Dennis, *J. Biol. Chem.* **250,** 9008 (1975).

[13] E. A. Dennis, *Arch. Biochem. Biophys.* **158,** 485 (1973).

[14] M. F. Roberts, A.-B. Otnaess, C. R. Kensil, and E. A. Dennis, *J. Biol. Chem.* **253,** 1252 (1978).

ing the amount of detergent in the micelle results in lower enzymatic activity due to surface dilution of the phospholipid substrate.[13]

When a single type of phospholipid is present in an assay, the enzyme shows a preference for phosphatidylcholine over phosphatidylethanolamine over phosphatidylserine.[14] In addition to this head group specificity, the enzyme is sensitive to the chain length of the fatty acid. The enzymatic rate increases as the chain length decreases from 16 carbons (palmitate) to 8 carbons.[14] This effect could represent the specificity of the enzyme. However, it may be due to slight changes in the lipid interface caused by the presence of the different fatty acids or to a decrease in product inhibition by the shorter fatty acids.[15]

The enzyme can hydrolyze thio ester bonds in the *sn*-2 position of phospholipids but will not hydrolyze amide bonds such as that in sphingomyelin.[14] The ability to hydrolyze thio ester bonds has been utilized to develop a thiol-based assay for the enzyme.[16]

Kinetic Properties. The rate of phospholipase A_2 hydrolysis depends on both the bulk concentration of phospholipid and on the concentration of phospholipid in the interface (mole ratio). Kinetic analysis of phospholipase A_2 activity is very complex and must vary both of these parameters.[17] When the mole ratio of Triton X-100 to egg phosphatidylcholine is held constant at 2 : 1, the apparent K_m is estimated to be between 2 and 5 mM with an apparent V_{max} of 2000 units/mg.[6] The rate of egg phosphatidylethanolamine hydrolysis is 15-fold slower than the rate of PC hydrolysis.[15,18] However, the rate of phosphatidylethanolamine hydrolysis increases up to 20-fold with the addition of choline-containing activators such as dodecylphosphorylcholine, sphingomyelin, or phosphatidylcholine.[15,18]

Inhibitors. Cobra venom phospholipase A_2 is inhibited by Ba^{2+} and Sr^{2+}, which are competitive inhibitors of the Ca^{2+} cofactor. The K_D for both these metals is about 0.6 mM.[10] The enzyme is subject to product inhibition by fatty acids.[15] This inhibition can be minimized by utilizing a shorter chain phospholipid, the fatty acid products of which can diffuse out of the micelle, or by adding bovine serum albumin to the assay to extract fatty acids from the micelles. The enzyme is inhibited by aromatic dyes such as Cibacron Blue, which binds the enzyme reversibly with a K_D about 2 μM and K_i of 3.5 μM.[19] Phospholipids which contain an amide,[20]

[15] A. Plückthun and E. A. Dennis, *J. Biol. Chem.* **260,** 11099 (1985).

[16] L. Yu and E. A. Dennis, this volume [5].

[17] R. A. Deems, B. R. Eaton, and E. A. Dennis, *J. Biol. Chem.* **250,** 9013 (1975).

[18] M. Adamich, M. F. Roberts, and E. A. Dennis, *Biochemistry* **18,** 3308 (1979).

[19] R. E. Barden, P. L. Darke, R. A. Deems, and E. A. Dennis, *Biochemistry* **19,** 1621 (1980).

[20] L. Yu, R. A. Deems, J. Hajdu, and E. A. Dennis, *J. Biol. Chem.* **265,** 2657 (1990).

a fluoroketone,[21] or phosphonate[22] instead of an ester bond at the *sn*-2 position are also reversible inhibitors. Cobra venom phospholipase A_2 is completely inactivated by *p*-bromophenacyl bromide, which covalently modifies the active site histidine.[23] The sponge metabolite manoalide[24] and its synthetic analog manoalogue[25] cause a partial irreversible inactivation of the enzyme by modification of lysine residues.

Acknowledgments

Support for this work was provided by the National Institutes of Health (GM-20,501) and the National Science Foundation (DMB 89-17392).

[21] W. Yuan, R. J. Berman, and M. H. Gelb, *J. Am. Chem. Soc.* **109,** 8071 (1987).
[22] W. Yuan and M. H. Gelb, *J. Am. Chem. Soc.* **110,** 2665 (1988).
[23] M. F. Roberts, R. A. Deems, T. C. Mincey, and E. A. Dennis, *J. Biol. Chem.* **252,** 2405 (1977).
[24] D. Lombardo and E. A. Dennis, *J. Biol. Chem.* **260,** 7234 (1985).
[25] L. J. Reynolds, B. P. Morgan, G. A. Hite, E. D. Mihelich, and E. A. Dennis, *J. Am. Chem. Soc.* **110,** 5172 (1988).

[34] Phospholipase A_2 from Rat Liver Mitochondria

By H. van den Bosch, J. G. N. de Jong, and A. J. Aarsman

Introduction

Phospholipases A_2 (EC 3.1.1.4) are abundantly present in pancreatic juice and snake venoms, and detailed insight into the structure and mechanism of these extracellular enzymes is available.[1] Intracellular forms of phospholipase A_2 have been reported for a great variety of cell types, with the enzymes occurring in both soluble form and in association with the membranes of different subcellular organelles.[2] A long standing question concerns the structural relationships between the enzymes that are present in different compartments of a given cell, between the intracellular phospholipases A_2 from different cells and tissues, and between these cellular phospholipases A_2 and the extracellular ones. One approach to these problems is to purify cellular phospholipases A_2 for structural and enzymological characterization. This chapter deals with various aspects of the

[1] H. M. Verheij, A. J. Slotboom, and G. H. de Haas, *Rev. Physiol. Biochem. Pharmacol.* **91,** 91 (1981).
[2] H. van den Bosch, *Biochim. Biophys. Acta* **604,** 191 (1980).

purification and properties of rat liver mitochondrial phospholipase A_2 as an example of a low abundancy cellular phospholipase A_2. Rat liver mitochondria are one of the first subcellular organelles reported to contain phospholipase A_2,[3] and partial purifications of the enzyme have been described.[4,5]

Assay Method

Principle. Enzyme activity can be assayed by measuring the release of a radioactive fatty acid from the *sn*-2 position of radiolabeled phospholipids. The released fatty acid can be conveniently extracted by a modified Dole extraction procedure,[6] thus avoiding cumbersome thin-layer chromatographic (TLC) techniques. However, depending on the nature and unsaturation of the phospholipid substrate, the heptane layer of the Dole extraction procedure may contain up to 25% of the substrate. This can easily be removed by chromatographing an aliquot of the heptane layer over a minicolumn of 300 mg silica gel in a cotton wool-plugged Pasteur pipette.

Numerous assay methods for cellular phospholipases A_2 have been described and reviewed.[6,7] Many of these, including the method described here, cannot be applied indiscriminately for the measurement of phospholipase A_2 activity in crude (sub)cellular fractions because the consecutive action of phospholipase A_1 and lysophospholipase may also release the fatty acid from the *sn*-2 position of phospholipids. However, rat liver mitochondria do not contain phospholipase A_1 or lysophospholipases to any appreciable extent.[8,9] Although different radiolabeled phospholipids are in use for measurement of cellular phospholipases A_2, we prefer phosphatidylethanolamine. This substrate, especially in the presence of Ca^{2+}, easily associates with biomembranes and can reach the membrane-associated phospholipases, presumably through lateral diffusion.[10] These properties make this phospholipid the substrate of choice to compare the activity

[3] P. Bjørnstad, *J. Lipid Res.* **7,** 612 (1960).

[4] M. Waite and P. Sisson, *Biochemistry* **10,** 2377 (1971).

[5] Y. Natori, K. Karasawa, H. Arai, Y. Tamori-Natori, and S. Nojima, *J. Biochem.* (*Tokyo*) **93,** 631 (1983).

[6] H. van den Bosch and A. J. Aarsman, *Agents Actions* **9,** 382 (1979).

[7] L. J. Reynolds, W. N. Washburn, R. A. Deems, and E. A. Dennis, this volume [1].

[8] G. L. Scherphof, M. Waite, and L. L. M. van Deenen, *Biochim. Biophys. Acta* **125,** 406 (1966).

[9] J. M. de Winter, G. M. Vianen, and H. van den Bosch, *Biochim. Biophys. Acta* **712,** 332 (1982).

[10] H. B. M. Lenting, F. W. Neys, and H. van den Bosch, *Biochim. Biophys. Acta* **917,** 178 (1987).

of membrane-associated phospholipases A_2 before and after solubilization, while avoiding the use of detergents.

Reagents

1-Acyl-2-[1-^{14}C]linoleoyl-*sn*-glycero-3-phosphoethanolamine, 1 m*M*, sonicate in water[11]

Tris-HCl buffer, 0.5 *M*, pH 8.5

$CaCl_2$, 0.1 *M*

Dole's extraction medium[12] (2-propanol, *n*-heptane, 1 *N* H_2SO_4, 40 : 10 : 1, v/v)

n-Heptane

Silica gel (Silic AR cc-4, Mallinckrodt, St. Louis, MO)[13]

Procedure. The standard incubation mixture contains 0.2 m*M* substrate, 10 m*M* $CaCl_2$, and varying amounts of enzyme preparations in 0.5 ml of 0.1 *M* Tris-HCl (pH 8.5). After a 30-min incubation at 37°, the reaction is stopped by addition of 2.5 ml of Dole's extraction medium, followed by addition of 1.5 ml each of *n*-heptane and distilled water with vortexing for 15 sec after each addition. An aliquot of 1.0 ml of the upper heptane phase (total volume 2.0 ml) is chromatographed over a 300-mg silica gel minicolumn in a Pasteur pipette. The heptane eluate and a 1.0-ml diethyl ether wash are collected directly in scintillation vials containing 3.7 ml emulsifier (Packard) scintillation fluid for radioactivity measurements. Blanks containing no enzyme are included in each series to correct for nonenzymatic hydrolysis.

Enzyme Purification

Isolation of Mitochondria. Crude mitochondrial fractions are prepared from 10 or 20% (w/v) rat liver homogenates in 0.25 *M* sucrose, 1 m*M* EDTA, 1 m*M* phenylmethanesulfonyl fluoride, and 20 m*M* Tris-HCl (pH 7.4). The homogenate is first spun for 10 min at 1000 *g* and then for 10 min at 8000 *g*. The final pellet is resuspended in 50 m*M* Tris-HCl (pH 8.0) at

[11] This substrate can be prepared biosynthetically from [1-^{14}C]linoleic acid and 1-acyllysophosphatidylethanolamine using rat liver microsomes [H. van den Bosch, A. J. Aarsman, and L. L. M. van Deenen, *Biochim. Biophys. Acta* **348,** 197 (1974)]. Alternatively, commercially available preparations can be used. Routinely, the labeled substrates are diluted with either rat liver or egg yolk phosphatidylethanolamine to give a final specific activity of 300 dpm/nmol. Stock sonicates can be stored frozen and can be reused after thawing and brief resonication.

[12] V. P. Dole, *J. Clin. Invest.* **35,** 150 (1956).

[13] This type of silica gel gives optimal results with respect to fatty acid recovery and removal of excess substrate from the heptane layer.

a protein concentration of approximately 20 mg/ml. All these and further operations are carried out at 0–4°.

Solubilization. The resuspended mitochondria are poured into 10 times their volume of ice-cold acetone containing 0.125 ml of concentrated ammonia per liter acetone.[4] After stirring for 5 min, the mixture is centrifuged for 5 min at 16,000 *g*. The supernatant is decanted, and the pellet is briefly dried under a stream of nitrogen and resuspended in extraction buffer consisting of 20 m*M* Tris-HCl buffer (pH 7.4) containing 2 m*M* EDTA, 1 *M* KCl, and 10% (v/v) glycerol,[14] using a volume equivalent to one-half that of the original mitochondrial suspension. The suspension is stirred for 1 hr and then centrifuged for 20 min at 25,000 *g* to yield a clear extract. This can be stored frozen without any appreciable loss of activity. Removal of the salt by dialysis leads to the formation of protein precipitates that contain all phospholipase A_2 activity. This excludes ion-exchange chromatography at this stage, and gel filtration in buffers containing 1 *M* KCl is used as the next purification step.[15] The drawback of the low capacity of gel filtration as a first purification step is compensated by an unusually high purification factor due to the low molecular weight of the enzyme.

Gel Filtration. The extract (24 ml, 215 mg protein, 1600 mU) is filtered over an AcA 54 column (3.5 × 145 cm), equilibrated and eluted with extraction buffer, at a flow rate of 40 ml/hr. Fractions of 20 ml are collected (Fig. 1). Enzyme after AcA 54 chromatography is stable when stored at 4° (less than 25% activity loss in 6 months) and forms a convenient starting material for further purification.

Hydroxyapatite Chromatography. Pooled fractions from the gel-filtration step (240 ml) are concentrated 2-fold in a dialysis bag put in 300 g Aquacide F (Calbiochem, La Jolla, CA) and dialyzed overnight against 10 volumes of 3 m*M* potassium phosphate (pH 7.4), 0.2 *M* KCl, and 10% (v/v) glycerol. The dialyzate is pumped onto a hydroxyapatite column (1.8 × 4.5 cm) at a flow rate of 15 ml/hr. The column is rinsed with 1 bed volume of buffer and then eluted with a gradient consisting of 4 bed volumes each of 3 and 300 m*M* potassium phosphate (pH 7.4) in 0.2 *M* KCl and 10% (v/v) glycerol. Fractions of 4 ml are collected. Phospholipase A_2 activity elutes at 100 m*M* phosphate, somewhat retarded with respect to the main protein peak.

Matrex Gel Blue A Chromatography. Pooled fractions from the previous step (48 ml) are dialyzed against 50 volumes of 20 m*M* Tris-HCl (pH 7.4), 0.2 *M* KCl, and 10% (v/v) glycerol. The dialyzate is pumped onto a

[14] Initially 10 m*M* 2-mercaptoethanol was included in this buffer, but this can be omitted.

[15] For unknown reasons AcA 54 gives considerably higher recoveries than Sephadex G-75. When the AcA column is eluted with a buffer containing 0.15 rather than 1 *M* KCl, phospholipase A_2 activity elutes in the void volume protein peak without much increase in specific activity.

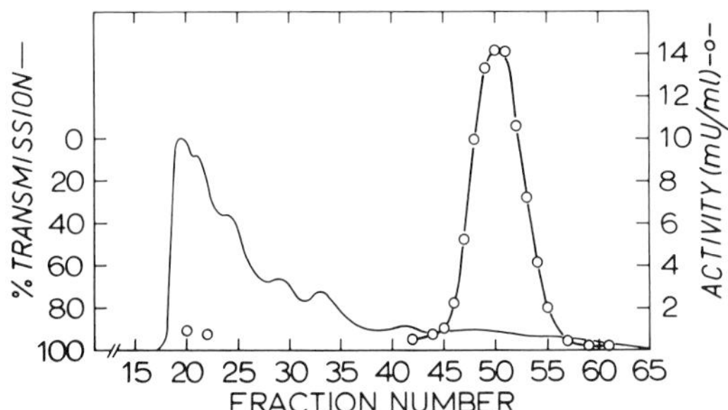

FIG. 1. AcA 54 chromatography of phospholipase A_2 solubilized from rat liver mitochondria. For details, see text. The eluent containing phospholipase A_2 represents a stable enzyme preparation that can be used for further purification by either hydroxyapatite and Matrex gel Blue A chromatography, ligand affinity chromatography, or immunoaffinity chromatography. [Reproduced with permission from J. G. N. de Jong, H. Amesz, A. J. Aarsman, H. B. M. Lenting, and H. van den Bosch, *Eur. J. Biochem.* **164,** 129 (1987).]

Matrex gel Blue A column (0.8 × 7 cm) at a flow rate of 10 ml/hr. The column is previously treated with 0.5 *N* NaOH/8 *M* urea according to the instructions of the supplier (Amicon, Danvers, MA) and equilibrated with the above dialysis buffer. After sample application the column is rinsed with 2 bed volumes of the same buffer and then eluted with a linear gradient consisting of 7 bed volumes each of 0.2 and 1.0 *M* KCl in 20 m*M* Tris-HCl (pH 7.4) and 10% (v/v) glycerol. Fractions of 2.3 ml are collected. Phospholipase A_2 elutes at 0.75 *M* KCl. Table I summarizes the purification procedure.

TABLE I
PURIFICATION OF PHOSPHOLIPASE A_2 FROM RAT LIVER MITOCHONDRIA

Step	Total protein[a] (mg)	Total activity[b] (mU)	Specific activity (mU/mg)	Recovery (%)	Purification (-fold)
Mitochondria	1343	2283	1.7	100	—
Extract	216	1598	7.4	71	4
AcA 54	4.5	1440	320	63	188
Hydroxyapatite	0.47	1233	2630	54	1550
Matrex gel Blue A	0.095	822	8650	36	5090

[a] Protein is measured according to O. H. Lowry, N. J. Rosebrough, A. L. Farr, and R. J. Randall, *J. Biol. Chem.* **193,** 265 (1951). Dilute samples, namely, those from AcA 54 fraction on, were first precipitated with trichloroacetic acid (7%, w/v, final) in the presence of 0.02% (w/v) sodium deoxycholate.

[b] One milliunit (mU) represents the release of 1 nmol fatty acid/min.

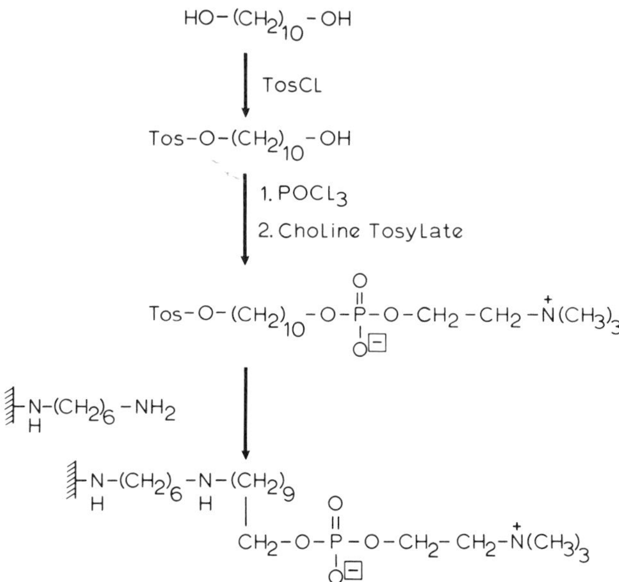

FIG. 2. Schematic representation of the synthesis of a ligand affinity adsorbent for phospholipase A_2. For details, see text. [Reproduced with permission from A. J. Aarsman, F. Neys, and H. van den Bosch, *Biochim. Biophys. Acta* **792,** 363 (1984).]

Affinity Purification of Phospholipase A_2

Alternative, single-step, purification procedures for rat liver mitochondrial phospholipase A_2 starting from the stable preparation obtained after AcA 54 gel filtration rely on affinity chromatography and can be distinguished in ligand affinity chromatography and immunoaffinity chromatography. The leading thought behind the development of the ligand affinity adsorbent represented in Fig. 2 has been that phospholipase A_2 is able to hydrolyze glycolphosphatidylcholines[16] and to bind *n*-alkylphosphocholines in a Ca^{2+}-dependent manner.[17] Thus, the enzyme binds to immobilized decane-1-*O*-phosphocholine in the presence of Ca^{2+} and can be eluted in the presence of EDTA.[18]

Synthesis of Ligand Affinity Adsorbent. The synthesis (Fig. 2) starts with the preparation of the monotosylated derivative of 1,10-decanediol.

[16] G. H. de Haas, N. M. Postema, W. Nieuwenhuizen, and L. L. M. van Deenen, *Biochim. Biophys. Acta* **159,** 103 (1968).

[17] M. C. E. van Dam-Mieras, A. J. Slotboom, W. A. Pieterson, and G. H. de Haas, *Biochemistry* **14,** 5387 (1975).

[18] A. J. Aarsman, F. Neys, and H. van den Bosch, *Biochim. Biophys. Acta* **792,** 363 (1984).

This is synthesized by adding *p*-toluenesulfonyl chloride (19.1 g, 100 mmol) in 20 ml pyridine dropwise to a solution of 17.4 g (100 mmol) of 1,10-decanediol in 40 ml pyridine at 0°. After 1 hr at 0° and 3 hr at room temperature, the reaction is stopped by addition of ice-cold 1 *N* HCl, and the products are extracted with 400 ml of diethyl ether. The ether layer is washed successively with 300 ml each of a 5% (w/v) $NaHCO_3$ solution and water. After drying on Na_2SO_4 the ether layer is concentrated *in vacuo,* and the product is purified on a column (4 × 50 cm) of kieselgel 60 (Merck, Darmstadt, FRG) using petroleum ether/diethyl ether (3:7, v/v) as eluent. The monotosyl derivative of 1,10-decanediol is obtained in a yield of 70%. R_f values on TLC [Schleicher and Schüll (Dassel, FRG) F 1500 (LS 254) in $CHCl_3$] are 0.07, 0.27, and 0.65 for 1,10-decanediol, monotosyl, and ditosyl derivatives, respectively. The tosylated derivatives can be detected under 256-nm UV light.

The monotosyl derivative (1.71 g, 5.2 mmol) together with 12.5 mmol pyridine in 30 ml dry dichloromethane is added dropwise to a stirred solution of 843 mg (5.5 mmol) of $POCl_3$ in 20 ml of dry dichloromethane at 0°. Stirring is continued for 1 hr at room temperature after which TLC analysis shows the complete disappearance of the monotosyl derivative of decanediol. Pyridine (1 ml, 12.5 mmol) and choline toluene sulfonate (1.9 g, 6.9 mmol) are added, and stirring is continued for 16 hr.[19] After addition of 1 ml each of pyridine and water and stirring for 2 hr, water (50 ml) and methanol (100 ml) are added, and the pH is adjusted to 3 with 1 *N* HCl. The product is extracted twice with 50 ml $CHCl_3$, evaporated to dryness, and percolated over a mixed ion-exchange column (Amberlite IRA-45 and IRC-50) in chloroform/methanol/water (5:4:1, v/v). The product is further purified on kieselgel 60 by elution with chloroform/methanol mixtures. The purified compound, namely, 10-*O*-*p*-toluenesulfonyldecane-1-*O*-phosphocholine, shows an R_f value of 0.33 on TLC (see above) with chloroform/methanol/water (60:40:10, v/v) as developing solvent and is UV-, phosphor-, and choline-positive.

The ligand is coupled to AH Sepharose 4B according to the procedure of Nilsson and Mosbach.[20] Ligand (245 mg, 0.5 mmol) is added to 2 g freeze-dried AH Sepharose 4B, swollen and washed as recommended by the supplier (Pharmacia, Uppsala, Sweden), in 10 ml of 0.2 *M* $NaHCO_3$ (pH 10.7) and allowed to react for 20 hr at 37° with gentle agitation. The gel is washed successively with 150-ml portions of water, 0.1 *M* sodium acetate buffer (pH 4.0), water, 0.5 *M* NaCl, water, 0.2 *M* sodium bicarbon-

[19] H. Brockerhoff and N. K. W. Ayengar, *Lipids* **14,** 88 (1979).
[20] K. Nilsson and K. Mosbach, *Biochem. Biophys. Res. Commun.* **102,** 449 (1981).

ate, and water. Phosphor determination[21] on destructed gels shows the coupling of 1.0–3.2 μmol ligand/ml gel for various preparations.

Ligand Affinity Chromatography. The ligand affinity adsorbent is packed into a column (bed 0.5 × 3.5 cm) and equilibrated with 50 m*M* Tris-HCl (pH 7.4) containing 0.2 *M* KCl, 10% (v/v) glycerol, 10 m*M* $CaCl_2$, and 1 m*M* EDTA. Pooled AcA 54 fractions[22] (see Fig. 1) are dialyzed against this equilibration buffer, and the dialyzate, containing up to 1 mg protein, is loaded on the column at a flow rate of 8 ml/hr. The column is rinsed with equilibration buffer until the eluate is free of material absorbing at 280 nm (approximately 10 bed volumes) and then eluted with the above buffer except that 0.2 *M* KCl and 10 m*M* $CaCl_2$ are replaced by 0.5 *M* KCl and 50 m*M* EDTA, respectively. Fractions of 0.67 ml are collected at a flow rate of 8 ml/hr throughout. Phospholipase A_2 starts to elute after 2 to 5 bed volumes and is recovered in 80 to 100% yield in 10 to 15 bed volumes.

Immunoaffinity Chromatography. Starting from pooled AcA 54 fractions the enzyme can be obtained in homogeneous and more concentrated form by immunoaffinity chromatography. Since this method depends on the availability of purified monoclonal antibodies to the enzyme[23] it is not described here in detail. A full account of this procedure has been published recently.[24]

Properties

Stability. In the presence of 10% (v/v) glycerol the enzyme is stable until AcA 54 purification. Thereafter, the enzyme preparations steadily lose activity on storage, presumably due to adsorption of the enzyme to glass and other surfaces at these low protein concentrations. Sodium dodecyl sulfate–polyacrylamide gel electrophoresis (SDS–PAGE) of stored or dialyzed enzyme solutions clearly shows a decrease of enzyme protein over time in solution. Rinsing of the tube or dialysis bag with SDS solutions shows the presence of adsorbed protein. The enzyme itself is fairly stable and can be treated for 24 hr with 5 *M* urea without loss of activity provided the urea concentration in the assay mixture remains less than 0.2 *M*.

Purity. Pooled fractions from Matrex gel Blue A,[9] ligand affinity chro-

[21] G. Rouser, S. Fleischer, and A. Yamamoto, *Lipids* **5,** 494 (1970).

[22] Direct application of the mitochondrial extract to the affinity column results in partial elution of the phospholipase A_2 activity in the break-through peak.

[23] J. G. N. de Jong, H. Amesz, A. J. Aarsman, H. B. M. Lenting, and H. van den Bosch, *Eur. J. Biochem.* **164,** 129 (1987).

[24] A. J. Aarsman, J. G. N. de Jong, E. Arnoldussen, F. W. Neys, P. D. van Wassenaar, and H. van den Bosch, *J. Biol. Chem.* **264,** 10008 (1989).

matography,[18] and immunoaffinity chromatography[24] all give a single band on SDS–PAGE corresponding to a molecular weight of 14,000. A single amino acid sequence is obtained[24] for the N-terminal 24 residues, which indicates that the enzyme belongs to group II phospholipases A$_2$, lacking Cys-11.

Enzymatic Properties. The enzyme is absolutely specific for the acyl ester bond at the *sn*-2 position of phospholipids, as could be deduced from the release of [^{14}C]linoleate and the production of [^{3}H]palmitoyllysophosphatidylethanolamine exclusively from doubly-labeled 1-[9,10-$^{3}H_2$]palmitoyl-2-[1-^{14}C]linoleoylphosphatidylethanolamine.[9] The enzyme requires Ca^{2+} for activity, although this can be replaced by Sr^{2+} to give about 30% of the activity measured in the presence of Ca^{2+}. The Ca^{2+} requirement of the enzyme is not mediated by calmodulin.[25] The specific activity of the enzyme obtained after rapid immunoaffinity purification amounts to 28 or 50 U/mg, depending on whether protein is measured by the Lowry procedure or by amino acid analysis, respectively.[24] These specific activities are obtained in routine assays using 0.2 m*M* phosphatidylethanolamine. Kinetic analyses with this substrate indicate a K_m of 1.2 m*M* and an apparent V_{max} of 350 U/mg.

Inhibitors. The enzyme is inhibited by *p*-bromophenacyl bromide, and Ca^{2+} protects against this inhibition.[25] Neither diisopropyl fluorophosphate, phenylmethylsulfonyl fluoride nor the thiol reagents *N*-ethylmaleimide, iodoacetamide, or 5,5′-dithiobis(2-nitrobenzoic acid) inhibit enzyme activity.[9]

[25] J. M. de Winter, J. Korpancova, and H. van den Bosch, *Arch. Biochem. Biophys.* **234,** 243 (1984).

[35] Assay and Purification of Phospholipase A$_2$ from Human Synovial Fluid in Rheumatoid Arthritis

By RUTH M. KRAMER and R. BLAKE PEPINSKY

Introduction

Phospholipases A$_2$ (PLA$_2$; phosphatide 2-acylhydrolase, EC 3.1.1.4) are a diverse family of enzymes that hydrolyze the *sn*-2 fatty acyl ester bond of phosphoglycerides, liberating free fatty acids and lysophospholipids. PLA$_2$ from mammalian pancreas and snake venoms are abundant and consequently have been purified and extensively studied.[1,2] PLA_2 from other cell and tissue sources, in contrast, are trace enzymes and not well defined. Related enzymes have been purified from bovine intestine,[3] rabbit and rat inflammatory exudate,[4,5] rat and human platelets,[6,7] rat spleen,[8] rat liver,[9] and from the joint fluid of patients with rheumatoid arthritis.[7,10–12] In rheumatoid arthritis the synovial fluid contains PLA_2 in a soluble form and in relatively high amounts. Several procedures have been developed for purifying PLA_2 from rheumatoid synovial fluid, but low yields are reported, with typical recoveries accounting for only 4–8% of the total enzyme activity.[10–12] This chapter describes a purification protocol that exploits the remarkable stability and solubility of this PLA_2 at low pH and results in yields of total enzymatic activity of greater than 50%. With this procedure approximately 10 μg of homogeneous PLA_2 is routinely recovered from only 50 ml of joint fluid. The same procedure has also been used to purify a secretory PLA_2 from human platelets[7] and thus provides a simple strategy for isolating acid-stable PLA_2.

Materials

S-Sepharose Fast Flow and Sephadex G-50 superfine are obtained from Pharmacia LKB Biotechnology Inc. (Piscataway, NJ). [9,10-^{3}H(N)]Oleic acid (2–10 Ci/mmol in ethanol) is obtained from NEN Research Products (Boston, MA). Bovine serum albumin (essentially fatty acid free) and Brij 35 are from Sigma (St. Louis, MO).

[1] H. M. Verheij, A. J. Slotboom, and G. H. de Haas, *Rev. Phys. Biochem. Pharmacol.* **91,** 91 (1981).

[2] E. A. Dennis, *in* "The Enzymes" (P. D. Boyer, ed.), 3rd ed., Vol. 16, p. 307. Academic Press, New York, 1983.

[3] R. Verger, F. Ferrato, C. M. Mansbach, and G. Pieroni, *Biochemistry* **21,** 6883 (1982).

[4] S. Forst, J. Weiss, P. Elsbach, J. M. Maraganore, I. Reardon, and R. L. Heinrikson, *Biochemistry* **25,** 8381 (1986).

[5] H. W. Chang, I. Kudo, M. Tomita, and K. Inoue, *J. Biochem.* (*Tokyo*) **102,** 147 (1987).

[6] K. Horigome, M. Hayakawa, K. Inoue, and S. Nojima, *J. Biochem.* (*Tokyo*) **101,** 625 (1987).

[7] R. M. Kramer, C. Hession, B. Johansen, G. Hayes, P. McGray, E. P. Chow, R. Tizard, and R. B. Pepinsky, *J. Biol. Chem.* **264,** 5768 (1989).

[8] T. Ono, H. Tojo, S. Kuramitsu, H. Kayamiyama, and M. Okamoto, *J. Biol. Chem.* **263,** 5732 (1988).

[9] A. J. Aarsman, J. G. N. de Jong, E. Arnoldussen, F. W. Neys, P. D. van Wassenaar, and H. van den Bosch, *J. Biol. Chem.* **264,** 10008 (1989).

[10] E. Stefanski, W. Pruzanski, B. Sternby, and P. Vadas, *J. Biochem.* (*Tokyo*) **100,** 1297 (1986).

[11] C.-Y. Lai and K. Wada, *Biochem. Biophys. Res. Commun.* **157,** 488 (1988).

[12] S. Hara, I. Kudo, H. W. Chang, K. Matsuta, T. Miyamoto, and K. Inoue, *J. Biochem.* (*Tokyo*) **105,** 395 (1989).

Preparation of [^{3}H]Oleate-Labeled *Escherichia coli* Substrate

An overnight culture of *E. coli* (MCΔ7 strain, available from R. B. Pepinsky) in 1% Bacto-tryptone, 0.5% NaCl is diluted 1 to 20 into fresh broth and regrown, monitoring cell growth with a Klett–Summerson colorimeter (at 670 nm). After 1 hr, at a Klett reading of 40, Brij 35 (100 mg/ml in water) and [^{3}H]oleic acid are added to final concentrations of 1 mg/ml and 50 μCi/ml, respectively. After 5 hr, when cell growth levels off (reaching a Klett reading of 150), the suspension is autoclaved for 20 min at 120° and the flask stored overnight at 4°. The bacteria are pelleted by centrifugation at 15,000 *g* for 30 min in a Sorvall RC-5B centrifuge using a SS34 rotor. The loose pellets are combined and washed several times with suspension buffer (10 m*M* $CaCl_2$, 0.7 *M* Tris-HCl, pH 8) containing 10 mg/ml bovine serum albumin until the ^{3}H radioactivity in the supernatant is low and remains constant. The washed bacteria are stored at 4° in suspension buffer [10^8 counts per minute (cpm)/ml] containing 0.2% (w/v) sodium azide. Typically, when a 400-ml culture is labeled with 20 mCi of [^{3}H]oleic acid, 5–10% of the added radioactivity is incorporated into phospholipids, yielding a specific radioactivity of approximately 200,000 cpm/nmol of phosphatidylethanolamine plus phosphatidylglycerol. A single preparation routinely can be used for up to 6 months, although the background radioactivity increases with storage.

Prior to use, an aliquot of the [^{3}H]oleate-labeled *E. coli* cells is washed twice with 10 volumes of 12 m*M* EGTA, 200 m*M* Tris-HCl, pH 8, and then with 10 volumes of assay buffer (250 m*M* Tris-HCl, pH 9). The bacteria are resuspended in assay buffer at 100,000 cpm per 10 μl.

Assay of Synovial Fluid Phospholipase A_2

The most convenient and sensitive assay for the synovial fluid PLA$_2$ uses [^{3}H]oleate-labeled *E. coli* cells as the substrate.

Reagents

250 m*M* Tris-HCl, pH 9 (assay buffer)
1 *M* $CaCl_2$
100 mg/ml bovine serum albumin
[^{3}H]Oleate-labeled *E. coli* substrate
Enzyme-containing fractions to be assayed
2 *N* HCl (stopping reagent I)
20 mg/ml bovine serum albumin (stopping reagent II)

Procedure. PLA$_2$ assays are carried out in 200-μl reactions containing 10 m*M* $CaCl_2$ and 1 mg/ml bovine serum albumin. Briefly, a reaction cocktail is prepared by mixing 100 μl of 1 *M* $CaCl_2$ and 100 μl of bovine serum albumin (100 mg/ml) with 7.3 ml of assay buffer. To 150 μl of the reaction mixture is added 10 μl of the *E. coli* substrate, and the volume is adjusted with assay buffer such that the volume including enzyme will be 200 μl. Aliquots of enzyme and corresponding controls are added, and the mixture is incubated for 15 min at 37° in 1.5-ml Eppendorf tubes. The reaction is stopped by adding 100 μl of 2 *N* HCl to each tube, followed by the addition of 100 μl of 20 mg/ml of bovine serum albumin. Tubes are vortexed and incubated on ice for 20 min. The undigested substrate is pelleted at 4° in an Eppendorf centrifuge for 5 min at 10,000 *g*. An aliquot of the supernatant (250 μl) containing the released [^{3}H]oleic acid is removed, mixed with 6 ml of compatible liquid scintillation fluid, and analyzed for radioactivity by liquid scintillation counting. Nonspecific release of radiolabel from the substrate is monitored by performing control incubations in the absence of added enzyme. The released ^{3}H radioactivity in the controls should be less than 2000 cpm and is subtracted from the total counts of each sample. A PLA_2 standard (e.g., purified porcine pancreatic PLA_2 from Sigma, ~600 units/mg) is also carried through each assay. The assay is linear with enzyme concentration up to 10,000 cpm net of released [^{3}H]oleic acid. Typically about 8000 cpm net of [^{3}H]oleic acid is liberated in the presence of 10 pg of purified porcine pancreatic PLA_2.

Purification of Human Rheumatoid Synovial Fluid Phospholipase A_2

All of the purification steps are carried out at 4° unless specified otherwise. Generally, the entire process requires about 5 days. The procedure may be interrupted, with fractions stored at 4° until further processing. In working with samples after the gel-filtration step, dilutions of column fractions should be made into acetate buffer containing 1 mg/ml bovine serum albumin to prevent loss of enzyme due to surface adsorption.

Step 1: Preparation of Rheumatoid Synovial Fluids. Synovial fluids are aspirated from patients diagnosed with classic rheumatoid arthritis. Fresh synovial fluid is clarified by centrifugation for 20 min at 3000 *g* and stored at −70°. Small aliquots (100 μl) of fluids are removed from each batch of fluid for later examination of PLA_2 activity. Prescreening samples for PLA_2 activity is important as large sample-to-sample variations are observed resulting from differences in disease state. For assay of PLA_2 activity in synovial fluids, aliquots are diluted 100-fold with phosphate-buffered saline (PBS) to bring the activity within the linear range of the assay.

Step 2: Acid Treatment of Rheumatoid Synovial Fluid. Synovial fluids are thawed, pooled to yield 50 ml (~2 g of protein), and mixed with

50 ml of 0.36 *N* sulfuric acid. One hundred milliliters of 0.18 *N* sulfuric acid in 150 m*M* NaCl is added to reduce the sample viscosity. The mixture is incubated on ice for 60 min and then dialyzed overnight against two changes of 4 liters of 200 m*M* NaCl, 50 m*M* sodium acetate, pH 4.5, using Spectra/Por membranes (MW cutoff 3500) (Spectrum Medical Industries, Los Angeles, CA). The dialyzed sample is centrifuged at 15,000 *g* for 40 min to remove precipitated material.

Step 3: Column Chromatography on S-Sepharose. The acid-soluble extract is applied to a column of S-Sepharose Fast Flow (1.6 × 27 cm) preequilibrated with 200 m*M* NaCl, 50 m*M* sodium acetate, pH 4.5. The column is washed with 150 ml of the same buffer and then developed with a 550-ml salt gradient from 200 m*M* to 2 *M* NaCl in 50 m*M* sodium acetate, pH 4.5. Loading, washing, and elution of the S-Sepharose column are performed at a flow rate of 90 ml/hr. Fractions are monitored for PLA_2 activity (*E. coli* assay) and for total protein content by measuring absorbance at 280 nm. The PLA_2 activity elutes from the column with approximately 1 *M* NaCl and is recovered in fractions 84–98 (Fig. 1A). The fractions are pooled and concentrated to 0.8 ml using an Amicon (Danvers, MA) ultrafiltration stirred cell with a YM5 membrane.

Step 4: Gel Filtration on Sephadex G-50. The concentrated peak from the S-Sepharose column is applied to a Sephadex G-50 superfine column (1 × 48 cm) preequilibrated in 500 m*M* NaCl, 50 m*M* sodium acetate, pH 4.5. The sample is chromatographed at a flow rate of 2 ml/hr, collecting 0.5-ml fractions. Fractions are monitored for PLA_2 activity and for protein content by measuring absorbance at 280 nm. The PLA_2 activity elutes as a single peak (fractions 45–56) with an apparent molecular weight of 13,000 (Fig. 1B).

Step 5: Reversed-Phase High-Performance Liquid Chromatography. The pooled peak fractions from the gel-filtration column (6 ml) are further purified by reversed-phase HPLC on a C_4 column (Vydac, 0.46 × 25 cm) equilibrated at 29° with 0.1% trifluoroacetic acid. The reversed-phase column is developed at a flow rate of 1 ml/min with a gradient of acetonitrile from 0 to 75% in 0.1% trifluoroacetic acid, collecting 0.5 ml fractions. The total duration of the chromatography is 45 min, and the elution of protein is monitored on-line by absorbance at 214 nm (range 0–0.2 AUF) and 280 nm (range 0–0.05 AUF). Aliquots of the fractions are diluted into 500 m*M* NaCl, 50 m*M* sodium acetate, pH 4.5, containing 1 mg/ml bovine serum albumin and assayed for PLA_2 activity. The PLA_2 activity is recovered as a single peak eluting at 35% acetonitrile (fraction 43, Fig. 1C). The activity in acetonitrile/TFA is stable for months when stored at 4°. For characterizing functional properties of the PLA_2, HPLC fractions are diluted into 500

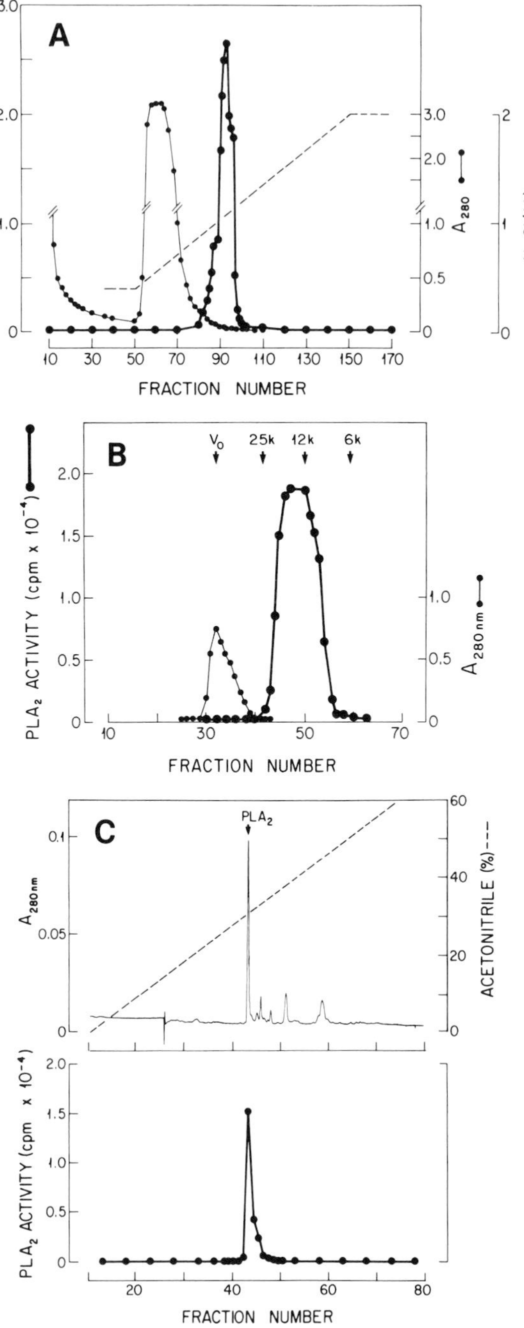
A
PLA₂ ACTIVITY (cpm x 10⁻⁴)
A₂₈₀
NaCl (M)
FRACTION NUMBER
B
V₀
25k
12k
6k
PLA₂ ACTIVITY (cpm x 10⁻⁴)
A₂₈₀ nm
FRACTION NUMBER
C
PLA₂
A₂₈₀ nm
ACETONITRILE (%)
PLA₂ ACTIVITY (cpm x 10⁻⁴)
FRACTION NUMBER

TABLE I
PURIFICATION OF PHOSPHOLIPASE A_2 FROM HUMAN RHEUMATOID ARTHRITIS SYNOVIAL FLUID

Purification step	Protein (mg)	PLA_2 activity: Units[a]	Units/mg	Yield (%)	Purification (-fold)
1. Extraction (pH 1)	1932	1.63	0.84×10^{-3}	100	
2. Dialysis (pH 4.5)	1582	1.60	1.01×10^{-3}	98	
3. S-Sepharose	4.2	0.12	0.029	7	
4. Sephadex G-50	~0.1	0.68	6.8	42	>8000
5. Reversed-phase HPLC	~0.005	0.94	188	58	>223,810

[a] One unit equals 1 μmol substrate (*E. coli* phosphatidylethanolamine + phosphatidylglycerol) hydrolyzed/min.

m*M* NaCl, 50 m*M* sodium acetate, pH 4.5, containing 1 mg/ml bovine serum albumin.

The results of a typical purification of PLA_2 from 50 ml of rheumatoid synovial fluid are summarized in Table I. The enzyme was purified over 200,000-fold with an overall yield of 58%. The final preparation exhibited a specific activity of 188 units/mg. The apparent activity of synovial fluid PLA_2 shows an unusual dependence on ionic strength that complicates quantitation during purification, as evidenced by the fluctuations observed in total activity units. Notably, there is an apparent loss of total PLA_2 activity after the cation-exchange chromatography that is restored on gel filtration. This transient drop in activity results from storage of the PLA_2 in the high-salt elution buffer.

The purity of the final preparation is assessed by sodium dodecyl sulfate (SDS)–polyacrylamide gel electrophoresis. SDS is added to reversed-phase HPLC fractions containing PLA_2 activity to a final concentration of 0.01%, and the samples are dried in a Speed-Vac concentrator (Savant, Farmingdale, NY). Enzyme preparations are dissolved in electrophoresis sample buffer, incubated for 10 min at 60°, and subjected to SDS–PAGE on 16% gels using the Laemmli discontinuous buffer system.[13]

[13] U. K. Laemmli, *Nature (London)* **227,** 680 (1970).

FIG. 1. Elution profile of PLA_2 activity (using *E. coli* substrate) and total protein (absorbance at 280 nm) from chromatography columns: (A) S-Sepharose, (B) Sephadex G-50, and (C) C_4 reversed-phase HPLC.

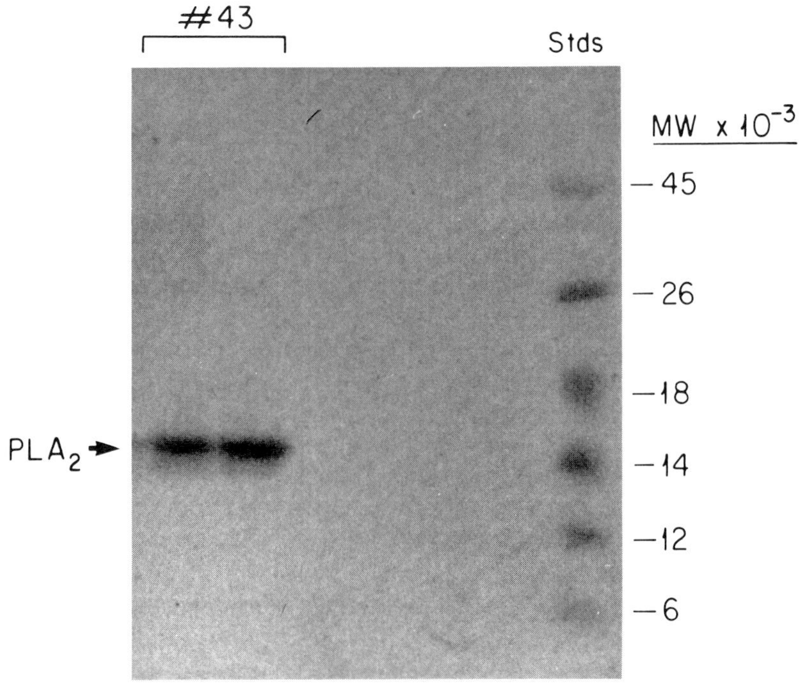

FIG. 2. Analysis of reversed-phase HPLC-purified rheumatoid synovial fluid PLA_2 (see Fig. 1C, fraction 43) by SDS-polyacrylamide gel electrophoresis (16% gel), followed by electroblotting onto PVDF (Immobilon) membrane paper and staining with Coomassie blue.

In the preparation shown in Fig. 2, we were interested in obtaining sequence information and therefore subjected the entire purified PLA_2 preparation described in Table I to gel electrophoresis. The protein was electroblotted onto PVDF membrane (Millipore, Bedford, MA) paper according to Matsudaira[14] modified to include 0.01% SDS in the blotting buffer, stained with Coomassie Blue R250, and directly subjected to N-terminal sequence analysis.

Properties of Synovial Fluid Phospholipase A_2

The synovial fluid PLA_2 is a basic protein with an isoelectric point exceeding 10.5 and has an apparent molecular weight of 13,000 by gel filtration and 14,000 by SDS–polyacrylamide gel electrophoresis. The enzyme exhibits a broad pH optimum from 8 to 10 and is Ca^{2+}-dependent,

[14] P. Matsudaira, *J. Biol. Chem.* **262,** 10035 (1987).

requiring 10 mM Ca^{2+} for optimal activity. The PLA_2 shows a great preference for the *E. coli* substrate and hydrolyzes phospholipid/deoxycholate dispersions and sonicated liposomes at rates that are 100- and 1000-fold slower, respectively. In deoxycholate dispersions phosphatidylethanolamine is the preferred substrate and is hydrolyzed at a 4 times faster rate than phosphatidylcholine. The corresponding gene of the PLA_2 has been isolated and sequenced.[7,15] The deduced amino acid sequence of the protein consists of 124 amino acids, contains structural features common to all known PLA_2, and has a half-cystine pattern that is characteristic for Group II snake venom PLA_2.[16] The sequence is preceded by a signal peptide, confirming that the PLA_2 is a secretory enzyme.

It should be noted that there are intracellular (nonsecretory) PLA_2 enzymes of higher molecular weight ($\geqslant$14,000) that are activated by submicromolar Ca^{2+} and prefer phosphatidylcholine substrates. These PLA_2 are likely to represent distinct enzymes.[17,18] The high molecular weight PLA_2 appear to be inactivated by acid treatment and thus are lost during the purification procedure described here.

[15] J. J. Seilhamer, W. Pruzanski, P. Vadas, S. Plant, J. A. Miller, J. Kloss, and L. K. Johnson, *J. Biol. Chem.* **264,** 5335 (1989).

[16] R. L. Heinrikson, E. T. Krueger, and P. S. Keim, *J. Biol. Chem.* **252,** 4913 (1977).

[17] R. M. Kramer, J. A. Jakubowski, and D. Deykin, *Biochim. Biophys. Acta* **959,** 269 (1988).

[18] L. Loeb and R. Gross, *J. Biol. Chem.* **261,** 10467 (1986).

[36] Purification of Mammalian Nonpancreatic Extracellular Phospholipases A_2

By SHUNTARO HARA, HYEUN WOOK CHANG, KAZUHIKO HORIGOME, ICHIRO KUDO, and KEIZO INOUE

Introduction

High levels of extracellular phospholipase A_2 have been found at inflamed sites in some experimental animals and humans with diseases. It has been reported that various kinds of inflammatory cells, such as platelets, macrophages, and polymorphonuclear leukocytes, secrete phospholipase A_2 extracellularly on stimulation. We have purified and characterized

TABLE I
CHARACTERISTICS OF PURIFIED MAMMALIAN PANCREATIC AND NONPANCREATIC EXTRACELLULAR PHOSPHOLIPASES A_2

Characteristic	Phospholipase A_2	
	Nonpancreatic[a]	Pancreatic[b]
Molecular weight	14K	14K
Group (pattern of S—S bonds)	II	I
Amino acid composition	Basic	Acidic
Affinity for heparin	High	Low
Substrate specificity[c]	PE > PC	PE = PC
Optimum pH	Alkaline	Alkaline
Ca^{2+} requirement	Yes	Yes
Stability at pH 4.0	Yes	Yes
Effect of *p*-bromophenacyl bromide	Inhibition	Inhibition
Effect of deoxycholate	Inhibition	Stimulation

[a] From rat platelets.
[b] From porcine pancreas.
[c] PE, Phosphatidylethanolamine; PC, phosphatidylcholine.

phospholipases A_2 found at inflamed sites in humans[1,2] and rats,[3] and those secreted from rat[4,5] and rabbit[6] platelets.

Our observations together with the results of Forst *et al.,* who purified an extracellular phospholipase A_2 from rabbit ascites fluid,[7] indicate that these nonpancreatic extracellular phospholipases A_2 share common properties, such as molecular weight, optimum pH, Ca^{2+} requirement, substrate specificity, detergent sensitivity, stability at acidic pH, and heparin affinity (Table I). These enzymes, however, show different substrate speci-

[1] S. Hara, I. Kudo, H. W. Chang, K. Matsuta, T. Miyamoto, and K. Inoue, *J. Biochem.* (*Tokyo*) **105,** 395 (1989).

[2] S. Hara, I. Kudo, K. Matsuta, T. Miyamoto, and K. Inoue, *J. Biochem.* (*Tokyo*) **104,** 326 (1988).

[3] H. W. Chang, I. Kudo, M. Tomita, and K. Inoue, *J. Biochem.* (*Tokyo*) **102,** 147 (1987).

[4] K. Horigome, M. Hayakawa, K. Inoue, and S. Nojima, *J. Biochem.* (*Tokyo*) **101,** 625 (1987).

[5] M. Hayakawa, K. Horigome, I. Kudo, M. Tomita, S. Nojima, and K. Inoue, *J. Biochem.* (*Tokyo*) **101,** 1311 (1987).

[6] H. Mizushima, I. Kudo, K. Horigome, M. Murakami, M. Hayakawa, D. K. Kim, E. Kondo, M. Tomita, and K. Inoue, *J. Biochem.* (*Tokyo*) **105,** 520 (1989).

[7] S. Forst, J. Weiss, P. Elsbach, J. M. Maraganore, I. Reardon, and R. L. Heinrikson, *Biochemistry* **25,** 8381 (1986).

ficities and detergent sensitivities from those of mammalian pancreatic phospholipase A_2. Analysis of the entire amino acid sequence of phospholipase A_2 purified from rat platelets[8] indicates that the nonpancreatic enzyme shows approximately 30% homology with the pancreatic enzyme and contains some well conserved residues also present in the pancreatic enzyme. Remarkable structural features commonly observed in the nonpancreatic enzymes are a high content of basic amino acids and the absence of Cys-11, which is strictly conserved in the group I family of phospholipases A_2 (those from mammalian pancreatic juice and from the venom of elapid and hydrophid snakes). The enzymes may therefore belong to the group II family, to which Crotalidae and Viperidae snake venom enzymes belong.

In this chapter, we describe methods for purification of phospholipases A_2 from supernatants of rat activated platelets,[4] human synovial fluid,[1] and rat peritoneal fluid.[3]

Purification of Extracellular Phospholipase A_2 Released from Rat Platelets

The most distinct property commonly observed in mammalian nonpancreatic extracellular phospholipases A_2 is a high affinity for heparin. Heparin-Sepharose affinity chromatography is therefore a powerful tool for the purification of these enzymes. Extracellular phospholipase A_2 released from rat platelets is purified about 770-fold to near homogeneity by the sequential use of column chromatography on heparin-Sepharose and TSK gel G2000SW [gel-filtration high-performance liquid chromatography (HPLC)]. Lysophosphatidylserine-specific lysophospholipase, which is also released from rat platelets, is separated from phospholipase A_2 by chromatography on heparin-Sepharose.

Reagents

Heparin-Sepharose CL-6B (Pharmacia Fine Chemicals, Uppsala, Sweden)
TSK gel G2000SW (Tosoh, Tokyo, Japan)
Heparin-Sepharose buffer: 10 m*M* Tris-HCl (pH 7.4) containing 0.15 or 1.5 *M* KCl
HPLC buffer: 0.2 *M* Na_2SO_4, 10 m*M* acetic acid, pH 3.0

[8] M. Hayakawa, I. Kudo, M. Tomita, S. Nojima, and K. Inoue, *J. Biochem.* (*Tokyo*) **104,** 767 (1988).

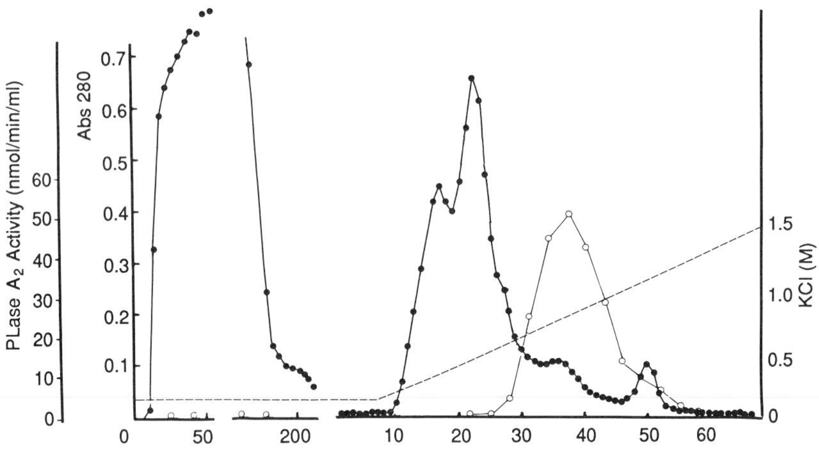

FIG. 1. Elution profile of rat platelet secretory phospholipase A$_2$ separated by heparin-Sepharose affinity chromatography. Proteins released from activated rat platelets (~170 mg) were applied to a column (1 × 9.5 cm) of heparin-Sepharose CL-6B preequilibrated with buffer (150 mM KCl, 10 mM Tris-HCl, pH 7.4) and eluted as described in the text. The dashed line indicates the concentration of KCl. ●, Protein; ○, phospholipase A$_2$ activity.

Procedures

Preparation of supernatants from activated rat platelets is described in detail elsewhere in this volume.[9]

Step 1: Heparin-Sepharose Affinity Chromatography. The supernatant from 3 × 10^{11} platelets is applied to a heparin-Sepharose CL-6B column (1 × 9.5 cm), which has been equilibrated previously with buffer [10 mM Tris-HCl (pH 7.4) containing 0.15 M KCl], at 4°. The column is washed extensively with the starting buffer, then subsequently eluted with 70 ml of a linear concentration gradient of KCl (0.15 to 1.5 M). Phospholipase A$_2$ activity is eluted at 0.6 to 1.1 M KCl (Fig. 1).

Step 2: Gel-Filtration HPLC. The fractions showing phospholipase A$_2$ activity recovered from the heparin-Sepharose column are concentrated to about 0.5 ml by ultrafiltration with Centricon 10 (Amicon, Danvers, MA). The resulting sample is applied to a TSK gel G2000SW column (0.75 × 60 cm) in the HPLC system [Shimadzu Model LC-5A liquid delivery modules (Shimadzu, Kyoto, Japan)], which has been equilibrated with HPLC buffer, at a flow rate of 0.2 ml/min. The single peak of phospholipase A$_2$ activity appears to correspond to a major protein peak, with a molecular weight estimated to be about 15,000 from the elution profile. The

[9] M. Murakami, K. Takayama, M. Umeda, I. Kudo, and K. Inoue, this volume [20].

final preparation gives a single protein band, with an estimated molecular weight of about 13,500, on sodium dodecyl sulfate (SDS)–polyacrylamide gel electrophoresis.

Purification of Extracellular Phospholipase A_2 from Synovial Fluid of Rheumatoid Arthritis Patients

Human rheumatoid synovial fluid phospholipase A_2 also shows high affinity for heparin. The enzyme is purified about 1.7×10^5-fold to near homogeneity by sequential use of heparin-Sepharose, butyl-Toyopearl, and reversed-phase HPLC. It should be noted that the amount of phospholipase A_2 in human rheumatoid synovial fluid is approximately 60 times lower than that in the medium of rat platelets stimulated with thrombin.

Reagents

Sepharose CL-4B (Pharmacia)
Heparin-Sepharose CL-6B (Pharmacia)
Butyl-Toyopearl 650C (Tosoh)
Capcell Pak C_8 (Shiseido, Tokyo, Japan)
Heparin-Sepharose buffer: 50 m*M* Tris-HCl (pH 7.4) containing 0.15 or 0.5 *M* NaCl
Hydrophobic chromatography buffer: 0.5 *M* NaCl, 50 m*M* Tris-HCl (pH 7.4) containing 0, 10, 20, or 30% ammonium sulfate
HPLC buffers: A, 5% acetonitrile in 0.1% trifluoroacetic acid; B, 55% acetonitrile in 0.1% trifluoroacetic acid

Procedures

Step 1: Heparin-Sepharose Affinity Chromatography. Pooled human synovial fluid from patients with rheumatoid arthritis (480 ml) is loaded onto a Sepharose CL-4B precolumn (2.5 × 4 cm) and eluted with 50 m*M* Tris-HCl (pH 7.4) containing 0.15 *M* NaCl. The eluate is loaded onto a heparin-Sepharose CL-6B column (2.5 × 8 cm) which has been equilibrated with 50 m*M* Tris-HCl (pH 7.4) containing 0.15 *M* NaCl. The phospholipase A_2 activity has affinity for the column under these conditions and is eluted as a single peak in the fractions eluted with the same buffer containing 0.5 *M* NaCl.

Step 2: Butyl-Toyopearl Hydrophobic Chromatography. The fractions with phospholipase A_2 activity that have been recovered from the heparin-Sepharose column are brought to 30% saturation with solid ammonium sulfate and applied to a butyl-Toyopearl column (1.5 × 25 cm), which has been equilibrated with 50 m*M* Tris-HCl (pH 7.4) containing 0.5 *M* NaCl

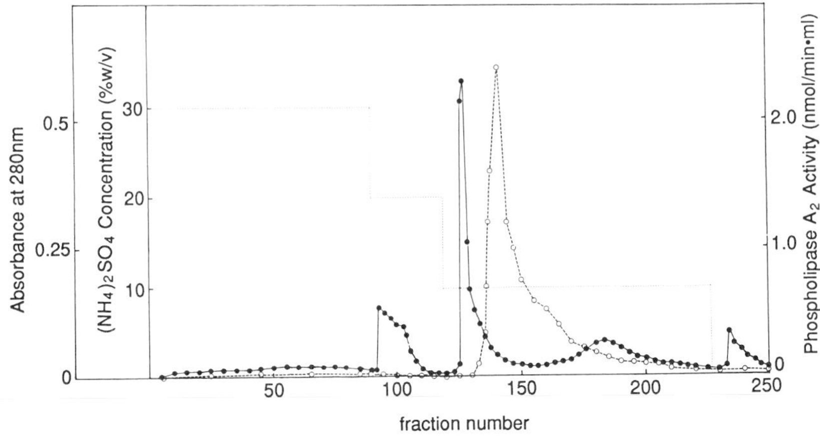

FIG. 2. Elution profile of human synovial fluid phospholipase A_2 separated by butyl-Toyopearl hydrophobic chromatography. The pooled fractions with phospholipase A_2 activity recovered from the heparin-Sepharose column were brought to 30% ammonium sulfate, applied to a column (1.5 × 25 cm) of butyl-Toyopearl 650C preequilibrated with buffer containing 30% ammonium sulfate, and eluted as described in the text. The dashed line indicates the concentration of ammonium sulfate. ●, Protein; ○, phospholipase A_2 activity.

and 30% ammonium sulfate. The column is washed with the same buffer, and it is eluted with a stepwise gradient of decreasing ammonium sulfate concentration from 30 to 0%. The phospholipase A_2 activity has affinity for the column in the presence of 30% ammonium sulfate, and the activity is eluted as a single peak in the fractions eluted with buffer containing 10% ammonium sulfate (Fig. 2).

Step 3: Reversed-Phase HPLC. The fractions showing phospholipase A_2 activity are concentrated to about 10 ml by ultrafiltration with a Centricon 10.The concentrated protein is applied to a Capcell Pak C_8 column (0.46 × 15 cm) in the HPLC system [two Gilson Model 302 liquid delivery modules and a Gilson Model 811 dynamic mixer (Gilson Medical Electronics, Villiers-le-Bel, France)], which has been equilibrated with buffer A. The protein is eluted with a linear gradient of increasing concentration of acetonitrile; the single peak of phospholipase A_2 activity appears to correspond to one of the protein peaks. The fraction with phospholipase A_2 activity is again subjected to the same reversed-phase HPLC and is eluted under the same conditions. The single peak of activity appears to correspond to a major protein peak. The final preparation gives a single protein band, with a molecular weight estimated to be about 13,700, on SDS–polyacrylamide gel electrophoresis.

Purification of Extracellular Phospholipase A_2 from Peritoneal Cavity of Caseinate-Treated Rats

Extracellular phospholipase A_2, which shares a common structure with the enzyme released from activated rat platelets, is detected in the peritoneal cavity of rats injected intraperitoneally with casein.[10] The cellular origin of this enzyme is most probably some inflammatory cells distinct from platelets. As mentioned above, heparin-Sepharose chromatography is a powerful tool for the purification of mammalian nonpancreatic phospholipase A_2. However, it was not applicable for rat peritoneal phospholipase A_2 because we collected the peritoneal fluid by washing with heparinized saline. The enzyme can be purified about 14,000-fold to near homogeneity from rat peritoneal fluid without using heparin-Sepharose chromatography.

Reagents

Sephadex G-75 superfine (Pharmacia)
Toyopearl HW65 (Tosoh)
TSK gel ODS 120-T (Tosoh)
Gel filtration buffer: 50 m*M* Tris-HCl (pH 7.4) containing 1 *M* NaCl
Hydrophobic chromatography buffer: 1 *M* NaCl, 50 m*M* Tris-HCl (pH 7.4) containing 0, 10, 20, or 30% ammonium sulfate
HPLC buffers: A, 5% acetonitrile in 0.02% HCl; B, 55% acetonitrile in 0.02% HCl

Procedures

Step 1: Sephadex G-75 Gel-Filtration Chromatography. Cell-free peritoneal fluid from 40 rats that have received an intraperitoneal injection of casein is collected by washing with heparinized saline. After removal of the bulk of the protein, which is precipitated by bringing the peritoneal fluid to 30% ammonium sulfate, the supernatents showing phospholipase A_2 activity are precipitated by bringing to 60% ammonium sulfate. In one set of experiments, one-eleventh of the protein (7.3 ml, 200 mg) from the ammonium sulfate precipitation is loaded onto a Sephadex G-75 column (2.64 × 100 cm). The column is then eluted with gel-filtration buffer. The phospholipase A_2 activity is eluted as a single peak.

Step 2: Toyopearl HW-65 Hydrophobic Chromatography. One-third of the fractions (200 ml, 5 mg) showing phospholipase A_2 activity recovered from the gel-filtration step are brought to 30% saturation with solid ammo-

[10] H. W. Chang, I. Kudo, S. Hara, K. Karasawa, and K. Inoue, *J. Biochem.* (*Tokyo*) **100,** 1099 (1986).

nium sulfate and applied to a Toyopearl HW-65 column (1.5 × 30 cm) equilibrated previously with 50 m*M* Tris-HCl (pH 7.4) containing 1 *M* NaCl and 30% ammonium sulfate. The phospholipase A_2 has affinity for the column in the presence of 30% ammonium sulfate. The column is eluted with a stepwise gradient of decreasing ammonium sulfate concentration from 30 to 0%. The activity is eluted as a single peak in the fractions eluted with buffer containing 10% ammonium sulfate.

Step 3: Reversed-Phase HPLC. The fractions showing phospholipase A_2 activity, which have been recovered from the Toyopearl HW-65 chromatography, are concentrated to about 1 ml by ultrafiltration with a Centricon 10. The concentrated protein is applied to a TSK ODS 120-T column (0.46 × 15 cm) in the Gilson HPLC system, which has been equilibrated with buffer A. The protein is eluted with a linear gradient of increasing concentration of acetonitrile from 5 to 55% in 50 min at a flow rate of 1 ml/min. The single peak of activity appears to correspond to a major protein peak. The final preparation shows a single protein band, with a molecular weight estimated to be about 13,500, on SDS-polyacrylamide gel electrophoresis.

Comments on Purification Procedures

(1) The most striking feature commonly observed in mammalian nonpancreatic extracellular phospholipases A_2 is a high affinity for heparin. Heparin-Sepharose affinity chromatography is, therefore, a powerful tool for purification of all of these enzymes. Rabbit platelet secretory phospholipase A_2 can be purified by the use of heparin-Sepharose as well as other nonpancreatic phospholipases A_2. This affinity column method is available in the presence of CHAPS.[11] It should also be emphasized that these mammalian nonpancreatic extracellular phospholipases A_2 are basic proteins. Extracellular phospholipases A_2 have been successfully purified from rabbit ascites fluid[7] and human synovial fluid[12] by the use of cation-exchange chromatography.

(2) We have performed gel-filtration HPLC on TSK G2000SW at acidic pH and gel-filtration chromatography on Sephadex G-75 with a buffer of high ionic strength for the following reasons. First, at neutral pH, the phospholipases A_2 interact with TSK G2000SW resin, and the recovery is very low. Second, under isotonic conditions, the enzymes aggregate or interact with other proteins. When we purified the enzyme from rat perito-

[11] M. Hayakawa, I. Kudo, M. Tomita, and K. Inoue, *J. Biochem. (Tokyo)* **103,** 263 (1988).

[12] R. M. Kramer, C. Hession, B. Johansen, G. Hayes, P. McGray, E. P. Chow, R. Tizzard, and R. B. Pepinsky, *J. Biol. Chem.* **264,** 5768 (1989).

neal fluid or human synovial fluid, the recovery of total activity from Sephadex G-75 gel-filtration chromatography in the presence of 1.0 *M* NaCl was approximately 200%; some endogenous inhibitory factors might have been removed during the gel-filtration step.

(3) Hydrophobic chromatography is useful for purification of the extracellular phospholipases A_2. We have successfully used a butyl-Toyopearl column for the purification of human nonpancreatic extracellular phospholipase A_2. The rat nonpancreatic extracellular enzyme was also applied to this column, but it could not be eluted under the same conditions. This observation suggests that the hydrophobicity of the enzyme may differ among various species and that the rat enzyme is more hydrophobic than the human one. A Toyopearl HW-65 column is usually used as a gel-filtration column, but it can also be used as a hydrophobic column when ammonium sulfate is present.[13] We applied this column with less hydrophobicity for the purification of rat nonpancreatic extracellular phospholipase A_2, which is thought to be more hydrophobic. The nonpancreatic extracellular enzymes are stable in acidic solvents containing acetonitrile. The stability of the enzyme enables us to use reversed-phase HPLC with an acid–acetonitrile system for the purification. Extracellular phospholipase A_2 can be purified from rat platelets by the use of reversed-phase HPLC instead of gel-filtration HPLC.

(4) Monoclonal antibodies against mammalian nonpancreatic extracellular phospholipases A_2 were used to develop a rapid immunoaffinity chromatography method for purification of the enzymes. These methods are described elsewhere in this volume.[9]

Conclusions

We have devised procedures for purifying mammalian extracellular nonpancreatic phospholipase A_2. These procedures utilize heparin-Sepharose affinity chromatography, hydrophobic chromatography, and gel-filtration chromatography. The most efficient step is chromatography over heparin-Sepharose. The sequential use of the chromatography steps should facilitate purification of mammalian extracellular nonpancreatic phospholipases A_2 from various other sources.

[13] N. Sakihama, H. Ohmori, N. Sugimoto, Y. Yamasaki, R. Oshino, and M. Shin, *J. Biochem. (Tokyo)* **93,** 129 (1983).

[37] Spleen Phospholipases A_2

By HIROMASA TOJO, TAKASHI ONO, and MITSUHIRO OKAMOTO

Introduction

Phospholipase A_2 (PLA_2; EC 3.1.1.4) stereospecifically hydrolyzes the fatty acyl ester bonds at the *sn*-2 position of glycerophospholipids. The enzymes exist in almost every type of cell studied so far and are distributed in diverse subcellular fractions. Detailed comparison of the molecular properties of PLA_2s purified to homogeneity from different subcellular locations may provide an insight into their functional differences. Rat or human spleen is a relatively rich source of PLA_2 among tissues other than pancreas, and it contains PLA_2s in both soluble and membrane-associated forms. Recently, the two forms of PLA_2 were purified from rat spleen and characterized in detail. They were demonstrated to be distinct from each other: the soluble form was identical to pancreatic PLA_2, whereas the membrane-associated form was of viperid/crotalid type.[1,2] This chapter describes the assay method, purification procedures, and properties of the enzymes.

Assay Method[1]

Principle. PLA_2 specifically catalyzes the hydrolysis of the fatty acyl bond at position 2 of 3-*sn*-phosphoglycerides. The fatty acids released are extracted by the method of Dole and Meinertz[3] and then derivatized with 9-anthryldiazomethane[4] (ADAM). The derivatized fatty acids are analyzed by reversed-phase high-performance liquid chromatography (HPLC).

Reagents for Standard Assay

0.05% (w/v) 9-Anthryldiazomethane (Funakoshi Co., Tokyo, Japan), in ethyl acetate/methanol (1 : 9, v/v)

[1] H. Tojo, T. Ono, S. Kuramitsu, H. Kagamiyama, and M. Okamoto, *J. Biol. Chem.* **263,** 5724 (1988).

[2] T. Ono, H. Tojo, S. Kuramitsu, H. Kagamiyama, and M. Okamoto, *J. Biol. Chem.* **263,** 5732 (1988).

[3] V. P. Dole and H. Meinertz, *J. Biol. Chem.* **235,** 2595 (1960).

[4] N. Nimura and T. Kinoshita, *Anal. Lett.* **13,** 191 (1980).

5 m*M* 1-Palmitoyl-2-oleoyl-*sn*-glycero-3-phosphoethanolamine in chloroform
100 m*M* Sodium deoxycholate in methanol
25 m*M* Calcium chloride
0.5 *M* Tris-HCl, 0.5 *M* NaCl, pH 8.5
0.5 m*M* Heptadecanoic acid in methanol
Dole's reagent (*n*-heptane/2-propanol/2 *N* sulfuric acid, 10:40:1, v/v) *n*-Heptane

Preparation of Substrate Solutions. Mixed micellar solutions of deoxycholate and phosphatidylethanolamine (PE) are prepared as reported.[5] Solutions of sodium deoxycholate in methanol and PE in chloroform are mixed to obtain the desired molar ratios. We have routinely employed a deoxycholate/PE ratio of 6; since an optimal molar ratio depends on the detergent–phospholipid combination, the ratio should be determined for each set of detergent and phospholipid. The solvents are evaporated under nitrogen, and then the samples are further dried *in vacuo* for about 30 min. An appropriate amount of the buffer is added, and the tubes are vortexed for 2 min at about 37°. In the case of assay without detergent the substrate is sonicated 3 times for 20 sec in the buffer.

Preparation of ADAM Solution. Since ADAM is relatively unstable in methanolic solution, the solution of ADAM used for the assay should be prepared immediately before use. For convenience, ADAM dissolved in a small volume of ethyl acetate (e.g., 25 mg/ml) is divided into small portions containing known amounts, then the solvent is evaporated. Store in darkness at −35°. Only the amount needed for the day's work is first dissolved in 1 volume of ethyl acetate, and then 9 volumes of methanol is added to make up the final concentration of ADAM to 0.05%.

Procedures. The assay mixtures contain 5 m*M* $CaCl_2$, 1 m*M* 1-palmitoyl-2-oleoyl-PE, 6 m*M* deoxycholate, 0.1 *M* NaCl, 0.1 *M* Tris-HCl, pH 8.5, and the enzyme sample, in a final volume of 50 μl. In a control tube $CaCl_2$ is replaced by EDTA (10 m*M*). The enzyme reaction is stopped by the addition of 200 μl of Dole's reagent. To the reaction mixture are added 120 μl of heptane and 70 μl of water, and then 5 nmol of heptadecanoic acid (10 μl) is added as an internal standard. The tubes are vortexed for about 20 sec. Fatty acids are extracted into the upper heptane phase.[3] Fifty microliters of the heptane layer is transferred to a 0.3-ml microvial, and the solvent is evaporated with an aspirator. Then 50 μl of the 0.05% ADAM solution is added to the vial. The vial is incubated for about 15 min at room temperature, and then the derivatized fatty acids are analyzed on a Superspher RP-18 (4 × 50 mm, Merck, Darmstadt, FRG) or a Cos-

[5] S. Yedgar, R. Hertz, and S. Gatt, *Chem. Phys. Lipids* **13,** 404 (1974).

mosil 3C18 (2.1 × 50 mm, Nacalai Tesque, Kyoto, Japan) column with a UV detector at 254 nm. The solvent system employed is 95–98% acetonitrile at 1 ml/min. The samples were automatically injected with a Gilson (Villiers-le-Bel, France) autosampler 231-401.

Comments

(1) This method is applicable to the assay with any phospholipid class containing long-chain fatty acids as a substrate. Many kinds of synthetic phospholipids with high purity are now commercially available (e.g., from Avanti Polar Lipids, Inc., Birmingham, AL). Hence, the substrate specificities of PLA_2s are easily evaluated by this method.[1,2] When 1-palmitoyl-2-oleoylphosphatidylglycerol was used as a substrate, the oleic acid released by PLA_2 reacted with ADAM significantly poorly at PG concentrations above 1 m*M*. Phosphatidylglycerol (PG) itself did not hinder the reaction. Although we did not further examine the reason for this, the problem could be circumvented by treatment of the heptane phase (100 μl) with silicic acid (~5 mg).

Some portions of phospholipids are extracted into the upper heptane layer together with fatty acids by the Dole procedure. When phosphatidylserine (PS) was used as a substrate, a carboxylate group of the extracted PS was derivatized with ADAM, and the PS ester was eluted faster than most of derivatized long-chain fatty acids on HPLC. The PS extracted in the heptane layer could, if necessary, be removed by the silicic acid treatment.

(2) The positional specificity of phospholipase A action can be determined by this method with a mixed-acyl phospholipid as a substrate: the time courses of the hydrolysis at the *sn*-1 position and of that at the *sn*-2 position are followed separately and compared with the time courses of the hydrolysis by pancreatic PLA_2, which exclusively hydrolyzes the ester bond at the *sn*-2 position.

(3) When ADAM–derivatized fatty acids are separated on a HPLC column of small size, it is essential to keep the pre- and postcolumn dead volumes associated with a HPLC system as small as possible. Otherwise, the resolving power of the small column is severely hampered.

Purification of Phospholipases A_2

From Rat Spleen Supernatant

Purification Strategy. The key step of the purification procedures is reversed-phase HPLC. The HPLC method is a powerful tool for enrichment and purification of trace amounts of cellular PLA_2s, which are stable in organic solvents.[6] Since we introduced the HPLC method to the purification of a soluble PLA_2 from rat spleen,[6] the method has been successfully applied to the purification of PLA_2s of various origins including rat pancreas,[7] rat gastric mucosa,[8] human spleen,[9,10] rabbit[11] and rat[12] ascites, human synovial fluid,[13–14a] human platelets,[13] and the macrophagelike cell line $P388D_1$.[15] The procedure described here includes some modifications to adapt the HPLC method reported previously for a relatively larger scale preparation.

HPLC System. The HPLC system consists of two Gilson Model 302 liquid delivery modules and a sample injector with a 8-ml sample loop. For microbore HPLC, the system equipped with a Gilson 811 dynamic mixer with a 65-μl mixing chamber is operated in the microflow mode. The effluent is monitored simultaneously at 280 and 210 nm with a Gilson 116 detector with a 1.6-μl flow cell. A well-established solvent system, trifluoroacetic acid (TFA)/acetonitrile, is effective for the purification of PLA_2s. The solvent system employs an acidic (pH ~2.1) mobile phase, which usually maximizes the separation for most peptides and small proteins.[16] No irreversible inactivation of the enzymes takes place during elution at the acidic pH in the presence of the organic solvent.[6] The eluants used are as follows: eluant A, 0.1% (v/v) TFA in water; eluant B, 95% (v/v) acetonitrile, 0.1% TFA.

Buffers

Buffer A, 20 m*M* Tris-HCl, pH 7.4

Buffer B, 10 m*M* 2-(*N*-morpholino)ethanesulfonic acid (MES), pH 6.0

[6] H. Tojo, T. Teramoto, T. Yamano, and M. Okamoto, *Anal. Biochem.* **137,** 533 (1984).

[7] T. Ono, H. Tojo, K. Inoue, H. Kagamiyama, T. Yamano, and M. Okamoto, *J. Biochem.* (*Tokyo*) **96,** 785 (1984).

[8] H. Tojo, T. Ono, and M. Okamoto, *Biochem. Biophys. Res. Commun.* **151,** 1188 (1988).

[9] K. Nakaguchi, J. Nishijima, M. Ogawa, T. Mori, H. Tojo, T. Yamano, and M. Okamoto, *Enzyme* (*Basel*) **35,** 2 (1986).

[10] A. Kanda, T. Ono, N. Yoshida, H. Tojo, and M. Okamoto, *Biochem. Biophys. Res. Commun.* **163,** 42 (1989).

[11] S. Forst, J. Weiss, P. Elsbach, J. M. Maraganore, I. Reardon, and R. L. Heinrikson, *Biochemistry* **25,** 8381 (1986).

[12] H. W. Chang, I. Kudo, M. Tomita, and K. Inoue, *J. Biochem.* (*Tokyo*) **102,** 147 (1987).

[13] R. M. Kramer, C. Hession, B. Johansen, G. Hayes, P. McGray, E. P. Chow, R. Tizard, and R. B. Pepinski, *J. Biol. Chem.* **264,** 5768 (1989).

[14] S. Hara, I. Kudo, H. W. Chang, K. Matsuta, T. Miyamoto, and K. Inoue, *J. Biochem.* (*Tokyo*) **105,** 395 (1989).

[14a] J. J. Seilhamer, S. Plant, W. Pruzanski, J. Schilling, E. Stefanski, P. Vadas, and L. K. Johnson, *J. Biochem.* (*Tokyo*) **106,** 38 (1989).

[15] R. J. Ulevitch, Y. Watanabe, M. Sano, M. D. Lister, R. A. Deems, and E. A. Dennis, *J. Biol. Chem.* **263,** 3079 (1988).

[16] W. C. Mahoney and M. A. Hermodson, *J. Biol. Chem.* **255,** 11199 (1980).

Buffer C, 20 m*M* 2-methyl-2-amino-1,3-propanediol, pH 9.5

Buffer D, 100 m*M* Tris-HCl, pH 7.4

Preparation of Tissue Extract. Rats are anesthetized by the intraperitoneal injection of pentobarbital. Spleens are removed, carefully prepared to avoid the involvement of pancreas tissues, and then stored at −35°. Frozen rat spleen tissue (500 g) is homogenized in 2 liters of buffer A containing 1 m*M* $CaCl_2$ with a blender. The homogenate is centrifuged at 108,000 *g* for 1 hr. The supernatant contains about 10% of the total PLA_2 activity, the pellet about 90%. The pellet is stored at −35° for further use.

DEAE–Cellulose Treatment. The supernatant is applied to a DEAE-cellulose column (8 × 30 cm) preequilibrated with buffer A. The resulting flow-through fractions which contain the PLA_2 activity are pooled. The pH of the pooled fractions is adjusted to 6.0 by the addition of 1 *N* HCl, and then the PLA_2 is concentrated with a short S-Sepharose column (5 × 6 cm) preequilibrated with buffer B. The PLA_2 activity is eluted with buffer D containing 0.5 *M* NaCl. *Note*: If the capacity of the DEAE-cellulose column is insufficient to remove contaminants, lowering pH of the flow-through fractions may form significant amounts of precipitates. This results in very low recovery of the enzyme activity.

TEAE–Cellulose Column Chromatography. The concentrated PLA_2 fraction is dialyzed against buffer C. The resultant solution is applied to a TEAE–cellulose column (3 × 10 cm) preequilibrated with buffer C. The major part (86%) of the PLA_2 activity is eluted under these conditions, while a small part (14%) is bound to the column. The PLA_2 contained in the major flow-through fraction, which is named PLA_2 S-1, is further purified. The enzyme activity bound to the column is eluted in a stepwise manner with buffer D containing 0.2 *M* NaCl.

BioGel P-30 Column Chromatography. The pooled PLA_2 S-1 fraction is dialyzed against buffer B, and then the resultant solution is concentrated with a short S-Sepharose column (2 × 1.5 cm) as described above. The concentrated PLA_2 fraction is then applied to a BioGel P-30 (minus 400 mesh) column (2 × 50 cm) preequilibrated with buffer A containing 0.5 *M* NaCl. A single peak of PLA_2 S-1 activity appears at an elution volume of 80 ml. The fractions containing PLA_2 activity are pooled.

Reversed-Phase HPLC. The pooled fraction is injected onto a Cosmosil 5C8-300 column (4 × 250 mm) initially equilibrated with eluant A. After the column is washed for 20 min with eluant A, the enzyme is eluted from the column with a linear gradient of increasing acetonitrile concentration, from 0 to 15% in eluant B in 5 min and then from 15 to 40% in eluant B in 60 min. The PLA_2 activity is eluted in a fraction corresponding to a single absorbance peak at about 27% acetonitrile in eluant B. A

TABLE I
PURIFICATION OF PHOSPHOLIPASE A_2 S-1 FROM RAT SPLEEN SUPERNATANT

Step	Total protein (mg)	Total activity (μmol/min)	Specific activity (nmol/min/mg)	Yield (%)
Homogenate	37,600	246	6.54	—
Supernatant	24,400	17.6	0.72	100
DEAE-cellulose	2340	9.33	3.99	53.0
TEAE-cellulose	752	5.60	7.45	31.8
BioGel P-30	9.66	3.96	410	22.5
Reversed-phase HPLC	0.042	3.2	76,200	18.2

small amount (~20 μl/ml) of 1 *M* Tris is added to the pooled fractions to neutralize the enzyme solution, and then the fractions are stored at −35°. The presence of acetonitrile does not affect the PLA_2 activity, but it somewhat prevents nonspecific adsorption of the enzyme to glass tubes or polypropylene tubes. Usually, this preparation is essentially homogeneous as judged by sodium dodecyl sulfate (SDS)-gel electrophoresis. However, when minor contaminants are present in the preparation, rechromatography of an aliquot (~5 μg) on a small Cosmosil 5C8-300 column (1 × 150 or 2.1 × 30 mm) is effective for further purification. A summary of a typical enzyme purification is shown in Table I.

From Rat Spleen Membrane Fraction[2]

Buffers

Buffer A, 10 m*M* Tris-HCl, pH 7.4
Buffer B, 10 m*M* Tris-HCl, 0.3% lithium dodecyl sulfate (LDS), pH 7.4
Buffer C, 10 m*M* Tris-HCl, 1 *M* NaCl, pH 7.4
Buffer D, 200 m*M* Tris-HCl, 0.5% 3-[(3-cholamidopropyl)dimethylammonio]-1-propane sulfonate (CHAPS), pH 7.4

Solubilization. Various salts and detergents are tested for the ability to extract PLA_2 from the splenic membrane fraction (108,000 *g* pellet). Extraction is carried out at 4° for 2 hr. A high concentration (up to 1 *M*) of KCl hardly solubilizes the enzyme activity, whereas treatment with 1 *M* KBr extracts about 20% of the PLA_2 activity in the membrane fraction. Interestingly, the PLA_2 activity in the membrane fraction of human spleen could be well solubilized (~70%) by the KBr treatment.[9] The reason for this species difference in the solubility of the enzymes is unknown at present. Since bromide is a chaotropic ion, the KBr treatment may perturb hydrophobic interactions as well as ionic interactions between the enzyme and the membrane. After being solubilized by 1 *M* KBr, the human splenic PLA_2 is purified without any detergent.[9,10] CHAPS and octylglucoside do not solubilize PLA_2 from the membrane fraction of rat spleen in the concentration range of 0.5 to 1%. In contrast, treatment with a much stronger detergent, LDS (0.3%), extracts about 90% of the PLA_2 activity. The PLA_2 activity shows no change in 0.3% LDS for at least several hours. Since LDS has a lower Kraft point than SDS, LDS does not precipitate at 4°. Hence, LDS is superior to SDS for the solubilization and further purification of PLA_2 at 4°.

The 108,000 *g* pellet is homogenized with 10 volumes of buffer B with a blender, and then the mixture is stirred for 2 hr at 4°. The resultant solution is centrifuged at 108,000 *g* for 1 hr.

DEAE–Cellulofine AM Treatment. The supernatant obtained is applied to a DEAE–Cellulofine AM column (6 × 10 cm, Seikagaku Kogyo, Tokyo, Japan) previously equilibrated with buffer A. Since the flow rate becomes gradually slower, during loading of the sample, because of the presence of LDS, the enzyme solution is thoroughly mixed with the gel by occasional stirring with a glass bar to gain a sufficient flow rate. The enzyme is adsorbed to the column under these conditions. The enzyme activity is eluted with buffer D containing 0.5 *M* Trizma sulfate, the pH of which is adjusted to 7.0 with 1 *M* Tris base. After being diluted 10-fold with buffer A to reduce the salt concentration, the enzyme fraction is applied to an S-Sepharose column (4 × 10 cm) preequilibrated with buffer A. The enzyme activity is eluted from the column with buffer C containing 5% CHAPS.

Octyl–Sepharose Chromatography. The pooled PLA_2 fraction is diluted 5-fold with buffer A to reduce the concentration of CHAPS. Then the appropriate amounts of lithium sulfate and Trizma sulfate are added to the enzyme solution to a final concentration of 0.5 *M* each, and the pH of the solution is adjusted to 7.0 with 1 *M* Tris base. The resultant solution is applied to an octyl-Sepharose column (2.5 × 20 cm) preequilibrated with buffer A containing 1 *M* lithium sulfate. Elution is performed with buffer A containing 1% CHAPS. The enzyme solution is diluted 10-fold with buffer A and concentrated with a small S–Sepharose column (1 × 3 cm) previously equilibrated with buffer A. The column is washed with buffer A containing 0.3 *M* NaCl and 0.5% CHAPS and then eluted with buffer C containing 0.5% CHAPS.

Gel Filtration on Cellulofine GCL 300-m Column. The concentrate is applied to a Cellulofine GCL 300-m (Seikagaku Kogyo) gel-filtration column (4.5 × 74 cm) preequilibrated with buffer D. A peak of PLA_2 activity is eluted at an elution volume of 780 ml.

S–Sepharose Chromatography. The pooled enzyme fraction is applied

TABLE II
PURIFICATION OF PHOSPHOLIPASE A_2 M FROM RAT SPLEEN MEMBRANE FRACTION[a]

Step	Total protein (mg)	Total activity (μmol/min)	Specific activity (nmol/min/mg)	Yield (%)
Crude membrane	2700	4.80	1.8	100
Extract	2360	4.40	1.9	91.7
DEAE-cellulose	385	3.51	9.1	73.1
Octyl-Sepharose	34.8	0.32[b]	9.2	6.7
Cellulofine GCL 300-m	3.0	0.96[b]	320	20.0
S-Sepharose	0.25	1.60	6400	33.3
BioGel P-30	0.01	0.30	30,000	6.3

[a] Data from T. Ono *et al.*[2]
[b] The reason for the apparent decrease in the total activity was not further examined.

to an S–Sepharose column (0.8 × 8 cm) preequilibrated with buffer A. Elution is performed with a linear gradient from 0.3 to 0.8 *M* NaCl in buffer A. The PLA_2 activity is eluted at 0.5–0.6 *M* NaCl. The fractions containing the enzyme activity are pooled and diluted 5-fold with buffer A. Then the solution is concentrated with a small S–Sepharose column (0.7 × 1 cm) as described at the step of octyl-Sepharose chromatography.

BioGel P-30 Gel Chromatography. The concentrate is applied to a BioGel P-30 column (2.2 × 45 cm) preequilibrated with buffer D. The PLA_2 activity is eluted, at an elution volume near the total available column volume, much later than the position expected from the molecular weight estimated on SDS–gel electrophoresis. This may suggest the hydrophobic interaction of the enzyme with the gels. The purified enzyme, named PLA_2 M, is concentrated with a small S–Sepharose column as described above and stored at −35°. Table II summarizes the results of a typical enzyme purification.

Properties

Purity and Molecular Weight. The PLA_2 preparations (PLA_2 S-1 and PLA_2 M) purified from the two sources were homogeneous as judged by SDS–gel electrophoresis and analytical HPLC.[1,2] Both enzymes migrated on SDS gels as a single protein band to the same position as rat pancreatic PLA_2: the apparent molecular weight of both enzymes was estimated to be 13,600, in agreement with the weights calculated from the amino acid sequences. PLA_2 S-1 was eluted as a sharp peak at the same retention time as pancreatic PLA_2 on reversed-phase HPLC, whereas PLA_2 M was eluted later, as a significantly broad peak, than PLA_2 S-1. This indicated that PLA_2 M has a molecular surface more hydrophobic than PLA_2 S-1.

Catalytic Properties. Purified PLA_2 S-1 and PLA_2 M absolutely required calcium ions for activity: the concentrations of calcium ion which give half-maximum velocity were 0.03 mM for PLA_2 S-1 and 0.5 mM for PLA_2 M.[1,2] Optimal activities were found in the range of pH 8.0–10.5 for PLA_2 S-1 and pH 8.0–9.5 for PLA_2 M. The positional specificity of PLA_2 S-1 and PLA_2 M regarding the hydrolysis of an acyl ester bond at the *sn*-2 position of phospholipids was confirmed by the HPLC method as described above.

The substrate specificities of PLA_2 S-1 and PLA_2 M were examined for sonicated phospholipid vesicles and for mixed micelles of phospholipids and bile salts (cholate or deoxycholate) at various molar ratios. In all cases, the substrate specificity of PLA_2 S-1 was practically identical to that of pancreatic PLA_2. The hydrolytic rate order for sonicated phospholipids in the case of PLA_2 M was PG $>$ PE $=$ PS $>$ phosphatidylcholine (PC), whereas that in the case of PLA_2S-1 was PG $>$ PE $>$ PC $>$ PS. The general trend in the specificity of PLA_2 M toward the mixed micelles was similar to that of PLA_2 S-1: PG was the best substrate, PS the poorest. There was, however, a clear difference in the dependence of the specificity on the cholate/phospholipid (PE or PC) molar ratio. The highest specific activities were obtained at a cholate/PG ratio of 6 (0.5 mmol/min/mg) for PLA_2 S-1 and at cholate/PG ratios ranging from 2.5 to 7 (0.15 mmol/min/mg) for PLA_2 M.

Immunochemical Properties. PLA_2 S-1 is immunochemically identical to rat pancreatic PLA_2, whereas there is no immunochemical similarity between PLA_2 M and pancreatic PLA_2.[1,2] An antipancreatic PLA_2 antibody did not recognize PLA_2 M, and conversely an anti-PLA_2 M antibody did not recognize the pancreatic PLA_2.

Structural Properties. The amino acid compositions of PLA_2 S-1 and pancreatic PLA_2 are very similar; in addition, the peptide maps and sequences of the amino-terminal 32 residues of the two enzymes are identical. The enzyme purified from rat spleen supernatant is, therefore, of the pancreatic type (group I).[1] When rat tissues were screened with the antipancreatic PLA_2 antibody[17] or rat pancreatic PLA_2 cDNA[18] as a probe, the pancreatic type PLA_2 was found in gastric mucosa and lung as well as in spleen. The gastric enzyme and its zymogen were purified and character-

[17] M. Okamoto, T. Ono, H. Tojo, and T. Yamano, *Biochem. Biophys. Res. Commun.* **128,** 788 (1985).

[18] O. Ohara, M. Tamaki, E. Nakamura, Y. Tsuruta, Y. Fujii, M. Shin, H. Teraoka, and M. Okamoto, *J. Biochem.* (*Tokyo*) **99,** 733 (1986).

ized.[8] The cDNAs encoding pancreatic PLA_2 were cloned from gastric mucosa and lung, and their nucleotide sequences were completely identical to that isolated from pancreas.[19] The pancreatic type PLA_2 was found to exist also in human spleen.[10]

The amino acid composition and peptide map of PLA_2 M were quite different from those of pancreatic PLA_2. The amino-terminal sequence of PLA_2 M revealed the absence of Cys-11, characteristic of group II (viperid/crotalid venom type) PLA_2s.[20] Group I (pancreatic and elapid type) PLA_2s are known to have a disulfide pair between Cys-11 and Cys-77.[20] More recently, the complete amino acid sequence of PLA_2 M was determined by nucleotide sequencing of PLA_2 M cDNA[21] as well as by protein sequencing[22] of PLA_2 M. The results from both the methods were completely identical and confirmed that PLA_2 M belongs to the group II PLA_2 category, that is, PLA_2 M has a Cys-50 and a carboxy-terminal extension of the seven residues terminating in a cysteine residue. The hydropathy profile of PLA_2 M did not show any plausible membrane binding site. The primary structure of PLA_2 M was different only by four residues from that of rat platelet secretory PLA_2.[23] The PLA_2 M cDNA contains a sequence similar to typical signal sequences for secretory proteins. The mechanism of anchoring of the enzyme on the membrane therefore remains to be clarified.

The primary structure of PLA_2 M purified from human spleen was also determined by protein sequencing.[10] The sequence was also homologous to that of group II enzymes and identical to that deduced from the nucleotide sequence of the corresponding gene cloned from a genomic library.[13,24]

[19] T. Sakata, E. Nakamura, Y. Tsuruta, M. Tamaki, H. Teraoka, H. Tojo, T. Ono, and M. Okamoto, *Biochim. Biophys. Acta* **1007,** 124 (1989).

[20] R. L. Heinrikson, E. T. Krueger, and P. S. Keim, *J. Biol. Chem.* **252,** 4913 (1977).

[21] J. Ishizaki, O. Ohara, E. Nakamura, M. Tamaki, T. Ono, A. Kanda, N. Yoshida, H. Teraoka, H. Tojo, and M. Okamoto, *Biochem. Biophys. Res. Commun.* **162,** 1030 (1989).

[22] T. Ono, H. Tojo, N. Yoshida, and M. Okamoto submitted for publication.

[23] M. Hayakawa, I. Kudo, M. Tomita, S. Nojima, and K. Inoue, *J. Biochem.* (*Tokyo*) **104,** 767 (1988).

[24] J. J. Seilhamer, W. Pruzanski, P. Vadas, S. Plant, J. A. Miller, J. Kloss, and L. K. Johnson, *J. Biol. Chem.* **264,** 5335 (1989).

[38] Purification and Characterization of Cytosolic Phospholipase A_2 Activities from Canine Myocardium and Sheep Platelets

By STANLEY L. HAZEN, LORI A. LOEB, and RICHARD W. GROSS

Introduction

Activation of intracellular phospholipases plays a critical role in the generation of biologically active metabolites such as eicosanoids and platelet-activating factor. Although many phospholipases are activated during signal transduction (e.g.,phospholipases A_2, C, and D), recent studies have demonstrated the importance of phospholipase A_2 activation in the generation of the majority of arachidonic acid mass released during cellular stimulation and in the initiation of platelet-activating factor synthesis.[1–4] Furthermore, the concomitant accumulation of unsaturated fatty acids and lysophospholipids during myocardial ischemia has underscored the importance of accelerated phospholipid metabolism mediated by phospholipase A_2 as a primary determinant of the pathophysiologic sequelae of myocardial ischemia.[5–7]

Although it has traditionally been assumed that intracellular phospholipase A_2 activities are mechanistically and structurally related to the classic low molecular weight phospholipases A_2, recent work has identified several new types of intracellular phospholipases A_2 which have kinetic characteristics (e.g., calcium sensitivity, substrate selectivity) which implicate them as the likely enzymic mediators of signal transduction in mammalian cells. In this chapter, we describe the purification and characterization of platelet[8] and myocardial[9,10] cytosolic phospholipases A_2 as examples of

[1] A. D. Purdon and J. B. Smith, *J. Biol. Chem.* **260,** 12700 (1985).
[2] M. J. Broekman, *J. Lipid Res.* **27,** 884 (1986).
[3] R. L. Wykle, B. Malone, and F. Snyder, *J. Biol. Chem.* **255,** 10256 (1980).
[4] J. D. Hanahan, *Annu. Rev. Biochem.* **55,** 483 (1986).
[5] K. R. Chien, A. Han, A. Sen, L. M. Buja, and J. T. Willerson, *Circ. Res.* **54,** 313 (1984).
[6] P. B. Corr, R. W. Gross, and B. E. Sobel, *Circ. Res.* **55,** 135 (1984).
[7] R. W. Gross, *in* "Pathobiology of Cardiovascular Injury" (H. L. Stone and W. B. Weglicki, eds.), p. 287. Martinus Nijhoff, Boston, Massachusetts, 1985.
[8] L. A. Loeb and R. W. Gross, *J. Biol. Chem.* **261,** 10467 (1986).
[9] R. A. Wolf and R. W. Gross, *J. Biol. Chem.* **260,** 7295 (1985).
[10] S. L. Hazen, R. J. Stuppy, and R. W. Gross, *J. Biol. Chem.* **265,** 10622 (1990).

intracellular phospholipases A_2 which have markedly different kinetic and physical characteristics in comparison to the venom and pancreatic enzymes.[11,12]

Phospholipase A_2 Assays

Platelet phospholipase A_2 is assayed in the presence of 65 mM Tris-Cl (pH 7.2), 10 mM $CaCl_2$, whereas myocardial phospholipase is assayed in the presence of 100 mM Tris-Cl (pH 7.0), 4 mM EGTA. Routine assays of platelet or myocardial phospholipase A_2 activities are performed by incubation of platelet sonicates, myocardial cytosol, or column chromatographic fractions (25–100 μl) with 2 μM radiolabeled lipid (600 Ci/mol) in the appropriate assay buffer at a final volume of 250 μl for 5 min at 37° in 10 × 75 mm borosilicate test tubes.

Reactions are initiated by injection of ethanolic stock solutions (5–10 μl) containing the appropriate amount of radiolabeled phospholipid. For routine assays, reactions are quenched with 100 μl butanol, vortexed, centrifuged at 2000 g_{max} for 2 min, and separated by thin-layer chromatography (TLC) on silica G channeled plates in a petroleum ether/ethyl ether/acetic acid (80 : 20 : 1, v/v) tank. To identify regions containing radiolabeled fatty acid, unlabeled oleic acid standard is spotted on each channel and, following chromatography, is visualized by brief I_2 staining. Regions corresponding to fatty acid (R_f 0.7) are scraped into vials and quantified by scintillation spectrometry after addition of fluor. This procedure yields near quantitative recovery of fatty acid.

When phospholipid, lysophospholipid, and fatty acid products are to be isolated and quantified simultaneously, reactions are quenched with 200 μl butanol, vortexed, and centrifuged, and products are isolated by TLC utilizing a chloroform/acetone/methanol/acetic acid/water (6 : 8 : 2 : 2 : 1, v/v) solvent system, which resolves lysophosphatidylcholine (R_f 0.15), phosphatidylcholine and lysophosphatidylethanolamine (R_f 0.40), phosphatidylethanolamine (R_f 0.65), and fatty acid (R_f 1.0). Under these conditions, analyses demonstrated that recoveries of 99% of fatty acid, 95% of phosphatidylcholine, 87% of phosphatidylethanolamine, 84% of lysophosphatidylcholine, and 74% of lysophosphatidylethanolamine are routinely achieved.

[11] H. M. Verheij, A. J. Slotboom, and G. H. de Haas, *Rev. Physiol. Biochem. Pharmacol.* **91,** 91 (1981).

[12] E. A. Dennis, *in* "The Enzymes" (P. D. Boyer, ed.), 3rd Ed., Vol. 16, p. 307. Academic Press, New York, 1983.

Miscellaneous Procedures and Sources of Materials

Synthesis of high specific activity radiolabeled plasmenylcholine molecular species (*sn*-2 radiolabeled with either palmitic, oleic, or arachidonic acids) is achieved by the dicyclohexylcarbodiimide-mediated synthesis of radiolabeled fatty acid anhydride and the subsequent *N,N*-dimethyl-4-aminopyridine-catalyzed condensation of radiolabeled anhydride with homogeneous 1-*O*-(*Z*)-hexadec-1′-enyl-*sn*-glycero-3-phosphocholine.[13] Syntheses of *sn*-2-radiolabeled diacyl- and 1-*O*-alkyl-2-acylglycerophosphocholine molecular species are achieved similarly utilizing either 1-hexadecanoyl-*sn*-glycero-3-phosphocholine or 1-*O*-hexadecyl-*sn*-glycero-3-phosphocholine and the appropriate radiolabeled fatty acid as starting materials. Protein content is determined utilizing the Bio-Rad (Richmond, CA) protein assay kit or QuantiGold (Diversified Biotech, Newton, MA) with bovine serum albumin (BSA) as standard.

Oleic, palmitic, and arachidonic acids are purchased from Nu Chek Prep (Elysian, MN). Bovine heart lecithin is obtained from Avanti Polar Lipids (Birmingham, AL). Radiolabeled reagents are purchased from New England Nuclear (Boston, MA). Chromatography resins and fast protein liquid chromatography (FPLC) columns are purchased from Pharmacia-LKB (Piscataway, NJ) and high-performance liquid chromatography (HPLC) columns are obtained from P. J. Cobert (St. Louis, MO). All other materials are obtained from Sigma (St. Louis, MO).

Purification of Canine Myocardial Cytosolic Phospholipase A_2

Early studies demonstrated that the cytosolic fraction possesses the major measurable phospholipase A_2 activity in canine myocardial homogenates and that this activity is calcium independent and selective for plasmalogen substrate.[9] To gain insight into the physical and kinetic characteristics of the enzyme catalyzing this novel phospholipase A_2 activity, the enzyme was purified to homogeneity utilizing sequential column chromatographies.[10]

Fresh ventricular tissue is obtained from mongrel dogs (25–35 kg), fed *ad libitum*) after intravenous injection of sodium pentothal (40 mg/kg). The heart is surgically removed and placed in homogenization buffer at 0° (0.25 *M* sucrose, 10 m*M* imidazole, 10 m*M* KCl, 10 m*M* potassium phosphate, pH 7.8, adjusted with HCl). All subsequent procedures are performed at 0°–4°. Ventricular tissue (100–150 g) is rapidly isolated, trimmed of fat, rinsed, weighed, placed in fresh homogenization buffer (3 : 1, v/w),

[13] R. A. Wolf and R. W. Gross, *J. Lipid Res.* **26,** 629 (1985).

and rapidly minced (0.2 × 0.4 cm pieces) with a sharp pair of scissors. Homogenization is performed with 3 strokes of a loose-fitting Potter–Elvehjem apparatus operated at 2000 rpm. Nuclei and cellular debris are removed from the homogenate by centrifugation at 1000 g_{max} for 10 min. The resultant supernatant is centrifuged at 10,000 g_{max} for 10 min, and the mitochondrial pellet (which does not contain substantial phospholipase activity) is discarded. The supernatant is subsequently centrifuged at 100,000 g_{max} for 60 min, yielding a microsomal fraction (pellet) and a cytosolic fraction (supernatant).

Myocardial cytosol is filtered through glass wool, dialyzed twice (8 hr each) against 50 volumes of buffer 1 (15 m*M* imidazole, 5 m*M* potassium phosphate, 10% glycerol, pH 7.8), and applied to a preequilibrated DEAE-Sephacel column (5 × 7 cm) at 3 ml/min. After washing with buffer 1 [containing 1 m*M* dithiothreitol (DTT)], phospholipase A_2 activity is eluted by direct application of 100 m*M* NaCl in elution buffer (10 m*M* imidazole, 10 m*M* KCl, 10% glycerol, 1 m*M* DTT, pH 8.0). Fractions possessing phospholipase A_2 activity are pooled and dialyzed against 20 liters of buffer 2 (10 m*M* imidazole, 10 m*M* KCl, 25% glycerol, 1 m*M* DTT, pH 8.0) overnight. The dialyzed DEAE-Sephacel eluate is subsequently loaded at 1.8 ml/min onto a Polybuffer Exchanger-94 chromatofocusing column (1.6 × 30 cm) previously equilibrated with buffer 2. Application of a polybuffer (PB) mixture (10% PB96, 5% PB74, 25% glycerol, 1 m*M* DTT, pH 7.1) results in elution of a sharply focused peak of phospholipase A_2 activity with an apparent isoelectric point of 7.55 (Fig. 1A).

Chromatofocusing fractions containing phospholipase A_2 activity are immediately applied to a 1 × 1 cm column of N^6-[(6-aminohexyl)carbamoylmethyl]ATP-agarose previously equilibrated with buffer 3 (10 m*M* imidazole, 25% glycerol, 1 m*M* DTT, pH 8.3) at 2 ml/min. The column is washed with 20 ml of buffer 3 alone, with 5 ml of buffer 3 containing 10 m*M* adenosine, with 15 ml of buffer 3 containing 10 m*M* AMP, and once again with 5 ml of buffer 3 alone (to remove all UV-absorbing AMP). Application of buffer 3 containing 1 m*M* ATP results in the quantitative elution of phospholipase A_2 activity from the ATP-agarose matrix (Fig. 1B). Phospholipase A_2 activity in the ATP-agarose eluate is purified 52,000-fold (from canine myocardial cytosol) and is moderately stable when stored at 0°–4° ($t_{1/2}$ 5–7 days) or stable indefinitely (6 months) in liquid nitrogen.

Phospholipase A_2 is further purified by the direct application of the ATP-agarose eluate to an HR 5/5 Mono Q column previously equilibrated with buffer 4 (20 m*M* imidazole, 25% glycerol, 1 m*M* DTT, pH 8.3). Application of a nonlinear salt gradient (0–450 m*M* NaCl) results in the elution of phospholipase A_2 activity (Fig. 2A). Mono Q active fractions are identified and immediately loaded onto a Koken hydroxylapatite HPLC

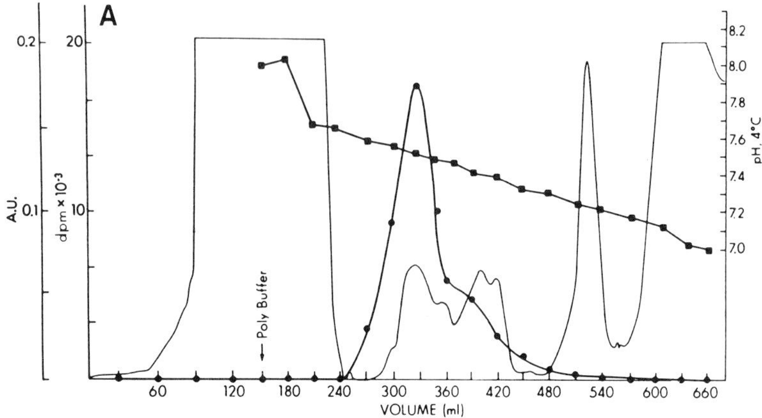

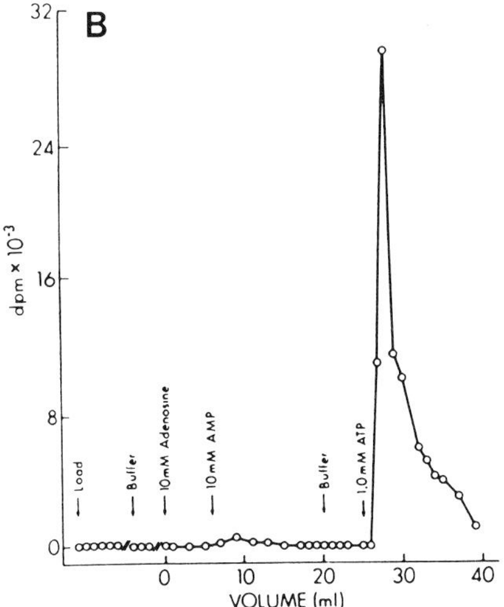

FIG. 1. Chromatofocusing and ATP affinity chromatographies of canine myocardial cytosolic phospholipase A_2. Eluate from the DEAE-Sephacel column was further purified utilizing sequential chromatofocusing (A) and ATP-agarose (B) chromatographies as described in the text. (A) Fatty acid release from 1-*O*-(*Z*)-hexadec-1′-enyl-2-[9,10-^{3}H]octadec-9′-enoyl-*sn*-glycero-3-phosphocholine (●); UV absorbance at 280 nm (—); pH (■). (B) Fatty acid release from 1-*O*-(*Z*)-hexadec-1′-enyl-2-[9,10-^{3}H]octadec-9′-enoyl-*sn*-glycero-3-phosphocholine (○).

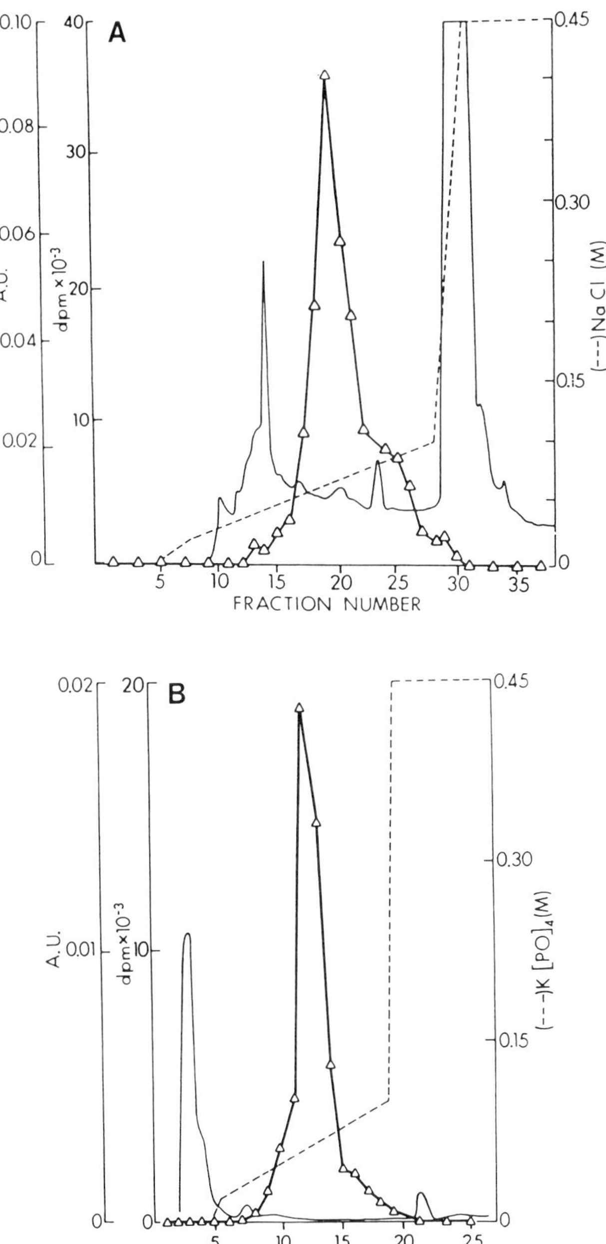

FIG. 2. Mono Q and HPLC hydroxylapatite chromatographies of canine myocardial phospholipase A_2. Phospholipase A_2 from the ATP affinity chromatography step was purified

TABLE I
PURIFICATION OF MYOCARDIAL CYTOSOLIC PHOSPHOLIPASE A_2[a]

Step	Protein (mg)	Total activity (nmol/min)	Specific activity (nmol/mg · min)	Purification (-fold)	Yield (%)
Cytosol	2430	3570	1.5	1	100
DEAE-Sephacel	540	3430	6.3	4.3	96
Chromatofocusing	6.0	3110	515	350	87
ATP-agarose	0.04	3070	76,500	52,040	86
Mono Q	0.006	1540	256,700	174,600	43
HPLC hydroxylapatite	0.003	680	226,700	154,200	19

[a] Myocardial cytosol and eluates from DEAE-Sephacel, chromatofocusing, ATP-agarose, Mono Q, and HPLC hydroxylapatite columns were incubated with 75 μM of 1-*O*-(*Z*)-hexadec-1′-enyl-2-[9,10-^{3}H]octadec-9′-enoyl-*sn*-glycero-3-phosphocholine in the presence of 4 m*M* EGTA. Fatty acid was extracted with butanol, separated by TLC, and quantified by scintillation spectrometry as described in the text.

column (4 mm × 10 cm) previously equilibrated with buffer 5 (10 m*M* potassium phosphate, 25% glycerol, 1 m*M* DTT, pH 7.4). Homogeneous phospholipase A_2 is obtained by application of a shallow, nonlinear potassium phosphate gradient (0–450 m*M*) (Fig. 2B).

Collectively, this series of column chromatographic steps results in a 154,000-fold purification of canine myocardial cytosolic phospholipase A_2 to a specific activity of 227 μmol/mg · min, with an overall yield of 19% (Table I). The purity of the preparation is substantiated by iodination of the active fractions from the hydroxylapatite column with Bolton–Hunter reagent, separation on sodium dodecyl sulfate-polyacrylamide gel electrophoresis (SDS-PAGE), and subsequent visualization by ^{125}I-autoradiography. In multiple preparations a 40-kDa polypeptide is the only protein whose intensity paralleled enzymatic activity ($n > 10$) and is the only band visualized after autoradiography of the most active hydroxylapatite fraction ($n = 3$).

to homogeneity by sequential Mono Q chromatography (A) and HPLC on hydroxylapatite (B) as described in the text. (A) Fatty acid release from 1-*O*-(*Z*)-hexadec-1′-enyl-2-[9,10-^{3}H]octadec-9′-enoyl-*sn*-glycero-3-phosphocholine (△); UV absorbance at 280 nm (—); NaCl gradient (- - -). (B) Fatty acid release from 1-*O*-(*Z*)-hexadec-1′-enyl-2-[9,10-^{3}H]octadec-9′-enoyl-*sn*-glycero-3-phosphocholine (△); UV absorbance at 280 nm (—); potassium phosphate gradient (- - -).

Kinetic Analysis of Purified Canine Myocardial Cytosolic Phospholipase A_2

Homogeneous myocardial phospholipase A_2 is a calcium-independent phospholipase with absolute regiospecificity for cleavage of the *sn*-2-acyl linkage in diradylglycerol phospholipids.[10] Myocardial phospholipase A_2 possesses a pH optimum of 6.4 (for each phospholipid substrate examined) and has an absolute requirement for DTT for expression of enzymatic activity.[10] The purified polypeptide is remarkable for its ability to selectively hydrolyze ether-containing glycerophospholipids in homogeneous vesicles (subclass rank order: plasmenylcholine > alkyl ether choline glycerophospholipid > phosphatidylcholine) or in mixed bilayers comprised of equimolar mixtures of plasmenylcholine and phosphatidylcholine.[10] Analysis of phospholipase A_2 activity utilizing phosphatidylcholine molecular species containing palmitate at the *sn*-1 position and a variety of fatty acids at the *sn*-2 position reveals a preference for hydrolysis of substrates with eicosatetraenoic fatty acids at the *sn*-2 position (preference for cleavage: arachidonate > oleate > palmitate ≫ acetate). The purified polypeptide also possesses intrinsic lysophospholipase and palmitoyl-CoA hydrolase activities, albeit at rates 2–3 orders of magnitude less than its phospholipase A_2 activity.[10]

Purification of Sheep Platelet Cytosolic Phospholipase A_2

Whole blood from three sheep (~5 liters) is collected directly into bottles containing acidified citrate–dextrose (ACD) to a final concentration of 15% (v/v) (ACD is 16 g citric acid, 44 g trisodium citrate, 49 g dextrose in 2 liters water). Fresh anticoagulated whole blood is poured into polyethylene tubes (~40 ml each) and centrifuged at 160 g_{max} for 45 min at 20°. The upper two-thirds to three-quarters of the platelet-rich supernatant is removed and transferred to polyethylene tubes containing 4 ml ACD (final volume, 40 ml),and each is subsequently centrifuged at 1500 g_{max} for 20 min at 20°. The white platelet pellets are washed by gentle resuspension in 2 ml of buffer A (50 m*M* Tris-Cl, 1 m*M* EGTA, 10% glycerol, pH 7.0 at 4°) containing 0.25 *M* dextrose, and the resultant mixtures are centrifuged at 1500 g_{max} for 20 min at 20°. Platelet pellets are gently resuspended in 2 ml of fresh buffer A containing, in addition, 0.25 *M* dextrose and are sonicated in 12.5-ml aliquots on ice using a Heat Systems sonicater (Farmingdale, NY) equipped with a microtip probe at a setting of 1 for three 15-sec bursts. All subsequent purification steps are performed at 4°. The sonicate is subsequently centrifuged at 100,000 g_{max} for 60 min, and the supernatant (cytosol) is dialyzed against 100 volumes of buffer A for 6 hr.

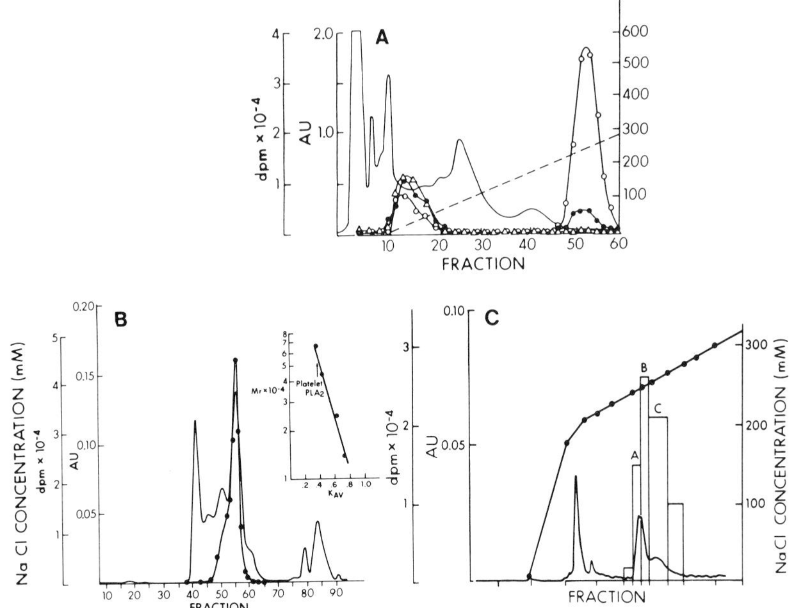

FIG. 3. Chromatography of platelet cytosolic phospholipase A_2. Sheep platelet phospholipase A_2 was purified utilizing sequential DEAE-cellulose (A), Superose 12 (B), and Mono Q (C) columns as described in the text. (A) Fatty acid release from 1-hexadecanoyl-2-[9,10-^{3}H]octadec-9′-enoyl-*sn*-glycero-3-phosphocholine (△), 1-*O*-hexadecyl-2-[9,10-^{3}H]octadec-9′-enoyl-*sn*-glycero-3-phosphocholine (●), or 1-*O*-(*Z*)-hexadec-1′-enyl-2-[9,10-^{3}H]octadec-9′-enoyl-*sn*-glycero-3-phosphocholine (○); UV absorbance at 280 nm (—); NaCl gradient (– – –). (B) Fatty acid release from 1-*O*-(*Z*)-hexadec-1′-enyl-2-[9,10-^{3}H]octadec-9′-enoyl-*sn*-glycero-3-phosphocholine (●); UV absorbance at 280 nm (—). (C) The vertical bars indicate fatty acid release from 5 μM of 1-*O*-(*Z*)-hexadec-1′-enyl-2-[9,10-^{3}H]octadec-9′-enoyl-*sn*-glycero-3-phosphocholine; UV absorbance at 280 nm (—); NaCl gradient (●—●).

Dialyzed platelet cytosol is loaded onto a DEAE-cellulose column (2.6 × 12 cm) previously equilibrated with buffer A, and phospholipase A_2 activities are eluted overnight at 1.6 ml/min utilizing a linear 2-liter salt gradient (0–410 mM NaCl) in buffer A. Assays of eluates from the DEAE-cellulose column reveal the presence of two peaks of phospholipase A_2 activity eluting at approximately 80 mM NaCl [phospholipase A_2 (α)] and at 230 mM NaCl [phospholipase A_2 (β)] (Fig. 3A). Phospholipase A_2 (α) hydrolyzes plasmenylcholine, phosphatidylcholine, and alkylether choline glycerophospholipids at similar rates. However, phospholipase A_2 (β), which contains the overwhelming majority of total observable phospholipase A_2 activity in platelet cytosol, preferentially hydrolyzes 1-*O*-(*Z*)-hexadec-1′-enyl-2-[9,10-^{3}H]octadec-9′-enoyl-*sn*-glycero-3-phospho-

TABLE II
PURIFICATION OF SHEEP PLATELET CYTOSOLIC PHOSPHOLIPASE A_2[a]

Step	Protein (mg)	Total activity (nmol/min)	Specific activity (nmol/mg · min)	Purification (-fold)	Yield (%)
Cytosol	510	8.7	0.017	1	100
DEAE-cellulose	5.7	11.4	2.0	118	117
Superose 12	0.4	2.8	7.0	410	29
Mono Q_A	0.05	0.4	8	710	
Mono Q_B	0.05	0.5	10	650	
Mono Q_C	0.03	1.8	60	3530	

[a] Purification of platelet phospholipase A_2 isoforms was achieved by sequential DEAE-cellulose, Superose 12, and Mono Q chromatographies. Aliquots of column eluates were assayed as described. The subscripts refer to the Mono Q fractions identified in Fig. 3C.

choline (plasmenylcholine) substrate in comparison to 1-*O*-hexadecyl-2-[9,10-^{3}H]-octadec-9′-enoyl-*sn*-glycero-3-phosphocholine (alkylacylglycerophosphorylcholine), whereas only diminutive amounts of hydrolysis are present utilizing 1-hexadecanoyl-2-[9,10-^{3}H]octadec-9′-enoyl-*sn*-glycero-3-phosphocholine as substrate (phosphatidylcholine).

The major observable peak of phospholipase A_2 activity in the cytosolic fraction, phospholipase A_2 (β), is subsequently pooled, dialyzed against 6 liters of buffer A at pH 8.0 for 3 hr and concentrated on a DEAE-cellulose column (1 × 2 cm). The column is washed with buffer A (pH 8.0) and then eluted with buffer A (adjusted to pH 7.0 with HCl) containing 410 m*M* NaCl. The concentrated phospholipase A_2 (β) activity is loaded onto tandem columns of Superose 12 (each 1.6 × 50 cm) and subsequently eluted utilizing buffer C (50 m*M* HEPES, 10% glycerol, pH 7.4). Phospholipase A_2 (β) activity elutes with an apparent molecular weight of 58 K (Fig. 3B). Active fractions are pooled and loaded onto an HR 5/5 Mono Q anion exchange column, and phospholipase A_2 (β) isoforms are eluted utilizing a nonlinear salt gradient (0–350 m*M* NaCl in buffer C) (Fig. 3C). Phospholipase A_2 activity is present in three resolvable UV-absorbing peaks. The purity of each peak is evaluated by SDS-PAGE and silver staining, which reveal a single intense band at 30 K in each UV-absorbing peak.

This series of steps collectively results in a 3000-fold purification of phospholipase A_2 (β) isoforms in 30% overall yield (combined activities from each isoform). Kinetic analysis of each resolvable isoform reveals that phospholipases A_2 (β_1) (Fig. 3C, fraction A) and (β_2) (fraction B) possess nearly identical specific activities, whereas the specific activity of phospholipase A_2 (β_3) (fraction C) is significantly greater (Table II). Taken

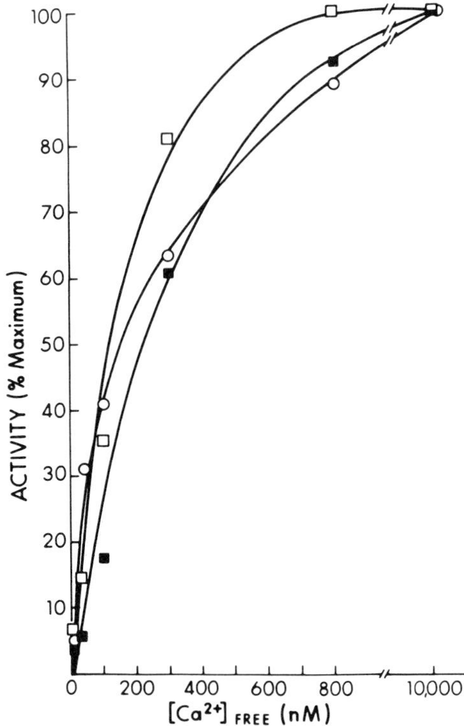

FIG. 4. Calcium ion sensitivity of sheep platelet phospholipase A_2 isoforms. Approximately 500 ng of purified phospholipase A_2 isoforms [phospholipase A_2 (β_1) (□), phospholipase A_2 (β_2) (○)] and 100 ng of phospholipase A_2 (β_3) (■) were incubated with appropriate concentrations of EGTA and $CaCl_2$ in 50 m*M* Tris-HCl (pH 7.5) at 37° to achieve the indicated concentrations of free calcium ion. Phospholipase A_2 isoforms were incubated for 5 min at 37° with 1-hexadec-1′-enyl-2-[9,10-^{3}H]octadec-9′-enoyl-*sn*-glycero-3-phosphocholine, and released fatty acid was quantified following TLC by scintillation spectrometry.

together, these results demonstrate that platelet phospholipase A_2 (β) is a dimeric polypeptide (58 kDa native mass) comprised of chromatographically resolvable isoforms of nearly identical monomeric molecular mass (30 kDa).

Characterization of Platelet Cytosolic Phospholipase A_2

Calcium Requirement

The calcium sensitivity of the purified phospholipase A_2 (β) isoforms is determined utilizing an EGTA/$CaCl_2$ buffering system.[14] Each of the

[14] H. Portzehl, P. C. Caldwell, and J. C. Reugg, *Biochim. Biophys. Acta* **79,** 581 (1964).

isoforms is fully activated by physiologic concentrations of calcium ion, exhibiting 30 and 90% of maximal enzymatic activity in the presence of 200 and 800 n*M* calcium, respectively (Fig. 4). In contrast, the less acidic phospholipase A_2 [cytosolic phospholipase A_2 (α)] requires over 1 m*M* $CaCl_2$ to reach half-maximal activity, exhibiting nominal activity at calcium concentrations less than 100 μM. Thus, phospholipase A_2 (β) is the major phospholipase A_2 activity in platelet cytosol, is comprised of dimeric polypeptides which are exquisitely sensitive to physiologic alterations in calcium ion concentration (200–800 n*M*), and demonstrates a substantial preference for ether-linked phospholipid substrates. Since the majority of platelet arachidonic acid mass is contained in plasmalogen molecular species, these results suggest that activation of this phospholipase A_2 (β) during platelet stimulation by physiologic alterations in calcium ion concentration (500–800 n*M*) results in the selective release of arachidonic acid and contributes to the synthesis of platelet-activating factor.

Acknowledgments

This research was supported by Grants HL34839 and HL35864. RWG is the recipient of an Established Investigator Award from the American Heart Association.

[39] Platelet-Activating Factor Acetylhydrolase in Human Erythrocytes

By DIANA M. STAFFORINI, STEPHEN M. PRESCOTT, and THOMAS M. MCINTYRE

Introduction

Platelet-activating factor (1-*O*-alkyl-2-acetyl-*sn*-glycero-3-phosphocholine, PAF) is a phospholipid that has been shown to possess potent biological activity. PAF induces hypotension, leukopenia, and thrombocytopenia[1] and increases vascular permeability.[2] PAF also activates platelets, neutrophils, and macrophages. A variety of cells, for example, neutrophils, macrophages, and endothelial cells, synthesize PAF on appropriate

[1] M. Halonen, J. D. Palmer, I. C. Lohman, L. M. McManus, and R. N. Pinckard, *Annu. Rev. Resp. Dis.* **122,** 915 (1980).

[2] J. Bjork and G. Smedegard, *J. Allergy Clin. Immunol.* **71,** 145 (1983).

stimulation.[3] Some inflammatory cells release most of the PAF that they synthesize (e.g., monocytes),[4] in contrast to vascular cells (e.g., endothelial cells)[5,6] which do not. PAF has been identified in human blood, saliva, urine, and amniotic fluid.[7–9]

The rate of removal of PAF is likely an important means of regulating the bioactivity of this compound. Farr and co-workers[10–12] were the first to demonstrate the occurrence, in mammalian plasma, of an enzyme that inactivates PAF by removal of the acetyl group esterifying the *sn*-2 position of glycerol, to produce lyso-PAF and acetate, which are biologically inactive. This reaction is catalyzed by a specific PAF acetylhydrolase (1-alkyl-2-acetylglycerophosphocholine esterase, EC 3.1.1.47, 1-alkyl-2-acetyl-*sn*-glycero-3-phosphocholine acetohydrolase). The plasma PAF acetylhydrolase efficiently hydrolyzes PAF released by inflammatory cells such as monocytes.[13] In addition, the plasma enzyme is capable of inactivating PAF synthesized (but not released) by stimulated endothelial cells.[14] Thus, the plasma PAF acetylhydrolase acts as a scavenger of circulating and cell-associated PAF.

In addition to the plasma enzyme, intracellular PAF acetylhydrolases have been described in a variety of tissues and blood cells.[15,16] Blank *et*

[3] S. M. Prescott, G. A. Zimmerman, and T. M. McIntyre, *Proc. Natl. Acad. Sci. U.S.A.* **81,** 3534 (1984).

[4] M. R. Elstad, S. M. Prescott, T. M. McIntyre, and G. A. Zimmerman, *J. Immunol.* **140,** 1618 (1988).

[5] S. M. Prescott, G. A. Zimmerman, and T. M. McIntyre, *Proc. Natl. Acad. Sci. U.S.A.* **81,** 3534 (1984).

[6] F. Bussolino, F. Breviario, C. Tetta, M. Aglietta, A. Mantovani, and E. Dejana, *J. Clin. Invest.* **77,** 2027 (1986).

[7] K. E. Grandel, R. S. Farr, A. A. Wanderer, T. C. Eisenstadt, and S. I. Wasserman, *N. Engl. J. Med.* **313,** 405 (1985).

[8] C. P. Cox, M. L. Wardlow, R. Jorgensen, and R. S. Farr, *J. Immunol.* **127,** 46 (1981).

[9] M. M. Billah and J. M. Johnston, *Biochem. Biophys. Res. Commun.* **113,** 51 (1983).

[10] R. S. Farr, C. P. Cox, M. L. Wardlow, and R. Jorgensen, *Clin. Immunol. Immunopathol.* **15,** 318 (1980).

[11] R. S. Farr, M. L. Wardlow, C. P. Cox, K. E. Meng, and D. E. Greene, *Fed. Proc., Fed. Am. Soc. Exp. Biol.* **42,** 3120 (1983).

[12] M. L. Wardlow, C. P. Cox, K. E. Meng, D. Greene, and R. S. Farr, *J. Immunol.* **136,** 3441 (1986).

[13] D. M. Stafforini, M. R. Elstad, T. M. McIntyre, G. A. Zimmerman, and S. M. Prescott, *J. Biol. Chem.* **265,** 9682 (1990).

[14] G. A. Zimmerman, T. M. McIntyre, M. Mehra, and S. M. Prescott, *J. Cell Biol.* **110,** 529 (1990).

[15] R. Yanoshita, I. Kudo, K. Ikizawa, H. W. Chang, S. Kobayashi, M. Ohno, S. Nojima, and K. Inoue, *J. Biochem. (Tokyo)* **103,** 815 (1988).

[16] T.-c. Lee, B. Malone, S. I. Wasserman, V. Fitzgerald, and F. Snyder, *Biochem. Biophys. Res. Commun.* **105,** 1303 (1982).

al.[17] described a PAF acetylhydrolase activity in the cytosolic fraction of mammalian tissues, Nijssen *et al.*[18] reported the presence of a similar activity in rat lung cytosol, and we have studied the properties of PAF acetylhydrolases in human tissues and blood cells.[19] These activities belong to a family of intracellular PAF acetylhydrolases that, like the plasma enzyme, are constitutively active, require a short-chain residue at the *sn*-2 position, and are calcium independent. However, the intracellular activities can be differentiated from the plasma enzyme by a variety of biochemical criteria.[19,20] The function(s) of these activities is yet to be determined, but they may play an important role in decreasing the accumulation of intracellular PAF. For example, human macrophages regulate PAF accumulation by increasing the levels of intracellular PAF acetylhydrolase activity[21]; this may limit the activation of cells that respond to the lipid. In addition, the plasma[22] and intracellular[23] PAF acetylhydrolases catalyze the degradation of oxidatively fragmented phospholipids. This may have major pathological significance since oxidatively fragmented phospholipids with toxic and biological activity have been shown to occur *in vivo*.[24,25] Thus, the PAF acetylhydrolases may play an important role as scavengers of PAF and toxic phospholipids.

Human erythrocytes contain a PAF acetylhydrolase activity which is highly specific for hydrolysis of phospholipids with short-chain acyl groups, much like the human plasma and tissue PAF acetylhydrolases.[19,26] However, the erythrocyte activity is due to a distinct protein with biochemical properties that clearly differentiate it from the plasma and other tissue activities.[19] This chapter describes in detail the procedure to assay this activity and a partial purification that allows the investigator to perform preliminary characterization studies. In addition, we describe the bio-

[17] M. L. Blank, T.-c. Lee, V. Fitzgerald, and F. Snyder, *J. Biol. Chem.* **256,** 175 (1981).

[18] J. G. Nijssen, C. F. P. Roosenbloom, and H. van den Bosch, *Biochim. Biophys. Acta* **876,** 611 (1986).

[19] D. M. Stafforini, S. M. Prescott, G. A. Zimmerman, and T. M. McIntyre, *Lipids,* in press (1990).

[20] M. L. Blank, M. N. Hall, E. A. Cress, and F. Snyder, *Biochem. Biophys. Res. Commun.* **113,** 666 (1983).

[21] M. R. Elstad, D. M. Stafforini, T. M. McIntyre, S. M. Prescott, and G. A. Zimmerman, *J. Biol. Chem.* **264,** 8467 (1989).

[22] K. E. Stremler, D. M. Stafforini, S. M. Prescott, G. A. Zimmerman, and T. M. McIntyre, *J. Biol. Chem.* **264,** 5331 (1989).

[23] D. M. Stafforini, S. M. Prescott, and T. M. McIntyre, unpublished results.

[24] A.Tokumura, K. Takauchi, T. Asai, K. Kamiyasu, T. Ogawa, and H. Tsukatani, *J. Lipid Res.* **30,** 219 (1989).

[25] A. Tokumura, T. Asai, K. Takauchi, K. Kamiyasu, T. Ogawa, and H. Tsukatani, *Biochem. Biophys. Res. Commun.* **155,** 863 (1988).

[26] D. M. Stafforini, S. M. Prescott, and T. M. McIntyre, *FASEB J.* **2,** A1375 (1988).

chemical characteristics of the partially purified enzyme and compare this activity to the plasma PAF acetylhydrolase.

Assay of Human Erythrocyte Platelet-Activating Factor Acetylhydrolase

Principle. To assay PAF acetylhydrolase activity, we use PAF labeled in the acetyl group. The amount of radioactivity released from the *sn*-2 position of the glycerol backbone is a direct measure of hydrolysis.[27,28] The released acetate can be easily separated from remaining substrate by reversed-phase column chromatography on disposable cartridges. The assay can be carried out easily and inexpensively. In addition, other short-chain glycerophospholipids labeled at the *sn*-2 position can be used as substrates. However, the only commercially available phospholipid with these features is PAF; thus, it is convenient to use this compound initially as a substrate for purification and characterization purposes.

Reagents

[*acetyl*-^{3}H]PAF: We usually prepare 4 ml of a 0.1 m*M* solution by first mixing 400 nmol of PAF (supplied in chloroform, Avanti Polar Lipids, Birmingham, AL) with 4.5 μCi of hexadecyl-2-acetyl-*sn*-glyceryl-3-phosphorycholine, 1-*O*-[*acetyl*-^{3}H(N)] (supplied in ethanol, New England Nuclear, Boston, MA). After evaporation of the solvents (by a stream of nitrogen) we add 4 ml of HEPES buffer, and the solution then is sonicated for 5 min at 4° and 100 W, using a 4-mm needle probe in a Braun Sonicator (Model 1510). The solution should be stored frozen to avoid nonenzymatic hydrolysis; it can be used for at least 1 week. We repeat the sonication step each time the substrate is thawed. Duplicate aliquots should be counted to determine the specific radioactivity of the substrate prepared each time; our working solutions are 100 μM with 10,000 counts per minute (cpm)/nmol.

HEPES buffer: 0.1 *M*, adjust the pH to 7.6 with 1 *M* potassium hydroxide. Prepare 100 ml.

Acetic acid: prepare 100 ml of a 10 *M* solution. Add 57.1 ml of glacial acetic acid slowly to 42.9 ml of distilled water.

Sodium acetate: 0.1 *M*. Prepare 500 ml.

Octadecylsilica gel cartridges: purchased from Baker Chemical Co. (Phillipsburg, NJ). Each assay requires an individual column. They can be reused several times, provided that they are properly regen-

[27] D. M. Stafforini, S. M. Prescott, and T. M. McIntyre, *J. Biol. Chem.* **262,** 4223 (1987).

[28] D. M. Stafforini, T. M. McIntyre, and S. M. Prescott, this series, Vol. 187, p. 344.

erated. Before each use, wash each column with 3 ml of chloroform/methanol (1 : 2, v/v), followed by 3 ml of 95% ethanol, and, finally, 3 ml of water.

Standard Assay Procedure. Human erythrocytes, obtained by drawing blood in EDTA or citrate, are lysed by dilution into NH_4Cl (final concentration 0.84%, w/v) for 10 min at room temperature. The erythrocyte lysate then can be separated from white blood cells by centrifugation at 2000 *g* for 20 min. Aliquots (5 μl) of the samples to be assayed (diluted in HEPES buffer, if necessary) are mixed with 5 μl of 400 m*M* dithioerythritol or dithiothreitol and 40 μl of 0.1 m*M* [*acetyl*-^{3}H]PAF in polypropylene tubes and then incubated for 15 min at 37°. Glass tubes should be avoided since substrates will bind to the glass surface. After incubation, 50 μl of acetic acid is added to stop the reaction, followed by 1.5 ml of sodium acetate solution. Each reaction mixture is then passed through an octadecylsilica gel cartridge, and the filtrates are collected in 15-ml scintillation vials. Each assay tube then is washed with an additional 1.5 ml of sodium acetate solution, and the wash is also passed through the cartridge and combined with the original effluent. Ten milliliters of Opti-Fluor (Packard Instruments Co.) is added to the vials, and the amount of radioactivity is determined in a liquid scintillation counter. When there are many samples, it is convenient to use a multiplace vacuum manifold to allow several samples to be processed simultaneously. If only a few samples need to be assayed, one can get satisfactory results by manually pushing the product of the reaction through a syringe attached to an octadecylsilica gel cartridge.

Expression of Results. The amount of enzymatic activity present is expressed after correction for quenching (see below), incubation time, dilution factors, and the amount of enzyme present in the assay. Using the procedure detailed above we found that the hydrolysis of PAF by a human erythrocyte lysate was linear with time (up to 15–30 min, Fig. 1A) and with protein concentration (up to 0.4 μl of packed erythrocytes per assay, Fig. 1B). The apparent K_m for PAF in lysed red blood cells was 32 μM (Fig. 1C); the K_m of a partially purified preparation of the enzyme (see below) was 10.2 μM. The optimal amount of reducing agent to be added varies slightly from preparation to preparation. Thus, it is advisable to first establish the optimal concentration of reducing agent for a given fraction before performing an experiment.

When assaying erythrocyte lysates or crude samples containing substantial amounts of hemoglobin, there can be artifacts due to quenching. Appropriate corrections should be made by comparison with a quenching curve where a constant amount of [^{3}H]acetate is counted in the absence or presence of the colored sample after the latter is passed through an octadecylsilica cartridge, as above.

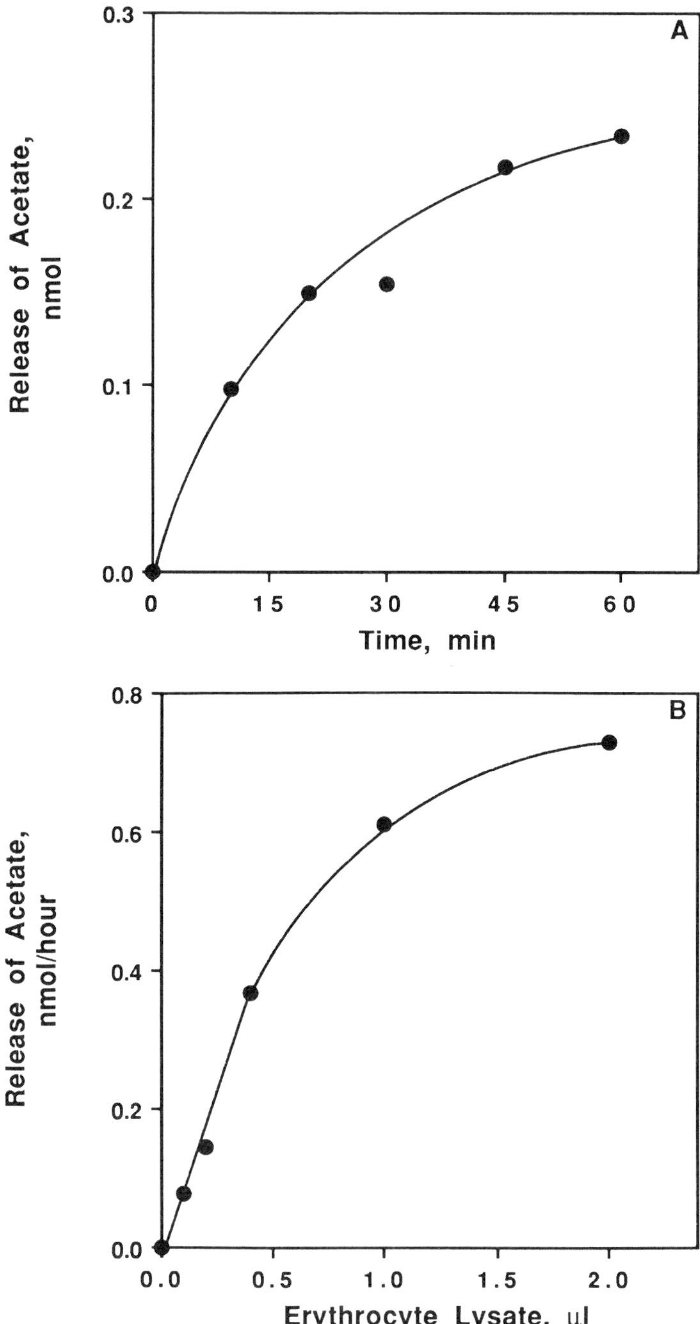

FIG. 1. Dependence of PAF hydrolysis on (A) time, (B) protein, and (C) substrate concentration.

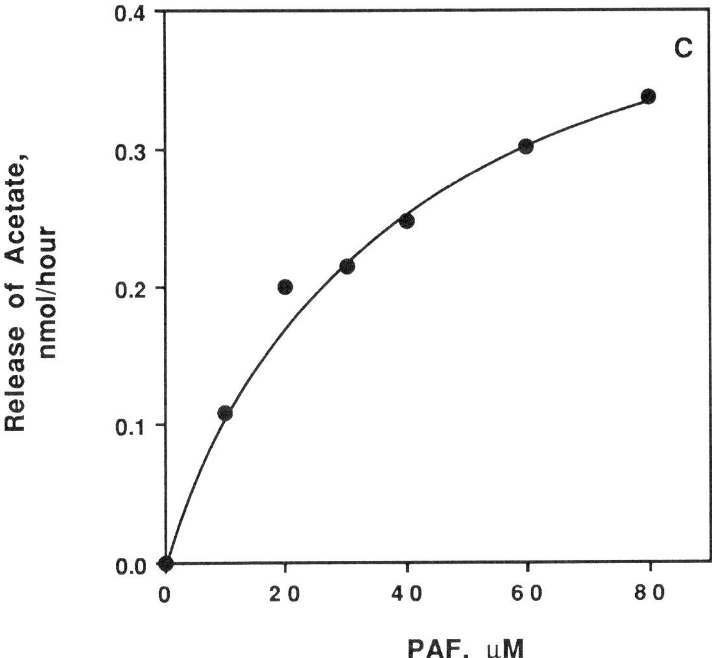

FIG. 1. (*continued*)

Comments. We have also used another octadecylsilica gel cartridge, the Waters (Milford, MA) Sep-Pak, but have found the Baker product to give less variable results. However, both types of cartridges can be reused at least 10 times without loss of binding capacity. The amount of protein permissible in the solution to be applied to the cartridges should be determined, since large amounts of protein will prevent binding of PAF to the resin. This results in what appears to be high activity since the product becomes contaminated with substrate. This should be examined by carrying out mock assays terminated at zero time. The Baker cartridge has a higher capacity than the Waters cartridge. Finally, following any change in procedure, one should verify that the apparent product is not contaminated with substrate. The effluent should be extracted[29] and examined by thin-layer chromatography (TLC)[30] or high-performance liquid chromatography (HPLC).[31]

[29] E. G. Bligh and W. F. Dyer, *Can. J. Biochem. Physiol.* **37,** 911 (1959).
[30] H. W. Mueller, J. T. O'Flaherty, and R. L. Wykle, *J. Biol. Chem.* **258,** 6213 (1983).
[31] A. R. Brash, C. D. Ingram, and T. M. Harris, *Biochemistry* **26,** 546 (1987).

Partial Purification of Human Erythrocyte Platelet-Activating Factor Acetylhydrolase

The PAF acetylhydrolase activity present in human erythrocytes can initially be purified from a human erythrocyte lysate by ion-exchange chromatography on DEAE-Sepharose CL-6B, as described below. This procedure separates the enzyme from hemoglobin and yields a preparation that is 485-fold purified from erythrocytes.

Reagents

Human blood cells: draw 200 ml of blood, using EDTA or sodium citrate as an anticoagulant. Keep the blood at 4° throughout the remaining steps.

EDTA: prepare 100 ml of a 0.25 *M* solution. Adjust the pH to 7.0 with NaOH.

Phosphate buffer: prepare 100 ml of a 1 *M* sodium phosphate stock by weighing equimolar amounts of NaH_2PO_4 and Na_2HPO_4. Make the necessary amount of 5 m*M* sodium phosphate buffer; if prepared as described above, the pH of this solution will be 6.85.

Saline: prepare 1 liter of 0.9% NaCl (9 g of solid NaCl in 1 liter of water).

DEAE-Sepharose CL-6B: 600 ml. Equilibrate the gel in 5 m*M* sodium phosphate buffer (pH 6.85) following the instructions provided by the manufacturer (Pharmacia, Piscataway, NJ).

Other reagents: KCl, NH_4Cl, aprotinin, and dithioerythritol. All of these reagents can be purchased from Sigma (St. Louis, MO).

Procedure. Centrifuge 200 ml of freshly drawn blood for 20 min at 2000 *g*. Discard the supernatant, suspend the blood cells in 0.9% NaCl (3 volumes of saline per volume of packed cells), and centrifuge as above. Discard the straw-colored supernatant and wash the cells twice more or until the washes are free of protein [determined by the Bio-Rad (Richmond, CA) protein assay] and free of plasma PAF acetylhydrolase activity. Discard the white buffy coat remaining on top of the red blood cells; to 100 ml of packed cells add 8.3 g of NH_4Cl, 38.5 mg of dithioerythritol, aprotinin (1.2 trypsin inhibitor units), 4 ml of 0.25 *M* EDTA, and water to a final volume of 1 liter. Stir the cells overnight at 4° and then dilute the preparation 5-fold by adding 4 liters of 5 m*M* sodium phosphate buffer (pH 6.85). Add 300 ml of DEAE-Sepharose CL-6B equilibrated in the same buffer and stir at 4° until an aliquot of the supernatant contains no PAF acetylhydrolase activity (3–4 hr). Decant the gel, discard the supernatant, and pack the slurry in a column of wide diameter (we use a 5 × 90 cm glass column). Wash the gel with 1 liter of the 5 m*M* sodium phosphate buffer; discard the reddish flow-through and the wash, which contain most of the hemoglobin present in the original cell lysate. Wash the column with 500 ml of 0.3 *M* KCl in buffer and collect five 100 ml-fractions; most of the activity is present in the third and fourth KCl washes. Pool the active fractions (~200 ml), dilute them 5-fold with sodium phosphate buffer, add dithioerythritol to a final concentration of 250 μM, and load the preparation (usually 1 liter) on a 2.5 × 60 cm DEAE-Sepharose CL-6B column at a flow rate of 100 ml/hr. Wash the column with 600 ml of buffer containing 250 μM dithioerythritol and elute the PAF acetylhydrolase with a 500-ml linear gradient of KCl (0–0.3 *M*) in buffer, at a flow rate of 35–40 ml/hr. If necessary, the column can be chased with buffer containing 0.3 *M* KCl. The active fractions can be concentrated by ultrafiltration using a PM10 Diaflo ultrafiltration membrane (Amicon, Danvers, MA). The preparation should be dialyzed to eliminate the KCl, and it is stable for at least several weeks if stored frozen in the presence of dithioerythritol (250 μM).

Comments. The specific activity in a fresh human erythrocyte lysate is 3.15 nmol/hr/mg protein. The effluent from the first DEAE step is 143-fold purified (specific activity 450 nmol/ml/hr). The second DEAE step yields a preparation that is 485-fold purified (specific activity 1530 nmol/ml/hr). The recovery of activity is close to 100%. The buffer and pH used during chromatography on DEAE is critical. In Tris-HCl (pH 7.5) we have found that hemoglobin binds to the resin and copurifies with the erythrocyte PAF acetylhydrolase. Thus, sodium phosphate buffer (pH 6.85) is the optimal choice for this initial separation.

Properties of Human Erythrocyte Platelet-Activating Factor Acetylhydrolase

The human erythrocyte PAF acetylhydrolase has a broad range of pH in which it is active. Maximal hydrolysis is observed at pH 7.6 with HEPES buffer, but the enzyme is also active when assayed in Tris-HCl (pH 7.5) and phosphate buffers (pH 6.8). The enzyme is stable when stored at −70° in phosphate and Tris-HCl buffers. It exhibits a half-life of 144 min at 37°.

Substrate Specificity. The human erythrocyte PAF acetylhydrolase has a high specificity for glycerophospholipids with short-chain acyl groups at the *sn*-2 position of glycerol. Phosphatidylcholines containing very long-chain acyl groups (i.e., arachidonoyl) at the *sn*-2 position are not inhibitors of PAF hydrolysis by the purified enzyme. In addition, 1-palmitoyl-2-hexanoylglycerophosphocholine is not a substrate of the purified enzyme. 1-Palmitoyl-2-glutaroylglycerophosphocholine was hydrolyzed by the partially purified enzyme, albeit at 50% of the rate at which PAF was hydrolyzed. It remains to be established whether the suitability of this compound as a substrate is due to the oxidized nature of the substituent at the *sn*-2

TABLE I
SULFHYDRYL GROUPS ESSENTIAL FOR ACTIVITY OF HUMAN ERYTHROCYTE PAF ACETYLHYDROLASE[a]

Additions[b]	Preincubation[c] (min)	Relative acetylhydrolase activity (%)	
		Erythrocyte	Plasma
Dithiothreitol (1 m*M*)	10	152	94
Dithioerythritol (1 m*M*)	10	136	N.D.[d]
DTNB (2 m*M*)	30	4	83
Iodoacetic acid (2 m*M*)	10	48.4	90.4
Iodoacetic acid (20 m*M*)	10	5.2	67.7

[a] Partially purified preparations of the acetylhydrolases were used for these studies.
[b] Final concentrations in the assay mixtures.
[c] At 37°.
[d] N.D., Not determined.

position, or whether the specificity of the enzyme decreases abruptly as the length of the *sn*-2 acyl groups increases from five to six carbons.

Sensitivity to Sulfhydryl Reagents. Three lines of evidence suggest the presence of sulfhydryl groups essential for activity of the human erythrocyte PAF acetylhydrolase. First, the addition of reducing agents is necessary for maximal hydrolysis (Table I). Second, the activity can be stabilized on storage by supplementing reducing agents (data not shown). In addition, PAF hydrolysis is inhibited by pretreatment with agents such as dithiobis(2-nitrobenzoic acid) (DTNB) and iodoacetic acid, which inactivate enzymes by reacting with free sulfhydryl groups (Table I). In contrast, the human plasma PAF acetylhydrolase does not require the addition of reducing agents for maximal activity, nor is the enzyme inhibited by DTNB or iodoacetic acid.

Metals and Chelators. A variety of heavy metals are inhibitors of the erythrocyte PAF acetylhydrolase activity (Fig. 2). The presence of sulfhydryl groups essential for activity likely accounts for the inhibition exerted by heavy metals such as cadmium, lead and copper. PAF did not protect the enzyme from copper inhibition, suggesting that heavy metals and phospholipid substrates bind to different sites.

The addition of calcium, an obligatory cofactor for activity of most phospholipases A_2, is not necessary for PAF hydrolysis by the human erythrocyte PAF acetylhydrolase (Table II). Further, the divalent metal chelator EDTA does not significantly inhibit PAF hydrolysis.

Histidine Inhibitors. The presence of essential histidine residues at the

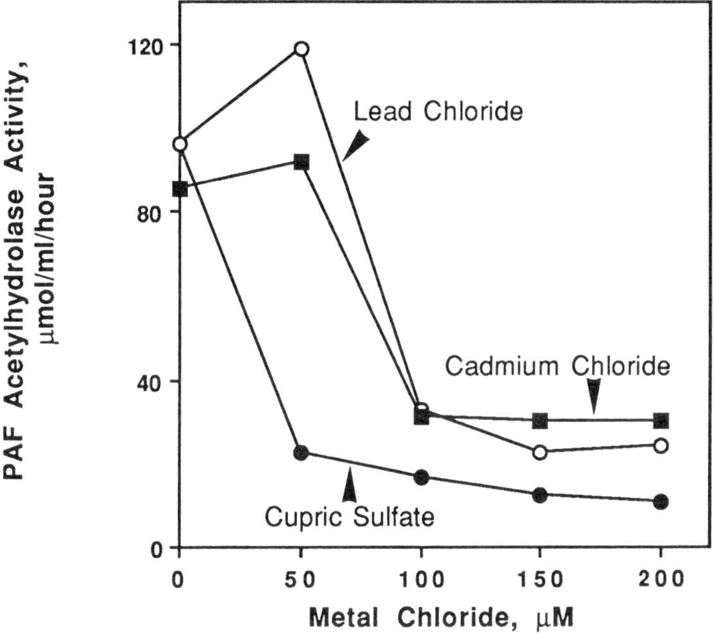

FIG. 2. Effect of metals on human erythrocyte PAF acetylhydrolase activity.

TABLE II
EFFECT OF CALCIUM IONS AND EDTA ON HYDROLYSIS OF PAF CATALYZED BY HUMAN ERYTHROCYTES[a]

Additions[b]	Relative activity (%)
$CaCl_2$ (0.5 m*M*)	64
$CaCl_2$ (1.0 m*M*)	42
EDTA (12.5 m*M*)	88
EDTA (25.0 m*M*)	75

[a] A washed erythrocyte lysate (5 μl of a 25-fold dilution) was preincubated for 30 min at 37° in the absence or presence of $CaCl_2$ or EDTA, in a total volume of 10 μl. Then 40 μl of 0.1 m*M* [*acetyl*-^{3}H]PAF was added, and incubations were continued for 30 min at 37°.

[b] Final concentrations in the assay mixtures.

active site of the human erythrocyte PAF acetylhydrolase was examined by testing the effects of diethyl pyrocarbonate (DEPC) and *p*-bromophenacyl bromide (pBPB), a nucleophilic reagent that derivatizes a histidine at the active site of the phospholipase A_2 purified from snake venom.[32] The PAF acetylhydrolase activity from human erythrocytes was inhibited by both compounds; in contrast, the activity present in human plasma was resistant (Fig. 3). This suggests the presence of an essential histidine residue(s) at the active site of the erythrocyte, but not the plasma, activity.

Serine Esterase Inhibitors. The presence of serine residues at the active site of the human erythrocyte PAF acetylhydrolase was examined by testing the effect of the serine esterase inhibitor diisopropyl fluorophosphate (DFP) on PAF hydrolysis. Preincubation with DFP (up to 1 m*M*) for 30 min at 37° almost completely inhibited PAF hydrolysis by the human erythrocyte PAF acetylhydrolase. The plasma activity was inhibited to a lesser extent (Fig. 4).

Susceptibility to Proteolysis. An additional line of evidence suggesting that the erythrocyte and plasma PAF acetylhydrolases are distinct enzymes is their different sensitivities to proteolysis. We incubated the erythrocyte and plasma PAF acetylhydrolases with various proteases and then examined the amount of activity remaining after protease treatment (Table III). The erythrocyte activity was sensitive to all the proteases tested, in contrast to the plasma activity which was resistant.

Effect of Detergents. Since the phospholipids that are substrates of the erythrocyte PAF acetylhydrolase are hydrophobic, it may be necessary and often advisable to add detergents to the incubation assay, especially when testing phospholipids with *sn*-2 acyl chains longer than acetate. The detergent of choice must provide adequate substrate solubilization and not inhibit enzymatic activity. We examined the effect of detergents on PAF hydrolysis by the human erythrocyte PAF acetylhydrolase (Table IV). Tween 20 and decyl-β-D-glucopyranoside are suitable choices since they do not significantly inhibit PAF hydrolysis.

Comparison between Erythrocyte and Other Intracellular PAF Acetylhydrolases. The human erythrocyte PAF acetylhydrolase can clearly be distinguished from other intracellular PAF acetylhydrolases by several criteria.[19] First, the erythrocyte activity is sensitive to sulfhydryl reagents, and it requires the addition of reducing agents for maximal activity, in contrast to other cellular PAF acetylhydrolase activities. Second, sodium fluoride inhibits only the erythrocyte activity. In addition, the erythrocyte activity is extremely sensitive to proteolysis, in contrast to other intracellular PAF acetylhydrolases which are only partially sensitive to treatment

[32] M. F. Roberts, R. A. Deems, T. C. Mincey, and E. A. Dennis, *J. Biol. Chem.* **252,** 2405 (1977).

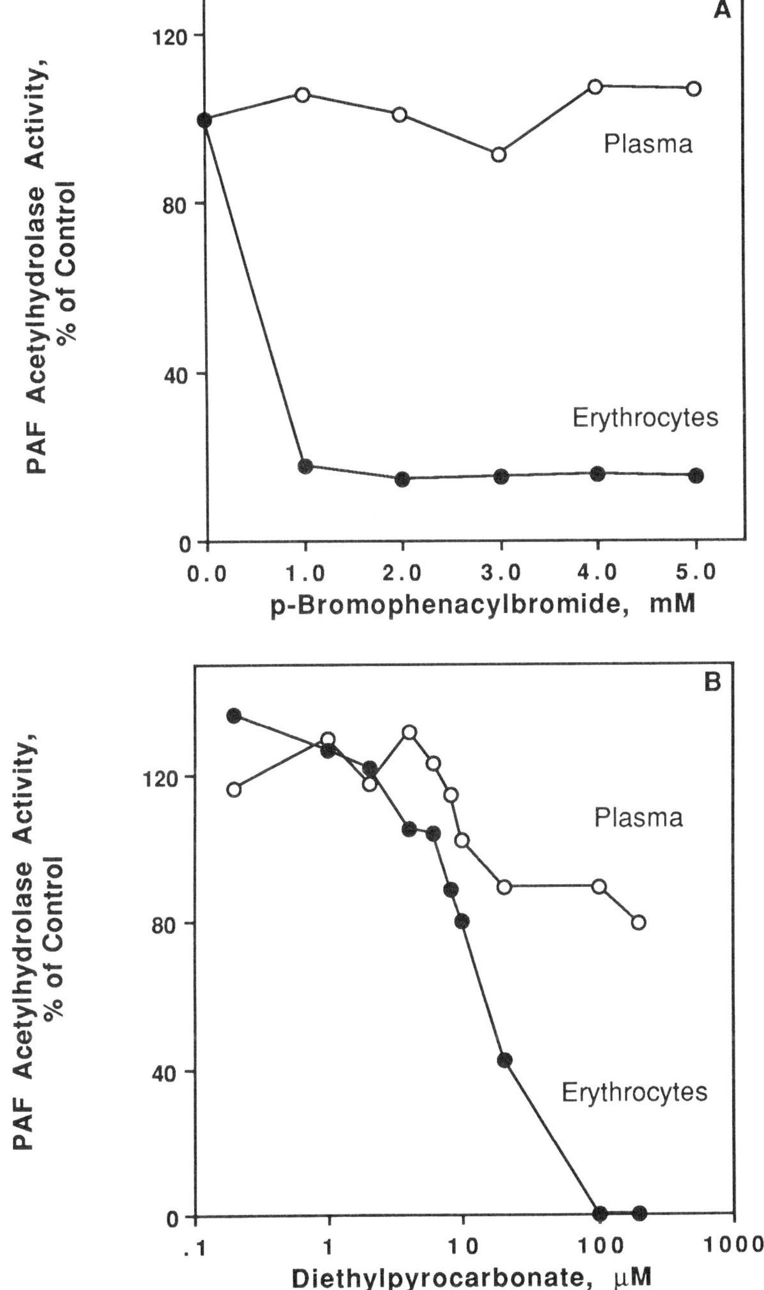

FIG. 3. Effect of histidine reagents (A, *p*-bromophenacyl bromide; B, diethyl pyrocarbonate) on human plasma and erythrocyte PAF acetylhydrolase activities.

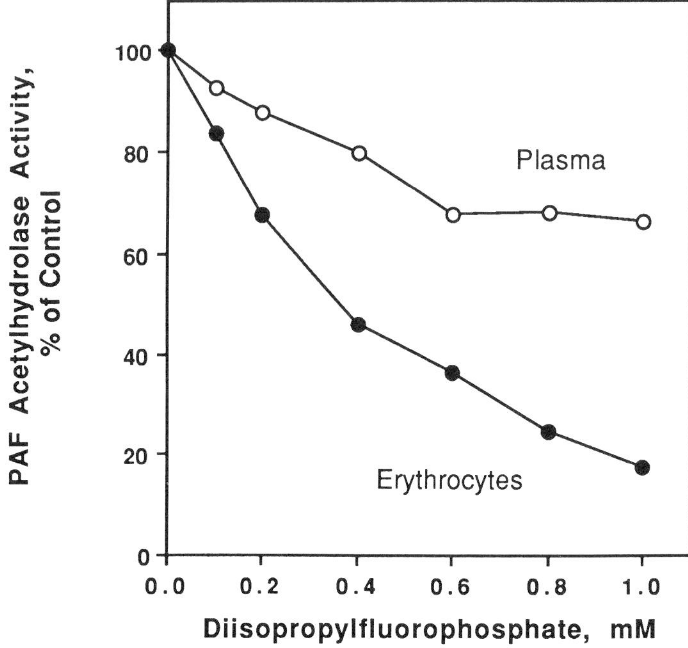

FIG. 4. Effect of diisopropyl fluorophosphate on human plasma and erythrocyte PAF acetylhydrolase activities.

TABLE III
SUSCEPTIBILITY OF HUMAN ERYTHROCYTE AND PLASMA PAF ACETYLHYDROLASES TO PROTEOLYSIS[a]

Additions	Preincubation time (min)	Relative activity (%)	
		Plasma	Erythrocyte
Trypsin (0.25 mg)	30	85	11
Protease K (4 μg)	30	104	13
	90	107	7
Papain (10^{-2} units)	30	98	55
	90	98	18
V_8 Protease (4 μg)	30	81	49
	90	109	24

[a] Preincubation with added proteases was carried out for 30 or 90 min at 37° in a total volume of 50 μl. Then 20-μl aliquots were assayed with 30 μl of 0.1 mM [*acetyl*-^{3}H]PAF, and the incubations were continued for 30 min at 37°.

TABLE IV
EFFECT OF DETERGENTS ON HUMAN ERYTHROCYTE PAF ACETYLHYDROLASE

Detergent added[a]	Critical micellar concentration[b]	PAF acetylhydrolase activity (% of control)
CHAPS	6–10 mM	13
Tween 20	59 μM	81
Triton X-100	250 μM	43
Deoxycholate	4–8 mM	0
Zwittergent 3-10	25–40 mM	0
Zwittergent 3-12	2–4 mM	1
Zwittergent 3-14	100–400 μM	8
Zwittergent 3-16	10–60 μM	57
Dodecyl-β-D-maltoside	100–600 μM	17
Heptyl-β-D-glucopyranoside	79 mM	8
Nonyl-β-D-glucopyranoside	6.5 mM	2
Decyl-β-D-glucopyranoside	2–3 mM	80

[a] Detergents were added to standard assays at the critical micellar concentration. A partially purified preparation of the erythrocyte PAF acetylhydrolase was used for these studies.

[b] Critical micellar concentrations reported by Calbiochem Corporation (La Jolla, CA).

with proteases. In conclusion, the human erythrocyte PAF acetylhydrolase is a distinct enzyme with properties that differentiate it from the other intracellular PAF acetylhydrolases and from the plasma activity.

Role of Human Erythrocyte PAF Acetylhydrolase. Human erythrocytes do not synthesize PAF when stimulated with calcium ionophore or agonists known to induce PAF production by other cells.[33] However, PAF is produced and released by inflammatory cells, and hemolysis of erythrocytes at sites of inflammation would result in the release of the intracellular fluid. The release of PAF acetylhydrolase activity could down-regulate the response(s) elicited by PAF by limiting further recruitment of cells that respond to the lipid.

A second potential role of the erythrocyte PAF acetylhydrolase is the degradation of toxic products of lipid peroxidation such as oxidatively fragmented phospholipids. The production of this class of compounds in erythrocytes is likely to occur since these cells are exposed to high oxygen tensions, and they contain large quantities of hemoglobin, a catalyst of lipid peroxidation. The erythrocyte PAF acetylhydrolase could provide a mechanism to protect the cell against tissue damage since it would recognize and degrade only toxic species without acting on structural membrane phospholipids.

[33] T. M. McIntyre, unpublished results, 1989.

[40] Phospholipase Activity of Lecithin–Cholesterol Acyltransferase

By CHRISTOPHER J. FIELDING and XAVIER COLLET

Introduction

Lecithin–cholesterol acyltransferase (LCAT; EC 2.3.1.43) in mammalian plasma reacts with high-density lipoprotein (HDL) to catalyze the transfer of an acyl group from lecithin to the 3-position of cholesterol with the production of lysolecithin and cholesteryl ester (Table I). Shortly after the first isolation of the enzyme protein,[1] it was determined that, in the absence of sterol, LCAT could also catalyze the hydrolysis of lecithin with the production of lysolecithin and unesterified fatty acid.[2,3] Subsequently it was shown that the LCAT protein was also responsible for the lecithin–lysolecithin acyltransferase (LLAT) exchange reaction which had been reported earlier to catalyze the exchange of acyl chains between lecithin and lysolecithin in plasma.[4,5] LCAT also shows esterase activity with short- and medium-chain *p*-nitrophenyl esters.[6]

The LCAT gene has been cloned and sequenced and the predicted amino acid sequence of the enzyme described.[7,8] Identical information was subsequently obtained by direct peptide sequencing.[9] The amino acid sequence of LCAT shows some local similarities to other lipases, particularly lipoprotein lipase and the heparin-released hepatic triglyceride lipase.[10,11] Like these, LCAT appears to be a classic serine hydrolase. The

[1] J. J. Albers, V. G. Cabana, and Y. D. B. Stahl, *Biochemistry* **15,** 1084 (1976).

[2] U. Piran and T. Nishida, *J. Biochem. (Tokyo)* **80,** 887 (1976).

[3] L. Aron, S. Jones, and C. J. Fielding, *J. Biol. Chem.* **253,** 7220 (1978)

[4] P. V. Subbaiah, this series, Vol. 129 [47].

[5] P. V. Subbaiah, J. J. Albers, C. H. Chen, and J. D. Bagdade, *J. Biol. Chem.* **255,** 9275 (1980).

[6] F. S. Bonelli and A. Jonas, *J. Biol. Chem.* **264,** 14723, (1989).

[7] J. McLean, C. Fielding, D. Drayna, H. Dieplinger, B. Baer, W. Kohr, W. Henzel, and R. Lawn, *Proc. Natl. Acad. Sci. U.S.A.* **83,** 2335 (1986).

[8] J. McLean, K. Wion, D. Drayna, C. Fielding, and R. Lawn, *Nucleic Acids Res.* **14,** 9397 (1986).

[9] C. Y. Yang, D. Manoogian, Q. Pao, F. S. Lee, R. D. Knapp, A. M. Gotto, and H. J. Pownall, *J. Biol. Chem.* **262,** 3086 (1987).

[10] K. Wion, T. D. Kirchgessner, A. J. Lusis, M. C. Schotz, and R. M. Lawn, *Science* **235,** 1638 (1987).

TABLE I
REACTIONS CATALYZED BY LECITHIN–CHOLESTEROL ACYLTRANSFERASE

Phospholipid substrate	-OH substrate	Phospholipid product	Acylated product
Lecithin	Alcohols (sterols, long-chain alcohols)	Lysolecithin	Cholesteryl esters, long-chain esters
Lecithin	Water	Lysolecithin	Unesterified fatty acid
Lecithin	Lysolecithin	Lysolecithin	Lecithin

presence of active site serine and histidine residues has been demonstrated by chemical modification,[12,13] and it is likely that the active site of LCAT involves the classic triad of serine, histidine, and aspartate residues.

The phospholipase activity of LCAT has some unusual features. It is Ca^{2+} independent, as shown by its undiminished activity in the presence of high concentrations of EGTA or EDTA.[3] LCAT phospholipase and transferase activities show predominant 2-position specificity[3,14] but not the absolute specificity shown by many phospholipases (see other chapters in this volume). Finally the phospholipase activity of LCAT, like its acyltransferase activity, is markedly dependent on the presence of lipoprotein apolipoproteins, most effectively apolipoprotein A-I (apoA-I), the major protein of the physiological substrate, HDL. Pure LCAT shows very little catalytic activity with phospholipid vesicles in the absence of apoA-I.[3]

As is the case with other phospholipases, the activity of LCAT is highly dependent on the physical structure of the lipid interface. Studies with various phospholipids dissolved in a nonhydrolyzable phosphatidyl ether matrix show that much of the substrate specificity observed when LCAT reacts with pure phospholipid dispersions reflects this type of effect, rather than true substrate specificity.[15]

Two recent reviews summarize current information about LCAT biochemistry. One summarizes the gene and protein structure and mechanism

[11] G. A. Martin, S. Busch, G. D. Meredith, A. D. Cardin, D. T. Blankenship, S. T. J. Mao, A. E. Rechlin, C. W. Woods, M. M. Racke, M. P. Schafer, M. C. Fitzgerald, D. M. Burke, M. A. Flanagan, and R. L. Jackson, *J. Biol. Chem.* **263,** 10907 (1988).

[12] M. Jauhiainen and P. J. Dolphin, *J. Biol. Chem.* **261,** 7032 (1986).

[13] M. Jauhiainen, N. D. Ridgeway, and P. J. Dolphin, *Biochim. Biophys. Acta* **918,** 175 (1987).

[14] G. Assmann, G. Schmitz, N. Donath, and D. Lekim, *Scand. J. Clin. Invest.* **38** (Suppl. 150), 16 (1978).

[15] H. J. Pownall, Q. Pao, and J. B. Massey, *J. Biol. Chem.* **260,** 2146 (1985).

of LCAT.[16] The other provides a comprehensive account of biophysical aspects of substrate structure and mechanism.[17] This chapter focuses on the phospholipase reactions of LCAT.

Isolation of Lecithin–Cholesterol Acyltransferase Protein

Human plasma contains LCAT protein at a concentration of about 6 μg/ml[18] and has been used as source for almost all purification procedures described. The available methods are referenced in an earlier chapter in this series.[19] The following method is suitable for routine use.

Step 1: Ultracentrifugal Flotation. Plasma (1000 ml) from blood collected in a final concentration of 10 m*M* sodium citrate (pH 7.0) is brought to 1 m*M* with disodium EDTA and to a solvent density of 1.21 g/ml with solid KBr (0.33 g/ml plasma) that has been maintained in a desiccator at 120°. Centrifugation is carried out for 48 hr at 40,000 rpm in a Beckman ultracentrifuge at 0°–4°. The floating colored lipoprotein layer is first removed by suction. The clear intermediate zone is then collected for further purification.

Step 2: Phenyl-Agarose Affinity Chromatography. The material from Step 1 is applied to a column (2.5 × 20 cm) of phenyl-agarose (Pharmacia-LKB, Uppsala, Sweden) equilibrated with 3 *M* NaCl, 1 m*M* EDTA. After washing with 500 ml of the same solution, the column is washed with 0.15 *M* NaCl, 1 m*M* EDTA, until the OD_{280} of the eluate is below 0.05. The remaining bound protein is then eluted with distilled water.

Step 3: DEAE–Cellulose Chromatography. The eluate from Step 2 is applied to a column (1 × 10 cm) of DEAE–cellulose [Whatman (Clifton, NJ) DE-52] equilibrated with 10 m*M* Tris-HCl (pH 7.4) and fractionated on a 250-ml linear gradient of 0–0.35 *M* NaCl in the same buffer. The enzyme activity is eluted in approximately 0.15 *M* NaCl.

Step 4: Hydroxylapatite Chromatography. The DEAE–cellulose fractions containing LCAT activity are pooled and passed on to a column (1.5 × 4 cm) of hydroxylapatite (BioGel HT, Bio-Rad, Richmond, CA) equilibrated with 10 m*M* Tris-HCl buffer, pH 7.0. The column is then eluted with a gradient (0–15 m*M*) of phosphate in the same buffer. Different batches of hydroxylapatite show considerably different binding properties

[16] C. J. Fielding, *in* "Advances in Cholesterol Research" (M. Esfahani and J. Swaney, eds.), p. 271. Telford Press, New Jersey, 1990.

[17] A. Jonas, *in* "Plasma Lipoproteins" (A. M. Gotto, ed.), p. 299. Elsevier, Amsterdam, 1989.

[18] J. J. Albers, J. L. Adolphson, and C. H. Chen, *J. Clin. Invest.* **67,** 141 (1981).

[19] J. J. Albers, C. H. Chen, and A. G. Lacko, this series, Vol. 129 [45].

and must be standardized. LCAT activity is typically eluted as a sharp peak at phosphate concentrations of about 5 m*M*.

The product, essentially pure by sensitive silver staining following sodium dodecyl sulfate (SDS)-gel electrophoresis, is recovered in a yield of about 10% purified about 20,000-fold from the original plasma.

Assay of Lecithin–Cholesterol Acyltransferase Phospholipase Activity

Principle. LCAT phospholipase is assayed as the production of unesterified fatty acid from synthetic lecithins labeled in the acyl moiety with ^{3}H or ^{14}C.

$$\text{Lecithin} + H_2O \rightarrow \text{lysolecithin} + \text{fatty acid}$$

Sources of Labeled Lecithins. A variety of synthetic labeled lecithins can now be purchased commercially. Double-labeled lecithins or unusual lecithin species unavailable commercially are readily synthesized. Labeled unesterified fatty acid is converted to its anhydride with cyclohexylcarbodiimide, and the labeled anhydride is then reacted with anhydrous glycerylphosphorylcholine–$CdCl_2$ complex or unlabeled 1-acyllysolecithin under vacuum at 70° for 72 hr.[3] The lecithin product is purified from fatty acid or unconverted lysolecithin by extraction into chloroform–methanol and thin-layer chromatography on silica gel layers developed in chloroform–methanol–water (65 : 35 : 5, v/v). Although pure dipalmitoyllecithin vesicles are a poor substrate for LCAT, the commercially available di[^{3}H]-palmitoyl)lecithin is readily hydrolyzed codispersed with a synthetic phosphatidylcholine ether or with egg lecithin and in this form is the most economical substrate for the routine phospholipase assay of LCAT.

Sources of Activator Apolipoprotein A-I. ApoA-I is routinely prepared in the laboratory from human plasma HDL by delipidation with ethanol and ether followed by molecular sieve and DEAE–cellulose chromatography in buffer solutions containing 8 *M* urea.[20] It is now also available commercially as a lyophilized powder (Sigma, St. Louis, MO) which dissolves readily at a concentration of 1 mg/ml in 2 m*M* phosphate buffer, pH 7.4, and gives an activation of LCAT with synthetic phospholipid vesicles that is comparable with that obtained with locally prepared flash-frozen apoA-I in the same buffer solution.

Preparation of Single-Walled Lecithin Vesicles. A convenient French Press method has been described in detail by Hamilton *et al.*,[21] who also

[20] C. Edelstein, C. T. Lim, and A. M. Scanu, *J. Biol. Chem.* **247,** 5842 (1972).

[21] R. L. Hamilton, J. Goerke, L. S. S. Guo, M. C. Williams, and R. J. Havel, *J. Lipid Res.* **21,** 981 (1980).

described the size and structure of the lecithin dispersions obtained. The instrument (Aminco, Silver Spring, MD) consists of a titanium cell with a small orifice and a press used to force the lecithin dispersion through the orifice at pressures of up to 2000 pounds per square inch. Lecithin (1–5 mg/ml, 0.5–5.0 ml) is mixed to an opalescent dispersion by hand vortexing in distilled water, then transferred to the cell and dispersed by 3 cycles of pressure and release. The clear product consists largely of uniform single-walled vesicles of diameter about 200 Å.[22] If required the product can be purified by subsequent molecular sieve chromatography on a column (0.9 × 20 cm) of Sepharose 4B (Pharmacia-LKB) in 0.1 *M* Tris-HCl buffer (pH 7.4), although (as discussed below) activation of these vesicles with apoA-I involves conversion of the single-walled vesicles to a characteristic discoidal structure.[22]

Single-walled vesicles can also be obtained with a sonifier (Heat Systems-Utrasonics, Plainview, NY)[23] The crude lecithin dispersion is sonicated for 3 min under nitrogen with cooling. It is usually necesary to remove titanium fragments from the probe by centrifugation (1000 *g*, 15 min) before use.

Lecithin in solvents or detergent solutions vesiculate spontaneously when dialyzed,[24,25] and such vesicles have been used as effective substrates for LCAT activity. Advantages of these methods include their utility, as specialized equipment is not needed. A potential disadvantage may be the retention of traces of dilute trapped solvent or detergent in the vesicles. This can be overcome by appropriate control procedures, for example by demonstrating complete removal of labeled detergent following dialysis.

Assay Media for LCAT Phospholipase Activity. The kinetic characteristics and cofactor requirements of LCAT phospholipase activity are very similar to those of the corresponding acyltransferase activity. Assay media contain the following components (volumes given are per assay): 50 μl 2-palmitoyl[9,10-^{3}H]phosphatidylcholine in egg lecithin [final concentration to μmol/ml; specific activity 0.5–1.0 × 10^6 disintegrations per minute (dpm)/μmol] dispersed in distilled water and 50 μl apolipoprotein A-I (0.25 mg/ml) reconstituted in 1 m*M* phosphate buffer (pH 7.4). These reagents are preincubated for 60 min at 37° to complete apoA-I–lecithin discoidal lipoprotein formation.[22] After incubation the following are added: 50 μl recrystallized human albumin solution (100 mg/ml) in 0.15 *M* NaCl, pH

[22] L. S. S. Guo, R. L. Hamilton, J. Goerke, J. N. Weinstein, and R. J. Havel, *J. Lipid Res.* **21,** 993 (1980).

[23] C. H. Huang, *Biochemistry* **8,** 344 (1969).

[24] C. H. Chen and J. J. Albers, *J. Lipid Res.* **23,** 680 (1982).

[25] S. Batzri and E. D. Korn, *Biochim. Biophys. Acta* **298,** 1015 (1973).

7.4 (to bind product unesterified fatty acid and lysolecithin), 50 μl of 50 mM Tris-HCl buffer (pH 7.4), and 200 μl of 0.15 M NaCl.

The reaction is started by addition of 50 μl of purified LCAT solution or 50 μl of the same buffer without enzyme. Incubation is normally linear for at least 60 min, if less than 5% of the initial substrate is hydrolyzed. At the end of the incubation, the assays are chilled in ice water, and an equal volume of methanol containing 0.5 M H_2SO_4 is added, then the same volume of chloroform. After vortexing to mix the phases, 250 μl of the lower chloroform phase is taken for thin-layer chromatography on silica gel G layers on plastic sheets (Merck, Darmstadt, FRG) developed in hexane–diethyl ether–acetic acid (83 : 16 : 1, v/v/v). The simultaneous and reproducible application of multiple samples to the silica gel layers is conveniently automated with an TLC Multispotter (AIS, Libertyville, IL) or similar instrument. After development and drying (5–10 min) in a fume hood, the unesterified fatty acid region of the silica gel (R_f ~0.4) is identified with minimal exposure in an iodine vapor tank. After the removal of visible iodine in the fume hood, the fatty acid region is cut out and counted with a liquid scintillation counter. In this system unhydrolyzed lecithin and lysolecithin both remain near the origin. The recovery of unesterified fatty acid is over 90%.

Enzymatic Properties of Lecithin–Cholesterol Acyltransferase-Dependent Phospholipase Activity

Substrate Specificity. As assayed in a nonhydrolyzable matrix of phosphatidyl ether, LCAT hydrolyzes lecithin and phosphatidylethanolamine at roughly equivalent rates. Other phospholipids are hydrolyzed more slowly. There is no detectable activity against the *N*-acyl residue of sphingomyelin.[26]

In reaction with pure substrates of long-chain lecithins LCAT shows the greatest reactivity with mono- and diunsaturated lecithins and less activity with the corresponding saturated species.[15,18,24] However, this probably relates mainly to the physical structure of the substrate since dipalmitoyllecithin, a very poor substrate as a pure lipid, is relatively reactive in a dimyristoylphosphatidyl ether matrix.[15]

Although LCAT shows a predominant 2-positional specificity, this is not absolute[3,15] (Table II). The generation of minor amounts of 1-position acylation products may be the result of acyl migration catalyzed by the LLAT reaction during the course of hydrolysis:

$$\text{1-Palmitoyl-2-oleyllecithin} \rightarrow \text{1-palmitoyllysolecithin} + \text{oleic acid} \qquad (1)$$

[26] C. J. Fielding, *Scand. J. Clin. Lab. Invest.* **33** (Suppl. 137), 15 (1974).

TABLE II
POSITIONAL SPECIFICITY OF LCAT PHOSPHOLIPASE AND ACYLTRANSFERASE ACTIVITIES[a]

	Percentage of fatty acids released			
	Unesterified fatty acids		Cholesteryl ester fatty acids	
Lecithin substrate	16:0	18:1	16:0	18:1
1-Palmitoyl-2-oleyl	13.4	86.6	23.3	76.7
1-Oleyl-2-palmitoyl	90.8	9.2	90.4	9.6

[a] Data from Ref. 3.

1-Palmitoyl-2-oleyllecithin +
1-palmitoyllecithin → 1,2-dipalmitoyllecithin + 1-oleyllecithin (2)
1,2-Dipalmitoyllecithin → 1-palmitoyllecithin + palmitic acid (3)

The relative proportions of 1- and 2-position fatty acids would then depend on the relative rates of the LCAT and LLAT reactions of the enzyme with a given substrate. As the rate of the LLAT reaction is increased about 10-fold in the presence of low-density lipoprotein (LDL),[5] this type of reaction may be more significant quantitatively in native plasma than in experiments with pure reconstituted reagents as shown in Table II.

Dependence of LCAT Phospholipase Activity on ApoA-I. Both acyltransferase and phospholipase activities of LCAT with long-chain lecithins

TABLE III
ApoA-I DEPENDENCE OF LCAT PHOSPHOLIPASE ACTIVITY[a]

ApoA-I (μg/ml assay)	Oleic acid released (nmol/ml/hr)
0	<0.02
2.5	0.18
5.0	0.56
10.0	3.58
15.0	4.40
20.0	4.69
25.0	4.38
50.0	3.03

[a] Substrate in these experiments was di[9,10-^{3}H]oleyllecithin (0.25 mM), and assays were conducted for 60 min at 37°.

TABLE IV
EFFECTS OF EGTA ON LCAT PHOSPHOLIPASE ACTIVITY[a]

EGTA concentration (m*M*)	Oleic acid released
0	100
1	97 ± 15
10	105 ± 16
50	185 ± 26

[a] Values given are expressed as the percentage of hydroysis rates in the absence of EGTA. The substrate in these experiments was di[9,10-^{3}H]oleyllecithin (0.25 m*M*) activated with 25 μg/ml apoA-I.

are highly dependent on the presence of lipoprotein apoproteins, of which the most effective is apolipoprotein A-I, the major proteins of the physiological substrate HDL (Table III). Maximal activation in the presence of near-saturating lecithin concentrations (0.25 m*M* lecithin) is obtained at 20–25 μg apoA-I/ml (a molar ratio of about 300 : 1), compatible with the presence of two apoA-I molecules on a disk of molecular weight about 450,000.

One major reason for this activating effect probably lies in the ability of apoA-I to induce such disk formation from unilamellar vesicles. Phospholipid–apoA-I disks are stable and as synthetic recombinants or as native discoidal lipoproteins are the most effective substrates for LCAT. It is of interest that the stimulation (fold) of LCAT activity by apoA-I is much less with medium- and short-chain than with long-chain lecithins (C. J. Fielding and X. Collet, unpublished data). LCAT activity with soluble esters is also apoA-I independent.[6]

Calcium Dependence of LCAT Phospholipase Activity. EDTA or EGTA at concentrations up to 50 m*M* were without inhibitory effect on LCAT activity, and the highest concentrations induced a modest rise in the rate of unesterified fatty acid production (Table IV) or (in the presence of cholesterol) a more modest increase in cholesteryl ester synthesis.[3]

Acknowledgments

The research of this laboratory was supported by the National Institutes of Health through Arteriosclerosis SCOR HL 14237.

Section V

Lysophospholipase

A. Lysophospholipase
Articles 41 through 45

B. Lysoplasmalogenase
Article 46

[41] Lysophospholipases from *Escherichia coli*

By KEN KARASAWA and SHOSHICHI NOJIMA

Introduction

$$\text{1-Acyl-}sn\text{-glycero-3-phosphoethanolamine} + H_2O \xrightarrow{\text{lysophospholipase } L_1} \text{glycerophosphoethanolamine} + \text{fatty acid}$$

$$\text{2-Acyl-}sn\text{-glycero-3-phosphoethanolamine} + H_2O \xrightarrow{\text{lysophospholipase } L_2} \text{glycerophosphoethanolamine} + \text{fatty acid}$$

A number of enzymes termed lysophospholipases have activity toward diacyl phospholipids and therefore fall into the category of phospholipase B.[1,2] In *Escherichia coli*, there are two kinds of lipolytic enzymes without activities toward the diacyl phospholipids. One is cytosolic, and the other is membrane-associated.[3,4] They are termed lysophospholipase L_1 and lysophospholipase L_2, respectively, since the enzymes preferentially hydrolyze 1- and 2-acyllysophospholipids, respectively. However, the enzymes purified from cytoplasm or membrane fractions are not absolutely specific for the acyl ester position. Nevertheless, several lines of genetic and biochemical evidence demonstrate that lysophospholipase L_1 and lysophospholipase L_2 are distinct proteins. Two kinds of mutants of lysophospholipase L_2 (one with an elevated level of the enzymes and another defective in it) were isolated, and the relative specific activities of lysophospholipase L_1 and lysophospholipase L_2 were found to be variable in a series of mutant colonies.[5] In addition, lysophospholipase L_1 and lysophospholipase L_2 were purified from *E. coli* strains each of which contained a hybrid plasmid bearing the structure genes and overproduced the enzymatic activities. The homogeneous lysophospholipase L_1 and lysophospholipase L_2 preparations exhibited distinct physical properties (e.g., molecular weight) and substrate specificity.[6,7]

[1] H. van den Bosch, *in* "Phospholipids" (J. N. Hawthorn and G. B. Ansell, eds.), p. 313. Elsevier, Amsterdam, New York, and Oxford, 1982.

[2] M. Waite. "The Phospholipases." Plenum, New York, 1987.

[3] F. D. Albright, D. A. White, and W. J. Lennarz, *J. Biol. Chem.* **248,** 3968 (1973).

[4] O. Doi and S. Nojima, *J. Biol. Chem.* **250,** 5208 (1975).

[5] T. Kobayashi, H. Homma, Y. Natori, I. Kudo, K. Inoue, and S. Nojima, *J. Biochem. (Tokyo)* **96,** 137 (1984).

[6] K. Karasawa, I. Kudo, T. Kobayashi, T. Sa-eki, K. Inoue, and S. Nojima, *J. Biochem. (Tokyo)* **98,** 1117 (1985).

Assay Method

Principle. The assay method is based on the conversion of 1- or 2-[^{14}C]acyl-*sn*-glycero-3-phosphoethanolamine to 1- or 2- ^{14}C-labeled fatty acid, extraction of the released labeled fatty acid by a modification of Dole's solvent extraction procedure,[8,9] and determination of the radioactivity of the upper heptane phase.

Reagents

- 1,2-[^{14}C]Diacylphosphatidylethanolamine [400 disintegrations per minute (dpm)/nmol], biosynthesized using *E. coli* (see [27] in this volume)
- Phospholipase A_2, prepared prior to use as follows: One milligram of Habu snake (*Trimeresurus flavoviridis* Hallowell[10]) venom is dissolved in acetate buffer (5 m*M*, pH 5.0), followed by heating for 5 min at 100° in a boiling water bath, and denatured proteins are removed by centrifugation for 15 min at 3000 rpm and 4°. The resulting supernatant is used as phospholipase A_2.
- *Rhizopus delemar* lipase (Seikagaku Kogyo, Tokyo, Japan)
- Tris-HCl buffer (pH 8.0), 100 m*M*
- Tris–maleate buffer (pH 5.6), 50 m*M*
- Petroleum ether/diethyl ether (1 : 1, v/v)
- Methanol
- Chloroform
- Citric acid solution, 0.5 *M*
- Tris-HCl buffer (pH 7.0), 100 m*M*
- Dole's extraction medium (2-propanol/*n*-heptane/1 *N* H_2SO_4, 20 : 78 : 7, v/v)
- Wakogel C-200 (Wako, Tokyo, Japan)
- EDTA, 50 m*M*
- ACS liquid scintillation fluid (Amersham Corp., Arlington Heights, IL)

Preparation of Substrate.[11] 1-[^{14}C]Acyl-*sn*-glycero-3-phosphoethanolamine, the substrate for the assay of lysophospholipase L_1, is prepared as follows. Four micromoles of *E. coli* phosphatidylethanolamine (400 dpm/

---

[7] K. Karasawa, I. Kudo, T. Kobayashi, H. Homma, N. Chiba, H. Mizushima, K. Inoue, and S. Nojima, *J. Biochem.* (*Tokyo*) in press (1991).

[8] V. P. Dole and H. Meinertz, *J. Biol. Chem.* **235,** 2595 (1960).

[9] G. S. Sundaram, K. M. M. Shakir, G. Barnes, and S. Margolis, *J. Biol. Chem.* **253,** 7703 (1978).

[10] M. Matsumoto and Y. Suzuki, *J. Biochem.* (*Tokyo*) **73,** 793 (1973).

[11] M. Nishijima, Y. Akamatsu, and S. Nojima, *J. Biol. Chem.* **249,** 5658 (1974).

nmol) in chloroform solution is added to a round-bottomed test tube (1 × 10.5 cm). Chloroform is evaporated under N_2, and the labeled phospholipid is redissolved in ethanol/diethyl ether (5 : 95, v/v), followed by the addition of 0.2 ml of 100 m*M* Tris-HCl (pH 8.0), 0.1 ml of 100 m*M* $CaCl_2$, and 0.2 ml of phospholipase A_2 in 5 m*M* acetate buffer (pH 5.0). The reaction mixture is incubated at room temperature for 2 hr, then shaken vigorously with 0.5 ml of water, 1 ml of methanol, and 4 ml of petroleum ether/diethyl ether (1 : 1, v/v). The supernatant is discarded, and the lower phase is extracted twice more with petroleum ether/diethyl ether. The lower phase is combined with 0.2 ml of 0.5 *M* citric acid, and the mixture is extracted with a mixture of 0.5 ml of methanol and 2.2 ml of chloroform. The resulting chloroform layer contained 1-[^{14}C]acyl-*sn*-glycero-3-phosphoethanolamine with a specific radioactivity of 200 dpm/nmol.

2-[^{14}C]Acyl-*sn*-glycero-3-phosphoethanolamine, the substrate for the assay of lysophospholipase L_2, is prepared as follows. Four micromoles of *E. coli* phosphatidylethanolamine (400 dpm/nmol) in chloroform solution is added to a round-bottomed test tube, and the chloroform is evaporated *in vacuo* under N_2; then 0.6 ml of 50 m*M* Tris–maleate (pH 5.6) and 0.1 ml of 100 m*M* $CaCl_2$ are added to the tube. The phospholipid is suspended in a bath-type sonicator, then 0.3 ml of *Rhizopus delemar* lipase (1 mg protein/ml) in a solution of 50 m*M* Tris–maleate (pH 5.6) and 0.1 ml of diethyl ether are added to the tube. The removal of fatty acid and extraction of lysophospholipid are accomplished by the same method as described above. At alkaline pH, 2-[^{14}C]acyl-*sn*-glycero-3-phosphoethanolamine rapidly undergoes migration of the fatty acyl group to the 1-position of glycerol, so it should be used soon after preparation.

Procedure. Lysophospholipase L_1 activity is determined as follows. For convenience, a 3 m*M* stock solution of 1-[^{14}C]acyl-*sn*-glycero-3-phosphoethanolamine in water is prepared by sonication for approximately 30 sec, and then 10 μl of the substrate sonicate is transferred to a test tube. After addition of 25 μl of 100 m*M* Tris-HCl (pH 7.0) and 5 μl of 50 m*M* EDTA, enzyme is added to give a final volume of 50 μl. After rapid mixing, the tube is incubated at 30 min for 5–10 min. The reaction is stopped by addition of 1.25 ml of Dole's reagent, and the mixture is incubated at 60° for 1 min. The mixture is cooled to room temperature, then 0.7 ml of water and 0.75 ml of *n*-heptane are added. The mixture is vortexed for 30 sec and centrifuged at 3000 rpm for 5 min. An aliquot of 0.8 ml of the upper heptane phase is transferred using a Pasteur pipette (Corning Glass Works, Corning, NY) into a tube containing 0.8 ml of *n*-heptane. Then approximately 100 mg of Wakogel C-200 is added, and the tube is vigorously vortexed for 30 sec and centrifuged for 5 min at 3000 rpm. The entire heptane phase is transferred to a scintillation vial containing 10 ml of ACS scintillation fluid.

For the assay of lysophospholipase L_2, 2-[^{14}C]acyl-*sn*-glycero-3-phosphoethanolamine is used as the substrate; otherwise the enzyme assay is performed according to the same procedure as described above.

Lysophospholipase L_2 from *Escherichia coli* Membranes

The structure gene, *pldB*, coding for lysophospholipase L_2 was cloned on the plasmid pKO1.[12] The *E. coli* strain transformed by pB1071,[13] a subclone of pKO1, overproduced lysophospholipase L_2 activity at an approximately 45 times higher level than that of the wild type.[6] For purification purposes it is advantageous to start with a hybrid plasmid-bearing strain, KL1699/pB1071, since relatively few steps are required to obtain a near-homogeneous preparation.

Purification Procedure[6]

Step 1: Cell-Free Extract. Escherichia coli strain KL1699/pB1071, which overproduces lysophospholipase L_2, is grown at 37° to the late log phase in 9 liters of Luria broth. The medium contains, per liter, Bactotryptone (10 g), yeast extract (5 g), and NaCl (5 g). Harvested cells (wet weight 21 g) are suspended in 90 ml of 10 m*M* Tris-HCl (pH 7.5)/5 m*M* $MgCl_2$ and disrupted at a pressure of 400 kg/cm^2 in a French press. Then, 1 mg/ml of pancreatic DNase I dissolved in 1 *M* $MgCl_2$ solution is added to give a final concentration of 15 μg/ml, and the suspension is incubated at room temperature for 1 hr.

Step 2: Membrane Extract. Lysophospholipase L_2 is a membrane-bound enzyme. In order to isolate the membrane fraction, the cell-free extract is centrifuged at 105,000 *g* for 1 hr at 4°. The precipitate fraction is resuspended in 94 ml of 50 m*M* Tris-HCl (pH 7.4) containing 1 *M* KCl and extracted for 1 hr at 4° with gentle stirring. Unextracted material is removed by centrifugation for 1 hr at 105,000 *g* and 4°.

Step 3: Ammonium Sulfate Fractionation. The KCl extract (84 ml) is brought to 40% saturation with solid ammonium sulfate (20.4 g). After 30 min of stirring, the precipitate is collected by centrifugation at 10,000 *g* for 15 min. The ammonium sulfate precipitate is redissolved in 18 ml of 50

[12] H. Homma, T. Kobayashi, Y. Ito, I. Kudo, K. Inoue, H. Ikeda, M. Sekiguchi, and S. Nojima, *J. Biochem.* (*Tokyo*) **94,** 2079 (1983).

[13] T. Kobayashi, I. Kudo, H. Homma, K. Karasawa, K. Inoue, H. Ikeda, and S. Nojima, *J. Biochem.* (*Tokyo*) **98,** 1007 (1985).

TABLE I
PURIFICATION OF LYSOPHOSPHOLIPASE L_2 FROM *Escherichia coli* KL1699 BEARING PLASMID pB1071

Step	Total protein (mg)	Total activity (nmol/min)	Yield (%)	Specific activity (nmol/min/mg)	Purification (-fold)
1. Cell-free homogenate	2140	71,000	100	33.2	1
2. Membrane fraction	1230	75,200	106	61.1	1.84
3. KCl extract	287	26,500	37.3	92.3	2.78
4. $(NH_4)_2SO_4$ (0–40%)	76.1	16,700	23.6	220	6.63
5. Chromatofocusing	ND[a]	30,900	43.5	—	—
6. Heparin-Sepharose CL-6B	0.615	14,500	20.3	23,500	708

[a] Not determined owing to the inhibitory effect of Polybuffer in the method of Lowry *et al.*[14]

m*M* Tris-HCl (pH 7.4) containing 0.5% (w/v) CHAPS and dialyzed against the same buffer for 3 hr at 4°.

Step 4: Chromatofocusing. The dialyzed fraction is applied to a column (1.5 × 22.5 cm) of Polybuffer 94 exchanger (Pharmacia Fine Chemicals, Piscataway, NJ) previously equilibrated with 50 m*M* Tris-HCl (pH 7.4) containing 0.5% CHAPS. The column is eluted with Polybuffer 74, generating a pH gradient from 7.4 to 5.0. The lysophospholipase L_2 activity is eluted as a single peak at approximately pH 7.2. The enzymatic activity is recovered at a volume of 42 ml.

Step 5: Heparin-Sepharose CL-6B Affinity Chromatography. The pooled fraction is applied to a column (1.0 × 27 cm) of heparin-Sepharose CL-6B (Pharmacia) previously equilibrated with 50 m*M* Tris-HCl (pH 7.4) containing 0.2 *M* NaCl. The column is washed with 3 column volumes of the same buffer and then eluted with an 80-ml linear gradient of NaCl from 0.2 to 1.0 *M* in 50 m*M* Tris-HCl (pH 7.4). The activity is recovered as a single peak at a volume of 31 ml.

Purity. The purification scheme summarized in Table I[14] indicates an overall purification of 708-fold from the original strain (KL1699/pB1071). The final preparation exhibits a single protein band on sodium dodecylsulfate (SDS)–polyacrylamide gel electrophoresis. The yield is approximately 20.3% (specific activity 23,500 nmol/min/mg).

[14] O. H. Lowry, N. J. Rosebrough, A. L. Farr, and R. J. Randall, *J. Biol. Chem.* **193,** 265 (1951).

Properties

Stability and Storage. The purified enzyme is stable for 1 month at $-80°$ in 50 mM Tris-HCl (pH 7.4) containing 0.5–1.0 NaCl. Repeated freezing and thawing of the enzyme solution, however, cause a significant loss of the activity. It is recommended that the enzyme solution be divided into small aliquots and stored at $-80°$.

Physical Properties. The molecular weight of the purified enzyme is estimated to be 38,500 by SDS–polyacrylamide gel electrophoresis, although the same preparation gives an apparent molecular weight of 14,000 in gel filtration on Sephacryl S-200. The abnormal behavior of the enzyme on gel filtration may be due to its specific interaction with the resin. The pI of the enzyme is 7.2.

Substrate Specificity. The reaction velocities (nmol/min/mg) with various substrates are as follows: 2-acyl-*sn*-glycero-3-phosphoethanolamine, 19,900; 2-acyl-*sn*-glycero-3-phosphocholine, 12,910; 2-acyl-*sn*-glycero-3-phosphoglycerol, 2123. The purified lysophospholipase L_2 hydrolyzes 2-acyl lysophospholipids approximately 2 to 3 times faster than the 1-acyl isomers.

Other Reactions. The purified enzyme catalyzes the transacylation reaction to yield acyl phosphatidylglycerol from phosphatidylglycerol and 2-acyl(or 1-acyl)lysophospholipid. As in the hydrolysis reaction, the 2-acyl isomer of lysophospholipid is utilized as an acyl donor to form acylphosphatidylglycerol more effectively than the 1-acyl isomer.

Kinetics. An apparent K_m of 142 μM and an apparent V_{max} of 16,000 nmol/min/mg are obtained from Lineweaver–Burk plots for the purified enzyme. No abrupt change in the activity is seen at concentration above and below the critical micellar concentration. This result indicates that *E. coli* lysophospholipase L_2 is active toward substrates in the monomer state as well as in the micellar state.

Inhibitors. The *E. coli* lysophospholipase L_2 is inhibited by such detergents as Triton X-100 and deoxycholate. This inhibition is presumably attributable to dilution of the substrate in the micelles. The purified enzyme is irreversibly inactivated by diisopropyl fluorophosphate. Ca^{2+} and EDTA do not affect the enzyme activity at all.

Primary Structure of Lysophospholipase L_2

The DNA sequence of the *pldB* gene coding for lysophospholipase L_2 of *E. coli* was determined, and the amino acid sequence was deduced[15] to

[15] T. Kobayashi, I. Kudo, K. Karasawa, H. Mizushima, K. Inoue, and S. Nojima, *J. Biochem.* (*Tokyo*) **98,** 1017 (1985).

FIG. 1. Amino acid sequence of lysophospholipase L_2 from *E. coli*.

be as shown in Fig. 1. The deduced amino acid sequence of lysophospholipase L_2 contains 340 amino acid residues, corresponding to a protein with a molecular weight of 38,934, and coincides at its NH_2-terminal region with the sequence determined for the 15 NH_2-terminal residues of the purified preparation. *Escherichia coli* lysophospholipase L_2 has a high content of arginine residues (36 out of 340 residues). It is postulated that detergent-resistant phospholipase A in the outer membrane of *E. coli* is synthesized as a precursor with a typical signal peptide composed of 20 amino acids. In contrast, no DNA sequence corresponding to a signal peptide has been found near the NH_2 terminus of lysophospholipase L_2. The hydropathy profile, calculated according to Kyte and Doolittle,[16] indicates that the hydrophobic segments of lysophospholipase L_2 are relatively short compared with those of other membrane-bound proteins. This is consistent with the fact that *E. coli* lysophospholipase L_2 is a peripheral protein which is easily solubilized from the inner membrane by 1 *M* KCl.

Lysophospholipase L_1 from *Escherichia coli* Cytoplasm

The structural gene, *pldC*, coding for lysophospholipase L_1 was cloned on the ColE1 hybrid plasmid after screening of the Clarke and Carbon *E. coli* collection for lysophospholipase L_1 activity. An *E. coli* strain transformed by a subcloned plasmid, pC124, overproduces the enzyme activity by approximately 11.4 times compared with the parental strain.

[16] J. Kyte and R. F. Doolittle, *J. Mol. Biol.* **157,** 105 (1982).

Purification Procedure[7]

Step 1: Preparation of Cytosol. Escherichia coli strain KL1699/pC124, which overproduces lysophospholipase L1, is grown at 37° to the late log phase in 9 liters of Luria broth. Harvested cells are resuspended in 90 ml of 10 m*M* Tris-HCl (pH 7.5)/5 m*M* $MgCl_2$ and disrupted at a pressure of 400 kg/cm^2 in a French press. Then 1 mg/ml of pancreatic DNase I dissolved in 1 *M* $MgCl_2$ is added to give a final concentration of 15 μg/ml, and the suspension is incubated at room temperature for 1 hr. The cytosol fraction is obtained by centrifugation at 105,000 *g* for 1 hr at 4°.

Step 2: Streptomycin Treatment and $(NH_4)_2SO_4$ Fractionation. An aqueous 40% (w/v) streptomycin sulfate solution (3.93 ml) is added dropwise to 105 ml of the cytosol fraction with gentle stirring, and stirring is continued for an additional 40 min. The mixture is centrifuged at 10,000 *g* for 50 min at 4°. The supernatant (100 ml) is mixed with 12.9 g of $(NH_4)_2SO_4$ (55% saturation) with gentle stirring at 4°. After centrifugation, the precipitate is dissolved in 10.5 ml of 50 m*M* potassium phosphate buffer (pH 7.5) and dialyzed against the same buffer for 3 hr at 4°.

Step 3: Sephacryl S-300 Column Chromatography. The dialyzed fraction is applied to a column (3.0 × 50 cm) of Sephacryl S-300 (Pharmacia) previously equilibrated with 50 m*M* potassium phosphate buffer (pH 7.5). Fractions with lysophospholipase L_1 activity (52 ml) are pooled, and the pooled fraction is dialyzed against 50 m*M* Tris-HCl buffer (pH 7.5).

Step 4: DEAE–cellulose Column Chromatography. The dialyzed fraction is applied to a column of DEAE–cellulose (1.5 × 16.9 cm) which is developed with a 140-ml salt gradient from 0 to 0.3 *M* KCl in 50 m*M* Tris-HCl (pH 7.5). Fractions with lysophospholipase L_1 activity are eluted with 0.1 *M* KCl and pooled. The pooled fraction (32 ml) is brought to 70% saturation with solid ammonium sulfate (15.1 g). After 30 min of stirring, the precipitate is collected by centrifugation for 15 min at 10,000 *g* and 4°. The ammonium sulfate precipitate is dissolved in 5 ml of 1 m*M* potassium phosphate buffer (pH 7.5) and dialyzed against the same buffer for 3 hr at 4°.

Step 5: Hydroxyapatite Column Chromatography. The dialyzed fraction is applied to a column (1.0 × 12.7 cm) of hydroxyapatite, previously equilibrated with 1 m*M* potassium phosphate buffer (pH 7.5). The column is eluted with a linear gradient of 1 to 100 m*M* phosphate buffer (pH 7.5). Fractions with lysophospholipase L_1 activity are eluted with 70 m*M* potassium phosphate and pooled. The pooled fraction (10.5 ml) is concentrated to 2 ml with a Centricon 10 (Amicon, Danvers, MA).

Step 6: Sephacryl S-200 Column Chromatography. The concentrated solution is applied to a column (1.5 × 45 cm) of Sephacryl S-200, equili-

TABLE II
PURIFICATION OF LYSOPHOSPHOLIPASE L_1 FROM *Escherichia coli* KL1699 BEARING PLASMID pC124

Step	Protein: Total (mg)	Protein: Recovery (%)	Lysophospholipase L_1 activity: Total (nmol/min)	Lysophospholipase L_1 activity: Recovery (%)	Lysophospholipase L_1 activity: Specific (nmol/min/mg)	Purification (-fold)
105,000 *g* supernatant	1180	100	11,100	100	9.4	1.0
Streptomycin fractionation supernatant	956	81	10,300	92.8	10.8	1.14
$(NH_4)_2SO_4$ (35–55%)	311	26	5380	48.4	17.3	1.84
Sephacryl S-300	340	28.8	5240	47.2	15.4	1.63
DEAE–cellulose	8	0.67	1980	17.8	247	26.3
Hydroxyapatite	1.05	0.089	873	7.87	831	88.4
Sephacryl S-200	0.059	0.005	612	5,51	10,380	1104

brated with 10 m*M* potassium phosphate buffer (pH 7.5). Fractions with lysophospholipase L_1 activity are pooled.

Purity. As shown by the summary of the purification in Table II, the overall purification is 1104-fold from the plasmid-containing strain. The final yield is approximately 5.51%, with a specific activity of 10,380 nmol/min/mg. The final enzyme preparation contains a major band with an apparent molecular weight of 20,500 and a minor band of 22,000 by SDS–polyacrylamide gel electrophoresis.

Properties

Stability and Storage. The purified enzyme is stable for at least 1 month at $-80°$.

Physical Properties. Gel-permeation chromatography on TSK G 3000 SW (Tosoh, Tokyo, Japan) indicates that the lysophospholipase L_1 has a molecular weight of approximately 21,000. This value is in good agreement with that estimated by SDS–polyacrylamide gel electrophoresis as described above, suggesting that lysophospholipase L_1 is a monomeric polypeptide.

Substrate Specificity. The purified lysophospholipase L_1 hydrolyzes 1-acyl-*sn*-glycero-3-phosphoethanolamine approximately 2 times faster than the 2-acyl isomer. The purified lysophospholipase does not hydrolyze the acyl linkage of diacylglycerophospholipid and shows no acylphosphatidylglycerol synthase activity.

[42] Phospholipase B from *Penicillium notatum*

By KUNIHIKO SAITO, JUNKO SUGATANI, and TADAYOSHI OKUMURA

Introduction

A pure enzyme isolated from aqueous extracts of *Penicillium notatum* is a glycoprotein with a molecular weight of about 95,000 that catalyzes reactions (1) and (2). The enzymatic activity concerned with reaction (1) is phospholipase B activity (B activity), as proposed by McMurry and Magee,[1] and that concerned with reaction (2) is lysophospholipase activity (lyso activity).

1,2-Diacyl-*sn*-glycero-3-phosphocholine → fatty acids + glycerophosphocholine (1)
1-(or 2-)Acyllysoglycerophosphocholine → fatty acid + glycerophosphocholine (2)

We propose that the active glycoprotein for phospholipase B has two phospholipase activities. When phospholipase B undergoes limited proteolysis by endogenous protease(s) in *P. notatum,* which often occurs at the initial stage of purification, the B activity decreases greatly, but all of the lyso activity remains.[2] In this chapter, the original phospholipase B is described as the native form of phospholipase B and the partially proteolyzed one considered as the modified form.

Similar phospholipases have been reported by others in microorganisms, plants, and animal tissues, some of which are described under different names, for example, calcium-independent phospholipase A_2 activity (A_2 activity),[3] lysophospholipase with phospholipase A_1 activity (A_1 activity), or a kind of phospholipase B.[4] Recent papers describe phospholipase B from *Torulaspora delbrueckii,*[5] *Saccharomyces cerevisiae,*[6] and intestinal brush border membranes.[7] However, the correlation between B activity and lyso activity on a protein chemistry level has not been reported. On the other hand, a lysophospholipase which does not catalyze the deacylation of diacylglycerophospholipids but only of monoacylglycero-

[1] W. C. McMurray and W. L. Magee, *Annu. Rev. Biochem.* **41,** 129 (1972).
[2] J. Sugatani, T. Okumura, K. Saito, K. Ikeda, and K. Hamaguchi, *J. Biochem.* (*Tokyo*) **95,** 1407 (1984).
[3] S. Pind and A. Kuksis, *Biochim. Biophys. Acta* **901,** 78 (1987).
[4] Y. Nishijima, Y. Akamatsu, and S. Nojima, *J. Biol. Chem.* **249,** 568 (1974).
[5] Y. Kuwabara, M. Maruyama, Y. Watanabe, and S. Tanaka, *J. Biochem.* (*Tokyo*) **104,** 236 (1988).
[6] W. Witt, M. E. Schweingruber, and A. Mertshing, *Biochim. Biophys. Acta* **795,** 108 (1984).
[7] A. Gassama-Diane, J. Fanvelt, and H. Chap, *J. Biol. Chem.* **264,** 9470 (1989).

phospholipids has been reported.[8] But it is still uncertain whether the lysophospholipase is an artifact of the phospholipase B produced by limited proteolysis as seen in *P. notatum*.

Assay Systems

Phospholipase B Activity

Standard System. The substrate (1–2 μmol) is dispersed in 0.4 ml of 0.1 *M* acetate buffer (pH 5.0) containing 2 m*M* EDTA; 0.1 ml of enzyme solution (1–4 μg protein) is added, and the mixture is incubated for various times at 30°. The substrate (3–4 m*M*) is dispersed in water ultrasonically using a Branson sonicator with a titanium probe at an output of 60 W for 10 min in an ice–salt bath. The titanium particles produced during dispersion are removed by filtration with a Millipore filter. After incubation, the reaction is stopped by adding 0.1 ml of 5% bovine serum albumin (BSA) followed by 0.4 ml of 10% perchloric acid on ice.[9] After vortexing, the mixture is centrifuged, and 1.0 ml of the supernatant is assayed for glycerophosphate esters by the methods of Wells and Dittmer[10] or Bartlett.[11]

Radioactive Substrates. The assay system is the same as the above except for the use of radioactive substrates. After various incubation times, the enzyme reaction is stopped by adding 1.0 ml methanol and 0.5 ml chloroform (one-phase system of Bligh and Dyer[12]), followed by a further 0.5 ml of chloroform and water. After centrifugation, the chloroform layer is removed and washed once with 1.0 ml of methanol–water (10 : 9, v/v); the combined methanol–water layers are washed with chloroform. Each of the combined chloroform and methanol–water phases is counted for ^{14}C and ^{32}P in a scintillation counter. Aliquots of the chloroform phase are then subjected to thin-layer chromatography (TLC) on silica gel H in the solvent system of chloroform–methanol–acetic acid–water (25 : 15 : 4 : 2, by volume). Radioactive spots on the TLC plates located by autoradiography are scraped directly into counting vials, and ^{32}P and ^{14}C activities are determined.

The methanol–water phase of the hydrolyzate is chromatographed on Whatman No. 1 paper in *n*-butanol–acetic acid–water (5 : 3 : 1, by volume). Radioactive spots are located by autoradiography, cut out and counted directly. Results show that ^{32}P- and ^{14}C-labeled 1,2-diacyl-*sn*-glycero-3-

[8] H. van den Bosch and J. G. De Jong, *Biochim. Biophys. Acta* **398,** 244 (1975).

[9] R. M. C. Dawson, *Biochem. J.* **70,** 559 (1958).

[10] M. A. Wells and J. C. Dittmer, *Biochemistry* **5,** 3405 (1966).

[11] D. R. Bartlett, *J. Biol. Chem.* **234,** 466 (1959).

[12] E. C. Bligh and W. F. Dyer, *Can. J. Biochem. Physiol.* **37,** 911 (1959).

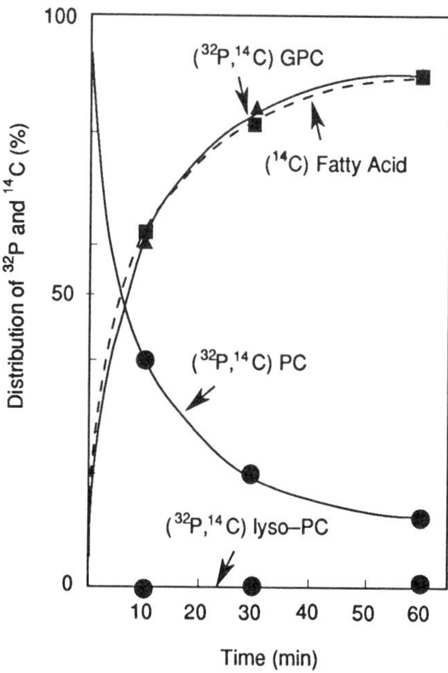

FIG. 1. Hydrolysis of ^{32}P- and ^{14}C-labeled 1,2-diacyl-*sn*-glycero-3-phosphocholine. GPC, Glycerophosphocholine; lyso-PC, 1-(or 2-)acyl-*sn*-glycero-3-phosphocholine; PC, 1,2-diacyl-*sn*-glycero-3-phosphocholine.

phosphocholine is hydrolyzed with the concomitant release of ^{14}C-labeled fatty acids and [^{14}C,^{32}P]glycerophosphocholine (Fig. 1). No labeled lyso compound is detected. Throughout the reaction, only one radioactive spot is determined on chromatograms of the water-soluble hydrolysis products, and it has the same R_f value as authentic glycerophosphocholine.

Assay in Presence of Detergent. In the presence of Triton X-100, the system consists of 5 m*M* substrate, 0.2 *M* acetate buffer (pH 5.0) containing 2 m*M* EDTA, 30 m*M* Triton X-100, and the enzyme protein in a final volume of 0.4 ml. After incubation, 1.5 ml of a chloroform–methanol mixture (2 : 1, v/v) is added, followed by 0.5 ml each of chloroform and water as described above. For assay systems with diethyl ether, 10% (v/v) diethyl ether is added to the substrate suspension or dispersion and the enzyme protein is added.

Lysophospholipase Activity

The substrate (2 μmol) is dissolved in 0.3 ml of 0.2 *M* acetate buffer (pH 5.0) containing 2 m*M* EDTA and 0.1 ml of enzyme protein. After incubation at 30° for 5 min, the reaction is stopped by adding 0.1 ml of 5%

(w/v) bovine serum albumin and 0.4 ml of 10% perchloric acid on ice. The glycerophosphocholine liberated is determined as described above.

Purification[13–17]

Penicillium notatum (FIO 4640) is grown aerobically in a culture medium (pH 5.4) containing 3.5% corn steep liquor, 5.5% lactose, 0.7% KH_2PO_4, 0.5% $CaCO_3$, 0.3% $MgSO_4$, and 0.25% soybean oil for 48 hr at 26° with continuous shaking in a 3 m^3 tank (2000 liters). The mycelia (51.2 kg) are obtained and stored at −20°. All purification steps thereafter are carried out at 4° unless otherwise stated. The mycelia are kindly supplied from Toyo Brewing Co., Ltd.

One kilogram of *P. notatum* cells is added to 5 volumes of 20 m*M* EDTA and 0.1 m*M* phenylmethylsulfonyl fluoride (PMSF), pH 7.0, homogenized with an Ultra Turrax disperser (Janke & Kunkel KG, FRG) set at maximum speed for 15 min and then centrifuged at 13,700 *g* for 25 min in the GS-3 rotor of a Sorvall centrifuge. The supernatant is brought to 90% saturation with solid ammonium sulfate and, after sitting overnight, is centrifuged at 13,700 *g* for 15 min. More than 90% of lysophospholipase activity, which is used as a measure of phospholipase B activity throughout the purification, appears in the supernatant. The supernatant is concentrated by dialysis against 20 m*M* phosphate–1 m*M* EDTA buffer, pH 7.4, using a hollow fiber H1P10 type cartridge (Amicon, Danvers, MA). The dialyzate is applied to a DEAE-Sephadex A-50 column (2 × 11 cm) equilibrated with the same buffer. Inactive proteins are washed out with approximately 300 ml of 30 m*M* phosphate buffer at a flow rate of 15 ml/hr, and the enzyme is eluted with 150 ml of 80 m*M* phosphate buffer at the same flow rate as above. Then the pooled active fraction is dialyzed against 1 m*M* EDTA (pH 7) and freeze-dried. The preparation is stable for several months at −20°.

The post DEAE-Sephadex A-50 fraction is dissolved in 20 m*M* phosphate–1 m*M* EDTA buffer, pH 7.4, adjusted to pH 4.0 with glacial acetic acid, and applied to a phosphatidylserine-AH Sepharose column (1.3 × 5 cm, 1.6 μmol of phosphatidylserine bound/ml of gel) equilibrated with 0.2 *M* acetate–2 m*M* EDTA buffer, pH 4.0. The column is washed successively with the same buffer (60 ml), 0.2 *M* phosphate–2 m*M* EDTA buffer (pH 7.4, 100 ml) and 0.2% Brij 58 in 20 m*M* phosphate–1 m*M* EDTA buffer

[13] K. Saito and K. Sato, *Biochim. Biophys. Acta* **151,** 706 (1968).
[14] N. Kawasaki and K. Saito, *Biochim. Biophys. Acta* **296,** 426 (1973).
[15] N. Kawasaki, J. Sugatani, and K. Saito, *J. Biochem. (Tokyo)* **77,** 1233 (1975).
[16] T. Okumura, J. Sugatani, and K. Saito, *Arch. Biochem. Biophys.* **211,** 419 (1981).
[17] T. Okumura, S. Kimura, and K. Saito, *Biochim. Biophys. Acta* **617,** 264 (1980).

TABLE I
PURIFICATION OF PHOSPHOLIPASE B FROM *Penicillium notatum*

Purification step	Total activity[a] (units/kg mycelium)	Specific activity (units/mg protein)	Recover (%)
Original extract	57,000	1.5	100
$(NH_4)_2SO_4$ fractionation	52,000	125	91
DEAE-Sephadex A-50 chromatography	41,000	381	72
Phosphatidylserine-AH affinity chromatography	34,000	5100	60
Hydroxylapatite chromatography	23,000	6900	41

[a] Enzyme activity is assayed using lysophosphatidylcholine as substrate.

(pH 7.4, 50 ml) at a flow rate of 30 ml/hr. Then the enzyme is eluted with 0.1% Triton X-100 in 30 m*M* phosphate–1 m*M* EDTA buffer (pH 7.4, 50 ml).

The affinity resin is prepared as follows: AH Sepharose 4B (Pharmacia, Piscataway, NJ) is swollen and washed as recommended by the supplier. Phosphatidylserine (200 mg) purified from bovine brain and *N,N'*-dicyclohexylcarbodiimide (1.33 g) are added to 24 ml of the packed gel in 35 ml of tetrahydrofuran, and the pH is adjusted to 5.0. The coupling reaction is allowed to proceed for 24 hr at room temperature with gentle shaking after which acetic acid (2.5 ml) is added to block the uncoupled AH Sepharose. The gel is washed successively with tetrahydrofuran (90%, 200 ml), NaCl (1 *M*, 500 ml), and is equilibrated and packed with the column buffer.

The postphosphatidylserine-AH Sepharose fraction containing Triton X-100 is dialyzed against 5 m*M* phosphate–1 m*M* EDTA buffer, pH 7.4, and applied to hydroxylapatite column (2 × 9 cm) equilibrated with the same buffer. The column is washed with 8 m*M* phosphate–1 m*M* EDTA buffer, pH 7.4, and the enzyme is eluted with 40 m*M* phosphate–1 m*M* EDTA buffer, pH 7.4, at a flow rate of 30 ml/hr. The active fraction is dialyzed against 0.5 m*M* EDTA, pH 7.0, lyophilized, and stored at −20° without a loss of activity for several months, but inactivated by freezing and thawing. The purification procedure is summarized in Table I.

Characterization of *Penicillium notatum* Phospholipase B[2,16–18]

The molecular size of phospholipase B (native form) is 95,000 Da in the presence and absence of 2-mercaptoethanol on slab sodium dodecyl

[18] Y. Takeuchi, T. Okumura, J. Sugatani, and K. Saito, *Arch. Biochem. Biophys.* **252,** 206 (1987).

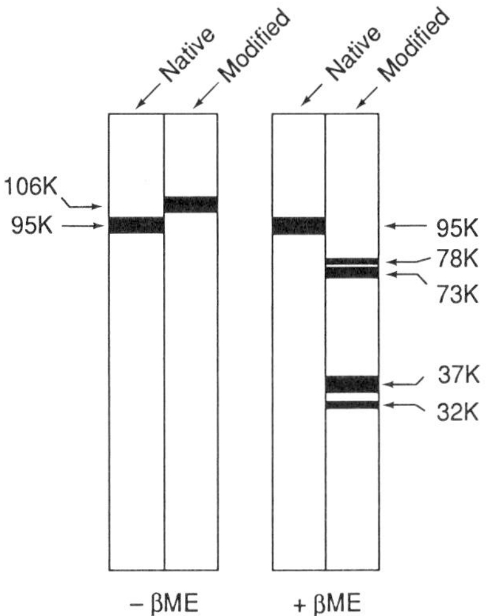

FIG. 2. Slab SDS–PAGE patterns of the native and modified forms of phospholipase B in the presence or absence of 2-mercaptoethanol (βME).

sulfate–polyacrylamide gel electrophoresis (SDS–PAGE). However, phospholipase B modified by protease(s) (modified form) gives a 106K band in the absence of 2-mercaptoethanol and is reductively cleaved to 78K, 73K, 37K, and 32K peptides by 2-mercaptoethanol. The protease(s), which attacks the phospholipase B molecule and is very sensitive to phenylmethylsulfonyl fluoride, exists intracellularly in *P. notatum* and increases at the late stationary phase of the growth. Figure 2 shows the typical pattern of the native and modified phospholipase B on slab SDS–PAGE in the presence and absence of 2-mercaptoethanol. The 37K plus 32K fragment and the 78K plus 73K fragment, respectively, are the N-terminal and C-terminal segments of the original 95K protein.

The native and modified forms are glycoproteins with 29% carbohydrate (23% glucose and mannose and 6% *N*-acetylglucosamine) and have the same isoelectric point of 4.0. The circular dichroism spectrum in the far-ultraviolet region of the modified enzyme is different from that of the native one, showing the conformational change between the native and modified forms of the enzyme. The conformational change in protein structure of phospholipase B results in a significant difference in pH dependence of enzyme activity (Fig. 3). The decrease in the B activity of the modified form is based on the decrease in both the A_1 and A_2 activities. Endoglycosidase H removes the carbohydrate from the native and modified enzymes to the same extent, but the B and lyso activities increase in

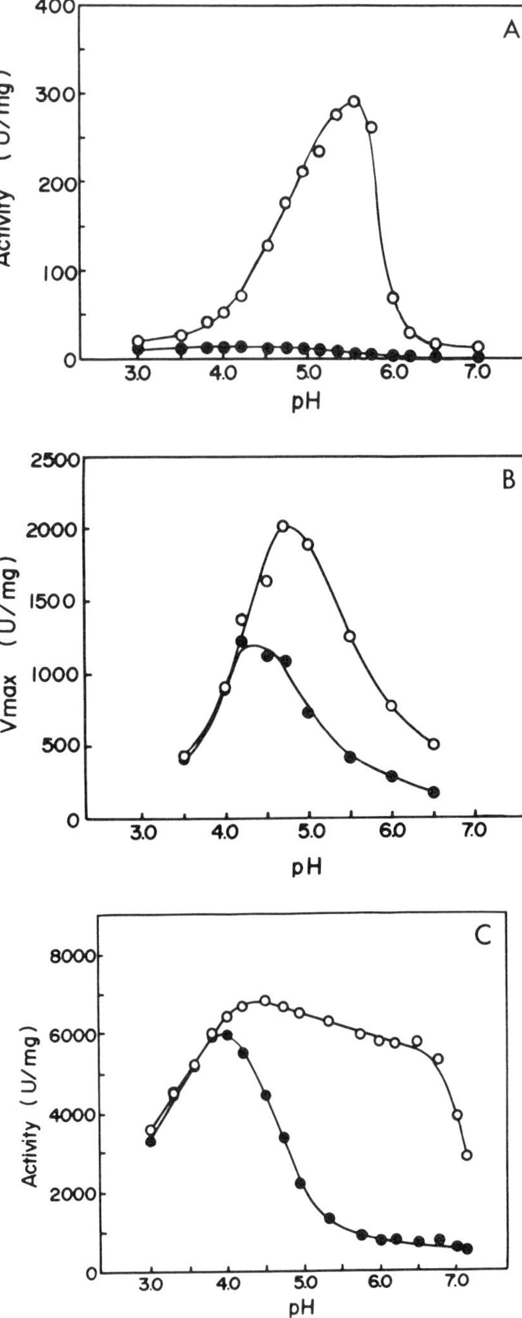
A
Activity (U/mg)
400
300
200
100
0
3.0
4.0
5.0
6.0
7.0
pH
B
Vmax (U/mg)
2500
2000
1500
1000
500
0
3.0
4.0
5.0
6.0
7.0
pH
C
Activity (U/mg)
8000
6000
4000
2000
0
3.0
4.0
5.0
6.0
7.0
pH

the modified enzyme and not in the native one. The carbohydrate moiety of phospholipase B may efficiently be involved in the enzymatic activity.

Substrate Specificity

Stereospecific A_2 Activity and Nonstereospecific A_1 Activity. Penicillium notatum phospholipase B catalyzes the hydrolysis of the 2-acyl ester bond of L isomers of choline glycerophospholipids (1,2-diacyl-, 1-*O*-alk-1′-enyl-2-acyl-, and 1-*O*-alkyl-2-acyl-*sn*-glycerophosphocholines), but not that of the D isomers. The enzyme hydrolyzes both the 1-acyl ester bonds of the L and D isomers in a nonstereospecific phospholipase A_1 activity.[19,20]

Dipalmitoylglycerophospholipids with Different Polar Groups. The phospholipase B hydrolyzes phospholipids in the following order: phosphatidylserine > phosphatidylinositol > phosphatidic acid > phosphatidylcholine > phosphatidylethanolamine acid > cardiolipin.[21]

Lysophospholipase Activity. The enzyme hydrolyzes 1-acyl-2-lyso-*sn*-glycero-3-phosphocholine more than the 2-acyl one. Lyso activity toward different fatty acyl chains shows the preference $C_{10} < C_{12} > C_{14} > C_{16} > C_{18}$.[19,21]

Lipase and Esterase Activities. Monoacyl(oleoyl or palmitoyl)glycerol is also hydrolyzed, especially in the presence of sodium taurocholate. But diacyl- and triacylglycerols including triacetoyl- and tributyroylglycerols, *p*-nitrophenyl acetate, and cholesteryl oleate are not hydrolyzed.[19]

A_1 and A_2 Activities of the Phospholipase B. In conventional measurements with isotope-labeled 1,2-diacyl-*sn*-glycerophospholipids, A_1, A_2, L_1, and L_2 activities are indistinguishable.[19,20] But with 1,2-dipalmitoyl-*sn*-glycero-3-phosphocholine, 2,3-[^{14}C]dipalmitoyl-*sn*-glycero-1-phosphocholine, and 1-*O*-alkyl-2-[^{14}C]palmitoyl-*sn*-glycero-3-phosphocholine, the B activity and its A_1 and A_2 activities are determined independently. The system consists of a 5 m*M* substrate mixture as mentioned above, 0.2 *M* acetate buffer (pH 5.0) containing 2 m*M* EDTA, 30 m*M* Triton X-100, and the enzyme protein in a final volume of 0.4 ml. After incubation at 30° for 30 min, the reaction products are partitioned in chloroform and metha-

[19] J. Sugatani, N. Kawasaki, and K. Saito, *Biochim. Biophys. Acta* **529,** 29 (1978).

[20] J. Sugatani, T. Okumura, and K. Saito, *Biochim. Biophys. Acta* **620,** 372 (1980).

[21] K. Saito and M. Kates, *Biochim. Biophys. Acta* **369,** 245 (1974).

FIG. 3. pH dependence of activities of native and modified forms of phospholipase B. (A) Egg yolk 1,2-diacyl-*sn*-glycero-3-phosphocholine; (B) 1,2-dioctanoyl-*sn*-glycero-3-phosphocholine; (C) egg yolk 1-acyl-2-lyso-*sn*-glycero-3-phosphocholine.

nol–water layers by the method of Bligh and Dyer. The lipid products and glycerophosphocholine liberated are analyzed as mentioned above.

Preparation of Substrates

1-O-Alkyl-2-[^{14}C]palmitoyl-sn-glycero-3-phosphocholine. The choline glycerophospholipid fraction is prepared from beef heart lipids by column chromatography on alumina and silicic acid and hydrolyzed with methanolic 0.25 *M* KOH at 37° for 30 min.[19] The resulting 1-*O*-ack-1′-enyl-2-lyso-*sn*-glycero-3-phosphocholine (lysoplasmalogen) fraction is hydrogenated over platinum oxide, and mixed with [^{14}C]palmitic acid (2 μmol) neutralized with tetraethylammonium hydroxide and palmitic anhydride (4 μmol) *in vacuo* at 80° overnight. The product is purified by column chromatography on silicic acid.

2,3-[^{14}C]Dipalmitoyl-sn-glycero-1-phosphocholine. 2,3-Dipalmitoyl-*sn*-glycero-1-phosphocholine (40 μmol), obtained from hydrolysis of *rac*-1,2-dipalmitoyl-*sn*-glycero-3-phosphocholine with phospholipase A_2 (*Crotalus adamanteus*), is incubated with lipase (*Rhizopus arrhizus var. Delemar,* 5 mg) in 10 ml of 19 m*M* sodium taurocholate, 5 m*M* $CaCl_2$, and 90 m*M* borate buffer (pH 5.7) at 37° for 3 hr.[20] The resulting monopalmitoyl-*sn*-glycero-1-phosphocholine is mixed with [^{14}C]palmitic acid neutralized with tetraethylammonium hydroxide and its fatty acid anhydride *in vacuo* at 80° overnight.

Uniformly Labeled [^{14}C,^{32}P]diacylglycerophosphocholine. Candida lipolytica, NRRLY 1094, is grown on standard yeast medium in the presence of [^{14}C] acetate and ortho[^{32}P]phosphate.[21] The cells are harvested, and total lipids are extracted. The lipids are separated into individual components by preparative TLC. The labeled glycerophosphocholine and glycerophosphoethanolamine are chromatographically pure. The specific activity [disintegrations per minute (dpm)/nmol] of labeled glycerophosphocholine is 805 for ^{32}P and 326 for ^{14}C. The constituent fatty acids are 16 : 0 (7%), 16 : 1 (13%), 18 : 1 (36%), and 18 : 2 (44%); therefore, the suspension of this highly unsaturated substrate in acetate buffer (above) is found to be hydrolyzed in the absence of any activator and without being subjected to sonication (Fig. 1).

1-[^{14}C]Palmitoyl-sn-glycero-3-phosphocholine and 2-[^{14}C]Palmitoyl-sn-glycero-3-phosphocholine. 1-[^{14}C]Palmitoyl-*sn*-glycero-3-phosphocholine is prepared from hydrolysis of 1,2-[^{14}C]palmitoyl-*sn*-glycero-3-phosphocholine with phospholipase A_2.[19] 2-[^{14}C]Palmitoyl-*sn*-glycero-3-phosphocholine was prepared from 1-*O*-alk-1′-enyl-2-[^{14}C]palmitoyl-*sn*-glycero-3-phosphocholine (2.5 μmol) dispersed in 1.5 ml petroleum ether and 1.5 ml of 0.1 *M* boric acid by the addition of iodine-saturated petroleum ether. These substrates should be used as soon as possible, because slow acyl migration occurs, particularly at acidic pH.[22]

Egg Yolk Glycerophosphocholine and Its Lyso Compound. These substrates are prepared as previously reported.[13]

Other Commercially Available Substrates. A homologous series of synthetic 1,2-diacyl-*sn*-glycero-3-phosphocholines containing two identical, saturated normal-chain fatty acids with even chain lengths from C_8 to C_{18}, as well as $C_{18:1}$, and 1,2-dipalmitoyl-*sn*-glycero-3-phosphoethanolamine are purchased from Serdary Research Laboratories (London, Ontario, Canada). When necessary, the diacyl compounds are purified by preparative TLC. The short-chain glycerophosphocholines are purified on small column of silicic acid, eluted first with chloroform to remove simple lipids and then with a mixture of chloroform–methanol (1 : 1, v/v).[21] A homologous series of lysoglycerophosphocholines with acyl chains from C_{14} to C_{18} is prepared from the corresponding diacylglycerophosphocholines by hydrolysis with phospholipase A_2 (*Crotalus adamanteus*) in an ether medium.[23] Lyso-C_{10} and -C_{12} glycerophosphocholine are purchased from Serdary Research Laboratories. All short-chain lysoglycerophosphocholines are unstable on storage in a mixture of chloroform–methanol (1 : 1, v/v) at 0° for longer than 2 weeks; therefore, they are always purified by silicic acid column immediately before use.

Stimulation and Inhibition

Penicillium notatum phospholipase B activity is stimulated by diethyl ether, Triton X-100, and chlorpromazine (local anesthetic). Fe^{2+} and Fe^{3+} inhibit the B activity, but Mg^{2+}, Mn^{2+}, Zn^{2+}, Cu^{2+} and Hg^{2+} do not. Ca^{2+} is quite independent. Lyso activity is inhibited by detergents but not affected by the other reagents tested. Diisopropyl fluorophosphate (DFP) inhibits the B and lyso activities at a relatively high concentration (50 m*M*). Both activities are rather heat-labile.[15,19–21]

Both activities are inhibited to the same extent by chemical modification of the enzyme protein with diethyl *p*-nitrophenyl phosphate [serine reagent, in 0.1 *M* NaCl and 0.1 *M* acetate buffer (pH 5.5) in the presence of sodium taurocholate], *N*-bromosuccinimide and 2-hydroxy-5-nitrobenzyl bromide (tryptophan reagent, in the dark in 0.1 *M* acetate buffer, pH 4.0), and phenylglyoxal (arginine reagent, in the dark in 0.1 *M* *N*-ethylmorpholine–acetate buffer, pH 8.0). These observations suggest that both activities share the same active site of the phospholipase B.

[22] A. Plückthun and E. A. Dennis, *Biochemistry* **21,** 1743 (1982).

[23] D. J. Hanahan, M. Rodbell, and L. D. Turner, *J. Biol. Chem.* **206,** 431 (1954).

Conclusion

Phospholipase B in *P. notatum,* at least, catalyzes the complete deacylation of all kinds of natural glycerophospholipids including lysoglycerophospholipids; therefore, the phospholipase B has both intrinsic B and lyso activities. Following proteolytic modification, the B activity is almost completely lost, but the lyso activity remains intact. The modified enzyme gives two large peptides and two small peptides by reductive cleavage with 2-mercaptoethanol. The large peptides are located at the C-terminal part, and the small peptides are at the N-terminal part of the native enzyme. In "Enzyme Nomenclature" published in 1984, phospholipase B is described as one of the other names for lysophospholipase (EC 3.1.1.5); however, as mentioned above, this may not be true for phospholipase B in *P. notatum.* How universally applicable the present model is to other lysophospholipases reported so far is the next problem to be solved.

Acknowledgments

This study is supported in part by a Grant-in-Aid for Scientific Research (63480132) from the Ministry of Education, Science, and Culture of Japan, and by the Science Research Promotion Fund from the Japan Private School Promotion Foundation (1987–1989).

[43] Lysophospholipases I and II from P388D$_1$ Macrophage-like Cell Line

By Ying Yi Zhang, Raymond A. Deems, and Edward A. Dennis

Introduction

1-Acyl-*sn*-glycero-3-phosphorylcholine + $H_2O \rightarrow$ glycero-3-phosphorylcholine + fatty acid

Until recently, membrane phospholipids were viewed as the inert building blocks of cellular membranes. We now appreciate that these "building blocks" are actively metabolized via a complex control system and that they dramatically affect numerous enzyme functions. In addition, it has been found that many of the products of phospholipid catabolism are also very potent cellular modulators. One of these products, lysophospholipid, is a strong detergent whose presence in the membrane can dramatically affect membrane fluidity, membrane integrity, and cell viability.[1] Lysophospholipases (EC 3.1.1.5) are the main enzymes responsible for controlling the level of this biologically active species. We have purified and characterized two of these enzymes from the P388D$_1$ macrophage-like cell line.[2]

Assay

Principle. Lysophospholipase hydrolyzes the fatty acid ester bond of lysophospholipids, liberating water-soluble glycerol phosphatides and fatty acids. Because of the low levels of lysophospholipase activity found in cells, a radioactive assay is the best method of determining their activity. The assay employs lysophosphatidylcholine (lyso-PC) that is radiolabeled in the fatty acid chain. Palmitoyllsyo-PC, the substrate used in the standard assay, forms micelles in aqueous solutions and has a critical micelle concentration (CMC) of 7 μM.[3] Since the standard assay contains 125 μM palmitoyllyso-PC, about 7 μM is present as monomers and about 118 μM is present as micelles. At the end of the reaction, the lipid components are extracted, separated, and counted.

Procedure. The standard assay mixture contains 0.1 mM Tris-HCl (pH 8.0), 125 μM [^{14}C]palmitoyllyso-PC (0.2 μCi/μmol) in a total volume of 0.5 ml. The assay is started by adding 5 to 10 μl (40 to 80 ng) of purified enzyme or a comparable amount of activity of a partially purified preparation. The incubation is carried out at 40° for 30 min. The extent of hydrolysis is determined by one of two methods. The first uses a modified Dole extraction,[4] and the other employs a modified Bligh and Dyer extraction,[5] followed by separation of the reactants and products by thin-layer chromatography (TLC).

Dole Assay. In the Dole assay, the reaction is stopped by the addition of 2.5 ml of 2-propanol/heptane/1 N H_2SO_4 (20 : 5 : 1, v/v). About 0.1 g of silica gel (Bio-Sil A, 100–200 mesh, Bio-Rad, Richmond, CA) is added and vortexed immediately. Then 1.5 ml of heptane and 1.5 ml of deionized water are added and vortexed for at least 10 sec. One milliliter of the upper heptane phase is removed, mixed with 5 ml of scintillation fluid, and counted.

TLC Assay. In the TLC assay, the reaction is stopped by the addition

[1] R. E. Stafford and E. A. Dennis, *Colloids Surf.* **30,** 47 (1988).
[2] Y. Zhang and E. A. Dennis, *J. Biol. Chem.* **263,** 9965 (1988).
[3] R. E. Stafford, T. Fanni, and E. A. Dennis, *Biochemistry* **28,** 5113 (1989).
[4] S. A. Ibrahim, *Biochim. Biophys. Acta* **137,** 413 (1967).
[5] E. G. Bligh and W. J. Dyer, *Can. J. Biochem. Physiol.* **31,** 911 (1959).

of 1 ml chloroform/methanol/glacial acetic acid (2 : 4 : 1, v/v), and the solution is vortexed. An additional 0.5 ml of deionized water and 0.5 ml of chloroform are added, and the solution is vortexed again. The chloroform phase is removed and dried under vacuum at 40°. The residue is dissolved in 30 μl of chloroform/methanol/glacial acetic acid (2 : 4 : 1, v/v). The entire sample is spotted onto silica gel G plates, and the lipid components are separated by elution with chloroform/methanol/acetic acid/water (25 : 15 : 4 : 2, v/v). The lipids are visualized with I_2 vapor, and the zones corresponding to fatty acid and lyso-PC are scraped directly into scintillation vials and counted with 6 ml of scintillation fluid.

The efficiencies of the various extractions and their effects on the accuracy of the assays have been determined.[2] The TLC assay overestimates the amount of free fatty acid in the assay by 2–4% while the Dole assay underestimates it by 5–9%. The average deviation of triplicate experimental points is always within 5% in both assays. Overall, the TLC assay is more accurate and reliable than the Dole assay, although it is much more laborious. We use the TLC assay for all kinetic studies and the Dole assay to monitor the purification.

Purification Procedures

Cell Culture. The $P388D_1$ cells were originally provided by Dr. H. S. Koren.[6] Similar cells can be obtained from the American Type Culture Collection (Rockville, MD). These cells are maintained in culture[7,8] at 37° and 5% CO_2 in RMPI 1640 medium supplemented with 10% fetal calf serum (Hyclone Laboratories, Logan, UT), 2 m*M* L-glutamine, 50 units/ml of penicillin, and 50 μg/ml of streptomycin. The cell culture is started with 1×10^5 cells/ml in 60 ml of culture medium in 150-cm^2 culture flasks. The cells generally are confluent in 2 days, after which the culture is inoculated into a 850-cm^3 roller bottle containing 450 ml of culture medium and incubated at 0.3 rpm on a bottle roller in a warm room without CO_2. After 3 days of incubation, any adherent cells are suspended into the medium by agitation, and all cells are harvested by low-speed centrifugation (600 *g*, 4°, 15 min). Routinely, 3.5–4.5 $\times$ 10^9 cells are harvested from eight roller bottles with over 90% cell viability.

[6] H. S. Koren, B. S. Handwerger, and J. R. Wunderlich, *J. Immunol.* **114,** 894 (1975).

[7] M. I. Ross, R. A. Deems, A. J. Jesaitis, E. A. Dennis, and R. J. Ulevitch, *Arch. Biochem. Biophys.* **238,** 247 (1985).

[8] R. J. Ulevitch, M. Sano, Y. Watanabe, M. D. Lister, R. A. Deems, and E. A. Dennis, *J. Biol. Chem.* **263,** 3079 (1988).

TABLE I
PURIFICATION OF LYSOPHOSPHOLIPASE I[a]

Step	Total protein (mg)	Total activity (nmol/min)	Yield (%)	Specific activity (nmol/min/mg)	Purification (-fold)
. Homogenate	15,300	2410	100	0.158	1
. LS-1	6530	1190	49	0.182	1.2
. DEAE–Sephacel	808	955	40	1.18	7.5
. Sephadex G-75	193	737	31	3.82	24
. Blue Sepharose					
Peak I	3.50	341	14	97.4	620
Peak II	5.40	149	6.2	27.6	180
. Chromatofocusing of peak I	0.321	288	12	897	5700
. Sephadex G-75	0.154	206	8.6	1340	8500

[a] Reprinted from Zhang and Dennis,[2] with permission.

Purification of Lysophospholipase I

Two purification schemes are presented. The first is designed to maximize the yield of lysophospholipase I, and the second is required to purify lysophospholipase II. Although lysophospholipase I can also be obtained from the second method, its yield is much lower than in the first. The results of a typical purification of lysophospholipase I, employing the first purification scheme, are summarized in Table I.[2] All of the following steps are carried out at 4° to 8°.

Step 1: Preparation of Cell Homogenates. The cells harvested from roller bottles are suspended in 0.34 *M* sucrose, 10 m*M* HEPES, pH 7.5, 1 m*M* EDTA, and 1 m*M* ATP. The suspended cells (1×10^9 cells/30 ml) are continuously mixed by a magnetic stirring bar in a plastic cylindrical chamber in a precooled Parr cell disruption bomb at 4° for 15 min under 600 psi N_2 and then lysed by releasing the pressure.

Step 2: Preparation of LS-1. The cell lysate is centrifuged (1000 *g*, 4°, 15 min), and the resultant supernatant (LS-1) is removed and stored at −20° for subsequent use as the starting material for both purifications. Generally, 1×10^{10} cells (from 20 roller bottles) yielded about 280 ml of LS-1.

Step 3: DEAE–Sephacel Chromatography. LS-1 is thawed and adjusted to pH 8.3 with 1 *N* NaOH and loaded onto a DEAE–Sephacel column which has been previously equilibrated with 10 m*M* Tris-HCl, pH 8.3, 2 m*M* EDTA, and 10 m*M* 2-mercaptoethanol (2-ME). The column (2.5×40 cm) is eluted with 600 ml equilibration buffer containing a 0–0.24 *M* NaCl gradient. Optimum column resolution is obtained with a column bed volume to sample volume ratio of 1 : 1.6. A slow linear flow rate (6 cm/hr) is required to efficiently remove the large amounts of turbid nucleic acid present in LS-1. Two or three active peaks are recovered from the DEAE–Sephacel column. The first and second activity peaks elute in the void volume, and the third peak elutes at the very beginning of the NaCl gradient (20 m*M*). The appearance of the first peak is not always reproducible, and, in some preparations, only one active peak appears in the void volume. We found that it is more efficient to pool the active peaks together at this stage since their separation on the Blue-Sepharose column is much cleaner and more reproducible.

Step 4: Sephadex G-75 Chromatography. To concentrate the protein, a single 80% saturation ammonium sulfate cut is taken by adding solid ammonium sulfate to the pooled DEAE fractions over a 20-min period. The mixture is stirred slowly for another 30–40 min, and the precipitate is pelleted by centrifugation (12,000 *g*, 4°, 30 min). The pellet is resuspended in 10 to 20 ml of 20 m*M* Tris-HCl (pH 8.0), 2 m*M* EDTA, and 10 m*M* 2-ME and immediately loaded onto a Sephadex G-75–50 column (2.5 × 100 cm) which has been equilibrated with 20 m*M* Tris-HCl (pH 8.0), 2 m*M* EDTA, and 5 m*M* 2-ME. The protein is eluted with equilibration buffer at a flow rate of 20 ml/hr. Although the purification with this column is only 3-fold, it serves to remove the ammonium sulfate. The final 2-ME concentration should be 5 m*M* since 10 m*M* 2-ME prevents the enzyme from binding to the Blue Sepharose column. This procedure is employed rather than dialysis because the latter resulted in a 30% loss of enzymatic activity.

Step 5: Blue Sepharose Chromatography. The active fractions pooled from the Sephadex G-75 column are adjusted to pH 6.0 with 2% acetic acid and immediately loaded onto a Blue Sepharose CL-6B (Pharmacia, Piscataway, NJ) column (1.5 × 20 cm) previously equilibrated with 10 m*M* Tris–acetate (pH 6.0), and 1 m*M* EDTA. Both lysophospholipases I and II bind to the column completely. However, the column equilibration conditions are more critical for lysophospholipase I than for lysophospholipase II. Increasing the buffer pH, substituting Tris–acetate with imidazole-HCl, or adding 2-ME to the equilibration buffer will prevent the lysophospholipase I from binding to the column. After loading the sample, the column is washed with 1 volume of the equilibration buffer, and the protein is eluted with a 300-ml pH gradient from pH 6.0 to 8.5 with 25 m*M* imidazole-HCl buffer containing 10 m*M* 2-ME, 10% glycerol, and 10 μM lyso-PC. The latter three components were required to maintain enzymatic activity during this and the next step. Two lysophospholipases are separated completely on the Blue Sepharose column. Lysophospholipase I elutes at the very beginning of the gradient (pH 6.0) in a sharp activity peak. The lysophospholipase II elutes at about pH 7 in a broad active peak. This column yields a 26-fold purification of lysophospholipase I and a 7-fold purification of lysophospholipase II.

Step 6: Chromatofocusing. The lysophospholipase I fractions are pooled and, after the pH is adjusted to 7.5 with 0.1 *M* NaOH, loaded onto a chromatofocusing column. The chromatofocusing column (1 × 20 cm) is packed with Polybuffer Exchange 94 resin (Pharmacia) and equilibrated with 25 m*M* imidazole-HCl buffer, pH 7.5, containing 10 m*M* 2-ME, 10% glycerol, and 10 μ*M* lyso-PC. After the sample is loaded, the column is washed with 1 volume of the equilibration buffer and eluted with a 1 to 8 dilution of Polybuffer 74 (Pharmacia) at pH 4.5 containing 10 m*M* 2-ME, 10% glycerol, and 10 μ*M* lyso-PC. Lysophospholipase I elutes at about pH 5.5. The fractions (5 ml) are collected into 1.5 ml of 0.15 *M* imidazole buffer, pH 7.5, containing 10 m*M* 2-ME and 8 m*M* EDTA. The active fractions are pooled, placed in Spectra/Por 2 dialysis tubing (12,000–14,000 molecular weight cutoff), and concentrated against 70–80% (w/v) polyethylene glycol in water (MW 15,000–20,000; Sigma, St. Louis, MO).

Step 7: Sephadex G-75 Chromatography. The concentrated active pool (~5 ml) from the chromatofocusing column is loaded onto a Sephadex G-75–50 column (2.5 × 50 cm) which was equilibrated with 25 m*M* imidazole-HCl (pH 7.5), containing 30 m*M* NaCl, 10% glycerol, 10 m*M* 2-ME, 2 m*M* EDTA, and 10 μ*M* lyso-PC. Lysophospholipase I elutes as a single active peak at about 30,000 daltons. The active fractions from this column are immediately placed into Spectra/Por 2 dialysis tubing and immersed in 100% glycerol containing 10 m*M* 2-ME at 4° for 2–3 hr. This purified enzyme solution is stored at 0.5-ml aliquots in siliconized vials at −20° and is stable for at least 1 year. This preparation gave a single band on sodium dodecyl sulfate-polyacrylamide gel electrophoresis (SDS–PAGE) using either silver staining or Coomassie blue staining. The final yield of the purification procedure is 8.6%, with a 8500-fold purification. Various preparations have yielded specific activities between 1.3 and 1.7 μmol/min/mg.

Purification of Lysophospholipase II

The lysophospholipase II preparation obtained from the Blue Sepharose column, Step 5 above, proved to be difficult to purify further. This is due to the instability of the enzyme (see Properties). The conditions required to stabilize the two enzymes are not identical, and, in fact, some conditions required to purify lysophospholipase I lead to the inactivation of lysophospholipase II. Therefore, the purification of lysophospholipase

TABLE II
PURIFICATION OF LYSOPHOSPHOLIPASE II

Step	Total protein (mg)	Total activity (nmol/min)	Yield (%)	Specific activity (nmol/min/mg)	Purificatio (-fold)
1. LS-1	8285	1559	100	0.187	1
2. Sephadex G-25	1800	1113	71	0.62	3.3
3. DEAE–Sephacel					
Peak I	133	578	37	4.34	23
Peak II	82	347	22	4.22	23
4. Sephadex G-75 of peak II	6.8	288	18	42.1	225
5. Blue Sepharose	0.1	120	7.7	1229	6555
6. Chromatofocusing	0.0092	32	2	3440	18,350

II is carried out separately, using the same columns but different conditions. We also found that the addition of two precipitation steps at the beginning of the purification greatly enhanced the efficiency of the DEAE–Sephacel column. The first was a streptomycin precipitation that removes nucleic acid. The second was an ammonium sulfate precipitation, used mainly to concentrate the sample.

Step 1: LS-1. The starting material (LS-1) is prepared as described above for lysophospholipase I. Again, all steps are carried out at 4°. Table II summarizes a typical purification of lysophospholipase II.[9]

Step 2: Streptomycin and Ammonium Sulfate Precipitations. The removal of nucleic acid is accomplished by streptomycin precipitation. Ten percent (w/v) streptomycin is added to LS-1 dropwise to a final concentration of 1%. The mixture is stirred slowly for another 10–30 min. It is then centrifuged (15,000 *g*, 4°, 10 min), and the pellet, containing the nucleic acids, is discarded. Thus, large amounts of nucleic acid are removed from LS-1 without affecting the lysophospholipase activity.

Solid ammonium sulfate is added to the streptomycin supernatant to a final concentration of 30% of saturation. The ammonium sulfate is added slowly over 30–60 min, and the solution is then stirred slowly for another 30–45 min and, finally, centrifuged (15,000 *g*, 4°, 30 min). The supernatant is collected and subjected to another ammonium sulfate precipitation at 55% saturation following the same procedures, except that the pellet is collected. The lysophospholipase is stable in the 55% saturated ammonium sulfate solution for at least 1 month. The 55% ammonium sulfate pellet is dissolved in 50 ml of buffer A (10 m*M* Tris, pH 8.0, and 10 m*M* 2-ME). A little precipitate is insoluble and is removed by centrifugation (2000 *g*, 4°, 15 min). The solution is desalted on a 2.5 × 90 cm Sephadex G-25–80 column, which is equilibrated and eluted with buffer A.

About 30% of the total lysophospholipase activity is lost during the ammonium sulfate precipitation while yielding a 3-fold purification. One of the major purposes of this step is to reduce the volume of the starting material. Generally, 1300 ml of LS-1 is reduced to 90 ml after the ammonium sulfate precipitation and desalting on the Sephadex G-25 column.

Step 3: DEAE–Sephacel Chromatography. The pooled active fractions from the Sephadex G-25 column are loaded onto a DEAE–Sephacel column (2.5 × 40 cm), preequilibrated with buffer A. The column is then washed with 100–150 ml of buffer A, and the protein is eluted with a 600-ml linear NaCl gradient (0–0.24 *M*) in buffer A. Two lysophospholipase peaks are eluted. The first peak is lysophospholipase II, eluting in 50 m*M* NaCl, and the second is lysophospholipase I, eluting in 90 m*M* NaCl. The pH of the collected lysophospholipse I pool is 8.9, and that of the lysophospholipase II pool is 8.7, although the pH of the elution buffer is only 8.0. Since the enzymes are not stable at such a high pH, the pH values are lowered to 7.0 with 1 *M* imidazole, pH 6.0, immediately after collection. Alternatively, the fractions can be collected directly into 0.2 *M* imidazole, pH 6.0, 1 m*M* EDTA, and 10 m*M* 2-ME.

In this purification procedure, all of the lysophospholipase activity binds to the column and elutes in two distinct, reproducible peaks. This is in contrast to the previous purification scheme in which a large portion of the activity eluted in the void volume and the two lysophospholipases could not be separated. This difference is probably due to the removal of the nucleic acids from LS-1 by the streptomycin precipitation in Step 2 and the elimination of the EDTA in the equilibration buffer.

Step 4: Sephadex G-75 Chromatography. The lysophospholipase II pool is concentrated to 10–20 ml by an Amicon (Danvers, MA) ultrafiltration device equipped with a YM10 membrane and then chromatographed on a Sephadex G-75–50 column (2.5 × 90 cm), which has been preequilibrated with 25 m*M* imidazole, pH 6.7, 1 m*M* EDTA, and 5 m*M* 2-ME at a flow rate of 20 ml/hr. This step produced a 10-fold purification, with an 80% yield of activity. In addition, the enzyme is obtained in a solution close to the equilibration buffer of the next column.

Step 5: Blue Sepharose Chromatography. The G-75 pool is loaded onto a Blue Sepharose column (1.5 × 6 cm) which is preequilibrated with 25 m*M* imidazole (pH 6.7) and 1 m*M* EDTA. After loading, the column is washed with 50 ml of equilibration buffer and eluted with 25 m*M* imidazole (pH 6.7), 60 m*M* NaCl, 10% glycerol, 1 m*M* EDTA, 10 m*M* 2-ME, and 10 μ*M* lyso-PC. The flow rate of the column is 15 ml/hr. To efficiently elute the bound enzyme with a small volume of the elution buffer, the column flow is stopped for 1 hr after the first 5 ml of elution buffer is applied.

The equilibration buffer for this column is 25 m*M* imidazole, whereas in the purification of lysophospholipase I it is 10 mM Tris. This change in buffer takes advantage of the fact that lysophospholipase I does not bind to Blue Sepharose in 25 mM imidazole, but lysophospholipase II does. Both the pH and concentration of the imidazole are also critical in allowing lysophospholipase II to bind to the Blue Sepharose column but preventing many other proteins from binding to the column, including lysophospholipase I. More than 95% of the contaminating protein in the Sephadex G-75 II pool is removed by this step. The activity yield from this column is 40%, with a 30-fold purification.

Step 6: Chromatofocusing Chromatography. The active fractions from the Blue Sepharose column are pooled and concentrated to 5 ml via Amicon ultrafiltration (described above). To change the buffer, the protein is passed through a Sephadex G-50–80 column (1.5 × 20 cm), which is preequilibrated and then eluted with CF elution buffer (7% Polybuffer 96, pH 7.0, 10 mM 2-ME, 10% glycerol, and 10 μM lyso-PC). It is then loaded onto a chromatofocusing column (1 × 13 cm) packed with Polybuffer Exchange 94 resin and equilibrated with 25 mM Tris, pH 8.5, 10 mM 2-ME. Since lysophospholipase II is not stable above pH 7.5, 2–3 ml of CF elution buffer is applied to the column before the loading of the sample. After loading, the protein is eluted with CF elution buffer. Lysophospholipase II elutes at about pH 7.8. Two-milliliter fractions are collected into 0.12 ml of 40% glycerol, 0.16 M imidazole, pH 6.0, and 10 mM 2-ME to protect the enzyme from inactivation at high pH.

Although the chromatofocusing column is used for both lysophospholipases I and II, the column operation is totally different for the two enzymes. Lysophospholipase II did not bind to the chromatofocusing resin at pH values below 8.5. Considering that the enzyme would be gradually denatured at pH values above 7.5, the lysophospholipase II is first transferred to the chromatofocusing column elution buffer, via the Sephadex G-50 column, which is at pH 7.0. It is then loaded onto the chromatofocusing column equilibrated at pH 8.5. In this way, lysophospholipid II binds to the column but avoids the direct contact with high-pH buffer.

This preparation of lysophospholipase II is homogeneous by SDS–PAGE. The enzyme is purified 18,350-fold, with a 2% yield. The purified lysophospholipase II is very dilute and is not stable for more than 1 day. The addition of 0.5 mg/ml of cytochrome *c* protects the enzyme activity completely. This solution can be frozen and kept at −20° without losing enzyme activity. It can also be stored by concentrating the sample in glycerol as described in the lysophospholipase I purification.

The lysophospholipase I, which is separated from lysophospholipase II by the DEAE step, can be further purified to homogeneity by using the same procedures described before. However, the yield is about 40% lower

TABLE III
COMPARISON OF LYSOPHOSPHOLIPASES I AND II

Characteristic	Lysophospholipase I	Lysophospholipase II
Molecular weight	27,000	28,000
Isoelectric point	4.4	6.1
Optimum pH	7.5–9	7.5–9
Kinetic model	Hill	Hill
V_{max} (nmol/min/mg)	1345	3516
Half-maximal saturation (μM)	5	4
Hill coefficient	2.1	1.8
EDTA, Ca^{2+}, Mg^{2+}, Mn^{2+}	No effect	No effect
Co^{2+}, Zn^{2+}, Cu^{2+}, Hg^{2+}	Inhibited	Inhibited
Stable pH range	4.5–8.5	6–7.5
Phosphorylcholine affinity column binding	Yes	No
React with lysophospholipase I antibody	Yes	No

than achieved with the first purification scheme. The lower yield is perhaps due to the loss of some of the enzyme during the ammonium sulfate precipitation step or during the chromatography steps due to the different elution conditions.

Properties

Purity and Physical Properties. Both lysophospholipases I and II gave single sharp bands on SDS–PAGE, whether detected by Coomassie blue or silver staining. From the SDS gels, we determined the molecular weight of lysophospholipase I to be 27,000 and lysophospholipase II to be 28,000.[2] Isoelectric focusing of the lysophospholipase peaks from the Blue Sepharose column showed that lysophospholipase I had an isoelectric point of 4.4 and lysophospholipase II an isoelectric point of 6.1.[2] These and other properties of lysophospholipase I and II are compared in Table III.[9]

Antibody Cross-Reactivity. Polyclonal antibodies were developed to SDS–denatured lysophospholipase I. Western blot analyses of lysophospholipase I and II were performed. The antibody bound only to the lysophospholipase I; no cross-reaction with lysophospholipase II was detected.[9] Although the enzymes are very similar in many kinetic characteristics, the lack of cross-reactivity indicated that there are major differences in the amino acid composition of these proteins. They appear to be quite distinct enzyme species.

Phosphorylcholine Affinity Column Binding. The hypothesis that the

[9] Y. Y. Zhang, Ph.D Dissertation, University of California at San Diego (1989).

two enzymes are distinct is bolstered by their interaction with a phosphorylcholine affinity column.[10] The functional group of this column is a 10-carbon methylene chain terminated at one end by a phosphorylcholine moiety. The other end of the lysophospholipid analog is linked to the Sepharose beads through an hexanediamine arm.[10] Lysophospholipase I bound to this column very tightly while lysophospholipase II did not bind at all.[9]

Stability. Enzyme instability is a major problem in the purification of the enzymes.[2,9] After the DEAE column, the enzyme solutions could not be stored at −20° in the absence of glycerol or cytochrome *c* (see below) since the enzymes lose all of their activity during the freezing and thawing process. It was possible, however, to store the crude enzyme solution at 4° for several weeks with little loss of activity.

The purified lysophospholipase I is extremely unstable. When kept in the elution buffer of the last column, the enzyme lost half of its activity within 24 hr and was completely inactive in 3 days. This instability can be overcome by concentrating the enzyme solution by dialyzing it against 100% glycerol (see Step 7 of lysophospholipase I purifcation). The purified enzyme solution was usually concentrated to 10–25 μg/ml protein and contained 60–70% glycerol. With this treatment, the enzyme retained its activity for at least 1 year at −20°. Dilution of the concentrated enzyme into aqueous buffer resulted in immediate loss of activity; however, it does not lose activity when added directly to an assay mixture containing more than 20 μM lysophospholipid. If the substrate concentration is below this, the enzyme again loses activity. For example, if lysophospholipase I is incubated with 3 μM lyso-PC, it loses 80% of its activity in 60 min. It appears that lysophospholipase II behaves similarly.[9]

This instability at low substrate concentrations is a serious obstacle to conducting kinetic studies because the substrate concentration at half-maximal activity for both enzymes is less than 10 μM. Therefore, a large portion of the substrate dependence curve falls within this region. To carry out kinetic studies, some means of stabilizing the enzyme under these assay conditions are required. After an extensive search, we found that increasing the protein concentration was the only method of stabilizing the lysophospholipase activities. Of those proteins tested, the only one that did not interfere with the assay was cytochrome *c*. The enzyme is stable when diluted into 0.5 mg/ml cytochrome *c*. It maintains its activity even when incubated at 40° for 60 min. Therefore, all kinetic studies, as well as other studies requiring low levels of lyso-PC, are carried out in the presence of 0.5 mg/ml cytochrome *c*.

[10] A. J. Aarsman, R. Neys, and H. van den Bosch, *Biochim. Biophys. Acta* **792,** 363 (1984).

Lysophospholipase II was found to be very unstable after Blue Sepharose chromatography, but it could be stabilized in the same ways as lysophospholipase I. In addition, it was also found that lysophospholipase II is only stable over a narrow pH range. While the lysophospholipase I is stable between pH 5.0 and 8.5, the lysophospholipase II is stable only between pH 6.0 and 7.5. Outside of this pH range, significant loss of activity usually occurred within 5 to 10 hr. The narrower region of stability requires the lysophospholipase II to be purified separately from lysophospholipase I and necessitated various changes in column conditions.

Kinetics. Both lysophospholipase I and II activities have broad pH optima that extend from pH 7.5 to 9. There is no apparent requirement for metal ions. In fact, four metal ions, Co^{2+}, Zn^{2+}, Cu^{2+}, Hg^{2+}, were found to inhibit both enzymes.[2,9]

Kinetic analysis[2,9] of the substrate dependence indicated that the enzymes did not follow simple Michaelis–Menten kinetics. The enzyme kinetics do fit a Hill model. Lysophospholipase I had a V_{max} of 1345 nmol/min/mg, a half-maximal saturation of 5 μM, and a Hill coefficient of 2.1. Lysophospholipase II had a V_{max} of 3516 nmol/min/mg, a half-maximal saturation of 4 μM, and a Hill coefficient of 1.8. Lysophospholipase II is 2.5 times more active than lysophospholipase I; however, both enzymes have about the same half-maximal saturation and Hill coefficients. It appears that both enzymes bind two lysophospholipid molecules in a cooperative manner.

Specificity. Many of the previously examined lysophospholipases have other esterase activities.[11–16] Lysophospholipase I did not have any phospholipase A_2, phospholipase A_1, acyltransferase, transacylase, cholesterol esterase, or nonspecific esterase activity. The only other ester that it could hydrolyze was the fatty acid ester of monoolein, and this was at a rate 20 times lower than the hydrolysis of palmitoyllyso-PC.

Lysophospholipase I showed no preference for fatty acid chain length and hydrolyzed *sn*-2 lyso-PC as well as it did *sn*-1 lyso-PC. It also hydrolyzed lyso-PG as well as it did lyso-PC. We have studied a series of

[11] O. Doi and S. Nojima, *J. Biol. Chem.* **250,** 5208 (1975).

[12] K. Karasawa, I. Kudo, T. Kobayashi, T. Sa-Eki, K. Inoue, and S. Nojima, *J. Biochem. (Tokyo)* **98,** 1117 (1985).

[13] G. Brumley and H. van den Bosch, *J. Lipid Res.* **18,** 523 (1977)

[14] H. van den Bosch, A. J. Aarsman, J. G. N. De Jong, and L. L. M. van Deenen, *Biochim. Biophys. Acta* **296,** 94 (1973).

[15] J. G. N. de Jong, H. van den Bosch, A. J. Aarsman, and L. L. M. van Deenen, *Biochim. Biophys. Acta* **296,** 105 (1973).

[16] J. G. N. de Jong, H. van den Bosch, D. Rijken, and L. L. M. van Deenen, *Biochim. Biophys. Acta* **369,** 50 (1974).

lysophospholipids at concentrations above, at, and below their CMC and have found that the aggregation of the substrate did not seem to affect the enzymatic rate in any way. This is in sharp contrast to the extracellular phospholipases A_2 whose activities dramatically increase when monomeric phospholipid substrates are converted to aggregated structures. Thus, these lysophospholipases seem to hydrolyze both monomeric and micellar lysophospholipid equally well[9] and do not exhibit interfacial activation.

Acknowledgments

Support for this work was provided by the National Institutes of Health (GM-20,501) and National Science Foundation (DMB 89-17392).

[44] Lysophospholipases from Bovine Liver

By H. van den Bosch, J. G. N. de Jong, and A. J. Aarsman

Introduction

Lysophospholipases occur widely in nature both in prokaryotic and eukaryotic cells.[1] In conjunction with phospholipases A_1 and A_2 the enzymes are thought to play essential roles in the catabolism of membrane phosphoglycerides by catalyzing deacylating processes and in preventing the accumulation of the lytic intermediary lysophosphoglycerides. Both membrane-associated and soluble forms of lysophospholipases have been described.[1] This chapter deals with the purification and brief characterization of a membrane-bound and a soluble lysophospholipase from bovine liver.[2,3]

Assay Method

Principle. A convenient and generally applicable assay procedure for lysophospholipases consists of measuring the release of radioactive fatty acid from acyl-labeled lysophospholipids by a modified Dole extraction procedure.[4,5] This procedure circumvents time-consuming thin-layer chro-

[1] H. van den Bosch, *in* "Phospholipids" (J. N. Hawthorne and G. B. Ansell, eds.), New Comprehensive Biochemistry, Vol. 4, p. 313. Elsevier, Amsterdam, 1982.

[2] J. G. N. de Jong, H. van den Bosch, D. Rijken, and L. L. M. van Deenen, *Biochim. Biophys. Acta* **369,** 50 (1974).

[3] H. van den Bosch and J. G. N. de Jong, *Biochim. Biophys. Acta* **398,** 244 (1975).

matographic methods. Continuous spectrophotometric assays for lysophospholipases have been developed utilizing substrate analogs in which the acyl group is linked to the backbone of the substrate by a thio ester linkage.[6–8] Since these substrates are not commercially available and the procedure can only be applied for enzyme solutions that are freed of 2-mercaptoethanol, this procedure is not further described.

Reagents

1-[1-^{14}C]Palmitoyl-*sn*-glycero-3-phosphocholine, 2 m*M*, sonicated in water[9]

Potassium phosphate buffer, 0.1 *M*, pH 7.0 or 7.5

Dole's extraction medium[10] (2-propanol, *n*-heptane, 1 *N* H_2SO_4, 40 : 10 : 1, by volume)

Silica gel [Merck (Darmstadt, FRG) Kieselgel 60, reinst]

Procedure. The incubation mixtures consist of varying amounts of enzyme protein in a total volume of 0.5 ml of 20 m*M* potassium phosphate buffer (pH 7.5, lysophospholipase I; pH 7.0, lysophospholipase II). The assay mixture for lysophospholipase I contains in addition 2 m*M* 2-mercaptoethanol. After all components are pipetted and distilled water is added to give a volume of 0.4 ml, the reaction is started by addition of 0.1 ml of the stock substrate solution to give a final substrate concentration of 0.4 m*M*. After 10 min of incubation at 37° the reaction is stopped by adding 2.5 ml of Dole's extraction mixture[10] and vortexing for 15 sec. Then approximately 100 mg of silica gel is added[11] by means of a glass spatula, followed by 1.5 ml of heptane and 1.5 ml of water. The mixture is vortexed for 15 sec, and the phases are allowed to separate. An aliquot of 1 ml of the upper heptane phase (total volume 2 ml) is transferred with an Eppendorf pipette to a scintillation vial for radioactivity measurement. A blank incubation without enzyme is included in each series to correct for nonenzy-

[4] S. A. Ibrahim, *Biochim. Biophys. Acta* **137,** 413 (1967).

[5] H. van den Bosch, G. M. Vianen, and G. P. H. van Heusden, this series, Vol. 71, p. 513.

[6] A. J. Aarsman, L. L. M. van Deenen, and H. van den Bosch, *Bioorg. Chem.* **5,** 241 (1976).

[7] H. van den Bosch and A. J. Aarsman, *Agents Actions* **9,** 382 (1979).

[8] L. Yu and E. A. Dennis, this volume [5].

[9] Commercially available 1-[1-^{14}C]palmitoyl-*sn*-glycero-3-phosphocholine is diluted to the required specific radioactivity (routinely 50 dpm/nmol) with unlabeled palmitoyllysophosphatidylcholine. A 2 m*M* stock solution in water is prepared by sonication for 30 sec. This sonicate can be stored indefinitely at −20°. After thawing at room temperature and resonication for 10 sec it is ready for use.

[10] V. P. Dole, *J. Clin. Invest.* **35,** 150 (1956).

[11] This addition is necessary to prevent partitioning of traces of lysophosphatidylcholine, especially when substrates with unsaturated fatty acids are used, into the wet heptane layer.

matic hydrolysis and traces of free fatty acid that are present in the substrate sonicate.

Purification of Lysophospholipase I

Homogenate. All steps are carried out at 0°–4°. Fresh bovine liver (1000 g) is homogenized in 2 liters of 20 m*M* Tris-HCl (pH 7.3) containing 0.15 *M* NaCl during 2 min in a Waring blendor and filtered through 2 layers of cheesecloth to give a homogenate.

Extract. Solubilization of lysophospholipase activity is achieved after delipidation of the homogenate by extraction with *n*-butanol. The homogenate is treated with an equal volume of ice-cold *n*-butanol saturated with water. The mixture is stirred vigorously with a glass rod for 2 min and centrifuged immediately for 20 min at 10,000 rpm in the GSA rotor of a Sorvall RC-2B centrifuge. The upper butanol layer is removed with a pipette connected to an aspirator. The water layer is filtered through cheesecloth to yield a clear extract. The extract is dialyzed against 3 batches of 4 volumes each of 20 m*M* Tris-HCl buffer (pH 7.3). The inactive precipitate that forms during dialysis is removed by centrifugation.

DEAE-Cellulose Chromatography. The clear supernatant from the previous step is applied to a DEAE-cellulose column [Whatman (Maidstone, UK) DE-23, 7 × 18 cm] equilibrated with the dialysis buffer. This step separates two lysophospholipases, denoted lysophospholipase I and II, respectively. Lysophospholipase I is somewhat retarded on this column and is eluted just after the break-through peak. Lysophospholipase II binds to the column and can only be eluted after application of a salt gradient (see the results of a comparable pilot experiment on DEAE-Sephadex, Fig. 1). The ratio of lysophospholipase I to lysophospholipase II varied from 0.17 to 0.24 in five experiments. Assuming that the recoveries for both enzymes until this stage of the purification are the same, the total recovery, specific activity, and purification factor for each enzyme can be calculated in homogenate and extract.

Ammonium Sulfate Precipitation. Pooled fractions from the previous step containing lysophospholipase I are brought to 60% saturation with solid $(NH_4)_2SO_4$ and stirred for 45 min. The precipitate is collected by centrifugation and dissolved in 50 ml of 20 m*M* Tris-HCl buffer (pH 7.3) containing 0.15 *M* NaCl.

Sephadex G-100 Gel Filtration. The solution obtained in the previous step is applied to a Sephadex G-100 column (3 × 150 cm) and eluted with the same buffer. Active fractions are pooled and dialyzed against 20 m*M* Tris-HCl buffer (pH 7.3) containing 10 m*M* 2-mercaptoethanol.

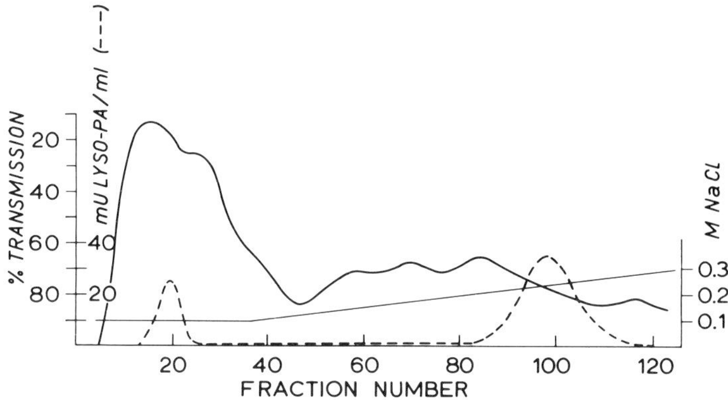

FIG. 1. Chromatography of lipid-free bovine liver extract on DEAE-Sephadex A-50 demonstrating the presence of two separate lysophospholipases. Lipid-free extract (2700 mg protein) was made 0.1 *M* with respect to NaCl and applied to a column of DEAE-Sephadex A-50 (4.5 × 10 cm) equilibrated with 20 m*M* Tris-HCl (pH 7.3) containing 0.1 *M* NaCl. The column is washed with 600 ml of this buffer and then eluted with a gradient of 650 ml each of 0.1 and 0.3 *M* NaCl in 20 m*M* Tris-HCl (pH 7.3). Fractions of 18 ml are collected at a flow rate of 54 ml/hr. Lyso-PA, Lysophospholipase activity. [Reproduced with permission from J. G. N. de Jong, H. van den Bosch, D. Rijken, and L. L. M. van Deenen, *Biochim. Biophys. Acta* **369,** 50 (1974).]

Second DEAE-Cellulose Chromatography. The dialyzate is chromatographed on DEAE-cellulose (Whatman DE-52 microgranular, 2.5 × 60 cm). The enzyme elutes from this column with the equilibration buffer while over 90% of the contaminating proteins are bound.

CM-Cellulose Chromatography. Initial attempts to purify lysophospholipase I further by cation-exchange column chromatography resulted in great losses of enzyme activity under conditions where the enzyme was bound to the column. The enzyme is therefore further purified in a CM-cellulose batch procedure. The pooled fractions from the previous step are dialyzed against 30 volumes of 80 m*M* sodium acetate (pH 5.0) containing 10 m*M* 2-mercaptoethanol. Ten grams of CM-cellulose (Whatman CM-52 microgranular) are added to the dialyzate, and the mixture is stirred for 30 min and then centrifuged. The supernatant is concentrated by ultrafiltration [Amicon (Danvers, MA) Diaflo, UM2 membrane] to one-seventh of the original volume. This procedure is necessary since lyophilization of the CM-cellulose supernatant causes great losses of activity. The solution obtained after ultrafiltration is brought to 90% saturation with solid $(NH_4)_2SO_4$, and the precipitate is collected by centrifugation. The pellet

TABLE I
PURIFICATION OF BOVINE LIVER LYSOPHOSPHOLIPASE I[a]

Fraction	Total protein[b] (mg)	Total activity[c] (U)	Specific activity (mU/mg)	Recovery (%)	Purification (-fold)
Homogenate	184,000	72[d]	0.39	100	—
Extract	37,900	47[d]	1.24	65	3.2
DEAE-cellulose	6300	40	6.3	55	16
$(NH_4)_2SO_4$	3400	39	11.5	54	30
Sephadex G-100	483	31	64	43	164
DEAE-cellulose	39	23.9	610	33	1564
CM-cellulose	25	17.9	720	25	1846
Sephadex G-50	8	11.2	1400	15	3590

[a] The purification starts from 1000 g of fresh bovine liver. Reproduced with permission from J. G. N. de Jong, H. van den Bosch, D. Rijken, and L. L. M. van Deenen, *Biochim. Biophys. Acta* **369,** 50 (1974).

[b] Protein is determined according to O. H. Lowry, N. J. Rosebrough, A. L. Farr, and R. J. Randall, *J. Biol. Chem.* **193,** 265 (1951). Samples containing 2-mercaptoethanol are first carboxymethylated with iodoacetate according to E. Ross and G. Schatz, *Anal. Biochem.* **54,** 304 (1973).

[c] One unit (U) of enzyme activity represents the release of 1 μmol fatty acid per minute.

[d] Total activity in homogenate and extract is calculated from the relative contribution of lysophospholipase I and II to the total activity in these fractions.

is dissolved in 2.5 ml of 20 m*M* Tris-HCl (pH 7.3) containing 10 m*M* 2-mercaptoethanol and 0.15 *M* NaCl.

Sephadex G-50 Gel Filtration. The enzyme solution is filtered through a Sephadex G-50 column (1.5 × 100 cm) equilibrated with the same buffer. Coinciding activity and protein peaks are pooled to yield a 3600-fold-purified enzyme. The purification is summarized in Table I.

Purification of Lysophospholipase II

Homogenate and delipidated extract are prepared as described for the purification of lysophospholipase I. The extract is dialyzed against 20 m*M* Tris-HCl (pH 7.3) containing 0.1 *M* NaCl.

DEAE-Sephadex Chromatography. The dialyzate is applied to a DEAE-Sephadex A-50 column (4 × 22 cm) equilibrated with dialysis buffer. The column is rinsed with this buffer until the eluate is free of material absorbing at 280 nm. A linear gradient of 5 bed volumes each of 0.1 and 0.3 *M* NaCl in 20 m*M* Tris-HCl (pH 7.3) is then applied to the column. Lysophospholipase II elutes at 0.18–0.23 *M* NaCl. The fractions

TABLE II
PURIFICATION OF BOVINE LIVER LYSOPHOSPHOLIPASE II[a]

Fraction	Total protein (mg)	Total activity (U)	Specific activity (mU/mg)	Recovery (%)	Purification (-fold)
Homogenate	166,000	294	1.77	100	—
Extract	23,300	144	6.2	49	3.5
DEAE-Sephadex	816	135	165	46	93
Hydroxylapatite	197	117	595	40	336
Sephadex G-100	55	73	1320	25	745
Sephadex G-100	48	65	1360	22	770

[a] See footnotes to Table I. Reproduced with permission from J. G. N. de Jong, H. van den Bosch, D. Rijken, and L. L. M. van Deenen, *Biochim. Biophys. Acta* **369,** 50 (1974).

containing enzyme activity are pooled and dialyzed against 10 m*M* potassium phosphate buffer (pH 7.0).

Hydroxylapatite Chromatography. The dialyzate from the previous step is pumped onto a hydroxylapatite column (BioGel HTP, 4 × 20 cm) equilibrated with dialysis buffer. After sample application the column is washed with 160 ml of this buffer, and then a linear gradient of 5 bed volumes each of 10 and 100 m*M* potassium phosphate buffer (pH 7.0) is applied. Lysophospholipase II elutes at 55–85 m*M* phosphate. Pooled fractions are dialyzed against distilled water and lyophilized.

Sephadex G-100 Gel Filtration. The lyophilized powder is dissolved in a minimal volume of 20 m*M* Tris-HCl (pH 7.3) containing 0.15 *M* NaCl and percolated through a Sephadex G-100 column (3 × 150 cm) at a flow rate of 25 ml/hr. The second of two, only partially resolved, protein peaks contains the lysophospholipase II. The active fractions are pooled, dialyzed against distilled water, and lyophilized to repeat the gel-filtration step. This results in elution of a symmetrical protein peak coinciding with lysophospholipase II activity. The purification is summarized in Table II.

Properties

Stability. Purified lysophospholipase I is unstable and loses activity on repeated freezing and thawing. The enzyme can be stored, however, at −20° in 10 m*M* Tris-HCl (pH 7.3) containing 5 m*M* 2-mercaptoethanol and 50% (v/v) glycerol for years without appreciable loss of activity. Lysophospholipase II is stable after lyophilization and can be stored in this form for years without loss of activity.

Purity. Both enzymes give a single band[2] in nondenaturing[12] and in sodium dodecyl sulfate (SDS)–polyacrylamide[13] gels.

Physical Properties. Molecular weights determined by SDS–PAGE and Sephadex G-100 gel filtration amount to 24,000 and 26,000, respectively, for lysophospholipase I. The corresponding values for lysophospholipase II are 63,000 and 57,000, respectively. These results indicate that both enzymes consist of single polypeptide chains. Isoelectric points, determined by isoelectric focusing,[14] are 5.2 for enzyme I and 4.5 for enzyme II. Antibodies raised against lysophospholipase I are not cross-reactive with enzyme II and vice versa,[15] indicating different surface structures for the two enzymes.

Subcellular Distribution. Lysophospholipase I is a soluble enzyme with a bimodal distribution over mitochondrial matrix and cytoplasm. Lysophospholipase II is a peripherally membrane-associated enzyme present at the luminal side of bovine liver endoplasmic reticulum.[3,16]

Enzymological Properties. The deacylation of 1-palmitoyllysophosphatidylcholine by lysophospholipase I proceeds optimally with a broad pH optimum between pH 6 and 8. The pH optimum for lysophospholipase II is 8.5. Neither enzyme requires Ca^{2+} for activity. Both enzymes are sensitive to detergents and thiol reagents, although to different degrees. Lysophospholipase II is more sensitive to detergents such as sodium deoxycholate and Triton X-100 than enzyme I, whereas lysophospholipase I is considerably more sensitive to inhibition by thiol reagents.[2] Both enzymes show general esterolytic properties and hydrolyze tributyrin and *p*-nitrophenyl acetate at rates that are in fact higher than observed with 1-palmitoyllysolecithin. Studies with a series of lysophosphatidylcholines varying in chain length from C_6 to C_{18} indicate that lysophospholipase II is active on substrate monomers as well as on substrate micelles, with no break in the activity profile at the critical micelle concentrations of the substrates.[17] The enzyme is also active toward short-chain diacylphosphatidylcholines bearing hexanoic, octanoic, decanoic, or dodecanoic acyl chains and essentially produces stoichiometric amounts of octanoyllysophosphatidylcholine from dioctanoylphosphatidylcholine due to the low activity of the enzyme toward short-chain lysophosphatidylcholines. The

[12] B. J. Davis, *Ann. N.Y. Acad. Sci.* **121,** 404 (1964).

[13] A. L. Shapiro, E. Vinnela, and J. V. Maizel, *Biochem. Biophys. Res. Commun.* **28,** 815 (1967).

[14] O. Vesterberg, this series, Vol. 22, p. 389.

[15] J. G. N. de Jong, A. M. H. P. van den Besselaar, and H. van den Bosch, *Biochim. Biophys. Acta* **441,** 221 (1976).

[16] J. H. E. Moonen and H. van den Bosch, *Biochim. Biophys. Acta* **573,** 114 (1979).

[17] J. G. N. de Jong, R. Dijkman, and H. van den Bosch, *Chem. Phys. Lipids* **15,** 125 (1975).

degradation of 1-octanoyl-2-pentanoylphosphatidylcholine produces nearly exclusively pentanoyllysophosphatidylcholine, indicating mainly phospholipase A_1 activity of the lysophospholipase II. However, comparative experiments with 1-hexyl-2-hexanoyl- and 1-hexanoyl-2-hexylphosphatidylcholine indicate that also fatty acids from the *sn*-2 position in these short-chain phosphatidylcholines can be removed, albeit at a rate less than 10% of that of reaction at the *sn*-1 position.[17]

[17] J. G. N. de Jong, R. Dijkman, and H. van den Bosch, *Chem. Phys. Lipids* **15,** 125 (1975).

[45] Purification of Lysophospholipase and Lysophospholipase-Transacylase from Rabbit Myocardium

By RICHARD W. GROSS

Introduction

Accumulation of diminutive amounts of lysophosphoglycerides in cellular membranes results in alterations of membrane molecular dynamics which are accompanied by changes in the kinetics of transmembrane proteins.[1-3] Thus, a vast excess of enzymatic activity which degrades lysophosphoglycerides (e.g., lysophospholipase) is present in most cell types to prevent accumulation of these potentially noxious metabolites. Lysophosphoglyceride catabolism is mediated either by enzyme-catalyzed nucleophilic attack of the *sn*-1 acyl group by "activated" water or by the formation of an acyl enzyme intermediate which can either be hydrolyzed by water (lysophospholipase activity) or transferred to a second molecule of lysophosphatidylcholine (transacylase activity).

Recently, lysophosphoglycerides have been implicated as the biochemical mediators of electrophysiologic dysfunction in ischemic myocardium.[4-6] Accordingly, the identification and purification of the enzymes

[1] H. U. Weltzien, *Biochim. Biophys. Acta* **559,** 259 (1979).

[2] R. W. Gross, P. B. Corr, B. I. Lee, J. E. Saffitz, W. A. Crafford, Jr., and B. E. Sobel, *Circ. Res.* **51,** 27 (1982).

[3] K. L. Fink and R. W. Gross, *Circ. Res.* **55,** 585 (1984).

[4] B. E. Sobel, P. B. Corr, A. K. Robison, R. A. Goldstein, F. X. Witkowski, and M. S. Klein, *J. Clin. Invest.* **62,** 546 (1978).

mediating lysophosphoglyceride catabolism in mammalian myocardium has assumed particular importance. The majority of metabolic activity capable of mediating lysophospholipid hydrolysis in mammalian myocardium is catalyzed by cytosolic lysophospholipase and cytosolic lysophospholipase-transacylase.[7] This chapter describes the chromatographic methods we have developed for the purification of rabbit myocardial lysophospholipase and lysophospholipase-transacylase to homogeneity.[8,9]

Lysophospholipase and Lysophospholipase-Transacylase Assays

Lysophospholipase and lysophospholipase-transacylase activities are quantified by incubating 1-[1-^{14}C]hexadecanoyl-*sn*-glycero-3-phosphocholine (palmitoyllysophosphatidylcholine) with 50 or 100 μl of column fractions in 75 m*M* potassium phosphate, pH 7.0 for 10 min at 37° in a final volume of 700 μl. Reaction products are extracted with butanol (700 μl) and separated utilizing silica OF thin-layer chromatography (TLC) plates developed with chloroform/acetone/methanol/acetic acid/water (3 : 4 : 1 : 1 : 0.5, v/v), and fatty acid release (R_f = 1) or phosphatidylcholine synthesis (R_f = 0.4) is assessed by scraping appropriate regions into scintillation vials, adding 10 ml of Aquasol II, and counting by scintillation spectrometry. It is important to note that phosphatidylcholine synthesized by lysophospholipase-transacylase has a specific activity twice that of substrate and released fatty acid.

Miscellaneous Procedures and Sources of Materials

Radiolabeled palmitoyllysophosphatidylcholine and Aquasol II are available from New England Nuclear (Boston, MA), and unlabeled lysophosphatidylcholine is obtained from Sigma (St. Louis, MO). Silica OF TLC plates are obtained from Analabs (Norwalk, CT). DEAE-Sephacel, polybuffer exchanger, Polybuffer 74, and Mono Q columns are obtained from Pharmacia (Piscataway, NJ). Hydroxylapatite is obtained from Bio-Rad (Richmond, CA), and AcA resin is obtained from LKB (Gaithersburg, MD). Most other reagents are obtained from Sigma.

[5] P. B. Corr, R. W. Gross, and B. E. Sobel, *Circ. Res.* **55,** (1984).

[6] L. A. Scherrer and R. W. Gross, *Mol. Cell. Biochem.* **88,** 97 (1989).

[7] R. W. Gross and B. E. Sobel, *J. Biol. Chem.* **257,** 6702 (1982).

[8] R. W. Gross and B. E. Sobel, *J. Biol. Chem.* **258,** 5221 (1983).

[9] R. W. Gross, R. C. Drisdel, and B. E. Sobel, *J. Biol. Chem.* **258,** 15165 (1983).

Purification of Rabbit Myocardial Lysophospholipase

Reagents

Buffer 1: 0.25 *M* sucrose, 10 m*M* potassium phosphate, 1 m*M* EDTA, 1 m*M* dithiothreitol (DTT), pH 7.4

Buffer 2: 28 m*M* potassium phosphate, 10% glycerol, 10 m*M* mercaptoethanol, 10 μM lysophosphatidylcholine, pH 7.55

Buffer 3: 28 m*M* potassium phosphate, 10% glycerol, 10 m*M* mercaptoethanol, 10 μM lysophosphatidylcholine, pH 7.0

Buffer 4: 25 m*M* imidazole, 10% glycerol, 10 μM mercaptoethanol, 10 μM lysophosphatidylcholine, pH 7.0

Buffer 5: Polybuffer 74 (diluted 1 : 5, v/v), 10% glycerol, 10 m*M* mercaptoethanol, 10 μM lysophosphatidylcholine, pH 4.5

Preparation of Rabbit Myocardial Cytosol. New Zealand White rabbits (typically 10–20) are sacrificed by cervical dislocation, a left thoracotomy is performed, and hearts (~6 g/heart) are rapidly placed in ice-cold buffer 1. All procedures are performed at 0°–4°. Ventricular myocardium is isolated, extensively rinsed in buffer 1, and homogenized utilizing three 30-sec bursts from a Polytron homogenizer at a setting of 5 to yield a 25% homogenate (w/v). Nuclei, mitochondria, and cellular debris are removed by low-speed centrifugation, and the resulting supernatant is further centrifuged at 100,000 g_{max} for 1 hr to obtain myocardial cytosol.

Myocardial cytosol is brought to 55% saturation by slow addition of solid ammonium sulfate (e.g., 51 g added to 170 ml over 2 min), the mixture is stirred at 4° for an additional 9 min, and the resulting precipitate is pelleted by centrifugation at 10,000 g_{max} for 5 min. It is essential that the pellet be quickly and gently resuspended in buffer 2 (1.4 ml/g wet weight tissue, typically accomplished utilizing a rubber policeman). The resuspended ammonium sulfate precipitate is dialyzed for 14 hr against 100 volumes of buffer 2, centrifuged at 10,000 g_{max} (to remove a precipitate which does not contain substantive lysophospholipase activity), and loaded onto a DEAE-Sephacel anion-exchange column (2.6 × 15 cm) previously equilibrated with buffer 2. Under these conditions, rabbit myocardial lysophospholipase elutes in the void volume whereas myocardial lysophospholipase-transacylase activity binds to the column matrix and requires high salt concentrations for elution (200 m*M* NaCl) (Fig. 1A).

Fractions containing over 80% of maximal lysophospholipase activity are pooled (void volume) and concentrated (utilizing an Amicon ultrafiltration device equipped with a YM10 membrane) to a final volume of 8 ml, and the concentrate is degassed for 5 min. Lysophospholipase is further purified utilizing a 1.6 × 90 cm gel column comprised of AcA 44 medium

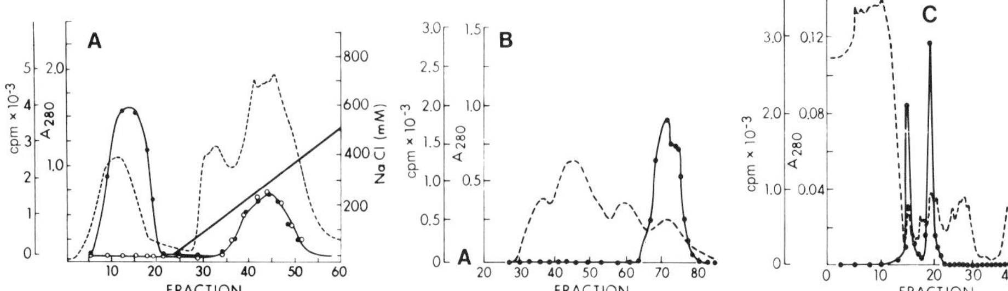

FIG. 1. Purification of rabbit myocardial lysophospholipase. Rabbit myocardial lysophospholipase activity was purified by sequential column chromatographies including DEAE-Sephacel (A), gel filtration (B), hydroxylapatite (not shown), and chromatofocusing (C), as described in the text. [^{14}C]Palmitic acid released from [^{14}C]palmitoyllysophosphatidylcholine (●); [^{14}C]phosphatidylcholine synthesized from [^{14}C]lysophosphatidylcholine (○); UV absorbance at 280 nm (– – –); NaCl gradient (—).

previously equilibrated with buffer 3, and chromatography is performed at a flow rate of 0.17 ml/min. Lysophospholipase activity typically elutes between 140 and 155 ml (K_{av} 0.6), corresponding to a molecular weight of 20 K (Fig. 1B).

Fractions containing greater than 80% maximal activity are pooled and directly loaded onto a hydroxylapatite column (1.6 × 15 cm) previously equilibrated with buffer 3. Lysophospholipase does not bind to hydroxylapatite resin, and active fractions in the void volume are combined, dialyzed against 100 volumes of buffer 4 for 12 hr, and loaded onto a 0.9 × 18 cm chromatofocusing column comprised of Polybuffer exchanger 94 previously equilibrated with buffer 4. After the active fractions are loaded, equilibration buffer is applied at 0.7 ml/min for 1 column volume prior to chromatofocusing by the generation of an *in situ* pH gradient utilizing buffer 5. It is imperative that the column eluate flow into neutralizing buffer [3 ml of 0.15 *M* potassium phosphate (pH 7.4) at a final ratio of buffer to eluate of 1 : 2, (v/v)] to preserve enzyme activity by minimizing exposure of enzyme to acidic conditions. Routinely, two peaks of lysophospholipase activity elute with apparent isoelectric points of 6.7 and 5.6, referred to as lysophospholipase A and lysophospholipase B, respectively (Fig. 1C). Polybuffer is removed from each lysophospholipase isoform by gel-filtration chromatography utilizing a 1.6 × 90 cm column comprised of AcA 54 resin run at a flow rate of 10 ml/hr with buffer 3. Lysophospholipases A and B possess equivalent specific activities of 7 μmol/mg · min (Table I), have an apparent pH optimum of 7.5, and appear as a single band at 23K after sodium dodecyl sulfate-polyacrylamide gel electrophoresis (SDS–PAGE) and silver staining. Remarkably, both lysophospholipases

TABLE I
PURIFICATION OF LYSOPHOSPHOLIPASE FROM RABBIT HEART

Step	Protein (mg)	Total activity[a]	Specific activity[a]	Purification (-fold)	Yield[b] (%)
Homogenate	7616	2132	0.28	1	100
Cytosol	867	1041	1.2	4	49
Ammonium sulfate	246	871	3.5	12	41
DEAE-Sephacel	55	589	10.7	38	28
AcA 44	5.5	548	99.6	356	26
Hydroxylapatite	0.84	303	361	1289	14
Chromatofocusing (A)[c]		55			2
Chromatofocusing (B)[c]		85			4
AcA 54 (A)[c]	0.003	23	7602	27,143	1
AcA 54 (B)[c]	0.008	62	7578	27,064	3

[a] One unit of activity is defined as the activity necessary to convert 1 nmol of lyso-PC to 1 nmol of fatty acid and 1 nmol of glycerophosphorylcholine in 1 min at 37°.

[b] No corrections for activity lost during preparation of cytosol or elution from DEAE-Sephacel owing to removal of other enzymes (i.e., microsomal lysophospholipase or cytosolic lysophospholipase-transacylase) have been made.

[c] A and B refer to the order of elution of activity peaks from the chromatofocusing column.

A and B are capable of hydrolyzing palmitoyl-CoA (specific activity 23 μmol/mg · min) but do not catalyze the hydrolysis of acylcarnitine, phosphatidylcholine, or triglycerides.

Purification of Rabbit Myocardial Lysophospholipase-Transacylase

Reagents

- Buffer 6: 0.25 *M* sucrose, 10 m*M* sodium phosphate, 1 m*M* dithiothreitol, 1 m*M* EGTA, pH 7.4
- Buffer 7: 50 m*M* sodium phosphate, 10% glycerol, 10 m*M* mercaptoethanol, pH 6.95
- Buffer 8: 20 m*M* potassium phosphate, 10% glycerol, 10 m*M* mercaptoethanol, pH 7.0
- Buffer 9: 20 m*M* sodium phosphate, 10% glycerol, 10 m*M* mercaptoethanol, pH 7.6

Procedure. New Zealand White rabbits are sacrificed by cervical dislocation, a left thoracotomy is performed, and hearts are placed in ice-cold buffer 6. All subsequent operations are performed at 0°–4°. Ventricular myocardium is obtained by dissecting left ventricular muscle from the

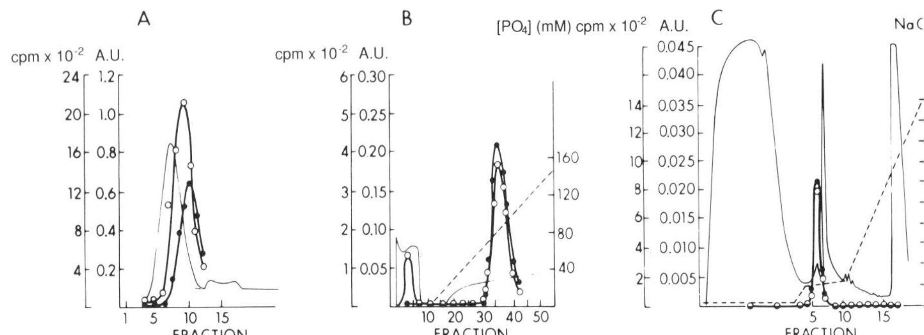

FIG. 2. Purification of rabbit myocardial lysophospholipase-transacylase. Rabbit myocardial lysophospholipase-transacylase was purified by sequential DEAE-Sephacel (A), hydroxylapatite (B), and Mono Q (C) chromatographies, as described in the text. [^{14}C]Palmitic acid released from [^{14}C]palmitoyllysophosphatidylcholine (●); [^{14}C]phosphatidylcholine synthesized from [^{14}C]palmitoyllysophosphatidylcholine (○); UV absorbance at 280 nm (—); NaCl or phosphate gradient, as indicated (– – –).

great vessels, atria, and connective tissue and is thoroughly rinsed in buffer 6. Myocardium is subsequently minced into small pieces utilizing a sharp pair of scissors (3 × 5 mm strips), and cells are disrupted utilizing three 30-sec bursts from a Polytron apparatus at a setting of 5.5. The homogenate is immediately centrifuged at 1100 g_{max} for 10 min, and the supernatant is placed in an ice-cold beaker. Next, the pellets are resuspended in buffer 6 (25%, w/v) and homogenized and centrifuged a second time utilizing the above protocol. The supernatants are combined and sequentially centrifuged at 7800, 24,000, and 105,000 g_{max} for 10, 10, and 60 min, respectively. To 115 ml of myocardial cytosol (105,000 g_{max} supernatant), 33 g of solid ammonium sulfate is added over 2 min (53% saturation), and, after stirring for an additional 10 min, lysophospholipase-transacylase is pelleted by centrifugation at 24,000 g_{max} for 5 min. After rapid resuspension in buffer 7 (1.4 ml/g wet), the resuspended precipitate is dialyzed against 50 volumes of buffer 7 for 15 hr. The precipitate which forms contains only minute amounts of lysophospholipase-transacylase activity and is removed by centrifugation.

Lysophospholipase-transacylase is purified by sequential column chromatographies. The resuspended dialyzed ammonium sulfate precipitate is applied to a 2.6 × 28 cm column comprised of DEAE-Sephacel resin previously equilibrated with buffer 7. After loading, 50 ml of buffer 7 is applied to the column followed by 200 ml of buffer 7 containing, in addition, 20 m*M* NaCl. Lysophospholipase-transacylase activity is present during the latter stages of elution of nonretained proteins and typically is partially

TABLE II

PURIFICATION OF LYSOPHOSPHOLIPASE-TRANSACYLASE FROM RABBIT HEART

Step	Protein (mg)	Activity		Specific activity		Purification (-fold)	
		Hydrolase[a,b]	Transacylase[c]	Hydrolase	Transacylase	Hydrolase	Transacylase
Homogenate	11,200	2206	336	0.20	0.03	1	1
Cytosol	1294	1178	233	0.91	0.18	5	6
Ammonium sulfate	243	520	134	2.1	0.55	11	18
DEAE-Sephacel	12	214	121	17.8	10.1	90	337
Hydroxylapatite	0.32	57	32	179	101	909	3367
Mono Q	0.002	8.2	4.4	3904	2,095	19,523	69,841

[a] One unit of activity is defined as the activity necessary to convert 1 nmol of lyso-PC to 1 nmol of fatty acid and 1 nmol of glycerophosphorylcholine in 1 min at 37°.

[b] Corrections for activity lost during preparation of cytosol owing to removal of other enzymes with lysophospholipase activity during purification have not been made.

[c] One unit of activity is defined as the activity necessary to convert 2 nmol of lyso-PC to 1 nmol of phosphatidylcholine and 1 nmol of glycerophosphorylcholine in 1 min at 37°.

resolved from lysophospholipase activity (Fig. 2A). Routinely, the initial 25–30% of lysophospholipase-transacylase activity is excluded from the pooled fractions because of its relatively greater contamination.

Further purification of lysophospholipase-transacylase is performed utilizing a 1.6 × 5 cm column of hydroxylapatite resin previously equilibrated with buffer 8. The active fractions from DEAE-Sephacel chromatography are directly applied to the equilibrated hydroxylapatite column. Residual lysophospholipase elutes in the void volume whereas lysophospholipase-transacylase is quantitatively adsorbed to the hydroxylapatite matrix and eluted utilizing a linear gradient from 20 to 150 m*M* potassium phosphate (Fig. 2B).

Fractions containing in excess of 30% peak activity are pooled and dialyzed against 100 volumes of buffer 9. Lysophospholipase-transacylase activity is loaded onto an HR 5/5 Mono Q FPLC anion-exchange column previously equilibrated with buffer 9 and subsequently eluted utilizing a discontinuous NaCl gradient (Fig. 2C). Active fractions elute at approximately 100 m*M* NaCl in a small sharp peak which precedes the major UV-absorbing peak (Fig. 2C). It should be noted that the pressure utilized during Mono Q column chromatography is critical since myocardial lysophospholipase-transacylase rapidly loses activity when pressure exceeds 2.0 MPa. Accordingly, flow rates utilized for this step are entirely dependent on the flow–pressure characteristics of the column employed. New columns typically are run at 1 ml/min, whereas the flow is reduced to 0.2–0.5 ml/min for older columns which have poorer flow–pressure characteristics. SDS–PAGE of the most active fraction followed by silver staining demonstrated a single band at 63K. This sequence of chromatographic steps results in the 69,000-fold purification of lysophospholipase-transacylase activity to homogeneity and a final specific activity of 8 μmol/mg · min (lysophospholipase activity), with typical yields of 1–2% (Table II).

Acknowledgments

This work was supported by National Institutes of Health Grant HL34839. R.W.G. is the recipient of an Established Investigator Award from the American Heart Association.

[46] Lysoplasmalogenase: Solubilization and Partial Purification from Liver Microsomes

By M. S. JURKOWITZ-ALEXANDER and L. A. HORROCKS

Introduction

The plasmalogens (alk-1′-enylglycerophosphoethanolamine and -choline) are ether-linked glycerophospholipids characterized by the presence of a vinyl ether substituent at the *sn*-1 position of the glycerol backbone.[1] Found in all mammalian cells,[2] plasmalogens are particularly abundant in heart[3] and neural[2] cell membranes, where they are highly enriched in arachidonic acid at the *sn*-2 position.[4]

Plasmalogens are probably essential for normal cell function. It was suggested that the severe reduction in cellular plasmalogen content is sufficient to create the lethal abnormalities present in Zellweger Syndrome.[5] However, the biological functions of the plasmalogens are not clearly elucidated. Plasmalogens are known to alter the physical properties of biological membranes.[6,7] Recent evidence from studies involving Chinese hamster ovary cells[8,9] suggests that the vinyl ether linkage of plasmalogens may function as an antioxidant and thus protect other membrane lipids and proteins from damage by reactive oxygen species. In this reaction the free radical cleaves the vinyl ether bond of plasmalogen and forms aldehydes and lysoglycerophospholipid products. Free radical oxidation of vinyl ether linkages occurs in cell-free model systems also.[10–12]

[1] L. A. Horrocks and M. Sharma, *in* "Phospholipids" (J. N. Hawthorne and G. B. Ansell, eds.), p. 51. Elsevier, Amsterdam, 1982.
[2] L. A. Horrocks, *in* "Ether Lipids: Chemistry and Biology" (F. Snyder, ed.), p. 177. Academic Press, New York, 1972.
[3] R. W. Gross, *Biochemistry* **24,** 1662 (1985).
[4] F. Snyder, T. Lee, and R. L. Wykle, *in* "Enzymes of Biological Membranes" (A. N. Martonosi, ed.), Vol. 2. Plenum, New York, 1985.
[5] N. S. A. Heymans, R. B. H. Schutgens, R. Ten, H. van den Bosch, and P. Borst, *Nature* **306,** 69 (1983).
[6] H. Goldfine, N. C. Johnstone, J. Mattai, and G. C. Shipley, *Biochemistry* **26,** 2814 (1987).
[7] J. H. Pak, V. P. Bork, R. E. Norberg, M. H. Greer, R. A. Wolf, and R. W. Gross, *Biochemistry* **26,** 4824 (1987).
[8] R. A. Zoeller, O. H. Morand, and C. R. H. Raetz, *J. Biol. Chem.* **263,** 11590 (1988).
[9] O. H. Morand, R. A. Zoeller, and C. R. H. Raetz, *J. Biol. Chem.* **263,** 11597 (1988).
[10] A. Frimer, *Chem. Rev.* **79,** 359 (1979).
[11] E. Yavin and S. Gatt, *Eur. J. Biochem.* **25,** 431 (1972).
[12] E. Yavin and S. Gatt, *Eur. J. Biochem.* **25,** 437 (1972).

Lysoplasmalogens (1-alk-1′-enyl-2-*sn*-glycero-3-phosphoethanolamine or -choline) are the *sn*-2-deacylated forms of plasmalogens. They are formed from plasmalogen following hydrolytic cleavage of the ester bond at the *sn*-2 position of the glycerol backbone.[13,14] The reaction is catalyzed by a phospholipase A_2 enzyme. Novel calcium-independent plasmalogen-selective phospholipases A_2 have recently been purified from heart.[15] The lysoplasmalogens are amphiphilic lipids and are normally maintained at very low levels within cells.

An enzyme involved in the catabolism of lysoplasmalogen is lysoplasmalogenase (EC 3.3.2.2; EC 3.3.2.5). This enzyme, first discovered in liver microsomes,[16] catalyzes the hydrolysis of the vinyl ether bond of choline-[16,17] and ethanolamine-[17,18] lysoplasmalogens, as well as 1-alk-1′-enylglycerol.[19] The products of the reaction are glycerophosphocholine, glycerophosphoethanolamine, and glycerol, respectively, and a long-chain fatty aldehyde. It is specific for the 2-deacylated form of plasmalogen. The enzyme is associated with the microsomes and is found in liver,[16–18] brain,[20] and heart.[21] The liver and brain enzymes have similar properties,[20,22] but heart lysoplasmalogenase differs significantly with respect to detergent and divalent cation effects and kinetic parameters.[21]

Lysoplasmalogenase may be important in regulation of the cell content of plasmalogen by determining the concentration of lysoplasmalogen available for reacylation or transacylation.[23] A reciprocal relationship exists between the tissue levels of plasmalogens and the activities of lysoplasmalogenase in brain and liver tissues. The brain contains high concentrations of plasmalogens and low lysoplasmalogenase activity, and liver has the opposite.[18,20] In order to study the physicochemical properties and the control mechanisms for this potentially important enzyme, we solubilized, purified, and characterized lysoplasmalogenase from rat liver micro-

[13] R. A. Wolf and R. W. Gross, *J. Biol. Chem.* **260,** 7295 (1985).

[14] L. A. Loeb and R. W. Gross, *J. Biol. Chem.* **261,** 10467 (1986).

[15] S. L. Hazen, R. J. Stuppy, and R. W. Gross, *J. Biol. Chem.* **265,** 10622 (1990).

[16] H. R. Warner and W. E. M. Lands, *J. Biol. Chem.* **236,** 2404 (1961).

[17] M. Jurkowitz-Alexander, H. Ebata, J. S. Mills, E. J. Murphy, and L. A. Horrocks, *Biochim. Biophys. Acta* **1002,** 203 (1989).

[18] J. Gunawan and H. Debuch, *Hoppe-Seyler's Z. Physiol. Chem.* **362,** 445 (1981).

[19] U. Franken, H. Debuch, J. Gunawan, and A. Harder, *in* "Enzymes of Lipid Metabolism II" (L. Freysz, H. Dreyfus, R. Massarelli, and S. Gatt, eds.), p. 165. Plenum, New York, 1986.

[20] J. Gunawan and H. Debuch, *J. Neurochem.* **39,** 693 (1982).

[21] G. Arthur, L. Page, T. Mock, and P. C. Choy, *Biochem. J.* **236,** 475 (1986).

[22] M. S. Jurkowitz, C. Flynn, E. S. Murphy, and L. A. Horrocks, *J. Neurochem.* **48,** S150 (1987).

[23] R. M. Kramer and D. Deykin, *J. Biol. Chem.* **258,** 13806 (1983).

somes.[17] Lysoplasmalogenase is the first enzyme to be purified which catalyzes the hydrolysis of the vinyl ether bond of the plasmalogens.

Following purification of lysoplasmalogenase, the enzyme was characterized.[17] A 300-fold purification resulted in a specific activity of 11 μmol/min/mg protein. In agreement with earlier studies using native microsomes,[16,18] the purified enzyme had no hydrolytic activity with intact plasmalogen or 1-acyl-*sn*-glycero-3-phosphoethanolamine but was specific for the lysoplasmalogen. Kinetic analyses demonstrated apparent K_m values of 6 and 40 μM for choline and ethanolamine lysoplasmalogen, respectively; the V_{max} values were 11.7 and 13.6 μmol/min/mg protein with the two substrates, respectively. The pH optima are between 6.6 and 7.1 with both substrates. The enzyme requires no cofactors and is not affected by low millimolar Ca^{2+}, Mg^{2+}, Mn^{2+}, or EDTA. It is inhibited by sulfhydryl-reacting reagents. The enzyme is inhibited by detergents, with K_i values for Triton X-100, deoxycholate, and octylglucoside of 0.09, 3.2, and 17 mM, respectively.[17] The solubilization and purification procedure is described below.

Solubilization and Purification of Lysoplasmalogenase

Assay Methods

The assay procedure is described in [7] of this volume[24] along with the substrate preparations. In this coupled enzyme assay, the aldehyde released during the hydrolysis of the vinyl ether bond of lysoplasmalogen is reduced to the alcohol by exogenously added alcohol dehydrogenase. The reaction is followed spectrophotometrically by observing the absorbance change at 340 nm when NADH is oxidized.[17,25]

Purification Procedure[17]

Materials. Bovine brain ethanolamine glycerophospholipid and bovine heart choline glycerophospholipid were obtained from Serdary Research Laboratories (London, Ontario, Canada). The detergent 1-*O*-*n*-octyl-β-D-glucopyranoside (octylglucoside) was from Boehringer Mannheim (Indianapolis, IN). The BioGel HPHT (hydroxylapatite) column was from Bio-Rad (Richmond, CA). Diethylaminoethyl-cellulose (DE-52) was obtained from Whatman Chemical Separation Ltd. (Maidstone, UK). All other

[24] M. S. Jurkowitz-Alexander, Y. Hirashima, and L. A. Horrocks, this volume [7].

[25] Y. Hirashima, M. S. Jurkowitz-Alexander, A. A. Farooqui, and L. A. Horrocks, *Anal. Biochem.* **176,** 180 (1989).

chemicals are of reagent grade and are from Sigma Chemical Co. (St. Louis, MO).

Microsomal Preparation. All procedures are carried out at 0°. Five male Sprague-Dawley rats (300–380 g) are sacrificed by decapitation. The livers are removed, chilled immediately in 0.25 *M* sucrose (medium 1), and the blood is rinsed off. Livers are minced and homogenized in 7 volumes of medium 1 using a power-drived Teflon pestle in a glass homogenizing vessel, with approximately 9 strokes at 600 rpm.

The homogenate is centrifuged at 600 *g* for 20 min. The supernatant is filtered through cheesecloth and centrifuged at 23,000 *g* for 20 min. This supernatant is centrifuged at 100,000 *g* for 60 min. The pellet containing the light microsomes is washed once by suspending in 25 ml of medium 1 with gentle homogenization, followed by centrifugation at 100,000 *g* for 60 min. A Sorvall RC-2 centrifuge with GSA rotor is used for the first two centrifugations and a Beckman ultracentrifuge with Ti 60 rotor for the last two centrifugations. Finally, the microsomes are suspended at 25 mg/ml in medium 1. Microsomes may be frozen at −20°, and lysoplasmalogenase activity is maintained; however, loss of activity occurs with repeated freezing and thawing. The yield of light microsomes is between 0.5 and 1.0% of the wet weight of liver tissue.

Solubilization of Microsomes with Octylglucoside. All procedures are carried out at 0°. Previously frozen or freshly prepared microsomes (300 mg) are suspended at 12 mg protein/ml in the following medium (medium S): 125 m*M* sucrose, 60 m*M* KCl, 80 m*M* glycylglycine (pH 7.0 with NaOH), and 1 m*M* dithiothreitol (DTT). Dithiothreitol solutions should be prepared fresh daily.

The detergent is added dropwise from a 10% aqueous solution of octylglucoside (342 m*M*) to a final concentration of 64 m*M*. The final concentration of protein is adjusted to 9 mg/ml with a small amount of 80 m*M* glycylglycine (pH 7.0) buffer. The detergent to protein ratio (weight/weight) should be between 1.8 and 2.2 and is critical for reproducibility in the solubilization procedure. The microsomes are stirred during detergent addition and are kept on ice for 15 min with occasional stirring. The addition of octylglucoside causes the turbid microsomes to clarify. The solubilized microsomes are then centrifuged at 100,000 *g* for 60 min at 4° using a Ti 60 rotor in a Beckman centrifuge. The clear amber supernatant containing the lysoplasmalogenase activity is removed by pipette from the insoluble protein pellet.

The protein present in the supernatant should contain approximately 70% of the protein present in the microsomes, or about 250 mg protein. It contains 110% of the units present in the microsomes. Liver lysoplasmalogenase can also be solubilized with the less expensive detergent sodium deoxycholate.[17] This detergent was not used because of its ionic nature.

Aliquots of the microsomes, solubilized microsomes, and solubilized microsomal supernatant are assayed for protein concentrations and for lysoplasmalogenase activity in order to quantitate recovery and specific activities. The solubilized enzyme requires the presence of detergent in order to remain in solution. If dialyzed against detergent-free buffer, the enzyme precipitates out of solution, with a loss of activity. All columns are equilibrated and developed in buffers containing 17 m*M* octylglucoside.

Purification of Lysoplasmalogenase by Anion-Exchange Chromatography on DEAE-Cellulose. The 100,000 *g* solubilized supernatant is applied directly, without dialysis or dilution, to the prepared DEAE-cellulose column. Little or no activity is lost if the preparation is stored overnight at 0° prior to application to the column. The column is prepared and developed at 4°–6°. Diethylaminoethyl (DEAE)-cellulose (Whatman DE-52) is suspended and washed in 20 m*M* MOPS buffer adjusted to pH 7.0 with Tris base and packed into a column (1 × 24 cm); the bed volume is 17 ml. The column is washed with 3 bed volumes (51 ml) of medium A (20 m*M* MOPS buffer adjusted to pH 7.0 with Tris base, 0.5 m*M* DTT, and 17 m*M* octylglucoside). The solubilized microsomes (150 mg protein in 22 ml), solubilized within 18 hr and maintained at 0°, are applied to the column at a flow rate of 0.5 ml/min. An isocratic elution with 55 ml of buffer A is followed by a 150 ml linear gradient (0–2.0 *M* KCl) made from 75 ml of buffer A and 75 ml of buffer B (buffer B is buffer A containing 2.0 *M* KCl, with the pH adjusted to 7.0 with HCl, if necessary). The flow rate is 0.25 ml/min. The eluate is collected in 1.0-ml fractions which are assayed for lysoplasmalogenase activity and for protein content.[26]

Lysoplasmalogenase is eluted very late in the KCl gradient between 1.45 and 1.65 *M* KCl. The enriched fractions are pooled. These contain 85% of the total activity and 4.5% of the total protein applied to the column (Table I). The specific activities of the enriched fractions range between 1000 and 1800 nmol/min/mg protein, representing 20 to 30-fold enhancements of activity. Chromatographic resolution on the DEAE column is highly reproducible (see Fig. 1A). The DEAE-purified enzyme is either maintained at 0° overnight for application on the HPHT column or stored at −20° in the presence of 50% glycerol for use in characterization studies or in the coupled enzyme assay for plasmalogen-specific phospholipase A_2 activity.[24]

Purification of Lysoplasmalogenase on Hydroxylapatite Column. The

[26] M. M. Bradford, *Anal. Biochem.* **72,** 248 (1976).

TABLE I
PURIFICATION OF ALKENYLHYDROLASE FROM LIVER MICROSOMES[a]

Purification step	Protein (mg)	Activity		Purification (-fold)	Yield (%)
		Units	Units/mg		
Microsomes	94	3780	40	—	100
Solubilized microsomes	66.0	3780	57	1.43	100
DEAE-cellulose	3.0	3300	1100	27.5	87.3
Hydroxylapatite	0.240	1920	8000	200	50.8

[a] Reactions were assayed in glycylglycine buffer under standard reaction conditions as described in [7], this volume. The pH was 6.9, and the substrate was 60 μM choline lysoplasmalogen. Reproduced with permission from Ref. 17.

enriched fraction from the DEAE-cellulose column is kept at 0° until it is applied to the HPHT column. The column procedure is carried out at 25°. A BioGel HPHT column connected to a Pharmacia fast protein liquid chromatography (FPLC) system is equilibrated with buffer C (20 mM potassium phosphate buffer, pH 6.9, containing 0.20 mM $CaCl_2$, 17 mM octylglucoside, 0.25 mM DTT, and 0.25 mM phenylmethylsulfonic acid). A portion (3 mg protein) of the enriched fraction from the DEAE column is filtered (0.2 μm pore size) and injected onto the column at a flow rate of 0.5 ml/min. An isocratic elution with 5 ml of buffer C is followed by a linear gradient made up of 15 ml of buffer C and 15 ml of buffer D (1.0 M potassium phosphate buffer, pH 6.9, containing 0.010 mM $CaCl_2$, 17 mM octylglucoside, 0.25 mM DTT, and 0.25 mM phenylmethylsulfonic acid) at a flow rate of 0.5 ml/min. The fractions are collected and maintained on ice and assayed for activity and protein concentration.

The enzyme elutes as a major peak between 0.55 and 0.70 M potassium phosphate. The enriched fractions are pooled. They contain 51% of the activity applied to the HPHT column (Table I). The specific activity is 8000 nmol/min/mg protein, an 8- to 10-fold increase in specific activity. The HPHT-purified enzyme is used within 24 hr or is stored at −20° in the presence of 50% glycerol. The HPHT-purified enzyme separates into approximately five bands on polyacrylamide gel electrophoresis.

With the HPHT column the major activity peak and the major protein peak elute in close proximity. This may be the reason why the degree of purification is not so great as with the DEAE-cellulose column. Also, the separation on the HPHT column is not so reproducible as with the DEAE column. The kinetic parameters and pH optima of the microsomes and purified fractions are shown in Table II.

Stability of Lysoplasmalogenase Activity. Lysoplasmalogenase activ-

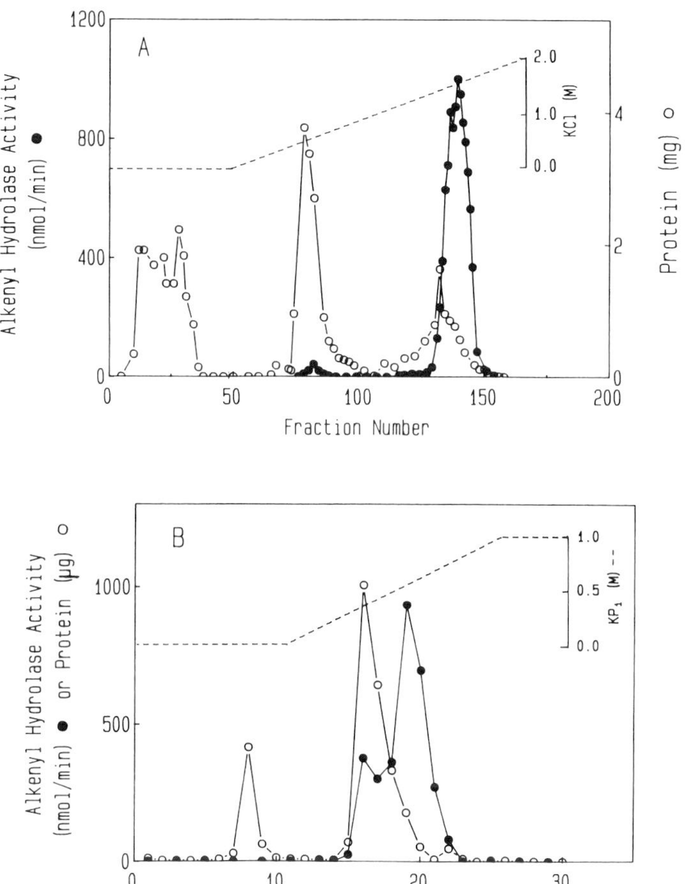

FIG. 1. (A) Separation of alkenylhydrolase by DEAE–cellulose chromatography. An aliquot (150 mg) of solubilized liver microsomes was applied to a 17-ml DEAE-cellulose column. An isocratic elution was followed by a 0–2.0 *M* KCl gradient elution.The flow rate was 0.25 ml/min; 1-ml aliquots were collected. The activities are shown as units/fraction. The protein values are shown as mg/fraction. Activities were determined under standard assay conditions with 60 μM choline lysoplasmalogen as described in [7], this volume. Protein concentrations were determined by the Bradford method. Fractions 134 through 145 were pooled and applied to the HPHT column. (B) Separation of alkenylhydrolase by hydroxylapatite chromatography. A portion of the pooled DEAE fractions (3.0 mg protein) was applied to the HPHT column. Following an isocratic elution, a gradient of 10–1000 m*M* potassium phosphate was applied. Fractions (1 ml) were assayed for activity and protein concentrations as for (A). Fractions 19–21 were pooled as the enriched fractions. (Reproduced with permission from Ref. 17.)

TABLE II
PROPERTIES OF MICROSOMAL AND PARTIALLY PURIFIED ALKENYLHYDROLASE[a]

Source	K_m (μM)		V_{max} (nmol/min/mg protein)		pH optima: Cho or Etn
	Cho	Etn	Cho	Etn	
Microsomes	50	100	50	55	6.8–7.0
DEA-cellulose	16	40	1700	1800	6.7–7.0
Hydroxylapatite	6	40	11,600	13,600	6.7–7.0

[a] Buffer systems: MES (Na^+), pH 5.8–6.5; MOPS (Na^+), pH 6.3–7.2, and glycylglycine (Na^+), pH 6.9–8.2. Cho and Etn designate the substrates 1-alkenyl-*sn*-glycerophosphocholine and 1-alkenyl-*sn*-glycerophosphoethanolamine, respectively.

ity in the microsomes is stable on storage at −20°, but the enzyme loses activity with repeated freeze–thaw cycles. The solubilized microsomes, 100,000 *g* supernatant, DEAE-purified enzyme, and HPHT-purified enzyme lose all activity if frozen at −20°. However, they are stable at this temperature for months in the presence of 50% glycerol. They are also stabilized at this temperature by the addition of an aqueous suspension of diradylglycerophosphocholine to 600 μM final concentration. The enzymes are stable for several days at 0°.

Conclusion

In summary, lysoplasmalogenase was solubilized with octylglucoside, a detergent which is relatively nondenaturing to proteins. The enzyme appears to be lipophilic and requires the presence of octylglucoside throughout the purification procedure. The enzyme is purified 300-fold by sequential DEAE-cellulose and hydroxylapatite chromatographies. Another column separation is needed for purification to homogeneity. Conditions are described which stabilize the enzyme. This purification may lead to an understanding of the role of the enzyme in plasmalogen metabolism in normal and pathological conditions.

Section VI

Phospholipase C

A. Phospholipase C
Articles 47 through 51

B. Sphingomyelinase
Articles 52 and 53

C. Phosphatidate Phosphatase
Articles 54 and 55

[47] Phosphatidylinositol-Specific Phospholipases C from *Bacillus cereus* and *Bacillus thuringiensis*

By O. HAYES GRIFFITH, JOHANNES J. VOLWERK, and ANDREAS KUPPE

Introduction

Phosphatidylinositol-specific phospholipase C (PI-PLC; EC 3.1.4.10, 1-phosphatidylinositol phosphodiesterase) from *Bacillus cereus* and *Bacillus thuringiensis* catalyzes the cleavage of the *sn*-3 phosphodiester bond of phosphatidylinositol (PI). Under some conditions the cyclic phosphate may then undergo hydrolysis to the D-*myo*-inositol phosphate.

RCOO RCOO H O HO P O OH O 2 1 6 HO 4 OH 3 HO OH 5 —PI-PLC→ RCOO RCOO H HO + O OH P O O HO OH HO OH

Phosphatidylinositol 1,2-diacyl-*sn*-glycerol D-*myo*-inositol 1,2-cyclic phosphate

Bacteria of the genus *Bacillus* secrete a variety of hydrolytic enzymes including proteases, amylases, glucanases, and lipases, depending on the specific strain.[1] *Bacillus cereus* secretes three phospholipases, one specific for PI (the PI-PLC), a second hydrolyzing sphingomyelin, and a third with a preference for hydrolyzing phosphatidylcholine, phosphatidylserine, and phosphatidylethanolamine.[2-4] *Bacillus cereus* and *B. thuringiensis* are closely related rod-shaped gram-positive bacteria that are nonpathogenic to man. The enzymes are excreted in relatively large quantities across the single cell membrane into the growth medium. This facilitates the purification of PI-PLC because intracellular proteins are removed with the cells during an initial centrifugation step. PI-PLC is isolated from the

[1] A. Krieg, *in* "The Prokaryotes: A Handbook on Habitats, Isolation, and Identification of Bacteria" (M. P. Starr, ed.), p. 1743. Springer-Verlag, New York, 1981.

[2] H. Ikezawa, M. Mori, T. Ohyabu, and R. Taguchi, *Biochim. Biophys. Acta* **528,** 247 (1978).

[3] C. Little, this series, Vol. 71 [83].

[4] J. J. Volwerk, P. B. Wetherwax, L. M. Evans, A. Kuppe, and O. H. Griffith, *J. Cell. Biochem.* **39,** 315 (1989).

supernatants of *B. cereus* and *B. thuringiensis* cultures in milligram quantities by an extension of the procedure of Ikezawa and Taguchi.[5] The final product is homogeneous on sodium dodecyl sulfate-polyacrylamide gel electrophoresis (SDS–PAGE), exhibits high specific activity, and was used to raise monoclonal antibodies and provide polypeptide sequence information for the cloning and DNA sequencing of PI-PLC from *B. cereus*.

Purification of Phosphatidylinositol-Specific Phospholipase C from Culture Media of *Bacillus cereus* and *Bacillus thuringiensis*

The same procedure is used to purify PI-PLC from *B. cereus* and *B. thuringiensis*.

Growth Conditions

The growth medium for *B. cereus* (ATCC 6464) and *B. thuringiensis* (ATCC 10792) contains 40 g Bacto-peptone, 40 g yeast extract, 20 g NaCl, 1 g Na_2HPO_4, and water to reach a final volume of 4 liters, adjusting the pH to 7.0 with 1 *M* NaOH. Stock cultures are prepared from single colony isolates by inoculating 100 ml of medium and growing overnight on a shaker at 37°. Stock cultures are stored in 4-ml aliquots containing 50% glycerol at −20°.

To generate starting material for the purification of PI-PLC, *B. cereus* or *B. thuringiensis* is grown in 4-liter batches in a jar fermentor. A preculture is prepared by inoculating 100 ml of medium with 4 ml of stock culture followed by incubation on a shaker at 37° for 12 hr. The preculture is then added to 4 liters medium in a jar fermentor and incubated at 37° with the stirrer set at 400 rpm and the air flow at 10 liters/min, while the pH is maintained at 7.0 by addition of NaOH and H_3PO_4. Incubation is continued for 3.5 hr until the culture is in the early stationary phase. The increase of PI-PLC activity in the culture medium closely parallels cell growth and is optimal in the late logarithmic/early stationary phase.[6]

Purification Procedure

Ammonium Sulfate Precipitation and Dialysis. The *B. cereus* or *B. thuringiensis* culture is cooled to 4°, and the cells are removed by centrifugation for 30 min at 13,000 *g*. Solid ammonium sulfate is added

[5] H. Ikezawa and R. Taguchi, this series, Vol. 71 [84].

[6] H. Ikezawa, M. Yamanegi, R. Taguchi, T. Miyashita, and T. Ohyabu, *Biochim. Biophys. Acta* **450,** 154 (1976).

slowly to the supernatant to 90% saturation (576 g/liter), and the solution is kept overnight at 4°. The dark-brown precipitate is collected by centrifugation for 30 min at 13,000 *g*, dissolved in a small volume of cold 5 m*M* Tris–maleate buffer (pH 6.5), and dialyzed against 3 changes of the same buffer at 4°. Subsequent chromatography and dialysis steps are carried out at 4°.

CM-Sephadex Column Chromatography. The dialyzed material is collected, and a small precipitate is removed by centrifugation for 30 min at 4500 *g*. The clear supernatant is applied to a column of CM-Sephadex (3.5 × 10 cm) and eluted with Tris–maleate buffer at a flow rate of 50 ml/hr. The dark-colored break-through fractions, which contain the PI-PLC activity, are pooled and dialyzed against 3 changes of 20 m*M* Tris-HCl (pH 8.5).

DEAE-Cellulose Column Chromatography. Following dialysis, the solution is applied to a DEAE-cellulose column (3.5 × 10 cm) equilibrated in 20 m*M* Tris-HCl (pH 8.5) and eluted with a linear gradient (2 × 750 ml) from 0 to 0.3 *M* NaCl in the same buffer. The fractions containing PI-PLC activity are pooled and concentrated to approximately 10 ml by ultrafiltration [Amicon (Danvers, MA) cell with a YM10 membrane].

Phenyl-Sepharose Column Chromatography. The concentrated DEAE-cellulose fractions are loaded onto a Phenyl-Sepharose column (2 × 7 cm) equilibrated in 20 m*M* Tris-HCl (pH 7.5). The column is rinsed exhaustively with the same buffer (at least 50 ml) until the UV monitor gives a baseline reading. PI-PLC activity is then eluted with the same buffer containing 50% ethylene glycol (by volume). Fractions containing PI-PLC activity are pooled, and the ethylene glycol is removed and the solution concentrated by ultrafiltration. Enzyme solutions are stored in 20 m*M* Tris-HCl (pH 7.5) at 1–2 mg protein/ml at −20° and are stable for at least 1 year, withstanding repeated freezing and thawing.

Comments. The results of a typical purification of the enzymes from *B. cereus* and *B. thuringiensis* are summarized and compared in Table I. Figure 1 shows an SDS–polyacrylamide gel[7] of the material obtained at the various stages of purification of both preparations. The purity of the final enzyme preparations is generally better than 95%, with similar specific activities for the *B. cereus* and *B. thuringiensis* enzymes in the range of 1200 to 1500 units/mg[4,8] using the PI/deoxycholate assay described below. During the early steps of the purification, the specific enzyme activities are affected by the presence of colored contaminants that con-

[7] U. K. Laemmli, *Nature (London)* **227,** 680 (1970).

[8] J. J. Volwerk, J. A. Koke, P. B. Wetherwax, and O. H. Griffith, *FEMS Microbiol. Lett.* **61,** 237 (1989).

TABLE I
PURIFICATION OF PHOSPHATIDYLINOSITOL-SPECIFIC PHOSPHOLIPASE C FROM *Bacillus thuringiensis*[a] AND *Bacillus cereus*[b]

Step	Total protein (mg)		Total activity (units)		Specific activity (units/mg)		Yield (%)	
	Bt	Bc	Bt	Bc	Bt	Bc	Bt	Bc
$(NH_4)_2SO_4$ (90% saturated)	145	194	6176	6900	43	36	100	100
CM-Sephadex	71	73	4132	5920	58	80	67	86
DEAE-cellulose	6.4	3.8	2106	2180	332	580	34	32
Phenyl-Sepharose	0.9	1.4	1420	1850	1515	1280	23	27

[a] Bt, ATCC 10792.
[b] Bc, ATCC 6464.

tribute strongly in the protein determinations. These contaminants are completely removed by the final Phenyl-Sepharose step. The Bradford protein assay[9] is preferred since it is the least sensitive to the colored material. In some of our *B. thuringiensis* preparations we observe a second minor polypeptide component of slightly higher molecular weight than the main PI-PLC band.[8] Western blot experiments using monoclonal antibodies specific for the *B. cereus* and *B. thuringiensis* PI-PLCs indicate that this polypeptide is immunologically related to the enzyme (A. Kuppe, unpublished observations).

The initial steps of the purification procedure described here are essentially the same as those reported earlier for *B. cereus*.[5] However, addition of the final Phenyl-Sepharose step greatly improves the purity of the preparation and is partly responsible for the significantly higher specific enzyme activities both of the *B. cereus* and *B. thuringiensis* preparations compared to those reported previously.[5] A different purification procedure for the *B. thuringiensis* enzyme has been reported more recently and also yields enzyme with a high specific activity.[10] We find that the procedure described here works well both for the *B. cereus* and *B. thuringiensis* PI-PLCs, yielding preparations of similar quality for these highly homologous enzymes.

[9] M. Bradford, *Anal. Biochem.* **72,** 248 (1976).
[10] M. G. Low, J. Stiernberg, G. L. Waneck, R. A. Flavell, and P. W. Kincade, *J. Immunol. Methods* **113,** 101 (1988).

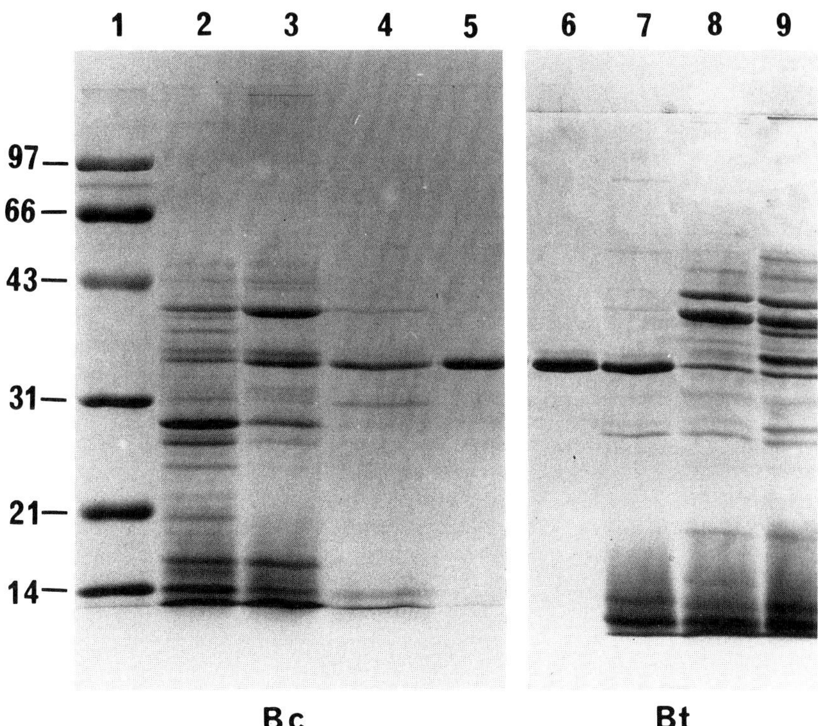

FIG. 1. SDS–polyacrylamide gel electrophoresis of *B. cereus* and *B. thuringiensis* PI-PLC (Bc and Bt, respectively) during the various stages of purification. Acrylamide gels (12%) were run with a 6% acrylamide stacking gel and the discontinuous buffer system of Laemmli.[7] Lane 1, molecular weight markers [molecular weights ($\times 10^{-3}$) are indicated at left]; lanes 2 and 9, after ammonium sulfate precipitation; lanes 3 and 8, after CM-Sephadex chromatography; lanes 4 and 7, after DEAE-cellulose chromatography; lanes 5 and 6, after Phenyl-Sepharose chromatography.

Assay Methods

Two different types of assays are currently available to measure enzyme activities of PI-PLC from *B. cereus* and *B. thuringiensis*. The first method is based on the enzyme-catalyzed cleavage of the substrate PI in the presence of detergent and quantitation of the water-soluble product, inositol phosphate. The procedure we describe below is an adaptation of an assay method reported previously[5] but employs radiolabeled PI instead of a colorimetric phosphorus determination. The second method is based on the ability of bacterial PI-PLC to release into the medium a number of enzymes tethered to membranes by means of a glycosylphosphatidylinositol(GPI)-containing membrane anchor.[11] Quantitation of the released enzyme by means of its own specific reaction thus provides a measure of the PI-PLC activity.

Method A: Cleavage of Phosphatidylinositol

Reagents

0.1 *M* sodium borate/HCl (pH 7.5)
0.8% (w/v) sodium deoxycholate
Chloroform/methanol/concentrated HCl (66 : 33 : 1, by volume)
10 m*M* radiolabeled PI [specific radioactivity 70,000 cpm (counts/min)/μmol] in aqueous suspension is prepared by mixing the appropriate amounts of cold PI (Avanti, Birmingham, AL, from bovine brain) and [*inositol*-2-^{3}H]PI (NEN, Wilmington, DE), in chloroform, followed by evaporation of the solvent and resuspension in water by bath sonication for 5 min. The lipid suspension is stored at $-20°$ in 1-ml aliquots.

Assay Procedure. The reaction mixture is prepared by mixing (in this order) 0.1 ml PI suspension, 0.1 ml deoxycholate solution, and 0.2 ml borate buffer. The reaction is initiated by addition of 0.1 ml enzyme solution containing up to 0.005 units of PI-PLC appropriately diluted in 0.1% bovine serum albumin (pH 7.5). Incubation is at 37° for 10 min, and the reaction is terminated by addition of 2.5 ml of the chloroform/methanol/HCl mixture and vigorous stirring. After brief centrifugation to separate the layers, 0.5 ml of the aqueous upper layer (~1 ml total) is transferred to a scintillation vial, cocktail is added, and the sample is counted with automatic quench correction. Counts are multiplied by a factor of 2 to estimate the total amount of substrate cleaved in the assay mixture. One unit is defined as the amount of enzyme converting 1 μmol of substrate per minute. Specific enzyme activities are expressed in units/milligram protein. Total protein is determined using the Bradford protein assay.[9]

Comments. A strict proportionality between the amount of enzyme added and the counts in the aqueous layer is observed for up to 0.005 units of PI-PLC. Use of radiolabeled PI significantly improves both the ease and sensitivity of the assay. A drawback is the high cost of cold and radiolabeled PI. However, for comparative purposes, for example, finding the active peak in column fractions, the assay can be performed with as little as 0.025 ml PI suspension.

[11] M. G. Low and A. R. Saltiel, *Science* **239,** 268 (1988).

Method B: Enzyme Release Assay

Detailed procedures for determining the PI-PLC-catalyzed release of alkaline phosphatase from rat kidney slices and acetylcholinesterase from bovine erythrocytes have been described by Ikezawa and Taguchi.[5] We use the latter procedure to compare the enzyme-release activities of the *B. cereus* and *B. thuringiensis* preparations.[8]

Properties of the Enzymes

The PI-PLCs purified from *B. cereus* and *B. thuringiensis* are nearly identical in their physicochemical properties examined thus far. We did not observe any (substantial) differences in systematic comparisons of several properties of the purified enzymes: specific activity toward substrate (PI, GPI), molecular weight and electrophoretic mobility, sensitivity to certain effectors, and interaction with inhibitory monoclonal antibodies.

Amino Acid Sequence. Inspection of the amino acid sequences of the enzymes, derived by translation of the sequenced genes,[12,13] has shown that there are only eight amino acid substitutions apparent between the proteins (Fig. 2). These amino acid substitutions do not seem to alter the enzymatic properties of the enzyme. There is an amino acid sequence similarity of the *B. cereus* enzyme with a GPI-specific phospholipase C from *Trypanosoma brucei*.[13] The bacterial enzyme shows limited sequence similarity with eukaryotic PI-PLCs presumed to be involved in the signal transduction of Ca^{2+}-mobilizing hormones.[13]

Molecular Weight and Isoelectric Point. The PI-PLCs purified by the above method comigrate in SDS–PAGE (on Laemmli-type mini slab gels of 12% acrylamide/0.32% *N,N'*-methylenebisacrylamide) and run at about 35,000 apparent molecular weight. This value agrees with the molecular weights calculated from the sequences of the enzymes (e.g., 34,466 for the *B. cereus* PI-PLC). The value is higher than that observed by gel filtration (23,000–29,000)[5]; preliminary results from our laboratory (J. J. Volwerk, unpublished observation) indicate that this discrepancy can be ascribed to interaction of the proteins with the column matrix. The isoelectric point of the *B. cereus* PI-PLC is reported to be 5.4 using ampholite (pH 5–8).[2]

pH Optimum. The pH optimum appears to be rather broad and does

[12] D. J. Henner, M. Yang, E. Chen, R. Hellmiss, H. Rodriguez, and M. G. Low, *Nucleic Acids Res.* **16,** 10383 (1988).

[13] A. Kuppe, L. M. Evans, D. A. McMillen, and O. H. Griffith, *J. Bacteriol.* **171,** 6077 (1989).

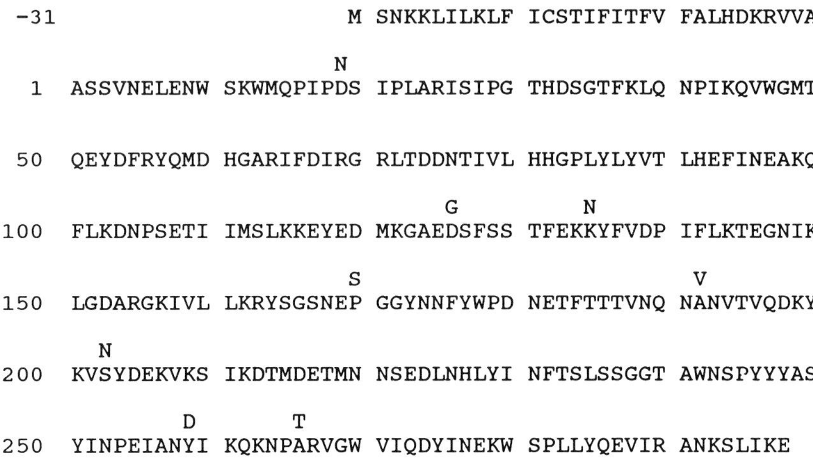

FIG. 2. Amino acid sequence of the PI-PLC from *B. cereus*. The protein is synthesized as a larger precursor carrying an N-terminal signal peptide (position -31 to -1) in addition to the mature enzyme of 298 residues. Amino acid replacements in the PI-PLC from *B. thuringiensis* are indicated above the sequence.

not show much variation between pH 5 and 8.5 in the PI-cleavage assay using Tris–acetate buffers. The enzyme remains active after short exposure to low pH (pH 2.5, <60 min).

Substrate Specificity. Enzyme preparations from *B. cereus* and *B. thuringiensis* show no proteolytic activity and cleave only lipids containing the *myo*-inositol group. The specific activities toward cleavage of PI and the GPI anchors are very similar for the two enzymes.[8] Diacylglycerol is the only product detectable in the organic phase after prolonged incubation of PI with the *B. cereus* enzyme. The enzymes do not recognize as substrate the more highly phosphorylated inositol phospholipids phosphatidylinositol phosphate (PIP) and phosphatidylinositol 4,5-bisphosphate (PIP_2). No hydrolysis of phosphatidylcholine (PC), phosphatidylethanolamine (PE), or sphingomyelin is observed.[4]

Influence of Effectors. Treatment of the enzyme with a reducing agent (dithiothreitol) or the thiol reagents iodoacetamide and *N*-ethylmaleimide has no effect on enzyme activity. The enzymes contain no disulfide bonds, and, in fact, the mature protein shows no Cys residue in the amino acid sequence. As with many phospholipases, the presence of detergent in the assay activates the enzyme. At the detergent/PI ratios of 2 : 1 used in the standard assay, deoxycholate activates most strongly. Under these conditions, several other detergents activate to a lesser degree: cholate

(30% of the maximal deoxycholate activation), Triton X-100 (18%), and *n*-octylglucoside (12%).[4]

Metal ions exert an inhibitory effect on enzyme activity, possibly due to interaction with the substrate rather than enzyme. Divalent cations inhibit at lower concentrations (1–10 m*M*: Ca, Mg < Mn, Zn) than NaCl (100 m*M*). The inhibitory effect of metal ions is pH dependent.[14]

Inhibitors. Several nonhydrolyzable enzyme inhibitors have been synthesized, based on the structure of the substrate PI.[15,16] In these molecules, the PI-PLC-sensitive P—O bond is replaced by a P—C bond (phosphonate). In place of the diacylglycerol moiety, single chains of varying lengths were introduced in the inhibitors. The best inhibitor in this series, 1-*myo*-inositol (4-palmitoyloxybutan-1)phosphonate, reduces PI-PLC activity to half-maximum at 5 m*M* concentration under standard assay conditions. PI-PLC-specific monoclonal antibodies almost completely inhibit the enzymes at low antibody/PI-PLC ratios.[17] This effect is observed in both of the commonly used assay systems, PI cleavage and release of acetylcholinesterase from the surface of bovine erythrocytes through cleavage of the membrane anchor. The PI-PLCs from *B. cereus* and *B. thuringiensis* are identical in their interaction with the inhibitory monoclonal antibodies. The lipid inhibitors and antibodies should prove useful in the study of structure–function relationships of the bacterial PI-PLCs and in the investigation and characterization of membrane proteins containing glycosyl-phosphatidylinositol anchors.

Enzyme Stability. As mentioned above, the purified PI-PLCs from *B. cereus* and *B. thuringiensis* remain active after many freeze–thaw cycles. This is observed for enzyme preparations stored in 20 m*M* Tris-HCl (pH 7.5) buffer at concentrations between 0.1 and 3 mg protein/ml. PI-PLC is also fairly stable during prolonged incubations at room temperature and at 37°. We suspect that PI-PLC is only partially denatured by SDS–PAGE dissociation buffer (3% SDS in stacking gel buffer, no mercaptoethanol) or renatures rapidly on removal of the SDS. This notion stems from results of experiments where enzyme activity is measured in gel slices after gel electrophoretic separation of the (ammonium sulfate-precipitated) secreted proteins of *B. cereus*.[4]

In summary, the PI-PLCs of *B. cereus* and *B. thuringiensis* are very nearly identical. These PI-PLCs are stable enzymes which can be purified

[14] R. Sundler, A. W. Alberts, and P. R. Vagelos, *J. Biol. Chem.* **253,** 4175 (1978).

[15] M. S. Shashidhar, J. F. W. Keana, J. J. Volwerk, and O. H. Griffith, *Chem. Phys. Lipids* **53,** 103 (1990).

[16] M. S. Shashidhar, J. J. Volwerk, J. F. W. Keana, and O. H. Griffith, *Biochim. Biophys. Acta* **1042,** 410 (1990).

[17] A. Kuppe, K. K. Hedberg, J. J. Volwerk, and O. H. Griffith, *Biochim. Biophys. Acta* **1047,** 47 (1990).

to high specific activity by the method described here without contamination by other phospholipases or proteases. The enzymes are used in the study of novel cell surface proteins containing the glycosylphosphatidylinositol anchors and may serve as model systems for the eukaryotic phospholipases involved in signal transduction.

Acknowledgments

We are pleased to acknowledge useful discussions with our colleagues Dr. M. S. Shashidhar, Dr. H. Stewart Hendrickson, and Mr. John A. Koke. This work was supported by U.S. Public Health Service Grant GM 25698.

[48] Assays of Phosphoinositide-Specific Phospholipase C and Purification of Isozymes from Bovine Brains

By SUE GOO RHEE, SUNG HO RYU, KEE YOUNG LEE, and KEY SEUNG CHO

Introduction

It is well established that in a variety of cells, receptor-mediated phosphoinositide-specific phospholipase C (PLC) hydrolyzes phosphatidylinositol 4,5-bisphosphate [PtdIns(4,5)P_2] to yield diacylglycerol and inositol 1,4,5-trisphosphate [Ins(1,4,5)P_3]. Both products act as intracellular messengers, the former by activating protein kinase C and the latter by mobilizing intracellular Ca^{2+} pools. PLC is present in most mammalian cells[1] as well as in plants[2] and various microorganisms.[3] Multiple forms of PLC enzymes have been purified from both particulate and soluble fractions of a variety of mammalian tissues. An examination of the molecular weights of PLC isozymes determined by sodium dodecyl sulfate (SDS)-polyacrylamide gel electrophoresis, the amino acid sequences deduced from cDNA, and the immunocross-reactivity indicates that PLC enzymes purified from mammalian tissues fall into four groups[1]: PLC-α with a molecular weight of 60,000–70,000,[4-6] PLC-β with 140,000–155,000,[7-10] PLC-γ with

[1] S. G. Rhee, P.-G. Suh, S. H. Ryu, and S. Y. Lee, *Science* **244,** 456 (1989).

[2] H. Pfaffmann, E. Hartmann, A. O. Brightman, and D. J. Morre, *Plant Physiol.* **85,** 1151 (1987).

[3] H. Ikezawa and R. Taguchi, this series, Vol. 71, p. 731.

[4] T. Takenawa and Y. Nagai, *J. Biol. Chem.* **256,** 6769 (1981).

[5] S. L. Hoffman and P. W. Majerus, *J. Biol. Chem.* **257,** 6461 (1982).

[6] C. F. Bennett and S. T. Crooke, *J. Biol. Chem.* **262,** 13789 (1987).

145,000–148,000,[7-9] and PLC-δ with 85,000–88,000.[9,11-13] These isoforms are expressed differently between tissues, between individual cells, and during development.[1]

Despite these differences, all the purified isozymes exhibit similar catalytic properties. (1) They are specific for inositol phospholipid and do not hydrolyze other phospholipids such as phosphatidylcholine (PC), phosphatidylethanolamine (PE), and phosphatidylserine (PS). (2) They hydrolyze three inositol-containing lipids, namely, phosphatidylinositol (PtdIns), phosphatidylinositol 4-phosphate [PtdIns(4)P], and PtdIns(4,5)P_2. (3) The hydrolyses of these three substrates are dependent on Ca^{2+}. (4) Optimal Ca^{2+} concentrations required for the hydrolysis of PtdIns(4)P and PtdIns(4,5)P_2 are significantly less than that for PtdIns. (5) Recently, new phosphoinositides, PtdIns(3)P and PtdIns(3,4)P_2, have been found in cells stimulated with growth factors and transformed by certain oncoproteins. None of the four types of PLC isozymes can hydrolyze these phosphoinositides containing phosphate at the 3-OH position of the inositol ring.[14,15]

In the past several years it has become evident that guanine nucleotide-binding proteins (G proteins) are involved in linking receptor activation to PLC. However, neither the nature of G protein nor whether all PLC isozymes require G protein for activation is known. In addition, various metal-chelating reagents,[16] nucleotides,[7] proteins,[9] and lipids[17-19] might differently affect PLC activity. Since all three substrates for PLC are

[7] S. H. Ryu, K. S. Cho, K.-Y. Lee, P.-G. Suh, and S. G. Rhee, *J. Biol. Chem.* **262,** 12511 (1987).

[8] K.-Y. Lee, S. H. Ryu, P.-G. Suh, W. C. Choi, and S. G. Rhee, *Proc. Natl. Acad. Sci. U.S.A.* **84,** 5540 (1987).

[9] S. H. Ryu, P.-G. Suh, K. S. Cho, K.-Y. Lee, and S. G. Rhee, *Proc. Natl. Acad. Sci. U.S.A.* **84,** 6649 (1987).

[10] M. Katan and P. J. Parker, *Eur. J. Biochem.* **168,** 413 (1987).

[11] M. J. Rebecchi and O. M. Rosen, *J. Biol. Chem.* **262,** 12526 (1987).

[12] Y. Homma, J. Imaki, O. Nakanish, and T. Takenawa, *J. Biol. Chem.* **263,** 6592 (1988).

[13] T. Fukui, R. J. Lutz, and J. M. Lowenstein, *J. Biol. Chem.* **263,** 17730 (1988).

[14] D. L. Lips, P. W. Majerus, F. R. Gorga, A. T. Young, and T. L. Benjamin, *J. Biol. Chem.* **264,** 8759 (1989).

[15] L. A. Serunian, M. T. Haber, T. Fukui, J. W. Kim, S. G. Rhee, J. M. Lowenstein, and L. C. Cantley, *J. Biol. Chem.* **264,** 17809 (1989).

[16] D. Bojanic, M. A. Wallace, R. J. H. Wojcikiewicz, and J. N. Fain, *Biochem. Biophys. Res. Commun.* **147,** 1088 (1987).

[17] S. L. Hofmann and P. W. Majerus, *J. Biol. Chem.* **257,** 14359 (1982).

[18] R. F. Irvine, A. J. Letcher, and R. M. C. Dawson, *Biochem. J.* **218,** 177 (1984).

[19] S. Jackowski and C. O. Rock, *Arch. Biochem. Biophys.* **268,** 516 (1989).

water-insoluble lipids, which substrate is used and how the homogeneous solution of a substrate is prepared can significantly affect the outcome of studies on effector molecules. Various assay procedures have been described in the literature. This chapter summarizes three representative PLC assay procedures and the procedures used in our laboratory for the purification of PLC-β, PLC-γ, and PLC-δ from bovine brain.

Assay Using ^{3}H-Labeled Phosphatidylinositol in Presence of Detergent

Reagents

Substrate solution: For 50 assays, 2.5 mg of PtdIns and 1 μCi of Ptd[^{3}H]Ins are mixed and dried under a stream of N_2. Five milliliters of buffer containing 100 m*M* HEPES, pH 7.0, 2% (w/v) octylglucoside or 0.2% (w/v) deoxycholate, and 0.2 m*M* dithiothreitol (DTT) is added, and the substrate mixture is sonicated briefly to disperse the lipid uniformly.

Ca^{2+}/EGTA solution: 4 m*M* EGTA and 12 m*M* $CaCl_2$

Procedure. To 100 μl substrate solution plus 50 μl Ca^{2+}/EGTA solution, 50–0 μl of water and 0–50 μl of PLC enzyme source are added to make a final volume of 200 μl. The final assay mixture contains 50 μg of PtdIns, approximately 20,000 counts per minute (cpm) of Ptd[^{3}H]Ins, 1% octylglucoside (or 0.1% deoxycholate), 0.1 m*M* DTT, 1 m*M* EGTA, and 3 m*M* $CaCl_2$ in 50 m*M* HEPES buffer, pH 7.0. After incubating the reaction mixture for the appropriate length of time (5–10 min) at 37°, the reaction is stopped by adding 1 ml of chloroform/methanol/HCl (100 : 100 : 0.6, by volume), followed by 0.3 ml of 1 *N* HCl containing 5 m*M* EGTA. The aqueous and organic phases are separated by centrifugation, and a 300-μl portion of the upper aqueous phase is removed for liquid scintillation counting.

Comments. The PtdIns-hydrolyzing activities of various PLC isozymes are maximal at millimolar concentrations of Ca^{2+}.[5,6,9–14] Because certain PLC enzymes are strongly inhibited by contaminating heavy metals such as Hg^{2+}, Cd^{2+}, and Zn^{2+} and appear to contain essential sulfhydryl groups,[7] EGTA and DTT are included. Under the assay conditions, the reaction rate is fairly linear with respect to enzyme concentration when less than 20% of the substrate is hydrolyzed during the reaction. In addition, PtdIns is inexpensive compared to polyphosphoinositides, and the homogeneous substrate solution is easy to prepare in the presence of detergent. Therefore, this assay is commonly used during purification and for the detection of PLC. However, the presence of detergent might entangle studies on the effects of potential regulators, such as lipids and protein factors, on PLC.

Assay Using Small Unilamellar Vesicles of ^{3}H-Labeled Phosphatidylinositol 4,5-Bisphosphate

Reagents

Substrate solution: For 50 assays, 0.75 mg PtdIns(4,5)P_2 and 1.5 μCi Ptd[^{3}H]Ins(4,5)P_2 are mixed and dried under a stream of N_2. Five milliliters of 65 m*M* HEPES, pH 7.0, 100 m*M* NaCl is added. The samples are sonicated extensively for six 3-min intervals interspersed with periods of cooling at ice temperature under a stream of N_2. Samples are then centrifuged at 130,000 *g* for 90 min at 4°, and the clear supernatant is removed. Usually 60–90% of the ^{3}H radioactivity is recovered in the supernatant.

$MgCl_2$ solution: 12 m*M* $MgCl_2$ in 50 m*M* HEPES, pH 7.0, containing 200 m*M* NaCl and 0.4 m*M* DTT

Ca^{2+}/EGTA solution: 10 m*M* EGTA and 10 m*M* $CaCl_2$ in 50 m*M* HEPES, pH 7.0. This ratio of EGTA to Ca^{2+} gives a free Ca^{2+} concentration of 45 μM under the assay conditions described here.

Procedure. To a mixture of 100 μl substrate solution, 50 μl $MgCl_2$ solution, and 20 μl Ca^{2+}/EGTA solution, 30–0 μl of water and 0–30 μl of PLC enzyme source are added to make a final volume of 200 μl. The final assay mixture contains approximately 10 μg of PtdIns(4,5)P_2, about 20,000 cpm of Ptd[^{3}H]Ins(4,5)P_2, 3 m*M* $MgCl_2$, 100 m*M* NaCl, 2 m*M* EGTA, 2 m*M* $CaCl_2$, and 0.1 m*M* DTT in 50 m*M* HEPES buffer, pH 7.0. Exact amounts of PtdIns(4,5)P_2 in the assay mixture are dependent on the recovery yield after centrifugation at 130,000 *g*. After incubating the reaction mixture for 2–5 min at 37°, the reaction is stopped, and the product is quantitated as described in the previous assay.

Comments. In living cells, only PtdIns(4,5)P_2, but not PtdIns and PtdIns(4)P, is considered to be hydrolyzed by PLC in the presence of submicromolar concentrations of Ca^{2+} in response to agonist stimulation.[20] This assay using small unilamellar vesicles of PtdIns(4,5)P_2 is designed to mimic the physiological condition as closely as possible. Various lipid components can be included in the substrate vesicles by adding stock solutions of individual lipids. It has been shown that PLC activity is inhibited by PC,[17,18] activated by diglycerides[17] and phosphatidic acid,[19] and not affected significantly by PE and PS.[17,18] The activity of PLC

[20] A. R. Hughes and J. W. Putney, Jr., *J. Biol. Chem.* **264,** 9400 (1989).

enzymes toward PtdIns(4,5)P_2 is dependent on Ca^{2+} and is maximal at 10–100 μM Ca^{2+}. At a higher concentration of Ca^{2+} reactivity decreases, partly because the surface area of lipid vesicles decreases as a consequence of fusion of unilamellar vesicles. The unilamellar vesicles are especially unstable at higher concentrations of lipids and divalent cations. The free Ca^{2+} concentration can be varied by changing the concentration of Ca^{2+} in the Ca^{2+}/EGTA solution. Free Ca^{2+} concentrations are calculated by using the following constants[21,22]: log K_1 = 9.22 and log K_2 = 8.60 for the protonation of EGTA, log K_1 = 10.01 and log K_2 = 3.79 for the association of Ca^{2+} and EGTA, and log K_1 = 5.42 for the assocation of Mg^{2+} and EGTA.

Assay to Demonstrate Agonist-Sensitive Phospholipase C Activity Using Exogenously Added Phosphatidylinositol 4,5-Bisphosphate

It is evident now that the breakdown of endogenously labeled phosphoinositides in membranes from a variety of cells is enhanced by GTP or hydrolysis-resistant GTP analogs such as GTPγS and by agonists.[23] However, the agonist-sensitive activation of PLC has been difficult to demonstrate with exogenously added substrate. Recently, Claro *et al.*[24] carefully optimized the assay conditions to demonstrate that the hydrolysis of exogenously added Ptd[^{3}H]Ins(4,5)P_2 is stimulated by the muscarinic cholinergic agonist carbachol in the presence of GTPγS.

Briefly, 100 μg of brain membranes is incubated in buffer containing 8 mM Tris–maleate (pH 6.8), 30 μM exogenous PtdIns(4,5)P_2, and 0.1 μM free Ca^{2+} buffered in the presence of 3 mM EGTA, 6 mM $MgCl_2$, 1 mM deoxycholate, 10 mM LiCl, 2 mM ATP, and 1 μM GTPγS plus 0.1 mM carbachol. Interestingly, the presence of 1 mM deoxycholate is critical. There is no detectable PtdIns(4,5)P_2 breakdown in the absence of deoxycholate.

Purification of Three Phospholipase C Isozymes from Bovine Brain

Previously, we separately reported three procedures for the purification of cytosolic forms of PLC-β and PLC-γ,[7,9] the particulate-associated form of PLC-β,[8] and the cytosolic PLC-δ.[9] The procedure described here

[21] A. E. Martell and R. M. Smith, *in* "Critical Stability Constants," Vol. 1, p. 269. Plenum, New York, 1974.

[22] C. Y. Huang, V. Chau, P. B. Chock, J. H. Wang, and R. K. Sharma, *Proc. Natl. Acad. Sci. U.S.A.* **78,** 871 (1981).

[23] J. N. Fain, M. A. Wallace, and R. J. H. Wojcikiewicz, *FASEB J.* **2,** 2569 (1988).

[24] E. Claro, M. A. Wallace, H.-M. Lee, and J. N. Fain, *J. Biol. Chem.* **264,** 18288 (1989).

is essentially a combination of the three procedures. Slight modifications were necessary to make it possible to obtain three isozymes from the same batch of brains.

In brain about 80–90% of PLC-γ and PLC-δ is in the cytosol while 70–80% of PLC-β is associated with the particulate fraction. Therefore, PLC-γ and PLC-δ are purified from the cytosolic fraction and PLC-β mainly from the particulate fraction. However, since the particulate and cytosolic forms of PLC-β are identical with respect to amino acid sequence,[1] PLC-β-containing fractions pooled during the purification of the cytosolic PLC-γ and PLC-δ are combined with the PLC-β fraction from the particulate fraction. A total of 36 bovine brains are used for the purification. Twelve brains are processed at one time.

Step 1: Separation of Cytosolic and Particulate Fractions. Twelve bovine brains are freshly obtained from a local slaughterhouse, and the cerebra (3.3 kg) are homogenized in a Waring blendor with 6.6 liters of buffer containing 20 m*M* Tris-HCl, pH 7.4, 5 m*M* EGTA, 2 m*M* phenylmethylsulfonyl fluoride (PMSF), and 0.1 m*M* DTT. The homogenate is centrifuged for 30 min at 13,000 *g* at 4°. Both the precipitate and supernatant are saved for Steps 2 and 3, respectively.

Step 2: Preparation of Extracts from Particulate Fractions. The precipitate from Step 1 is resuspended in the same homogenization buffer (6.6 liters) and homogenized again to ensure complete breakage of cells. The homogenate is centrifuged for 30 min at 13,000 *g*. The washed pellet is suspended in 2 *M* KCl in homogenization buffer and stirred for 2 hr at 4°. The suspension is then centrifuged for 90 min at 13,000 *g*. The supernatant is brought to 60% $(NH_4)_2SO_4$ saturation by adding solid salt. This suspension is centrifuged for 30 min at 13,000 *g*, and the pellet is suspended in 500 ml of homogenization buffer; the suspension is dialyzed overnight against the homogenization buffer. Dialyzed solution is centrifuged for 30 min at 13,000 *g* to remove insoluble particles, and the supernatant, which is still very turbid, is kept at −20° to be combined with the dialyzed solutions from two other identical preparations.

Step 3: Preparation of Cytosolic Extracts. The supernatant from Step 1 is adjusted to pH 4.8 with 1 *M* acetic acid. After 30 min at 4°, precipitates are collected by centrifugation and dissolved in 1 liter of homogenization buffer. Insoluble materials are pelleted by centrifuging for 30 min at 13,000 *g*, and the turbid supernatant is removed for Step 4.

Step 4: Ion-Exchange Chromatography on DEAE-Cellulose. The supernatant from Step 3 is applied to a DE-52 DEAE-cellulose (Whatman Biosystems, Maidstone, UK) column (8 × 40 cm), which has been equilibrated with 20 m*M* Tris-HCl, pH 7.6, 1 m*M* EGTA, 0.1 m*M* DTT. The column is eluted with a 6-liter linear KCl gradient from 0 to 225 m*M* KCl

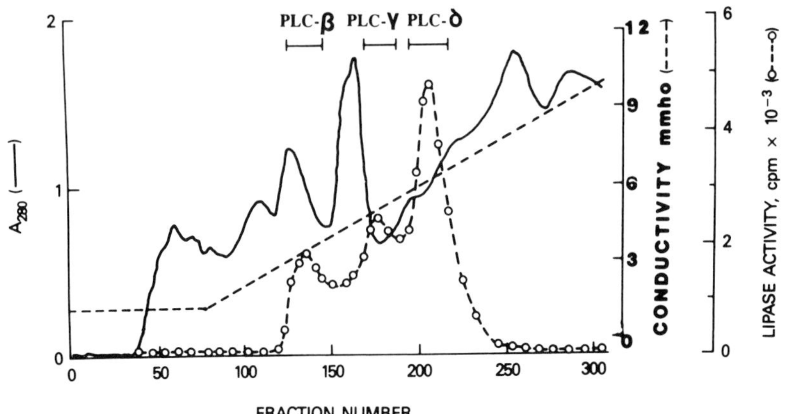

FIG. 1. Ion-exchange chromatography on DEAE-cellulose. Detailed procedures are described in Step 4.

in 50 m*M* Tris-HCl, pH 7.6, 1 m*M* EGTA, and 0.1 m*M* DTT. Three PLC activity peaks are eluted as shown in Fig. 1. The peak fractions are pooled separately. The first peak, which contains PLC-δ, is further purified immediately in the next step. The second peak fractions containing PLC-β are concentrated to about 100 ml and combined with the extracts of particulate fractions from Step 2. The third peak fractions containing PLC-γ are concentrated to about 200 ml and stored frozen to be combined with concentrated fractions of PLC-γ from two other identical preparations.

Purification of PLC-δ

Step 5: Heparin-Agarose Chromatography of PLC-δ. The PLC-δ fractions pooled from the previous step (750 ml) are directly applied to a heparin-agarose column (5 × 15 cm) equilibrated with 20 m*M* HEPES, pH 7.0, 0.1 m*M* DTT, and 1 m*M* EGTA. The column is eluted with a 1.8-liter linear gradient of NaCl from 100 to 700 m*M* NaCl in equilibration buffer. The peak fractions (240 ml) are pooled, concentrated to approximately 10 ml in an Amicon (Danvers, MA) filtration apparatus, and stored frozen to be combined with concentrated fractions of PLC-δ from two other identical preparations.

Step 6: Reversed-Phase Chromatography of PLC-δ on TSK Phenyl-5-PW. Solid KCl is added to the combined concentrated fractions (35 ml) from Step 5 to give a final concentration of 3 *M*, and the mixtures are centrifuged to remove denatured proteins. The supernatants are applied at a flow rate of 5.0 ml/min to a high-performance liquid chromatography

(HPLC) preparative TSK phenyl-5-PW column (21.5 × 150 mm; Bio-Rad, Richmond, CA) equilibrated with 20 m*M* HEPES, pH 7.0, 3 *M* KCl, 1 m*M* EGTA, and 0.1 m*M* DTT. Elution is continued at 5.0 ml/min with a decreasing KCl gradient from 3 to 1.2 *M* KCl for 10 min and with a decreasing KCl gradient from 1.2 to 0 *M* KCl for 20 min. Fractions (25 ml) containing PLC activity are pooled and washed in an Amicon filtration apparatus with 20 m*M* MOPS buffer, pH 5.7, 0.1 m*M* DTT, 1 m*M* EGTA, and finally concentrated to about 10 ml.

Step 7: Ion-Exchange Chromatography of PLC-δ on a Mono S Column. The washed protein solution (~10 ml) from Step 6 is applied at a flow rate of 1.0 ml/min to a Mono S column (70 × 6 mm, Pharmacia, Piscataway, NJ) equilibrated with 20 m*M* MOPS, pH 5.7, 0.1 m*M* DTT, and 1 m*M* EGTA. Elution is continued at 1.0 ml/min with a NaCl gradient from 0 to 300 m*M* NaCl for 20 min and from 300 m*M* to 1 *M* for 10 min. Peak fractions (1.2 ml) are collected manually, diluted with 2 ml of 20 m*M* HEPES (pH 7.0), concentrated in a Centricon microconcentrator (Amicon) to approximately 0.5 ml, separated into aliquots, and stored at −20°. A total of 0.3–0.6 mg of homogeneous PLC-δ is obtained, with a yield of 2–4%.

Purification of PLC-β

Step 8: Ion-Exchange Chromatography of PLC-β on DEAE-Cellulose. Because of turbidity, the combined protein solution from Steps 2 and 3 cannot be chromatographed on a DEAE-cellulose column directly. Therefore, two stages of DEAE-cellulose chromatography, a batch procedure followed by a column step, are employed. In the batch step, all of the combined proteins are adsorbed on 2 liters of DEAE-cellulose equilibrated with 20 m*M* Tris-HCl, pH 7.6, containing 5 m*M* EGTA and 0.1 m*M* DTT. The DEAE-cellulose slurry is stirred and then collected in a 4-liter sintered glass (coarse) filter funnel. The DEAE-cellulose is washed with the equilibration buffer until it is free of turbid lipid materials and unbound protein. For the column procedure, the washed DEAE-cellulose is removed from the filter funnel, mixed with the equilibration buffer, and poured onto a column already containing a 10 cm high bed of equilibrated DEAE-cellulose (final dimension, 8 × 45 cm). The column is eluted at a flow rate of 8 ml/min with an 8-liter linear gradient from 0 to 300 m*M* KCl buffer containing 50 m*M* Tris-HCl, pH 7.6, 1 m*M* EGTA, and 0.1 m*M* DTT. The activity peak is eluted at a KCl concentration of 110 m*M*. The peak fractions (600 ml) are pooled.

Step 9: Heparin-Agarose Chromatography of PLC-β. The pooled fraction from Step 8 (600 ml) is applied to a heparin-agarose column (5 × 25 cm) equilibrated with 20 m*M* HEPES, pH 7.0, 100 m*M* NaCl, 0.1 m*M*

DTT, and 1 m*M* EGTA. The column is eluted with a linear gradient from 100 to 500 m*M* NaCl in 1.5 liters of equilibrium buffer. Peak fractions (310 ml) are pooled and concentrated on an Amicon filter to 27 ml.

Step 10: Reversed-Phase Chromatography of PLC-β on TSK Phenyl-5-PW. Solid KCl is added to the concentrated fractions from Step 9 to give a concentration of 3 *M*, and the mixtures are centrifuged to remove denatured proteins. The supernatants are applied at a flow rate of 5 ml/min to an HPLC preparative phenyl-5-PW column (150 × 215 mm) equilibrated with 20 m*M* HEPES, pH 7.0, 3 *M* KCl, 1 m*M* EGTA, and 0.1 m*M* DTT. Elution is continued at 5 ml/min with a decreasing KCl gradient from 3 to 1.2 *M* for 15 min and with a decreasing gradient from 1.2 to 0 *M* for 20 min. Then the column is washed with a KCl-free buffer. Fractions containing each of the two peaks of PLC activity (15 ml for fraction M1 and 13 ml for fraction M2) are collected separately. The pooled solutions are washed with a KCl-free 20 m*M* HEPES, pH 7.0, and are concentrated to 5 ml in an Amicon filter concentrating procedure. Analysis on SDS-polyacrylamide gels indicates that fractions M1 and M2 contain 150-kDa (PLC-β1) and 140-kDa (PLC-β2) forms of PLC, respectively. The two forms are immunologically indistinguishable. Whether PLC-β2 is a proteolytic fragment of PLC-β1 or a product of alternately spliced mRNA is not known. About 15 mg of PLC-β1 and 8 mg of PLC-β2 are obtained.

Purification of PLC-γ

Step 11: Affinity Chromatography of PLC-γ on Matrex Green Gel. The combined PLC-γ fractions (800 ml) from Step 4 are applied to a Matrex green gel column (5 × 17 cm) equilibrated with 50 m*M* Tris-HCl, pH 7.4, containing 1 m*M* EGTA, and the column is eluted with a 600-ml KCl gradient (0.15–1 *M*). Peak fractions (200 ml) are pooled and concentrated to 50 ml on Amicon filtration apparatus.

Step 12: Heparin-Agarose Chromatography of PLC-γ. Fractions from Step 11 are diluted with 20 m*M* HEPES buffer, pH 7.0, containing no NaCl, to a conductivity equivalent to that of 100 m*M* NaCl buffer and applied to heparin-agarose columns (5 × 17 cm) equilibrated with 20 m*M* HEPES, pH 7.0, containing 1 m*M* EGTA and 0.1 m*M* DTT. The columns are eluted with an 800-ml linear gradient from 100 to 500 m*M* NaCl present in the equilibration buffer. Peak fractions (150 ml) are pooled and concentrated to 5 ml.

Step 13. Reversed-Phase Chromatography of PLC-γ on TSK Phenyl-5-PW. Solid KCl is added to the concentrated fractions from Step 12 to give a concentration of 3 *M*, and the mixtures are centrifuged to remove the precipitate formed. The supernatants are applied at a flow rate of 1 ml/

min to an HPLC phenyl-5-PW column (7.5 × 75 mm) equilibrated with 20 mM HEPES, pH 7.0, 3 M KCl, 1 mM EGTA, and 0.1 mM DT. The proteins are eluted at a flow rate of 1 ml/min, by successive applications of (1) the equilibration buffer for 5 min, (2) a decreasing KCl gradient from 3 to 1.2 M for 10 min, and (3) a decreasing KCl gradient from 1.2 to 0 M for 25 min. Then the column is washed with a KCl-free buffer. Fractions of 1.0 ml are collected. PLC-γ is eluted in a peak centered at 30 min. Peak fractions are pooled and washed with 50 mM Tris-HCl buffer, pH 7.6, containing 1 mM EGTA and 0.1 mM DTT.

Step 14: Ion-Exchange Chromatography of PLC-γ on a Mono-Q Column. The washed PLC-γ samples (5 ml) are applied at a flow rate of 1 ml/min to a Mono Q column (70 × 6 mm) equilibrated with 20 mM Tris-HCl, pH 7.6, 1 mM EGTA, and 0.1 mM DTT. The proteins are eluted at a flow rate of 1.0 ml/min by successive applications of KCl gradients from 0 to 0.3 M for 20 min and from 0.3 to 1.0 M for 10 min. A major protein peak coinciding with PLC activity emerged at 20 min. Peak fractions (3 ml) are pooled, washed with 20 mM Tris-HCl buffer, pH 7.6, containing 0.1 mM DTT and 1 mM EGTA, concentrated to approximately 1 ml, separated into aliquots, and stored at −20°. This procedure yields about 4 mg of PLC-γ (>95% pure).

[49] Properties of Phospholipase C Isozymes

By TADAOMI TAKENAWA, YOSHIMI HOMMA, and YASUFUMI EMORI

Introduction

Phospholipase C (PLC)-catalyzed hydrolysis of phosphatidylinositol 4,5-bisphosphate (PIP_2) has been proposed to be a crucial step in the cellular response to calcium-mobilizing hormones and neurotransmitters.[1–3] Numerous attempts to isolate PLC from various tissues have been made. As a result, many kinds of PLC have been isolated and characterized from various mammalian tissues. We have also purified a PLC from rat liver.[4–14]

[1] M. J. Berridge and R. F. Irvine, *Nature (London)* **312,** 315 (1984).
[2] Y. Nishizuka, *Science* **225,** 1365 (1984).
[3] P. W. Majerus, T. M. Connolly, H. Dechmyn, T. S. Ross, T. E. Bross, H. Ishii, V. S. Bansal, and D. B. Willson, *Science* **234,** 1519 (1986).
[4] T. Takenawa and Y. Nagai, *J. Biol. Chem.* **256,** 6769 (1981).
[5] S. L. Hofmann and P. W. Majerus, *J. Biol. Chem.* **257,** 6461 (1982).
[6] H. Hakata, J. Kobayashi, and G. Kosaki, *J. Biochem. (Tokyo)* **92,** 929 (1982).

Phospholipase C in Rat Brain

Ryu *et al.*[7] isolated from bovine brain three types of PLC, having molecular weights of 150K, 145K, and 85K, respectively, when measured by sodium dodecyl sulfate (SDS)-gel electrophoresis. Moreover, the three PLCs are immunologically distinct as evidenced by the fact that monoclonal antibodies directed against each enzyme do not cross-react, demonstrating the presence of PLC isozymes in brain. We also isolated from rat brain two types of PLC with molecular weights for both of 85K, although their immunological reactivity was different.[8] One was found to be same as the PLC (85K) reported by Ryu *et al.* previously. Therefore, at least four types of PLC may exist in rat brain. According to Rhee's classification,[15] these PLCs are divided to types β, γ, δ, and ε, respectively. We purified these types of PLC from rat brain and examined their properties[16] (summarized in Table I).

PLC-β was purified from rat brain membrane fractions, and PLC-γ, PLC-δ, and PLC-ε were from rat brain cytosol fractions. The molecular weight of PLC-β was found to be 150K. The contribution of PLC-β to total PLC activity was found to be about 50%, suggesting that PLC-β shares the largest activity in rat brain. PLC-β required fairly high Ca^{2+} concentrations (10 μM) for hydrolysis of PIP_2, but a much higher concentration of Ca^{2+} was needed to hydrolyze PI (1 mM or more). PLC-γ, which corresponds to the 145K PLC purified by Ryu *et al.* from bovine brain, showed similar properties to those of PLC-β on the basis of substrate specificity and Ca^{2+} dependence. PLC-γ accounted for around 20% of the total PLC activity. PLC-δ and PLC-ε were purified 2810-fold and 4010-fold, respectively, to homogeneity.[8,16] The molecular weight of these PLCs was estimated to be 85K. PLC-δ accounted for about 20–25% of the total activity in rat brain, but the activity of PLC-ε was less than 5%. Like

[7] S. H. Ryu, P. G. Suh, K. S. Cho, K. Y. Lee, and S. G. Rhee, *Proc. Natl. Acad. Sci. U.S.A.* **84,** 6649 (1987).

[8] Y. Homma, J. Imaki, O. Nakanishi, and T. Takenawa, *J. Biol. Chem.* **263,** 6592 (1988).

[9] O. Nakanishi, Y. Homma, H. Kawasaki, Y. Emori, K. Suzuki, and T. Takenawa, *Biochem. J.* **256,** 453 (1988).

[10] Y. Banno, Y. Yada, and Y. Nozawa, *J. Biol. Chem.* **263,** 11459 (1988).

[11] C. F. Bennett and S. T. Crooke, *J. Biol. Chem.* **262,** 13789 (1987).

[12] M. Katan and P. J. Parker, *Eur. J. Biochem.* **168,** 413 (1987).

[13] M. J. Rebecchi and O. M. Rosen, *J. Biol. Chem.* **262,** 12526 (1987).

[14] T. Fukui, R. J. Lutz, and J. M. Lowenstein, *J. Biol. Chem.* **263,** 17730 (1988).

[15] S. G. Rhee, P.-G. Suh, S.-H. Ryu, and S. Y. Lee, *Science* **244,** 546 (1989).

[16] T. Takenawa, Y, Homma, and O. Nakanishi, *in* "Physiology and Pharmacology of Transmembrane Signalling" (T. Segawa, M. Endo, M. Ui, and K. Kurihara, eds.), p. 207. Elsevier, Amsterdam, 1989.

TABLE I
PROPERTIES OF PHOSPHOLIPASE C ISOZYMES IN RAT BRAIN[a]

Type of PLC	Substrate	V_{max} (μmol/min/ng)	K_m (μM)	Optimal Ca^{2+} (M)
β	PIP_2	25.3	120	10^{-5}
	PIP	12.9	—	10^{-5}–10^{-4}
	PI	20.1	90	$>10^{-3}$
γ	PIP_2	16.2	140	10^{-5}
	PIP	12.3	—	10^{-5}–10^{-4}
	PI	18.3	100	$>10^{-3}$
δ	PIP_2	15.3	200	5×10^{-5}
	PIP	8.4	125	10^{-4}
	PI	19.2	135	$>10^{-3}$
ε	PIP_2	12.9	130	10^{-6}
	PIP	6.5	80	5×10^{-6}
	PI	1.3	>200	$>10^{-3}$

[a] PIP_2, Phosphatidylinositol 4,5-biphosphate; PIP, phosphatidylinositol 4-phosphate; PI, phosphatidylinositol.

PLC-β and PLC-γ, these enzyme required high Ca^{2+} concentrations to hydrolyze PI. PLC-ε especially needed much higher Ca^{2+} concentrations and did not show any activity for PI hydrolyis at a Ca^{2+} level lower than 1 mM. When PIP_2 was used as substrate, PLC activity could be detected at lower Ca^{2+} concentrations. Optimal Ca^{2+} concentrations for PLC-β, PLC-γ, and PLC-δ were around 10 μM, whereas that for PLC-ε was found to be very low (1 μM or less). However, K_m and V_{max} values were almost identical among PLC isozymes when PIP_2 was used as substrate.

Recently, cDNA clones corresponding to PLC-α, -β, -γ, and -δ have also been isolated.[17–20] Three PLCs (β, γ, and δ) had two conserved regions considered to be catalytic domains for PLC activity in common. However, another PLC (α) had a totally different amino acid sequence showing similarity to thioredoxin of *Escherichia coli*.[17] These results suggest that there are several genes encoding PLCs having completely different structures.

[17] C. F. Bennett, J. M. Malcareck, A. Varrichio, and S. T. Crook, *Nature (London)* **334,** 268 (1988).

[18] M. Katan, R. W. Kriz, N. Totty, R. Philip, E. Meldrum, R. A. Aldalpe, J. L. Knopf, and P. J. Parker, *Cell (Cambridge, Mass.)* **54,** 171 (1988).

[19] M. L. Stalh, C. R. Ferenz, K. L. Kellehe, R. W. Kriz, and J. L. Knopf, *Nature (London)* **332,** 209 (1988).

[20] P. G. Suh, S. H. Ryu, K. H. Moon, H. W. Suh, and S. G. Rhee, *Cell (Cambridge, Mass.)* **54,** 161 (1988).

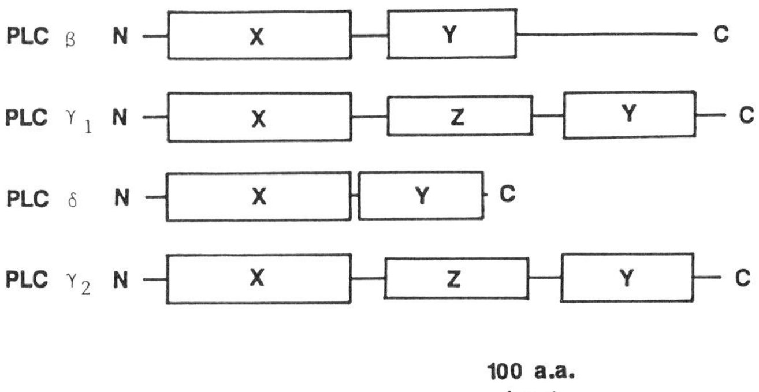

FIG. 1. Schematic structure of four isozymes of PLC. Homologous domains common to PLC-β, -γ_1, -γ_2, and -δ are represented by X and Y, and the *src*-related domain is represented by Z. a.a., Amino acid.

Phospholipase C γ_2

In addition to cDNA clones for various PLCs (α, β, γ, and δ) we have isolated a cDNA clone encoded a new type of PLC from a rat muscle cDNA library that contains a single open reading frame encoding 1265 amino acid residues.[17] The deduced amino acid sequence differs from those of PLC-α, -β, -γ, and -δ, though it contains two homologous regions common to PLC-β, -γ, and -δ (Fig. 1). These two regions were previously designated as domains X and Y by Rhee *et al.*[15] Like PLC-γ, this PLC could be structually divided into three domains (X, Y, and Z).[17] The first and third domains (domains X and Y) are common to PLC-β, -γ, and -δ. The second domain (domain Z), also found in PLC-γ, is related to the N-terminal regulatory domains of oncogenes in the *src* family. Therefore, the structure of this PLC is most similar to that of PLC-γ, and 50.2% of the amino acid residues were identical. As reported previously,[15] domain Z contains three subdomains, A, B-C1, and B-C2, which are also present in the new PLC (Fig. 2). Although the functions of the homologous sequences are not clear, a regulatory role of domain Z on PLC activity is suggested because of homology with the regulatory portion of *src* family tyrosine kinases, *crk*, and the GTP-activating protein of *ras p*21[19,21–23] Amino acid residues of domain Z in the new PLC are 60% identical to those of PLC-γ. Thus, this PLC should be named PLC-γ_2, with the PLC-γ previously reported designated as PLC-γ_1.

[21] Y. Emori, Y. Homma, H. Sorimachi, H. Kawasaki, O. Nakanishi, K. Suzuki, and T. Takenawa, *J. Biol. Chem.* **261,** 21886, (1989).

[22] B. J. Mayer, M. Hamaguchi, and H. Hanafusa, *Nature (London)* **332,** 272 (1988).

[23] U. S. Vogel, R. A. Dixon, M. D. Schaber, R. F. Diehl, M. S. Marshall, E. M. Scolnick, I. S. Sigal, and J. B. Gibbs, *Nature (London)* **335,** 90 (1988).

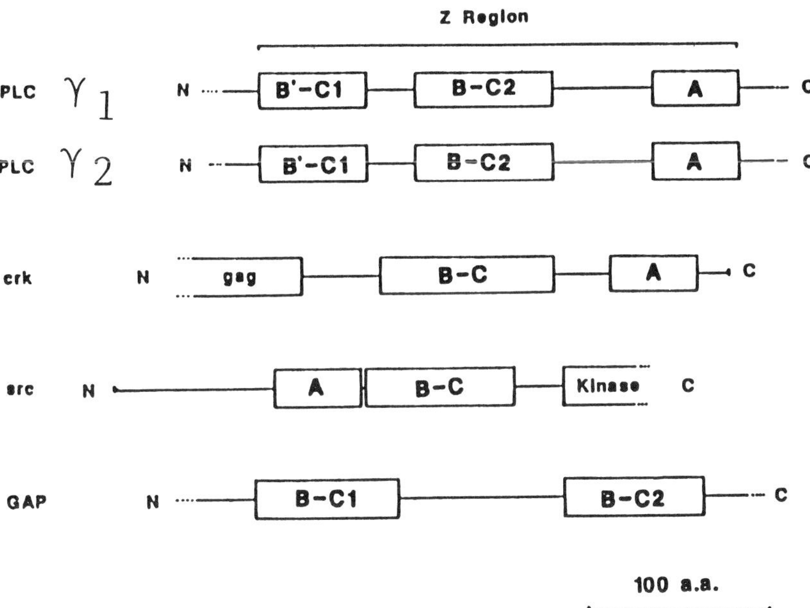

FIG. 2. Schematic representation of relevant regions of PLC-γ_1 and PLC-γ_2, oncogenes (*src* and *crk*), and the GTP-activating protein (GAP). Regions that exhibit similarity are represented by A and B-C. In cases where two homologous regions were observed, they are designated B-C1 and B-C2.

We have established an expression system for PLC in *E. coli* and determined the essential region for PLC activity. Analysis revealed that even small deletions in domain X result in a complete loss of PLC activity and that deletion of the N-terminal 220 residues, which are not contained in previously described homologous regions, also cause the loss of activity. In addition, a precise homology search showed that the X and Y regions are longer than previously reported. Thus, domain X contains about 400 amino acid residues (from 50 to 450 of PLC-γ_2) and domain Y contains about 270 residues (from 930 to 1200) (Fig. 1). These domains are essential for enzymatic activity. Therefore, PLC-δ is considered to have the minimal structure for PLC, since a region not included in domains X and Y is very short in PLC-δ. On the other hand, PLC-γ_1 and -γ_2 contain a long intervening region (domain Z) between the X and Y domains. PLC-β has a unique long C-terminal sequence rich in basic amino acid residues and a unique sequence with a cluster of Glu residues between the X and Y domains. These regions may play important roles in the regulation of PLC activity through the interaction with other molecules. However, even after

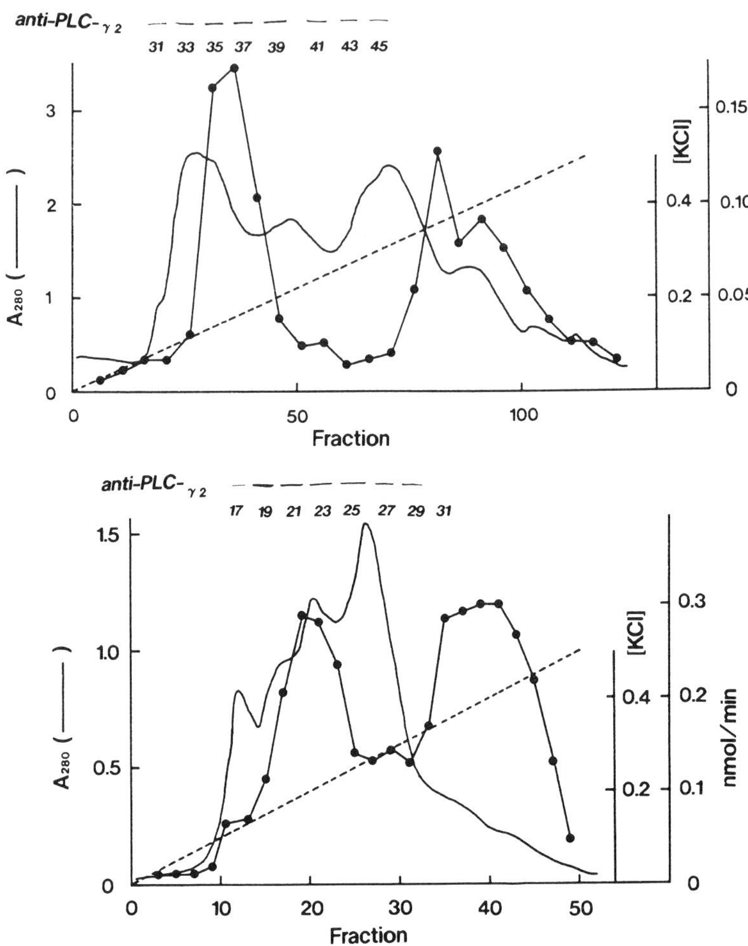

FIG. 3. Purification of PLC-γ_2 from bovine spleen. The supernatant fractions obtained from spleen homogenates were centrifuged at 100,000 *g* for 1 hr and applied to a Q Sepharose fast-flow (top) and then a heparin-Sepharose column (bottom). PLC activity was measured using PIP_2 as a substrate (●), and PLC-γ_2 was detected using anti-PLC-γ_2 antibody. Western blot patterns produced by anti-PLC-γ_2 antibody are shown above the graphs.

domain Z was deleted from PLC-γ_2, most of PLC activity remained. The results suggest that this region is not necessary for PLC activity.

Next we examined the mRNA levels of four types of PLC (β, γ_1, γ_2, and δ) in various rat tissues by Northern analysis.[24] All four types were

[24] Y. Homma, T. Takenawa, Y. Emori, H. Sorimachi, and K. Suzuki, *Biochim. Biophys. Res. Commun.* **164,** 406 (1988).

ubiquitously expressed in rat tissues, but levels of mRNAs for the four PLC isozymes varied in tissues and cells. The level of PLC-β mRNA was extremely high in brain. PLC-γ_1 and PLC-δ had a more widespread distribution. The level of PLC-γ_1 mRNA was relatively high in brain, lung, spleen, thymus, and testis, and PLC-δ mRNA was high in skeletal muscle, spleen, and testis. On the other hand, PLC-γ_2 mRNA was found to be expressed abundantly in spleen.

We therefore tried to isolate PLC-γ_2 from bovine spleen.[25] Spleen cells were homogenized with a Polytron homogenizer in 5 volumes of homogenization buffer consisting of 50 m*M* Tris-HCl (pH 7.4), 0.25 *M* sucrose, 2 m*M* EDTA, 0.5 m*M* EGTA, 1 m*M* dithiothreitol, 0.1 m*M* phenylmethylsulfonyl fluoride, 0.1 m*M* diisopropyl fluoride, 10 μg/ml leupeptin, 10 μg/ml aprotinin, 10 μg/ml bestatin, and 10 μg/ml pepstatin A. The homogenate was centrifuged at 100,000 *g* for 1 hr. The supernatant was applied to a Q Sepharose fast-flow column equilibrated with 20 m*M* Tris-HCl (pH 7.4) and 1 m*M* EDTA. PLCs were eluted with a linear gradient of KCl from 0 to 0.5 *M*. Two PLC activity peaks appeared (Fig. 3). The first peak contained two PLCs, PLC-γ_2 and PLC-δ; the second peak was PLC-γ_1 as evidenced by reaction of each type with specific antibody. The first peak fractions were applied to a heparin-Sepharose column with 10 m*M* HEPES–NaOH (pH 7.0) and 1 m*M* EDTA after dialysis against HEPES buffer. The column was eluted with a linear gradient of KCl from 0 to 0.5 *M* (Fig. 3). PLC activities were eluted separately as two peaks. The first peak (eluted at 0.3 *M* KCl) consisted of PLC-γ_2, and the second peak contained PLC-δ. The first peak fractions were applied to an S Sepharose fast-flow column after dialysis against 20 m*M* HEPES–NaOH (pH 7.0) and 1 m*M* EDTA. The column was eluted with a linear gradient from 0 to 0.6 *M* NaCl. PLC activity was eluted as a single peak around 0.25 *M* NaCl. The active fractions were next applied to a heparin 5PW HPLC column. The column was eluted with a linear gradient of NaCl from 0 to 0.5 *M*. PLC-γ_2 activity was eluted as a single band and was verified with 7.5% SDS–polyacrylamide gel electrophoresis. The molecular weight was estimated as 145,000, comparable to the molecular weight estimated from the cDNA.

The biochemical properties of PLC-γ_2 were similar to those of PLC-γ_1. When PIP_2 was used as a substrate, the optimal Ca^{2+} concentration was found to be 10 μM. On the other hand, PLC-γ_2 hydrolyzed PI at an optimal Ca^{2+} concentration of 10 m*M*. The optimal pH for PLC-γ_2 activity was found to be 5.5 and 6.5 for hydrolysis of PI and PIP_2, respectively.

[25] Y. Homma, Y. Emori, F. Shibasaki, K. Suzuki, and T. Takenawa, *Biochem. J.* **269,** 13 (1990).

At pH 7.0 and 100 μM Ca^{2+}, apparent K_m values for PI and PIP_2 were 110 and 160 μM, respectively. Since the tissue distribution of PLC-γ_1 and PLC-γ_2 is different, PLC-γ_2 may play roles similar to those of PLC-γ_1 in different cells.

[50] Phosphatidylinositol-Specific Phospholipase C from Human Platelets

By YOSHINORI NOZAWA and YOSHIKO BANNO

Introduction

Phosphoinositide (PI)-specific phospholipase C (PLC) plays a crucial role in transmembrane signaling in cells exposed to various extracellular agonists, such as hormones, neurotransmitters, autacoids, and growth factors. The activated PI-PLC generates the second messenger molecules inositol 1,4,5-trisphosphate and 1,2-diacylglycerol, which mobilize calcium ions from intracellular stores and activates protein kinase C, respectively. These initial biochemical events eventually lead to metabolism, secretion, nerve excitation, muscle contraction, and cell proliferation. Despite numerous data supporting the stimulus–response coupling theory, the precise mechanism at the molecular level for induction of cell functions by occupancy of calcium-mobilizing receptors involving PI-PLC is poorly understood.

Recent substantial evidence indicates existence of several types of PI-PLC isozymes in mammalian cells.[1] The physiological significance of such a multiplicity of PI-PLC remains to be explored. Platelets also contain multiple forms of PI-PLC in both membrane and cytosol fractions. In this chapter, we describe the purification and partial characterization of PI-PLCs of human platelets.

Preparation of Membrane and Cytosol Fractions

Human platelets are isolated from outdated blood.[2] Whole human blood in plastic bags are centrifuged at 1400 g for 5 min to remove erythrocytes and leukocytes. The supernatant (platelet-rich plasma) is then centrifuged

[1] S. G. Rhee, P.-G. Suh, S.-H. Ryu, and S. Y. Lee, *Science* **244,** 546 (1989).
[2] N. L. Baenziger and P. W. Majerus, this series, Vol. 31, p. 149.

at 2500 *g* for 5 min at 4°, and the resulting platelet pellet is suspended in plasma. The platelets thus obtained are washed twice with Tris–citrate–bicarbonate buffer,[3] pH 7.0, containing 5 m*M* EGTA by centrifugation for 15 min at 2000 *g*. Washed platelets are resuspended to a final concentration of 5×10^9 cells/ml in lysis buffer [20 m*M* Tris-HCl buffer, pH 7.4, 20 m*M* EGTA, 2 m*M* EDTA, and 1 m*M* phenylmethylsulfonyl fluoride (PMSF)] and allowed to stand on ice for 1 hr. The platelet suspensions are disrupted by sonication on ice for a total of 5 min with 15-sec bursts of a probe-type sonicator (Branson sonifier, B-12). After removing unbroken platelets by centrifugation at 2400 *g* for 10 min, the supernatant is subjected to centrifugation at 105,000 *g* for 60 min. The supernatant (cytosol fraction) is withdrawn, and the pellet (membrane fraction) is resuspended in a small volume of buffer A [20 m*M* Tris-HCl, pH 7.4, 20 m*M* EGTA, 1 m*M* EDTA, 1 m*M* dithiothreitol (DTT), and 0.5 m*M* PMSF] and stored at −80°.

Assay for Phosphoinositide-Specific Phospholipase C Activity

PI-PLC activity is assayed by measuring the formation of radioactive inositol phosphates from 250 μM [^{3}H]PI [15,000 disintegrations per minute (dpm)] or 200 μM [^{3}H]PIP_2 (20,000 dpm) prepared from [*inositol*-2-^{3}H(N)]-phosphatidylinositol (16.6 Ci/mmol) or [*inositol*-2-^{3}H(N)]phosphatidylinositol 4,5-bisphosphate (6.7 Ci/mmol) (New England Nuclear, Boston, MA), as described previously.[4]

Assay Mixture. The final concentrations of components of the mixture (total volume 50 μl) are as follows:

1. 250 μM PI containing [^{3}H]PI (15,000 dpm)
   $CaCl_2$, 2 m*M*
   Tris–maleate buffer, 25 m*M*, pH 5.5
   KCl, 80 m*M*
2. 250 μM PI containing [^{3}H]PI (15,000 dpm)
   $CaCl_2$, 2 m*M*
   Tris–maleate buffer, 25 m*M*, pH 7.0
   Sodium deoxycholate, 1 mg/ml
   KCl, 80 m*M*
3. 200 μM PIP_2 containing [^{3}H]PIP_2 (20,000 dpm) or PIP_2/phosphatidylethanolamine (PE), 40/200 μM containing [^{3}H]PIP_2 (15,000 dpm)
   $CaCl_2$/EGTA, 10 μM (free Ca^{2+} concentration)
   Tris–maleate buffer, 25 m*M*, pH 6.5

[3] S. E. Rittenhouse-Simmons, *J. Clin. Invest.* **8,** 580 (1979).
[4] Y. Banno and Y. Nozawa, *Biochem. J.* **248,** 95 (1987).

Sodium deoxycholate, 1 mg/ml
KCl, 80 mM

Procedure. Free Ca^{2+} concentrations are adjusted to the desired levels using Ca^{2+}/EGTA buffers containing 2 mM EGTA final concentration and the appropriate amount of $CaCl_2$.[5] To prepare substrates for kinetics study, [^{3}H]PI or [^{3}H]PIP_2 dissolved in chloroform, enough for several assays, is kept under a stream of nitrogen, and the residues are dissolved in distilled water in a sonicator bath. Over the course of purification steps on various columns, PLC activity is assayed using [^{3}H]PI as substrate at pH 5.5 (assay mixture 1) or at pH 7.0 (assay mixture 2) and with [^{3}H]PIP_2/PE (1 : 5 molar ratio) as substrate (assay mixture 3). Small unilamellar vesicles of [^{3}H]PIP_2/PE are prepared as follows: the lipids are dispersed into the assay buffer by vigorous vortexing and then sonicated for 2 min to yield 2 nmol of [^{3}H]PIP_2 (15,000 dpm) with 10 nmol of PE/assay.

The assay is initiated by the addition of enzyme solution. After incubation for 10 min at 37°, the reaction is terminated by addition of 0.25 ml of chloroform/methanol/concentrated HCl (100 : 100 : 0.6, by volume) and 0.1 ml of a 5 mM EGTA, 1 N HCl solution. The mixtures are vortex-mixed and then centrifuged for 10 min at 2000 g at 25°. A 0.2-ml portion of the upper aqueous phase is carefully transferred to a vial, mixed with 6 ml of scintillation fluid, and the radioactivity determined in a liquid scintillation counter (Beckman LS-9000). The water-soluble reaction products from incubation of the purified PLCs with [^{3}H]PI and [^{3}H]PIP_2 are analyzed by chromatography on a column of Dowex AG1-X8 according to the method of Downes and Michell.[6] More than 90% of the product derived from [^{3}H]PI and [^{3}H]PIP_2 hydrolysis corresponds to the carrier inositol monophosphate and inositol 1,4,5-trisphosphate, respectively.

Purification of Membrane-Bound Phospholipase C

The study of distribution of PIP_2-hydrolyzing activity in human platelets has shown that 20% of the total activity of the homogenate is associated with the particulate fraction.[4] The frozen platelet membrane fractions are thawed, resuspended in buffer A, and spun down at 105,000 g for 60 min. The washed pellet is resuspended in 2 M KCl in buffer A and stirred for 2 hr at 4°. Buffer A is then added to adjust the KCl concentration to 1 M, and the suspension is centrifuged at 105,000 g for 60 min. The residual pellet is suspended in buffer A and extracted with an equal volume of

[5] J. Raaflaub, *Methods Biochem. Anal.* **3,** 301 (1960).
[6] C. P. Downes and R. H. Michell, *Biochem. J.* **198,** 133 (1981).

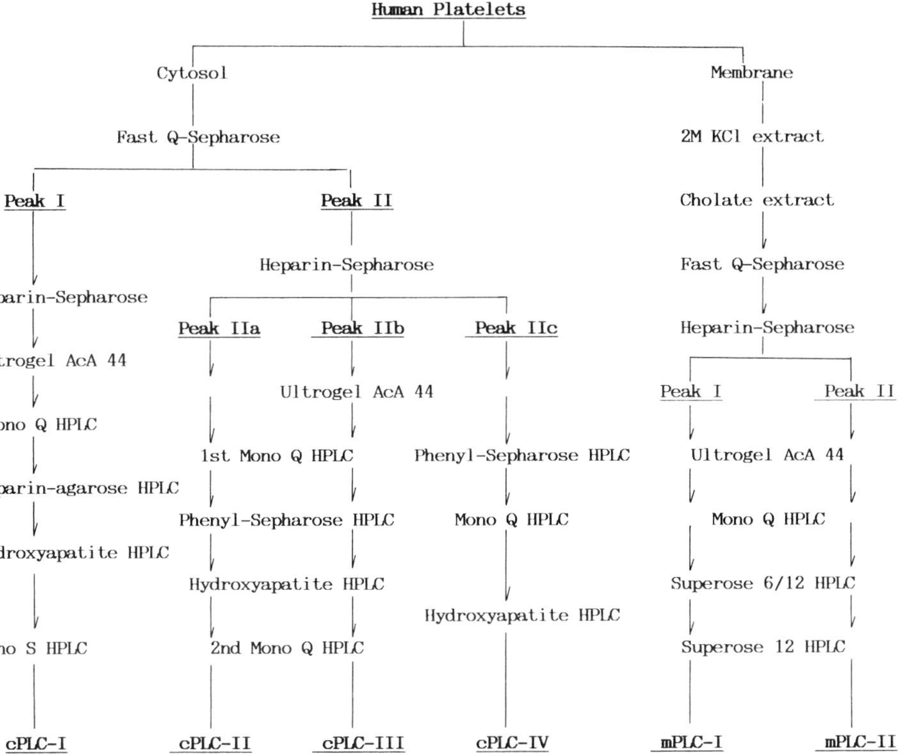

FIG. 1. Purification of phosphatidylinositol-specific phospholipases C from human platelets.

buffer A containing 2% sodium cholate and 200 m*M* NaCl, pH 7.4. After stirring for 2 hr at 4°, the suspension is sedimented by centrifugation at 105,000 *g* for 60 min to obtain the cholate extract for the subsequent purification process (Fig. 1). About one-fourth of the particulate-associated [^{3}H]PIP$_2$-hydrolyzing activity is released by extraction with 2 *M* KCl. When the residual pellet obtained after the repeated KCl extraction is treated with 1% sodium cholate and 0.1 *M* NaCl, nearly 80% of the [^{3}H]PIP$_2$-hydrolyzing activity and about 40% of the total membrane protein are released.

Fast Q-Sepharose Column Chromatography. The cholate extract is dialyzed overnight against buffer B containing 20 m*M* Tris-HCl, pH 7.4, 1 m*M* DTT, 1 m*M* EGTA, 0.5 m*M* PMSF, 0.1 *M* NaCl, and 0.5% sodium cholate. The dialyzed solution is applied onto a Fast Q-Sepharose column

(6.0 × 25 cm) equilibrated with buffer B. After washing the column with 500 ml of the same buffer to remove unbound protein, elution is performed with a linear concentration gradient of NaCl from 0.1 to 0.5 *M* in 2000 ml of buffer B. The majority of PLC activity for [^{3}H]PI and [^{3}H]PIP$_2$ is eluted between 0.2 and 0.3 *M* NaCl. The active fractions are pooled and dialyzed overnight against buffer C (20 m*M* Tris-HCl, pH 7.4, 1 m*M* EDTA, 0.1 *M* NaCl, 0.5 *M* PMSF, 1 m*M* DTT, and 0.5% sodium cholate).

Heparin-Sepharose Column Chromatography. The dialyzate from the Fast Q-Sepharose step is applied to a heparin-Sepharose column (3 × 10 cm) equilibrated with buffer C, and the column is first washed with the same buffer and then eluted with a 1000-ml NaCl linear gradient (0.1–0.7 *M*) in buffer C. The two activity peaks, mPLC-I and mPLC-II, are eluted at 0.35–0.45 and 0.5–0.6 *M* NaCl, respectively. The fractions containing mPLC-I and mPLC-II are pooled separately and concentrated in an Amicon (Danvers, MA) filtration apparatus (YM10 membrane).

Ultrogel AcA 44 Column Chromatography. The concentrated samples (mPLC-I and mPLC-II) obtained as above are separately applied to Ultrogel AcA 44 columns (2.5 × 90 cm) equilibrated with buffer C containing 0.3 *M* NaCl. The active fractions are pooled and concentrated as above. The concentrated mPLC-I and mPLC-II solutions are diluted one-tenth with salt-free buffer D (20 m*M* Tris-HCl, pH 7.4, 1 m*M* DTT, 1 m*M* EGTA, and 10% glycerol).

Mono Q HPLC. Aliquots of mPLC-I and mPLC-II obtained from the previous step are applied separately to Mono Q columns (HR 10/10) equilibrated with buffer D. The proteins are eluted at a flow rate of 2 ml/min by successive application of increasing NaCl gradients from 0 to 0.3 *M* for 5 min, from 0.3 to 0.5 *M* for 30 min, and from 0.5 to 0.6 *M* for 5 min using a fast liquid chromatography. The fractions exhibiting PLC activity are pooled and concentrated in a Centricon microconcentrator (Amicon).

Superose 6-12 Combination HPLC. The mPLC-I and mPLC-II (0.1 ml) solutions pooled from the Mono Q column step are separately applied to Superose 6-12 combination columns (HR 10/30) connected in series and equilibrated with buffer C containing 0.3 *M* NaCl. Elution is carried out with the same buffer at a flow rate of 0.4 ml/min, and 0.2-ml fractions are collected for assay of activity.

Superose 12 HPLC. The mPLC-II solution obtained from the Superose 6-12 combination column is applied to a Superose 12 column (HR 10/30) equilibrated with buffer C containing 0.15 *M* NaCl. The column is eluted with the same buffer at a flow rate of 0.4 ml/min, and 0.1-ml fractions are collected. The active fractions are pooled and concentrated for rechromatography on the same column, producing a nearly symmetric activity peak.

Purification of Cytosolic Phospholipase C

PI-hydrolyzing activity (98% at pH 5.5 and 90% at pH 7.0) and PIP_2-hydrolyzing activity (80%) are found to locate in the cytosolic fraction of human platelets. The cytosolic fraction (1500 ml) is dialyzed against buffer A (20 m*M* Tris-HCl, pH 7.4, 5 m*M* EGTA, 1 m*M* EDTA, 0.5 m*M* PMSF, 1 m*M* DTT, and 10% glycerol) for subsequent purification (Fig. 1).

Fast Q-Sepharose. The supernatant obtained after centrifugation of the dialyzate is loaded onto a Fast Q-Sepharose column (6.0 × 20 cm) equilibrated with buffer A. After washing with the same buffer, elution is performed with a linear concentration gradient from 0.1 to 0.4 *M* NaCl in buffer A. When assayed for [^{3}H]PI hydrolysis at pH 5.5 and [^{3}H]PIP_2 hydrolysis at pH 6.5, the two activity peaks are eluted at between 0.15 and 0.28 and between 0.3 and 0.4 *M* NaCl. The first peak fraction (peak I) exhibits major [^{3}H]PI-hydrolyzing activity at pH 5.5, and the second activity peak (peak II) shows preferential hydrolysis for [^{3}H]PIP_2. The activity for [^{3}H]PI hydrolysis at pH 7.0 coincides with the peak II fraction. The fractions containing peak I and peak II are pooled separately and concentrated in an Amicon concentrator (YM10 membrane). The concentrated enzyme solutions are dialyzed against buffer B (20 m*M* Tris-HCl, pH 7.4, 1 m*M* EDTA, 0.1 m*M* PMSF, 1 m*M* DTT, and 10% glycerol).

Heparin-Sepharose Column Chromatography. The dialyzates of peak I and peak II from the Fast Q-Sepharose column steps are separately applied to heparin-Sepharose columns (3 × 15 cm) equilibrated with buffer B. The column is first washed with the same buffer, and then elution is performed with a linear gradient ranging from 0.1 to 0.7 *M* NaCl in buffer B. The [^{3}H]PI-hydrolyzing activity of peak I is eluted in a single peak between 0.2 and 0.3 *M* NaCl. The [^{3}H]PIP_2-hydrolyzing activity of peak II is resolved into three activity peaks (IIa, IIb, IIc) with linear gradient of NaCl (0.1–0.7 *M*). The first activity peak (IIa) elutes at between 0.2 and 0.35 *M*, the second (IIb) between 0.35 and 0.55 *M*, and the third (IIc) between 0.55 and 0.65 *M* are pooled separately. The ratios of total [^{3}H]PIP_2-hydrolyzing activity of peaks IIa, IIb, and IIc are 24.8, 49.6, and 25.6%, respectively. When the PLC activity of the three peaks is measured using [^{3}H]PI as substrate at pH 7.0, activity is detected in both peak IIa and peak IIb, whereas [^{3}H]PI-hydrolyzing activity is hardly detected in peak IIc.

Peak I, designated cPLC-I, with high [^{3}H]PI-hydrolyzing activity at 5.5 from the heparin-Sepharose column chromatography step is further purified by Mono Q HPLC, heparin-agarose HPLC, hydroxyapatite HPLC, and Mono S HPLC. The cPLC-I is eluted with 0.15–0.18 *M* NaCl from Mono Q, with 0.3 *M* NaCl from heparin-agarose, with 0.2 *M* NaCl from Mono S, and with 0.23 *M* potassium phosphate buffer from the hydroxyapatite column. Peak IIc obtained by heparin-Sepharose column chromatography, which exhibits a preferential specificity for [^{3}H]PIP$_2$ hydrolysis, is further purified as follows. The activity is measured using [^{3}H]PIP$_2$/PE as substrate.

Ultrogel AcA 44 Column Chromatography. The concentrated peak IIc solution is applied to an Ultrogel AcA 44 column (4.0 × 90 cm) equilibrated with buffer B containing 0.3 *M* NaCl. Active fractions are pooled.

Phenyl-Sepharose HPLC. Solid KCl is added to the peak IIc solution from the Ultrogel AcA 44 column to give a concentration of 3 *M*. The enzyme solution is applied to an HPLC Phenyl-Sepharose column (HR 5/5) equilibrated with buffer B containing 3 *M* KCl. The proteins are eluted at a flow rate of 0.5 ml/min for 5 min, with a decreasing KCl gradient from 3.0 to 1.8 *M* for 5 min, and with decreasing KCl followed by a linear cholate gradient of 0–0.4% in buffer B for 30 min. The active fractions eluted between 0.15 and 0.25% cholate are pooled and diluted to one-fifth with buffer B.

Mono Q HPLC. The enzyme solution from the Phenyl-Sepharose column step is applied to a Mono Q column equilibrated with buffer B. The proteins are eluted at a flow rate of 1 ml/min with increasing NaCl from 0 to 0.2 *M* for 5 min, from 0.2 to 0.35 *M* for 30 min, and from 0.35 to 0.7 *M* for 5 min. The fractions coinciding with PLC activity elute at about 0.3 *M* NaCl and are pooled and diluted to one-tenth with buffer C (20 m*M* Tris-HCl, pH 7.4, 1 m*M* DTT, 0.1 m*M* EDTA and 10% glycerol).

Hydroxyapatite HCA-100 HPLC. An aliquot of peak IIc after the Mono Q column step is applied to a hydroxyapatite HCA-100 column (5 × 10 cm) equilibrated with buffer C. The proteins are eluted with a 30-ml linear gradient ranging from 50 to 400 m*M* potassium phosphate buffer, pH 7.4, at a flow rate of 0.5 ml/min. The activity fractions are pooled and concentrated in a Centricon microconcentrator (cPLC-IV).

Peak IIa and peak IIb obtained from the heparin-Sepharose column chromatography step are further purified on Ultrogel AcA 44, the first Mono Q column, phenyl-Sepharose, hydroxyapatite, and the second Mono Q column. By the first mono Q HPLC, IIa is eluted with 0.2–0.28 *M* NaCl, and IIb is eluted with 0.3–0.4 *M* NaCl. Peak IIa is eluted with 0.3–0.37% cholate on Phenyl-Sepharose and with 0.25 *M* potassium phosphate buffer on hydroxyapatite. The elution profiles of peak IIb on these two columns are similar to those of peak IIa. By the second Mono Q column chromatography, IIa (designated as cPLC-II) and IIb (designated as cPLC-III) are eluted with 0.25 and 0.32 *M* NaCl, respectively.

Properties

There are more than four different forms of PLC in the cytosol[7–9] and two forms in the membrane fraction[10,11] of human platelets. The results of identification of human platelet PLC isoforms by antibodies against various types of soluble PLCs (brain PLC-β, -γ_1, -δ,[12,13] uterus PLC-α,[14] and PLC-γ_2[15,16]) indicate that human platelet cPLC-I is recognized by the antibody against PLC-γ_2. On the other hand, platelet cPLC-II, -III, and mPLC-II preparations do not react with any other PLC antibodies.

The catalytic properties of the platelet PLC isoforms are studied by using either [^{3}H]PI or [^{3}H]PIP$_2$ as substrates. Hydrolysis of both PI and PIP$_2$ by platelet PLC isoforms is dependent on Ca^{2+}. However, at low Ca^{2+} concentrations (10 μM) [^{3}H]PIP$_2$ is the preferred substrate for platelet PLC isoforms. When assayed with [^{3}H]PI as substrate at high calcium concentration (2 mM), cPLC-I activity is most active at pH 5.5; however, at neutral pH the order of specific activity is cPLC-II > cPLC-I > cPLC-III. On the other hand, cPLC-IV has very low activity for PI hydrolysis at both pH values. The order of specific activity is cPLC-IV > cPLC-III > cPLC-II > cPLC-I for [^{3}H]PIP$_2$ hydrolysis, indicating that cPLC-IV is the most specific for [^{3}H]PIP$_2$. The preference for [^{3}H]PIP$_2$ hydrolysis is also observed for the mPLC-I and mPLC-II preparations. The K_m values for [^{3}H]PI hydrolysis of cPLC-IV or mPLC-II are higher than those of cPLC-I or cPLC-II, namely, 0.7 and 0.5 mM for cPLC-IV and mPLC-II, and 0.1 and 0.08 mM for cPLC-I and cPLC-II, respectively.

The effects of various metal ions on the PIP$_2$-hydrolyzing activities of platelet PLC isoforms have been examined. Platelet cytosolic and membrane PLCs are inhibited by addition of 2 mM EGTA or 2 mM EDTA. The addition of Ca^{2+} (10 μM free) causes activation whereas Mg^{2+} does

[7] Y. Banno, S. Nakashima, and Y. Nozawa, *Biochem. Biophys. Res. Commun.* **136,** 713 (1986).

[8] M. G. Low, R. C. Carroll, and A. C. Cox, *Biochem. J.* **243,** 763 (1987).

[9] V. Manne and H. F. Kung, *Biochem. J.* **243,** 763 (1987).

[10] Y. Banno, Y. Yada, and Y. Nozawa, *J. Biol. Chem.* **263,** 11459 (1988).

[11] J. J. Baldassare, P. A. Henderson, and G. J. Fisher, *Biochemistry* **28,** 6010 (1989).

[12] S. H. Ryu, P.-G. Suh, K. S. Cho, K. Y. Lee, and S. G. Rhee, *J. Biol. Chem.* **262,** 12511 (1987).

[13] S. H. Ryu, P.-G. Suh, K. S. Cho, K. Y. Lee, and S. G. Rhee, *Proc. Natl. Acad. Sci. U.S.A.* **84,** 6649 (1987).

[14] C. F. Bennett and S. T. Crooke, *J. Biol. Chem.* **262,** 13789 (1987).

[15] S. Ohta, A. Matsui, Y. Nozawa, and Y. Kagawa, *FEBS Lett.* **242,** 31 (1988).

[16] Y. Homma, T. Takenawa, Y. Emori, H. Sorimachi, and K. Suzuki, *Biochem. Biophys. Res. Commun.* **164,** 406 (1989).

not. At low Ca^{2+} concentrations (10^{-7} *M*) corresponding to the level in resting platelets, cPLC-IV and mPLC-II act to hydrolyze PIP_2 at 20–26% of maximum hydrolysis, but cPLC-I does not. Other metal ions (50 μM) such as Mn^{2+}, Fe^{2+}, and Cu^{2+} show some inhibition (15–20%). The inhibitory potency of Hg^{2+} differs among PLC isoforms; cPLC-I and mPLC-II are extremely sensitive to Hg^{2+} ($I_{0.5}$ <1 μM), whereas cPLC-IV is less sensitive to the ion ($I_{0.5}$ >5 μM).

[51] Purification of Guinea Pig Uterus Phosphoinositide-Specific Phospholipase C

By C. FRANK BENNETT, MICHAEL P. ANGIOLI, and STANLEY T. CROOKE

Introduction

Hydrolysis of membrane phospholipids following occupancy of cell surface receptors is a common signal transduction pathway for a variety of agonists. Hydrolysis of phosphoinositides by phospholipase C (PLC) is well documented to be a major transduction mechanism for calcium-mobilizing agonists. Phospholipase C-mediated hydrolysis of phosphatidylinositol 4,5-bisphosphate (PIP_2) generates two second messenger molecules, inositol 1,4,5-trisphosphate and diacylglycerol.[1,2] Purification of the enzymes which hydrolyze the phosphodiester bond of phosphatidylinositol 4,5-bisphosphate (PI-PLC) from a variety of tissues has recently been accomplished. These studies have demonstrated that multiple forms of PI-PLC exist within the same tissue or cell type.[3,4] Isolation of the cDNA clones for four PI-PLC isoenzymes demonstrated that the different forms of PI-PLC are distinct gene products with similar enzymatic activity (reviewed in Refs. 3 and 4).

In a study of phospholipases in tissues which respond to peptidoleukotrienes, as part of our long-term studies on these agents, we determined that guinea pig uterus was a relatively rich source of phosphoinositide-specific phospholipase C and exhibited moderate levels of a phosphatidylethanolamine-preferring phospholipase A_2 activity and low levels of phosphatidylcholine-specific phospholipase C activity. We subsequently puri-

[1] M. J. Berridge, *Annu. Rev. Biochem.* **56,** 159 (1987).
[2] U. Kikkawa and Y. Nishizuka, *Annu. Rev. Cell Biol.* **2,** 149 (1986).
[3] S. G. Rhee, P.-G. Suh, S.-H. Ryu, and S. Y. Lee, *Science* **244,** 546 (1989).
[4] S. T. Crooke and C. F. Bennett, *Cell Calcium* **10,** 309 (1989).

fied a phosphoinositide-specific phospholipase C to apparent homogeneity from uterine tissue[5] that exhibited biochemical characteristics similar to a PI-PLC previously purified from sheep seminal vesicles.[6]

Materials

Female guinea pigs are obtained from Hazelton Research. Rat basophilic leukemia cells (RBL-1) and the human promonocytic leukemia cell line U-937 are obtained from the American Type Culture Collection (Rockville, MD). DEAE-Sepharose CL-6B, AH-Sepharose, heparin-Sepharose, and Sephacryl S-200 are obtained from Pharmacia (Uppsala, Sweden). Affi-Gel blue is purchased from Bio-Rad Laboratories (Richmond, CA). 1-Stearoyl-2-[1-^{14}C]arachidonylphosphatidylinositol (specific activity 60 mCi/mmol) is purchased from Amersham (Arlington Heights, IL). L-[*myo-inositol*-2-^{3}H(N)]Phosphatidylinositol (PI, specific activity 8.4 Ci/mmol), L-[*myo-inositol*-2-^{3}H(N)]phosphatidylinositol 4-phosphate (PIP, specific activity 1.5 Ci/mmol), and L-[*myo-inositol*-2-^{3}H(N)]phosphatidylinositol 4,5-bisphosphate (PIP_2, specific activity 2.0 Ci/mmol) are purchased from New England Nuclear (Boston, MA). Silica gel G thin-layer chromatography plates are purchased from Analtech (Newark, DE).

Enzyme Assays

Phospholipase C activity is measured by either the formation of [^{14}C]diacylglycerol from 1-stearoyl-2[1-^{14}C]arachidonylphosphatidylinositol or, alternatively, by the formation of water-soluble inositol products from L-[*myo-inositol*-2-^{3}H]phosphatidylinositol.[5,6] Using semipurified material, similar results are obtained by measuring either product of phospholipase C activity. Measurement of organic soluble products by thin-layer chromatography reveals other enzyme activities which hydrolyzed phosphatidylinositol present in the crude guinea pig uterus cellular fractions, such as a phosphatidylinositol-preferring phospholipase A_2 activity (C. F. Bennett, unpublished observations, 1986). The phospholipase A_2 activity is resolved from the PI-PLC activity by the first ion-exchange chromatography step. Measurement of water-soluble products is an easier assay to perform and is less time consuming; therefore, it is used to analyze column fractions during enzyme purification and characterization of the purified enzyme.

The standard reaction mixtures contain 50 m*M* Bis-Tris (pH 7.0), 50 m*M* KCl, 2.4 m*M* deoxycholate, 10 μM labeled substrate (diluted with

[5] C. F. Bennett and S. T. Crooke, *J. Biol. Chem.* **262,** 13789 (1987).

[6] S. L. Hoffman and P. W. Majerus, *J. Biol. Chem.* **257,** 6461 (1982).

unlabeled substrate to a specific activity of 45.4 mCi/mmol), and 1 mM $CaCl_2$ for PI or 0.5 mM $CaCl_2$ for PIP and PIP_2 as substrates. Between 5 and 10 μl of the column fractions is analyzed for activity in a final volume of 50 μl. The reaction mixture is incubated at 37° for 10 min. Enzyme reactions are stopped by the addition of 250 μl of chloroform/methanol/HCl (50 : 50 : 0.3, by volume) followed by 75 μl of 1 N HCl containing 5 mM EGTA for quantitation of water-soluble inositol phosphates. The radioactivity in a 150-μl aliquot of the aqueous layer is determined. Quantitation of [^{14}C]diacylglycerol is performed by stopping the enzyme reaction and extracting the lipids by sequential addition of 50 μl chloroform/methanol (1 : 2), 50 μl chloroform, and 50 μl of 4 M KCl. The organic layer is spotted onto silica gel G thin-layer chromatography plates, and the plates are developed using petroleum ether/diethyl ether/acetic acid (70 : 30 : 1) as the solvent system. Plates are stained with iodine vapors, and the region comigrating with unlabeled diacylglycerol and arachidonic acid is scraped into scintillation vials. Methanol (0.5 ml) is added to the vials followed by 10 ml scintillation cocktail. Assays are found to be linear with respect to time and protein when less than 30% of the substrate is hydrolyzed.

Purification of Phosphoinositide-Specific Phospholipase C from Guinea Pig Uterus

Preparation of Cell Cytosol. Uteruses are collected from mature (400–600 g) asynchronous guinea pigs and washed in ice-cold phosphate-buffered saline. Fat and connective tissue are removed. No attempt is made to selectively isolate endometrium or myometrium. Frozen uteruses could be used for the isolation of PI-PLC; however, the final specific activity is typically 30–50% of the activity obtained from fresh tissue. All subsequent manipulations are performed at 4°. Uteruses are homogenized in 10 mM Tris-HCl (pH 7.4), 5 mM $MgCl_2$, 1 mM EDTA, 0.25 M sucrose, 0.5 mM phenylmethylsulfonyl fluoride, 0.1 mM leupeptin, and 10 μg/ml aprotinin with a Polytron homogenizer. The homogenate is centrifuged at 1000 g for 15 min. The subsequent supernatant is centrifuged at 15,000 g for 20 min, then at 100,000 g for 60 min. Typically 90% of the PI-PLC enzyme activity is found in the cytosolic fractions.

Column Chromatography of Cytosolic Fractions. Cell cytosol is applied to a DEAE–Sepharose CL-6B column (5 × 19 cm) equilibrated with 10 mM Tris-HCl (pH 7.6), 0.2 mM EDTA, and 20% glycerol. Glycerol is added to all subsequent buffers to help stabilize enzyme activity. No enzyme activity is detected in the unbound material. Bound protein is eluted with a linear 2-liter gradient of 0 to 500 mM KCl at a flow rate of

40 ml/hr. Two peaks of PI-PLC activity are resolved on the DEAE-Sepharose column and are pooled separately (peaks I and II). The ratio of the two peaks varies from preparation to preparation and is thought to reflect the estrus status of the animals.

Pooled fractions from the DEAE-Sepharose column are dialyzed against 20 m*M* HEPES (pH 6.8), 20 m*M* NaCl, 2 m*M* EGTA, 20% glycerol and applied separately to an AH-Sepharose column (2.6 × 17 cm) equilibrated with the same buffer. No activity is detected in the flow-through fractions. Bound protein is eluted with a linear 600-ml 20 to 500 m*M* NaCl gradient at 20 ml/hr. Peak I partially resolves into two peaks (peaks Ia and Ib), which are pooled separately. Peak II elutes from the AH-Sepharose column between 350 and 400 m*M* NaCl. Samples are dialyzed against 20 m*M* HEPES (pH 6.7), 0.5 m*M* EGTA, and 20% glycerol and applied separately to a heparin-Sepharose column (1.6 × 27 cm). All three forms of PI-PLC activity bound to the heparin-Sepharose column. Enzyme activity is eluted with a linear 200 ml 0 to 750 m*M* KCl gradient at a flow rate of 40 ml/hr.

Heparin-Sepharose fractions containing PI-PLC enzyme activity are pooled, concentrated to 2 ml by an Amicon (Danvers, MA) concentrator, and applied to a Sephacryl S-200 column (1.6 × 88 cm) equilibrated with 20 m*M* Tris-HCl (pH 7.0), 75 m*M* KCl, 0.2 m*M* EGTA, and 20% glycerol at a flow rate of 20 ml/hr. Both PI-PLC Ia and Ib elute from the Sephacryl S-200 column with an apparent molecular weight of 58,000 (Fig. 1A), while PI-PLC II elutes from the column at a position corresponding to a molecular weight of 72,000. Subsequent studies revealed that better resolution of PI-PLC II from contaminating proteins could be achieved on a Sephacryl S-300 column (2.6 × 105 cm). Analysis of the PL-PLC Ib fractions from the Sephacryl S-200 column by sodium dodecyl sulfate (SDS)–polyacrylamide gel electrophoresis and staining with Coomassie Brilliant blue R-250 demonstrated the presence of a 60K protein which coeluted with enzyme activity (Fig. 1B). Fractions on the descending side of the major protein peak, containing the peak of PI-PLC enzyme activity, are selectively pooled. Following Sephacryl S-200 chromatography PI-PLC Ib appears to be 80% pure, whereas PI-PLC Ia is approximately 50% pure. Further purification of PI-PLC Ib to greater than 95% purity is achieved by step elution from an Affi-Gel blue column. Protein is applied to an Affi-Gel column (0.6 × 3 cm) equilibrated with 20 m*M* Bis-Tris (pH 6.8), 200 m*M* KCl, 0.5 m*M* EGTA, and 20% glycerol. Proteins are eluted stepwise with 400, 600, and 750 m*M* KCl. PI-PLC Ib is eluted from the column by 600 m*M* KCl.

Following Affi-Gel blue chromatography, PI-PLC Ib is purified greater

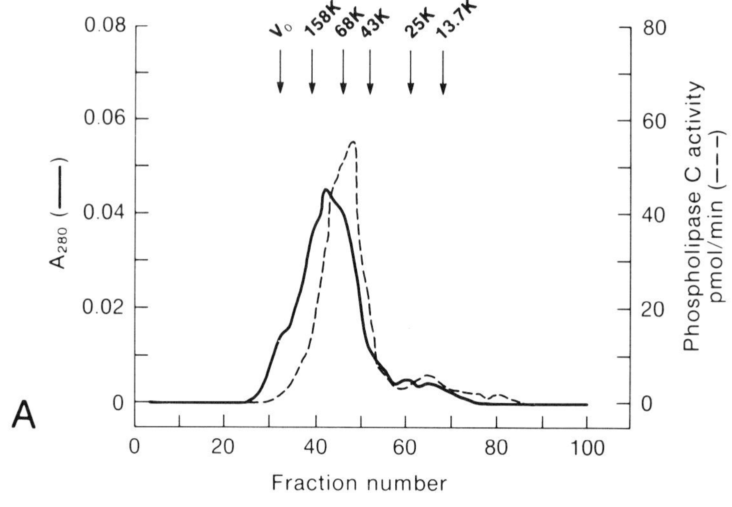

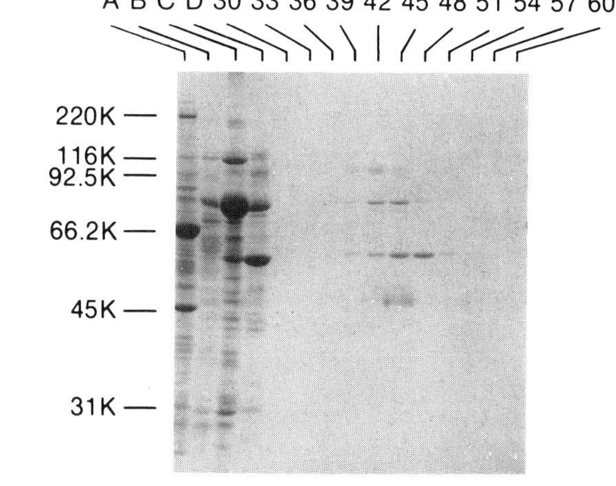

FIG. 1. Sephacryl S-200 chromatography of PLC-α. Phospholipase C activity corresponding to peak Ib after chromatography on a heparin-Sepharose column was pooled, concentrated to 2.0 ml, and loaded onto a Sephacryl S-200 column. (A) Enzyme activity was determined in 5-μl aliquots from each fraction (1.9 ml) as described in the text. The molecular weight reference proteins correspond to aldolase, 158K; bovine serum albumin, 68K; ovalbumin, 43K; chymotrypsinogen, 25K; and RNase A, 13.7K. (B) Fraction at each stage of purification and from the Sephacryl S-200 column were separated on an 8% polyacrylamide gel and stained with Coomassie Brilliant blue R-250. Lane A, 75 μg protein from the 100,000 *g* supernatant fraction of guinea pig uterus; lane B, 50 μg protein from peak I of the DEAE-Sepharose column; lane C, 50 μg protein from the AH-Sepharose column; lane D, 25 μg protein from the heparin-Sepharose column. Lanes 30–60 correspond to 150-μl aliquots of the indicated fractions from the Sephacryl S-200 column shown in (A).

TABLE I
PURIFICATION OF PHOSPHOLIPASE C-α FROM GUINEA PIG UTERUS[a]

Fraction	Total protein (mg)	Specific activity (nmol/mg/min)	Total activity (nmol/min)	Enrichment (-fold)	Yield (%)
Homogenate	2356.0	0.5	1178	—	100
100,000 *g* supernatant	792	1.8	1425.6	3.6	121.0
DEAE Fraction I	106	7.4	784.4	14.8	66.6
AH-Sepharose	5.9	71.0	418.9	142	35.6
Heparin-Sepharose	1.4	281.4	394	562.8	33.4
Sephacryl S-200	0.4	569.6	227.8	1139.2	19.3
Affi-Gel Blue	0.06	717.8	43.1	1435.6	3.6

[a] Enzyme activity was determined using 10 μM phosphatidylinositol as a substrate. The reaction mixture contained 2.4 mM deoxycholate, 50 mM Bis-Tris, 50 mM KCl, 1 mM $CaCl_2$ (pH 7.0).

than 1400-fold, giving a final yield of 3.6% (Table I). In keeping with the nomenclature recently established by Rhee *et al.*,[3] we refer to this enzyme as PLC-α. Neither PI-PLC Ia nor PI-PLC II are purified to homogeneity. Antibodies prepared against PI-PLC Ib cross-react with a 60K protein in the PI-PLC Ia fractions, suggesting that PI-PLC Ia and Ib are immunologically related. Several attempts were made to purify PI-PLC II to apparent homogeneity, including fast protein liquid chromatography (FPLC) ion-exchange columns, hydrophobic interaction columns, and HPLC gel filtration. At the stage of greatest purity (~500-fold), PI-PLC II consisted of three major proteins exhibiting apparent molecular weights of 56,000, 73,000, and 87,000 as demonstrated by SDS–polyacrylamide gel electrophoresis.

Purification of Phosphoinositide-Specific Phospholipase CI from Other Tissues

Using similar protocols as established for guinea pig uterus we were able to obtain PLC-α to greater than 50% purity from the rat basophilic cell line (RBL-1) and greater than 80% from the human promonocytic cell line U-937 (starting with 40–60 g of cells each). The rat and the guinea pig enzyme were indistinguishable by electrophoretic mobility and immunological properties.[5] The human enzyme migrated approximately 2K faster on SDS–polyacrylamide gels, giving an apparent molecular weight of 58K.

TABLE II
PROPERTIES OF GUINEA PIG UTERUS PHOSPHOLIPASE C-α[a]

Property	Value
Protein length	504 amino acids
Molecular weight	
SDS–PAGE	60,000
Gel filtration	58,000
mRNA	2.0 kilobases
Isoelectric point	6.7
Subcellular localization	Cytoplasmic and membrane bound
Processed amino terminus	Yes
Postranslational modifications	
Phosphorylation	Yes
Glycosylation	No
Substrate preference	
PI	K_m 11 μM; V_{max} 0.7 μmol/mg/min
PIP	K_m 40 μM; V_{max} 3.2 μmol/mg/min
PIP_2	K_m 100 μM; V_{max} 7.1 μmol/mg/min
Sensitivity to chemical modification	
N-ethylmaleimide	No
5,5-Dithiobis(2-nitrobenzoic acid)	Yes
p-Hydroxymercuribenzoate	Yes
Diethyl pyrocarbonate	Yes
Manoalide	Yes

[a] Data are compiled from Refs. 5, 7, and 8 and from C. F. Bennett and M. P. Angioli, unpublished studies, 1987.

Characterization of Purified Phosphoinositide-Specific Phospholipase CI

The biochemical properties of PLC-α are summarized in Table II.[7,8] PLC-α purified from guinea pig uterus exhibits an apparent molecular weight of 60K on SDS–polyacrylamide gels and a p*I* value of 6.7. Like other PI-PLC isoenzymes which have been described, PLC-α will hydrolyze all three phosphoinositide substrates.[5] The enzyme exhibited a higher rate of hydrolysis for PIP_2 followed by PIP then PI. The K_m for the different substrates was reversed. Hydrolysis of PI by PLC-α generates both cyclic inositol 1,2-phosphate and inositol 1-phosphate in approximately equal ratios.[9] Anionic detergents markedly enhance enzyme activity, whereas nonionic detergents inhibit activity. The enzyme is sensitive to high monovalent cation concentrations, being irreversibly inactivated by 2 *M* NaCl.

[7] C. F. Bennett, J. M. Balcarek, A. Varrichio, and S. T. Crooke, *Nature* (*London*) **334,** 268 (1988).

[8] C. F. Bennett, S. Mong, M. A. Clark, L. Wheeler, and S. T. Crooke, *Mol. Pharmacol.* **32,** 587 (1988).

[9] G. Lin, C. F. Bennett, and M.-D. Tsai, *Biochemistry* **29,** 2747 (1990).

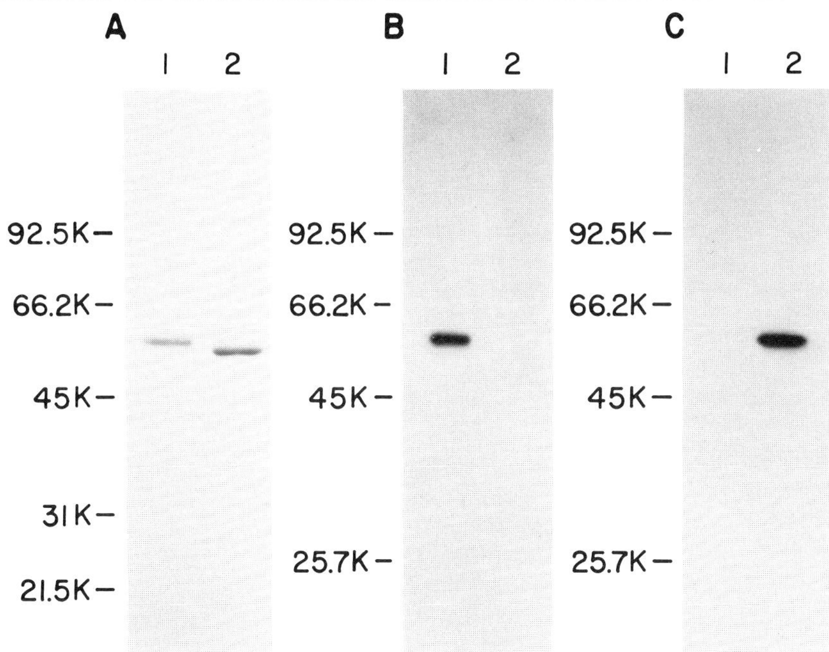

FIG. 2. SDS–PAGE and Immunoblots of PLC-α and PDI. Purified guinea pig uterus PLC-α (lane 1) and purified rat liver PDI (lane 2) were separated on an SDS–polyacrylamide gel containing 10% acrylamide. Protein was stained with Coomassie brillant blue R-250 (A) or transferred to nitrocellulose paper and probed with either PLC-α antibodies (B) or PDI antibodies (C) followed by ^{125}I-labeled goat anti-rabbit IgG. Each lane contains 1 μg protein.

Treatment of PLC-α with *N*-ethylmaleimide at concentrations as high as 30 m*M* had no effect on enzyme activity. PLC-α was inhibited by 5,5-dithiobis(2-nitrobenzoic acid), and mercurial compounds were very potent inhibitors of enzyme activity. The marine natural product manoalide, which selectively modifies lysine residues, inhibited PLC-α at concentration of 2 to 5 μM.[8]

The enzyme is localized in cytosolic fractions and associated with cellular membranes. The cellular membrane compartment with which PLC-α is associated has not been characterized. PLC-α is a substrate for protein kinase C *in vivo* and *in vitro*.[5] Phosphorylation by protein kinase C *in vivo* caused an apparent loss of activity from cell membranes. Optimal phosphorylation by protein kinase C *in vitro* was obtained in the presence of 100 n*M* calcium and PIP_2, in addition to phosphatidylserine and diolein normally used in protein kinase C assays (M. A. Angioli and C. F. Bennett, unpublished observations, 1988). The major phosphorylation site for

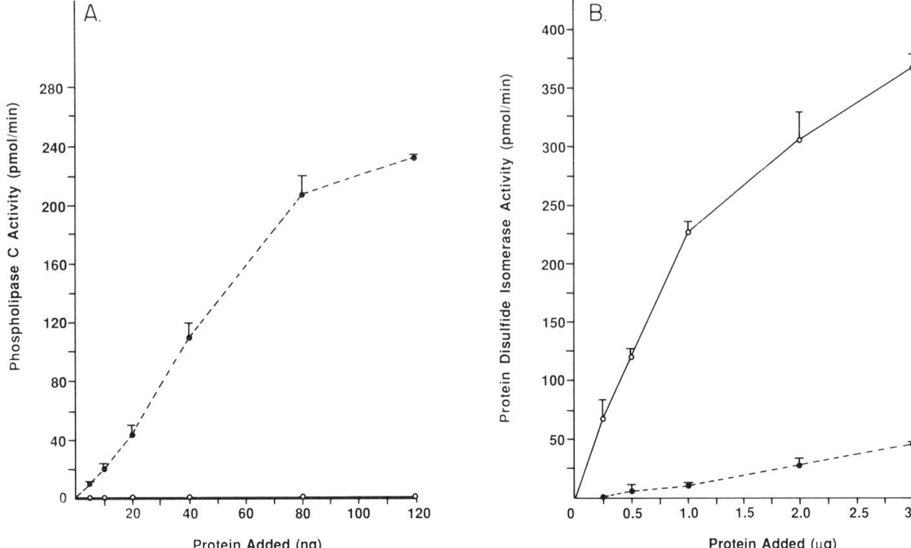

FIG. 3. Comparison of PLC-α and PDI Enzymatic Activities. (A), Phosphoinositide-specific phospholipase C activity was measured using increasing concentrations of PLC-α (●---●) or PDI (○---○). Enzyme assays were performed using 20 μM [^{3}H]PI as a substrate as described in text. (B) Protein disulfide isomerase activity was determined using increasing concentrations of PLC-α (●---●) or PDI (○—○) by measuring glutathione-insulin transhydrogenase activity as described. Values represent mean ± standard deviation (n = 3). Similar results were obtained in two separate experiments.

PLC-α by protein kinase C *in vitro* was serine-126 (M. A. Angioli, M. Strohsacker, and C. F. Bennett, unpublished data). Examination of the effect of protein phosphorylation revealed no change in catalytic rate or calcium sensitivity. Because phosphorylation by protein kinase C has no demonstrable effect on the catalytic activity of PLC-α, yet activation of protein kinase C by phorbol esters abrogate agonist-mediated phosphoinositide hydrolysis, we have proposed that phosphorylation may interrupt the ability of PLC-α to interact with either the guanine nucleotide-binding protein or some unidentified membrane component.

Molecular cloning and sequence analysis of the rat PLC-α cDNA demonstrated that PLC-α exhibited only limited sequence similarity to the other mammalian enzymes which have been cloned.[3,4,7] Remarkably, PLC-α exhibited more extensive sequence similarity with another class of intracellular proteins, namely protein disulfide isomerases.[10] Both proteins

[10] J. C. Edman, L. Ellis, R. W. Blacher, R. A. Roth, and W. J. Rutter, *Nature* (*London*) **317,** 267 (1985).

contain two domains with extensive sequence homology to thioredoxin, one at the carboxyl-terminal portion of the protein and the second at the amino terminus.[4,7,10] Examination of the relationship between PLC-α and protein disulfide isomerase revealed that antibodies prepared against rat PLC-α did not recognize protein disulfide isomerase purified from rat liver and, conversely, that antibodies prepared against protein disulfide isomerase did not recognize PLC-α (Fig. 2). Purified protein disulfide isomerase did not hydrolyze any of the phosphoinositides. However, purified PLC-α did exhibit low levels of insulin–glutathione transhydrogenase activity,[11] approximately 5% of the activity exhibited by protein disulfide isomerase (Fig. 3). The function of the thioredoxin domains in PLC-α is open to speculation at this time. It was originally thought that the cysteine residues corresponding to the active site of thioredoxin could participate in the catalytic mechanism by forming an inositol thiophosphate intermediate.[4] However, experiments examining the stereochemical course of the enzyme-catalyzed reaction with PLC-α rule out a mechanism in which the enzyme forms a covalent intermediate with the substrate.[9]

The role played by PLC-α in receptor-mediated phosphoinositide hydrolysis is currently unknown. Preliminary data suggest that PLC-α is coupled to cell surface receptors known to promote phosphoinositide hydrolysis.[12] Further characterization of PLC-α and other PI-PLC enzymes will provide a better understanding for the need for multiple distinct enzymes which apparently perform the same enzyme reaction.

[11] R. A. Roth, *Biochem. Biophys. Res. Comm.* **98,** 431 (1981).

[12] N. Aiyar, C. F. Bennett, P. Nambi, W. Valinski, M. Angioli, M. Minnich, and S. T. Crooke, *Biochem. J.* **261,** 63 (1989).

[52] Human Acid Sphingomyelinase from Human Urine

By L. E. QUINTERN and K. SANDHOFF

Introduction

Acid sphingomyelinase (EC 3.1.4.12, sphingomyelin phosphodiesterase) is a lysosomal enzyme catalyzing the hydrolysis of sphingomyelin to ceramide and phosphorylcholine.[1] In addition to the lysosomal enzyme, a neutral sphingomyelinase and other sphingomyelinases have also been described.[2–4] Deficiency of acid sphingomyelinase activity results in the accumulation of sphingomyelin as seen in Niemann–Pick disease, types A and B, which are severe lysosomal lipid storage disorders.[5,6] In order to characterize the enzyme, several purification procedures have been developed using different human tissues as starting material. The presence of acid sphingomyelinase in human urine offers the possibility of purifying the enzyme from a source which is easily obtained. The purified enzyme has a low hydrolytic activity if the assay contains pure sphingomyelin liposomes; the addition of detergents stimulates activity. The natural stimulating agent may be a small heat-stable glycoprotein as discussed by Morimoto *et al.*[7]

Assay Methods

Reagents. The incubation mixture contains (in a volume of 200 μl) 100 μM [*choline-methyl*-^{3}H]sphingomyelin from bovine brain (37×10^{10} Bq/mol); 200 mM sodium/acetate buffer (pH 4.5); 0.05% (w/v) Nonidet P-40 (NP-40) as detergent; and an aliquot of the enzyme preparation [diluted to an extent giving not more than 10% substrate utilization during the time of incubation (see below)] in 10 mM Tris-HCl buffer (pH 7.2), 0.1% (w/w) Nonidet P-40 or crude enzyme solution up to a volume of 20 μl.

[1] J. N. Kanfer, O. Young, D. Shapiro, and R. O. Brady, *J. Biol. Chem.* **241,** 1081 (1966).

[2] S. Chatterjee and N. Gosh, *J. Biol. Chem.* **264,** 12554 (1989).

[3] M. W. Spence, D. M. Byers, F. B. St.C. Palmer, and H. Cook, *J. Biol. Chem.* **264,** 5358 (1989).

[4] T. Vanha-Perttula, *FEBS Lett.* **233,** 263 (1988).

[5] R. O. Brady, J. N. Kanfer, M. B. Mock, and D. S. Fredrickson, *Proc. Natl. Acad. Sci. U.S.A.* **55,** 366 (1966).

[6] P. B. Schneider and E. P. Kennedy, *J. Lipid Res.* **8,** 202 (1967).

[7] S. Morimoto, B. M. Martin, Y. Kishimoto, and J. S. O'Brien, *Biochim. Biophys. Res. Commun.* **156,** 403 (1988).

METHODS IN ENZYMOLOGY, VOL. 197

Procedure. The assay should be carried out as follows. [^{3}H]Sphingomyelin dissolved in 20 μl toluene/ethanol (2 : 1, v/v), or in another organic solvent, should be added to each test tube and the solvent evaporated in a stream of nitrogen at room temperature. After adding the buffer and detergent, the solution should be mixed (e.g., for 5 sec with a Vortex Genie Mixer) and then cooled to 0° on ice; the enzyme can now be added. The reaction (e.g., 60 min at 37° in a shaking water bath) can be stopped by the addition of 800 μl chloroform/methanol (2 : 1, v/v) and 100 μl of 0.1% (w/w) unlabeled phosphorylcholine in water. After mixing the two phases can be separated by centrifugation (i.e., 5 min in an uncooled tabletop centrifuge). The upper water phase contains [^{3}H]phosphorylcholine released by the enzyme. In order to measure enzyme activity 100 μl of the upper water layer is removed and mixed with scintillation fluid, and the radioactivity measured in a scintillation counter. Appropriate blanks should be carried out with denatured enzyme (15 min, 95°).

Purification Procedure

The purification steps are carried out at 4° unless otherwise stated. The urine of certain male or female hospital patients contains much higher amounts of acid sphingomyelinase than normal controls. Apart from an increase in the excretion of the enzyme into the urine of patients with peritonitis and some other diseases,[8] there is often a greatly enhanced excretion of other proteins as well. For successful enzyme purification only urine with slightly elevated protein content but a highly elevated level of sphingomyelinase activity should be used. Urine should be collected daily. After the addition of sodium acetate buffer (20 m*M*; pH 5.0) and sodium azide (0.01%), the urine can be stored for several days; pH 5.0 is chosen because the enzyme in the undialyzed urine is stable at this pH.

Table I summarizes the data for a purification of the enzyme from the urine of patients with peritonitis. The purification is done as follows: the clear supernatant from stored urine is concentrated by ultrafiltration [i.e., 50 liters is reduced to 1 liter with a Millipore (Bedford, MA) ultrafiltration system using a cassette filter Type PTGC MG 10,000]. After dialysis (3 × 24 hr, 3 × 10 liters buffer) against 20 m*M* sodium acetate (pH 5.0)/50 m*M* NaCl/0.01% NaN_3 and dialysis for 24 hr against 10 liters 10 m*M* Tris-HCl buffer (pH 7.2), 100 m*M* NaCl/0.01% NaN_3 (buffer A), the solution is clarified by centrifugation (1 hr, 10,000 *g*).

Step 1: Octyl-Sepharose Chromatography. The supernatant is then loaded onto an octyl-Sepharose column previously equilibrated with buffer

[8] L. E. Quintern, T. S. Zenk, and K. Sandhoff, *Biochim. Biophys. Acta* **1003,** 121 (1989).

TABLE I
PURIFICATION OF ACID SPHINGOMYELINASE FROM URINE OF PATIENTS WITH PERITONITIS[a]

Step	Volume (ml)	Protein (mg)	Activity (μmol/hr)	Specific activity (μmol/hr/mg)	Yield (%)	Enrichment (-fold)
Urine concentrate (from 49 liters)	1150	19,750	61,000	2	100	—
Octyl-Sepharose	400	760	24,900	33	41	15
Concanavalin A-Sepharose	460	37	28,600	777	47	353
Blue-Sepharose	63	3	16,000	5550	26	2523
DEAE-cellulose	68	2	13,500	6900	22	3136

[a] Acid sphingomyelinase activity was measured in a test volume of 200 μl which contained 200 m*M* sodium acetate buffer, pH 4.5, 50 μM [^{3}H]sphingomyelin, and 0.05% Nonidet P-40. Data from L. E. Quintern, T. S. Zenk, and K. Sandhoff, *Biochim. Biophys. Acta* **1003,** 121 (1989).

A. Unbound proteins are removed with at least 800 ml of buffer A. The bound sphingomyelinase elutes in high yield (~40%) with 300 ml of 1% (w/v) Nonidet P-40 in a buffer containing 10 m*M* Tris-HCl (pH 7.2)/100 m*M* NaCl/1 m*M* $CaCl_2$/1 m*M* $MgCl_2$/1 m*M* $MnCl_2$/0.01% NaN_3. The flow rate should not be higher than 30 ml/hr during loading, washing, and elution; pH 7.2 is chosen because in the presence of Nonidet P-40 the enzyme is most stable at this pH. During loading, washing, and elution the column should always be run in such a way that solutions of higher density enter the column from the bottom and solutions of lower density from the top; these precautions should be followed during every chromatographic step.

Step 2: Concanavalin A-Sepharose Chromatography. The octyl-Sepharose eluate containing the enzyme is loaded onto a concanavalin A-Sepharose column (1.6 × 24 cm) equilibrated with 10 m*M* Tris-HCl buffer (pH 7.2), 100 m*M* NaCl, 1 m*M* $CaCl_2$, 1 m*M* $MgCl_2$, 1 m*M* $MnCl_2$, 0.01%, NaN_3, and 0.1% Nonidet P-40 (buffer B); the column should be washed with at least 200 ml of buffer B. We have had good results by performing a further washing step with 200 ml of buffer B containing 0.2% (w/w) α-methyl-D-glucopyranoside, followed by at least another 200 ml of buffer B. The enzyme is eluted in good yield at room temperature with 15% (w/w) α-methyl-D-glucopyranoside in buffer B using a fraction collector; the flow rate should not exceed 10, 30, and 2 ml/hr during loading, washing, and elution, respectively. It is important to maintain the flow direction in the column. In this way the first fractions of the eluate contain the bulk of contaminating proteins and the acid sphingomyelinase elutes later. Fractions with the highest specific activity are pooled and dialyzed against 6 changes (6 × 12 hr) of 4 liters each of 30 m*M* Tris-HCl buffer (pH 7.2), 0.01% NaN_3, 0.1% Nonidet P-40 (buffer C).

Step 3: Blue-Sepharose Chromatography. The dialyzed fractions are loaded onto a freshly prepared Blue-Sepharose column (1.6 × 8 cm) equilibrated with buffer C. After washing with 200 ml buffer C the bound enzyme is eluted with a linear sodium chloride gradient in buffer C (0–1 *M* NaCl; total volume 50 + 50 ml); the flow rate during loading, washing, and elution should be 3, 30, and 2 ml/hr, respectively. Most of the enzyme elutes in the presence of around 500 m*M* NaCl. Fractions with the highest specific activities are pooled and then dialyzed against 6 changes of 2 liters of 40 m*M* Tris-HCl (pH 7.6), 0.1% NP-40 (buffer D).

Step 4: DEAE-Cellulose Chromatography. The dialyzed fractions of the Blue-Sepharose eluate are applied to a DEAE-cellulose column (1 × 2 cm) equilibrated with buffer D. Under these conditions most of the enzyme does not bind to the column, but most of the contaminating proteins do.

Additional Remarks

In order to obtain pure acid sphingomyelinase protein, the enzyme (post-DEAE-cellulose) should be concentrated and separated from the detergent Nonidet P-40 by ethanol precipitation, alkylated with iodoacetamide,[9] and purified on a hydrophobic HPLC column (e.g., HIBAR Lichrospher, 500 CH-2/10 μm) with a linear gradient of acetonitrile (0–70%, v/v) in 0.05% trifluoroacetic acid. This procedure gives denatured but pure enzyme protein (e.g., for immunochemical experiments).

Properties

The purified glycosylated enzyme has an apparent molecular mass of 72 kDa determined with SDS–gel electrophoresis, which was reduced to 61 kDa after deglycosylation. It has an isoelectric point (p*I*) of 7.5, and the optimum for enzyme activity occurs in the range pH 4.5–5.0. Enzyme activity can be stimulated with Nonidet P-40, various cholates, or by the addition of certain lipids to a detergent-free assay system.[10] Under some conditions the freshly prepared enzyme hydrolyzes phosphatidylcholine and phosphatidylglycerol with phospholipase C-like activity.[10] The en-

[9] J. Hempel, H. v. Bahr-Lindström, and H. Jörnvall, *Eur. J. Biochem.* **141,** 21 (1984).

[10] L. E. Quintern, G. Weitz, H. Nehrkorn, J. M. Tager, A. W. Schram, and K. Sandhoff, *Biochim. Biophys. Acta* **922,** 323 (1987).

zyme has an apparent K_m value of around 60 μM toward the three lipids sphingomyelin, phosphatidylcholine, and phosphatidylglycerol in the presence of Nonidet P-40, sodium taurodeoxycholate, and Nonidet P-40, respectively. Reducing agents (e.g., dithiothreitol) inactivate the enzyme. Divalent cations (Ca^{2+}, Co^{2+}, Mg^{2+}, Mn^{2+}) or the cation-chelating agent EDTA up to a concentration of 20 mM have no effect on enzyme activity. AMP (K_i 45–74 μM), 9-β-D-arabinofuranosyladenine 5′-monophosphate (K_i ~4 μM) and phosphatidylinositol 4′,5′-bisphosphate (K_i ~1 μM) are potent inhibitors of acid sphingomyelinase. In the presence of 0.1% Nonidet P-40 and greater than 100 mM NaCl the enzyme can be stored for several weeks with only a small loss in total hydrolytic activity.

Additional Remarks

The cDNA of the enzyme has been cloned.[11] The occurrence of alternatively processed transcripts indicates the presence of isoenzymes of this enzyme.

Acknowledgments

We are grateful to Roger Klein for proofreading the manuscript and the Deutsche Forschungsgemeinschaft for financial support (Grant Sa 257/12-1).

[11] L. E. Quintern, E. H. Schuchmann, O. Levran, M. Suchi, K. Ferlinz, H. Reinke, K. Sandhoff, and R. J. Desnick, *EMBO J.* **8,** 2469 (1989).

[53] Purification of Neutral Sphingomyelinase from Human Urine

By SUBROTO CHATTERJEE and NUPUR GHOSH

Introduction

Sphingomyelinase (EC 3.1.4.12) catalyzes the hydrolytic cleavage of sphingomyelin via reaction (1).[1] It has been shown that sphingomyelin may be cleaved at both acid pH optima and neutral pH optima.[2–4] The lack

[1] R. O. Brady, J. N. Kanfer, M. B. Moek, and D. S. Fredrickson, *Proc. Natl. Acad. Sci. U.S.A.* **55,** 366 (1966).
[2] P. Ghosh and S. Chatterjee, *J. Biol. Chem.* **262,** 12550 (1987).
[3] S. Gatt, *Biochem. Biophys. Res. Commun.* **68,** 235 (1976).
[4] B. G. Rao and M. W. Spence, *J. Lipid Res.* **17,** 506 (1976).

$$\text{Sphingomyelin} \rightarrow \text{ceramide} + \text{phosphocholine} \quad (1)$$

or deficiency of the acid sphingomyelinase (a lysosomal enzyme) leading to the storage of sphingomyelin in Niemann–Pick disease is well established.[1] Indirect evidence suggests that a sphingomyelinase (having a neutral pH optima) may be involved in catabolizing cell surface sphingomyelin.[5] Ceramide released from such a reaction is speculated to be involved in the generation of sphingosine (via the action of ceramidase), and the sphingosine may be subsequently involved in the cascade of reactions leading to the regulation of protein kinase activity.[6,7] We have shown that decreased activity of neutral sphingomyelinase (N–SMase) and acid sphingomyelinase *in vitro* in cultured human proximal tubular cells incubated with gentamicin and *in vivo* in patients receiving gentamicin, leads to the accumulation of sphingomyelin.[8,9]

Assay Method

The activity of neutral sphingomyelinase is measured, using [^{14}C]sphingomyelin as the substrate, by a modification of the procedure of Brady *et al.*[1] The water-soluble product phosphocholine is separated from [^{14}C] sphingomyelin following trichloroacetic acid (TCA) precipitation of the mixture and extraction with ether.

Reagents

[^{14}C]Sphingomyelin, purchased from Amersham Searle (Arlington Heights, IL), (specific activity 57 mCi/mmol)
0.2 *M* Tris–glycine buffer (pH 7.4)
Human serum albumin (fatty acid-free)
$MgCl_2$
Cutscum (a detergent available through Fisher Scientific Company, Pittsburgh, PA)
10% Trichloroacetic acid (TCA)
Bovine serum albumin (BSA)
Ether (anhydrous)
Aquasol II (New England Nuclear, Boston, MA)

[5] C. W. Slife, E. Wang, R. Hunter, S. Wang, C. Burgess, D. C. Liotta, and A. H. Merrill, Jr., *J. Biol. Chem.* **264,** 10371 (1989).

[6] Y. A. Hannun, C. R. Loomis, A. H. Merrill, Jr., and R. M. Bell, *J. Biol. Chem.* **261,** 12604 (1986).

[7] Y. A. Hannun and R. M. Bell, *Science* **2,** 500 (1989).

[8] S. Chatterjee, *J. Biochem. Toxicol.* **2,** 181 (1987).

[9] S. Chatterjee and S. Bose, *J. Biochem. Toxicol.* **3,** 47 (1988).

Unless otherwise mentioned all other chemicals described here are available from Sigma Chemical Company (St. Louis, MO).

Procedure. The assay mixture for the measurement of neutral sphingomyelinase activity consists of the following: 25 nmol of Tris–glycine buffer, pH 7.4, 2.5 nmol of Mg^{2+}, 50 nmol of [^{14}C]sphingomyelin [20,000 disintegrations per minute (dpm)], 0.5 mg of human serum albumin, 50 μg of Cutscum, and 10–50 μg of enzyme protein. The final volume of the assay mixture is adjusted to 200 μl with water, and incubation is carried out for 1 hr at 37°. The assay is terminated with 1 ml of ice-cold TCA. The contents of the tubes are mixed, allowed to settle for 5 min at room temperature, and then centrifuged for 5 min at 2000 rpm in a swinging bucket (IEC, bench top centrifuge) at 10°. One milliliter of the supernatant is carefully withdrawn using disposable plastic Eppendorf tips and transferred to a separate set of glass test tubes (13 × 100 mm).

One milliliter of ether is added to the samples mixed, allowed to settle for 5 min on ice, and centrifuged as above. The aqueous layer (~800 μl, maximum) is withdrawn and transferred into glass scintillation vials. Such samples are dried overnight, or the radioactivity is measured directly after mixing with 10 ml of Aquasol II in a scintillation spectrometer (Beckman LS-3801).

Product Identification. Following enzyme assay, the products are analyzed by thin-layer chromatography. The aqueous phase containing [^{14}C]phosphocholine is mixed with unlabeled phosphocholine and separated by thin-layer chromatography using methanol, 0.5% NaCl, and NH_4OH (100 : 100 : 2, v/v). The gel area corresponding to phosphocholine is scraped and the radioactivity measured in a scintillation spectrometer.

Neutral Sphingomyelinase Purification

We have developed the followign steps to purify N-SMase from urine concentrates.[10] To isolate N-SMase from tissues,[10] step I below is replaced by subcellular fractionation and isolation of plasma membranes[2] that serve as the starting material.

Reagents

Sephadex G-75
Biolyte (Bio-Rad, Richmond, CA)
CH-Sepharose
Sodium acetate buffer: 25 m*M* NaCl, 1 m*M* EDTA, 20 m*M* sodium acetate, pH 6.5

[10] S. Chatterjee and N. Ghosh, *J. Biol. Chem.* **264,** 12554 (1989).

Phenylmethylsulfonyl fluoride
Sodium azide
Glycine
Sphingosylphosphocholine
Phenyl-Sepharose CL-4B

Step I: Preparation of Urine Concentrate. We collect about 2–3 liters of normal urine from male subjects daily. Male volunteers are preferred over females because of lack of contamination by blood cells, excessive number of uroepithelial cells, and urogenital secretions. For best results, protease and other inhibitors should be added to the container and to all the purification buffers, namely 10 μM phenylmethylsulfonyl fluoride (PMSF) and 0.001% sodium azide. Without such inhibitors, the enzyme is susceptible to extensive proteolytic cleavage. The urine sample is immediately centrifuged at 16,270 *g* (30 min at 4°) in a Sorvall RC-2B refrigerated centrifuge. The cell pellet, a transparent layer at the bottom, is carefully removed, the supernatant is concentrated in a Millipore Minitan ultrafiltration device (Bedford, MA), using a filter (PTGC-OMT-05) with a nominal cutoff of M_r 10,000. Next, the urine concentrate is further concentrated to 5–10 ml in an Amicon (Danvers MA) YM10 membrane filter with a nominal cutoff of M_r 10,000. The filters are washed with 1% NaOH and water and reused up to several weeks. Fresh filters are used daily for ultrafiltration purposes. The urine concentrate is dialyzed against 2 liters of sodium acetate buffer (25 mM NaCl, 1 mM EDTA, 20 mM sodium acetate, pH 6.5) for 24 hr. Urine concentrate from at least 10 liters of urine is used for the isolation of N-SMase.

Step II: Chromatography on Sephadex G-75. The urine concentrate is applied to a Sephadex G-75 column (2.5 × 34 cm) preequilibrated with the acetate buffer above and eluted. First, absorbance at 280 nm is measured in all the fractions. The positive samples are used for protein and enzyme activity determinations. For example, fractions 8–10 in Fig. 1 containing high N-SMase activity are pooled and concentrated by ultrafiltration.

Step III: Preparative Isoelectric Focusing. The concentrate (30 mg protein) obtained following Sephadex G-75 column chromatography is dialyzed against glycine and placed in a Biolyte gel bed containing carrier ampholytes of pH range 4–10 using a sample applicator provided by the manufacturer (LKB Instrument Co.). Electrofocusing is carried out at constant power of 8 W for 14–16 hr. To visualize the focused zones a print is made of the gel on a nitrocellulose paper and stained with Coomassie blue. The focused zones are cut out, suspended in water, and the pH measured. Since the p*I* value of N-SMase is approximately 6.5,[10] gel areas corresponding to this pH range (fractions 12–15 in Fig. 2) are transferred to a Pasteur pipette column and eluted with 0.2 M Tris-glycine buffer, pH

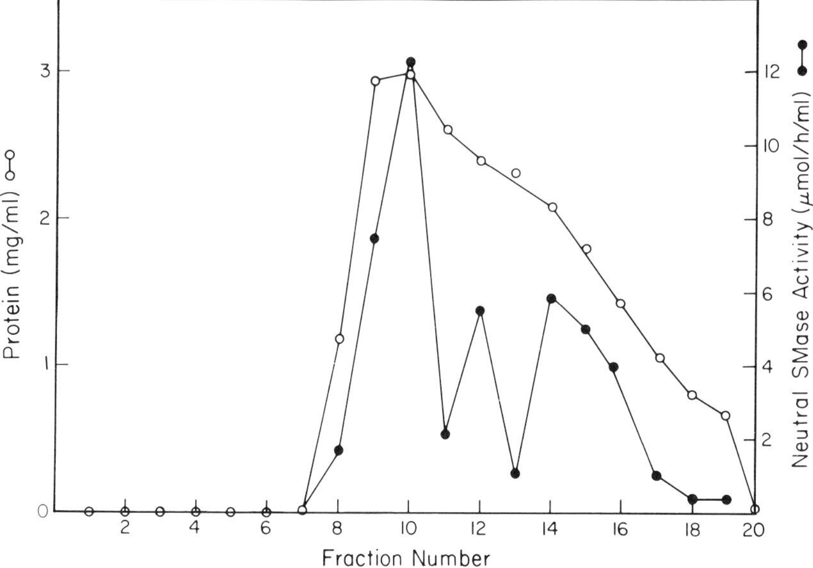

FIG. 1. Sephadex G-75 chromatography of neutral sphingomyelinase from human urine. Approximately 246 mg of urine concentrate was applied to a Sephadex G-75 column (2.5 × 34 cm) preequilibrated with a buffer containing 20 m*M* sodium acetate, 25 m*M* NaCl, 1 m*M* EDTA (pH 6.5). The column was washed with the same buffer at a flow rate of 25 ml/hr, and 5-ml fractions were collected in a Pharmacia (Piscataway, NJ) fraction collector (FRAC-100). All operations were carried out at 4°. The content of protein and neutral sphingomyelinase activity were measured in the fractions. [Reproduced from S. Chatterjee and N. Ghosh, *J. Biol. Chem.* **264,** 12554 (1989).]

7.4. The sample is concentrated and the ampholytes removed on columns loaded with Phenyl-Sepharose CL-4B (1 × 4 cm) by sequential elution with 0.2 *M* Tris–glycine buffer and 0.1–1 *M* phosphate. Samples are withdrawn for the measurement of protein and enzyme activity. Alternatively, we have developed a relatively simple method to pursue preparative isoelectric focusing of N-SMase using the Rotofor (Bio-Rad, Richmond, CA) equipment. The concentrate (~50 ml) obtained following ultrafiltration (step II above) is mixed with ampholytes (2%, w/v), pH range 4–10, and subjected to isoelectric focusing at a constant power of 12 W with initial voltage of 400 V. The total time required is approximately 4–5 hr.

The latter procedure of preparative isoelectric focusing has several advantages over the Biolyte gel isoelectric focusing method. (1) The samples are in a liquid state; accordingly, the step involving elution of protein from the Biolyte gel bed is not required. (2) pH measurements and ampho-

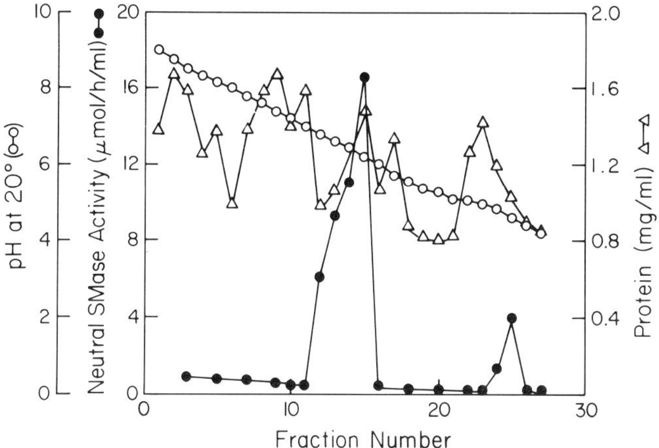

FIG. 2. Preparative isoelectric focusing of neutral sphingomyelinase from human urine. A portion of the pooled post-Sephadex G-75 fractions was dialyzed against 1% glycine, placed (30 mg protein) in a gel bed of Biolyte containing carrier ampholytes of pH range 4–10, and subjected to electrofocusing as described in the text. The protein content and N-SMase activity were measured. [Reproduced from S. Chatterjee and N. Ghosh, *J. Biol. Chem.* **264,** 12554 (1989).]

lyte removal are easy. (3) Increased yields of protein are obtained. (4) The method is highly reproducible and (5) less time consuming.

Step IV: Chromatography on Sphingosylphosphocholine CH-Sepharose. Partially purified N-SMase, obtained following the above procedure, is dialyzed against 0.2 *M* Tris–glycine buffer, pH 7.4, and applied (5–10 mg protein) to a column of sphingosylphosphocholine-CH–Sepharose[11] preequilibrated with 0.2 *M* Tris–glycine buffer, pH 7.4. The column is sequentially washed with 5 bed volumes of Tris–glycine buffer and 10 bed volumes of Tris–glycine buffer containing 0.1 *M* NaCl. Next, the column is eluted with the above buffer containing 0.1% Cutscum. Protein and enzyme activity are measured in all the fractions. Fractions 21–24 in Fig. 3 containing most of the N-SMase activity are pooled, dialyzed against the 0.2 *M* Tris–glycine buffer, pH 7.4, and the activity of N-SMase and protein are measured. Such enzyme preparations are further characterized and used for various biochemical studies. By this procedure, the enzyme is purified approximately 440-fold and is almost pure as judged by polyacrylamide gel electrophoresis (Fig. 4, lane 2). It resolves as a single peak on reversed-phase HPLC.

[11] C. S. Jones, P. Shankaran, and J. W. Callahan, *Biochem. J.* **195,** 373 (1981).

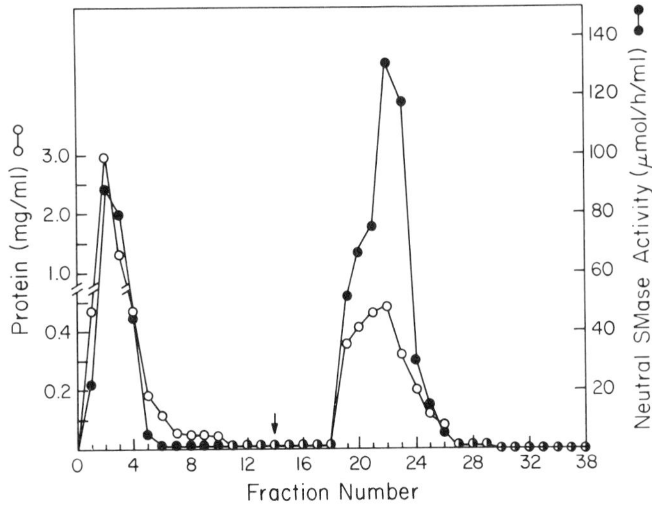

FIG. 3. Sphingosylphosphocholine CH-Sepharose chromatography of neutral sphingomyelinase from human urine. Pooled postpreparative isoelectric focusing fractions suspended in 0.2 *M* Tris–glycine buffer (pH 7.4) were applied to a column (1.0 × 7.0 cm) of sphingosylphosphocholine CH-Sepharose. The column was preequilibrated with the same buffer. Next, the column was washed with 5 bed volumes of 0.2 *M* Tris–glycine buffer (pH 7.4) containing 0.1 *M* NaCl, and 1-ml fractions were collected. The column was then eluted with the above buffer containing 0.1% Cutscum (indicated by arrow). Samples were dialyzed extensively, and the protein content and N-SMase activity were measured. [Reproduced from S. Chatterjee and N. Ghosh, *J. Biol. Chem.* **264,** 12554 (1989).]

General Properties

pH Optima, pI Value, and Size. Using [^{14}C]sphingomyelin as substrate, the maximal enzyme activity is at pH 7–9 and optimum pH activity at 7.4. The enzyme is composed of a single polypeptide whose apparent molecular weight is 92,000 (Fig. 4), with a p*I* of 6.5.

Metal Ion and Detergent Requirements. N-SMase requires Mg^{2+} and detergents (Cutscum but not Triton X-100) and 37° for optimum activity.

Inhibitors and Activitors. The activity of the enzyme is unaffected by adenosine 5′-monophosphate. It is susceptible to treatment with mild acids such as trifluoroacetic acid. In addition, heating at 50° for 1 hr inactivates the enzyme.

Stability and Storage. The enzyme is stable when stored in a liquid state for over 6 weeks and in a frozen state for several months.

Other Enzyme Activities Associated with N-SMase. The purified N-SMase is free of contaminating enzymes such as β-galactosidase, β-*N*-acetylglucosaminidase, and acid sphingomyelinase. It has low levels of

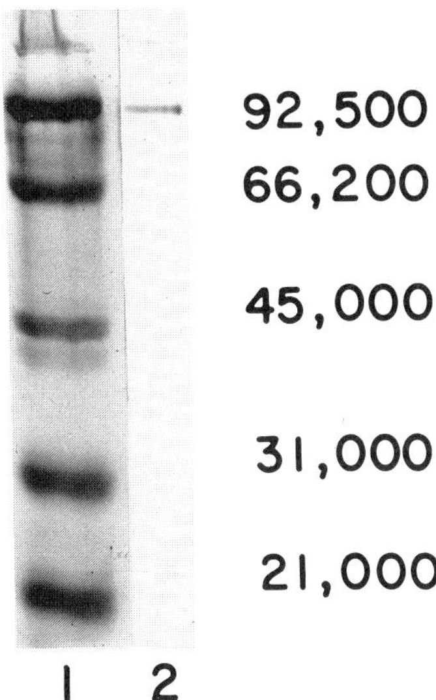

FIG. 4. Polyacrylamide gel electrophoresis of purified neutral sphingomyelinase from human urine. The samples were reduced, alkylated, loaded on the gel, and subjected to electrophoresis for 2 hr at room temperature; standard proteins (Lane 1) coelectrophoresed were phosphorylase *b* (M_r 92,500), bovine serum albumin (M_r 66,200), ovalbumin (M_r 45,000), carbonic anahydrase (M_r 31,000), soybean trypsin inhibitor (M_r 21,000), and lysozyme (M_r 14,400). Lane 2 contained 3 μg of purified neutral sphingomyelinase. [Reproduced from S. Chatterjee and N. Ghosh, *J. Biol. Chem.* **264,** 12554 (1989).]

phospholipase A_1 and A_2 activity when [^{14}C]phosphatidylcholine is used as a substrate and Nonidet P-40 and sodium taurodeoxycholate were added to the assay mixture. The enzyme is inactive toward phosphatidylglycerol and sphingomyelin at acid pH.

Acknowledgments

This work was supported by a National Institutes of Health grant (RO-AM-34657 to S.C.) and a Virginia L. Luette Memorial Fellowship of the American Heart Association, Maryland Affiliate (N.G.).

[54] Phosphatidate Phosphatase from Yeast

By GEORGE M. CARMAN and YI-PING LIN

Introduction

Phosphatidate phosphatase (EC 3.1.3.4; 3-*sn*-phosphatidate phosphohydrolase) catalyzes reaction (1).[1] The enzyme was first identified in the

$$\text{Phosphatidate} \rightarrow \text{diacylglycerol} + P_i \quad (1)$$

yeast *Saccharomyces cerevisiae* by Hosaka and Yamashita.[2] Phosphatidate phosphatase activity is associated with the membrane and cytosolic fractions of the cell.[2,3] The membrane-associated phosphatidate phosphatase has been purified to apparent homogeneity by standard protein purification procedures.[4] We describe here the purification and properties of the enzyme.

Preparation of Substrates

[γ-^{32}P]Phosphatidate is synthesized enzymatically from [γ-^{32}P]ATP and diacylglycerol using *Escherichia coli* diacylglycerol kinase (Lipidex, Inc., Westfield, NJ) under the assay conditions described elsewhere.[5] [2-^{3}H]Phosphatidate is prepared from [2-^{3}H]glycerol 3-phosphate and palmitoyl-CoA using yeast microsomal-associated glycerol-3-phosphate acyltransferase under the assay conditions described previously.[6] Labeled phosphatidate is purified by thin-layer chromatography using the solvent system chloroform/methanol/water (65 : 25 : 4, v/v).

Assay Methods

Phosphatidate phosphatase activity is routinely measured by following the release of water-soluble [^{32}P]P_i from 0.5 m*M* [γ-^{32}P]phosphatidate [1000–2000 counts per minute (cpm)/nmol] in 50 m*M* Tris–maleate buffer (pH 7.0) containing 5 m*M* Triton X-100, 10 m*M* 2-mercaptoethanol, 2 m*M* $MgCl_2$, and enzyme protein in a total volume of 0.1 ml at 30°.[3] The

[1] M. Kates, *Can. J. Biochem.* **35,** 575 (1955).
[2] K. Hosaka and S. Yamashita, *Biochim. Biophys. Acta* **796,** 102 (1984).
[3] K. R. Morlock, Y.-P. Lin, and G. M. Carman, *J. Bacteriol.* **170,** 3561 (1988).
[4] Y.-P. Lin and G. M. Carman, *J. Biol. Chem.* **264,** 8641 (1989).
[5] J. P. Walsh and R. M. Bell, *J. Biol. Chem.* **261,** 6239 (1986).
[6] T. S. Tillman and R. M. Bell, *J. Biol. Chem.* **261,** 9144 (1986).

reaction is terminated by the addition of 0.5 ml of 0.1 *N* HCl in methanol. Chloroform (1 ml) and 1 *M* $MgCl_2$ (1 ml) are added, the system is mixed, and the phases are separated by a 2-min centrifugation at 100 *g*. Four milliliters of Liquiscint (National Diagnostics, Manville, NJ) is added to a 0.5-ml aliquot of the aqueous phase, and radioactivity is determined by scintillation counting. Alternatively, phosphatidate phosphatase activity is measured by following the formation of [2-^{3}H]diacylglycerol from [2-^{3}H]phosphatidate (1000 cpm/nmol) under the assay conditions described above. The chloroform-soluble lipid product of the reaction diacylglycerol is analyzed with standard diacylglycerol by one-dimensional paper chromatography on EDTA-treated SG81 (Whatman, Clifton, NJ) paper[7] using the solvent system hexane–diethyl ether–glacial acetic acid (30 : 70 : 1, v/v). Carrier standard diacylglycerol is added to the chloroform phase prior to separation, and the position of the labeled diacylglycerol on the chromatograms is determined after exposure to iodine vapor. The amount of labeled diacylglycerol is determined by liquid scintillation counting. One unit of enzymatic activity is defined as the amount of enzyme that catalyzes the formation of 1 μmol of product per minute.

Growth of Yeast

Saccharomyces cerevisiae strain *ade5 MAT***a** is used as a representative wild-type strain[8,9] for enzyme purification. Cells are grown in 1% yeast extract, 2% peptone, and 2% glucose at 28° to late exponential phase, harvested by centrifugation, and stored at $-80°$ as described previously.[10]

Purification Procedure

All steps are performed at 5°–8°.

Step 1: Preparation of Cell Extract. Cells (85 g) are disrupted with glass beads (diameter 0.5 mm) with a BioSpec Products (Bartlesville, OK) Bead-Beater in 50 m*M* Tris–maleate buffer (pH 7.0) containing 1 m*M* disodium EDTA, 0.3 *M* sucrose, and 10 m*M* 2-mercaptoethanol as described previously.[10] Glass beads and unbroken cells are removed by centrifugation at 1500 *g* for 5 min to obtain the cell extract.

Step 2: Preparation of Total Membranes. Total membranes are collected from the cell extract by centrifugation at 100,000 *g* for 90 min.

[7] S. Steiner and R. L. Lester, *J. Bacteriol.* **109,** 81 (1972).

[8] L. S. Klig, M. J. Homann, G. M. Carman, and S. A. Henry, *J. Bacteriol.* **162,** 1135 (1985).

[9] M. L. Greenberg, L. S. Klig, V. A. Letts, B. S. Loewy, and S. A. Henry, *J. Bacteriol.* **153,** 791 (1983).

[10] A. S. Fischl and G. M. Carman, *J. Bacteriol.* **154,** 304 (1983).

Membrane pellets are washed with 50 m*M* Tris–maleate buffer (pH 7.0) containing 10 m*M* $MgCl_2$, 10 m*M* 2-mercaptoethanol, and 20% glycerol. Membranes are routinely frozen at −80° until the enzyme is purified.

Step 3: Preparation of Sodium Cholate Extract. Membranes are suspended in 50 m*M* Tris–maleate buffer (pH 7.0) containing 10 m*M* $MgCl_2$, 10 m*M* 2-mercaptoethanol, 20% glycerol, and 1% sodium cholate at a final protein concentration of 10 mg/ml. The suspension is incubated for 1 hr on a rotary shaker at 150 rpm. After the incubation, the suspension is centrifuged at 100,000 *g* for 90 min to obtain the sodium cholate extract (supernatant).

Step 4: DE-53 Chromatography. A DE-53 (Whatman) column (2.5 × 15 cm) is equilibrated with 5 column volumes of 50 m*M* Tris–maleate buffer (pH 7.0) containing 10 m*M* $MgCl_2$, 10 m*M* 2-mercaptoethanol, and 20% glycerol followed by 1 column volume of the same buffer containing 1% sodium cholate. It is important not to saturate the top portion of the column with sodium cholate since this prevents the binding of phosphatidate phosphatase. The sodium cholate extract is applied to the column at a flow rate of 30 ml/hr. The column is washed with 4 column volumes of equilibration buffer containing 1% sodium cholate followed by elution of the enzyme with 10 column volumes of a linear NaCl gradient (0–0.3 *M*) in the same buffer. The peak of phosphatidate phosphatase activity elutes from the column at a NaCl concentration of about 0.1 *M*. Fractions containing the most activity are pooled and used for the next step in the purification scheme.

Step 5: Affi-Gel Blue Chromatography. An Affi-Gel Blue (Bio-Rad, Richmond, CA) column (1.5 × 10 cm) is equilibrated with 5 column volumes of 50 m*M* Tris–maleate buffer (pH 7.0) containing 10 m*M* $MgCl_2$, 10 m*M* 2-mercaptoethanol, 20% glycerol, and 0.1 *M* NaCl followed by equilibration with 1 column volume of the same buffer containing 1% sodium cholate. The equilibration of the column with buffer containing 0.1 *M* NaCl eliminates the need to desalt the enzyme preparation from the previous step. The DE-53-purified enzyme is applied to the column at a flow rate of 30 ml/hr. The column is washed with 3.5 column volumes of equilibration buffer containing 0.3 *M* NaCl and 1% sodium cholate. Phosphatidate phosphatase is then eluted from the column with 9 column volumes of a linear NaCl gradient (0.3–1.0 *M*) in the same buffer at a flow rate of 30 ml/hr. The peak of phosphatidate phosphatase activity elutes from the column at a NaCl concentration of about 0.6 *M*. The most active fractions are pooled, and the enzyme preparation is desalted by dialysis against equilibration buffer.

Step 6: Hydroxylapatite Chromatography. A hydroxylapatite (BioGel HT, Bio-Rad) column (1.5 × 7 cm) is equilibrated with 10 m*M* potassium phosphate buffer (pH 7.0) containing 5 m*M* $MgCl_2$, 10 m*M* 2-mercaptoethanol, 20% glycerol, and 1% sodium cholate. Dialyzed enzyme from the previous step is applied to the column at a flow rate of 15 ml/hr. The column is washed with 2 column volumes of equilibration buffer followed by elution of phosphatidate phosphatase with 8 column volumes of a linear potassium phosphate gradient (0.01–0.15 *M*) in equilibration buffer. The concentration of $MgCl_2$ is increased to 10 m*M* in the elution buffer. The peak of activity elutes from the column at a potassium phosphate concentration of about 70 m*M*. The most active fractions are pooled and used for the next step in the purification.

Step 7: Mono Q Chromatography. A anion-exchange Mono Q (Pharmacia LKB Biotechnology, Piscataway, NJ) column (0.5 × 5 cm) is equilibrated with 6 column volumes of 50 m*M* Tris–maleate buffer (pH 7.0) containing 10 m*M* $MgCl_2$, 10 m*M* 2-mercaptoethanol, 20% glycerol, 1% sodium cholate, and 0.1 *M* NaCl. The hydroxylapatite-purified enzyme is applied to the column at a flow rate of 30 ml/hr. The column is washed with 8 column volumes of equilibration buffer followed by 2 column volumes of a linear NaCl gradient (0.1–0.17 *M*) in equilibration buffer. The enzyme is then eluted from the column with 10 column volumes of a linear NaCl gradient (0.17–0.4 *M*) in equilibration buffer. The peak of enzyme activity elutes at a NaCl concentration of about 0.22 *M*. Fractions containing activity are pooled and used for the next step.

Step 8: Superose 12 Chromatography. A gel-filtration Superose 12 (Pharmacia LKB Biotechnology) column (1 × 30 cm) is equilibrated with 50 m*M* Tris–maleate buffer (pH 7.0) containing 10 m*M* $MgCl_2$, 10 m*M* 2-mercaptoethanol, 20% glycerol, and 1% sodium cholate. The enzyme from the previous Mono Q column is applied to the column at a flow rate of 24 ml/hr followed by elution of the enzyme from the column with equilibration buffer. Phosphatidate phosphatase activity and protein elutes from the column as a single peak. Fractions containing phosphatidate phosphatase activity are pooled and stored at −80°. The purified enzyme is completely stable for at least 3 months of storage.

Enzyme Purity. The eight-step purification scheme summarized in Table I results in a phosphatidate phosphatase preparation that is apparently homogeneous, as evidenced by sodium dodecyl sulfate-polyacrylamide gel electrophoresis. The enzyme is purified 9833-fold relative to the activity in the cell extract, with a final specific activity of 30 μmol/min/mg.

Properties of Phosphatidate Phosphatase

Physical Properties. Phosphatidate phosphatase and its associated activity has an apparent minimum subunit molecular weight of 91,000.[4] The native molecular weight of the pure enzyme is estimated to be 93,000 by gel-filtration chromatography with Superose 12 in the presence of sodium

TABLE I
PURIFICATION OF PHOSPHATIDATE PHOSPHATASE FROM *Saccharomyces cerevisiae*[a]

Purification step	Total units (μmol/min)	Protein (mg)	Specific activity (units/mg)	Yield (%)	Purification (-fold)
1. Cell extract	14	4060	0.003	100	1
2. Total membranes	9.9	1115	0.009	71	3
3. Sodium cholate extract	7.9	333	0.024	56	8
4. DE-53	4.8	49	0.098	34	33
5. Affi-Gel Blue	2.4	6.8	0.353	17	117
6. Hydroxylapatite	1.5	0.34	4.41	11	1470
7. Mono Q	0.72	0.06	12.0	5	4000
8. Superose 12	0.59	0.02	29.5	4	9833

[a] Data from Lin and Carman.[4]

cholate.[4] Since the micellar molecular weight of sodium cholate ranges from 900 to 1800,[11] the molecular weight of phosphatidate phosphatase by gel-filtration chromatography is in close agreement with the molecular weight estimated by sodium dodecyl sulfate-polyacrylamide gel electrophoresis. It appears that phosphatidate phosphatase exists as a monomer. Phosphatidate phosphatase associates with Triton X-100 micelles in the absence of phosphatidate; however, the enzyme is more tightly associated with micelles when its substrate is present.[12]

Enzymological Properties. The pH optimum for the phosphatidate phosphatase reaction is 7. Maximum phosphatidate phosphatase activity at pH 7 is dependent on 1 m*M* magnesium ion. The magnesium requirement cannot be substituted by manganese. The addition of Triton X-100 to the assay system stimulates phosphatidate phosphatase activity (3-fold) to a maximum at 5 m*M* (molar ratio of Triton X-100 to phosphatidate of 10 : 1) followed by an apparent inhibition of activity at higher concentrations. These results are characteristic of surface dilution kinetics.[13] Monovalent ions do not affect phosphatidate phosphatase activity. Thioreactive compounds such as *p*-chloromercuriphenylsulfonic acid, *N*-ethylmaleimide, and Hg^{2+} ion inhibit phosphatidate phosphatase activity. The activation energy for the reaction is 11.9 kcal/mol, and the enzyme is labile above 30°. The turnover number (molecular activity) for the enzyme is 2.7×10^3 min^{-1} at pH 7 and 30°.

[11] A. Helenius and K. Simons, *Biochim. Biophys. Acta* **415,** 29 (1975).
[12] Y.-P. Lin and G. M. Carman, *J. Biol. Chem.* **265,** 166 (1990).
[13] R. A. Deems, B. R. Eaton, and E. A. Dennis, *J. Biol. Chem.* **250,** 9013 (1975).

Kinetic Properties toward Triton X-100–Phosphatidate Mixed Micelles. A detailed kinetic analysis of purified phosphatidate phosphatase has been performed using Triton X-100–phosphatidate mixed micelles.[12] Enzyme activity is dependent on the bulk and surface concentrations of phosphatidate. These results are consistent with the "surface dilution" kinetic scheme of Dennis and co-workers[13] where phosphatidate phosphatase binds to the mixed micelle surface before binding to its substrate and catalysis occur. The enzyme has 5- to 6-fold greater affinity (relected in the dissociation constant nK_s^A/x) for Triton X-100 micelles containing dioleoyl phosphatidate and dipalmitoyl phosphatidate when compared to micelles containing dicaproyl phosphatidate. The V_{max} for dioleoyl phosphatidate is 3.8-fold higher than the V_{max} for dipalmitoyl phosphatidate, whereas the interfacial Michaelis constant xK_m^B for dipalmitoyl phosphatidate is 3-fold lower than the xK_m^B for dioleoyl phosphatidate. The specificity constants (V_{max}/xK_m^B) of both substrates are similar, indicating that dioleoyl phosphatidate and dipalmitoyl phosphatidate are equally good substrates. Based on catalytic constants (V_{max} and xK_m^B), dicaproyl phosphatidate is the best substrate with an 11- and 14-fold greater specificity constant when compared to dioleoyl phosphatidate and dipalmitoyl phosphatidate, respectively.

Acknowledgments

This work was supported by Public Health Service Grant GM-28140 from the National Institutes of Health, New Jersey State funds, and the Charles and Johanna Busch Memorial Fund.

[55] Characterization and Assay of Phosphatidate Phosphatase

By Ashley Martin, Antonio Gomez-Muñoz, Zahirali Jamal, and David N. Brindley

Introduction

Phosphatidate phosphatase (phosphatidate phosphohydrolase, EC 3.1.3.4) is involved in the synthesis *de novo* of triacylglycerols, phosphatidylcholine, and phosphatidylethanolamine. It lies at a branch point in the pathway and could help to regulate the relative rates of flux of substrates passing through CDP-diacylglycerol, and to diacylglycerol. This form of the enzyme is thought to be Mg^{2+}-dependent. It is known that this phosphatase is under both long-term and short-term regulation. For example, in the case of liver glucocorticoids, glucagon (via cyclic AMP) and growth hormone appear to increase its rate of synthesis, whereas insulin opposes these effects. The half-life of the activity is also increased by glucagon and decreased by insulin. In the short-term, the activity can be increased by vasopressin within 5 min, and fatty acids cause a redistribution of the phosphatase from the cytosol to the endoplasmic reticulum. The latter event is thought to represent enzyme activation.[1]

More recently it has also become evident that phosphatidate phosphatase is involved in signal transduction following the stimulation of a phospholipase D.[2] The latter enzyme degrades phosphatidylcholine in the plasma membrane in response to agonists and produces phosphatidate that is converted to diacylglycerol so as to activate protein kinase C. Theoretically phosphatidate itself may also be a second messenger; therefore, the phosphatase might be involved in destroying one signal as well as in creating another. A phosphatidate phosphatase has been identified in plasma membrane fractions (see below). This enzyme is not stimulated by Mg^{2+} and is probably involved in this signal transduction.

In order to be able to investigate these events further, it is important to be able to assay phosphatidate phosphatases reliably and to distinguish between the respective activities that are involved in signal transduction and glycerolipid synthesis. There are two major problems in determining the activity of phosphatidate phosphatase. The first, as with other phospholipases, is to present the substrate in a suitable form for measuring enzyme activities in aqueous solutions. The second problem is that phosphatidate can be degraded by phospholipase A-type activities to yield glycerol phosphate (Fig. 1). The latter can then be converted to glycerol by acid or alkaline phosphatase activities. This would invalidate the analysis of inorganic phosphate or the liberation of water-soluble ^{32}P from *sn*-1,2-diacylglycerol glycerol [^{32}P]phosphate.[3] Equally, significant hydrolysis of diacylglycerol by lipase action could invalidate the determination of the phosphatase through measuring the concentration of diacylglycerol. This breakdown of diacylglycerol can be minimized by allowing only relatively low amounts of phosphatidate to be converted to diacylglycerol,[3] or by adding a lipase inhibitor.[4] It is therefore essential to characterize the tissue and subcellular fractions in question to decide on the appropriate way in which to determine phosphatase activity.

[1] D. N. Brindley, *in* "Phosphatidate Phosphohydrolase" (D. N. Brindley, ed.), Vol. 1, Chap. 1 and 2. CRC Series in Enzyme Biology, CRC Press, Boca Raton, Florida, 1988.

[2] K. Löffelholz, *Biochem. Pharmacol.* **38,** 1543 (1989).

[3] R. G. Sturton and D. N. Brindley, *Biochem. J.* **171,** 263 (1978).

[4] H. Ide and Y. Nakazawa, *Arch. Biochem. Biophys.* **271,** 177 (1989).

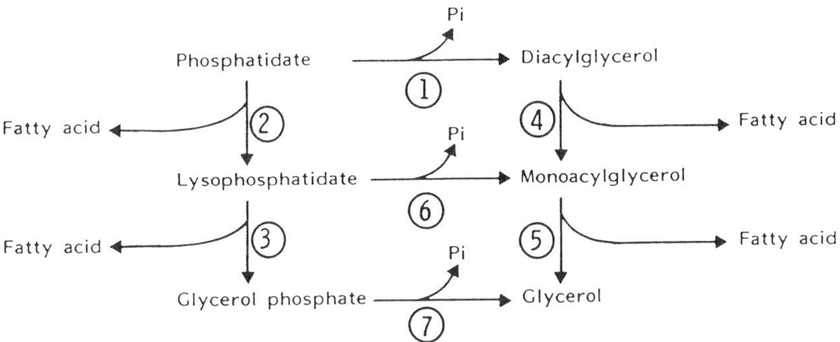

FIG. 1. Possible routes for the degradation of phosphatidate in the phosphatidate phosphatase assay. The reactions are catalyzed as follows: (1) phosphatidate phosphatase; (2) phospholipase A activity; (3) lysophospholipase; and (7) acid or alkaline phosphatase. (Reproduced from Ref. 1 with permission.)

Rationale for Measurement of Phosphatidate Phosphatase Activities in Rat Liver

The technique for assaying phosphatidate phosphatase activity is illustrated by taking rat liver as a source of enzyme activities. Rat liver contains two distinct phosphatidate phosphatases: one has an absolute requirement for Mg^{2+} and is inhibited by *N*-ethylmaleimide,[5] whereas the other is insensitive to inhibition by *N*-ethylmaleimide and does not require Mg^{2+} (Figs. 2 and 3[5]). In fact, the *N*-ethylmaleimide-insensitive phosphatase was inhibited by concentrations of Mg^{2+} above about 3.5 m*M*. It was therefore decided to use this differential effect of *N*-ethylmaleimide as a means of distinguishing between the two activities.[5] This is preferable to determining routinely a Mg^{2+} dependency since it is very difficult to eliminate bivalent cations from the substrate and from enzyme preparations.[6] The reaction rates were determined through the production of diacylglycerol as decided from previous work.[3,6]

Preparation of Radioactive Phosphatidate

[^{3}H]Phosphatidate is prepared enzymatically from glycerol phosphate and [^{3}H]palmitate.[6] An incubation volume of about 167 ml is used which contains 16.6 m*M* potassium phosphate buffer, pH 7.4, 1 m*M* dithiothreitol, 50 m*M* NaF (to prevent phosphatidate hydrolysis[1]), 13.4 m*M* $MgCl_2$,

[5] Z. Jamal, A. Martin, A. Gomez-Muñoz, and D. N. Brindley, *J. Biol. Chem.* **266,** in press (1991).

[6] A. Martin, P. Hales, and D. N. Brindley, *Biochem. J.* **245,** 347 (1987).

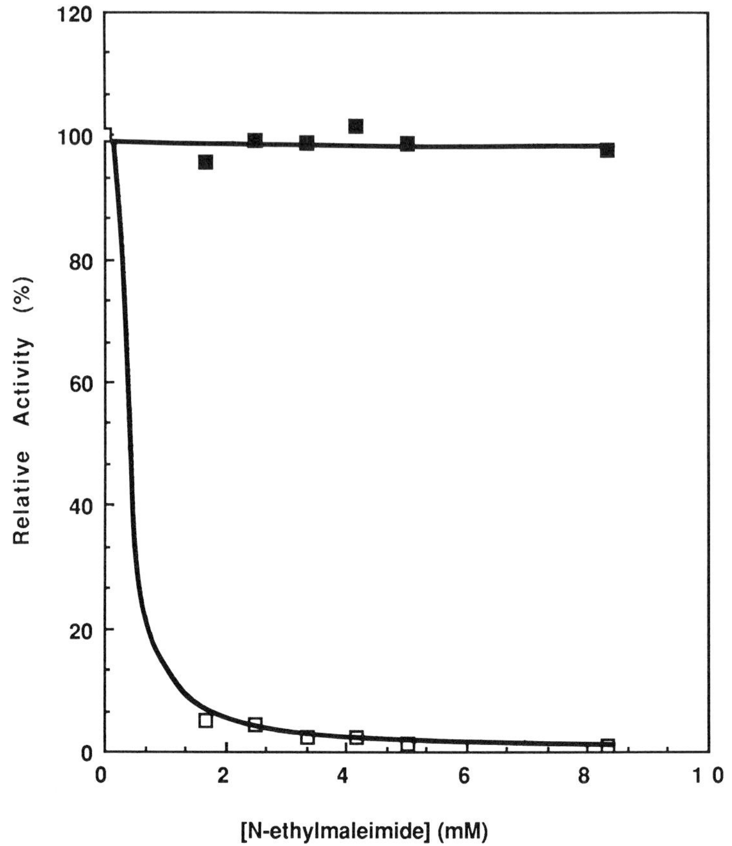

FIG. 2. Effects of *N*-ethylmaleimide on activities of the Mg^{2+}-dependent and non-Mg^{2+}-stimulated phosphatidate phosphatases of rat liver. The soluble and plasma membrane fractions of rat liver were preincubated for 10 min at 37° in the presence of *N*-ethylmaleimide. Measurements of the Mg^{2+}-dependent (□) and *N*-ethylmaleimide-insensitive phosphatidate phosphatases (■) were performed as described in the text. The fraction enriched with endoplasmic reticulum displayed the same characteristics as the Mg^{2+}-dependent phosphatidate phosphatase activity of the soluble fraction.[5]

8 m*M* ATP, 5 m*M* *rac*-glycerol 3-phosphate, 25 μM coenzyme A (CoA), 0.3 m*M* [^{3}H]palmitate (500 Ci/mol, prepared by warming with a 20% molar excess of KOH), 3 mg/ml of fatty acid-poor bovine serum albumin, and 3 mg/ml of microsomal protein prepared from rat liver. The incubation is for 30 min at 37°. The reactions are stopped by adding 3.75 volumes of chloroform/methanol (1 : 2, v/v). Two phases are then generated by adding 1.25 volumes of chloroform and 1.25 volumes of 2 *M* KCl containing 0.2 *M* H_3PO_4. The bottom phase is then washed 3 times with synthetic top

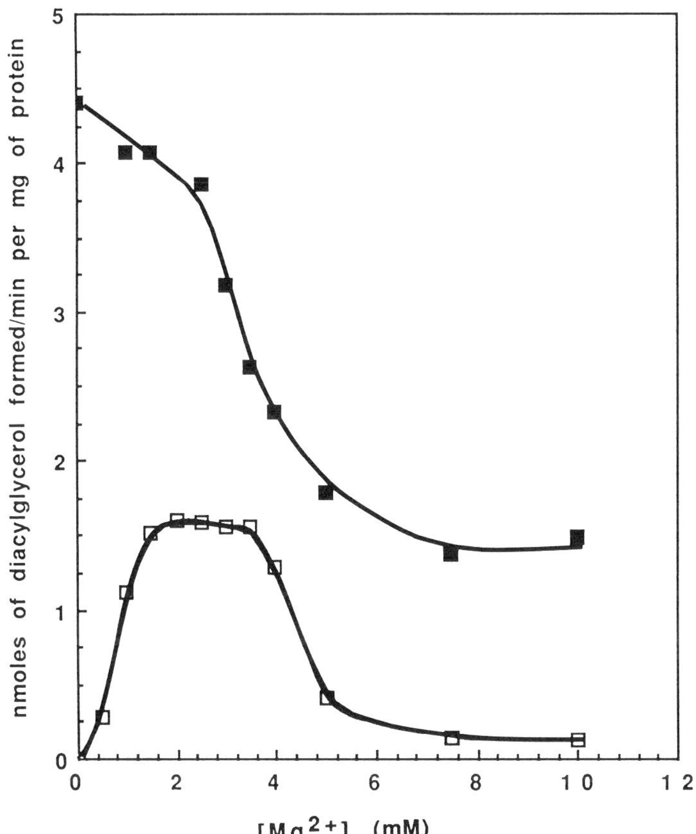

FIG. 3. Mg^{2+}-dependent and *N*-ethylmaleimide-insensitive phosphatidate phosphatase from rat liver. Rat liver was separated into fractions enriched in soluble components, plasma membranes, and endoplasmic reticulum.[5] Results are shown for the soluble and plasma membrane fractions. The endoplasmic reticulum also exhibited a Mg^{2+}-dependent phosphatidate phosphatase activity similar to that in the soluble fraction (see Ref. 6). Assay conditions are given in the text. The Mg^{2+}-dependent activity (□) in the soluble fraction was the difference in activities in the presence and absence of *N*-ethylmaleimide. The Mg^{2+}-independent activity that was not stimulated by Mg^{2+} was measured after preincubation with *N*-ethylmaleimide (■).

phase prepared by mixing equivalent volumes of solvent. The final bottom phase is dried down, and the lipids are dissolved in chloroform that had been treated with granular $CaCl_2$ to remove ethanol and then applied to a column of 60 g of SilicAR CC-7 60 (Mallinckrodt, St. Louis, MO) prepared in ethanol-free chloroform.[6] Neutral lipids are eluted with chloroform and phospholipids with chloroform containing 5, 10, 15, 20, and 50% methanol.

Fractions that contain ^3H are combined, and the identity of the [^3H]phosphatidate is confirmed by thin-layer chromatography on plates of silica gel 60 using chloroform/methanol/acetic acid/acetone/water (10 : 2 : 2 : 4 : 1, by volume) for development. Complete purification from unlabeled phospholipids is not considered to be necessary since the [^3H]phosphatidate is subsequently diluted more than 500-fold with nonradioactive phosphatidate. Radioactive phosphatidate is routinely stored under an atmosphere of N_2 at $-20°$ in chloroform solution.

sn-Diacylglycerol 3-[^{32}P]phosphate is prepared with diacylglycerol kinase, ATP, and *sn*-1,2-diacylglycerol using a method based on that of Pieringer and Kunnes.[7] Diacylglycerol kinase is prepared from *Escherichia coli* membranes,[7] except that the cells are broken in a French press and membranes are stored at $-70°$ before heat treatment in a boiling water bath for 10 min. The enzyme preparation is then cooled on ice and used in the following incubation (total volume 1 ml): 1.2 m*M* diolein (Sigma, St. Louis, MO), 50 m*M* sodium phosphate buffer, pH 7, 100 m*M* $MgCl_2$, 1% Triton X-100, 105 mCi [γ-^{32}P]ATP, 1 m*M* ATP, and 0.5 mg of heat-treated membrane protein from *E. coli*. The mixture is shaken to disperse the diacylglycerol and then incubated with shaking overnight (12–16 hr). Lipids are extracted with 3.75 ml of chloroform/methanol (1 : 2, v/v), and the phases are then separated with 1.25 ml of chloroform and 1.25 ml of 2 *M* KCl containing 0.2 *M* potassium phosphate, pH 7.0. The chloroform phase is washed 3 times with synthetic top phase prepared from equivalent volumes of solvents. This chloroform phase contains 25–33% of the original ^{32}P. The [^{32}P]phosphatidate is then purified on a 1-ml column of SilicAR CC-7 (100–200 mesh) as described above. It is then diluted to the appropriate specific radioactivity with nonradioactive phosphatidate and converted to the K^+ salt form as described below.

Methods for preparing phosphatidate labeled in the glycerol moiety are given elsewhere.[6] Nonradioactive phosphatidate can be prepared from egg yolk lecithin as described previously,[8] but it is now routinely purchased from Sigma.

Assay of Phosphatidate Phosphatase

Preparation of Phosphatidate as Enzyme Substrate

The [^3H]phosphatidate is diluted with nonradioactive phosphatidate to give a specific radioactivity of about 0.4 Ci/mol. The solvent is removed with a stream of N_2, and the lipid is dissolved in chloroform/methanol/

[7] R. A. Pieringer and R. A. Kunnes, *J. Biol. Chem.* **240,** 2833 (1965).

[8] R. G. Sturton and D. N. Brindley, *Biochem. J.* **162,** 25 (1977).

water (5 : 4 : 1, by volume) at a concentration of about 3.6 mg/ml. Chelex resin (K^+ form) is added at the rate of 50 ml[8] of resin per gram of phosphatidate. The mixture is stirred, and extra methanol is added as required to completely dissolve the phosphadidate in the solvent above the resin. After 30 min at room temperature the solvent mixture is recovered by filtration, the resin washed with more solvent mixture, and the composition of the filtrates adjusted to chloroform/methanol/2 *M* KCl (10 : 10 : 9, by volume). Phosphatidate is recovered in the bottom phase which is then taken to dryness by rotary evaporation. The product is dissolved in chloroform, and the concentration of phosphatidate is calculated from the recovery of radioactivity on the assumption that this is equivalent to that of nonradioactive phosphatidate. The Chelex treatment ensures a relatively complete removal of bivalent cations from the phosphatidate preparations. This facilitates the dispersion of the substrate in water and also means that the requirement for these cations could be subsequently determined with minimum interference.[6]

The K^+-phosphatidate is stored in chloroform and under N_2 at $-20°$. When the substrate is required for the assay of Mg^{2+}-dependent phosphatidate phosphatase, samples are taken and mixed with phosphatidylcholine (99% pure, prepared from egg yolk lecithin purchased from the Sigma). Phosphatidylcholine stimulates the Mg^{2+}-dependent phosphatase.[3] The solvent is removed with a stream of N_2, and 5.56 m*M* EDTA plus 5.56 m*M* EGTA, adjusted to pH 7 with KOH, are added to give a final concentration of 3.33 m*M* phosphatidate. The mixture is sonicated at 22 kHz with an amplitude of 8 μm peak to peak for 10–15 sec at 22°. To every 9 volumes of this mixture is then added 1 volume of 100 mg/ml of fatty acid-poor bovine serum albumin.

Alternatively, the phosphatidate substrate for the *N*-ethylmaleimide-insensitive phosphatidate phosphatase, which is not stimulated by Mg^{2+}, can be prepared in a similar way except that phosphatidylcholine is omitted since it decreases this activity.

Incubation Conditions for N-Ethylmaleimide-Insensitive Phosphatidate Phosphatase Activity That is Mg^{2+}-Independent

The enzyme activity is concentrated in a membrane fraction that can be purified from rat liver and which is enriched in 5′-nucleotidase and alkaline phosphodiesterase activities. This is therefore compatible with a plasma membrane location.[5] Subcellular fractions are prepared in 0.25 *M* sucrose, adjusted to pH 7.4 with $KHCO_3$ and containing 0.5 m*M* dithiothreitol. The fractions are incubated with 4.2 m*M* *N*-ethylmaleimide at 37° for 10 min and then cooled on ice. This concentration of *N*-ethylmaleimide is sufficient to remove the dithiothreitol from the sample and is capable of inhibiting the Mg^{2+}-dependent activity but preserving the non-Mg^{2+}-requiring phosphatase activity (Fig. 2).

The phosphatase is then assayed in the absence of Mg^{2+} in an incubation that contains (in 0.1 ml) 100 m*M* Tris–maleate buffer, pH 6.5, 1 m*M* dithiothreitol, 0.6 m*M* [^{3}H]phosphatidate (0.4 Ci/mol), and the 0.2 mg of albumin, 1 m*M* EDTA, plus 1 m*M* EGTA carried through with the substrate. Up to 50 μg of protein from the subcellular fractions is used, and the reaction is started with the phosphatidate substrate. The incubation time at 37° is 10–60 min depending on the activity of the fraction. The time is adjusted so that less than approximately 15% of the substrate is consumed during the incubation. The reaction rate is constant with time and proportional to protein concentration when this is kept below about 100 μg (Fig. 4). In later work, we discovered that the addition of 0.5% Triton X-100 to the incubations effectively extended the linear portion of the response curve up to about 200 μg of plasma membrane protein.[5] The use of Triton X-100 is now recommended for the routine assay of the *N*-ethylmaleimide-insensitive phosphohydrolase activity.

The reactions are stopped by adding 2 ml of chloroform/methanol (19 : 1, v/v) containing 0.08% olive oil as a carrier. One gram of aluminum oxide (alumina absorption 80–200 mesh from Fisher Scientific, Pittsburgh, PA) is added, and the mixture is shaken. When all the reactions have been stopped, the tubes are shaken for a further 2 min and the alumina pelleted in a benchtop centrifuge. Samples (1 ml) of the solvent are taken and dried in counting vials. ^{3}H is determined by scintillation counting after a delay of 24 hr so as to attain maximum counting efficiencies. The alumina is used to adsorb unreacted phosphatidate and any [^{3}H]palmitate that may have been released by phospholipase A-type activities acting on the phosphatidate.[6] This procedure is much more rapid than the use of thin-layer or column chromatography to purify the diacylglycerol.

When [^{32}P]phosphatidate is used the reactions are stopped by adding 2.28 ml of chloroform/methanol/water (1 : 2 : 0.64, by volume). Phases are separated after 5 min by adding 0.625 ml of chloroform and 0.625 ml of 0.4 *M* KCl in 0.2 *M* HCl. After shaking and centrifugation, 2-ml samples of the top phase are added to 4 ml of Aqueous Counting Scintillant (Amersham), and ^{32}P is determined by scintillation counting. If required, this method can be used to determine simultaneously the rate of diacylglycerol formed if [^{3}H]phosphatidate is also added to the substrate. In this case, carrier olive oil is added to the chloroform used to separate the phases, and samples of the bottom phase are treated with alumina (see above). The release of inorganic phosphate from nonradioactive phosphatidate can also be determined if the reactions are stopped with trichloroacetic acid.[3]

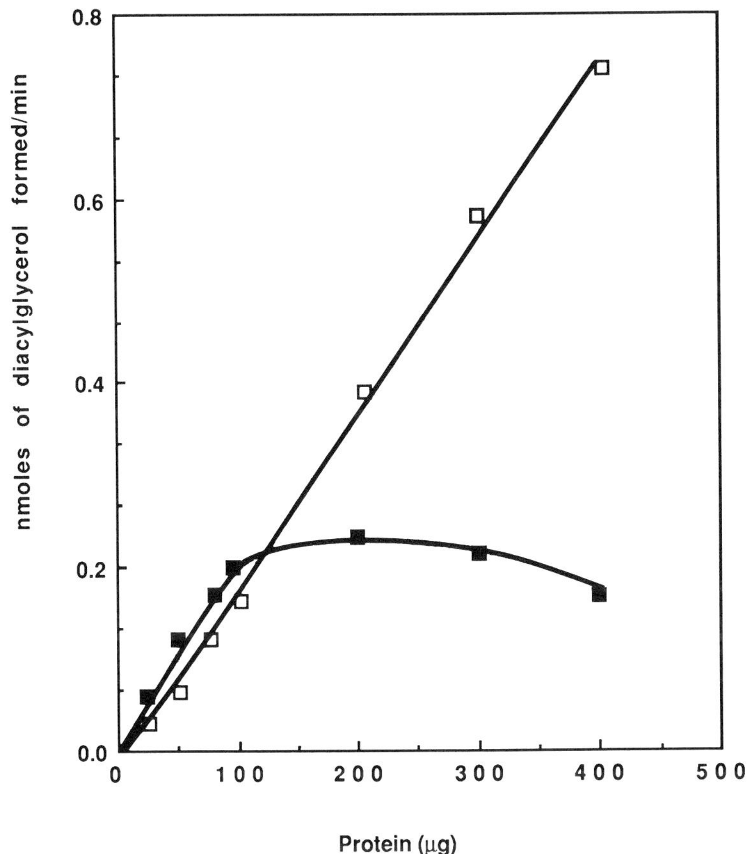

FIG. 4. Effect of protein concentration on the reaction rate obtained for the Mg^{2+}-dependent and *N*-ethylmaleimide-insensitive phosphatidate phosphatase activities. Activities of the Mg^{2+}-dependent (□) and *N*-ethylmaleimide-insensitive phosphatases (■) were measured in the soluble and plasma membrane fractions of rat liver as described in the text, and the effects of varying the amounts of cell fractions are shown.

When phosphatidate phosphatase is determined by the release of water-soluble ^{32}P, or by measurement of inorganic phosphate, it is necessary to prove that this is not being caused through the action of phospholipase A activities (Fig. 1). One way to do this is to use the [^{32}P]phosphatidate and to add *sn*-glycerol 2-phosphate to the assays to decrease the breakdown of *sn*-glycerol 3-[^{32}P]phosphate by acid or alkaline phosphatases. The ^{32}P products can then be identified by thin-layer chromatography on microgranular cellulose.[3] If significant quantities of glycerol [^{32}P]phosphate are detected then this method should not be used. Ideally the reaction rates calculated from the release of inorganic phosphate and diacylglycerol should be equal. Otherwise, it is essential to determine whether the results of one of these measurements is invalidated by phospholipase A type activities, or by diacylglycerol lipase.

Incubation Conditions for the Mg^{2+}-Dependent Phosphatidate Phosphatase

The enzyme activity is located in the soluble and microsomal fractions of rat liver. The activity is determined as the *N*-ethylmaleimide-sensitive phosphatidate phosphatase activity that is measured in the presence of Mg^{2+}.

Each assay contains (final volume 0.1 ml) 100 m*M* Tris–maleate buffer, pH 6.5, 1 m*M* dithiothreitol, 0.2 mg of fatty acid-poor bovine serum albumin, the mixed emulsion of 0.6 m*M* [^{3}H]phosphatidate (0.4 Ci/mol) with 0.4 m*M* phosphatidylcholine, the 0.2 mg of albumin, 1 m*M* EDTA, and 1 m*M* EGTA carried through with the substrate, 3 m*M* $MgCl_2$, and up to 200 μg of protein from the particle-free supernatant, or microsomal fraction of rat liver. Parallel incubations are performed with untreated subcellular fractions and those preincubated with *N*-ethylmaleimide as described above. The reactions are started by addition of the mixed micelles, and the incubations are at 37° for 10–60 min. The time of the incubation and the quantity of enzyme protein used are adjusted so that less than 15% of the phosphatidate is consumed during the reaction. This ensures that reaction rates are constant with time and proportional to the amount of enzyme added (Fig. 4).

The reaction rate can be calculated from the difference between the total phosphatase activity measured with untreated fraction and the activity obtained from the *N*-ethylmaleimide-insensitive phosphatase that is determined under the same assay conditions. This compensates for the presence of any *N*-ethylmaleimide-insensitive phosphatase activity. In these incubations the activity of the Mg^{2+}-dependent enzyme is favored by the presence of Mg^{2+} and phosphatidylcholine in the medium.[5]

It is again necessary to test the products of the reaction to decide whether to use the liberation of phosphate or the formation of diacylglycerol for the analysis.

Concluding Remarks

We have found these assays to give excellent reproducibility. However, it is very important that positive displacement pipettes be used to dispense the chloroform in which radioactive lipids are dissolved. The radiochemical assays based on the formation of [^{3}H]diacylglycerol or the liberation of ^{32}P give high sensitivity. However, this is limited by the blank value obtained in the absence of phosphatase activity, which in the case of the [^{3}H]phosphatidate can be up to 1% of the total radioactivity.

It is also important to be sure, especially for the *N*-ethylmaleimide-insensitive phosphatase, that the reaction rate is proportional to the amount of enzyme protein added throughout the range used (see Fig. 4). We normally do this by performing three assays at different protein concentrations instead of using replicates at one protein concentration. The mean specific activity is then calculated as the slope of the regression line, which also employs the blank value as the point of zero protein concentration.

Finally, it should again be emphasized that the assays for phosphatidate phosphatase should be characterized in each different tissue and subcellular fraction employed. This is necessary because of the phospholipase A-type activities that can also degrade phosphatidate and because of lipase activities that can destroy diacylglycerol. For example, because of the lipase activities in adipose tissue it is better to analyze for the release of phosphate or water-soluble ^{32}P from phosphatidate rather than assay for diacylglycerol.[9] Alternatively, diacylglycerol production can be determined for adipose tissue samples if 200 μM tetrahydrolipstatin (kindly donated by M. K. Meier, Hoffmann-LaRoche Ltd., Basel, Switzerland) is used to inhibit acylglycerol lipases.[5] However, with liver fractions no significant increase in diacylglycerol production was observed with tetrahydrolipstatin. We therefore conclude that lipase activity is not invalidating the determination of diacylglycerol formation in the latter case.[5]

The existence of at least two distinct phosphatidate phosphatases has been established here for liver. These activities have also been identified in adipose tissue and heart (A. Martin, A. Gomez-Muñoz, Z. Jamal, and D. N. Brindley, unpublished observations), and in lung tissue.[10] These different phosphatidate phosphatases need to be distinguished in future work on these enzymes.

[9] N. Lawson, A. D. Pollard, R. J. Jennings, M. I. Gurr, and D. N. Brindley, *Biochem. J.* **200,** 285 (1981).

[10] P. A. Walton and F. Possmayer, *Biochem. J.* **261,** 673 (1989).

Section VII

Phospholipase D

[56] Glycosylphosphatidylinositol-Specific Phospholipase D

By Kuo-Sen Huang, Shirley Li, and Martin G. Low

Introduction

The glycosylphosphatidylinositol-specific phospholipase D (GPI-PLD) was first observed as a result of its ability to degrade the GPI anchor of alkaline phosphatase during extraction from mammalian tissues with butanol.[1] This anchor-degrading activiy was initially thought to be due to the action of inositol phospholipid-specific phospholipases C, which are very active in most mammalian tissues. However, more detailed studies distinguished these two activities and indicated that a novel and highly specific phospholipase D was involved.[2] The activity was subsequently found to be abundant in mammalian plasma and serum, suggesting that the activity observed in the tissues may be, to a large extent, the result of blood contamination.[3–5] The GPI-PLD has recently been purified from human[6] and bovine[7] serum. The physiological function of this enzyme has not been determined, but it is proposed to play a role in the regulation of cell surface expression of GPI-anchored proteins (reviewed in Refs. 8–10).

Assay Methods

The GPI-PLD hydrolyzes the phosphodiester linkage of the phosphatidylinositol moiety in the GPI anchor of several membrane proteins. The products of this reaction are phosphatidic acid and an inositol residue which is attached to the C terminus of the protein through the glycan (for details of the structure of the GPI anchor, see Refs. 9 and 10). The assay

[1] M. G. Low and D. B. Zilversmit, *Biochemistry* **19,** 3913 (1980).
[2] A. S. Malik and M. G. Low, *Biochem. J.* **240,** 519 (1986).
[3] M. A. Davitz, D. Hereld, S. Shak, J. Krakow, P. T. Englund, and V. Nussenzweig, *Science* **238,** 81 (1987).
[4] M. G. Low and A. R. S. Prasad, *Proc. Natl. Acad. Sci.U.S.A.* **85,** 980 (1988).
[5] M. L. Cardoso de Almeida, M. J. Turner, B. B. Stambuk, and S. Schenkman, *Biochem. Biophys. Res. Commun.* **150,** 476 (1988).
[6] M. A. Davitz, J. Hom, and S. Schenkman, *J. Biol. Chem.* **264,** 13760 (1989).
[7] K.-S. Huang, S. Li, W.-J.C. Fung, J. D. Hulmes, L. Reik, Y.E. Pan, and M. G. Low, *J. Biol. Chem.* **265,** 17738 (1990).
[8] M. G. Low and A. R. Saltiel, *Science* **239,** 268 (1988).
[9] M. A. J. Ferguson and A. F. Williams, *Annu. Rev. Biochem.* **57,** 285 (1988).
[10] M. G. Low, *Biochim. Biophys. Acta,* **988,** 427 (1989).

methods that have been used in the purification procedure described here monitor production of either of these products.

Assay I: Alkaline Phosphatase as Substrate

In the following assay the loss of the hydrophobic part of the anchor from the protein is determined by partitioning in Triton X-114[11] and measuring the distribution of the protein between the hydrophilic, detergent-poor phase and the hydrophobic, detergent-rich phase. In principle, any GPI-anchored protein whose distribution could be determined easily would be an appropriate substrate. In the assay to be described here the substrate is human placental alkaline phosphatase, and its distribution in Triton X-114 is determined by measuring its ability to hydrolyze *p*-nitrophenyl phosphate in a subsequent incubation. The major disadvantage of this type of assay is that it does not distinguish between GPI-PLD-mediated cleavage and those catalyzed by other hydrolases especially GPI-PLC. It is also more time consuming since it involves a second incubation to assay for the degraded alkaline phosphatase. However, an advantage over the other assay procedure is the ready availability of the starting material for preparation of the substrate.

Purification of Human Placental Alkaline Phosphatase. Alkaline phosphatase, with the phosphatidylinositol anchor intact, is purified from human placenta by a procedure based on previous studies of anchor degradation in human placenta.[2] Except where specified otherwise, the following procedure is carried out at 0°–4°. Term human placenta frozen after delivery is thawed, membranes and fibrous material discarded, and the remaining tissue (typically about 300 g) homogenized in 100 m*M* Tris-HCl, pH 8.5 (3 ml/g tissue) containing benzamidine (1 m*M* final concentration) and phenylmethylsulfonyl fluoride (PMSF, 0.01 mg/ml). The homogenate is mixed with ice-cold butanol (3 ml/g tissue) and shaken vigorously at intervals for 5 min. The mixture is centrifuged at 5000 *g* for 45 min, and the lower aqueous phase is collected and dialyzed against 3 changes of 12 volumes of distilled water over a period of 48 hr. An alkaline pH is essential during the initial butanol extraction since the endogenous GPI-PLD is activated below pH 7.0 by the butanol. The dialyzate is concentrated 4-fold in an Amicon (Danvers, MA) ultrafiltration cell using a YM30 membrane. Triton X-114 [1% (w/v) final concentration], Tris-HCl, pH 8.5 (10 m*M*), $MgCl_2$ (0.1 m*M*), and zinc acetate (10 μM) are added as 10× or 100× solutions and the mixture incubated at 37° for 10 min. The phases are separated by centrifugation at 1500 *g* for 5 min and the upper, detergent-poor phase removed by aspiration.

[11] C. Bordier, *J. Biol. Chem.* **256,** 1604 (1981).

The lower, detergent-rich phase (containing alkaline phosphatase with an intact GPI anchor) is mixed with 1 volume of 10 m*M* Tris-HCl/0.1 m*M* $MgCl_2$/10 μ*M* zinc acetate, pH 8.0 (buffer A) and applied to a column (2.5 × 40 cm) of SM-2 BioBeads (Bio-Rad, Richmond, CA) or Amberlite XAD-2 (Sigma, St. Louis, MO) equilibrated and eluted with buffer A at a flow rate of 25 ml/h. The detergent-depleted eluate (100–170 ml eluted), containing most of the alkaline phosphatase activity, is concentrated in an Amicon ultrafiltration cell using a YM10 membrane to a volume of approximately 20 ml and centrifuged at 10,000 *g* for 15 min to remove insoluble material. The supernatant is applied to a column (two 2.5 × 118 cm columns connected together) of Sephacryl S-300 (Pharmacia, Piscataway, NJ) equilibrated with 150 m*M* NaCl/0.1 m*M* $MgCl_2$/10 μ*M* zinc acetate/10 m*M* HEPES–NaOH, pH 7.4, and eluted at a flow rate of 12 ml/hr. The high molecular weight, hydrophobic form of alkaline phosphatase (approximate elution volume 450–570 ml) is pooled and stored in aliquots at −20°. This material is stable for at least 12 months; one placenta produces sufficient purified alkaline phosphatase for approximately 10,000 of the anchor degradation assays described below. Sodium dodecyl sulfate (SDS)–polyacrylamide gel electrophoresis of the alkaline phosphatase shows it to be approximately 50% pure with alkaline phosphatase being the single major protein species present.

Assay Procedure. The alkaline phosphatase substrate (50 μl containing 1 volume of alkaline phosphatase, purified as described above, 2 volumes of 1% (w/v) Nonidet P-40 (NP-40), and 2 volumes of 200 m*M* Tris–maleate, pH 7) is incubated with aliquots of supernatant fractions, etc., in a total volume of 0.2 ml for 30 min at 37°. The incubation mixture is then diluted with 0.8 ml of ice-cold 150 m*M* NaCl/0.1 m*M* $MgCl_2$/10 μ*M* zinc acetate/10 m*M* HEPES–NaOH, pH 7.0, and a 50-μl aliquot is removed and mixed with 0.2 ml of the same buffer and 0.25 ml 2% precondensed Triton X-114. After sampling a 0.1-ml aliquot for assay of total alkaline phosphatase activity, the mixture is incubated at 37° for 10 min, then centrifuged (1500 *g*) immediately at room temperature* for 2 min to separate the phases, and a 0.1-ml aliquot of the upper phase is sampled. Alkaline phosphatase activity in the samples is determined as described previously[2] except that 0.2% Triton X-100 is included in the alkaline phosphatase assay incubation and 50 μl of 1% sodium deoxycholate is added after the enzyme reaction has been stopped.

Anchor degradation is measured by comparing the activity in the upper phase (i.e., the degraded form) with that in the total incubation mixture

* When large numbers of samples are assayed, it is advisable to carry out the centrifugation at 37° to avoid the high background due to partial phase mixing at room temperature.

before phase separation at 37°. One unit is the amount of enzyme hydrolyzing 1% of the alkaline phosphatase per minute. Under the conditions utilized here, approximately 5–10% of the total alkaline phosphatase is found in the upper phase in the absence of added anchor-degrading activity. This does not change substantially during incubations of up to 2 hr, indicating that the alkaline phosphatase purification procedure removes most of the GPI-PLD present in human placenta.[2]

Assay II: Variant Surface Glycoprotein as Substrate

In the following assay the GPI anchor of the protein substrate is biosynthetically labeled with a ^{3}H-labeled fatty acid, and the released phosphatidic acid can then readily be detected by liquid scintillation counting following extraction with aqueous butanol. Although the routine assay to be described below does not distinguish between GPI-PLD and GPI-PLC, this can be checked by taking the extracted ^{3}H-labeled product and analyzing it by thin-layer chromatography. For most GPI-anchored proteins, biosynthetic labeling is too low for use as substrates since they are of relatively low abundance. However the variant surface glycoprotein (VSG) of the parasitic protozoan *Trypanosoma brucei* is abundant and can be labelled with [^{3}H]myristic acid quite readily.

Preparation of Substrate. [^{3}H]Myristic acid-labeled VSG is prepared by a modification of the procedure of Hereld *et al.*[12] A rat is infected with *T. brucei* MITat 117 or 118 until the parasite density reaches approximately 10^9/ml. The blood is removed by cardiac puncture with 3.8% sodium citrate (1 ml/9 ml of blood) as anticoagulant. The blood is centrifuged at approximately 1000 *g* for 5 min. The majority of the plasma is removed and discarded. The trypanosomes form a white layer on top of the erythrocytes which is removed into a clean tube and washed at room temperature with phosphate-buffered saline containing 1% (w/v) glucose (PBSG; approximately 1 ml/ml of blood). The pellet of contaminating erythrocytes is removed and discarded after each centrifugation by transferring the trypanosomes into a clean tube. The washing is repeated (3–5 times) until erythrocytes are no longer visible; the yield from one rat is typically about 10^{10} trypanosomes.

Labeling medium is prepared in advance by drying 2 mCi of [9,10-^{3}H]myristic acid under nitrogen to remove toluene and redissolving it in 40 μl of 95% (v/v) ethanol. One-half milliliter of defatted bovine serum albumin (BSA) (Sigma, St. Louis, MO, Cat. No. A-6003), 20 mg/ml in 150 m*M* NaCl, 10 m*M* sodium phosphate buffer (pH 7.0), is added and mixed

[12] D. Hereld, J. L. Krakow, J. D. Bangs, G. W. Hart, and P. T. Englund, *J. Biol. Chem.* **261,** 13813 (1986).

vigorously to form a BSA–fatty acid complex. The mixture is then added to 30 ml minimal essential medium (MEM) buffered with 25 m*M* HEPES (pH 7.4) and containing 0.5 mg/ml defatted BSA. The washed trypanosomes are incubated in this medium at 37° for 60 min in a tissue culture flask with occasional gentle shaking to maintain suspension. An additional 20 ml of MEM/HEPES/BSA is added after about 40 min. Incubation can be continued for longer periods (up to 90 min), but in this situation, it is vital to monitor viability (i.e., maintenance of slender shape and rapid motility) of the trypanosomes frequently since lysis leads to activation of the endogenous GPI-specific phospholipase C (GPI-PLC) and rapid degradation of the GPI anchor of VSG. After incubation the flask is placed on ice for approximately 5 min, and the cell suspension is transferred to centrifuge tubes, centrifuged, and washed with ice-cold PBSG.

The washed cells are lysed by resuspension in 8 ml ice-cold hypotonic lysis buffer (10 m*M* sodium phosphate buffer, 1 μg/ml leupeptin, 0.1 m*M* tosyllysine chloromethyl ketone, pH 7.0). After 5 min, 0.42 ml of 100 m*M* *p*-chloromercuriphenylsulfonic acid (the sodium salt dissolved in 0.1 *M* NaOH) is added to inhibit the GPI-PLC, and the suspension is transferred to a 30-ml Corex tube and centrifuged at approximately 12,000 *g* for 15 min at 4°. The pellet is washed in an additional 8 ml of hypotonic lysis buffer and 0.42 ml of *p*-chloromercuriphenylsulfonic acid. The washed pellet is dissolved in 2 ml of 1% (w/v) SDS by heating in a 100° water bath with intermittent shaking. The solution is cooled to room temperature, 20 ml of water-saturated 1-butanol is added, and the solution is mixed thoroughly and centrifuged at 12,000 *g* for 20 min at room temperature. The upper phase is removed and discarded, and the remaining material (interface and lower phase) is reextracted with additional 20-ml portions of water-saturated butanol until the radioactivity in the aqueous phase is less than 10^4 counts/min (cpm)/ml (usually requires about 10 extractions). The lower aqueous phase is finally eliminated and the VSG precipitated by adding 20 ml anhydrous butanol. The precipitate is extracted 3 times with 5 ml of diethyl ether for 10 min to remove the butanol, and the diethyl ether is removed by evaporation. The dry precipitate is finally redissolved in 2–3 ml of 1% SDS by heating in a boiling water bath.

The procedure described above produces sufficient substrate for approximately 5000 of the GPI-PLD assays described below; it is stable at −20° for at least 1 year. The procedure can be interrupted after the membranes are dissolved in SDS; however, in order to minimize degradation of the VSG by endogenous GPI-PLC, it is advisable to carry out the first part of the procedure with minimal delay.

Assay Procedure. The assay is based on the procedures described by Hereld *et al.*[12] and Davitz *et al.*[3] [^{3}H]Myristate-labeled VSG (2000–3500 cpm, 2 μg; approximately 0.4 μl of the VSG solution prepared as described above) is mixed with 20 μl of 200 m*M* Tris–maleate, pH 7.0, 20 μl of 1% (w/v) NP-40, and 60 μl of water. The substrate–detergent mixture is incubated with aliquots of GPI-PLD or buffer (0.1 ml) in microcentrifuge tubes for 30 min at 37°. The reaction is stopped by the addition of 0.5 ml of butanol that has been saturated with water. After vortexing, the phases are separated by centrifugation at 1500 *g* for 3 min. The upper phase (0.3 ml) is sampled, mixed with scintillation fluid, and counted. One unit of GPI-PLD activity is arbitrarily defined as the amount of enzyme hydrolyzing 1% of the [^{3}H]myristate-labeled VSG per minute.

During the course of these studies it was observed that blanks containing GPI-PLD but not incubated often showed substantial hydrolysis. Preliminary investigations indicate that this may be due to a transient activation of the GPI-PLD by the butanol used for the extraction of the [^{3}H]phosphatidic acid. This may be related to the pronounced activation of GPI-PLD by butanol previously observed in several mammalian tissues.[1,2] This problem can be prevented by using butanol saturated with 1 *M* NH_4OH for extraction of the [^{3}H]phosphatidic acid. Under the conditions described here blank values are typically approximately 5% of the total counts.

Purification of GPI-PLD

The procedure given below is essentially as described by Huang *et al.*[7]

PEG-5000 Precipitation. Two and one-half liters of bovine serum are thawed at 4° in the presence of 0.5 m*M* PMSF and 0.02% NaN_3. With stirring at 4°, PEG-5000 is gradually added to a final concentration of 9%. The mixture is stirred for an additional 1 hr and centrifuged at 10,000 *g* for 25 min. The supernatant is collected and diluted with an equal volume of 50 m*M* Tris, pH 7.5, 100 m*M* NaCl, 0.02% NaN_3 (buffer B) containing 0.5 m*M* PMSF. All subsequent purification steps are performed at 4° except where noted.

Q Sepharose Chromatography. The diluted supernatant is loaded at a flow rate of 30 ml/min onto a Q Sepharose (Pharmacia, Piscataway, NJ) column (9 × 10 cm) equilibrated in buffer B plus 0.5 m*M* PMSF. Following a 10-bed volume wash with the equilibration buffer, GPI-PLD activity is eluted with a linear gradient of 0.1–1.0 *M* NaCl in 4 liters of 50 m*M* Tris, pH 7.5, 0.02% NaN_3, and 0.5 m*M* PMSF. The activity-containing fractions, which are eluted after approximately 1.1 liters, are pooled and concentrated by Amicon YM10 filtration to about 200 ml.

Sephacryl S-300 Chromatography. The YM10 concentrate is loaded onto two (10 × 53 cm) S-300 Sephacryl columns, equilibrated in buffer B, and linked in tandem. Proteins are eluted at a flow rate of 3.8 ml/min, and fractions of 23 ml are collected. After 3.9 liters active fractions are eluted and pooled.

Wheat Germ Lectin Sepharose Chromatography. NaCl and CHAPS are added to the Sephacryl S-300 pool to give final concentrations of 0.2 *M* and 0.6% (w/v), respectively. The sample is divided in half for two runs on a 40-ml (2.5 cm diameter) wheat germ lectin-Sepharose (Pharmacia) column equilibrated in 50 m*M* Tris, pH 7.5, 0.2 *M* NaCl, 0.02% NaN_3, and 0.6% CHAPS. The sample is loaded overnight at 4° at a flow rate of 17 ml/hr. The flow rate is then increased to 26 ml/hr, the column washed with 5 volumes of equilibration buffer, and the GPI-PLD activity eluted with 0.3 *M* *N*-acetylglucosamine.

Hydroxyapatite Ultrogel Chromatography. The wheat germ lectin eluates from two runs are combined (75 ml) and concentrated to approximately 10 ml. Nine volumes of 5 m*M* sodium phosphate, pH 6.8, 0.4% CHAPS, 0.02% NaN_3 is added, and the sample is loaded at room temperature (flow rate 3 ml/min) onto a 4.2 × 22 cm column of hydroxyapatite Ultrogel (IBF Biotechnics, Savage, MD) equilibrated in 5 m*M* sodium phosphate, pH 6.8, 0.6% CHAPS, and 0.02% NaN_3. GPI-PLD activity is collected in the unbound fractions, and the bound, contaminating proteins are eluted with 500 m*M* sodium phosphate, pH 6.8, 0.6% CHAPS, and 0.02% NaN_3.

Zinc Chelate Matrix Chromatography. GPI-PLD active fractions from hydroxyapatite agarose chromatography are pooled, concentrated by YM10 filtration to 21 ml, and the pH adjusted with the addition of a 20-fold dilution of 1 *M* Tris-HCl, pH 7.5. The sample is loaded onto a column (1.5 × 5.0 cm) of iminodiacetic acid on Fractogel TSK HW-65F (Pierce, Rockford, IL) chelated with zinc and equilibrated in 50 m*M* Tris, pH 7.5, 100 m*M* NaCl, 0.6% CHAPS (buffer C) containing 0.02% NaN_3. The first peak of activity is collected in 10–15 bed volumes of wash with equilibration buffer, and a sharper second peak of activity is eluted with 10 m*M* histidine in equilibration buffer.

Mono Q Chromatography. The two zinc chelate pools of activity are concentrated by YM10 filtration. Each sample (5 ml) is injected onto a Mono Q (HR5/5, Pharmacia) column equilibrated in buffer C at room temperature. GPI-PLD activities are eluted at a flow rate of 1 ml/min with a gradient of 0.1–0.19 *M* NaCl in 50 m*M* Tris, pH 7.5, and 0.6% CHAPS in 6 min, followed by isocratic elution at 0.19 *M* NaCl for 5 min and a gradient of 0.19–0.4 *M* NaCl in 14 min. Under these conditions, the first zinc chelate pool eluted as a single peak of activity at 0.19 *M* NaCl, and the second Zn-chelate pool is resolved into two peaks of activity at 0.19 and 0.25 *M* NaCl.

The procedure results in an overall purification in excess of 2000-fold. Since multiple peaks (due to the presence of aggregates with low specific activities) are revealed by the last two steps it is difficult to estimate recovery precisely. However, the combined activity recovered in the two peaks from the zinc chelate column is about 1% overall. The final Mono Q column is useful for reducing the levels of higher molecular weight contaminants from the aggregated GPI-PLD but gives relatively little increase in specific activity.

Properties of GPI-PLD

The GPI-PLD purified from bovine serum by the procedure described above has a molecular weight of approximately 100,000 according to SDS–gel electrophoresis[7] and a p*I* of about 5.6 as determined by two-dimensional electrophoresis (K.-S. Huang and S. Li, unpublished work, 1989). The purified enzyme also shows a unique amino-terminal sequence for 15 residues [H_2N-X-G-I-S-T-(H)-I-E-I-G-X-(R)-A-L-E-F-L] with no strong homology to that of any other known protein. GPI-PLD has also been purified from human serum by a different procedure and has a molecular weight of approximately 110,000.[6] To further characterize the bovine GPI-PLD and confirm that the 100K protein is GPI-PLD, the purified enzyme was immunized in mice. The antiserum was shown to neutralize GPI-PLD activity in both assay systems and to react with the 100K protein on Western blots.[7] Monoclonal antibodies were generated, and several of these precipitated both GPI-PLD activity and the 100K protein in the presence of anti-mouse IgG. One of these antibodies was used to develop an immunopurification procedure which also yielded a 100K protein.[7]

The bovine GPI-PLD has a molecular weight of approximately 200K when analyzed by gel-filtration high-performance liquid chromatography under nondenaturing conditions, suggesting that it is a dimer. A 400K form (presumably tetrameric) and other higher molecular weight aggregates are also resolved by the final Mono Q column. Amino acid composition, N-terminal sequence analysis, two-dimensional gel electrophoresis, and Western blotting analysis of the various forms of GPI-PLD isolated during this purification procedure also support the conclusion that they consist of aggregates of a common 100K subunit.[7] Aggregation of the GPI-PLD may also account for the previous estimate of a molecular weight of approximately 500K (by gel filtration) for the enzyme in plasma.[4]

The purified bovine enzyme is inhibited by EGTA and 1,10-phenanthroline, suggesting a requirement for divalent metal ions.[7] The purified enzyme from human plasma is also sensitive to inhibition by chelators, the inhibitory effect of EGTA being blocked by addition of Ca^{2+}.[6] The GPI-

PLD does not hydrolyze phosphatidylinositol or phosphatidylcholine and produces [^{3}H]phosphatidic acid as the only radiolabeled product when [^{3}H]myristate-labeled VSG is used as the substrate.[6,7] The properties of the purified GPI-PLDs correspond closely to those established previously from studies with plasma, serum, or partially purified GPI-PLD from a number of mammalian sources.[3-5]

[57] Solubilization and Purification of Rat Tissue Phospholipase D

By MUTSUHIRO KOBAYASHI and JULIAN N. KANFER

Introduction

A role for phospholipase D (EC 3.1.4.4) in signal transduction has emerged, but the mechanisms of its activation are still unknown. Two types of phospholipase D have been reported. There is a membrane-bound type which is rich in microsome and plasma membrane fractions and utilizes phosphatidylcholine and phosphatidylethanolamine as substrates. The other is phosphatidylinositol–glycan-specific and has been detected in serum. Mammalian phosphatidylcholine-specific phospholipase D exhibits both hydrolytic activity and transphosphatidylation activity like the plant phospholipase D,[1] as shown in Fig. 1. Transphosphatidylation activity is a characteristic of the enzyme, and the activity is easily detected because it produces unusual lipids (e.g., phosphatidylethanol) in the presence of primary alcohols (e.g., ethanol).[2]

Phospholipase D activity is barely detectable in the absence of appropriate activators *in vitro* measurement. Miranol H2M and taurodeoxycholate activate phospholipase D to some degree. Phospholipase D is best activated by unsaturated free fatty acids such as sodium oleate, arachidonate, linoleate, and linolenate.[3]

Taki and Kanfer[4] first solubilized and partially purified mammalian phospholipase D from rat brain with the detergent Miranol H2M. At that time it was not apparent that unsaturated free fatty acids such as oleic acid

[1] J. N. Kanfer, *Can. J. Biochem.* **58,** 1370 (1980).
[2] M. Kobayashi and J. N. Kanfer, *J. Neurochem.* **48,** 1597 (1987).
[3] R. Chalifour J. N. Kanfer, *J. Neurochem.* **39,** 299 (1982).
[4] T. Taki and J. N. Kanfer, *J. Biol. Chem.* **254,** 9761 (1979).

Hydrolytic Activity

Phosphatidylcholine + H_2O - - -> phosphatidic acid + choline

Transphosphatidylation Activity

Phosphatidylcholine + *n*–alcohol - -> phosphatidyl alcohol + choline

FIG. 1. Catalytic schema for hydrolytic and transphosphatidylation activity of phospholipase D.

are the most potent activator of phospholipase D. Miranol H2M is now commercially unavailable; therefore, we examined other detergents to solubilize mammalian phospholipase D.

Assay for Phospholipase D Activity

Assay Principle

The assay of phospholipase D is based on [^{14}C]phosphatidic acid formation from [^{14}C]phosphatidylcholine. The phosphatidic acid produced can be degraded to diacylglycerol by phosphatidic acid phosphatase. Therefore, the assay mixture contains sodium fluoride. Inhibition of phosphatidic acid phosphatase activity by fluoride is incomplete. Transphosphatidylation activity is easier to measure, but in this case we determined the hydrolytic activity since it is unequivocal.

Reagents

Mixed micelles of 12.5 m*M* [^{14}C]phosphatidylcholine and 25 m*M* sodium oleate: Egg yolk phosphatidylcholine (Serdary Research Laboratories, London, Ontario, Canada), 9.375 mg, and 5 μCi 1-palmitoyl-2 [^{14}C]oleoyl-*sn*-glycerol-3-phosphocholine (54.5 mCi/mmol, New England Nuclear, Boston, MA) are placed in a glass tube to adjust the specific activity to 0.42 Ci/mol, and 1 ml of 25 m*M* sodium oleate solution is added. The tube and contents are sonicated in a bath-type sonicator for 30 min. The resulting clear solution is used as the substrate.

500 m*M* β-dimethylglutaric acid buffer, pH 6.5

200 m*M* EDTA, pH 6.5

250 m*M* NaF

1 mg/ml phosphatidic acid solution in chloroform

Assay Procedure

The incubation mixture contains 2.5 m*M* [^{14}C]phosphatidylcholine (1.67 Ci/mol) and 5 m*M* sodium oleate microdispersion, 50 m*M* β-dimethylglutaric acid buffer, pH 6.5, 10 m*M* EDTA, 25 m*M* NaF, and enzyme preparation in a total volume of 120 μl. The reaction is initiated by adding enzyme protein. When the enzyme preparation contains Triton X-100, the final concentration of the detergent is adjusted to 0.01% or less. The incubations are carried out at 30° for 90 min in a shaking bath. After the incubation, reactions are terminated by the addition of 2 ml of chloroform followed by 1 ml of methanol. Fifteen microliters of the phosphatidic acid solution in chloroform is added to each tube as a carrier. The samples are processed by the method of Folch *et al.*[5] Lipids are separated by thin-layer chromatography on 20 × 20 cm silica gel 60 thin-layer chromatography plates (Merck, Darmstadt, FRG) with a solvent system consisting of chloroform/methanol/acetone/acetic acid/water (50 : 15 : 15 : 10 : 5, by volume). The dried plates are exposed to iodine vapor, and the areas cochromatographing with standard phosphatidic acid and at the solvent front are scraped off and collected in scintillation vials. Ten milliliters of Scinti-Verse I (Fisher Scientific Co., Pittsburgh, PA) is added, and the radioactivity is determined.

Preparation of Enzyme

Rat brain and lung are used as tissues for solubilizing phospholipase D. Twenty rats are decapitated, and the forebrains and lungs are quickly removed and weighed. Tissues are minced with a pair of scissors and homogenized with 7 strokes in 0.32 *M* sucrose, 5 m*M* HEPES, and 1 m*M* dithiothreitol (SMD) to make a 20% homogenate using a Teflon homogenizer. The homogenate is diluted with SMD to a final tissue weight concentration of 10%. The homogenate is centrifuged at 1000 *g* for 10 min at 4°, the supernatant is centrifuged at 12,000 *g* for 20 min, and the resultant supernatant is removed and centrifuged at 100,000 *g* for 60 min. The 100,000 *g* pellet is suspended in SMD, washed once with SMD, suspended in a minimal volume of SMD, and used as the microsomal fraction. A cruder particulate fraction is obtained by centrifugation of the 1000 *g* at 100,000 *g* for 60 min at 4°. The pellet recovered is washed once with SMD, suspended in a minimal volume of SMD, and used as the crude particulate fraction.

[5] J. Folch, M. Lees, and G. H. Sloane-Stanley, *J. Biol. Chem.* **226,** 497 (1957).

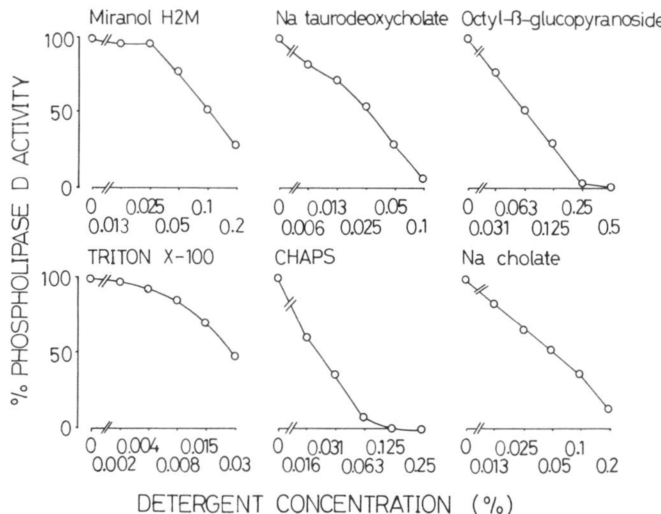

FIG. 2. Effect of various detergents on phospholipase D activity in the presence of the optimal concentration of sodium oleate. Phospholipase D activity was assayed with 2.5 m*M* [^{14}C]phosphatidylcholine and 5 m*M* sodium oleate. The phospholipase D activity was 89.3 nmol/mg/hr in the absence of detergent.

Effects of Various Detergents on Phospholipase D Activity

Effects of several concentrations of various detergents on rat brain microsomal phospholipase D activity are examined in the standard assay method. All the detergents examined inhibited the phospholipase D activity in a concentration-dependent manner (Fig. 2).

Effects of Various Treatments of Rat Brain Microsome on Phospholipase D Activity

The rat brain microsome fraction is mixed with 0.1 *M* $NaHCO_3$, 5 m*M* HEPES, 1 m*M* DTT, pH 7.3 (HD buffer), with or without 1 *M* NaCl or with or without 10 m*M* EGTA at 4° for 60 min with constant stirring. The final protein concentration is 4 mg/ml. The mixtures are centrifuged at 150,000 *g* for 60 min, and the phospholipase D activity is measured in both the supernatant and the pellet. The specific activity in the pellet is increased by approximately 25% with these treatments. The presence of $NaHCO_3$ increases the specific activity slightly better than the others. Phospholipase D is inactivated when this fraction is treated with acetone.

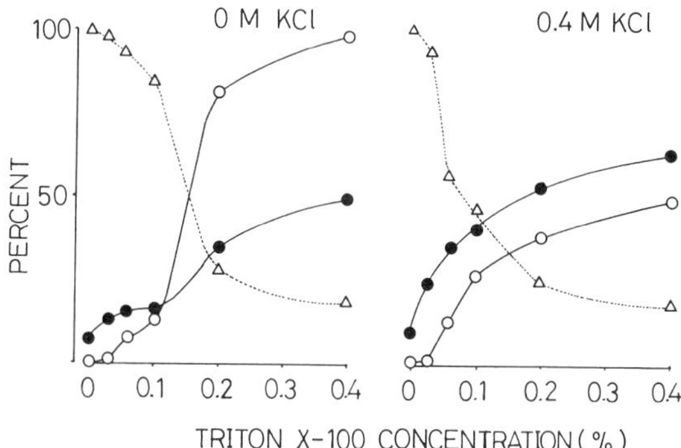

FIG. 3. Solubilization of phospholipase D activity with Triton X-100 in the absence and presence of 0.4 *M* NaCl. The total phospholipase D activity is shown in the curve with open circles, total protein recovered in the supernatant with the filled circles. The total phospholipase D activity remaining in the pellet is represented with open triangle. Solubilization was performed in the absence (left) or presence (right) of 0.4 *M* NaCl.

Solubilization with Triton X-100 in Presence of Various KCl Concentrations

The rat brain microsome fraction at a protein concentration of 2 mg/ml is suspended in HD buffer, and then varying amounts of 10% Triton X-100 solution and 1 *M* KCl are added to achieve required concentrations. The samples are mixed for 90 min in an ice bath with gentle stirring and centrifuged at 150,000 *g* for 60 min at 5°. The protein concentrations and enzyme activity in both the supernatant and the pellet are determined.

As shown in Fig. 3, 35–50% of protein and more than 80% of the phospholipase D activity are solubilized by 0.2–0.4% of Triton X-100 in the absence of KCl. The specific activity of solubilized fraction is 160–190 nmol/mg/hr and is increased about 2-fold above the original microsome. The total activity present in the supernatant and pellet is decreased in the presence of KCl.

Solubilization of Phospholipase D Activity with Other Detergents

Miranol H2M, sodium taurodeoxycholate, and CHAPS are also tested for the ability to solubilize phospholipase D. Miranol H2M successfully solubilizes phospholipase D; however, it is not employed because it is no longer commercially accessible. Sodium taurodeoxycholate solubilizes unrelated proteins rather than the enzyme from microsomal fraction and, therefore, is used for enriching the activity in the pellet. CHAPS solubilizes phospholipase D activity but does not increase the specific activity and might be useful for solubilizing phospholipase D.

TABLE I
PURIFICATION OF BRAIN PHOSPHOLIPASE D[a]

Step	Total protein (mg)	Specific activity (nmol/mg/hr)	Total activity (nmol/hr)	Yield (%)	Pur- cati- (-fo
1. Washed membranes	1427	99.6	142,129	100	1.0
2. 0.1 *M* $NaHCO_3$ treatment	1304	104.3	136,007	96	1.0
3. 0.2% Sodium taurodeoxycholate treatment	872	137.5	119,900	84	1.3
4. 0.4% Triton X-100 solubilization	153	464.1	71,007	50	4.6
5. DEAE-Sephacel column fractions	20	901.0	18,200	13	9.[illegible]

[a] Protein concentration was determined by the method of G. L. Peterson, this series, Vol. p. 95.

Solubilization of Phospholipase D and DEAE-Sephacel Column Chromatography

Reagents

5 m*M* HEPES, 1 m*M* dithiothreitol (DTT), pH 7.3 (HD buffer)
0.1 *M* $NaHCO_3$
1% sodium taurodeoxycholate solution, pH 7.4
10% Triton X-100

Procedure

Sodium Bicarbonate Treatment. Nine volumes of 0.1 *M* $NaHCO_3$ is added to 1 volume of the membrane fraction at a final protein concentration of 2–4 mg/ml. The mixture is stirred for 60 min at 4° and centrifuged at

TABLE II
PURIFICATION OF LUNG PHOSPHOLIPASE D[a]

Step	Total protein (mg)	Specific activity (nmol/mg/hr)	Total activity (nmol/hr)	Yield (%)	Pu- cat- (-fc
1. Washed membranes	156.0	305.9	45,932	100	1.[illegible]
2. 0.1 *M* $NaHCO_3$ treatment	120.6	401.3	48,397	105	1.[illegible]
3. 0.2% Sodium taurodeoxycholate treatment	67.1	375.5	25,196	55	1.[illegible]
4. Triton X-100 solubilization	31.7	541.7	17,172	37	1.[illegible]
5. DEAE-Sephacel column fractions	5.7	1151.1	6591	14	3.[illegible]

[a] Protein concentration was determined by the method of G. L. Peterson, this series, Vol. p. 95.

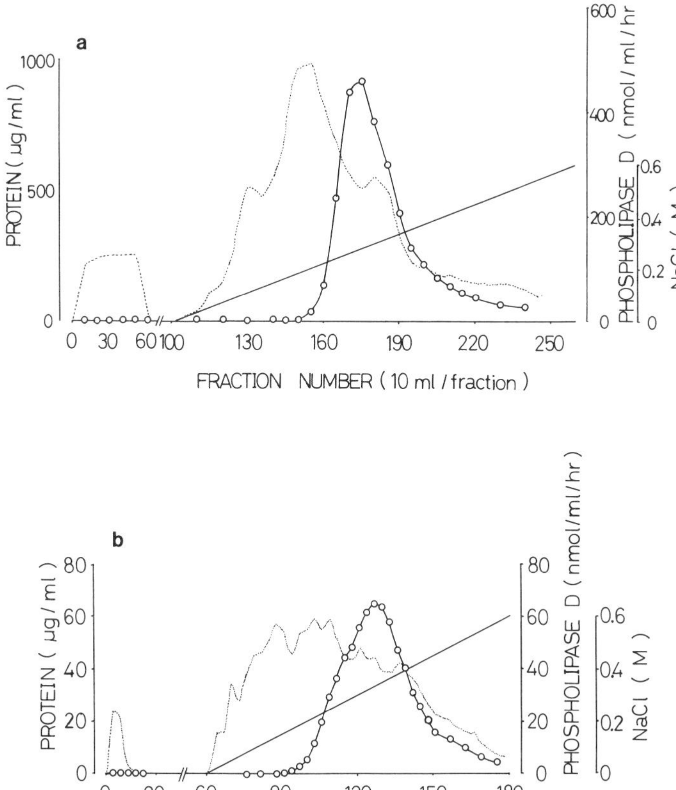

FIG. 4. DEAE-Sephacel column chromatography of Triton X-100 solubilized protein. Open circles indicate phospholipase D activity, and the dotted line indicates protein concentration. Phospholipase D activity appears at around 0.2 *M* NaCl in both brain (a) and lung (b).

150,000 *g* for 60 min, and the pellet is suspended in minimal volume of HD buffer.

Sodium Taurodeoxycholate Treatment. Sodium taurodeoxycholate (1%) and an appropriate amount of HD buffer are added to the $NaHCO_3$-treated pellet to give a final concentration of sodium taurodeoxycholate of 0.2%. The final protein concentration is 4–6 mg/ml. The mixture is stirred for 60 min at 4° and centrifuged at 150,000 *g* for 60 min, after which the pellet is suspended in a minimal volume of HD buffer.

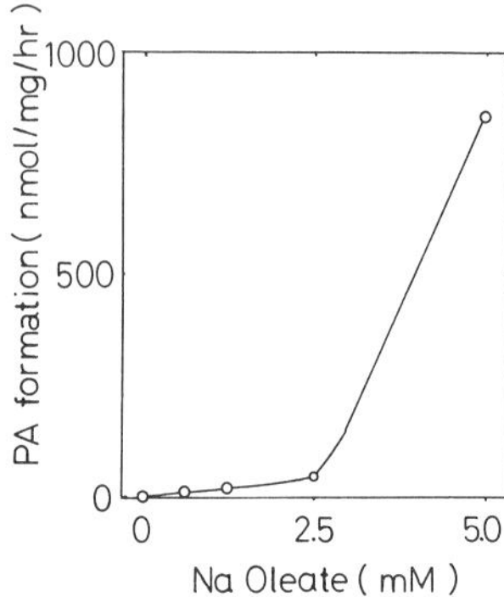

FIG. 5. Effect of sodium oleate on phospholipase D activity solubilized from brain. The concentration of [^{14}C]phosphatidylcholine was constant at 2.5 mM. Phospholipase D partially purified with DEAE-Sephacel was used for the assay in the absence and presence of the different concentrations of sodium oleate.

Solubilization with Triton X-100. Triton X-100 (10%) and appropriate amount of HD buffer are added to taurodeoxycholate-treated pellet to give a final concentration of Triton X-100 of 0.4%. The final protein concentration is 2–4 mg/ml. The mixture is stirred at 4° for 60 min, then centrifuged at 150,000 g for 60 min, and the supernatant is saved.

DEAE-Sephacel Column Chromatography. Solubilized proteins are applied to DEAE-Sephacel column (2.6 × 20 cm for brain and 1.6 × 11 cm for lung) preequilibrated with 5 mM HEPES, 1 mM DTT, pH 7.3, including 0.1% Triton X-100. After washing with 5 column volumes of equilibration buffer, proteins are eluted with a linear gradient of 400 ml each of equilibration buffer and 0.6 M NaCl in the equilibration buffer. Phospholipase D activity appears as a single peak at around 0.2 M NaCl as shown in Fig. 4a for brain and in Fig. 4b for lung. The fractions which had the specific activity of more than 500 nmol/mg protein/hr for brain or 1,000 nmol/mg protein/hr for lung are pooled.

The purification is summarized in Tables I and II. The specific activity was 901 nmol/mg protein/hr for brain or 1151 nmol/mg protein/hr for lung in the pooled fraction obtained from DEAE-Sephacel column chromatography with yields of 12.8 and 14% for brain and lung, respectively. Triton X-100 is removed from the pooled fraction using Extractigel-D (Pierce, Rockford, IL). Further attempts at purification with column chromatography including hydroxyapatite, Con A-Sepharose, AffiGel blue, octyl-Sepharose, phenyl-Sepharose, butyl-Sepharose, and phosphatidylcholine-Sepharose were unsuccessful.

Dependency on Sodium Oleate of Solubilized Phospholipase D

The phospholipase D activity obtained from the column was measured in the presence of various concentrations of sodium oleate. It was seen that the activity was still dependent on the presence of oleate (Fig. 5).

Phosphatidic Acid Phosphatase Activity in the Partially Purified Enzyme Fraction

The phospholipase D assay was performed in the absence or presence of the phosphatidic acid phosphatase inhibitor, sodium fluoride. Very little neutral lipid products were detected even in the absence of sodium fluoride, suggesting that phosphatidic acid phosphatase activity was absent or inactivated in the fraction.

Summary

Rat phospholipase D was successfully solubilized with Triton X-100. Solubilized phospholipase D is clearly separated from phosphatidic acid phosphatase and still dependent on sodium oleate for activity.

Acknowledgments

This work was supported by the grants from Medical Research Council of Canada and the U.S. Association for Dementia and Alzheimer's Disease.

[58] Lysophospholipase D

By ROBERT L. WYKLE and JAY C. STRUM

Introduction

Lysophospholipase D (EC 3.1.4.39) was discovered[1] during studies of plasmalogen biosynthesis in rat brain. When 1-*O*-[1-^{14}C]alkyl-2-lyso-*sn*-glycero-3-phosphoethanolamine (1-*O*-alkyl-GPE) was used as a substrate with microsomal preparations from rat brain, it was noted that recovery of the label in phospholipids was low and that a significant portion of the added 1-*O*-alkyl-GPE had been converted to 1-*O*-alkyl-*sn*-glycerol. The conversion to 1-*O*-alkyl-*sn*-glycerol could be blocked by washing the microsomal preparations with 5 m*M* EDTA. The studies revealed that the

[1] R. L. Wykle and J. M. Schremmer, *J. Biol. Chem.* **249,** 1742 (1974).

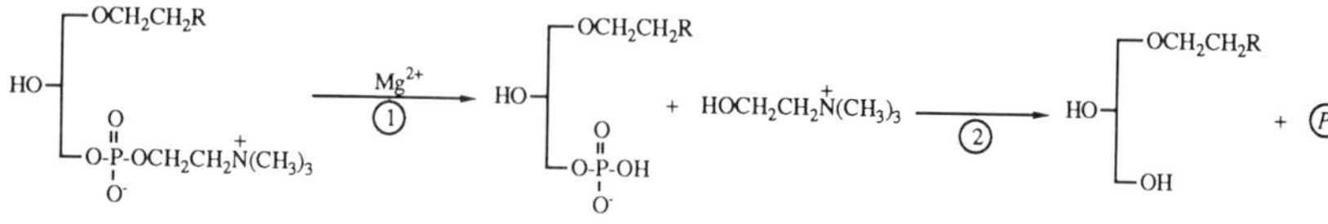

FIG. 1. Pathways catalyzed by (1) lysophospholipase D and (2) phosphohydrolase.

initial hydrolysis of 1-*O*-alkyl-GPE was catalyzed by a phospholipase D-type reaction that yields 1-*O*-alkyl-2-lyso-*sn*-glycero-3-phosphate which is subsequently dephosphorylated by a phosphohydrolase to yield 1-*O*-alkyl-*sn*-glycerol. Evidence for this sequence of reactions was obtained by adding NaF or Na_3VO_4 to the reaction mixture, which resulted in accumulation of 1-*O*-alkyl-*sn*-glycero-3-phosphate accompanied by a decrease in 1-*O*-alkyl-*sn*-glycerol. The rat brain enzyme responsible for removing the base was found to require Mg^{2+} for activity and to act only on substrates containing no acyl chain in the *sn*-2 position, hence the name lysophospholipase D.

Lysophospholipase D was found to act on 1-*O*-alkyl-2-lyso-*sn*-glycero-3-phosphocholine (1-*O*-alkyl-GPC) and the plasmalogen species 1-*O*-alk-1′-enyl-2-lyso-GPE as well as on 1-*O*-alkyl-GPE. The hydrolysis of the 1-*O*-alk-1′-enyl-GPE was studied in a mixture of 87% 1-*O*-alkyl-GPE and 13% 1-*O*-alk-1′-enyl-GPE and was not examined in detail. Other studies[2] revealed that lysophospholipase D is present in a number of tissues of the rat including liver, testes, lung, intestine, and kidney. Further studies of the substrate specificity of the rat liver lysophospholipase D[3] indicated the enzyme surprisingly does not hydrolyze acyl-linked substrates such as 1-acyl-2-lyso-GPC. Platelet-activating factor (PAF; 1-*O*-alkyl-2-acetyl-GPC) was not hydrolyzed by lysophospholipase D unless the 2-acetyl group was first removed by an acetylhydrolase.[3] The pathway catalyzed by lysophospholipase D is illustrated for 1-*O*-alkyl-GPC in Fig. 1; 1-*O*-alkyl-GPE and 1-*O*-alk-1′-enyl-GPE are hydrolyzed in the same manner.

Vierbuchen and co-workers[4] investigated the hydrolysis of 1-*O*-alkyl-GPE in rat brain microsomal preparations and demonstrated the release of ethanolamine as the major water-soluble product; however, some of the cleaved base was recovered in the absence of ATP as phosphoethanolamine, leading these workers to conclude that approximately 20% of

[2] R. L. Wykle, W. F. Kraemer, and J. M. Schremmer, *Arch. Biochem. Biophys.* **184,** 149 (1977).

[3] R. L. Wykle, W. F. Kraemer, and J. M. Schremmer, *Biochim. Biophys. Acta* **619,** 58 (1980).

[4] M. Vierbuchen, J. Gunawan, and H. Debuch, *Hoppe-Seyler's Z. Physiol. Chem.* **360,** 1091 (1979).

the hydrolysis may have been due to a phospholipase C. Since a significant fraction of the added substrate was converted to 1-*O*-alkyl-2-acyl-GPE, it is possible that hydrolysis of the latter species may also have occurred.

The hydrolysis of 1-*O*-alk-1′-enyl-GPE by rat brain preparations was further investigated by Gunawan and co-workers.[5] These workers concluded that although the plasmalogen is hydrolyzed by lysophospholipase D, the alkyl-linked species is a much better substrate. Increasing concentrations of 1-*O*-alkyl-GPE inhibited the hydrolysis of 1-*O*-alk-1′-enyl-GPE in an apparently noncompetitive manner.

Kawasaki and Snyder[6] found a novel lysophospholipase D in microsomal preparations of rabbit kidney medulla that requires Ca^{2+} for optimal activity. The rabbit kidney enzyme had a higher pH optimum (pH 8.4) than did the Mg^{2+}-dependent enzyme of the rat (pH 7.2). The hydrolysis of 1-*O*-alkyl-GPC by the Ca^{2+}-requiring enzyme yielded 1-*O*-alkyl-*sn*-glycero-3-phosphate and 1-*O*-alkyl-*sn*-glycerol in the same manner as the Mg^{2+}-dependent enzyme. The discovery of the Ca^{2+}-dependent lysophospholipase D opens the possibility that significant species differences may exist in the metal ion requirements and other characteristics of lysophospholipase D.

The physiological role of lysophospholipase D is unclear. One of the most remarkable characteristics of the enzyme is its apparent high selectivity for ether-linked species, and in particular the 1-*O*-alkyl species. No hydrolysis of 1-acyl-2-lyso-GPC was observed when it was codispersed in equimolar concentration with 1-alkyl-2-lyso-GPC, even though up to 50% of the alkyl-linked substrate was hydrolyzed.[3] Selective acylation as well as deacylation of the 1-acyl-GPC was ruled out as an explanation for the selectivity. Neither 1-acyl-*sn*-glycerol nor 1-acyl-*sn*-glycero-3-phosphate was detected in the studies.[3] The high specificity for alkyl-linked species suggests a special role for the enzyme in the metabolism of these lipids. Kawasaki and Snyder[6] and Qian *et al.*[7] have suggested that the enzyme may play a role in the metabolism of PAF and could possibly provide a pathway for the production of 1-alkyl-2-acetyl-*sn*-glycero-3-phosphate via acetylation of the 1-alkyl-2-lyso-*sn*-glycero-3-phosphate, thus providing a substrate for the putative *de novo* pathway of PAF biosynthesis.[8,9] In most cells, such as neutrophils, PAF appears to be catabolized by deacetylation

[5] J. Gunawan, M. Vierbuchen, and H. Debuch, *Hoppe-Seyler's Z. Physiol. Chem.* **360,** 971 (1979).

[6] T. Kawasaki and F. Snyder, *Biochim. Biophys. Acta* **920,** 85 (1987).

[7] C. Qian, T-c. Lee, and F. Snyder, *J. Lipid Med.* **1,** 113 (1989).

[8] T-c. Lee, B. Malone, and F. Snyder, *J. Biol. Chem.* **261,** 5373 (1986).

[9] F. Snyder, *in* "Platelet Activating Factor and Related Lipid Mediators" (F. Snyder, ed.), Chap. 4, p. 89. Plenum, New York, 1987.

followed by reacylation; we have been unable to detect lysophospholipase D in neutrophils (R. L. Wykle and J. C. Strum, unpublished results, 1990).

Another possible role for lysophospholipase D could be to provide a mechanism for interconversion of choline- and ethanolamine-linked ether species. Acylation of the 1-alkyl-GP to yield 1-alkyl-2-acyl-GP followed by dephosphorylation would yield the diglyceride which could then be reincorporated into choline- or ethanolamine-containing species by the cytidine 5′-diphosphoethanolamine-ethanolamine (choline) pathway.[10] The removal of ether-linked lysophosphoglycerides cannot be accomplished by phospholipase A-type acylhydrolases, but rather requires special mechanisms. The alkyl bonds can be cleaved by the alkyl cleavage enzyme, which requires O_2 and tetrahydropteridine, whereas the alk-1′-enyl chains can be hydrolyzed by plasmalogenase.[10] Since both systems can act directly on the lysophosphoglycerides,[10,11] it does not appear that lysophospholipase D is essential for removal of these species, but it could nevertheless play such a role.

Assay

Principle and Potential Problems

The formation of labeled 1-*O*-alkyl-*sn*-glycerol and 1-*O*-alkyl-2-lyso-*sn*-glycero-3-phosphate from 1-*O*-alkyl-2-lyso-GPC or 1-*O*-alkyl-2-lyso-GPE containing a radiolabeled 1-*O*-alkyl chain has most often been employed to detect lysophospholipase D activity. Since 1-*O*-[^{3}H]alkyl-2-lyso-GPC (lyso-PAF) containing 16 : 0 and 18 : 0 chains is commercially available and has been employed in many of the earlier studies, it will be used here as the model substrate. The 1-*O*-[^{3}H]alkyl-2-lyso-GPC is incubated with the microsomal fraction of interest in buffer-containing Mg^{2+} (Ca^{2+} for the Ca^{2+}-dependent enzyme). At the end of the incubation time, the lipids are extracted, separated by thin-layer chromatography (TLC), and radioassayed to determine the amount of products formed. In the absence of phosphatase inhibitors (NaF or Na_3VO_4) most of the 1-*O*-[^{3}H]alkyl-2-lyso-*sn*-glycero-3-phosphate formed is rapidly dephosphorylated, resulting in 1-*O*-[^{3}H]alkyl-*sn*-glycerol as the major product observed. However, it is important to monitor both products in the assay.

In most microsomal preparations a significant fraction of the added 1-*O*-[^{3}H]alkyl-2-lyso-GPC (or -GPE) will be acylated by a CoA-independent reaction to form 1-*O*-alkyl-2-acyl-GPC (or -GPE), which does not

[10] F. Snyder, T.-C. Lee, and R. L. Wykle, *in* "The Enzymes of Biological Membranes" (A. Martonosi, ed.), 2nd Ed. Vol. 2, Chap. 14, p. 1. Plenum, New York, 1985.

[11] F. Snyder, B. Malone, and C. Piantadosi, *Biochim. Biophys. Acta* **316**, 259 (1973).

serve as a substrate for lysophospholipase D. Removal of substrate by this competing reaction can be largely overcome by using higher concentrations of the lyso substrate. In earlier studies *p*-bromophenacyl bromide was found to strongly inhibit the acylation reaction but also weakly inhibit lysophospholipase D.[3] In view of this competing reaction, it is important to make certain the 1-*O*-[^{3}H]alkyl-2-lyso-GPC remains available as a substrate.

In homogenates or microsomes untreated by EDTA, no metal ion requirement is likely to be observed, but in microsomal preparations washed with 5 or 10 m*M* EDTA a Mg^{2+} or Ca^{2+} requirement is readily observed. It is possible that some microsomal preparations, or other cell fractions, might contain endogenous lysophospholipid substrates that would reduce the specific radioactivity of the added substrates and lead to an underestimation of activity. The assays described below are based on the studies of the Mg^{2+}-requiring enzyme of rat liver[1-3] and the Ca^{2+}-requiring enzyme of rabbit kidney medulla[6] described earlier.

Reagents

Substrates: 1-*O*-[^{3}H]Alkyl-2-lyso-*sn*-glycero-3-phosphocholine (lyso platelet-activating factor) is available from New England Nuclear (Boston, MA) and other sources. Unlabeled 1-*O*-alkyl-2-lyso-*sn*-glycero-3-phosphocholine (Bachem, Torrance, CA, and other sources) is used for dilution of radiolabeled substrate to the desired specific radioactivity. The 1-*O*-hexadecyl and 1-*O*-octadecyl species are most often available.

Buffer A: Tris-HCl (0.1 *M*, pH 7.2) containing 1 m*M* dithiothreitol and 5 m*M* $MgCl_2$ for the Mg^{2+}-requiring enzyme

Buffer B: Tris-HCl (0.1 *M*, pH 8.4) containing 5 m*M* $CaCl_2$ for the Ca^{2+}-requiring enzyme

Enzyme source: In all tissues examined the activity has been primarily associated with the microsomal fraction.

Procedure

The radiolabeled 1-*O*-[^{3}H]alkyl-2-lyso-GPC is adjusted to the desired specific activity [50,000 disintegrations/min (dpm)/nmol is generally a good range] by mixing with unlabeled 1-*O*-alkyl-2-lyso-GPC in chloroform; ideally the labeled and unlabeled substrates should have the same alkyl chain composition, but this is not critical for detection of activity. Immediately before use an aliquot of the diluted substrate is evaporated under a stream of nitrogen and resuspended in 95% ethanol at 2 nmol/μl. An aliquot of the ethanol solution should be counted to make certain the substrate is all dissolved.

The microsomal preparation (~100 μg protein suspended in buffer A or B, respectively, for the Mg^{2+}- or Ca^{2+}-requiring enzymes) is added to the assay buffer (A or B, depending on which activity is under investigation) to give a final volume of 0.5 ml. The labeled substrate is then added (20 nmol in 10 μl ethanol), and the mixture is shaken well to initiate the reaction. The incubation mixtures are shaken in a water bath at 37° for 10 min. The reaction is stopped by adding 150 μl of 2 *N* HCl, along with 25 μg lysophosphatidic acid and 150 μg phosphatidylcholine as carrier phospholipids[6]; the mixture is then extracted by a modified Bligh and Dyer procedure[12] in which the methanol contains 2% glacial acetic acid. The following volumes are added: 1.3 ml methanol/2% acetic acid, 1.2 ml chloroform, and 460 μl of 2.0 *M* KCl.[6] The lower organic phase is evaporated under a stream of nitrogen, and the extracted products are redissolved in a known volume of chloroform (100–200 μl).

The labeled products are separated on silica gel H layers developed in chloroform/methanol/concentrated NH_4OH (65 : 35 : 8, v/v). The separated products are visualized by exposure to iodine vapor, and the areas corresponding to 1-*O*-alkyl-2-lyso-*sn*-glycero-3-phosphate, 1-*O*-alkyl-2-lyso-GPC, 1-*O*-alkyl-2-acyl-GPC, and 1-*O*-alkyl-*sn*-glycerol are scraped and counted by liquid scintillation counting. Alternatively, the distribution of label along the plate can be determined by scraping and counting smaller zones, or by using a radiochromatogram imaging system (Bio-Scan, Inc., Washington, DC). Since the thin-layer chromatography system does not resolve ether-linked species from acyl-linked species, the corresponding acyl-linked phospholipids can be used as standards. A typical separation is shown in Fig. 2. The 1-*O*-[^{3}H]alkyl-*sn*-glycerol can be determined independently in the same manner by separating the products on silica gel G layers developed in ethyl ether/water (100 : 0.5, v/v). In the latter system, all the phospholipids remain at the origin. The percentage of label converted to 1-*O*-[^{3}H]alkyl-*sn*-glycerol and 1-*O*-[^{3}H]alkyl-*sn*-glycero-3-phosphate is a measure of the lysophospholipase D activity. Rat liver microsomes hydrolyzed approximately 1.9 nmol/min/mg protein, whereas microsomes of rabbit kidney medulla hydrolyzed 2.7 nmol/min/mg protein.

A number of points should be noted. The assay systems described may not be optimal for other enzyme preparations. Tracer levels of undiluted high specific activity substrates can be employed to detect lysophospholipase D when the enzyme source is limited, but much of the substrate may be acylated and become unavailable as a substrate. Neither labeled plasmalogen substrates nor labeled 1-*O*-alkyl-2-acyl-GPE are commer-

[12] E. G. Bligh and W. J. Dyer, *Can. J. Biochem. Physiol.* **37,** 911 (1959).

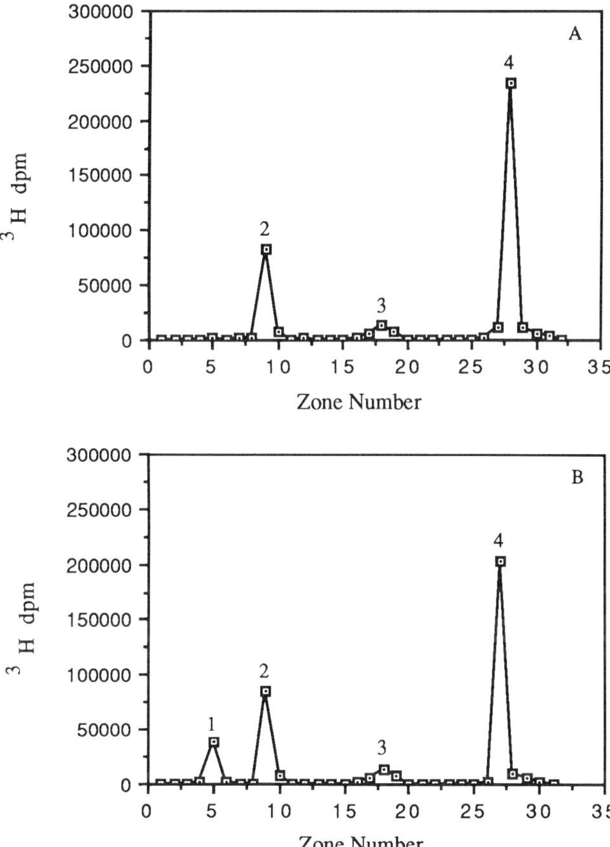

FIG. 2. Zonal profile of products obtained from a cell-free assay of lysophospholipase D. Microsomes from primary rat neurons (205 μg protein), in 5 mM $MgCl_2$, 0.1 M Tris-HCl (pH 7.2), and 1-*O*-hexadecyl-2-lyso-*sn*-glycero-3-phosphocholine 0.15 μCi; 56 Ci/mmol), were incubated at 37° for 30 min. Following incubations, lipids were extracted as described above and resolved by TLC in a solvent system of chloroform/methanol/ammonium hydroxide (65 : 35 : 8, v/v). Each lane was scraped into 5 mm zones and counted by liquid scintillation counting. (A) No addition; (B) 0.5 mM Na_3VO_4 added. Peaks: 1, 1-*O*-Alkyl-2-lyso-*sn*-glycerophosphate; 2, 1-*O*-alkyl-2-lyso-GPC; 3, 1-*O*-alkyl-2-acyl-GPC; and 4, 1-*O*-alkyl-*sn*-glycerol.

cially available to the knowledge of authors. These substrates can be prepared biosynthetically in the manner described earlier[1] by incubating labeled hexadecanol with appropriate cell preparations that contain significant levels of the ether lipids. 1-*O*-[^{3}H]Hexadecyl-2-lyso-GPE of high specific activity can also be prepared by a base-exchange reaction of ethanolamine with 1-*O*-[^{3}H]hexadecyl-2-acetyl-GPC catalyzed by cabbage phospholipase D[13–15] (R. L. Wykle and J. C. Strum, unpublished results, 1990) followed by hydrolysis of the 2-acetyl residue.[16] The 1-alkyl-2-lyso-GPE substrate can be added to incubation mixtures in ethyl ether/ethanol (2 : 1, v/v). NaF (20 m*M*) or sodium orthovanadate (1 m*M*) can be added to the incubation mixture to inhibit the phosphohydrolase and formation of 1-*O*-alkyl-*sn*-glycerol.[1,6]

Acknowledgments

Thanks to Connie McArthur for manuscript preparation. This work was supported by National Institutes of Health Grant HL-26818 and a grant from R. J. Reynolds Tobacco Co. J.C.S. was supported by a National Research Service Award (T32 CA-09422) from the National Cancer Institute.

[13] S. F. Yang, S. Freer, and A. A. Benson, *J. Biol. Chem.* **242,** 477 (1967).

[14] M. Saito, E. Bourque, and J. Kanfer, *Arch. Biochem. Biophys.* **164,** 420 (1974).

[15] K. Satouchi, R. N. Pinckard, L. M. McManus, and D. J. Hanahan, *J. Biol. Chem.* **256,** 4425 (1981).

[16] N. G. Clarke and R. M. C. Dawson, *Biochem. J.* **195,** 301 (1981).

Author Index

Numbers in parentheses are footnote reference numbers and indicate that an author's work is referred to although the name is not cited in the text.

A

B

C

D

E

F

G

H

I

J

K

L

M

N

O

P

Q

R

S

T

U

V

W

Y

Z

Subject Index

A

B

C

E

F

G

M

N

O

P

R